# Topological Vector Spaces

## Second Edition

# MONOGRAPHS AND TEXTBOOKS IN PURE AND APPLIED MATHEMATICS

## Recent Titles

*M. M. Rao*, Conditional Measures and Applications, Second Edition (2005)

*A. B. Kharazishvili*, Strange Functions in Real Analysis, Second Edition (2006)

*Vincenzo Ancona and Bernard Gaveau*, Differential Forms on Singular Varieties: De Rham and Hodge Theory Simplified (2005)

*Santiago Alves Tavares*, Generation of Multivariate Hermite Interpolating Polynomials (2005)

*Sergio Macías*, Topics on Continua (2005)

*Mircea Sofonea, Weimin Han, and Meir Shillor*, Analysis and Approximation of Contact Problems with Adhesion or Damage (2006)

*Marwan Moubachir and Jean-Paul Zolésio*, Moving Shape Analysis and Control: Applications to Fluid Structure Interactions (2006)

*Alfred Geroldinger and Franz Halter-Koch*, Non-Unique Factorizations: Algebraic, Combinatorial and Analytic Theory (2006)

*Kevin J. Hastings*, Introduction to the Mathematics of Operations Research with *Mathematica*®, Second Edition (2006)

*Robert Carlson*, A Concrete Introduction to Real Analysis (2006)

*John Dauns and Yiqiang Zhou*, Classes of Modules (2006)

*N. K. Govil, H. N. Mhaskar, Ram N. Mohapatra, Zuhair Nashed, and J. Szabados*, Frontiers in Interpolation and Approximation (2006)

*Luca Lorenzi and Marcello Bertoldi*, Analytical Methods for Markov Semigroups (2006)

*M. A. Al-Gwaiz and S. A. Elsanousi*, Elements of Real Analysis (2006)

*Theodore G. Faticoni*, Direct Sum Decompositions of Torsion-Free Finite Rank Groups (2007)

*R. Sivaramakrishnan*, Certain Number-Theoretic Episodes in Algebra (2006)

*Aderemi Kuku*, Representation Theory and Higher Algebraic K-Theory (2006)

*Robert Piziak and P. L. Odell*, Matrix Theory: From Generalized Inverses to Jordan Form (2007)

*Norman L. Johnson, Vikram Jha, and Mauro Biliotti*, Handbook of Finite Translation Planes (2007)

*Lieven Le Bruyn*, Noncommutative Geometry and Cayley-smooth Orders (2008)

*Fritz Schwarz*, Algorithmic Lie Theory for Solving Ordinary Differential Equations (2008)

*Jane Cronin*, Ordinary Differential Equations: Introduction and Qualitative Theory, Third Edition (2008)

*Su Gao*, Invariant Descriptive Set Theory (2009)

*Christopher Apelian and Steve Surace*, Real and Complex Analysis (2010)

*Norman L. Johnson*, Combinatorics of Spreads and Parallelisms (2010)

*Lawrence Narici and Edward Beckenstein*, Topological Vector Spaces, Second Edition (2010)

# Topological Vector Spaces

## Second Edition

**Lawrence Narici**
St. John's University
New York

**Edward Beckenstein**
St. John's University
New York

CRC Press
Taylor & Francis Group
Boca Raton London New York

CRC Press is an imprint of the
Taylor & Francis Group, an **informa** business
A CHAPMAN & HALL BOOK

Cover image by Jos Leys — www.josleys.com

Chapman & Hall/CRC
Taylor & Francis Group
6000 Broken Sound Parkway NW, Suite 300
Boca Raton, FL 33487-2742



Printed in the United States of America on acid-free paper
10 9 8 7 6 5 4 3 2 1

International Standard Book Number: 978-1-58488-866-6 (Hardback)



**Library of Congress Cataloging-in-Publication Data**

Narici, Lawrence.
Topological vector spaces / Lawrence Narici and Edward Beckenstein. -- 2nd ed.
p. cm.
Includes bibliographical references and index.
ISBN 978-1-58488-866-6
1. Linear topological spaces. I. Beckenstein, Edward, 1940- II. Title.

QA322.N375 2011
515'.73--dc22 2010007966

**Visit the Taylor & Francis Web site at**
**http://www.taylorandfrancis.com**

**and the CRC Press Web site at**
**http://www.crcpress.com**

# Contents

# Preface to This Edition

How is this edition different from the first? Aside from a great deal of rewriting and rearrangement, we have included much more on the Hahn–Banach theorem, e.g., its connection with the axiom of choice, uniqueness of Hahn–Banach extensions, things like that. There is a whole new chapter on vector-valued Hahn–Banach theorems and an enlarged presentation of the Banach–Stone theorem. We discuss different approaches to Banach–Stone as well as such variants as vector-valued versions. Since mathematics is done by mathematicians, we have included a couple of stories, some heart-throbbing, about them—Eduard Helly, for example. Helly proved the first Hahn–Banach theorem and he did it in a more flexible manner than Hahn or Banach. (And each of them relied on what Helly had done.) Instead of reducing the complex case to the real one, Helly's geometric approach treats the real and complex cases simultaneously. The technique was resuscitated in the 1970s and proved very effective in the investigation of vector-valued Hahn–Banach theorems. In Section 11.1 we discuss the Scottish Café, the café in Lwów (Poland, then, Ukraine now) where Banach and many others regularly gathered in the twilight interval between the conflagrations. When these young mathematicians first gathered there, there were no normed spaces, only collections of functions, sequences, etc., with various specific metrics that we call "norms"today. They (and others) recognized the features that united them and forged functional analysis. In so doing, they eliminated a lot of context-specific clutter. The perspicacity of that hardy group was commensurate with their curiosity but they were in the wrong place at a terrible time. Amidst the slaughter of many of his closest friends, including his thesis advisor, Banach survived the war, but barely. The university was closed, as were all universities in Poland, during the Nazi occupation. Officially reclassified as "subhuman," he spent the lean years feeding lice with his own blood at a bacteriological institute while living with the constant terror, including a stint in jail. The privations took their lethal toll and Banach died in August 1945.

We must thank Albrecht Pietsch [2007] for his sparkling history of Banach spaces. We have generously helped ourselves to this trove of remarkable scholarship.

We assume that you, the reader, will not read this book from page 1, that you will dip into it from time to time. Therefore we do not assume familiarity with our (pretty standard) notation and often repeat it to minimize visits to the index. The index is quite detailed—see the entry for continuity, for example—and should simplify the searching process.

We dedicate the volume to you, our fellow mathematicians, with a little extra for the many, many good friends we've made at conferences over the years. We raise our glasses (not $H_2O$) to you.

Lawrence Narici
Edward Beckenstein

# Preface to First Edition

Functional analysis started as a fusion of algebra and metric topology. With the accretion of general topology in the 1930s, it evolved from being the theory of Banach spaces to the theory of topological vector spaces (TVS)—not arbitrary TVS really, but a special kind introduced by von Neumann called locally convex spaces (LCS). (von Neumann, incidentally, called them convex spaces.) They are mostly what this book is about. The feature that makes them relatively more attractive is their good supply of continuous linear functionals. It becomes possible to study a LCS $X$ with the aid of its "dual space" $X'$ of continuous linear functionals. Some things—the class of closed convex sets, for example—can be freed from the topology on $X$ and viewed strictly as attributes of the "dual pair" $(X, X')$. These and others of the pioneering discoveries of Mackey are the subject matter of Chapter 8. Interestingly enough, some things whose discovery was strongly tied to duality theory, hence to LCS, are proving to be true in general TVS (compare Exercise 11.201 with the results of Chapter 11, for example), although they must be argued differently. In any case, we hardly touch on that delightful turn of events; a source that does is Adasch et al [1978].

This book is addressed to a mythical creature, known as The Beginner: he knows and likes general topology rather well, likes to see it used to prove things, and is delighted by the ingenuity of its quirky counterexamples; he knows and likes linear algebra after deducting those hideous calculations with matrices, but does not know anything about TVS; he likes concrete examples and do-able exercises. The detail presented should make for easy digestibility though, of course, some chewing will be required.

The exercises are divided into 100- and 200-level categories. The 100-level ones are meant to be fun. The 200-level ones possess some combination of the following attributes; harder, longer, tangential. Some of them are mini-subjects. Hints for selected exercises are given on the page right after the 200-level ones. What is a hint? Sometimes a word, sometimes "Use theorem so-and-so," sometimes not a hint at all, but the whole argument, occasionally reference to another source. We separated them from the exercises proper to discourage premature peeking, but of course we mean them to be used. We are undoubtedly guilty at times of (a) saying too much, (b) saying too little, and (rarely, we hope) (c) pointing you in the wrong direction. In no case (well, maybe one) does a subsequent result depend on an exercise. Most theorems and exercises have a "headline" to facilitate scanning.

A TVS is a topologized vector space in which the vector operations of addition and scalar multiplication are continuous. Before looking at them *per se*, we consider commutative topological groups. In other words, we postpone investigation of the effect of scalar multiplication. We can dispatch metrizability and completeness in this context since neither of them depends on scalar multiplication. (No knowledge of uniform spaces is assumed or used, incidentally.)

Topological groups are very localized things. If you know the topological contours around 0, you know them everywhere: there is a homeomorphism mapping any point of a topological group into any other point. For this reason we spend considerable time developing the properties of the neighborhoods of 0.

We get to TVS proper in Chapter 4. A significant difference between topological groups and TVS (over $\mathbf{R}$ or $\mathbf{C}$) is that TVS must always be connected topological spaces. Thus the discrete topology is a group topology but is never (except for the vector space consisting only of 0) a vector topology. While touching on the subject of separation, let us mention that we do not routinely require Hausdorff separation. We have tried to assume it only when needed, although it can definitely be removed in some spots.

By Chapter 5 we have arrived in the locally convex space, the territory in which we remain for the rest of the book. A locally convex space is a TVS in which each neighborhood of 0 contains a convex subneighborhood of 0. Even though that is correct, and the way a topologist might define it, locally convex spaces do not usually arise by specifying their neighborhoods; rather, they come equipped with a collection of seminorms. They are then endowed with the weakest topology that makes all the seminorms continuous.

A *seminormed space* is a TVS whose topology is defined by one seminorm. A subset of a seminormed space is *bounded* if it is contained in all sufficiently large multiples of the unit ball $U$. We say that a subset of a TVS is *bounded* if it is contained in all sufficiently large multiples of *any* neighborhood $U$ of 0. (As we show in Chapter 6, this is a stronger notion than that of metric boundedness.) If one has some acquaintance with normed spaces, there may be an expectation that neighborhoods of 0 in a general TVS are bounded. Not so. If there is so much as one bounded neighborhood of 0, the space is pseudometrizable; if there is a bounded convex neighborhood of 0, the space is seminormable.

As the four basic principles of functional analysis, we take the Hahn–Banach, Banach–Steinhaus, Krein–Milman, and closed graph theorems. In presenting the Hahn–Banach theorem we proceed by way of sublinear functionals. The basic idea is that a sublinear functional squashes down to a linear one. More precisely, a sublinear functional is linear if and only if it is a minimal element in the class of sublinear functionals [with respect to the ordering $f \leq g$ iff $f(x) \leq g(x)$ for all $x$] The problem of continuously extending a continuous linear map on a normed space into another normed space, rather than the underlying field, is probed in Chapter 10 where we present the solution obtained by Nachbin and Goodner. Kelley's finishing touch appears in Sec. 10.7, after extreme points have been introduced.

The Banach–Steinhaus theorem or principle of uniform boundedness asserts that, under certain conditions, a family of continuous linear maps which is bounded at each point of the (common) domain is uniformly bounded. The class of domain LCS for which the principle obtains is the class of barreled spaces, the subject of Chapter 11. For the sake not of The Beginner, but the Journeyman, we mention that Webb's recent proof of the barreledness of subspaces of countable codimension of barreled spaces appears in Sec. 11.12. (The result was discovered independently by Saxon and Levin and Valdivia earlier; a word of thanks to Steve Saxon here for apprising us of Webb's proof.)

The Krein–Milman theorem appears in Chapter 9. The theorem extends an old result of Minkowski's. An "extreme point" is a generalization of the notion of vertex of a convex polygon. Minkowski proved that a convex polygon in euclidean $n$-space can always be recovered as the convex hull of its vertices. The Krein–Milman theorem is an infinite-dimensional version. It says that a convex *compact* subset of a locally convex Hausdorff space may always be reconstituted as the *closure* of the convex hull of its extreme points. Thus the points of the original set may be *approximated* by the points of the convex hull of the extreme points in this case. An area opened up by the Krein–Milman theorem is Choquet theory. We give a brief introduction to it in Sec. 9.5.

Conditions under which a linear map with a closed graph is continuous are the focus of Chapter 14. We introduce webbed spaces (a slightly simpler version of them than de Wilde used) for the sake of proving that a closed linear map of an inductive limit of Banach spaces into a webbed space is continuous. Aside from its uses in solving some very practical problems, webbed spaces are already as permanent a fixture in functional analysis as barreled and bornological spaces are.

We started this book in 1969 and it would have been done much sooner if the chapter on topological algebras had not become so bloated that it became a separate book. There have been many people who have been helpful along this long road. Charles Suffel, although he has left this lovely land for the far shores of graph theory (something about connecting dots by lines, we believe), was a willing listener to many a late-night paradox. Donald McCarthy, despite the trauma induced by computer miasma, helped a lot in the daytime. Jean Schmets read chunks of the book, as did Marc de Wilde (another casualty in our ranks). Seth Warner gave us the benefit of his mighty erudition and so did Aaron Todd. Thanks are also due to Milos Dostal. More generally, thanks are due to Bourbaki, Köthe, Kelley and Namioka, and many, many others whose great works created and communicated this magnificent subject.

Lawrence Narici
Edward Beckenstein

# Chapter 1

# Background



With the exception of Secs. 1.4 and 1.5 on linear functionals, we list some terms and notations here for the sake of definiteness.

**R** and **C** denote the real and complex numbers with their usual topologies. **F** stands for either, without specifying which. $\mathbf{R}^+$ denotes the positive real numbers. **Q**, **N**, and **Z** denote the rationals, positive integers, and integers, respectively. "Iff" stands for if and only if. "Almost all" means "for all but a finite number." The end of proofs, examples, etc. is marked with a ■.

## 1.1 TOPOLOGY

Let $T$ be a set. A map $d : T \times T \mapsto [0, \infty)$ such that $d(s,t) = d(t,s)$ and $d(s,u) \leq d(s,t) + d(t,u)$ (the *triangle inequality*) for all $s,t,u \in T$ is a *pseudometric.* $(T,d)$ is a *pseudometric space.* A topological space whose topology is determined by a pseudometric is called *pseudometrizable.* If $d(s,t) = 0$

implies that $s = t$ then $d$ is a *metric.* If $(T, d)$ is a pseudometric space then the relation $s \sim t$ iff $d(s,t) = 0$ is an equivalence relation on $T$ and the set of equivalence classes $T/d$ is a metric space with respect to $d^{\sim}$ where, for $s^{\sim}, t^{\sim} \in T/d$, $d^{\sim}(s^{\sim}, t^{\sim}) = d(s,t)$ for any $s \in s^{\sim}$,and $t \in t^{\sim}$. If the pseudometric $d$ satisfies $d(s,u) \leq \max[d(s,t), d(t,u)]$ (the *strong* or *ultrametric triangle inequality*) for all $s,t,u \in T$, $d$ is an *ultrapseudometric* and $(T,d)$ an *ultrapseudometric space;* similar meanings attach to *ultrametric* and *ultrametric space.* If $d$ is a pseudometric on a set $T$, the *closed* and *open balls* of radius $r > 0$ about $t \in T$ are denoted, respectively, as

$$C(t,r) = \{s \in T : d(s,t) \leq r\} \quad \text{and} \quad B(t,r) = \{s \in T : d(s,t) < r\}$$

Most of the topologies encountered in the theory of topological vector spaces are defined by specifying neighborhoods of 0 rather than open sets. We do not require that a neighborhood of a point be an open set: we say that $V$ is a *neighborhood* of $t$ if $V$ contains an open set to which $t$ belongs.

The *discrete* topology on $T$ is the one in which all subsets are open; the *trivial* topology is $\{\varnothing, T\}$.

### 1.1.1 Closure and Interior

An *adherence point* $t$ of a subset $A$ of a topological space $T$ is a point whose every neighborhood meets (i.e., has nonempty intersection with) $A$. The set $\operatorname{cl} A$ of adherence points of $A$ is the *closure* of $A$. If $S$ is a topological subspace of $T$ and $A$ is a subset of $S$, then $\operatorname{cl}_s A$ denotes the closure of $A$ computed in $S$. We say that $t$ is an *interior point* of $A$ if $A$ contains a neighborhood of $t$. The set $\operatorname{int} A$ of interior points of $A$ is called the *interior* of $A$; for a subspace $S$ of $T$, $\operatorname{int}_s A$ denotes interior points of $A$ computed in $S$.

### 1.1.2 Filterbases and Nets

We deal with convergence through the use of nets and filterbases (rather than filters). We discuss some facts and properties of nets and filterbases in this section; for more information, Dugundji [1966] is an excellent source.

A collection $\mathcal{B}$ of nonempty sets such that the intersection of any two of them contains another from $\mathcal{B}$ is called a *filterbase.*

**Definition 1.1.1** CLUSTER POINTS AND LIMITS Let $\mathcal{B}$ be a filterbase of sets from a topological space $T$. (a) LIMITS We say that $t$ is a *limit* of $\mathcal{B}$ ($\mathcal{B}$ *converges* to $t$), $\mathcal{B} \to t$, if for any neighborhood $V$ of $t$ there is some $B \in \mathcal{B}$ such that $B \subset V$. The set of limits of $\mathcal{B}$ is denoted $\lim \mathcal{B}$.

(b) CLUSTER POINTS We say that $t$ is a *cluster point* of $\mathcal{B}$ if each neighborhood of $t$ meets each $B$ in $\mathcal{B}$. We denote the (closed) set of cluster points by $\operatorname{cl} \mathcal{B}$. ■

Evidently:

**Theorem 1.1.2** Given a filterbase $\mathcal{B}$ (a) $\operatorname{cl}\mathcal{B} = \cap \{\operatorname{cl} B : B \in \mathcal{B}\}$;
(b) limits are cluster points;
(c) if $T$ is Hausdorff, then limits of filterbases are unique.

**Example 1.1.3** FILTERBASES (a) Given a sequence $(t_n)$, the "tails" $B_n = \{t_j : j \geq n\}$, $n \in \mathbf{N}$, of the sequence $(t_n)$ form a filterbase called the *Fréchet* or *elementary* filterbase associated with $(t_n)$. Clearly, $(B_n) \to t$ iff $t_n \to t$.

(b) For any filterbase $\mathcal{B}$ and any map $f$, $f(\mathcal{B})$ is a filterbase; $f^{-1}(\mathcal{B})$ is a filterbase iff $\varnothing \notin f^{-1}(\mathcal{B})$

(c) The collection $V(t)$ of all neighborhoods of an element $t$ of a topological space forms a filterbase called the *neighborhood filter of* $t$. Generally, a filterbase $\mathcal{B}$ with the property that any superset of a member of $\mathcal{B}$ is also in $\mathcal{B}$ is called a *filter*—hence neighborhood *filter* rather than *filterbase.* The collection $\mathcal{F}(\mathcal{B})$ of supersets of the sets of a filterbase $\mathcal{B}$ is a filter called the *filter generated by* $\mathcal{B}$. The neighborhood filter $V(t)$ of a point $t$ obviously converges to $t$.

(d) TRACE If $\mathcal{B}$ is a filterbase and $A$ is a set such that $A \cap B \neq \varnothing$ for each $B \in \mathcal{B}$ then $\{B \cap A : B \in \mathcal{B}\}$ is a filterbase called the *trace of* $\mathcal{B}$ on $A$.

(e) If $\mathcal{A}$ and $\mathcal{B}$ are filterbases such that $A \cap B \neq \varnothing$ for each $A \in \mathcal{A}$ and $B \in \mathcal{B}$, then $\mathcal{A} \cap \mathcal{B} = \{A \cap B : A \in \mathcal{A}, B \in \mathcal{B}\}$ is a filterbase.

(f) In the complex plane, the collection of disks $C(0,n) = \{z \in \mathbf{C} : |z| \leq n\}$, $n \in \mathbf{N}$, is a filterbase which has no limit. ■

If $\mathcal{S}$ is a collection of sets that satisfies the finite intersection condition (i.e., the intersection of any finite number of them is nonempty), then the collection forms a *filter subbase;* the collection of finite intersections of sets from a filter subbase is the *filterbase generated* by $\mathcal{S}$.

Subordinate filterbases are the analog of subsequences. Given filterbases $\mathcal{A}$ and $\mathcal{B}$, we say that $\mathcal{B}$ is *subordinate to* $\mathcal{A}$, $\mathcal{B} < \mathcal{A}$, if each $A$ in $\mathcal{A}$ contains some $B$ in $\mathcal{B}$. Given a subsequence $(t_{n_k})$ of a sequence $(t_n)$, the Fréchet filterbase $\mathcal{B}$ of $(t_{n_k})$ is subordinate to the Fréchet filterbase $\mathcal{A}$ of $(t_n)$.

Convergence and continuity may be described by subordination as follows:

**Theorem 1.1.4** CONVERGENCE AND CONTINUITY Let $S$ and $T$ be topological spaces and let $V(p)$ denote the neighborhood filter of a point $p$. Let $f : S \to T$. Then: (a) $\mathcal{B} \to s$ iff $\mathcal{B} < V(s)$;
(b) $f$ is continuous at $s \in S$ iff $f(V(s)) < V(f(s))$;
(c) $f$ is continuous at $s \in S$ iff for any filterbase $\mathcal{B}$, $\mathcal{B} \to s \Rightarrow f(\mathcal{B}) \to f(s)$.

We note that subordination is *not* antisymmetric: $\mathcal{A} < \mathcal{B}$ and $\mathcal{B} < A$ does not imply that $\mathcal{A} = \mathcal{B}$. If $\mathcal{B}$ is such that for any filterbase $\mathcal{A}$, $\mathcal{A} < \mathcal{B}$ implies that $\mathcal{B} < \mathcal{A}$, then $\mathcal{B}$ is called a *maximal filterbase.* A straightforward Zorn's lemma argument shows that given any filterbase, there is a maximal filterbase subordinate to it. A useful characterization of maximal filterbases is the complement condition of Theorem 1.1.5.

**Theorem 1.1.5** COMPLEMENT CONDITION A filterbase $\mathcal{B}$ in a set $T$ is maximal iff for any subset $A$ of $T$ either $A$ contains an element of $\mathcal{B}$ or the complement $\complement A$ of $A$ contains an element of $\mathcal{B}$.

**Proof.** Let $\mathcal{B}$ be a maximal filterbase. Given a set $A$, either $B \subset A$ for some $B \in \mathcal{B}$ or $B \cap \complement A \neq \varnothing$ for all $B \in \mathcal{B}$. In the latter case, the trace $\mathcal{A} = \{B \cap \complement A : B \in \mathcal{B}\}$ of $\mathcal{B}$ on $\complement A$ is a filterbase which is clearly subordinate to $\mathcal{B}$. By the maximality of $\mathcal{B}$, $\mathcal{B} < \mathcal{A}$. Hence for any $B \in \mathcal{B}$, there must be a $D \in \mathcal{B}$ such that $D \subset B \cap \complement A \subset \complement A$.

Conversely, suppose that the condition holds on $\mathcal{B}$ and that $\mathcal{A} < \mathcal{B}$. Let $A \in \mathcal{A}$. By the condition there exists $B \in \mathcal{B}$ such that $B \subset A$ or $B \subset \complement A$. If $B \subset \complement A$ then, since $\mathcal{A} < \mathcal{B}$, there must be some $A' \in \mathcal{A}$ such that $A' \subset B$, hence the contradictory result that two elements of $\mathcal{A}$ do not meet. Therefore there must exist $B \in \mathcal{B}$ such that $B \subset A$, i.e., $\mathcal{B} < \mathcal{A}$. ■

**Theorem 1.1.6** CONVERGENCE IN PRODUCTS Let $\{T_s : s \in S\}$ be a family of topological spaces and let $T = \prod_{s \in S} T_s$ denote their Cartesian product endowed with the product topology. Let $pr_s$ denote the projection of $T$ onto $T_s$. For any filterbase $\mathcal{B}$ in $T$, $\mathcal{B} \to t$ iff $pr_s(\mathcal{B}) \to pr_s(t)$ for every $s \in S$.

**Proof.** Let $t = (t_s) \in T$. If $\mathcal{B} \to t$ then $pr_s(\mathcal{B}) \to pr_s(t)$ for every $s \in S$ by the continuity of projections and Theorem 1.1.4(c). Conversely, suppose that for all $s \in S$, $pr_s(\mathcal{B}) \to pr_s(t) = t_s$. We show that $\mathcal{B} \to t$. A basic neighborhood of $t$ is of the form $U = \cap_{i=1}^{n} pr_{s_i}^{-1}(V_{s_i})$ where $V_{s_i}$ is a neighborhood of $t_{s_i}$ in $T_{s_i}$, $i = 1, 2, \ldots, n$. Since each $pr_s(\mathcal{B}) \to t_s$, each filterbase $pr_s(\mathcal{B})$ is subordinate to the neighborhood filter $\mathcal{V}(t_s)$ in $T_s$. Therefore each $V_{s_i}$ contains some $pr_{s_i}(B_i)$, $B_i \in \mathcal{B}$. It follows that $B_i \subset pr_{s_i}^{-1}(V_{s_i})$ for $i = 1, 2, \ldots, n$. Since $\mathcal{B}$ is a filterbase, there is some $B \in \mathcal{B}$ such that $B \subset \cap_{i=1}^{n} B_i \subset \cap_{i=1}^{n} pr_{s_i}^{-1}(V_{s_i})$.
■

**Definition 1.1.7** DIRECTED SETS AND NETS A set $M$ together with a reflexive, transitive ordering relation $\leq$ such that finite subsets of $M$ have upper bounds in $M$ is called a *directed set*. A *net* in a topological space $T$ is a mapping $m \mapsto t_m$ from the directed set $M$ into $T$; we usually write $(t_m)$ or $\{t_m : m \in M\}$. We say that $t_m$ converges to $t$, $t_m \to t$, ($t$ is a *limit* of $(t_m)$) if for any neighborhood $V$ of $t$ there is an $m_0 \in M$ such that $t_n \in V$ for $n \geq m_0$. This is also expressed by saying that $t_m$ is *eventually* in any neighborhood of $t$. The set of limits of $(t_m)$ is denoted $\lim t_m$. We say that $t$ is a *cluster point* of $(t_m)$ if $t$ is frequently in any neighborhood $V$ of $t$, i.e., for any $m \in M$ there exists $n \geq m$ such that $t_n \in V$. The set of cluster points of $(t_m)$ is denoted $\mathrm{cl}(t_m)$. ■

In a first countable space, $t$ is a cluster point of a sequence $(t_n)$ iff a subsequence of $(t_n)$ converges to $t$.

As the proofs are widely available, we omit the details of showing that nets are topologically equivalent to filterbases in the sense that, given any net

$(t_m)$, there is a filterbase $\mathcal{B}$ with the same cluster points and limits as $(t_m)$ and vice-versa.

**Theorem 1.1.8** EQUIVALENCE OF NETS AND FILTERBASES Let $T$ be a topological space. (a) NET TO FILTERBASE Let $\{t_m : m \in M\}$ be a net in $T$. For each $m \in M$, let $B_m = \{t_n : n \geq m\}$. The collection $\mathcal{B} = \{B_m : m \in M\}$ is a filterbase for which $\operatorname{cl} \mathcal{B} = \operatorname{cl}(t_m)$ and $\lim \mathcal{B} = \lim t_m$.

(b) FILTERBASE TO NET Let $\mathcal{B}$ be a filterbase in $T$. The set $M$ of ordered pairs $(b, B)$ with $B \in \mathcal{B}$ and $b \in B$ ordered by taking $(a, A) \leq (b, B)$ if $B \subset A$ $(a \in A \in \mathcal{B})$ is a directed set and the map $M \to T$, $(b, B) \mapsto b$, is net with the same cluster points and limits as $\mathcal{B}$.

We will frequently use the net characterization of continuity of Theorem 1.1.9.

**Theorem 1.1.9** NETS AND CONTINUITY Let $S$ and $T$ be topological spaces. A map $f : S \to T$ is continuous at $s \in S$ iff for any net $s_m \to s$, $f(s_m) \to f(s)$.

### 1.1.3 Compactness

A topological space $T$ is *compact* if every open cover has a finite subcover. $T$ is:

(a) *locally compact* if each point in $T$ has a neighborhood whose closure is compact;

(b) *sequentially compact* if each sequence has a convergent subsequence;

(c) *countably compact* if every sequence has a cluster point (meaning that the sequence frequently enters any neighborhood of the point); this is equivalent to requiring that every countable open cover has a finite subcover or that every infinite subset $S \subset T$ has a limit point $t \in T$ (neighborhoods of $t$ contain infinitely many points of $S$). Note also that although sequential compactness implies countable compactness, the converse is false. Indeed, the Stone-Cech compactification $\beta\mathbf{N}$ of $\mathbf{N}$ is compact but not sequentially compact: there are no nontrivial convergent sequences in $\beta\mathbf{N}$ [Engelking 1977, p. 229].

(d) *$\sigma$-compact* if it can be written as a countable union of compact sets;

(e) *Lindelöf* if every open cover contains a countable subcover;

(f) *pseudocompact* if every real-valued continuous function on $T$ is bounded;

(g) *hemicompact* if there is a countable family $(K_n)$ of compact subsets of $T$ such that each compact subset of $T$ is contained in one of them;

(h) a subset of $T$ is *relatively compact* if its closure is compact.

If a set is both open and closed, we call it *clopen*. If $T$ has a base consisting of clopen sets, we call it *0-dimensional.* If there is a countable base for the topology, $T$ is *second countable*; if there is a countable neighborhood base at each point, then $T$ (or its topology) is *first countable.* If $T$ has a countable dense subset, $T$ is *separable.*

Our conventions about separation are:

(a) *Hausdorff* if distinct points are *separated by open sets,* i.e., for distinct $s, t \in T$, there exist disjoint neighborhoods $U$ and $V$ of $s$ and $t$, respectively.

(b) *regular* if **it is Hausdorff** and points and closed sets are separated by open sets, i.e., for a point $s \in T$ and a closed subset $F \subset T$ to which $s$ does not belong, there exist disjoint open subsets $U, V$ of $T$ with $s \in U$ and $F \subset V$. $T$ is *ultraregular* if points and closed sets are separated by clopen sets.

(c) *completely regular* if **it is Hausdorff** and for any point $t \in T$ and closed subset $F \subset T$ to which $t$ does not belong, there exists a continuous map $f : T \to [0,1]$ such that $f(t) = 0$ and $f(F) = \{1\}$.

(d) *normal* if **it is Hausdorff** and disjoint closed sets are separated by open sets. $T$ is *ultranormal* if disjoint closed sets are separated by clopen sets.

**Theorem 1.1.10** COMPACTNESS Let $T$ be a topological space.

(a) A cluster point of a maximal filterbase $\mathcal{M}$ is a limit.

$T$ is compact iff either of the following conditions holds:

(b) every filterbase has a cluster point (equivalently, each net has a convergent subnet);

(c) every maximal filterbase converges.

**Proof.** (a) Let $t$ be a cluster point of the maximal filterbase $\mathcal{M}$ and let $V$ be a neighborhood of $t$. Since $\mathcal{M}$ is maximal, there is some $M \in \mathcal{M}$ such that $M \subset V$ or $M \subset \complement V$. Since $t$ is a cluster point of $\mathcal{M}$, $M$ must meet $V$ so $M \subset V$.

(b) Let $T$ be compact and let $\mathcal{B}$ be a filterbase on $T$. The collection $\{\mathrm{cl}\, B : B \in \mathcal{B}\}$ is a collection of closed sets that satisfies the finite intersection condition. Hence, by the compactness of $T$ and Theorem 1.1.2(a), $\cap \{\mathrm{cl}\, B : B \in \mathcal{B}\} = \mathrm{cl}\, \mathcal{B} \neq \varnothing$. Conversely, let $\mathcal{B}$ be a collection of closed sets that satisfies the finite intersection condition so that they form a filter subbase. The filterbase $\mathcal{F}$ generated by $\mathcal{B}$ has a cluster point by hypothesis so $\varnothing \neq \cap \mathcal{F} \subset \cap \mathcal{B}$ and $T$ is compact. For the assertion about nets, we refer to Kelley 1976, p.136.

(c) Suppose that $T$ is compact and $\mathcal{M}$ is a maximal filterbase. $\mathcal{M}$ has a cluster point by (b) which is a limit by (a). Conversely, suppose that $\mathcal{B}$ is a filterbase. Using a Zorn's lemma argument, there exists a maximal filterbase

$\mathcal{A}$ subordinate to $\mathcal{B}$ [Dugundji 1966, p. 219]. By hypothesis, $\mathcal{A}$ has a limit $t$ so any neighborhood $V$ of $t$ must contain some $A \in \mathcal{A}$. For $B \in \mathcal{B}$ there exists $A' \in \mathcal{A}$ such that $A' \subset B$ so $\varnothing \neq A \cap A' \subset V \cap B$ and $t$ is a cluster point of $\mathcal{B}$. The compactness of $T$ now follows from (b). ■

# 1.2 VALUATION THEORY

In the body of the text all vector spaces are real or complex but we sometimes indicate what happens if the underlying field is not **R** or **C** but a field $K$ with an absolute value defined on it.

A map $|\cdot|$ of a field $K$ into $[0, \infty)$ such that $|ab| = |a|\,|b|$ and $|a+b| \leq |a|+|b|$ for all $a, b \in K$ and which is 0 only at 0 is called an *absolute value* or *valuation* on $K$. If, in addition, $|a+b| \leq \max(|a|, |b|)$ for all $a, b \in K$ (the *strong* or *ultrametric* triangle inequality), then the valuation is called *non-Archimedean* or an *ultravalue*. (The reason for the term *non-Archimedean* is that a valuation is non-Archimedean iff there is some real number $M$ such that $|n| \leq M$ for all "integers" $n$ in $K$; compare that to the "Archimedean" ordering of the real numbers.) The pair $(K, |\cdot|)$ is called a *valued field* or *ultravalued field* if $|\cdot|$ is an ultravalue. A valued field is always assumed to carry the topology induced by the metric $d(a, b) = |a - b|$, with respect to which, incidentally, addition and multiplication are continuous operations. If the valuation is non-Archimedean, then $K$ is 0-dimensional, as we now show.

**Theorem 1.2.1** NON-ARCHIMEDEAN PROPERTIES Let $(K, |\cdot|)$ be an ultravalued field. Then:

(a) If $|a| > |b|$, then $|a+b| = |a|$.

(b) Any point $b$ in a closed ball $C(a, r) = \{c \in K : |a - c| \leq r\}$, $r > 0$, is a center, i.e., $C(a, r) = C(b, r)$; the same is true for open balls.

(c) If two balls, open or closed, meet, then the one of smaller radius is contained in the one of larger radius.

(d) $K$ is 0-dimensional.

**Proof.** (a) If $|a| > |b|$, then $|a+b| \leq |a|$. If the latter inequality is strict then $|a| = |a+b-b| \leq \max(|a+b|, |b|) < |a|$; this contradiction yields the result.

(b) Suppose $b \in C(a, r)$. If $c \in C(b, r)$ then $|c - a| = |c - b + b - a| \leq \max(|c - b|, |b - a|) \leq r$, so $c \in C(a, r)$; hence $C(b, r) \subset C(a, r)$. The same argument establishes the reverse inclusion.

The result for open balls is proved in the same way.

(c) We prove only the statement for open balls. Suppose that $0 < r < t$ and that $c \in B(a, r) \cap B(b, t)$. By (b), then $B(c, r) \subset B(c, t) = B(b, t)$.

(d) We show that each open ball $B(a, r)$, $r > 0$, is closed. We effect this by showing that if $b \notin B(a, r)$ and $0 < t < r$, then $B(b, t) \cap B(a, r) = \varnothing$. Indeed, by (c), if $B(b, t) \cap B(a, r) \neq \varnothing$, then $B(b, t) \subset B(a, r)$ which contradicts $b \notin B(a, r)$. To see that closed balls $C(a, r)$ are clopen, note that $C(a, r) = \cup \{B(b, r) : b \in C(a, r)\}$ by (b). ■

**Example 1.2.2** VALUATIONS (a) [Bachman, 1964, p. 127] The usual absolute value $|\cdot|$ on **R** or **C** is a valuation which is *Archimedean*, i.e., not non-Archimedean. Any field $K$ with an Archimedean valuation is field-isomorphic to a subfield of **C** and the valuation on $K$, viewed as a subfield of **C**, is a power $|\cdot|^r$, $r > 0$, of the usual absolute value $|\cdot|$ on **C**.

(b) TRIVIAL VALUATION On any field $K$ the map sending each nonzero element into 1 and 0 into 0 is a non-Archimedean valuation called the *trivial valuation.* It induces the discrete topology. Its exclusion is marked by an expression such as "let $K$ be a nontrivially valued field."

(c) $p$-ADIC VALUATION Let $p$ be a positive prime. Any rational number $x$ can be written in the form $p^k (a/b)$ $(a, b, k \in \mathbf{Z})$ where $p$ is not a factor of $a$ or $b$. The *p-adic valuation* $|x|_p$ of $x$ is $p^{-k}$ and is non-Archimedean. The metric completion $\mathbf{Q}_p$. of the rationals $\left(\mathbf{Q}, |\cdot|_p\right)$ is called the *p-adic numbers.* ■

For more on valuation theory, see Bachman [1964] and Mahler [1973].

# 1.3 ALGEBRA

Except for a few exercises, when we say *vector space* (or *linear space*) we mean real or complex vector space. When we say *subspace* of a vector space, we mean *linear* subspace, i.e., one which is closed under addition and scalar multiplication. A linear map $A : X \to Y$ of a vector space $X$ into a vector space $Y$ is such that for all $a, b \in \mathbf{F} = \mathbf{R}$ or **C** and $x, y \in X$, $A(ax + by) = aAx + bAy$; if $Y = \mathbf{F}$, we call $A$ a linear functional. A *linear* or *vector isomorphism* is a 1-1 linear map—it does not have to be onto; if $A$ is onto, then $X$ and $Y$ are (*linearly*) *isomorphic.*

We deal with vector spaces of **F**-valued functions $x, y, \ldots$ on some set $T$ with respect to the *pointwise operations*, i.e., for each $t \in T$, $(x + y)(t) = x(t) + y(t)$ and $(ax)(t) = ax(t)$ for any scalar $a$.

The (*linear*) *span* of a set $E$ of vectors is denoted $[E]$. To say that $E$ *spans* the vector space $X$ means that $[E] = X$. A *Hamel base* for a vector space $X$ is a maximal (with respect to set inclusion) linearly independent set $B$ of vectors (or a minimal spanning set). A Hamel base (or *basis)* must exist in any vector space $X$ (Zorn's lemma) and necessarily spans the space. The dimension of a vector space is the cardinality of any Hamel base. The *algebraic dual* $X^*$ of $X$ is the linear space (pointwise operations) of all linear functionals on $X$. Concerning the size of $X^*$:

**Theorem 1.3.1** DIMENSION OF $X^*$ [Jacobson, 1953, p. 247]. If the vector space $X$ is finite-dimensional, then $\dim X = \dim X^*$. If $X$ is infinite-dimensional, $\dim X = b$ say, then $\dim X^* = c^b$, where $c$ denotes the cardinality of $\mathbf{F} = \mathbf{R}$ or **C**, i.e., the power of the continuum. Hence, for infinite-dimensional spaces $X$, $\dim X^* > \dim X$.

The space described in Example 1.3.2 is of denumerable dimension and is useful in some counterexamples in the text.

**Example 1.3.2** Space $\varphi$ of Finite Sequences The linear (pointwise operations) space of sequences $(a_n)$, $a_n \in \mathbf{F}$, which are 0 eventually is denoted $\varphi$. We often call it the space of *finite sequences*. It is isomorphic to the space $\mathbf{F}[x]$ of polynomials in one indeterminate $x$ with coefficients from $\mathbf{F}$. That it is of denumerable dimension can be seen by considering the Hamel base of *standard basis vectors* $\{e_n : n \in \mathbf{N}\}$ where $e_n$ is the sequence with 1 at the $n$th position and 0's everywhere else. ■

**Definition 1.3.3** Algebra Let $X$ be a vector space over $\mathbf{F}$. If there is a multiplication (indicated by juxtaposition) defined between elements of $X$ which is associative and distributive ($x(y+z) = xy + xz$) then $X$ is a an *algebra*; if $xy = yx$ for all $x, y \in X$, then $X$ is *commutative*. An element $e \in X$ such that $ex = xe = x$ for all $x \in X$ is an *identity* for $X$. If $X$ and $Y$ are algebras over $\mathbf{F}$ and $A : X \to Y$ is a linear map for which, for all $x, y \in X$, $A(xy) = AxAy$ then $A$ is an *algebra homomorphism* or just a *homomorphism* or a *multiplicative linear map;* if $A$ is 1-1, then it is an *algebra isomorphism*; if $A$ is bijective then $X$ and $Y$ are *algebra isomorphic.* ■

An important algebra is the commutative algebra $C(T, \mathbf{F})$ of continuous maps of the completely regular space $T$ into $\mathbf{F}$; the function $e : C(T, \mathbf{F}) \to \mathbf{F}$, $t \mapsto 1$, is an identity. There is no algebraic reason to consider apparently more general topological spaces because of Theorem 1.3.4; the proof can be found in Gillman and Jerison 1960 [3.9].

**Theorem 1.3.4** For every topological space $S$ there exists a completely regular space $T$ and a continuous mapping $h : S \to T$ such that for every $x \in C(S, \mathbf{F})$ the composition map $x \mapsto x \circ h$ is an algebra isomorphism onto $C(T, \mathbf{F})$.

## 1.4 LINEAR FUNCTIONALS

We collect some basic facts about linear functionals here. The subspace $N(f) = f^{-1}(0)$ of vectors on which a linear functional $f$ vanishes is called the *null space* or *kernel* of $f$. Theorem 1.4.1(b) shows that null spaces are generally very large, dimensionally speaking.

**Theorem 1.4.1** The Null Space Let $X$ be a vector space over $\mathbf{F} = \mathbf{R}$ or $\mathbf{C}$ and let $f$ be a nontrivial linear functional on $X$ with null space $N$. Then:

(a) $f$ is surjective;

(b) $X/N$ is linearly isomorphic to $\mathbf{F}$; hence the *codimension* $\dim X/N$ of $N$ is 1;

(c) for any $x \notin N$, $X = N \oplus \mathbf{F}x = \{n + ax \in X : n \in N, a \in \mathbf{F}\}$;

(d) for any scalar $a$ and any $x \in H = f^{-1}(a)$, $N = H - x = \{y - x : y \in H\}$;

(e) for any nontrivial linear functional $g$ on $X$, $N(f) \subset N(g)$ iff there is some scalar $a$ such that $g = af$ and $N(f) = N(g)$;

(f) Let $f$ be a linear functional on $X$. If $g_1, \ldots, g_n$ are nontrivial linear functionals on $X$, then $\cap_{i=1}^n N(g_i) \subset N(f)$ iff there are scalars $a_1, \ldots, a_n$ such that $f = \sum_{i=1}^n a_i g_i$. Hence if $\cap_{i=1}^n N(g_i) = \{0\}$ then, since $\{0\} \subset N(f)$ for any linear functional $f$, $\{g_1, \ldots, g_n\}$ spans the algebraic dual $X^*$.

**Proof.** (a) Choose a vector $x$ such that $f(x) = 1$. Then, for any scalar $a$, $f(ax) = a$.

(b) The map $x + N \mapsto f(x)$ is an isomorphism of $X/N$ onto $\mathbf{F}$.

(c) For any $x \notin N$ and $y \in X$, $y - (f(y)/f(x))x \in N$.

(d) Suppose $x \in H = f^{-1}(1)$. Then $H - x \subset N$. Conversely, if $w \in N$, then $w + x \in H$ and $w = w + x - x$.

(e) Suppose $N(f) \subset N(g)$. If $N(f)$ is a proper subset of $N(g)$, there is some $x \in N(g)$ such that $x \notin N(f)$. By (c), therefore, $X = N(f) + \mathbf{F}x \subset N(g)$ and $g$ would be 0. Hence $N(f) = N(g) = N$. For $x \notin N$ and $y \in X = N + \mathbf{F}x$, there exist $a \in \mathbf{F}$ and $n \in N$ such that $y = n + ax$; hence $g(y) = ag(x)$ and $f(y) = af(x)$. Therefore $g(y) = (f(y)/f(x))g(x)$ and it follows that $g = (g(x)/f(x))f$. The sufficiency of the condition is obvious.

(f) If $f = \sum_{i=1}^n a_i g_i$ then clearly $\cap_{i=1}^n N(g_i) \subset N(f)$. We prove the converse by induction on $n$. The result for $n = 1$ is (e) so we now assume that the result holds for $n - 1$ and suppose that $\cap_{i=1}^n N(g_i) \subset N(f)$. There is no loss of generality in assuming that $g_1, \ldots, g_n$ are linearly independent so we assume that no $g_k$ can be written as a linear combination of the remaining ones; hence, by the induction hypothesis, for each $1 \leq k \leq n$, $\cap_{i \neq k} N(g_i) \not\subset N(g_k)$ and there exist $x_k \in X$ $(1 \leq k \leq n)$ such that $g_j(x_k) = \delta_{jk}$, $1 \leq j \leq n$ (Kronecker delta). For any $x \in X$, each $g_j$ vanishes on $x - \sum_{k=1}^n g_k(x)x_k$, i.e., $x - \sum_{k=1}^n g_k(x)x_k \in \cap_{i=1}^n N(g_i) \subset N(f)$. Therefore $f = \sum_{j=1}^n f(x_j)g_j$. ■

On any complex vector space $X$, there is an intimate connection between the real and imaginary parts of a linear functional $f$ on $X$, namely that

$$\operatorname{Re} f(x) = \operatorname{Im} f(ix) \quad (x \in X)$$

Although usually credited to F. Murray [1936], H. Löwig discovered this in 1934. Any complex vector space $X$ can, of course, be viewed as a real one. A linear map of $X$ into $\mathbf{R}$ is then called a *real* linear functional or $\mathbf{R}$-*linear* functional on $X$. For emphasis we often refer to linear functionals $f$ on $X$ as *complex* linear functionals. For each $x \in X$, $f(x) = \operatorname{Re} f(x) + i \operatorname{Im} f(x)$ and $\operatorname{Re} f$ and $\operatorname{Im} f$ are each $\mathbf{R}$-linear functionals as is trivial to verify. A surprising fact about $\operatorname{Re} f$ and $\operatorname{Im} f$ is that $\operatorname{Re} f(ix) = -\operatorname{Im} f(x)$ and $\operatorname{Re} f(x) = \operatorname{Im} f(ix)$ for every $x \in X$ since $f(ix) = \operatorname{Re} f(ix) + i \operatorname{Im} f(ix) = if(x) = i \operatorname{Re} f(x) - \operatorname{Im} f(x)$. In summary:

**Theorem 1.4.2** REAL VERSUS COMPLEX Let $X$ be a complex vector space, let $X^*$ denote the collection of all linear functionals on $X$ and let $X_R^*$ denote the class of real linear functionals on $X$. Then for any $f \in X^*$ and any $x \in X$, $\operatorname{Im} f(x) = -\operatorname{Re} f(ix) =$ and the map $f \mapsto \operatorname{Re} f$ establishes the following 1-1 correspondence between $X^*$ and $X_R^*$:

$$\begin{array}{ccc} X^* & \leftrightarrow & X_R^* \\ f(x) = \operatorname{Re} f(x) - i \operatorname{Re} f(ix) & \leftrightarrow & \operatorname{Re} f(x) \end{array}$$

**Proof.** For $f, g \in X^*$, if $\operatorname{Re} f = \operatorname{Re} g$ then clearly $f = g$. To see that the correspondence is surjective, we show that if $r \in X_R^*$, then (its preimage) $f(x) = r(x) - ir(ix)$ $(x \in X)$ is a complex linear functional. Since $f$ is obviously a real linear functional, it only remains to show that $f(ix) = if(x)$. To that end, note that $f(ix) = r(ix) - ir(-x) = r(ix) + ir(x) = i(r(x) - ir(ix)) = if(x)$. ■

# 1.5 HYPERPLANES

A *maximal* subspace $M$ of a vector space $X$ is a proper subspace not properly contained in any proper subspace of $X$. Planes through the origin in $\mathbf{R}^3$, for example, are maximal subspaces of $\mathbf{R}^3$. As shown in Theorem 1.5.1, maximality of $M$ is equivalent to $M$ missing the dimension of $X$ by just 1 in the sense that $\dim X/M = 1$. A *linear variety* (*affine subspace, linear manifold*) is a translate $x + M$ of a subspace $M$. A *hyperplane* is a translate of a maximal subspace.

In Sec. 1.4 we saw the close connection between real and complex linear functionals. The development of analogous "geometric" statements about subspaces is the subject of this section.

**Theorem 1.5.1** MAXIMAL SUBSPACES AND HYPERPLANES Let $M$ be a subspace of a vector space $X$ over $\mathbf{F} = \mathbf{R}$ or $\mathbf{C}$. Then:

(a) $M$ is maximal iff $\dim X/M = 1$;

(b) $M$ is maximal iff $M = N(f)$, the null space of some nontrivial linear functional $f$ on $X$;

(c) a linear variety $H$ is a hyperplane iff $H = \{x \in X : f(x) = a\}$ for some nontrivial linear functional $f$ on $X$ and scalar $a$. For $a \neq 0$ we may replace $f$ by $a^{-1}f$ and say that $H = \{x \in X : f(x) = 1\}$.

**Proof.** (a) If $\dim X/M \geq 2$, there are $x, y \in X$ such that $x + M$ and $y + M$ are linearly independent in $X/M$. Moreover, letting $[\ ]$ denote linear span, we have the proper inclusions $M \subsetneq [M, x] \subsetneq [M, x, y]$ and $M$ is not maximal. Conversely, if $M$ is not maximal, there is a proper subspace $N$ of $X$ which contains $M$ properly; hence $\dim X/M > \dim N/M > 1$.

(b) Let $f$ be a linear functional and let $M = N(f)$. By Theorem 1.4.1(b), $\dim X/M = 1$; hence $M$ is maximal by (a). Conversely, if $M$ is maximal and

$x \notin M$, then $[M, x] = M + \mathbf{F}x = X$. Since $M \cap \mathbf{F}x = \{0\}$, if $y = m + ax$ ($m \in M$, $a \in \mathbf{F}$), then $m$ and $a$ are unique. Defining $f(y) = f(m + ax) = a$ for each $y \in X$, $f$ is a linear functional with null space $M$.

(c) If $H$ is a hyperplane, then $H = x + M$ for some $x \in X$ and maximal subspace $M \subset X$. Since $M$ is maximal, there is a nontrivial linear functional $f$ on $X$ such that $M = N(f)$ by (b). Let $f(x) = a$. Clearly $H \subset f^{-1}(a)$. As to the reverse inclusion, if $f(y) = a$ then $y - x \in M$ and $y \in H$, i.e., $H = \{y \in X : f(y) = a\}$.

Conversely, suppose $H = \{x \in X : f(x) = a\}$ for some nontrivial linear functional $f$ and scalar $a$. By Theorem 1.4.1(d), for any $x \in H$, $H - x$ is the null space of $f$ by Theorem 1.4.1(d); $H - x$ is maximal by (b). ■

If $X$ is a complex vector space, we say that $M \subset X$ is an **R**-*subspace* if it is closed under addition and multiplication by real scalars. An **R**-*variety* is a translate of an **R**-subspace; an **R**-*hyperplane*, a translate of a maximal **R**-subspace. As observed in Sec. 1.4, for any complex linear functional $f$ on $X$, $f(x) = r(x) - ir(ix)$ ($x \in X$) where $r = \operatorname{Re} f$ is a real linear functional on $X$. By Theorem 1.5.1(b), the null space $N(r)$ of $r$ is a maximal subspace of the real linear space $X$—in other words, a maximal **R**-subspace. As follows from Theorem 1.5.2(a) below, $N(f) = N(r) \cap iN(r)$.

**Theorem 1.5.2** Real and Complex Hyperplanes Let $X$ be a complex vector space. Then

(a) a subspace $M$ of $X$ is maximal iff there is a maximal **R**-subspace $N$ such that $M = N \cap iN$;

(b) a linear variety $H$ is a hyperplane iff there is a maximal **R**-subspace $M$ and $x, y \in X$ such that $H = (x + M) \cap (y + iM)$.

**Proof.** (a) If $M$ is a maximal subspace of $X$, then there is a nontrivial linear functional $f$ on $X$ such that $M = N(f)$ [Theorem 1.5.1(b)]. Moreover, with $r = \operatorname{Re} f$, $f(x) = r(x) - ir(ix)$ for every $x \in X$ [Theorem 1.4.2]. Hence

$$\begin{aligned} x \in M \quad &\Leftrightarrow \quad f(x) = 0 \Leftrightarrow r(x) = 0 \text{ and } r(ix) = 0 \\ &\Leftrightarrow \quad x \in N(r) \cap (-iN(r)) = N(r) \cap iN(r) \end{aligned}$$

Conversely, if $N$ is a maximal **R**-subspace, there is a real linear functional $r$ whose null space is $N$. With $f(x) = r(x) - ir(ix)$ ($x \in X$), $N \cap iN = N(f)$.

(b) If $H$ is a hyperplane, then $H = x + M$ for some $x \in X$ and maximal subspace $M$. There exists a maximal **R**-subspace $N$ such that $M = N \cap iN$ by (a). Thus $H = x + N \cap iN = (x + N) \cap (x + iN)$.

Conversely, if $N$ is a maximal **R**-subspace and $x, y \in X$, we want to show that $(x + N) \cap (y + iN) = H$ is a hyperplane. To do this, it suffices to show that there is a complex linear functional $f$ on $X$ and $a \in \mathbf{C}$ such that $H = \{w \in X : f(w) = a\}$. Since $N$ is a maximal **R**-subspace, there is a real linear functional $r$ on $X$ such that $N = N(r)$ by Theorem 1.5.1(b).

Let $f$ be the associated complex linear functional of Theorem 1.4.2, namely, $f(w) = r(w) - ir(iw)$ $(w \in X)$. With $a = r(x) - ir(iy)$, then

$$\begin{aligned} f(w) = a &\Leftrightarrow r(w) = r(x) \text{ and } r(iw) = r(iy) \\ &\Leftrightarrow w - x \in N \text{ and } w - y \in -iN = iN \\ &\Leftrightarrow w \in (x + N) \cap (y + iN) \quad \blacksquare \end{aligned}$$

In summary, we have the 1-1 correspondences of the following diagram where $N$ denotes null space.

| $\operatorname{Re} f(x) - i \operatorname{Re} f(ix) = f(x)$<br>complex linear functional | $\leftrightarrow$ | $N(f) = N(\operatorname{Re} f) \cap iN(\operatorname{Re} f)$<br>maximal subspace |
|---|---|---|
| $\updownarrow$ | | $\updownarrow$ |
| real linear functional<br>$\operatorname{Re} f(x)$ | $\leftrightarrow$ | maximal $\mathbf{R}$-subspace<br>$N(\operatorname{Re} f)$ |

## 1.6 MEASURE THEORY

A nonempty collection $\mathcal{S}$ of subsets of a set $S$ is an *algebra* (of sets) if $\mathcal{S}$ is closed with respect to the formation of finite unions and contains the complement of each of its members. If, in addition, it is closed with respect to the formation of countable unions, then $\mathcal{S}$ is a *$\sigma$-algebra* . A *measure* $m$ is a nonnegative function defined on a $\sigma$-algebra which is countably additive and vanishes on the null set. We call $(S, \mathcal{S}, m)$ a *measure space.* A property which holds at each point of $S$ except a set of a measure 0 is said to hold *almost everywhere* (a.e.). If there is some $s \in S$ such that $m(E) = 1$ if $s \in E$ and $m(E) = 0$ if $s \notin E$, then $m$ is called a *point mass.*

If $S$ is a locally compact Hausdorff space, the smallest $\sigma$-algebra containing the compact $G_\delta$-sets (countable intersections of open sets) is called the *Baire sets*. A measure $m$ on the Baire sets which is finite on compact $G_\delta$-sets is called a *Baire measure*. A Baire measure is *regular* in the sense that if $B$ is a Baire set,

$$\begin{aligned} m(B) &= \sup\{m(A) : B \supset A,\ A \text{ compact } G_\delta\} \\ &= \inf\{m(U) : B \subset U,\ U \text{ open Baire set}\} \end{aligned}$$

[Halmos, 1974a, p. 228]. The *Borel sets* are the $\sigma$-algebra generated by the compact (or open) subsets of $S$. A *Borel measure* $m$ is a measure defined on the Borel sets which is finite on compact sets. A Borel measure is *regular* if, given any Borel set $E$,

$$m(E) = \sup\{m(K) : E \supset K,\ K \text{ compact}\} = \inf\{m(U) : E \subset U,\ U \text{ open}\}$$

Unlike Baire measures, Borel measures do not have to be regular. A regular Borel measure $m$ on $S$ such that $m(S) = 1$ is called a *probability measure.* For more on measure theory, see Halmos [1974a] and Dunford and Schwartz [1958].

## 1.7 NORMED SPACES

Let $X$ be a vector space over a valued field $(K, |\cdot|)$. A *seminorm* $p$ on $X$ is a nonnegative map of $X$ such that $p(ax) = |a|p(x)$ and $p(x+y) \le p(x) + p(y)$ for all $x, y \in K$ and all $a \in K$. $(X, p)$ is called a *seminormed space*. If $p(x) = 0$ only when $x = 0$, then $p$ is a *norm* and $(X, p)$ is a normed space. The symbol $\|\cdot\|$ is reserved for norms. If $K$ is ultravalued and the norm satisfies $\|x+y\| \le \max[\|x\|, \|y\|]$ for all $x, y \in X$, then $\|\cdot\|$ is called a *non-Archimedean norm* or an *ultranorm* and $(X, \|\cdot\|)$ is called a *non-Archimedean* normed space or *ultranormed* space. If we say simply *normed space* we always mean that $K = \mathbf{R}$ or $\mathbf{C}$. Seminormed spaces $(X, p)$ are always assumed to carry the topology determined by the pseudometric $d(x, y) = p(x - y)$. It is trivial to verify that $N = p^{-1}(0)$ is a linear subspace. If a normed space is complete in the sense that every Cauchy sequence converges, then $X$ is called a *Banach space*. A linear map $A : X \to Y$ between normed spaces $X$ and $Y$ such that $\|Ax\| = \|x\|$ for each $x \in X$ is called a *linear isometry* or *norm isomorphism*; If $A$ is bijective, then $X$ and $Y$ are *linearly isometric* or *norm isomorphic* and $X$ and $Y$ are *norm isomorphs* of each other. Although we do not do it, the reader should be aware that in Banach space theory it is commonly said that two Banach spaces $X$ and $Y$ are "isomorphic" when they are linearly *homeomorphic* and that *isometry* usually means *linear* isometry.

Except in a few exercises, "vector space" means real or complex vector space from this point on.

Two norms $\|\cdot\|$ and $\|\cdot\|^*$ on a vector space $X$ are *equivalent* if they determine the same topology.

**Theorem 1.7.1** EQUIVALENT NORMS The norms $\|\cdot\|$ and $\|\cdot\|^*$ on the vector space $X$ are equivalent iff there exist positive numbers $a$ and $b$ such that $a\|x\| \le \|x\|^* \le b\|x\|$ for every $x$ in $X$.

**Proof.** Let $U$ and $U^*$ denote the closed solid unit balls $\{x \in X : \|x\| \le 1\}$, etc., determined by $\|\cdot\|$ and $\|\cdot\|^*$, respectively. If the condition is satisfied and $x \in U$, then $x/b \in U^*$, i.e., $(1/b)U \subset U^*$, so $U^*$ is a $\|\cdot\|$-neighborhood of 0. Similarly, $U$ is a $\|\cdot\|^*$-neighborhood of 0. Conversely, suppose that $\|\cdot\|$ and $\|\cdot\|^*$ determine the same topology. Then $U^*$ is a $\|\cdot\|$-neighborhood of 0, i.e., for some $a > 0$, $aU \subset U^*$. Hence, for any nonzero $x$, $\|ax/\|x\|\|^* \le 1$ or $\|x\|^* \le (1/a)\|x\|$. The other inequality is obtained similarly. ■

A result subsumed by Theorem 4.10.3(b) is the following:

**Theorem 1.7.2** All norms are equivalent on a finite-dimensional space.

If $X$ is a normed space, its *continuous* or *topological dual* (or *conjugate*) is the linear space $X'$ (pointwise operations) of all continuous linear functionals on $X$. $X'$ is a Banach space with respect to the *canonical norm*

$$\|f\| = \sup\{|f(x)| : \|x\| \le 1\}$$

Whenever we speak of $X'$ as a normed space, it is always with respect to this norm. Likewise, when we consider $X'' = (X')'$, we assume it carries its norm topology and call it the *bidual* of $X$. Every element $x \in X$ determines a continuous linear functional on $X'$, namely $Jx : X' \to \mathbf{F}$, $f \mapsto f(x)$. The map $J : X \to X''$, $x \mapsto Jx$, is called the *canonical embedding* and is a linear isometry of $X$ into $X''$ [Theorem 8.4.5(b)]. If $J(X) = X''$, then $X$ is said to be *reflexive*, a topic we take up at length in Chapter 15.

**Notation.** For any normed space $X$, $U(X)$ denotes the (closed, solid) *unit ball* $\{x \in X : \|x\| \leq 1\}$ of $X$; $SU(X)$ denotes $\{x \in X : \|x\| = 1\}$, the *surface of the unit ball.* ■

Some Banach spaces (see Dunford and Schwartz 1958 for the verifications) are listed below.

**Example 1.7.3** $\ell_p(n)$, $1 \leq p < \infty$, $n \in \mathbf{N}$, denotes the Banach space $\mathbf{F}^n$ of $n$-tuples $(x(i))$ normed by the *p-norm*

$$\|(x(i))\|_p = \left[\sum_{i=1}^{n} |x(i)|^p\right]^{1/p}$$

We call $\|\cdot\|_2$ the *Euclidean norm.* To restrict consideration to real $n$-tuples $(x(i))$, we write $\ell_p^{\mathbf{R}}(n)$. ■

**Example 1.7.4** $\ell_p$, $1 \leq p < \infty$, is the Banach space of all $p$th power summable sequences $x = (x(n))$ (i.e. $\sum_{n\in\mathbf{N}} |x(n)|^p < \infty$) of elements of $\mathbf{F}$, normed by $\|x\|_p = (\sum_{n\in\mathbf{N}} |x(n)|^p)^{1/p} < \infty$. We denote real $p$th power summable sequences by $\ell_p^{\mathbf{R}}$. ■

**Example 1.7.5** BOUNDED FUNCTIONS AND SEQUENCES $\ell_\infty(T)$ denotes the Banach space of all bounded $\mathbf{F}$-valued function $x$ on a set $T$ with *sup norm*: $\|x\|_\infty = \sup_{t\in T} |x(t)|$; $\ell_\infty^{\mathbf{R}}(T)$ or $\ell_\infty(T, \mathbf{R})$ denotes bounded real-valued functions. With $T = \{1, \ldots, n\}$ or $\mathbf{N}$ we write $\ell_\infty(n)$ or $\ell_\infty$, respectively, the spaces of $n$-tuples or bounded scalar sequences. ■

**Example 1.7.6** CONVERGENT AND NULL SEQUENCES The symbol $(c, \|\cdot\|_\infty) = c$ represents the Banach subspace of $\ell_\infty$ [Example 1.7.5] of all convergent sequences $(x(n))$ from $\mathbf{F}$; $c_0 = \{(x(n)) \in c : x(n) \to 0\}$ is the Banach subspace of all null sequences. ■

**Example 1.7.7** CONTINUOUS FUNCTIONS $C([a,b], \mathbf{F}, \|\cdot\|_\infty)$ or just $C[a,b]$ stands for the linear space of all continuous $\mathbf{F}$-valued maps of the closed interval $[a,b]$. It becomes a Banach space when endowed with the sup norm: $\|x\|_\infty = \sup|x[a,b]|$ for $x \in C[a,b]$. $C(T, \mathbf{F})$ denotes the linear space of continuous functions on the topological space $T$. If $T$ is a compact Hausdorff space and $C(T,\mathbf{F})$ is normed by $\|x\|_\infty = \sup|x(T)|$, it is a Banach space; we denote it by $C(T,\mathbf{F},\|\cdot\|_\infty)$. For an arbitrary topological space $T$, if $x \in C(T,\mathbf{F})$ is such that for any $r > 0$ there is a compact subset $K$ of $T$ such that $|x(t)| \leq r$ for all $t \in \complement K$, we say that $x$ *vanishes at infinity.* We denote the linear subspace of $C(T,\mathbf{F})$ of such functions by $C_\infty(T,\mathbf{F})$. With $\|\cdot\|_\infty$ as above, $C_\infty(T,\mathbf{F},\|\cdot\|_\infty)$ is a Banach space. For arbitrary $T$, $C_b(T,\mathbf{F},\|\cdot\|_\infty)$ denotes the Banach space of *bounded* continuous functions on $T$.

**Example 1.7.8** $L_p(E,m)$, $1 \leq p \leq \infty$, AND $L_\infty(E,m)$ Let $\mathcal{S}$ be a $\sigma$-algebra of subsets of a set $T$, $m$, a measure on $\mathcal{S}$, $E$ a fixed member of $\mathcal{S}$, and $1 \leq p < \infty$; $L_p(E,m)\hat{}$ denotes the collection of $\mathbf{F}$-valued measurable functions $x$ on $E$ such that $\int_E |x|^p \; dm < \infty$ seminormed by

$$q_p(x) = \left(\int_E |x|^p \; dm\right)^{1/p} < \infty$$

The collection $N = \{x \in L_p(E,m)\hat{} : q_p(x) = 0\}$ is a closed linear subspace of $L_p(E,m)\hat{}$ and $L_p(E,m)\hat{}/N = L_p(E,m)$ is a Banach space when normed by taking $\|(x+N)\|_p = \inf\{q_p(x+n) : n \in N\}$. In other words, we identify functions in $L_p(E,m)$ that are equal almost everywhere. As in Examples 1.7.3 and 1.7.4, that $q_p$ satisfies the triangle inequality follows from the Minkowski inequality, namely, for $x, y \in L_p(E,m)\hat{}$

$$\left(\int_E |x+y|^p \; dm\right)^{1/p} \leq \left(\int_E |x|^p \; dm\right)^{1/p} + \left(\int_E |y|^p \; dm\right)^{1/p}$$

and, for $p > 1$, equality holds iff $y$ is a scalar multiple of $x$ [Köthe 1983, pp. 135–140].

$L_\infty(E,m)\hat{}$ denotes the Banach space of *essentially bounded* measurable $\mathbf{F}$-valued functions—those functions $x$ for which there exists $M$ such that $|x(t)| \leq M$ almost everywhere on $E$, normed by the *essential sup:*

$$q_\infty(x) = \inf\{M : |x(t)| \leq M \text{ a.e. on } E\}$$

For $N = \{x \in L_\infty(E,m)\hat{} : q_\infty(x) = 0\}$, we take $L_\infty(E,m) = L_\infty(E,m)\hat{}/N$, normed by $\|(x+N)\|_\infty = \inf\{q_\infty(x+n) : n \in N\}$, i.e., we identify functions that are equal almost everywhere. Some important special cases are $L_p(T,m)$, $1 \leq p \leq \infty$, where $T$ is a subset of $\mathbf{R}$ such as $\mathbf{R}$ itself or a closed interval $[a,b]$ and $m$ is Lebesgue measure. In these cases we write simply $L_p(\mathbf{R})$ or $L_p[a,b]$, respectively.

**Example 1.7.9** $BV[a,b]$ The *variation* $v(x)$ of a function $x$ mapping $[a,b]$ into $\mathbf{F}$ is $\sup\{\sum_{i=0}^n |x(t_i) - x(t_{i-1})| : n \in \mathbf{N},\, a = t_0 < t_t < ... < t_n = b\}$ with the possibility that $v(x) = \infty$. $BV[a,b]$ denotes the functions of *bounded variation*, those $x$ for which $v(x) < \infty$. As norm on $BV[a,b]$ we take $\|x\| = v(x) + |x(a)|$ [without the $|x(a)|$ term, it would only be a seminorm, as nonzero constant functions would then have "norm" 0]. The vector subspace $NBV[a,b]$ of *normalized* functions of bounded variation consists of those $x$ for which $x(a) = 0$ and are continuous from the right everywhere on $[a,b]$. ■

A list of some Banach spaces and (norm isomorphs of) their continuous duals is given in Table 1.7.1. We refer to Dunford and Schwartz [1958] for the proofs not done later in the book. We do not specify a norm for $\mathbf{F}^n$ because all norms are equivalent on a finite-dimensional space [Theorem 4.10.3(b)].

| $X$ | $X'$ |
|---|---|
| $\mathbf{F}^n$ | $\mathbf{F}^n$ |
| $\ell_p,\ 1<p<\infty$ | $\ell_q,\ 1/p+1/q=1$ [Example 15.7.2] |
| $\ell_1$ | $\ell_\infty$ [Example 15.7.3] |
| $L_p(E,m),\ 1<p<\infty$ | $L_q(E,m),\ 1/p+1/q=1$ |
| $L_1(E,m)$ | $L_\infty(E,m)$ |
| $c$ | $\ell_1$ |
| $c_0$ | $\ell_1$ [Example 15.7.3] |
| $C[a,b]$ | $NBV[a,b]$ [Theorem 7.4.4] |

**Table 1.7.1** Continuous Duals of Banach Spaces

## 1.7.1 Inner Product Spaces

**Definition 1.7.10** INNER PRODUCT An *inner product* on a vector space $X$ is a map $(x,y) \mapsto \langle x,y\rangle$ of the Cartesian product $X \times X$ into $\mathbf{F}$ such that for all $x, y$, and $z$ in $X$ and all $a, b$ in $\mathbf{F}$ :

(a) $\overline{\langle x,y\rangle} = \langle y,x\rangle$, where the bar denotes complex conjugate;
(b) $\langle ax+by, z\rangle = a\langle x,z\rangle + b\langle y,z\rangle$ ; and
(c) $\langle x,x\rangle \geq 0$ and $= 0$ iff $x = 0$.

We call $(X, \langle\cdot,\cdot\rangle)$ an *inner product space.* If $\langle x,y\rangle = 0$, we say that $x$ and $y$ are *orthogonal* and we write $x \perp y$. If $x$ is orthogonal to each vector in a subset $S \subset X$, we write $x \perp S$. An *orthogonal set* is a subset of an inner product space consisting of mutually orthogonal vectors. ■

If $\langle\cdot,\cdot\rangle$ is an inner product on $X$, then $x \mapsto \langle x,x\rangle^{1/2}$ $(x \in X)$ is a norm. Whenever we speak of an inner product space as a normed space, this is the norm we mean. Although every inner product space is a normed space, the converse is false: A norm derived from an inner product satisfies the *parallelogram law*

$$||x+y||^2 + ||x-y||^2 = 2||x||^2 + 2||y||^2 \quad (x, y \in X)$$

The Banach space $L_1[0,1]$ of Example 1.7.8, violates the parallelogram law: The characteristic functions $x$ and $y$ of $[0,1/2]$ and $[1/2,1]$ satisfy $||x+y||_1^2 = ||x-y||_1^2 = 1$ while $2||x||_1^2 + 2||y||_1^2 = 2(1/4) + 2(1/4) = 1$. Therefore, the $L_1$-norm $\|\cdot\|_1$ is not derivable from an inner product.

**Definition 1.7.11** HILBERT SPACE An inner product space which is complete with respect to the inner product-derived norm is called a *Hilbert space.* ■

The spaces $\mathbf{F}^n$, $\ell_2$, and $L_2(E,m)$ are Hilbert spaces with inner products given by, respectively, for $x = (x(j))$, $y = (y(j))$ from $\mathbf{F}^n$ or $\ell_2$ or $x, y \in L_2(E,m)$,

$$\langle (x(j)), (y(j)) \rangle = \sum_{j=1}^{n} x(j)\, \overline{y(j)} \tag{1}$$

$$\langle (x(j)), (y(j)) \rangle = \sum_{j \in \mathbf{N}} x(j)\, \overline{y(j)} \tag{2}$$

$$\langle x, y \rangle = \int_E x\overline{y}\, dm \tag{3}$$

**Definition 1.7.12** ORTHONORMAL BASES An orthogonal subset is called *orthonormal* if it consists of unit vectors, i.e., vectors of norm 1. An orthonormal set which is not properly contained in any orthonormal set is called an *orthonormal basis* or *complete orthonormal set.* ■

Simple applications of Zorn's lemma—quoted in Sec. 7.3—reveal (1) that orthonormal bases exist in any inner product space, and (2) that any orthonormal set can be extended to an orthonormal basis. Some essential features of orthonormal bases are listed below [Bachman and Narici 2000, p. 155].

**Theorem 1.7.13** ORTHONORMAL BASES If $E$ is an orthonormal subset of a Hilbert space $X$ then the following statements are equivalent:

(a) $E$ is an orthonormal basis;
(b) if $x$ is orthogonal to each vector in $E$, then $x = 0$;
(c) $x = \Sigma_{e \in E} \langle x, e \rangle\, e$ for any $x$ in $X$;
(d) $\operatorname{cl}[E] = X$, where $[E]$ denotes the linear span of $E$;
(e) PARSEVAL'S IDENTITY: for any $x$ in $X$, $\|x\|^2 = \Sigma_{e \in E} |\langle x, e \rangle|^2$;
(f) $\langle x, y \rangle = \Sigma_{e \in E} \langle x, e \rangle \langle e, y \rangle$ for all $x$ and $y$ in $X$.

# Chapter 2

# Commutative Topological Groups



One of the principal reasons for superimposing a topological structure on an algebraic object is to permit the tools and insights of analysis (i.e., limits) to bear. Generally, in "topological algebra," a topology is introduced to an algebraic structure which makes the basic algebraic operations continuous. The idea of the topologic-algebraic fusion is fairly venerable and has numerous applications. It enabled Krull [1928], for example, to develop an infinite-dimensional Galois theory in which there exists a 1-1 correspondence between the intermediate fields and *closed* subgroups of the Galois group of a separable normal extension, after suitably topologizing the Galois group. In this connection—topologizing the Galois group of a separable normal field extension—it seems that a rudimentary grasp of the notion of topological group had already occurred to Dedekind in the nineteenth century, judging from his remark: "The set of these permutations forms a continuous multiplication in a certain sense, a question which we shall not address ourselves to any further here" [Dedekind 1932]. And certainly, Sophus Lie's "continuous groups," presently called "Lie groups," were topological groups whose theory was well developed in the second half of the nineteenth century. The notion of topological group to be introduced in this chapter was enunciated by Schreier [1925].

In this chapter we develop the most elementary properties of commutative topological groups. We are actually obtaining the basic properties of topological vector spaces, those that do not depend on scalar multiplication. We will see that topological groups are quite uniform in structure ("homogeneous"): the filter $V(0)$ of neighborhoods of 0 completely determines the topology in the sense that the filter $V(x)$ of neighborhoods of any point $x$ is $x + V(0) = \{x + V : V \in V(0)\}$.

After determining several properties of the neighborhoods of 0 in a topological group, we turn the question around in Sec. 2.3. We obtain conditions under which a collection of sets in a group $X$ will be a base at 0 for a topology which makes $X$ a topological group. These results enable us to efficiently topologize quotients and products of topological groups and groups of functions [Sec.2.6].

## 2.1 ELEMENTARY CONSIDERATIONS

**Notation.** The filter of neighborhoods of a point $x$ is denoted $V(x)$. ■

We define a topological group $X$ in this section and show that they would be a census taker's delight: if you know what is happening near 0, you know what is happening everywhere. They are *homogeneous* in that given any two points $x$ and $y$, there is a homeomorphism of $X$ onto itself mapping $x$ onto $y$. Thus, 0 serves as an "everypoint."

Let $X$ be an additive group. For any $n \in \mathbf{N}$ and $x \in X$, $nx$ denotes the sum obtained by adding $x$ to itself $n$ times and $-nx$ denotes $-x$ added to itself $n$ times. If $W$ is a subset of $X$, $-W = \{-w : w \in W\}$ and $W + W = \{x + y : x, y \in W\}$. If $X$ is also a topological space in which the algebraic operations are linked to the topology by

$(G_1)$ Continuity of Inversion: the map $x \mapsto -x$ of $X$ into $X$ is continuous;

$(G_2)$ Continuity of Addition: the map $(x, y) \mapsto x + y$ of $X \times X$ (product topology) into $X$ is continuous,

then $X$ is called a *topological group*. Equivalently, one says that the topology on $X$ is a *group topology* or is *compatible* with the group structure. As an alternate phrasing of $(G_1)$ and $(G_2)$, we have:

$(G_1)$ Continuity of Inversion: For any $x \in X$ and neighborhood $V$ of $-x$, there is a neighborhood $W$ of $x$ such that $-W \subset V$.

$(G_2)$ Continuity of Addition: For any $x, y \in X$ and any neighborhood $V$ of $x + y$ there exist neighborhoods $U$ of $x$ and $W$ of $y$ such that $U + W \subset V$

$(G_1)$ implies by induction that for any positive integer $n$ and any $W \in V(0)$, there is a $U \in V(0)$ such that $U + U + \cdots + U$ ($n$ terms) $\subset W$. We can combine $(G_1)$ and $(G_2)$ into:

> $(G_{12})$ The map $(x, y) \mapsto x - y$ of $X \times X$ into $X$ is continuous. Equivalently, for any $x, y \in X$ and any neighborhood $V$ of $x - y$, there exist neighborhoods $U$ of $x$ and $W$ of $y$ such that $U - W \subset V$.

Clearly, $(G_{12})$ is a consequence of $(G_1)$ and $(G_2)$. Conversely, $(G_{12})$ implies the continuity of the map $y \mapsto 0 - y$, so $(G_1)$ is implied by it. $(G_{12})$ and $(G_1)$ together imply $(G_2)$ for they imply the continuity of the map $(x, y) \mapsto x - (-y)$.

Any group with the discrete topology is a topological group, the continuity conditions being trivially satisfied; such *discrete groups* provide many counterexamples. (Although the discrete topology is a group topology for any group, it is never a vector topology because topological vector spaces are connected [Theorem 4.3.4].) More interesting examples of topological groups are given here, at the end of Sec. 2.3, and in Sec. 2.6.

**Example 2.1.1** $(\mathbf{R}, +)$, $(\mathbf{C}, +)$ AND $(X, \|\cdot\|, +)$ The additive groups $\mathbf{R}$ and $\mathbf{C}$ of real and complex numbers with their usual topologies are topological groups. More generally, so is any normed (or seminormed) space $(X, \|\cdot\|)$ viewed as an additive group. As we verify next, the triangle inequality yields continuity of addition and the fact that $\|x\| = \|-x\|$ makes for continuity of inversion. For $x \in X$, let $B(x, r) = \{z \in X : \|z - x\| < r\}$, $r > 0$. Given $x, y \in X$, consider a basic neighborhood $B(x-y, r)$, $r > 0$, of $x-y$. For $v \in B(x, r/2)$ and $w \in B(y, r/2)$, then $\|(v - w) - (x - y)\| \leq \|v - x\| + \|y - w\| < r$. Thus $B(x, r/2) - B(y, r/2) \subset B(x - y, r)$ and $(G_{12})$ is satisfied. ■

We say more about metric topological groups in Sec. 2.7.

**Example 2.1.2** $(\mathbf{R}^*, \cdot)$ AND $(\mathbf{C}^*, \cdot)$ The multiplicative groups $\mathbf{R}^*$ and $\mathbf{C}^*$ of nonzero real and complex numbers in their usual topologies are topological groups. To see this, consider nonzero (real or complex) numbers $a_0$ and $b_0$. For any nonzero numbers $a$ and $b$,

$$\begin{aligned} |ab - a_0 b_0| &= |(a - a_0)(b - b_0) + a_0(b - b_0) + (a - a_0)b_0| \\ &\leq |a - a_0|\,|b - b_0| + |a_0|\,|b - b_0| + |a - a_0||b_0| \end{aligned}$$

It follows that if $a$ is sufficiently close to $a_0$ and $b$ is sufficiently close to $b_0$, then $ab$ is close to $a_0 b_0$, i.e., multiplication is continuous. The continuity of the map $a \mapsto a^{-1}$ is verified by observing that the convergence of $a_n$ to $a$ implies the convergence of $1/a_n$ to $1/a$. ■

Another immediate consequence of the axioms is that translation and inversion are homeomorphisms:

**Theorem 2.1.3** TRANSLATION AND INVERSION HOMEOMORPHISMS For any topological group $X$ and any $z \in X$, the maps $x \mapsto -x$ and $x \mapsto x + z$ are homeomorphisms of $X$ onto $X$.

**Proof.** The inversion map $x \mapsto -x$ is clearly bijective. Continuity of inversion follows from $(G_1)$; its inverse is $x \mapsto -x$ (same map) so it is a homeomorphism. To see that translation $x \mapsto x+z$ is a homeomorphism, first observe that it is bijective. Its inverse, the map $x \mapsto x - z$, is also a translation, so it suffices to prove that translation is continuous to prove that it is a homeomorphism. If $W$ is a neighborhood of $z+x$, there must be neighborhoods $U$ and $V$ of $z$ and $x$, respectively, such that $U+V \subset W$ by $(G_2)$. Hence $V \mapsto z+V \subset U+V \subset W$ and translation is continuous. ■

We show next that translates $x + V$ of neighborhoods $V$ of 0 constitute the neighborhoods of $x$.

**Theorem 2.1.4** NEIGHBORHOOD BASES If $\mathcal{B}$ is a neighborhood base at 0 in a topological group $X$ then, for any $x \in X$, $x + \mathcal{B} = \{x + B : B \in \mathcal{B}\}$ is a neighborhood base for $V(x)$.

**Proof.** We use the notation of the statement. The map $y \mapsto y + x$ is a homeomorphism [Theorem 2.1.3] taking 0 into $x$. Its inverse is the map $y \mapsto y - x$. Thus if $V \in V(x)$ then $-x + V$ is a neighborhood of 0. Since $\mathcal{B}$ is a base, there exists $B \in \mathcal{B}$ such that $B \subset -x + V$. Hence $x + B \subset V$ and $x + \mathcal{B}$ is a neighborhood base at $x$. ■

**Definition 2.1.5** HOMOGENEOUS SPACES A topological space $X$ is *homogeneous* if, given any two points $x, y \in X$, there is a homeomorphism of $X$ onto $X$ mapping $x$ into $y$. ■

If $X$ is a topological group and $x, y \in X$, the translation map $z \mapsto z-x+y$ is a homeomorphism [Theorem 2.1.3] which takes $x$ into $y$; hence:

**Theorem 2.1.6** Any topological group is homogeneous.

It follows that for many important purposes, it suffices to examine the contours around 0 in a topological group. For example, if there is a compact neighborhood of 0, $X$ is locally compact; if $\{0\}$ is an open set, any one-point set is open—in other words, $X$ is discrete; $X$ is a $T_1$-space iff $\{0\}$ is closed.

**Definition 2.1.7** HOMOMORPHISM A (group) *homomorphism* is a map $f$ from a group $X$ into a group $Y$ such that $f(x + y) = f(x) + f(y)$ for all $x$ and $y$ in $X$. If, in addition, $f$ is 1-1, then $f$ is a (group) *isomorphism*. Continuous homomorphisms mapping one topological group into another are called *morphisms*. Group isomorphisms of one topological group into another which are homeomorphisms are called *group homeomorphisms* or *topological group isomorphisms*.

**Theorem 2.1.8** Continuity at One Point A homomorphism mapping one topological group into another continuous at one point is continuous.

**Proof.** Suppose that $X$ and $Y$ are topological groups and that $f : X \to Y$ is a homomorphism. Suppose that $f$ is continuous at $y \in X$. To see that $f$ is continuous everywhere, let $x \in X$ and suppose that the net $x_m \to x$. Then $y + x_m - x \to y$. Since $f$ is continuous at $y$,

$$f(y + x_m - x) = f(y) + f(x_m) - f(x) \to f(y)$$

which implies that $f(x_m) \to f(x)$ and proves the continuity of $f$ at $x$. ■

The use of "symmetric" (defined below) neighborhoods of 0 simplifies many arguments.

**Definition 2.1.9** Symmetric Sets A subset $E$ of a group $X$ is called *symmetric* if $E = -E$. ■

If $E$ is any subset of a topological group, then $E \cup -E$, $E \cap -E$, and $E - E$ are symmetric; moreover, if $E$ is symmetric, so is $\operatorname{cl} E$.

**Theorem 2.1.10** Base of Symmetric Neighborhoods of 0 In any topological group: (a) every neighborhood of 0 contains a symmetric neighborhood of 0; (b) given any neighborhood $V$ of 0, there is a symmetric neighborhood $W$ of 0 such that $W - W \subset V$.

**Proof.** (a) Since the map $x \mapsto -x$ is a homeomorphism, given any neighborhood $V$ of $0$ $-V$ is a neighborhood of 0; $U = V \cap -V$ is a symmetric neighborhood of 0 contained in $V$.

(b) By $(G_{12})$ and (a), given $V \in \mathcal{V}(0)$, there are symmetric neighborhoods $U, U' \in \mathcal{V}(0)$ such that $U - U' \subset V$. Now let $W = U \cap U'$. ■

As a simple but useful application, we have:

**Corollary 2.1.11** Let $X$ be topological group $X$, let $\mathcal{U}(0)$ denote the symmetric neighborhoods of 0. Then $\operatorname{cl}\{0\} = \cap \mathcal{U}(0) = \cap \mathcal{V}(0)$.

**Proof.** Since $\mathcal{U}(0)$ is a base for $\mathcal{V}(0)$, it suffices to prove that $\operatorname{cl}\{0\} = \cap \mathcal{U}(0)$. If $x \in \operatorname{cl}\{0\}$ then, for each $U \in \mathcal{U}(0)$, $x + U$ meets $\{0\}$, i.e., $0 \in x + U$. Since $U$ is symmetric, this means that $x \in U$ for all $U \in \mathcal{U}(0)$. Conversely, if $x \in \cap \mathcal{U}(0)$ then $-x \in \cap \mathcal{U}(0)$. Hence $0 \in x + U$ for all $U \in \mathcal{U}(0)$ which implies that $x \in \operatorname{cl}\{0\}$. ■

## 2.2 SEPARATION AND COMPACTNESS

**Notation.** $X$ denotes a commutative topological group with neighborhood filter $\mathcal{V}(0)$ at 0. ■

We show in Theorem 2.2.1 that every topological group is a regular topological space, i.e., given any neighborhood $V$ of a point $x$, there is a neighborhood $U$ of $x$ such that $\operatorname{cl} U \subset V$. Consequently, a nonregular topology such as the cofinite topology ($U$ is open iff $\complement U$ is finite) on any infinite group $X$ is not a group topology—in the case of the cofinite topology, the closure of any neighborhood is $X$.

**Theorem 2.2.1** REGULARITY Every topological group is regular. (Although this does not imply that every topological group is Hausdorff, it does if the group is $T_0$.)*

**Proof.** Let $V \in \mathcal{V}(0)$. Since $0 - 0 = 0$, it follows by ($G_{12}$) of Sec. 2.1 that there is some $U \in \mathcal{V}(0)$ such that $U - U \subset V$. We show that $\operatorname{cl} U \subset V$. If $x \in \operatorname{cl} U$ then every neighborhood $x + U$ of $x$ must meet $U$ so there must be $u, y \in U$ such that $x + u = y \in U$. Therefore, $x = y - u \in U - U \subset V$. ■

Concerning the separation properties of a topological group, we have:

**Theorem 2.2.2** SEPARATION In any topological group $X$ the following statements are equivalent:

(a) $X$ is a $T_0$-space.

(b) $X$ is a $T_2$-space.

(c) $\cap \mathcal{V}(0) = \{0\}$.

(d) For any nonzero element $x \in X$, there exists $V \in \mathcal{V}(0)$ such that $x \notin V$.

**Proof.** That (a) $\Rightarrow$ (b) is clear from the regularity of $X$ [Theorem 2.2.1]. To see that (b) $\Rightarrow$ (c), note that if $y \neq 0$ and $y \in \cap \mathcal{V}(0)$, then $X$ is not Hausdorff for $y$ cannot be separated from 0. To prove that (c) $\Rightarrow$ (a), we need only prove that $\{0\}$ is closed by Theorem 2.1.6 (homogeneity). To show this, we note that $\operatorname{cl}\{0\} = \cap \mathcal{V}(0) = \{0\}$ [Corollary 2.1.11]. Clearly (b) $\Rightarrow$ (d). Conversely, we show that (d) $\Rightarrow$ (c). If $x \neq 0$ and $x \notin V \in \mathcal{V}(0)$ then $x \notin \cap \mathcal{V}(0)$; hence $\cap \mathcal{V}(0) = \{0\}$. ■

By Theorem 2.1.3, translates of open sets or closed sets are open or closed, respectively. Moreover:

**Theorem 2.2.3** SUMS OF OPEN SETS In any topological group $X$, if $G$ is a nonempty open set and $B$ is any subset, then $G + B$ is open since we can write $G + B = \cup_{x \in B}(G + x)$. In particular, sums of open sets are open.

Sums of closed sets need not be closed, however, as the following example shows.

*Every topological group is completely regular [Bushaw 1963, p. 123, Th. 52] but generally not normal [Husain 1966, p. 51].

**Example 2.2.4** SUMS OF CLOSED SETS Let $X = (\mathbf{R}, +)$, the additive topological group of real numbers with the usual topology. The sets $A = \{n \in \mathbf{N} : n \geq 2\}$ and $B = \{-n - 1/n : n \geq 2\}$ are closed but $A + B$ contains $\{-1/n : n \geq 2\}$ which has 0 as an adherence point so $A + B$ is not closed. ■

The sum of a *compact* set and a closed set is generally closed, however. We prove this in Theorem 2.2.6 after establishing the following result.

**Theorem 2.2.5** NEIGHBORHOODS OF COMPACT SETS In any topological group, if the compact set $K$ is contained in the open set $U$ then there is a neighborhood $V$ of 0 such that $K + V \subset U$. (A *neighborhood of* $K$ is contained in $U$.)

**Proof.** Let $K$ and $U$ be as above. For each $x \in K$, there exists $V_x \in V(0)$ such that $x + V_x \subset U$. By continuity of addition, there exists a neighborhood $W_x$ of 0 such that $W_x + W_x \subset V_x$. Since $K$ is compact, a finite number $x_1 + W_{x_1}, \ldots, x_n + W_{x_n}$, of these sets covers $K$. Let $V = \cap_{j=1}^{n} W_{x_j}$. To show that $K + V \subset U$, consider any $x \in K$. There must be some integer $i$ such that $x \in x_i + W_{x_i}$. Hence

$$x + V \subset x_i + W_{x_i} + V \subset x_i + W_{x_i} + W_{x_i} \subset x_i + V_{x_i} \subset U$$

Therefore $K + V \subset U$. ■

**Theorem 2.2.6** COMPACT + CLOSED IS CLOSED In any topological group $X$, if $K$ is compact and $F$ is closed, then $K + F$ is closed.

**Proof.** Suppose $K$ is compact, $F$ is closed and let $x \in \operatorname{cl}(K + F)$. Then for all neighborhoods $V \in V(0)$,

$$(x + V) \cap (K + F) \neq \varnothing \Leftrightarrow (-K + x + V) \cap F \neq \varnothing \qquad (\star)$$

To show that $x \in K + F$, first observe that if $(-K + x) \cap F = \varnothing$, then the compact set $-K + x$ is contained in the open set $\complement F$. By Theorem 2.2.5 it follows that there is some $V \in V(0)$ such that $(-K + x + V) \subset \complement F$ which contradicts $(\star)$. ■

As a final word on closure and addition, we prove:

**Theorem 2.2.7** $\operatorname{cl} A + \operatorname{cl} B \subset \operatorname{cl}(A + B)$ If $A$ and $B$ are subsets of a topological group $X$, then $\operatorname{cl} A + \operatorname{cl} B \subset \operatorname{cl}(A + B)$.

**Proof.** Let $x \in \operatorname{cl} A$, $y \in \operatorname{cl} B$, and $U \in V(0)$. By the continuity of addition, there are neighborhoods $V, W \in V(0)$ such that $(x+V)+(y+W) \subset x+y+U$. Since $x \in \operatorname{cl} A$ and $y \in \operatorname{cl} B$, there exist $a \in A \cap (x + V)$ and $b \in B \cap (y + W)$. Hence $a + b \in x + y + U$ and therefore $x + y \in \operatorname{cl}(A + B)$. ■

As mentioned earlier, for a topological group to be locally compact it suffices for 0 to possess a compact neighborhood. A little more is true, however, namely:

**Theorem 2.2.8** LOCALLY COMPACT GROUPS For each neighborhood $V$ of 0 in a locally compact topological group, there is a relatively compact symmetric neighborhood $U$ of 0 such that $\mathrm{cl}\, U \subset V$ ($\mathrm{cl}\, U$ is symmetric as well).

**Proof.** Since any topological group is regular [Theorem 2.2.1], given $V \in V(0)$ there exists $W \in V(0)$ such that $\mathrm{cl}\, W \subset V$. Since $X$ is locally compact, there exists $W' \in V(0)$ such that $\mathrm{cl}\, W'$ is compact. Since $\mathrm{cl}\,(W \cap W') \subset \mathrm{cl}\, W'$, $\mathrm{cl}\,(W \cap W')$ is compact. Since $W \cap W'$ is a neighborhood of 0, there is a symmetric neighborhood $U$ of 0 such that $U \subset \mathrm{cl}\,(W \cap W')$ [Theorem 2.1.10]. As $\mathrm{cl}\, U \subset \mathrm{cl}\,(W \cap W') \subset V$, $\mathrm{cl}\, U$ is compact. Thus, $U$ is the desired neighborhood. Since the map $x \mapsto -x$ is a homeomorphism, $-\,\mathrm{cl}\, U = \mathrm{cl}\,(-U) = \mathrm{cl}\, U$ so $\mathrm{cl}\, U$ is symmetric. ■

**Definition 2.2.9** TOPOLOGICAL RINGS A ring $X$ endowed with a group topology and such that the map $(x, y) \mapsto xy$ of $X \times X$ into $X$ is continuous is called a *topological ring*. In this case the topology is said to be a *ring topology* or *compatible with the ring structure.* Fields $K$ topologized in such a way as to be topological rings and in which the map $x \mapsto x^{-1}$ from the nonzero elements $K^*$ of $K$ into $K^*$ is continuous are called *topological fields.* ■

To guarantee that a topologized field $K$ be a topological field, it is not enough to demand that $K$ be an additive topological group and $K^*$ a multiplicative topological group, for this would not require multiplications involving 0 be continuous. **R** and **C** with their usual topologies are topological fields, and more generally any valued field [Sec. 1.2] is a topological field.

## 2.3 BASES AT 0 FOR GROUP TOPOLOGIES

**Notation.** $X$ denotes a commutative typological group with neighborhood filter $V\,(0)$ at 0. ■

In the last section we noted some properties of bases for $V\,(0)$ in a topological group. In this section, we reverse direction and develop conditions for a filterbase $\mathcal{B}$ of subsets of a group $X$ to determine a group topology for which $\mathcal{B}$ is a basis at 0. The analogous discussion for topological vector spaces is in Sec. 4.5.

We have already seen [Theorem 2.1.10(b)] that:

($F$) For any $V \in V\,(0)$ there is a symmetric $W \in V\,(0)$ such that $W - W \subset V$

This condition on a filter $\mathcal{F}$ of sets from a group $X$ suffices to determine a compatible topology for $X$ in which $\mathcal{F}$ is the filter of neighborhoods of 0 as we prove next. $(F)$ immediately implies that for any $W \in \mathcal{F}$, there are $U, V \in \mathcal{F}$ such that $(U \cap V) - (U \cap V) \subset U - V \subset W$; therefore 0 belongs to each $W \in \mathcal{F}$.

**Theorem 2.3.1** BASE AT 0 FOR GROUP TOPOLOGY Let $\mathcal{B}$ be a filterbase in a group $X$. $\mathcal{B}$ is a neighborhood base at 0 for a group topology for $X$ if $\mathcal{B}$ satisfies:

$(FB)$ For each $B \in \mathcal{B}$ there is a symmetric $V \in \mathcal{B}$ such that $V - V \subset B$.

$(OB)$ If for each $B \in \mathcal{B}$ and any $x \in B$ there exists $V \in \mathcal{B}$ such that $x + V \subset B$ then $\mathcal{B}$ is an open base at 0.

**Proof.** Let $\mathcal{B}$ be a filterbase in a group $X$ and let $\mathcal{F}$ denote the filter of all supersets of sets in $\mathcal{B}$. For any $x \in X$ we show that $x + \mathcal{F}$ satisfies the neighborhood axioms at $x$.

Since, as noted above, 0 belongs to every $F \in \mathcal{F}$, it is clear that $x$ belongs to every $x + F \in x + \mathcal{F}$. We illustrate the argument for finite intersections by considering two sets. Given $F, G \in \mathcal{F}$, there exist $A, B \in \mathcal{B}$ such that $A \subset F$, $B \subset G$. Since $\mathcal{B}$ is a filterbase, there exists $C \in \mathcal{B}$ such that $C \subset A \cap B$. Therefore $x + C \subset (x + A) \cap (x + B) = x + A \cap B \subset x + F \cap G \in x + \mathcal{F}$. Now consider $F \in \mathcal{F}$ and $B \in \mathcal{B}$ such that $B \subset F$. By $(FB)$ there exists a symmetric $W \in \mathcal{B}$ such that $W - W \subset B$. For any $y \in x + W$, $y + W \subset x + W - W \subset x + B \subset x + F$ which implies that $x + F$ is a neighborhood of $y$. It follows that $x + \mathcal{F}$ is a system of neighborhoods of $x$.

To see that this is a group topology, we show that $(FB)$ implies that the neighborhoods $x + \mathcal{F}$ $(x \in X)$ satisfy $(G_{12})$ of Sec. 2.1. For $x, y \in X$ and $F \in \mathcal{F}$, consider the neighborhood $x - y + F$ of $x - y$. Choose $B \in \mathcal{B}$ such that $B \subset F$ and (by $(FB)$) $V \in \mathcal{B}$ such that $V - V \subset B$. Then

$$(x + V) - (y + V) \subset x - y + (V - V) \subset x - y + B \subset x - y + F$$

Condition $(OB)$ clearly makes each $B$ in $\mathcal{B}$ a neighborhood of each of its points, hence an open set. ∎

As an immediate corollary we have:

**Theorem 2.3.2** SUBBASES If $\mathcal{S}$ is a family of subsets of a group $X$ that satisfies the finite intersection condition (finite intersections of sets from $\mathcal{S}$ are not empty) and $(FB)$ then $\mathcal{S}$ is a neighborhood subbase at 0 for a group topology on $X$. If $\mathcal{S}$ also satisfies $(OB)$, then it is an open subbase.

**Example 2.3.3** FINITE TOPOLOGY Let $X$ and $Y$ be linear spaces over the same field and let $\mathcal{L}(X, Y)$ denote the commutative additive group of all linear maps of $X$ into $Y$ with respect to $(A + B)x = Ax + Bx$ for any $x \in X$ and $A, B \in \mathcal{L}(X, Y)$. For any $x \in X$ let $S(0, x) = \{A \in \mathcal{L}(X, Y) : Ax = 0\}$. It is a simple matter to verify that the sets $\{S(0, x) : x \in X\}$ satisfy the finite intersection condition (they all contain the trivial linear map $x \mapsto 0$) and

$(FB)$ of Theorem 2.3.1. Consequently, by Theorem 2.3.2, $\{S(0,x) : x \in X\}$ is a neighborhood *subbase* at 0 for a compatible topology on $\mathcal{L}(X,Y)$. This topology is called the *finite topology.*

If $A \neq 0$, then there is some $x \in X$ such that $Ax \neq 0$, so $A \notin S(0,x)$; the finite topology is therefore Hausdorff. Actually, its separation properties are even stronger, as we now show. The sets $S(0,x)$ also satisfy $(OB)$ of Theorem 2.3.1 for, for any $A \in S(0,x)$, $A + S(0,x) \subset S(0,x)$. Hence the sets $S(0,x)$ are an open subbase at 0. But the sets $S(0,x)$ are *subgroups* of $\mathcal{L}(X,Y)$; hence they must also be closed [Theorem 2.4.4(b)]. Consequently, the finite topology has a basis of clopen sets. This, together with the fact that it is Hausdorff, means that the finite topology is totally disconnected: The only connected subsets are points. As we shall see later [Theorem 4.3.4], this property prevents $\mathcal{L}(X,Y)$ with the finite topology from being a topological vector space. We further remark that the finite topology is discrete iff $X$ is finite-dimensional and that, by defining $(AB)x$ to be $A(Bx)$ for each $x \in X$, $\mathcal{L}(X,X)$ becomes a topological ring [Definition 2.2.9] with respect to the finite topology.

If $E$ is a Galois extension (= separable + normal) field of the field $L$ and $X$ denotes the set of all field automorphisms of $E$ which leave $L$ fixed, then $X$ is a group with respect to composition. $X$ is called the *Galois group* of $E$ over $L$. Viewing $X$ as a subset of $\mathcal{L}(E,E)$ endowed with the finite topology, then $X$ in its relative topology is also a topological group. In this context, it is often denominated the *Krull* topology after W. Krull, who proved [Krull 1928] that there is a 1-1 correspondence between the intermediate fields of $E$ over $L$ and the closed subgroups of $X$, thus developing a "Galois theory" in the infinite-dimensional case. For further discussion of this aspect of Galois theory, see Artin [1967, pp. 103–108] and Jacobson [1964, pp. 147-151]; for a more general discussion of the finite topology, see Jacobson [1953]. ■

## 2.4 SUBGROUPS AND PRODUCTS

**Notation.** $X$ denotes a commutative typological group with neighborhood filter $\mathcal{V}(0)$ at 0. ■

Consider a map $f : S \to T$ (or family of maps) where only one of $S, T$ is a topological space. We topologize the other in such a way as to make $f$ continuous. We topologize the domain $S$ in this section to get an *inverse image* (*weak*, *initial*) topology; in Sec. 2.5 we topologize the codomain $T$ to obtain a direct image (*final, identification*) topology. Inverse and direct image topologies determined by homomorphisms are group topologies.

**Definition 2.4.1** Inverse Image Topology Let $\{(Y_s, \mathcal{T}_s) : s \in S\}$ be a family of topological spaces and let $X$ be a set. Let $f_s : X \to Y_s$ $(s \in S)$ be a family of maps. The coarsest topology for $X$ with respect to which each

$f_s$ is continuous has $\{f_s^{-1}(\mathcal{T}_s) : s \in S\}$ as a subbase. It is called the *inverse image* (*weak, initial*) topology *induced by* $\{f_s : s \in S\}$ on $X$. (Cf. Example 4.7.1). ■

Our principal interest is when $X$ is a group, the $f_s$ are homomorphisms and the $Y_s$ are topological groups in which case we get a group topology on $X$ as we show next.

**Example 2.4.2** INVERSE IMAGE TOPOLOGY AND GROUPS Let $X$ be a group and let $\{X_s : s \in S\}$ be a family of topological groups such that for each $s \in S$ there is a homomorphism $h_s : X \to X_s$. For each $s \in S$ let $\mathcal{V}_s(0)$ denote the filter (or a base for the filter) of neighborhoods of 0 in $X_s$. A base $\mathcal{V}_{\mathcal{T}}(0)$ at 0 for the inverse image topology $\mathcal{T}$ for $X$ determined by the family $\{h_s : s \in S\}$ is given by intersections of the form $\cap_{k \in K} h_k^{-1}(V_k)$, where $K$ is a finite subset of $S$ and $V_k \in \mathcal{V}_k(0)$ for each $k \in K$. It is a group topology: To illustrate how to verify $(FB)$ of the basis theorem [2.3.1], consider $s, k \in S$ and suppose $V_s \in \mathcal{V}_s(0)$, and $V_k \in \mathcal{V}_k(0)$. Choose symmetric neighborhoods $B_s \in \mathcal{V}_s(0)$ and $B_k \in \mathcal{V}_k(0)$ such that $B_s + B_s \subset V_s$ and $B_k + B_k \subset V_k$, so that $h_s^{-1}(B_s) + h_s^{-1}(B_s) \subset h_s^{-1}(B_s + B_s) \subset h_s^{-1}(V_s)$ and $h_k^{-1}(B_k) + h_k^{-1}(B_k) \subset h_k^{-1}(V_k)$. Then $h_s^{-1}(B_s) \cap h_k^{-1}(B_k) + h_s^{-1}(B_s) \cap h_k^{-1}(B_k) \subset h_s^{-1}(V_s) \cap h_k^{-1}(V_k)$. ■

If $M$ is a subgroup of a topological group $X$ then the injection $I : M \to X$, $x \mapsto x$, is a homomorphism. The inverse image topology determined by $I$—i.e., $M$ in its relative topology—is therefore a group topology.

A subgroup of a topological group is said to be *closed*, *open*, etc., if it is a closed set, and so on. For example, the additive group of real numbers is a closed subgroup of the additive topological group of complex numbers.

**Theorem 2.4.3** CLOSURES OF SUBGROUPS If $M$ is a subgroup of a topological group $X$, then so is the closure $\operatorname{cl} M$ of $M$.

**Proof.** The continuous map $f : X \times X \to X$, $(x, y) \mapsto x - y$, maps $M \times M$ onto $M$. Since $f$ is continuous, $f(\operatorname{cl}(M \times M)) = f(\operatorname{cl} M \times \operatorname{cl} M) = \operatorname{cl} M - \operatorname{cl} M \subset \operatorname{cl} f((M \times M)) = \operatorname{cl} M$ and it follows that $\operatorname{cl} M$ is a subgroup. ■

As a result of Theorem 2.4.3, it is clear that $\operatorname{cl}\{0\}$ is a subgroup in any topological group.

**Theorem 2.4.4** OPEN AND CLOSED SUBGROUPS A subgroup $M$ of a topological group $X$ is:

(a) open iff its interior $\operatorname{int} M \neq \varnothing$;

(b) open only if $M$ is closed;

(c) discrete iff $M$ has an isolated point.

**Proof.** Let $M$ be as above. (a) If $M$ is open then $0 \in \operatorname{int} M$. Conversely, suppose that $M$ has an interior point $x$. For any $y$ in $M$, the map $z \mapsto z + (y - x)$ maps $M$ homeomorphically onto $M$ and maps $x$ into $y$. Thus $y$ is also an interior point of $M$.

(b) Suppose that $M$ is open. Then if $x \in \operatorname{cl} M$, $(x + M) \cap M \neq \varnothing$, so there exist $y, z \in M$ such that $x + y = z$; hence $x = z - y \in M$.

(c) If $x$ is an isolated point of $M$ then by translation as in (a), each point $y$ of $M$ is also isolated and therefore $M$ is discrete. The converse is clear. ■

We discuss products of topological groups next.

**Example 2.4.5** PRODUCTS Let $\{X_s : s \in S\}$ be a family of topological groups. For elements $x = (x_s)$ and $y = (y_s)$ from the product $X = \Pi_{s \in S} X_s$, define $x + y$ to be $(x_s + y_s)$. It is easy to verify that $X$ is a group with respect to this pointwise operation and that each projection $\operatorname{pr}_k : \Pi_{s \in S} X_s \to X_k$ is a homomorphism. The product topology $\mathcal{T}$, the coarsest topology with respect to which each projection is continuous, is the inverse image topology determined by $\{\operatorname{pr}_s : s \in S\}$. Since projections are homomorphisms, it is a group topology [Example 2.4.2]. Let $\mathcal{V}_s(0)$ denote the filter of neighborhoods of 0 in $X_s$ $(s \in S)$. A base at 0 for $\mathcal{T}$ is given by

$$\begin{gathered} \Pi_{s \in S} V_s, \text{ where each } V_s \in \mathcal{V}_s(0) \\ \text{and } V_s = X_s \text{ for almost all indices } s \end{gathered}$$

Whenever we speak of a product of topological groups it is always with respect to the product topology. ■

## 2.5 QUOTIENTS

Suppose $X$ is a topological space, $Y$ a set and $f : X \to Y$ a map. We want to topologize $Y$ in such a way that $f$ is continuous. The *direct image* (*final, identification*) topology $\mathcal{T}_f$ for $Y$ *induced by* $f$ has as its open sets those $U \subset Y$ such that $f^{-1}(U)$ is open in $X$; that this is a topology (as opposed to *generates* a topology) is clear since $f^{-1}$ preserves set operations. We further note that $\mathcal{T}_f$ reduces to the discrete topology on $\complement f(X)$ since, for any $A \subset \complement f(X)$, $f^{-1}(A) = \varnothing$, which is open and closed in $X$. If $f$ is not surjective, since $f(X)$ is $\mathcal{T}_f$-open $(X = f^{-1} f(X))$ then $Y = f(X) \cup \complement f(X)$ is $\mathcal{T}_f$-disconnected. In terms of neighborhoods, $V$ is a neighborhood of $y = f(x) \in Y$ iff $f^{-1}(V)$ is a neighborhood of $x$.

The direct image topology on $Y$ clearly makes $f$ continuous. Moreover, it is the finest topology for $Y$ with respect to which $f$ is continuous—if $\mathcal{T}$ is any other such topology, then $U \in \mathcal{T}$ implies that $f^{-1}(U)$ is open in $X$, i.e., $U \in \mathcal{T}_f$, so that $\mathcal{T} \subset \mathcal{T}_f$.

**Direct Image Topologies and Groups** Suppose $X$ is a topological group, $Y$ a group, and $h : X \to Y$ a homomorphism. The direct image topology induced by $h$ on $Y$ has as a base at 0 the collection $\mathcal{B}$ of all subsets $B \subset Y$ such that $h^{-1}(B)$ is a neighborhood of 0 in $X$. To see that it is a group topology, first note that $\mathcal{B}$ is clearly a filterbase. For any $B \in \mathcal{B}$ there exists a symmetric neighborhood $V$ of 0 in $X$ such that $V - V \subset h^{-1}(B)$ [Theorem 2.1.10(b)]. Hence

$$h(V - V) = h(V) - h(V) \subset B$$

Since $V \subset h^{-1}(h(V))$, it follows that $h(V) \in \mathcal{B}$; therefore $\mathcal{B}$ satisfies condition $(FB)$ of Theorem 2.3.1 and is a base a 0 for a group topology $\mathcal{T}$ for $X$. In summary:

**Theorem 2.5.1** DIRECT IMAGE TOPOLOGY ON A GROUP Suppose $X$ is a topological group, $Y$ a group, and $h : X \to Y$ a homomorphism. Then:

(a) the direct image topology for $Y$ is a group topology;

(b) the filterbase $\mathcal{B}$ of all subsets $B \subset Y$ such that $h^{-1}(B)$ is a neighborhood of 0 in $X$ is a base at 0 for the direct image topology.

**Definition 2.5.2** QUOTIENT TOPOLOGY Suppose $M$ is a subgroup of a topological group $X$ and let $k : X \to X/M$, $x \mapsto \dot{x} = x + M$ denote the canonical homomorphism. The direct image topology on $X/M$ determined by $k$ is called the *quotient topology* for $X/M$. Since it is determined by a homomorphism, it is a group topology on $X/M$. Whenever we discuss a quotient group, we assume that it carries its quotient topology. ■

We mention that $k^{-1}(\dot{x}) = x + M$ $(x \in X)$. By our general comments above on direct image topologies, $k$ is continuous.

**Theorem 2.5.3** QUOTIENT TOPOLOGY; $k$ IS OPEN BUT NOT CLOSED Let $M$ be a subgroup of a topological group $X$. The canonical map $k : X \to X/M$ is:

(a) open—indeed, every open subset of $X/M$ is of the form $G + M$ where $G$ is open in $X$; in particular, $k(\mathcal{V}(0))$ is the neighborhood filter at 0 for the quotient topology;

(b) $k$ generally does not map closed sets into closed sets.

**Proof.** We use the notation of the statement. (a) Let $G \subset X$ be open. Then $k^{-1}(k(G)) = G + M = \cup_{m \in M}(G + m)$ is open; therefore $k(G)$ is open. (b) Let $\mathbf{Z}$ denote the subgroup of integers of the additive group $\mathbf{R}$ of real numbers in its usual topology and consider the canonical map $k : \mathbf{R} \to \mathbf{R}/\mathbf{Z}$. For $x \in \mathbf{R}$, let $[x]$ denote the greatest integer $\leq x$. Every coset $x + \mathbf{Z}$ contains $x - [x]$ and no other number in $[0, 1)$. Hence we may map $\mathbf{R}/\mathbf{Z}$ in a bijective fashion onto $[0, 1)$ and reflect the quotient topology of $\mathbf{R}/\mathbf{Z}$ onto $[0, 1)$. A basis for $[0, 1)$ in this topology is then given by the sets $(a, b)$ and $[0, a) \cup (b, 1)$, $0 < a < b < 1$. The set $A = \{3/2, 9/4, \ldots, n + 2^{-n}, \ldots\}$ is closed in $\mathbf{R}$ but $k(A)$, under the identification, is $\{1/2, 1/4, \ldots, 2^{-n}, \ldots\}$ which has 0 as an adherence point; thus $k(A)$ is not closed. ■

**Theorem 2.5.4** HAUSDORFF QUOTIENT GROUPS Let $M$ be a subgroup of the topological group $X$. Then $X/M$ is Hausdorff iff $M$ is closed.

**Proof.** We use the notation of the statement. Let $k : X \to X/M$ denote the canonical map. If $X/M$ is Hausdorff, then $\{\dot{0}\}$ is closed; hence $k^{-1}(\{\dot{0}\}) = M$ is closed. Conversely, suppose that $M$ is closed. By Theorems 2.2.2 and 2.1.6, to show that $X/M$ is Hausdorff, it suffices to show that $\{\dot{0}\}$ is closed. To this end suppose that $\dot{x} \in \operatorname{cl}\{\dot{0}\}$. Then for every neighborhood $V$ of $0$, $\dot{0} \in k(x+V)$; equivalently, $(x+V) \cap M \neq \varnothing$ for each $V \in V(0)$. Thus $x \in \operatorname{cl} M = M$ and $\dot{x} = \dot{0}$. ■

It is often convenient to split a homomorphism $f : X \to Y$ as in the accompanying diagram.

$$\begin{array}{ccc} X & \overset{f}{\to} & Y \\ \downarrow k & & \uparrow I \\ X/N & \overset{\dot{f}}{\to} & f(X) \end{array}$$

Here $k$ denotes the canonical homomorphism $x \mapsto x + N = \dot{x}$ where $N = f^{-1}(0)$ denotes the kernel (subgroup) of $f$; $\dot{f}$ represents the bijective homomorphism $\dot{x} \mapsto f(x)$ and $I$ represents the canonical injection map $f(x) \mapsto f(x)$. We observe that $\dot{f}$ is an isomorphism of $X/N$ onto $f(X)$ called the *canonical isomorphism associated with f*.

**Theorem 2.5.5** Let $X$ and $Y$ be topological groups, $f : X \to Y$ a homomorphism and everything else as in the diagram above. Then

(a) The canonical injection $I$ is a group homeomorphism of $f(X)$ into $Y$;

(b) $f$ is continuous iff $\dot{f}$ is continuous;

(c) $\dot{f}$ is open iff $f$ is relatively open [i.e., $f$ maps open subsets of $X$ onto open subsets of $f(X)$].

**Proof.** We use the notation of the statement. (a) Clear.

(b) If $\dot{f}$ is continuous, then $f$ is the composite of three continuous maps and is therefore continuous. Conversely, suppose that $f$ is continuous and let $U$ be an open subset of $Y$. Since $k$ is an open map and $(I\dot{f})^{-1}(U) = k(f^{-1}(U))$, $I\dot{f}$ is continuous, from which the continuity of $I^{-1}I\dot{f} = \dot{f}$ follows.

(c) For each open subset $U$ of $X$, $f(U) = I\dot{f}(k(U))$. Since $k$ is open and $I$ is a homeomorphism, the result follows. ■

By Theorem 2.5.5, it is easy to see that:

**Theorem 2.5.6** $f$ is a relatively open continuous map iff $\dot{f}$ is a group homeomorphism.

## 2.6 $\mathcal{S}$-TOPOLOGIES

**Notation.** Throughout the section $T$ is a set and $X$ a commutative topological group with neighborhood filter $V(0)$ at 0. $F(T,X)$ denotes the set $X^T$ of all functions mapping $T$ into $X$. For $x, y \in F(T,X)$ and $t \in T$, define $(x+y)(t) = x(t)+y(t)$. With respect to these pointwise operations, $(F(T,X),+)$ is a commutative topological group. ■

We topologize $F(T,X)$ in various ways that make it a topological group. Various different notions of convergence of functions—uniform, pointwise, etc.—all fit under this umbrella. This was very important, historically, for it shows that general topology is a broad enough tool to encompass these diverse types of convergence. Our general remarks below apply just as well to any subgroup of $F(T,X)$. When $T$ is a topological space, we can consider the subgroup $C(T,X)$ of continuous maps of $T$ into $X$, as we do in Secs. 5.8–5.10.

The building blocks for the topologies for $F(T,X)$ to come are the sets

$$W(S,V) = \{x \in F(T,X) : x(S) \subset V\}, \ S \subset T, \ V \in V(0) \qquad (*)$$

Their vital properties follow. The proofs of all parts are immediate.

**Theorem 2.6.1** PROPERTIES OF $W(S,V)$ Let $T$ be a set and $X$ a commutative topological group with neighborhood filter $V(0)$ at 0. For any $S, S' \subset T$ and $U, V \in V(0)$ :

(a) $U \subset V \Rightarrow W(S,U) \subset W(S,V)$;

(b) $W(S,-V) = -W(S,V)$; thus if $V$ is symmetric, so is $W(S,V)$;

(c) $W(S,U) - W(S,U) \subset W(S,U-U)$;

(d) $S \subset S' \Rightarrow W(S,V) \supset W(S',V)$;

(e) for any family $\{S_i : i \in I\}$ of subsets of $T$, $W(\cup_{i\in I} S_i, V) = \cap_{i\in I} W(S_i, V)$;

(f) for any family $\{V_i : i \in I\}$ of neighborhoods of 0 in $X$, $W(S, \cap_{i\in I} V_i) = \cap_{i\in I} W(S, V_i)$.

**Example 2.6.2** $\mathcal{S}$-TOPOLOGIES Let $\mathcal{S}$ be a collection of subsets of $T$, and consider

$$B_{\mathcal{S}}(0) = \{W(S,V) : (S,V) \in \mathcal{S} \times V(0)\}$$

Since the 0 function, $t \mapsto 0$, of $T$ into $X$, belongs to each $W(S,V)$, finite intersections of sets from $B_{\mathcal{S}}(0)$ are nonempty. Let $S \in \mathcal{S}$ and $V \in V(0)$. Choose a symmetric neighborhood $U \in V(0)$ such that $U - U \subset V$. By Theorem 2.6.1(a,c), it follows that

$$W(S,U) - W(S,U) \subset W(S,U-U) \subset W(S,V) \qquad (**)$$

thus showing that $B_{\mathcal{S}}(0)$ satisfies condition $(FB)$ of Theorem 2.3.1. Thus:

$\mathcal{B}_{\mathcal{S}}(0)$ forms a neighborhood subbase at 0 for a group topology $\mathcal{T}_{\mathcal{S}}$ for $F(T,X)$ called the *topology of uniform convergence on* $\mathcal{S}$, or the $\mathcal{S}$*-topology*. We denote the neighborhood filter it determines at 0 by $\mathcal{V}_{\mathcal{S}}(0)$ and $F(T,X)$ equipped with $\mathcal{T}_{\mathcal{S}}$ by $F_{\mathcal{S}}(T,X)$. ■

Suppose the topological group $(X,d)$ is metrizable and and that $\mathcal{B}$ consists of the sets $B(0,r) = \{x \in X : d(x,0) < r\}$, $r > 0$. Then, for any collection $\mathcal{S}$ of subsets of $T$, the sets $W(S, B(0,r))$, $r > 0$, $S \in \mathcal{S}$, is a base at 0 for $\mathcal{T}_{\mathcal{S}}$.

Since $W(\cup_{i\in I} S_i, V) = \cap_{i\in I} W(S_i, V)$ [Theorem 2.6.1(e)], if $\mathcal{S}$ is closed under the formation of finite unions, then the $W(S,V)$ form a base at 0 rather than just a subbase. Actually, we can always assume that $\mathcal{S}$ is closed with respect to the formation of finite unions in view of the following result.

**Theorem 2.6.3** Subsets of Finite Unions Let $\mathcal{S}$ be a collection of subsets of the set $T$. The $\mathcal{S}$-topology $\mathcal{T}_{\mathcal{S}}$ is unaffected by replacing $\mathcal{S}$ by the collection $\mathcal{S}^*$ of all subsets of finite unions of sets of $\mathcal{S}$.

**Proof.** We use the notation of the statement. Since $\mathcal{S} \subset \mathcal{S}^*$, clearly $\mathcal{T}_{\mathcal{S}} \subset \mathcal{T}_{\mathcal{S}^*}$ so it only remains to show that $\mathcal{V}_{\mathcal{S}^*}(0) \subset \mathcal{V}_{\mathcal{S}}(0)$. To this end, suppose that $S_1, \ldots, S_n \in \mathcal{S}$ $(n \in \mathbf{N})$ and that $S \subset \cup_{i=1}^n S_i$. By Theorem 2.6.1(d), for any neighborhood $V$ of 0 in the topological group $X$,

$$W(\cup_{i=1}^n S_i, V) \subset W(S,V) \qquad [V \in \mathcal{V}(0)]$$

Since $W(\cup_{i=1}^n S_i, V) = \cap_{i=1}^n W(S_i, V)$ [Theorem 2.6.1(e)], $W(\cup_{i=1}^n S_i, V) \in \mathcal{V}_{\mathcal{S}}(0)$ and the result follows. ■

**Theorem 2.6.4** Hausdorff $\mathcal{S}$-Topologies Let $\mathcal{S}$ be a collection of subsets of the set $T$. If the topological group $X$ is Hausdorff and $\mathcal{S}$ covers $T$ then $F_{\mathcal{S}}(T,X)$ is Hausdorff.

**Proof.** We use the notation of the statement. By Theorem 2.2.2(c), it suffices to show that $\cap \mathcal{V}_{\mathcal{S}}(0) = \{0\}$. To this end let $x \in \cap \mathcal{V}_{\mathcal{S}}(0)$ and choose any $t \in T$. Now select $S \in \mathcal{S}$ such that $t \in S$. Since $x \in \cap \mathcal{V}_{\mathcal{S}}(0)$, $x \in \cap_{V\in\mathcal{V}(0)} W(S,V) = W(S, \cap \mathcal{V}(0)) = W(S, \{0\})$ by Theorem 2.6.1(f) and the fact that $X$ is Hausdorff. It follows that $x(t) = 0$. Since $t$ is arbitrary, $x = 0$. ■

Theorem 2.6.4 is one instance of a somewhat more general phenomenon: namely, that the behavior of a function $x \in F(T,X)$ on $\cup\mathcal{S}$ determines the position $x$ will occupy in the $\mathcal{S}$-topology. For example, if $x$ vanishes on $\cup\mathcal{S}$, then $x$ is an adherence point of the filter $\mathcal{V}_{\mathcal{S}}(0)$.

We consider some particular $\mathcal{S}$-topologies next.

**Example 2.6.5** Topology of Uniform Convergence Let $\mathcal{S} = \{T\}$. The sets $\{W(T,V) : V \in \mathcal{V}(0)\} = \mathcal{V}_T(0)$ form a filterbase in $F(T,X)$ by Theorem 2.6.1(f). The topology $\mathcal{T}_u$ obtained this way is called the *topology of uniform convergence*. We denote $F(T,X)$ endowed with $\mathcal{T}_u$ by $F(T,X,u)$. ■

The following remark explains why the preceding topology is called the topology of *uniform* convergence.

**Remark 2.6.6** "UNIFORM" CONVERGENCE Suppose a net $(x_m)$ from $F(T, X, u)$ converges to $x \in F(T, X, u)$. Then for any $V \in \mathcal{V}(0)$, $x_m(t) - x(t) \in V$ eventually for all $t \in T$. ■

**Example 2.6.7** UNIFORM CONVERGENCE ON A SUBSET Rather than $T$ as in Example 2.6.5, consider any subset $S$ of $T$. As in the preceding example, the sets $\mathcal{V}_S(0) = \{W(S, V) : V \in \mathcal{V}(0)\}$ are a base at 0 for a group topology on $F(T, X)$ called the *topology of uniform convergence on* $S$. In this case if a net from $F(T, X)$ converges, the convergence is "uniform" on the set $S$. If $X$ is metrizable, since the open balls $\{B(0, 1/n) : n \in \mathbf{N}\}$ are a base at 0 in $X$, the sets $\{W(S, B(0, 1/n))) : n \in \mathbf{N}\}$ are a base at 0 for $\mathcal{V}_S(0)$ and $\mathcal{T}_S$ is first countable. Hence, sequences are adequate to describe closures and continuity. ■

**Example 2.6.8** POINTWISE CONVERGENCE Let $\mathcal{S}$ denote the class of all singletons of $T$. In view of Theorem 2.6.3 we get the same $\mathcal{S}$-topology by taking $\mathcal{S}$ to be the set of all finite subsets of $T$. This $\mathcal{S}$-topology is called the *topology of pointwise convergence*; we denote $\mathcal{T}_\mathcal{S}$, $\mathcal{V}_\mathcal{S}(0)$, and $F_\mathcal{S}(T, X)$, respectively, by $\mathcal{T}_p$, $\mathcal{V}_p(0)$, and $F(T, X, p)$. It is clear that a net $(x_m)$ from $F(T, X, p)$ converges to $x \in F(T, X, p)$ iff the nets $(x_m(t))$ in $X$ converge to $x(t)$ for each $t \in T$. Another name for $\mathcal{T}_p$ is the *point-open topology*. If we replace $T$ by a subset $S$ of $T$, we call the resulting topology the *topology of pointwise convergence on* $S$.

Since the collection of all finite subsets of $T$ is generally a rather small collection, the topology of pointwise convergence is a fairly weak topology. A subbasic $\mathcal{T}_p$-neighborhood of 0 is a set of the form $\{x \in F(T, X) : x(t) \in V\}$ for some $t \in T$ and neighborhood $V$ of 0 in $X$ which is $V \times \Pi_{s \in T, s \neq t} X_s$ where $X_s = X$ for all $s \neq t$, i.e., a subbasic neighborhood of 0 for the product topology on $\Pi_{t \in T} X_t$, where $X_t = X$ for each $t \in T$; in other words *the product and point-open topologies coincide* on $F(T, X)$. By Theorem 2.6.3, given any collection $\mathcal{S}$ of subsets of $T$, the point-open topology on $\cup\mathcal{S}$ is coarser than the $\mathcal{S}$-topology. ■

**Example 2.6.9** COMPACT-OPEN TOPOLOGY Let $\mathcal{S}$ be the collection of all compact subsets of a topological space $T$. We denote this $\mathcal{S}$-topology by $\mathcal{T}_c$. It is called variously the *compact-open topology*, *topology of uniform convergence on compact sets* and the *topology of compact convergence*. $F_s(T, X)$ is denoted by $F(T, X, c)$ in this case. An open base at 0 for $\mathcal{T}_c$ is given by the sets $W(K, U)$, where $K$ is a compact subset of $T$ and $U$ an open neighborhood of 0 in $X$. Since there are generally more compact sets than finite sets, $\mathcal{T}_c$ is generally finer than the topology of pointwise convergence. ■

We consider next a topology which generalizes the topology of pointwise convergence on a one-point set. Let $\mathcal{B}$ be a filterbase in the set $T$ and $V$ a neighborhood of 0 in $X$. Let

$$Q(\mathcal{B}, V) = \{f \in F(T, X)\colon \text{for some } B \in \mathcal{B},\ f(B) \subset V\}$$

The sets $Q(\mathcal{B}, V)$ are a base at 0 for a compatible topology for $F(T, X)$ called the *topology of convergence on* $\mathcal{B}$ [Example 2.6.11]. First, we note some properties of the $Q(\mathcal{B}, V)$.

**Theorem 2.6.10** PROPERTIES OF $Q(\mathcal{B}, V)$ For any $U, V \in V(0)$,

(a) $U \subset V \Rightarrow Q(\mathcal{B}, U) \subset Q(\mathcal{B}, V)$

(b) $Q(\mathcal{B}, -V) = -Q(\mathcal{B}, V)$;

(c) $Q(\mathcal{B}, U) - Q(\mathcal{B}, U) \subset Q(\mathcal{B}, U - U)$;

(d) If $\mathcal{B}_1$ is subordinate to $\mathcal{B}_2$ (i.e., for each $B_2 \in \mathcal{B}_2$ there is a $B_1 \in \mathcal{B}_1$ such that $B_1 \subset B_2$), then $Q(\mathcal{B}_2, V) \subset Q(\mathcal{B}_1, V)$; and

(e) for any family $\{V_m : m \in M\}$ of neighborhoods of 0 in $X$, $Q(\mathcal{B}, \cap_{m \in M} V_m) = \cap_{m \in M} Q(\mathcal{B}, V_m)$.

**Proof.** In the notation of the statement, we prove only (c). For $f, g \in Q(\mathcal{B}, U)$, choose $B_f$ and $B_g$ from $\mathcal{B}$ such that $f(B_f) \subset U$ and $g(B_g) \subset U$. Choose $B \in \mathcal{B}$ such that $B \subset B_f \cap B_g$. Clearly, $f(B) - g(B) \subset U - U$. ■

**Example 2.6.11** TOPOLOGY OF CONVERGENCE ON A FILTERBASE Clearly, $0 \in Q(\mathcal{B}, V)$ for each $V \in V(0)$. This, in conjunction with Theorem 2.6.10(e), shows that $\{Q(\mathcal{B}, V) : V \in V(0)\}$ is a filterbase. Parts (a)–(c) of Theorem 2.6.10 show that the condition $(FB)$ of the basis Theorem 2.3.1 is satisfied so the sets $Q(\mathcal{B}, V)$ form a base at 0 for a group topology $\mathcal{T}_{\mathcal{B}}$ for $F(T, X)$; $\mathcal{T}_{\mathcal{B}}$ is called the *topology of convergence on the filterbase* $\mathcal{B}$.

To get an intuitive grasp of what convergence in this topology is like, suppose that $T$ is a topological space and that $\mathcal{B} \to t$ in $T$. Then a filterbase $\mathcal{F}$ from $F(T, X)$ "converges to $x$ on $\mathcal{B}$" (i.e., in the topology of convergence on $\mathcal{B}$) iff for each $V \in V(0)$ there is an $F \in \mathcal{F}$ such that $F - x \subset Q(\mathcal{B}, V)$, i.e., iff $\mathcal{F}$ is "close" to $x$ in the vicinity of $t$.

If $\mathcal{B}$ is a filterbase consisting of just one nonempty set $B$, then $Q(\mathcal{B}, V) = W(B, V)$ and $\mathcal{T}_{\mathcal{B}}$ is just the topology of uniform convergence on $B$, i.e., convergence on a filterbase subsumes uniform convergence on a subset.

As another illustration of a topology of convergence on a filterbase, suppose that $T = X = \mathbf{R}$. Let the filterbase in $T$ be the collection $\mathcal{B}$ of open balls $B(0, r)$, $r > 0$, and consider a sequence $(x_n)$ from $F(\mathbf{R}, \mathbf{R})$ converging to $x$ on $\mathcal{B}$. This means that, for any $r > 0$, there is some $N \in \mathbf{N}$ such that for each $n > N$, there is a $d_n > 0$ such that $(x_n - x)(B(0, d_n)) \subset B(0, r)$. Hence, if $x_n \to x$ uniformly on some $B(0, d)$, $d > 0$, in the usual function-theoretic sense, $x_n \to x$ in the topology of convergence on $\mathcal{B}$; hence uniform convergence is stronger than convergence on $\mathcal{B}$. Since convergence on $\mathcal{B}$ obviously does not imply uniform convergence on some neighborhood $B(0, d)$ of 0, uniform convergence on a neighborhood of zero is is strictly stronger than convergence on $\mathcal{B}$.

In virtue of Theorem 2.6.10(e), it follows that if $X$ is Hausdorff, then the topology of convergence on the filterbase $\mathcal{B}$ is Hausdorff iff the only function that vanishes on each $B \in \mathcal{B}$ is the zero function [Theorem 2.6.4]. ■

**Example 2.6.12** $\Phi$-TOPOLOGIES We use the same notation as Example 2.6.11. Extending the notion of topology of convergence on a filterbase to the topology of convergence on a *family* $\Phi$ of filterbases involves only an application of the subbase theorem [Theorem 2.3.2]: $B_\Phi(0) = \{Q(\mathcal{B}, V) : \mathcal{B} \in \Phi,\ V \in V(0)\}$ satisfies the finite intersection condition because the zero function is in each $Q(\mathcal{B}, V)$. The resulting topology $\mathcal{T}_\Phi$ is called the *topology of convergence on the family of filterbases* $\Phi$ or the $\Phi$-*topology*; the neighborhood filter at 0 is denoted by $V_\Phi(0)$ and the topological group by $F_\Phi(T, X)$.

The $\Phi$-topology yields convergence of a filterbase $\mathcal{F}$ from $F_\Phi(T, X)$ to $x \in F_\Phi(T, X)$ if $\mathcal{F}$ is close to $x$ at several sites in $T$—namely, iff $\mathcal{F} \to x$ on each $\mathcal{B} \in \Phi$ [see Example 2.6.11].

For the class $\Phi$ of all maximal filterbases in $T$ the resulting $\Phi$-topology is called the *topology of almost uniform convergence*. The topology of almost uniform convergence is quite similar to the topology of pointwise convergence [Example 2.6.8] and actually coincides with it in certain cases (for example, when $T$ is compact). ■

## 2.7 METRIZABILITY

As usual, $X$ denotes a commutative group. If $X$ is pseudometrizable and $d$ generates its topology, $B(0, r) = \{x \in X : d(0, x) < r\}$, $r > 0$.

We show in Theorem 2.7.3 that a topological group is pseudometrizable iff it has a countable base at 0, in which case the topology is generated by an *invariant* pseudometric. After showing that quotients of pseudometrizable groups are pseudometrizable, we show that the topology of any topological group is generated by a family of invariant pseudometrics.

First, we exhibit a pseudometric that does not yield a group topology.

**Example 2.7.1** INCOMPATIBLE PSEUDOMETRIC TOPOLOGY We define a pseudometric $d$ on the integers $\mathbf{Z}$ as follows: For $j, k \in \mathbf{Z}$, $d(j, k) = d(k, j) = 0$ if $j = k$ or $j < k \leq 0$; take $d(j, k) = 1 = d(k, j)$ if $j > k \geq 0$ and $d(j, k) = d(k, j) = d(0, k) = 1$ if $j \leq 0 \leq k$. It is trivial to verify that:

(a) $d$ is a pseudometric;

(b) any neighborhood of 0 must contain the negative integers; and

(c) the open ball $B(1, \frac{1}{2})$ of radius $\frac{1}{2}$ about 1 is $\{1\}$.

Writing 1 as a $1+0$, there can be no neighborhoods $U$ of 0 and $V$ of 1 such that $U + V \subset B(1, \frac{1}{2})$. This, then, is an incompatible pseudometric topology. (Is there an incompatible *metric* topology?) ■

A sufficient condition for a pseudometric $d$ to induce a group topology is that it be *invariant* (or *translation-invariant*), namely, that $d(x+z, y+z) = d(x, y)$ for all $x, y$ and $z$ in the group. Associated with an invariant pseudometric is a *value.*

**Definition 2.7.2** VALUES A *value* is a map $|\cdot|$ of a group $X$ into the nonnegative reals such that

(a) $|0| = 0$,
(b) $|x| = |-x|$, and
(c) $|x+y| \le |x| + |y|$. ■

The usual absolute value function on the additive groups of integers or real or complex numbers is a value. If $d$ is an invariant pseudometric, then $|\cdot| = d(\cdot, 0)$ is a value, called the *value associated with d.* Conversely, given a value $|\cdot|$, $|x-y|$ is an invariant pseudometric so there is a $1-1$ correspondence between invariant pseudometrics and values.

For any value $|\cdot|$, the open balls $B(0,r) = \{x \in X : |x| < r\}$, $r > 0$, are a filterbase of symmetric sets by (b). Moreover, by (c), $B(0,r/2) + B(0,r/2) \subset B(0,r)$ for any $r > 0$. Thus the sets $B(0,r)$, $r > 0$, are a base at 0 for a group topology by the basis theorem [2.3.1]. In summary: invariant pseudometrics determine group topologies. Nevertheless it is not *necessary* for a pseudometric to be invariant for it to generate a group topology, as the metric of Example 3.1.3 shows. Nevertheless, if a group topology is obtainable from a pseudometric then there is an invariant pseudometric that generates the topology [Theorem 2.7.3].

Metrizability of a topology depends on its separation and countability properties. Since group topologies have a certain amount of inherent separation (namely, regularity [Theorem 2.2.1]) metrizability reduces to just a countability condition. Theorem 2.7.3 was discovered independently by Birkhoff [1936] and Kakutani [1936].

**Theorem 2.7.3** PSEUDOMETRIZABLE $\Leftrightarrow$ COUNTABLE BASE AT 0 A topological group $X$ is pseudometrizable iff it has a countable base at 0. If $X$ is pseudometrizable then there exists an invariant pseudometric $d$ that generates its topology. $(X, d)$ is metrizable iff $(X, d)$ is Hausdorff.

**Proof.** We use the notation of the statement. The countability condition is obviously necessary. Conversely, suppose that $(V_n)$ is a denumerable base of neighborhoods of 0. By Theorem 2.1.10(b), we can define a sequence $(U_n)$ of symmetric neighborhoods of 0 such that $U_1 = V_1$ and $U_{n+1} + U_{n+1} + U_{n+1} \subset U_n \cap V_n$ for each $n \in \mathbf{N}$. Since $(V_n)$ is a base at 0 so is $(U_n)$. As a first step toward getting an invariant pseudometric that generates the topology, consider the function $f$ defined as follows: $f(x,y) = 1$ if $x - y \notin U_1$; $f(x,y) = 0$ if $x - y \in U_n$ for every $n$; otherwise, $f(x,y) = 2^{-k}$ if $x - y \in U_k$ but $x - y \notin U_{k+1}$. For future reference we note that, for any $n$,

$$f(x,y) \le 2^{-n} \quad \text{iff} \quad x - y \in U_n \tag{1}$$

Obviously, $f$ is nonnegative, symmetric, and $f(x+z,\ y+z) = f(x,y)$ for all $x, y$ and $z$ in $X$. It is almost a pseudometric. The function $d$, defined next, is: $d(x,y) = \inf \sum_{i=0}^{p-1} f(z_i, z_{i+1})$ where the infimum is taken over all finite sequences $z_0, \ldots, z_p$ for which $z_0 = x$ and $z_p = y$. In effect, we are seeking a path of shortest distance from $x$ to $y$. We show that $d$ is an invariant pseudometric such that

$$(1/2)f \leq d \leq f \tag{2}$$

Assuming that this has already been demonstrated, it follows that

$$U_n \subset B(0, 2^{-n}) = \{x \in X : d(x,0) < 2^{-n}\} \subset U_{n-1} \tag{3}$$

which shows that the metric topology coincides with the original. Therefore, let us verify that $d$ is an invariant pseudometric which satisfies (2).

Clearly, $d$ is nonnegative and $d(x,x) = 0$ for all $x \in X$. It is symmetric and invariant because $f$ is. That $d$ satisfies the triangle inequality is a routine verification and obviously $d \leq f$. The central part of the proof is to show that $(1/2)f \leq d$, the inequality which yields the right-hand inclusion in (3). We prove $(1/2)f \leq d$ by showing that

$$\sum_{i=0}^{p-1} f(z_i, z_{i+1}) \geq (1/2)f(x,y) \tag{4}$$

for any finite sequence $z_0 = x, z_1, \ldots, z_p = y$. We establish this by induction on $p$. Clearly, this holds for $p = 1$, so suppose that $p > 1$ and that (4) holds for all integers $q < p$ and $x, y$ in $X$. Let $a$ denote the sum on the left side of (4). Consider two possibilities for $a : a \geq 1/2$ and $a < 1/2$. If $a \geq 1/2$, obviously (4) holds because $f(x,y) \leq 1$. If $a < 1/2$, let $h$ be the largest integer such that $\sum_{i<h} f(z_i, z_{i+1}) \leq a/2$; hence $\sum_{i<h+1} f(z_i, z_{i+1}) > a/2$. This implies that $\sum_{i>h} f(z_i, z_{i+1}) < a/2$. By the induction hypothesis, $(1/2)f(x, z_h) \leq \sum_{i<h} f(z_i, z_{i+1})$, so $f(x, z_h) \leq a$. Similarly, $f(z_{h+1}, y) \leq a$, while, obviously, $f(z_h, z_{h+1}) \leq a$. For $a > 0$, let $k$ be the least positive integer such that $(1/2^k) \leq a$. Since $a < 1/2$, $k$ must be at least 2. Thus, by the choice of $k$, if follows that $f(x, z_h) \leq 2^{-k}$, $f(z_h, z_{h+1}) \leq 2^{-k}$, and $f(z_{h+1}, y) \leq 2^{-k}$. By (1), then $x - z_h$, $z_h - z_{h+1}$ and $z_{h+1} - y$ must each belong to $U_k$. Therefore, $x - y \in U_k + U_k + U_k \subset U_{k-1}$, so $f(x,y) \leq 2^{-(k-1)} \leq 2a$, which is what we wanted to show.

If $a = 0$, then $f(z_i, z_{i+1}) = 0$ for $i = 0, \ldots, p-1$, which implies that $z_i - z_{i+1} \in U_n$ for every $n$. Thus $x - y \in U_n + U_n + \cdots + U_n$ ($p$ summands) for every $n$. It follows that $x - y \in \cap_{n \in \mathbf{N}} U_n$ so $f(x,y) = 0$. Thus (4) holds when $a = 0$ as well and the proof of the pseudometrizability of $X$ is complete. ■

**From now on when we say a group is pseudometrizable, we assume that the pseudometric is translation invariant.**

**Corollary 2.7.4** QUOTIENTS If the topological group $X$ is pseudometrizable then, for any subspace $M$ of $X$, so is the quotient group $X/M$.

**Proof.** We use the notation of the statement. It suffices to note that if $X$ has a countable base $(V_n)$ at 0, the canonical images $(\dot{V}_n)$ are a countable base at $\dot{0}$ in $X/M$ [Theorem 2.5.3(a)]. ■

**Example 2.7.5** QUOTIENTS OF METRIZABLE GROUPS By Corollary 2.7.4 we know that if the topological group $(X, d)$ is pseudometrizable then so is any quotient group $X/M$. The following procedure produces a specific pseudometric for $X/M$. Let $|\cdot| = d(\cdot, 0)$ be the value associated with $d$. For $\dot{x} = x + M \in X/M$ we define the *quotient value* to be $|\dot{x}|_q = \inf\{|x+m| : m \in M\}$; the map $d_q(\dot{x}, \dot{y}) = |\dot{x} - \dot{y}|_q$ is called the (associated) *quotient pseudometric.* We show now that:

(a) $|\cdot|_q$ is a value on $X/M$;

(b) the quotient topology $\mathcal{T}_q$ and the topology $\mathcal{T}_{|\cdot|}$ induced by $|\cdot|_q$ on $X/M$ coincide; and

(c) if $d$ is a metric and $M$ is closed, then $X/M$ is metrizable. ■

**Proof.** We use the notation of the statement. (a) It is evident that $|\cdot|_q$ satisfies $|0|_q = 0$, part (a) of Definition 2.7.2. In regard to (b), for any $x \in X$ and $m \in M$, $|-\dot{x}|_q \leq |-x-m| = |x+m|$. Hence $|-\dot{x}|_q \leq |\dot{x}|_q$. Replacing $\dot{x}$ by $-\dot{x}$, it follows that $|-\dot{x}|_q = |\dot{x}|_q$. As for (c), the triangle inequality, note that for any $x, y \in X$ and $m_1, m_2 \in M$, $|\dot{x} + \dot{y}|_q \leq |x + m_1 + y + m_2| \leq |x + m_1| + |y + m_2|$. The desired result now follows from the arbitrariness of $m_1$ and $m_2$.

(b) Let $k : X \to X/M$ denote the canonical map. We show that $k(B(0,r)) = B(\dot{0}, r) = \{\dot{x} \in X/M : |\dot{x}|_q < r\}$ for each $r > 0$. If $x \in B(0,r)$, $r > 0$, then $|\dot{x}|_q \leq |x| < r$; hence $k(B(0,r)) \subset B(\dot{0}, r)$. Conversely, if $|\dot{x}|_q < r$ then there exists $m \in M$ such that $|x+m| < r$; hence $x+m \in B(0,r)$ and $k(x+m) = k(x) = \dot{x}$.

(c) In view of (a) and (b), it suffices to show that $|\dot{x}|_q = 0$ implies $\dot{x} = 0$. If $0 = |\dot{x}|_q = \inf\{|x+m| : m \in M\}$, there is a sequence $(m_n)$ of elements of $M$ such that $|x + m_n| \to 0$, i.e., such that $m_n \to -x$ which implies that $x \in M$ and $\dot{x} = 0$. ■

Suppose that $D$ is a family of invariant pseudometrics on a group $X$. For each $d \in D$, consider the associated value $|\cdot|_d = d(\cdot, 0)$. The symmetric sets $\{x \in X : |x|_d < r\}$ where $d \in D$ and $r > 0$ form a filter subbase satisfying the conditions of Theorem 2.3.2, including $(OB)$, and therefore are a filter subbase of open sets for a group topology $\mathcal{T}_D$ on $X$. We call $\mathcal{T}_D$ the *topology determined by the family of pseudometrics $D$*. It is trivial to verify that $\mathcal{T}_D$ is the weakest topology with respect to which each $|\cdot|_d$ is continuous. We show next that the topology of any topological group is generated by a family of invariant pseudometrics.

**Theorem 2.7.6** GROUP TOPOLOGIES ARE GENERATED BY PSEUDOMETRICS The topology of any topological group $(X, \mathcal{T})$ is determined by a family of invariant pseudometrics.

**Proof.** Let $U(0)$ denote the collection of symmetric neighborhoods of 0 in the topological group $(X, \mathcal{T})$. For each $U \in U(0)$ we may inductively define a sequence $(U_n)$ of symmetric neighborhoods of 0 such that $U_1 = U$ and $U_{n+1} + U_{n+1} + U_{n+1} \subset U_n$ for $n \geq 1$ as in the proof of Theorem 2.7.3. Also as in the proof of Theorem 2.7.3, there is an invariant pseudometric $d_U$ associated with each $U \in U(0)$ such that $x \in U$ iff $|x|_U \leq 1$ where $|\cdot|_U = d_U(\cdot, 0)$. In other words, the closed unit ball determined by $|\cdot|_U$ is just $U$ so the topology generated by $\{d_U : U \in U(0)\}$ is finer than $\mathcal{T}$. By (3) of the proof of Theorem 2.7.3, $|\cdot|_U$ is continuous; therefore each $\{x \in X : |x|_U < r\}$, $r > 0$, is a neighborhood of 0 in $X$ and the topology generated by $\{d_U : U \in U(0)\}$ is coarser than $\mathcal{T}$. ■

## 2.8 EXERCISES

**Notation.** $X$ is a commutative topological group with neighborhood filter $V(0)$ at 0.

2.101 Give an example of two topological groups which are algebraically isomorphic but not topologically isomorphic.

2.102 If $x$ is a nonzero element of a Hausdorff topological group, then there is a neighborhood $U$ of 0 such that $U \cap (x + U) = \varnothing$.

2.103 For any element $x$ of a topological group, $\text{cl}\{x\} = x + \text{cl}\{0\}$.

2.104 Identify the open subgroups of a connected topological group.

2.105 Homogeneous Topologies Let $X$ be an untopologized commutative group. A topology $\mathcal{T}$ for $X$ which makes all translations $x \mapsto x + w$ ($w \in X$) homeomorphisms is called a *homogeneous* topology. A homogeneous topology is a group topology iff for each neighborhood $U$ of 0 (a) there exists a neighborhood $V$ of 0 such that $V \subset -U$ and (b) there exist neighborhoods $V$ and $W$ of 0 such that $V + W \subset U$.

2.106 Closed Subgroups If $M$ is a subgroup of a topological group and $U$ a neighborhood of 0 such that $(\text{cl}\, U) \cap M$ is closed then $M$ is closed.

2.107 Show that a discrete subgroup of a $T_0$ topological group is closed.

2.108 Connectedness For any subset $S$ of $X$, let $[S]$ denote the subgroup generated by $S$.

(a) Show that if $X$ is connected, then $X = [V]$ for any neighborhood $V$ of 0—i.e., a connected group is generated by any neighborhood of 0.

(b) COMPONENTS The *component* $K(x)$ of a point $x$ in any topological space is the union of all connected sets containing $\{x\}$; it is the largest connected set containing $\{x\}$. In any topological group $X$, show that $K(0)$ is a closed subgroup and that, for any $x \in X$, $K(x) = x + K(0)$.

(c) A subgroup $I$ in a ring $E$ such that $aI \subset I$ ($Ia \subset I$) for all $a \in E$ is a left (right) *ideal*. Show that the component $K(0)$ in a topological ring $X$ is a closed 2-sided (i.e., left and right) ideal.

2.109 If $M$ is a dense subgroup of $X$ which is generated by every neighborhood of 0 in $M$ in the sense used in Exercise 2.108(a), then $X$ is generated by every neighborhood of 0 in $X$.

2.110 HOMOMORPHISMS Let $Y$ be a commutative topological group and let $f : X \to Y$ be a homomorphism. Show that:

(a) The direct and inverse images under $f$ of a symmetric set are symmetric.

(b) For any subsets $A$ and $B$ of $X$, $\operatorname{cl} f(A) + \operatorname{cl} f(B) \subset \operatorname{cl} f(A+B)$. The analogous statement holds for $f^{-1}$.

(c) ALMOST CONTINUOUS We say $f$ is *almost* (*nearly*) *continuous* if for any neighborhood $V$ of 0 in $Y$, $\operatorname{cl} f^{-1}(V)$ is a neighborhood of 0 in $X$. We say that $f$ is *almost* (*nearly*) *open* if for any neighborhood $U$ of 0 in $X$, $\operatorname{cl} f(U)$ is a neighborhood of 0 in $Y$. Let $K = f^{-1}(0)$, let $k : X \to X/K$ denote the canonical map $x \mapsto x + K$ and let $g : X/K \to Y$ be the map $x + K \mapsto g(x)$, i.e., $f = g \circ k$. Show that $g$ is almost continuous if $f$ is and that $g$ is almost relatively open iff $f$ is, i.e., if $U$ is any neighborhood in $X$, $\operatorname{cl} f(U)$ is a neighborhood in $f(X)$.

2.111 QUOTIENTS Let $M$ be a subgroup of $X$, equip $X/M$ with the quotient topology [Sec. 2.4], and let $k$ denote the canonical map $x \mapsto x + M$. Show that:

(a) If $X$ is compact or locally compact then $X/M$ is compact or locally compact, respectively.

(b) $X/M$ is discrete iff $M$ is open; $X/M$ carries the trivial topology iff $\operatorname{cl} M = X$.

(c) If $M$ and $X/M$ are compact or locally compact then so is $X$.

(d) Suppose that $M$ is closed and contains the subgroup $N$. Let $g : X \to X/N$ be the map $x \mapsto x + N$. Show that $g(M)$ is a closed subgroup of $X/N$.

(e) If $M$ is closed and $X$ is locally compact and 0-dimensional, then $X/M$ is 0-dimensional.

2.112 FINITE TOPOLOGIES Let $Z$ and $Y$ be vector spaces over the same field and let $\mathcal{L}(Z,Y)$ denote the additive topological group of linear maps from $Z$ into $Y$ with finite topology [Example 2.3.3]. Show that:

(a) $\mathcal{L}(Z,Y)$ is discrete iff $Z$ is finite-dimensional.

(b) Any finite-dimensional subspace of $\mathcal{L}(Z,Y)$ is closed.

2.113 FUNCTION GROUPS Let $\mathcal{S}$ be any collection of subsets of a set $T$ and let $F(T,X)$ be the topological group of all functions mapping $T$ into $X$ with pointwise operations endowed with the $\mathcal{S}$-topology of Example 2.6.2.

(a) Show that $F(T,X)$ may fail to be Hausdorff even though $X$ is.

(b) Let $W(S,V) = \{x \in F(T,X) : x(S) \subset V\}$ be as in $(*)$ of Sec. 2.6. Show that if $V$ is a clopen neighborhood of 0 in $X$ then $W(S,V)$ is clopen for any $S \in \mathcal{S}$. Thus, conclude that if $X$ is zero-dimensional, then so is $F(T,X)$.

(c) Let $T$ be a topological space. If $X$ is Hausdorff and $\cup\mathcal{S}$ is dense in $T$, show that the subspace $C(T,X)$ of continuous maps of $T$ into $X$ is a Hausdorff space.

2.114 MATRIX GROUPS Let $\mathbf{F}^n$ denote the additive topological group of $n$-tuples from $\mathbf{F} = \mathbf{R}$ or $\mathbf{C}$ with euclidean topology; $M_n(\mathbf{F})$ denotes the additive topological group of $n \times n$ matrices with entries from $\mathbf{F}$. Map $(a_{ij}) \in M_n(\mathbf{F})$ into $(a_{11}, \ldots, a_{1n}, a_{21}, \ldots, a_{nn})$ in $\mathbf{F}^{n^2}$ and endow $M_n(\mathbf{F})$ with the inverse image topology (Sec. 2.4) determined by this map.

(a) $M_n(\mathbf{F})$ is a Hausdorff topological group which is locally compact, not compact and second countable.

(b) GENERAL (FULL) LINEAR GROUP $GL_n(\mathbf{F})$ Let $GL_n(\mathbf{F})$ denote the nonsingular matrices of $M_n(\mathbf{F})$. $GL_n(\mathbf{F})$ is a noncommutative Hausdorff multiplicative topological group called the *general or full linear group of degree n*. It is an open subset of $M_n(\mathbf{F})$. Closed subgroups of $GL_n(\mathbf{F})$ are called *matrix groups* and are an important class of Lie groups.

(c) Each of the maps (i)–(iv) are homeomorphisms of $GL_n(F)$ onto itself. (i) and (iii) are antiautomorphsims while (ii) and (iv) are automorphisms.

   i. $A \mapsto A^{-1}$;
   ii. $A \mapsto \overline{A} = (\overline{a}_{ij})$, the bar denoting complex conjugate;
   iii. $A \mapsto A'$ where $A'$ denotes the transpose of $A$; and
   iv. $A \mapsto (A')^{-1}$.

(d) A matrix $A \in GL_n(\mathbf{F})$ is called *orthogonal* if.$A = \overline{A} = (A')^{-1}$. The subgroup of orthogonal matrices is designated $O_n(\mathbf{R})$. $A \in GL_n(\mathbf{C})$ is called *complex orthogonal* if $A = (A')^{-1}$. The subgroup of complex orthogonal matrices is designated $O_n(\mathbf{C})$. A matrix $A \in GL_n(\mathbf{C})$ is *unitary* if $A = (A')^{-1}$. The subgroup of unitary matrices is designated $U_n$. Show that $O_n(\mathbf{R})$, $O_n(\mathbf{C})$, and $U_n$ are closed subgroups of $GL_n(\mathbf{C})$.

## Hints

2.102 Choose a neighborhood $V$ of 0 to which $x$ does not belong and a neighborhood $U$ of 0 such that $U - U \subset V$.

2.103 If $F$ is any closed set to which $x$ belongs, then $F - x$ is a closed set containing 0.

2.104 See Theorem 2.4.4.

2.106 Let $x \in \operatorname{cl} M$ and let $(m_s)$ be a net of elements from $M$ that converges to $x$. Choose a symmetric neighborhood $V$ of 0 such that $V + V \subset U$. Since $\operatorname{cl} M$ is a subgroup, $-x \in \operatorname{cl} M$ so there exists $y \in (-x + V) \cap M$. Thus, eventually, $m_s + y \in (x+V)+(-x+V) = V+V \subset U$. Therefore $m_s + y \in U \cap M$ eventually. This and the fact that $(\operatorname{cl} U) \cap M$ is closed imply that $x + y \in \operatorname{cl} U \cap M$.

2.107 Choose a neighborhood $U$ of 0 such that $U \cap M = \{0\}$ and a neighborhood $V$ of 0 such that $\operatorname{cl} V \subset U$; $\operatorname{cl} V \cap M = \{0\}$ which is closed in $X$. Now use Exercise 2.106.

2.108 (a) Show that $[V]$ is open and use Theorem 2.4.4.

(b) $K(0)$ must be closed because the closure of a connected set is connected. To see that $K(0)$ is a subgroup, let $y \in K(0)$ and show that $K(0) - y$ is connected and contains 0. Hence $K(0) - K(0) \subset K(0)$. For the second statement, for any $x$ in $X$, consider the map $w \mapsto x + w$.

2.109 If $V$ is a symmetric neighborhood of 0 in $X$, then $V \cap M$ is a symmetric neighborhood of 0 in $M$. $[V]$ is open, therefore closed by Theorem 2.4.4. Since $[V] \subset M$, $[V] \subset \operatorname{cl} M$.

2.112 (a) For $x_1, \ldots, x_k \in Z$ and $y_1, \ldots, y_k \in Y$, show that sets of the form $\{A \in \mathcal{L}(Z,Y) : Ax_i = y_i,\ i = 1, \ldots, k\} = O(x_i, y_i)$ are open. Now if $\{x_1, \ldots, x_k\}$ is a basis for $Z$, then for any $A$ in $\mathcal{L}(Z,Y)$, $O(x_i, Ax_i) = \{A\}$ is open, i.e., singletons are open.

Conversely, if $\dim Z = \infty$, given any linearly independent subset $\{x_1, \ldots, x_k\} \subset Z$, it can be extended to a basis for $Z$. As linear maps exist mapping basis elements of $Z$ into arbitrary elements of $Y$, the open subsets of the finite topology must be infinite (nondenumerable, actually), so the topology cannot be discrete.

# Chapter 3

# Completeness



Some of the seminal results of early (i.e., metric) functional analysis, such as the closed graph and Banach-Steinhaus theorems [cf. Theorems 14.3.4 and 11.9.2], require completeness because completeness in a metric setting permits invocation of the Baire category theorem, namely, that a complete metric space is nonmeager [Theorem 11.7.2].

In order to investigate the validity of the above-mentioned results in the broader context of topological vector spaces, an analog of completeness is needed. For that, we use an analog of Cauchy sequence, namely, the *Cauchy net,* a net $(x_s)$ in a topological group such that, given any neighborhood $V$ of 0, $x_s - x_t \in V$ eventually. Certainly, Cauchy sequences in normed spaces are Cauchy nets, so it is accurate to say that Cauchy nets generalize Cauchy sequences. But do other more or less vital properties of Cauchy sequences transfer to Cauchy nets? For example, must a convergent net be Cauchy? Must a cluster point of a Cauchy net be a limit? We show here, easily, that both of these properties survive. Heartened by the knowledge that Cauchy nets share these properties of Cauchy sequences, we define a topological group to be *complete* if each Cauchy net converges. If topological vector spaces which are complete in this sense must also be nonmeager, then many classical proofs could painlessly be adapted to the more general setting. This is not the case, however. There are plenty of complete-but-meager topological vector spaces. The LF-spaces of Chap. 11, for example, are meager but complete.

Although completeness is of some importance in the theory of topological vector spaces, it has nothing like its former clout. We must turn to other settings to validate analogs of some of the classical theorems, in particular to "barreled spaces" for the Banach-Steinhaus theorem [Chap. 9] and to webbed spaces for the closed graph theorem [Chap. 13].

THROUGHOUT THE CHAPTER "GROUP" MEANS "COMMUTATIVE GROUP."

## 3.1 COMPLETENESS

**Notation.** $X$ denotes a commutative topological group with neighborhood filter $V(0)$ of 0. ■

As motivation for the notion of $U$-small defined below, consider a subset $S$ of a normed space. We could say that $S$ is small "of order $r$" if the diameter of $S$ is less than $r$: For all $x, y \in S$, $\|x - y\| < r$ or, equivalently, $x - y \in B(0, r)$. We use neighborhoods of 0 to characterize closeness in topological groups as well:

**Definition 3.1.1** $U$-SMALL SETS Let $U$ be a neighborhood of 0 in the topological group $X$. A subset $E$ is said to be *small of order $U$ ($U$-small)* if $x - y \in U$ for all $x, y \in E$. ■

A *Cauchy net* $(x_s)$ in a topological group $X$ is a net whose elements are arbitrarily "close" for sufficiently "large" indices: Given a neighborhood $U$ of 0, there exists an index $r$ such that $x_s - x_t \in U$ for $s, t \geq r$. Clearly a Cauchy sequence is a Cauchy net. It is also straightforward to verify that the continuous homomorphic image of a Cauchy net is a Cauchy net.

**Definition 3.1.2** COMPLETENESS A topological group $X$ is *complete* if every Cauchy net converges; if only each Cauchy sequence converges, then $X$ is said to be *sequentially complete.* A subset $E$ of $X$ is complete or sequentially complete if each Cauchy net or sequence in $E$ converges to a point of $E$, respectively. ■

The notion of completeness in a topological group is different from the notion of completeness in a pseudometrizable space $(X, d)$. In the latter case a net $(x_s)$ is Cauchy if $d(x_s, x_t)$ eventually becomes arbitrarily small. A situation in which the two notions differ is the subject of the next example.

**Example 3.1.3** METRIC VERSUS GROUP COMPLETENESS Consider the additive group $\mathbf{R}$ of real numbers. A metric for $\mathbf{R}$ is given by $d(x, y) = |\tan^{-1} x - \tan^{-1} y|$. As a metric space ($\mathbf{R}$,$d$) is not complete for $(n)$ is a nonconvergent

Cauchy sequence. Since, for $r > 0$, $B(0,r) = \{b \in \mathbf{R} : |\tan^{-1} b| < r\} = (-\tan^{-1} b, \tan^{-1} b)$, $d$ induces the usual Euclidean topology on $\mathbf{R}$, $(\mathbf{R},d)$ is a complete topological group. Note that $(n)$ is a Cauchy sequence in $(\mathbf{R}, d)$ but not in the topological group $\mathbf{R}$ with its usual topology. ■

The metric of Example 3.1.3 is not invariant [Sec. 2.7], yet it induces a group topology on $\mathbf{R}$. Invariance, or rather the lack of it, is really the key to the situation of the example. If $d$ is an invariant pseudometric, then $d(x_s, x_t)$ goes to 0 iff $d(x_s - x_t, 0)$ goes to 0, so what occurred in Example 3.1.3 cannot happen. In summary:

**Theorem 3.1.4** Invariant Metrics and Completeness If $d$ is an invariant pseudometric on the group $X$ then $(X, d)$ is complete as a pseudometric space iff it is complete as a topological group.

We can also describe completeness by means of filterbases. In a metric space $(X, d)$ the *diameter* $d(B)$ of a subset $B$ is $\sup\{d(x,y) : x, y \in B\}$. Hence, a sequence $(x_n)$ in $(X,d)$ is Cauchy iff the diameter of $B_n = \{x_n, x_{n+1}, \ldots\}$ goes to 0 with $n$. The sets $B_n$ constitute the Fréchet filterbase associated with $(x_n)$; evidently $d(B_n - B_n) \to 0$. This suggests the following way to define a Cauchy filterbase.

**Definition 3.1.5** Cauchy Filterbase A filterbase $\mathcal{B}$ in a topological group $X$ is Cauchy if, given any neighborhood $V$ of 0, there exists an element $B$ of $\mathcal{B}$ such that $B - B \subset V$. ■

**Theorem 3.1.6** Cauchy Nets and Filterbases Under the correspondence between nets $(x_s)$ and filterbases $\mathcal{B}$ noted in Theorem 1.1.8, $(x_s)$ is Cauchy iff $\mathcal{B}$ is Cauchy.

**Proof.** If $(x_s)_{s \in S}$ is a Cauchy net, the corresponding filterbase $\mathcal{B}$ is given by the sets $B_s = \{x_t : t \geq s\}$, $s \in S$. If $(x_s)$ is Cauchy, then given a neighborhood $V$ of 0, there exists an index $r$ such that $B_s - B_t \subset V$ for $s, t \geq r$.

Conversely, suppose $\mathcal{B}$ is a Cauchy filterbase. The set $S = \{(a, A) : a \in A \text{ and } A \in \mathcal{B}\}$ ordered by $(a, A) \leq (b, B)$ iff $B \subset A$ is a directed set. The net corresponding to $\mathcal{B}$ is the map $(b, B) \mapsto x_{(b,B)} = b$. Since $\mathcal{B}$ is Cauchy then, given a neighborhood $V$ of 0, there exists $C \in \mathcal{B}$ such that $C - C \subset V$. Let $c \in C$ and suppose $(a, A), (b, B) \geq (c, C)$, i.e., that $A, B \subset C$. Then $x_{(a,A)} - x_{(b,B)} = a - b \in A - B \subset C - C \subset V$ and the net $(x_{(b,B)})$ is seen to be Cauchy. ■

Another way to characterize Cauchy filterbases $\mathcal{B}$ which shows that $\mathcal{B}$ contains arbitrarily small sets $B$ (rather than $B - B$) is given in the following result.

**Theorem 3.1.7** CAUCHY FILTERBASES CONTAIN SMALL SETS In a topological group $X$, a filterbase $\mathcal{B}$ is a Cauchy filterbase iff for each neighborhood $V$ of 0, there exists $B \in \mathcal{B}$ and an element $x_B \in X$ such that $B \subset x_B + V$.

**Proof.** It is clear that Cauchy filterbases satisfy the condition. Conversely, suppose that the condition holds and let $V$ be a neighborhood of 0. Choose a neighborhood $U$ of 0 such that $U - U \subset V$. By the condition, there exists $B \in \mathcal{B}$ and an element $x_B \in X$ such that $B \subset x_B + U$. Thus $B - B \subset (x_B + U) - (x_B + U) = U - U \subset V$. ■

By Theorem 3.1.7, it is clear that convergent filterbases are Cauchy. It is also easy to see that a filterbase subordinate to a Cauchy filterbase is Cauchy.

**Theorem 3.1.8** CLUSTER POINTS OF CAUCHY NETS In a topological group $X$, if $x$ is a cluster point of a Cauchy net $(x_s)$ or a Cauchy filterbase $\mathcal{B}$, then $x_s \to x$ or $\mathcal{B} \to x$, respectively.

**Proof.** Let $V$ be a neighborhood of 0 and let $U$ be a neighborhood of 0 such that $U + U \subset V$. Suppose $(x_s)$ is a Cauchy net with cluster point $x$ and choose an index $t$ such that $x_r - x_s \in U$ for $r, s \geq t$. Choose $p \geq t$ such that $x_p \in x + U$. For any $r \geq t$, $x_r - x_p \in U$, so $x_r \in x_p + U \subset x + U + U \subset x + V$ and $(x_s)$ is seen to converge to $x$.

Now suppose that $x$ is a cluster point of the Cauchy filterbase $\mathcal{B}$ and $U \in \mathcal{V}(0)$. Since $x$ is a cluster point of $\mathcal{B}$, $x + U$ meets each $B \in \mathcal{B}$. Since $\mathcal{B}$ is Cauchy, there exists $B$ such that $B - B \subset U$. For any $b$ in $(x+U) \cap B$, $b - x \in U$. Hence, for any $b'$ in $B$,

$$b' - x = (b' - b) + (b - x) \in (B - B) + U \subset U + U \subset V$$

Thus $B \subset x + V$ and $\mathcal{B} \to x$. ■

**Theorem 3.1.9** CLOSEDNESS AND COMPLETENESS Let $E$ be a complete subset of a topological group $X$. Then:

(a) a closed subset $B$ of $E$ is complete;

(b) if $X$ is Hausdorff then $E$ is closed.

**Proof.** Let $E$ be a complete subset of a topological group $X$. (a) Let $(x_s)$ be a Cauchy net in $B$. Since $(x_s)$ must also be a Cauchy net in $E$, it must converge to a point $x$ of $E$. Hence $x \in \operatorname{cl} B = B$.

(b) Suppose that $x \in \operatorname{cl} E$ so that there exists a net $(x_s)$ in $E$ which converges to $x$. Since $E$ is complete, the Cauchy net $(x_s)$ converges to a point of $E$. As limits are unique in a Hausdorff space, $x \in E$. ■

We show next that the existence of a complete neighborhood of 0 implies that the group is complete.

**Theorem 3.1.10** COMPLETE NEIGHBORHOOD OF 0 If the topological group $X$ has a complete neighborhood $V$ of 0 then $X$ is complete.

**Proof.** Let $V$ be a complete neighborhood of 0. If $(x_s)$ is a Cauchy net, there is some index $t$ such that $x_s \in x_t + V$ for $s \geq t$. As a translate of a complete set, $x_t + V$ is complete. Since $\{x_s : s \geq t\}$ is also a Cauchy net, it converges to some $x \in x_t + V$; clearly, $x$ is also a limit of $(x_s)$. ■

**Theorem 3.1.11** COMPACT IMPLIES COMPLETE A compact subset $E$ of a topological group $X$ is complete. Hence any locally compact topological group $X$ is complete.

**Proof.** We use the notation of the statement. Let $V$ be a complete neighborhood of 0. If $(x_s)$ is a Cauchy net in $E$, then $(x_s)$ must have a cluster point $x$ in $E$ by Theorems 1.1.10(b) and 1.1.8. Since $(x_s)$ is Cauchy, it must converge to $x$ by Theorem 3.1.8.

As for the second statement, it suffices to note that a locally compact topological group has a compact neighborhood of 0, hence a complete neighborhood of 0. $X$ is therefore complete by Theorem 3.1.10. ■

Since discrete groups are locally compact, it follows that discrete groups are complete. The following result is also a consequence of Theorem 3.1.10.

**Theorem 3.1.12** A CRITERION FOR COMPLETENESS Let $\mathcal{T}_s$ and $\mathcal{T}_w$ be Hausdorff group topologies for a group $X$. Let $\mathcal{V}_s(0)$ denote the filter of $\mathcal{T}_s$-neighborhoods of 0. If $\mathcal{T}_s$ is stronger than $\mathcal{T}_w$ and there exists a base $\mathcal{B}_w$ of $\mathcal{T}_w$-complete sets for $\mathcal{V}_s(0)$, then $X$ is $\mathcal{T}_s$-complete.

**Proof.** We use the notation of the statement. To prove than $X$ is $\mathcal{T}_s$-complete, we show that it has a $\mathcal{T}_s$-complete neighborhood of 0 and use Theorem 3.1.10. Specifically, we show that any $B \in \mathcal{B}_w$ is $\mathcal{T}_s$-complete.

Let $(x_s)$ be a net from $B \in \mathcal{B}_w$ which is $\mathcal{T}_s$-Cauchy. Note that $(x_s)$ is also a Cauchy net with respect to the weaker topology $\mathcal{T}_w$. Since $B$ is $\mathcal{T}_w$-complete, there exists $x \in B$ such that $(x_s)$ is $\mathcal{T}_w$-convergent to $x$. Let $V$ be a $\mathcal{T}_s$-neighborhood of 0 and choose $U \in \mathcal{B}_w$ such that $U \subset V$. Since $U$ is $\mathcal{T}_w$-complete and $\mathcal{T}_w$ is Hausdorff, it follows that $U$ is $\mathcal{T}_w$-closed [Theorem 3.1.9]. Since $(x_s)$ is $\mathcal{T}_s$-Cauchy, there exists an index $t$ such that $x_s \in x_t + U$ for $s \geq t$. Thus $x \in \operatorname{cl}_{\mathcal{T}_w}(x_s + U) = x_s + U$ for any $s \geq t$. Therefore, $x \in x_s + V$ eventually and $(x_s)$ is $\mathcal{T}_s$-convergent to $x$. ■

## 3.2 FUNCTION GROUPS

**Notation.** $X$ denotes a commutative topological group and $T$ a set throughout. ■

If $T$ is any set, the set $F(T, X)$ of all maps of $T$ into $X$ forms a group with respect to pointwise addition of functions as discussed in Sec. 2.6. In Sec. 2.6 we also discussed a class of group topologies for $F(T, X)$ called $\mathcal{S}$-topologies:

$\mathcal{S}$ is any family of subsets of $T$ and the basic neighborhoods of 0 for the $\mathcal{S}$-topology are the sets $W(S,V) = \{x \in F(T,X) : x(S) \subset V\}$ where $S \in \mathcal{S}$ and $V$ is a neighborhood of 0 in $X$. By Theorem 2.6.3, we may suppose that $\mathcal{S}$ is closed with respect to the formation of subsets of finite unions. We denote $F(T,X)$ endowed with the $\mathcal{S}$-topology $\mathcal{T}_{\mathcal{S}}$ by $F_{\mathcal{S}}(T,X)$.

The result below provides a necessary and sufficient condition for $\mathcal{S}$-convergence and also relates $\mathcal{S}$-convergence and pointwise convergence [Example 2.6.8].

**Theorem 3.2.1** $\mathcal{S}$-CONVERGENCE AND POINTWISE CONVERGENCE With notation as above, a net $(x_m)$ from $F_{\mathcal{S}}(T,X)$ is $\mathcal{T}_{\mathcal{S}}$-convergent to $x \in F_{\mathcal{S}}(T,X)$ iff $(x_m)$ is a $\mathcal{T}_{\mathcal{S}}$-Cauchy net and $x_m(s) \to x(s)$ for each $s \in \cup\mathcal{S}$; this latter property is called *pointwise convergence on* $\cup\mathcal{S}$ and is the same as convergence with respect to the topology of pointwise convergence on $S = \cup\mathcal{S}$ discussed in Example 2.6.8.

**Proof.** We use the notation of the statement. Suppose that $x_m \to x$ with respect to $\mathcal{T}_{\mathcal{S}}$. Then $(x_m)$ is a $\mathcal{T}_{\mathcal{S}}$-Cauchy net. Since $\mathcal{S}$ may be supposed to contain all subsets of finite unions of sets of $\mathcal{S}$ [Theorem 2.6.3], given any neighborhood $V$ of 0 in $X$ and any $s \in S$, $x_m(s) - x(s) \in V$ eventually and the conditions are seen to be necessary.

Conversely, suppose that the $\mathcal{T}_{\mathcal{S}}$-Cauchy net $x_m \to x$ pointwise on $S \in \mathcal{S}$. Since $(x_m)$ is $\mathcal{T}_{\mathcal{S}}$-Cauchy, given any neighborhood $V$ of 0, there exists an index $r$ such that $x_m - x_n \in W(S,V)$ for $m, n \geq r$. Since $x_m \to x$ pointwise on $S$, for any $s \in S$, there is an index $a(s) \geq r$ such that $x_{a(s)}(s) - x(s) \in V$. Hence, for any $s \in S$ and $u \geq r$,

$$x_u(s) - x(s) = x_u(s) - x_{a(s)}(s) + x_{a(s)}(s) - x(s) \in V + V$$

Since $V$ is arbitrary, the desired result follows. ■

The next result provides a sufficient condition for the completeness of a function group $F_s(T,X)$.

**Theorem 3.2.2** $X$ COMPLETE $\Rightarrow F_{\mathcal{S}}(T,X)$ COMPLETE If the topological group $X$ is complete, then so is $F_{\mathcal{S}}(T,X)$ for any collection $\mathcal{S}$ of subsets of $T$.

**Proof.** We use the notation of the statement. Let $(x_m)$ be a Cauchy net in $F_{\mathcal{S}}(T,X)$ and let $H = \cup\mathcal{S}$. Since $\mathcal{S}$ may be assumed to contain singletons [Theorem 2.6.3], given any $t \in H = \cup\mathcal{S}$ and neighborhood $V$ of 0 in $X$, $x_m - x_n \in W(\{t\},V)$ eventually, i.e., $x_m(t) - x_n(t) \in V$ eventually. For each $t \in H$, then $(x_m(t))$ is a Cauchy net in $X$. Since $X$ is complete, each $(x_m(t))$ converges; we call the limit $x(t)$. For $t \notin H$, take $x(t) = 0$. Thus $x_m \to x$ pointwise on $H$ so, by Theorem 3.2.1, $x_m \to x$ in the $\mathcal{S}$-topology. ■

**Corollary 3.2.3** ANY POWER OF A COMPLETE GROUP IS COMPLETE If $X$ is a complete topological group and $T$ any set then $X^T = \Pi_{t \in T} X_t$ (product topology) is complete.

**Proof.** We use the notation of the statement. Let $\mathcal{S}$ be the collection of finite subsets of $T$ and consider $F_{\mathcal{S}}(T, X)$. As noted in Example 2.6.8, $X^T = F(T, X, p) = \Pi_{t \in T} X_t$ and the result follows from the theorem. ■

# 3.3 TOTAL BOUNDEDNESS

**Notation.** $X$ denotes a commutative topological group with neighborhood filter $V(0)$ of 0. ■

For $r > 0$, a *finite* $r$*-net* for a subset $S$ of a metric space $(T, d)$ is a finite collection $x_1, \ldots, x_n$ of elements from $T$ such that $S$ is covered by $r$-neighborhoods $B(x_i, r) = \{y : d(x_i, y) < r\}$: $S \subset \cup_{i=1}^n B(x_i, r)$. In a normed space $B(x, r) = x + B(0, r)$ for each $x$ and each $r > 0$ so the covering condition becomes $S \subset \cup_{i=1}^n (x_i + B(0, r))$. In an arbitrary metric space $(T, d)$, if a finite $r$-net exists for $S$ for every $r > 0$ then $S$ is said to be *totally bounded.* Moreover, by the triangle inequality, the $r$-net may actually be assumed to be a subset of $S$ for each $r$. The notion of total boundedness of Definition 3.3.1 is due to von Neumann [1935, pp. 7–8].

**Definition 3.3.1** TOTALLY BOUNDED A subset $B$ of a topological group $X$ is said to be *totally bounded* or *precompact* if for all neighborhoods $V$ of 0 there exists a finite number of elements $x_1, \ldots, x_n$ from $X$ such that the sets $E \subset \cup_{i=1}^n (x_i + V)$. ■

Clearly any subset of a totally bounded set is totally bounded. Total boundedness in a topological group $X$ is closely related to compactness. In Sec. 3.3.1, we show that relative compactness implies total boundedness; in a *complete* topological group, a closed and totally bounded subset is compact [Theorem 3.4.1].

**Theorem 3.3.2** COVERINGS BY SMALL SETS A subset $B$ of a topological group $X$ is totally bounded iff it can be covered by a finite number of small sets no matter what order of smallness is specified.

**Proof.** Suppose that $B$ is totally bounded and let $U \in V(0)$. Choose $V \in V(0)$ such that $V - V \subset U$. Since $B$ is totally bounded, there are elements $x_1, \ldots, x_n$ in $X$ such that $B$ is covered by the sets $\{x_i + V\}$. For any $x$ and $y$ from $x_i + V$,

$$x - y = x - x_i + x_i - y \in V - V \subset U$$

so each $x_i + V$ is small of order $U$.

Conversely, suppose that for each $U \in V(0)$, $B$ can be covered by a finite number of sets $\{B_i\}$, each of which is small of order $U$. Choose $x_i \in B_i$ for $i = 1, 2, \ldots, n$. Then for all $y \in B_i$, $y - x_i \in U$ or $y \in x_i + U$; hence $B_i \subset x_i + U$ and the total boundedness of $B$ follows. ■

### 3.3.1 Total Boundedness and Subbases

The next result shows that Theorem 3.3.2 can be refined to the extent that orders of smallness need only be investigated for subbases of the neighborhood filter $V(0)$ of 0.

**Theorem 3.3.3** FINITE SUBBASIC COVERS Let $\mathcal{B}$ be a subbase for the filter $V(0)$ in the topological group $X$ and suppose that $K \subset X$. If, for each $B \in \mathcal{B}$, there is a finite cover for $K$ by sets small of order $B$, then $K$ is totally bounded.

**Proof.** We use the notation of the statement. To show $K \subset X$ to be totally bounded, it suffices to show that $K$ has a finite cover by sets small of order $U$, where $U$ is an arbitrary member of a filterbase for $V(0)$—in particular where $U$ is a finite intersection of sets from $\mathcal{B}$. To simplify the notation, we suppose that $U = B_1 \cap B_2$ where $B_1, B_2 \in \mathcal{B}$. By hypothesis there exist finite covers $\{E_i\}$ and $\{F_j\}$ of $K$ where each $E_i$ is small of order $B_1$ and each $F_j$ is small of order $B_2$. Clearly, $\{E_i \cap F_j\}$ covers $K$ and it only remains to note that each $E_i \cap F_j$ is small of order $U$. ■

### 3.3.2 Cauchy Boundedness

The notion introduced next, Cauchy boundedness, is equivalent to total boundedness.

**Definition 3.3.4** CAUCHY BOUNDED Let $B$ be a subset of a topological group $X$. If for any $U \in V(0)$ and each denumerable subset $I$ of $B$ there are distinct points $x, y \in I$ such that $x - y \in U$, then $B$ is said to be *Cauchy bounded*. ■

**Theorem 3.3.5** CAUCHY BOUNDED = TOTALLY BOUNDED In any topological group $X$, Cauchy boundedness is equivalent to total boundedness.

**Proof.** Let $I$ be an infinite subset of the totally bounded set $B$ and let $U$ be a neighborhood of 0. Choose a neighborhood $V$ of 0 such that $V - V \subset U$. Since $B$ is totally bounded, there are elements $x_1, \ldots, x_n \in X$ such that $I \subset B \subset \cup_{i=1}^{n}(x_i + V)$. Since $I$ is infinite, for some $i$ there exist distinct $x, y \in I$ such that $x, y \in (x_i + V)$. Thus $x - y \in (x_i + V) - (x_i + V) = V - V \subset U$ and $B$ is Cauchy bounded.

Conversely, suppose that $B$ is not totally bounded so that there must be some symmetric neighborhood $U$ of 0 such that no finite number of translates of $U$ covers $B$. Thus there must be some $e_1 \in B$ such that $e_1 \notin U$. By induction we select a sequence $(e_n)$ of distinct points from $B$ such that for every $n \in \mathbf{N}$, $e_n \notin \cup_{j=1}^{n-1}(e_j + U)$. We contend that no two points of $I = \{e_1, \ldots\}$ can be $U$-close. To see this, suppose that $i < j$. Since $U = -U$, if $e_j - e_i \in \pm U$, this contradicts the way in which the $e$'s were chosen. Hence $B$ is not Cauchy bounded and the proof is complete. ■

**Theorem 3.3.6** CLOSURES OF TOTALLY BOUNDED SETS If $B$ is a totally bounded subset of a topological group $X$, then so is its closure.

**Proof.** Given a neighborhood $V$ of 0, choose a closed neighborhood $U$ of 0 such that $U \subset V$. For a totally bounded set $B \subset X$, there exist $x_i, \ldots, x_n \in B$ such that $B \subset \cup_{i=1}^{n}(x_i + U)$. Hence $\operatorname{cl} B \subset \operatorname{cl} \cup_{i=1}^{n}(x_i + U) = \cup_{i=1}^{n}(x_i + U) \subset \cup_{i=1}^{n}(x_i + V)$. ■

## 3.4 COMPACTNESS

**Notation.** $X$ denotes a commutative topological group with neighborhood filter $\mathcal{V}(0)$ of 0. ■

Suppose that $B$ is a relatively compact subset of $X$. If $V$ is any neighborhood of 0, the collection $\{x + V : x \in B\}$ certainly covers $\operatorname{cl} B$. Since $\operatorname{cl} B$ is compact, a finite subcollection $x_1 + V, \ldots, x_n + V$ covers $B$. Hence, if $B$ is relatively compact, $B$ is totally bounded. For $T_0$-topological groups—hence Hausdorff by Theorem 2.2.2—compact subsets are closed. In complete topological groups we have a converse statement, which explains why "precompact" is a synonym for "totally bounded."

**Theorem 3.4.1** A complete totally bounded subset of a topological group $X$ is compact; in particular [Theorem 3.1.9(a)], a closed and totally bounded subset of a complete group is compact.

**Proof.** We show that the complete totally bounded set $B$ is compact by showing that every maximal filterbase $\mathcal{B}$ on $B$ converges to a point of $B$ [Theorem 1.1.10]. Since $B$ is complete, it suffices to show that $\mathcal{B}$ is a Cauchy filterbase. To that end, let $V$ be a neighborhood of 0. Since $B$ is totally bounded, there exist $\{x_1, \ldots, x_n\} \subset X$ such that $B \subset \cup_{i=1}^{n}(x_i + V)$. We want to show that, for some $i$, there exists $E \in \mathcal{B}$ such that $E \subset x_i + V$. Since $\mathcal{B}$ is maximal, if $(x_i + V) \cap B$ does not contain any element of $\mathcal{B}$, then its complement $\complement(x_i + V)$ must contain some $E_i \in \mathcal{B}$ [Theorem 1.1.5(a)]. Hence, if none of the $x_i + V$ contain an element of $\mathcal{B}$, there exist $E_i$, $i = 1, 2, \ldots, n$, such that $E_i \subset \complement(x_i + V)$. Since $\mathcal{B}$ is a filterbase, there exists $E \in \mathcal{B}$ such that $E \subset \cap_{i=1}^{n} E_i \subset \cap_{i=1}^{n} \complement[(x_i + V)] \subset \complement B$, a contradiction. ■

If $K$ is a compact subset of a topological group $X$ then $K$ is totally bounded by the discussion preceding Theorem 3.4.1. By Theorem 3.1.11,

$K$ is complete. Conversely, by Theorem 3.4.1, a complete, totally bounded subset is compact. In summary:

**Theorem 3.4.2** COMPACT = COMPLETE + TOTALLY BOUNDED A subset of a topological group is compact iff it is complete and totally bounded.

## 3.5 UNIFORM CONTINUITY

**Notation.** $X$ denotes a commutative topological group with neighborhood filter $V(0)$ of 0. As usual, all groups are commutative. ∎

Let $f$ map a subset $D$ of the topological group $X$ into the topological group $Y$. We say that $f$ is *uniformly continuous* if to each neighborhood $V$ of 0 in $Y$ there exists a neighborhood $U$ of 0 in $X$ such that when $x, y \in D$ and $x - y \in U$, then $f(x) - f(y) \in V$. In other words, closeness anywhere in the domain—$x$ and $y$ can be anywhere—yields closeness in the range. Clearly, uniform continuity implies continuity. In Theorem 2.1.8 we proved that a homomorphism is continuous iff it is continuous at a single point. We can say more:

**Theorem 3.5.1** CONTINUOUS HOMOMORPHISMS ARE UNIFORMLY CONTINUOUS A continuous homomorphism $f$ mapping a topological group $X$ into the topological group $Y$ is uniformly continuous.

**Proof.** With notation as above, for any neighborhood $V$ of 0 in $Y$, there exists a neighborhood $U$ of 0 in $X$ such that $f(U) \subset V$ by the continuity of $f$. Therefore, if $x, y \in X$ and $x - y \in U$, then $f(x - y) = f(x) - f(y) \in V$. ∎

In any topological group the inversion map $x \mapsto -x$ is a continuous homomorphism, hence is uniformly continuous. Translation, although not a homomorphism, is seen to be uniformly continuous by a direct argument.

**Theorem 3.5.2** UNIFORM CONTINUOUS IMAGE OF A CAUCHY NET Let $f$ be a uniformly continuous map of a subset $D$ of a topological group $X$ into a topological group $Y$. If $(x_s)$ is a Cauchy net in $D$, then $(f(x_s))$ is a Cauchy net.

**Proof.** We use the notation above. Given the neighborhood $V$ of 0 in $Y$, choose a neighborhood $U$ of 0 in $X$ such that for $x, y \in D$ and $x - y \in U$, $f(x) - f(y) \in V$. Since $(x_s)$ is Cauchy, $x_s - x_t \in U$ for sufficiently large $s$ and $t$; hence $f(x_s) - f(y_t) \in V$. ∎

**Theorem 3.5.3** UNIFORM CONTINUOUS IMAGE OF A TOTALLY BOUNDED SET Let $X$ and $Y$ be topological groups and let $f : X \to Y$ be uniformly continuous. Then if $B$ is a totally bounded subset of $X$, $f(B)$ is totally bounded.

**Proof.** We use the notation above. Let $V$ by any neighborhood of 0 in $Y$. Choose a neighborhood $U$ of 0 in $X$ such that $x - y \in U$ implies $f(x) - f(y) \in V$. Since $B$ is totally bounded, there exist $x_1, \ldots, x_n \in B$ such that $B \subset \cup_{i=1}^n (x_i + U)$. To complete the proof we show that $f(x_i + U) \subset f(x_i) + V$ $(i = 1, 2, \ldots, n)$. To do that, suppose that $y \in x_i + U$. Then $y - x_i \in U$ and $f(y) - f(x_i) \in V$. Thus $f(B) \subset \cup_{i=1}^n f(x_i + U) \subset \cup_{i=1}^n (f(x_i) + V)$. ■

As an immediate application of Theorem 3.5.2 we show the product of complete sets to be complete.

**Theorem 3.5.4** PRODUCTS OF COMPLETE SETS Let $(X_m)_{m\in M}$ be a collection of topological groups and, for each $m \in M$, let $E_m$ be a complete subset of $X_m$. Then their product $\Pi_{m\in M} E_m$ is a complete subset of the topological group $X = \Pi_{m\in M} X_m$ in its product topology.

**Proof.** We use the notation above. Let $(x_s)$ be a Cauchy net in $\Pi_{m\in M} E_m$. Each projection $pr_m$ of $X$ onto $X_m$ is a continuous homomorphism, hence is uniformly continuous by Theorem 3.5.1. Thus each $(pr_m(x_s))$ is a Cauchy net in $E_m$ by Theorem 3.5.2. As each $E_m$ is complete, there exists $x_m \in E_m$ such that $\lim_s pr_m(x_s) = x_m$. It follows that $x_s \to x = (x_m)$ by the net version of Theorem 1.1.6. ■

As in Sec. 2.6, let $F(T, X)$ denote the group of all maps of the set $T$ into the topological group $X$. For each $t$ in $T$ the *evaluation map* (*Dirac measure*) $t'$ *determined by* $t$ is the map

$$\begin{array}{rccc} t': & F(T,X) & \rightarrow & X \\ & x & \mapsto & x(t) \end{array}$$

Evidently, $t'$ is a homomorphism. Suppose $\mathcal{S}$ is a collection of subsets of $T$ and we endow $F(T, X)$ with the $\mathcal{S}$-topology of Example 2.6.2. For $t \in S \in \mathcal{S}$ and $V$ a neighborhood of 0 in $X$, then $t'(W(S, V))) \subset V$ by the definition of $W(S, V)$. Therefore $t'$ is continuous at 0, hence uniformly continuous by Theorem 3.5.1.

**Definition 3.5.5** UNIMORPHISM An injective map $f$ of a subset $D$ of a topological group $X$ into a topological group $Y$ is called a *unimorphism* if $f$ and $f^{-1}$ are uniformly continuous. If $f : X \to Y$ is a bijective unimorphism, we say that $X$ and $Y$ are *unimorphic*. ■

Unimorphisms preserve properties of a topological group related to its uniform structure, such as completeness, as we now show.

**Theorem 3.5.6** UNIMORPHIC IMAGE OF A COMPLETE GROUP The unimorphic image $Y = f(X)$ of a complete topological group $X$ is complete.

**Proof.** We use the notation of the statement. If $(y_s)$ is a Cauchy net in $Y$, then $(f^{-1}(y_s))$ is a Cauchy net in $X$ by Theorem 3.5.2. Since $X$ is complete, $f^{-1}(y_s)$ converges to some $x$ in $X$. By the continuity of $f$, $f(f^{-1}(y_s)) = y_s \to f(x)$. ■

## 3.6 UNIFORMLY CONTINUOUS MAPS

**Notation.** $X$ denotes a commutative topological group with neighborhood filter $V(0)$ of 0. All groups are commutative. ■

Addition of nets in a topological group is a little more complicated than addition of sequences because the directed sets on which they are defined may differ. Even so, it does not present a serious problem.

**Example 3.6.1** ADDITION OF NETS If $S$ and $T$ are directed sets, it is easy to verify that $S \times T$ is directed by the *pointwise ordering*: $(a, b) \leq (c, d)$ iff $a \leq c$ and $b \leq d$. If $\{x_s : s \in S\}$ and $\{y_t : t \in T\}$ are nets in a topological group $X$, then we define their *sum* $(x_s) + (y_t)$ to be the net $(x_s + y_t)$ defined by $(s, t) \to x_s + y_t$. It is straightforward to verify that $x_s \to x$ and $y_t \to y$ implies that $x_s + y_t \to x + y$; we also write $\lim(x_s + y_t) = \lim x_s + \lim y_t$. ■

**Theorem 3.6.2** EXTENDING UNIFORMLY CONTINUOUS MAPS If $f$ is a uniformly continuous map of a dense subset $D$ of a topological group $X$ into a complete Hausdorff topological group $Y$, then there is a unique uniformly continuous extension $\overline{f}$ of $f$ to all of $X$.

**Proof.** We use the notation of the statement. Let $V_X(0)$ and $V_Y(0)$ denote the neighborhoods of 0 in $X$ and $Y$, respectively. Given $x \in X$, choose a net $(x_s)$ in $D$ which converges to $x$. Since $f$ is uniformly continuous $(f(x_s))$ is a Cauchy net in $Y$ by Theorem 3.5.2. Since $Y$ is complete and Hausdorff, $(f(x_s))$ has a unique limit which we define to be $\overline{f}(x)$. Extending a function in this way is called *extension by continuity;* $\overline{f}$ is called the *continuous extension of* $f$. We verify next that $\overline{f}$ is well defined.

Let $(x_s)$ and $(y_t)$ be nets in $D$ which converge to $x$. Let $a = \lim f(x_s)$ and $b = \lim f(y_t)$. To show that $a = b$, we show that $b \in \text{cl}\{a\} = \cap(a + V_Y(0)) = \{a\}$ [since $X$ is Hausdorff]. Let $V \in V_Y(0)$ and choose a symmetric neighborhood $U \in V_Y(0)$ such that $U + U + U \subset V$. Since $f$ is uniformly continuous, we can choose a neighborhood $W \in V_X(0)$ such that $f(W) \subset U$. Since $(x_s)$ and $(y_t)$ converge to $x$, $x_s - y_t \in W$ eventually; hence $f(y_t) - f(x_s) \in U$ eventually (addition of nets as in Example 3.6.1). Since $f(x_s) \to a$ and $f(y_t) \to b$, $f(x_s) \in a + U$ and $f(y_t) \in b + U$ eventually; hence, for sufficiently large $s$ and $t$,

$$b \in f(y_t) + U \subset (f(x_s) + U) + U \subset a + U + U + U \subset a + V$$

Therefore $b \in \text{cl}\{a\} = \{a\}$ and $\overline{f}$ is well defined.

To see that $\overline{f}$ is uniformly continuous, let $W \in V_Y(0)$ and let $V \in V_Y(0)$ be symmetric and such that $V + V + V \subset W$. Let $U \in V_X(0)$ be such that if $u, v \in D$ and $u - v \in U$, then $f(u) - f(v) \in V$. Suppose $x, y \in X$, $x - y \in U$ and $(x_s)$ and $(y_t)$ are nets in $D$ which converge to $x$ and $y$, respectively. For sufficiently large $s$ and $t$, $x_s - y_t \in U$; consequently, $f(x_s) - f(y_t) \in V$. Since $V$ is symmetric and $f(x_s) \to \overline{f}(x)$, for sufficiently large $s$ and $t$,

$$\overline{f}(x) \in f(x_s) + V \subset (f(y_t) + V) + V \subset ((\overline{f}(y) + V) + V) + V \subset \overline{f}(y) + W$$

As for uniqueness of $\overline{f}$, suppose that $g$ is also a continuous extension of $f$, $x \in X$, and $(x_s)$ is a net from $D$ converging to $x$. By continuity, $g(x) = \lim g(x_s) = \lim f(x_s) = \overline{f}(x)$. ■

The uniform continuity of $f$ in the previous result cannot be replaced by continuity as shown by Example 3.6.3.

**Example 3.6.3** NONEXTENDIBILITY Let $\mathbf{R}$ be the reals and $\mathbf{Q}$ the rationals and consider the continuous map $f : \mathbf{Q} \to \mathbf{R}$, $q \mapsto 1/(\sqrt{2} - q)$. As $q \to \sqrt{2}$, $1/\left|\sqrt{2} - q\right| \to \infty$ so there can be no continuous extension of $f$ to $\mathbf{R}$. ■

**Theorem 3.6.4** EXTENDING HOMOMORPHISMS If $f$ is a uniformly continuous homomorphism of a dense subset $D$ of a topological group $X$ into a complete Hausdorff topological group $Y$ then its continuous extension $\overline{f}$ to $X$ is a homomorphism.

**Proof.** We use the notation of the statement. If $x_s \to x$ and $y_t \to y$ then $x_s + y_t \to x + y$ [Example 3.6.1]. Hence $\overline{f}(x+y) = \lim f(x_s + y_t) = \lim(f(x_s) + f(y_t)) = \lim f(x_s) + \lim f(y_t) = \overline{f}(x) + \overline{f}(y)$. ■

**Theorem 3.6.5** DENSE ISOMORPHIC SUBGROUPS If $A$ and $B$ are dense topologically isomorphic subgroups of the Hausdorff topological groups $X$ and $Y$, respectively, then $X$ and $Y$ are topologically isomorphic.

**Proof.** We use the notation of the statement. Let $\overline{f}$ be the continuous extension of the topological isomorphism $f : A \to B$ to $X$; $\overline{f}$ is a continuous homomorphism by Theorem 3.6.4. Let $\overline{f^{-1}}$ be the continuous extension of $f^{-1}$ to $Y$. We now argue that $h = \overline{f^{-1}} \circ \overline{f}$ is the identity map $x \mapsto x$ of $X$ onto $X$. The restriction of $h$ to $A$ is clearly the identity map $i_A$ of $A$. As $i_X$ is a continuous extension of $i_A$ to $X$, it follows that $h = i_X$ by the uniqueness statement of Theorem 3.6.2. Similarly, the identity map $y \mapsto y$ of $Y$ onto itself is $\overline{f} \circ \overline{f^{-1}}$. These two equalities imply that $\overline{f}$ and $\overline{f^{-1}}$ are inverses of each other, from which we conclude that $\overline{f}$ is bijective. That $\overline{f}$ is an open map follows from the continuity of $\overline{f^{-1}} = (\overline{f})^{-1}$. ■

## 3.7 COMPLETION

**Notation.** $X$ denotes a commutative topological group with neighborhood filter $V(0)$ of 0. All groups are commutative. ■

A *completion* of a topological group $X$ is a complete topological group which contains a dense topologically isomorphic copy of $X$. We show in Theorem 3.7.1 that a pseudometrizable group $(X, d)$ has a completion. Since the topology of any topological group $X$ is generated by a family $\{d_s : s \in S\}$ of invariant pseudometrics [Theorem 2.7.6], we can complete each $(X_s, d_s)$. The product $\prod_{s \in S} \widehat{X_s}$ of these complete groups $\widehat{X_s}$ is complete by Theorem 3.5.4 and we can embed $X$ in the product by mapping $x \in X$ into the "tuple" $(x)$ consisting solely of $x$'s. The closure $\widehat{X}$ of this image is the desired completion of $X$ [Theorem 3.7.2]. Even though any *commutative* topological group possesses a completion, noncommutative groups generally do not. A noncommutative group $X$ possesses a completion $\widehat{X}$ *as a uniform space* but it may not be possible to extend the group operations from $X$ to $\widehat{X}$ [Bourbaki 1966a, p. 248; 1966b, p. 331].

**Theorem 3.7.1** Completion of a Pseudometrizable Group Any pseudometrizable topological group $(X, d)$, where $d$ is an invariant pseudometric, possesses a completion $(\widehat{X}, \widehat{d})$ as a pseudometrizable topological group where $\widehat{d}$ is an invariant pseudometric that extends $d$.

**Proof.** We use the notation of the statement. Let $\widehat{X}$ denote the collection of all Cauchy sequences $(x_n)$ from $X$. The class $\widehat{X}$ of these Cauchy sequences is a commutative group when we add sequences pointwise as $(x_n)+(y_n) = (x_n + y_n)$. Furthermore, $\widehat{d}((x_n), (y_n)) = \lim_n d(x_n, y_n)$ is an invariant pseudometric on $\widehat{X}$ so $\left(\widehat{X}, \widehat{d}\right)$ is a topological group. For $x \in X$, let $(x)$ denote the sequence whose every entry is $x$ and consider the map $f : X \to \widehat{X}$, $x \mapsto (x)$. Clearly $\widehat{d}$ extends $d$ on $f(X)$ so the isometric isomorphism $f$ embeds $X$ in $\widehat{X}$. Given $x = (x_n) \in \widehat{X}$, $f(x_k)$ is $\widehat{d}$-close to $x$ for sufficiently large $k$ because $(x_n)$ is Cauchy; hence $f(X)$ is dense in $\widehat{X}$. If $(f(x_n))$ is a Cauchy sequence in $f(X)$, it is straightforward to verify that $(x_n)$ is a Cauchy sequence in $X$ and that $(f(x_n))$ converges to $(x_n)$ in $\widehat{X}$. In other words, Cauchy sequences from $f(X)$ have limits in $\widehat{X}$. For an arbitrary Cauchy sequence $(w_n)$ from $\widehat{X}$, choose elements $(y_n)$ from $f(X)$ such that $\widehat{d}(w_n, y_n) < 1/n$ for every $n \in \mathbf{N}$. Since $(w_n)$ is Cauchy, so is $(y_n)$. Since each $(y_n)$ is a Cauchy sequence in $f(X)$, it must converge to a point $y \in \widehat{X}$; since $\widehat{d}(w_n, y_n) < 1/n$ for every $n \in \mathbf{N}$, it follows that $w_n \to y$ as well and establishes the completeness of $\widehat{X}$. Finally, for any $x, y \in X$, it is evident that $d(x, y) = \widehat{d}((x), (y))$.

For a more detailed approach to the proof of Theorem 3.7.1, see Bachman and Narici 2000, pp. 48–53. ■

**Theorem 3.7.2** COMPLETION OF A TOPOLOGICAL GROUP Any commutative topological group $X$ possesses a completion.

**Proof.** Let $U(0)$ denote the symmetric neighborhoods of 0 in the topological group $X$. As shown in the proof of Theorem 2.7.6, there is an invariant pseudometric $d_U$ associated with each $U \in U(0)$. Let $\widehat{X_U}$ denote the completion of $(X_U, d_U)$ and let $f_U : X \to \widehat{X_U}$, $x \mapsto (x)$, be the isometric isomorphism embedding $X_U$ in $\widehat{X_U}$ mentioned in the proof of Theorem 3.7.1. $\widehat{X} = \Pi_{U \in U(0)} \widehat{X_U}$ is complete by Theorem 3.5.4. Consider the map $f : X \to \widehat{X}$, $x \mapsto (f_U(x))_{U \in U(0)}$. It is routine to verify that $f$ is a topological isomorphism. Since the closure $\widehat{X}$ of $f(X)$ in $\widehat{X}$ is complete [Theorem 3.1.9(a)], this is the desired completion of $X$. ■

Are completions unique? No, not even to within topological isomorphism as shown in Example 3.7.3 which also demonstrates that completions of Hausdorff groups need not be Hausdorff.

**Example 3.7.3** NONUNIQUENESS OF COMPLETION Let $\mathbf{R}_t$ denote $\mathbf{R}$ endowed with the trivial topology. The nonempty basic open sets in $\mathbf{R} \times \mathbf{R}_t$ are of the form $(a, b) \times \mathbf{R}$ where $(a, b)$ is an open interval in $\mathbf{R}$. Each $(a, b) \times \mathbf{R}$ intersects $\mathbf{R} \times \{0\}$ in $(a, b) \times \{0\}$ so $\mathbf{R} \times \{0\}$ is dense in $\mathbf{R} \times \mathbf{R}_t$. Since $\mathbf{R}_t$ is obviously complete, so is $\mathbf{R} \times \mathbf{R}_t$ by Theorem 3.5.4. Thus, even though $\mathbf{R} \times \{0\}$ is complete, $\mathbf{R} \times \mathbf{R}_t$ is a completion of $\mathbf{R} \times \{0\}$ which is not topologically isomorphic to the Hausdorff space $\mathbf{R} \times \{0\}$ since $\mathbf{R} \times \mathbf{R}_t$ is not Hausdorff. ■

Obviously, if a topological group $X$ has a Hausdorff completion $\widehat{X}$ then $X$ must be Hausdorff. As the proof is widely available, we do not prove here that a Hausdorff topological group has a Hausdorff completion. Theorem 3.6.5 implies that Hausdorff completions are unique. If $(X, d)$ is a metric topological group it has a completion $\widehat{X}$ as a pseudometrizable group. With $N = \text{cl}\{0\} = \cap \widehat{V}(0)$, $\widehat{X}/N$ is a completion of $X$ as a metric topological group.

Completeness of a topological group $X$ does not generally imply completeness of $X/M$ even when $M$ is closed subgroup of $X$. If $X$ is metrizable, it does, however, as we show next.

**Theorem 3.7.4** COMPLETE QUOTIENT If $M$ is a closed subgroup of a complete pseudometrizable topological group $(X, d)$ then $X/M$ is complete.

**Proof.** We use the notation of the statement. Let $|\cdot| = d(\cdot, 0)$ be the value associated with $d$ and $|\dot{x}|_q = \inf\{|x + m| : m \in M\}$ be the associated quotient value of Example 2.7.5 for $X/M$. By Corollary 2.7.4 we know that $X/M$ is pseudometrizable so to demonstrate the completeness of $X/M$, it suffices to show that Cauchy sequences converge. To this end, let $(\dot{x}_n)$ be a Cauchy sequence from $X/M$. By choosing a subsequence if necessary, we may assume that $|\dot{x}_{n+1} - \dot{x}_n| < 1/2^n$ for each $n \in \mathbf{N}$. For $n = 1$, choose $m_1 \in M$ and let $w_1 = x_1 + m_1$. Now choose $w_2 \in \dot{x}_2 - \dot{x}_1$, $w_2 = (x_2 - x_1) + m_2$ for some $m_2 \in M$, such that $|w_2| < 1/2$. Similarly, choose $m_n \in M$ such that $w_n = x_n - x_{n-1} + m_n \in \dot{x}_n - \dot{x}_{n-1}$ with $|w_n| < 1/2^{n-1}$. Now form the sums $y_n = w_1 + \cdots + w_n$ and note that these sums telescope: $y_n = (x_1 + m_1) + (x_2 - x_1 + m_2) + \cdots + (x_n - x_{n-1} + m_n) = x_n + (m_1 + \cdots + m_n)$ so that the canonical image $k(y_n) = \dot{x}_n$. Since, for each $n \in \mathbf{N}$, $|y_{n+1} - y_n| = |w_{n+1}| < 1/2^n$, it follows that, for each $p \in \mathbf{N}$, $|y_{n+p} - y_n| \leq 1/2^{n+p} + \cdots + 1/2^n < 1/2^{n-1}$; hence $(y_n)$ is a Cauchy sequence in $X$ and therefore has a limit $y$. By the continuity of $k$, $k(y_n) = \dot{x}_n \to k(y)$ and $X/M$ is complete. ■

Two remarks are in order: (1) There is a converse to Theorem 3.7.4 which says that if $M$ and $X/M$ are complete, then so is the metrizable group $X$ [Exercise 3.201]. (2) For a counterexample to Theorem 3.7.4 in the absence of metrizability, a case of a complete topological group with an incomplete quotient, see Köthe [1983, p. 434] or Bourbaki [1955, Chap. IV, Sec. 4, Ex. 10b].

## 3.8 EXERCISES

**Notation.** $X$ denotes a commutative topological group with neighborhood filter $V(0)$ of 0. If $(X, d)$ is a pseudometrizable group, the pseudometric $d$ is invariant.

3.101 COMPLETENESS

(a) Let $\mathcal{B}$ be a filterbase in $X$. Show that $\mathcal{B} \to x$ iff $x \in \operatorname{cl} \mathcal{B}$ and $\mathcal{B}$ is a Cauchy filterbase.

(b) Prove that a discrete topological group is complete.

(c) Suppose that $X$ is not Hausdorff and that $x_0 \neq 0$ belongs to $\operatorname{cl}\{0\}$. Show that $E = \operatorname{cl}\{0\} \setminus \{x_0\}$ is compact (hence complete) but not closed.

3.102 COMPLETENESS OF $(\mathbf{R}^*, \cdot)$

(a) If $(X, d)$ is pseudometrizable, then $D(x, y) = d(x, y) + d(-x, -y)$ is a pseudometric which generates the same topology.

(b) Let $(\mathbf{R}^*, \cdot)$ denote the topological group of nonzero real numbers. For $x, y \in \mathbf{R}^*$, let $d(x, y) = |x - y|$, and $D(x, y) = |x - y| + |1/x - 1/y|$. Show that $(\mathbf{R}^*, \cdot)$ is complete with respect to $d$ and $D$. [Cf. Exercise 3.204.] The point is that the incomplete subset $\mathbf{R}^* \subset \mathbf{R}$ is complete as a multiplicative topological group.

3.103 If $|\cdot|$ is a value on a group $X$ [Def. 2.7.2], then $|\cdot|$ is uniformly continuous.

3.104 ABSOLUTE CONVERGENCE IN COMPLETE GROUPS Suppose that the pseudometrizable group $(X, d)$ is complete. Let $|\cdot| = d(\cdot, 0)$ be the value associated with $d$. Show that $\sum_{n\in\mathbf{N}} |x_n| < \infty$ implies that $\sum_{n\in\mathbf{N}} x_n < \infty$. In other words, absolute convergence implies convergence in a complete pseudometrizable group.

3.105 BASE FOR THE COMPLETION If $\mathcal{B}$ is a neighborhood base at 0 in the Hausdorff group $X$, then $\{\mathrm{cl}_{\widehat{X}}\, B : B \in \mathcal{B}\}$ is a base at 0 in the completion $\widehat{X}$.

3.106 A locally compact subgroup of a Hausdorff group is closed.

3.107 $M$-ADIC TOPOLOGY Let $M$ be an ideal in the commutative ring $A$ with identity. For each positive integer $n$, $M^n$ denotes the collection of finite sums of products of elements from $M$. $M^n$ is an ideal. Show that the collection $\{M^n : n \in \mathbf{N}\}$ forms a base at 0 for a topology which is compatible with the ring structure of $A$; it is called the *$M$-adic topology*. When $A$ carries the $M$-adic topology, we write $(A, M)$. By the *radical* $\sqrt{M}$ of $M$ is meant the ideal of all elements $x$ in $A$ such that $x^n \in M$ for some $n \in \mathbf{N}$. Show that if $A$ is Noetherian (i.e., every ascending chain $I_1 \subset I_2 \subset \cdots$ of ideals $I_n$ must terminate), then $(A, M) = (A, \sqrt{M})$.

3.108 Let $F(T, X)$ be the function group of all maps of the set $T$ into $X$ [Example 2.6.2]. Let $\mathcal{S}_1$ and $\mathcal{S}_2$ be families of subsets of $T$ such that $\mathcal{S}_1 \subset \mathcal{S}_2$. In reference to the $\mathcal{S}$-topologies of Example 2.6.2, show that if $F_{\mathcal{S}_1}(T, X)$ is complete, then $F_{\mathcal{S}_2}(T, X)$ is complete.

3.201 $M$, $X/M$ COMPLETE IMPLIES $X$ COMPLETE If $M$ is a complete subgroup of the metrizable group $(X, d)$ and $X/M$ is complete then so is $X$.

3.202 SUMMABILITY Let $\{x_i : i \in I\}$ be a collection of elements of the Hausdorff group $X$. For each finite subset $J \subset I$, let $s_J = \Sigma_{j\in J} x_j$. The set $\mathcal{F}$ of all finite subsets of $I$ is a directed set with respect to $A \leq B$ iff $B \subset A$ where $A, B \in \mathcal{F}$. Consider the net

$$\begin{aligned} s:\ \mathcal{F} &\to X \\ J &\mapsto s_J \end{aligned}$$

The collection $\{x_i\}$ is said to be *summable* if the net $s$ has a limit $x$ in $X$; $x$ is called the *sum* of the family $\{x_i\}$ and we write $x = \Sigma_{i\in I} x_i$.

(a) Prove that $\Sigma_{i\in I} x_i = x$ if, for each neighborhood $U$ of 0, there is a finite subset $L$ of $I$ such that $s_J \in x + U$ for each finite subset $J \subset I$ which contains $L$.

(b) CAUCHY CRITERION If $\{x_i\}$ is summable to $x$ then, for each neighborhood $U$ of 0, there is a finite subset $L$ of $I$ such that $\Sigma_{i \in J} x_i \in U$ for all finite subsets $J$ of $I$ which do not meet $L$. Prove that the condition is sufficient if $X$ is complete.

(c) $N$TH TERM GOES TO 0 If $\Sigma_{i \in I} x_i = x$ then any neighborhood $U$ of 0 contains all but finitely many of the $x_i$.

(d) Let $X$ be first countable and let $\Sigma_{i \in I} x_i = x$. Prove that the set $\{i \in I : x_i \neq 0\}$ is countable.

(e) Prove that if $X$ is not first countable then the conclusion of (d) no longer holds.

(f) If $X$ is complete, show that every subcollection of a convergent collection $\{x_i\}$ is summable.

(g) If $f$ is continuous homomorphism of $X$ into the Hausdorff group $Y$, show that if $\{x_i\}$ is summable in $X$ then $\{f(x_i)\}$ is summable in $Y$ and $\Sigma_{i \in I} f(x_i) = f(\Sigma_{i \in I} x_i)$.

3.203 COMPLETION OF A TOPOLOGICAL RING To complete a Hausdorff topological ring $A$ [Definition 2.2.9] to $\widehat{A}$, it is necessary to extend multiplication by continuity to $\widehat{A} \times \widehat{A}$. The following result is helpful in doing that.

(a) Let $E, F$ and $X$ be complete Hausdorff topological groups. Let $A$ be a dense subgroup of $E$ and $B$ a dense subgroup of $F$. A function $f : E \times F \to X$ is *$Z$-bilinear* if it is additive in each argument when the other is held fixed. If $f$ is a continuous $Z$-bilinear map of $A \times B$ into $X$, then show that $f$ can be extended by continuity to a continuous $Z$-bilinear map of $E \times F$ into $X$.

(b) Show that a Hausdorff topological ring $A$ is topologically ring isomorphic to a dense subring of a complete Hausdorff ring $\widehat{A}$.

3.204 PROPERTY OF A COMPLETE FIELD Let $E$ be a Hausdorff topological field [Definition 2.2.9] and suppose that $E$, viewed additively, is a complete topological group. Show that the multiplicative group $(E^*, \cdot)$ of nonzero elements of $E$ is also a complete topological group.

3.205 SPHERICAL COMPLETENESS If $T$ is a metric space, then $T$ is complete iff every decreasing sequence $(C_n)$ of closed balls whose diameters shrink to 0 has nonempty intersection. A stronger version of this condition is obtained when we drop the requirement that the diameters go to 0: A metric space is called *spherically complete* if every decreasing sequence of closed balls has nonempty intersection. The notion arises in Hahn-Banach theorems for continuous linear maps of an ultranormed space into another [Exercise 10.201]. Recall that an *ultrametric* is a metric $d$ such that $d(x, y) \leq \max[d(x, z), d(z, y)]$ for all $x, y$ and $z$.

(a) In an ultrametric space $(T, d)$ show that any collection $\mathcal{F}$ of closed balls is totally ordered by set inclusion (i.e., for all $U, V \in \mathcal{F}, U \subset V$ or $V \subset U$) iff they satisfy the binary intersection condition (any two of them meet).

(b) Show that an arbitrary ultrametric space is spherically complete iff any totally ordered collection of closed balls has nonempty intersection.

(c) Give an example of a complete ultrametric space which is not spherically complete.

(d) If $(T, d)$ is a complete ultrametric space, show that $D(x, y) = \inf\{2^{-n} : d(x, y) \leq 2^{-n},\ n \in \mathbf{Z}\}$ is an ultrametric with the properties that $d \leq D \leq 2d$ and $(T, D)$ is spherically complete. Conclude that spherical completeness is not a topological property in the sense that it is not preserved by homeomorphism.

## Hints

3.101 (c) Show that $E$ is a subset of any neighborhood of 0.

3.102 (b) Show that $D$-Cauchy sequences $(x_n)$ are $d$-Cauchy and bounded away from 0 in the sense that for some $r > 0$, $|x_n| \geq r$ for every $n$; cf. hint to Exercise 204.

3.103 Show that $||x| - |y|| \leq |x - y|$.

3.106 Use Theorems 3.1.11, 3.1.9 and Exercise 3.105.

3.201 Consider the canonical image $(\dot{x}_s)$ in $X/M$ of a Cauchy net $(x_s)$ in $X$.

3.202 (e) Consider the additive topological group $F(\mathbf{R}, \mathbf{R}, p)$ of real-valued functions on $\mathbf{R}$ with the topology of pointwise convergence defined in Example 2.6.8. For each $a \in \mathbf{R}$, let $k_a$ denote the characteristic function of $\{a\}$ and show that $\{k_a : a \in \mathbf{R}\}$ is summable to the function that is identically 1.

(g) Use the net analog of Theorem 1.1.4(c).

3.203 (a) Let $(x, y) \in E \times F$ and let $(x_s)$ and $(y_t)$ be Cauchy nets in $A$ and $B$ which converge to $x$ and $y$, respectively. Show that $f(x_s, y_t)$ is a Cauchy net by verifying the identity $f(x_1, y_1) - f(x, y) = f(x_1 - x, y_2) + f(x_2, y_1 - y) + f(x_1 - x, y_1 - y_2)$.

3.204 Show that if $(x_s)$ is a Cauchy net in $E^*$ (i.e., given a neighborhood $W$ of 1, $x_s x_t^{-1} \in W$ eventually) then $(x_s)$ is a Cauchy net with respect to the additive structure which does not converge to 0. To do this, let $U$ be a neighborhood of 0 in $E$ and let $V$ be a closed neighborhood of 0 such that $V \subset U$, $VV \subset U$, and $-1 \notin V$.

3.205 (c) On $\mathbf{N}$ define $d(m, n) = 0$ if $m = n$ and $\max(1 + 1/n, 1 + 1/m)$ otherwise.

# Chapter 4

# Topological Vector Spaces



**Definition** TOPOLOGICAL VECTOR SPACE A topological vector space (TVS) is a vector space $X$ over $\mathbf{F} = \mathbf{R}$ or $\mathbf{C}$ endowed with a group topology that makes scalar multiplication a continuous map of $\mathbf{F} \times X$ into $X$. ■

In one sense we have arrived; in another, we have already been here since Chap. 2. We investigate topological vector spaces *per se* in this chapter, by which we mean that we examine the implications of the continuous scalar multiplication superimposed on the topological group structure.

Topological vector spaces are, naturally, as localized as topological groups—knowledge of the neighborhoods of 0 constitutes knowledge of the neighborhoods of any point. In this case, however, the neighborhoods of 0 have more texture, more character, so to speak. For example, in a topological group, there is a base at 0 of *symmetric* neighborhoods $V$ of 0, i.e., $V = -V$. In a topological vector space there is a base of *balanced* neighborhoods $V$ of 0: If $x \in V$ then so does $ax$ for any scalar $a$ of magnitude less than or equal to 1.

There are other things too. Of special importance is a property that does not arise automatically but is usually hypothesized. That property is that there be a base of convex neighborhoods of 0. Everywhere but in this chapter we deal almost exclusively with spaces of this type.

The properties of bases at 0 are the subject of Secs. 4.1 to 4.5. We then discuss products, quotients, metrizability, and completion. These are little more than grafts onto what has already been established for topological groups.

In Sec. 4.9 we examine the idea of a topological direct sum. The idea is to split a TVS $X$ into two linear subspaces $M$ and $N$ in such a way that $X$ can be recovered by adding $M$ and $N$ back together again. The hard part is reconstituting $X$ topologically. It is not always possible. In Example 4.9.7 we show that the subspace $c_0$ (null sequences) of $\ell_\infty$ (bounded sequences with sup norm) has no "complement"—there is no subspace of $\ell_\infty$ which can be "added" to $c_0$ that will yield $\ell_\infty$.

Finite-dimensional topological vector spaces are characterized in two ways in Sec. 4.10: A Hausdorff topological vector space $X$ is $n$-dimensional over $\mathbf{F}$ iff $X$ is linearly homeomorphic to $\ell_\infty(n)$; moreover, $X$ is finite-dimensional iff it is locally compact. It is worth mentioning that linearly homeomorphic topological vector spaces can differ in significant ways; one such way is discussed before Theorem 4.10.4.

## 4.1 ABSORBENT AND BALANCED SETS

A *topological vector space* (*TVS*) is a vector space $X$ over $\mathbf{F} = \mathbf{R}$ or $\mathbf{C}$ equipped with a topology that makes vector addition and scalar multiplication continuous, i.e., the maps

$$\begin{array}{ccc} X \times X \to X & \text{and} & \mathbf{F} \times X \to X \\ (x, y) \mapsto x + y & & (a, x) \mapsto ax \end{array},$$

continuous. (If $\mathbf{E}$ is any topological field, the same definition defines a TVS over $\mathbf{E}$.) The topology is said to be a *linear* or *vector* topology. We sometimes abbreviate Hausdorff TVS as *HTVS*.

**Definition 4.1.1** LINEAR HOMEOMORPHISM A linear map $A$ of the TVS $X$ into the TVS $Y$ which is also a homeomorphism is called a *linear homeomorphism* or a *topological isomorphism*. If there is a *linear homeomorphism* mapping $X$ onto $Y$, $X$ is said to be *linearly homeomorphic* or *topologically isomorphic* to $Y$. If $A$ is a *linear homeomorphism* mapping $X$ into $Y$, we sometimes say $X$ is *embedded* in $Y$. ■

If $X$ is a TVS, it is a topological group. Hence, with $V(0)$ denoting the filter of neighborhoods of 0:

(a) The neighborhoods of any point $x \in X$ are $x + V(0)$ [Theorem 2.1.4].

(b) Any neighborhood $V$ of $0$ contains a closed neighborhood of $0$ [Theorem 2.2.1].

(c) For any neighborhood $V$ of $0$ there exists a neighborhood $U$ of $0$ such that $U + U \subset V$ [Theorem 2.1.10].

(d) There is a base of symmetric neighborhoods of $0$ [Theorem 2.1.10].

**Definition 4.1.2** ABSORBENT AND BALANCED Let $C(0,1) = \{a \in \mathbf{F} : |a| \leq 1\}$. A subset $E$ of a vector space $X$ is:

(a) *balanced* (*circled, equilibrée*) if $ax$ belongs to $E$ for any scalar $|a| \leq 1$ and $x \in E$, i.e., $C(0,1)E = \{ax : a \in C(0,1),\ x \in E\} \subset E$;

(b) *absorbent* (*absorbing, radial at 0*) if for any $x \in X$, there is a positive number $r$ such that $ax$ belongs to $E$ for all $|a| \leq r$. Thus, an absorbent set is one such that any vector can be shrunk into it. Equivalently, $E$ is absorbent if it can be inflated to cover any vector: Given $x \in X$ there exists $r > 0$ such that $x \in aE$ for $|a| \geq r$; consequently $X = \cup_{r>0} rE$. ■

Clearly, $0$ must belong to any balanced or absorbent set. The underlying field can make a substantial difference. The closed interval $[-1,1]$ is balanced in the real space $\mathbf{R}^2$, for example, but it is not a balanced subset of the complex space $\mathbf{C}$ because $i1 = i \notin [-1,1]$. Similarly, the union of two unit circles centered at $(1,0)$ and $(-1,0)$ together with $\{0\} \times [-1,1]$ is balanced in the real space $\mathbf{R}^2$ but not in the complex space $\mathbf{C}$.

**Example 4.1.3** ABSORBENT AND BALANCED SETS (*a*) In $\mathbf{R}^3$ the closed unit ball $C(0,1)$ centered at the origin is absorbent and balanced. The disk $C(0,1) \cap \mathbf{R}^2$ is balanced but not absorbent in $\mathbf{R}^3$.

(b) The unit circle $C(1/2,1)$, centered at $(1/2,0)$ in the real space $\mathbf{R}^2$, is absorbent but not balanced. (Having $0$ as an interior point suffices to make a set absorbent [Theorem 4.3.6(b)]. Can you think of an absorbent set in the real space $\mathbf{R}^2$ which does *not* have $0$ as an interior point?)

(c) The collection $P$ of polynomials is not an absorbent subset of the linear space $C([0,1], \mathbf{F})$ of continuous maps of $[0,1]$ into $\mathbf{F}$, but $P$ is clearly balanced.

(d) Obviously, any subspace $M$ of a vector space $X$ is balanced. If $E \subset X$ is absorbent then its linear span $[E]$ must be all of $X$. Thus, proper subspaces are never absorbent.

(e) Arbitrary unions and intersections of balanced sets are balanced.

(f) Arbitrary unions of absorbent sets are absorbent; infinite intersections of absorbent sets need not be absorbent: Consider a decreasing sequence of circles about the origin in $\mathbf{R}^2$ whose diameters shrink to $0$. Absorbency survives finite intersections, however [Theorem 4.1.7].

(g) In the more general situation of a TVS defined over a topological field $K$, neither *absorbent* nor *balanced* make sense. They do if $K$'s topology is defined by an absolute value. ■

**Definition 4.1.4** BALANCED HULL The *balanced hull* (*balanced envelope*) $E_b$ of a subset $E$ of a linear space $X$ is the intersection of all balanced sets containing it, $X$ itself being one such set. ■

**Theorem 4.1.5** BALANCED HULL Let $C(0,1)$ denote the closed unit disk in $\mathbf{F} = \mathbf{R}$ or $\mathbf{C}$ and let $E$ be a subset of the vector space $X$ over $\mathbf{F}$. Then $E_b = C(0,1)E = \cup\{aE : |a| \leq 1\}$.

**Proof.** We use the notation of the statement. Since $C(0,1)[C(0,1)E] \subset C(0,1)E$, $C(0,1)E$ is balanced. Therefore $C(0,1)E \supset E_b$. If $S$ is any balanced set containing $E$ then $S \supset C(0,1)E$. Consequently, $E_b \supset C(0,1)E$. ■

The *balanced core* $B(E)$ of a subset $E$ of a vector space $X$ is the largest balanced subset of $E$. It is the union of all balanced subsets of $E$. As $E$ may not have any balanced subsets, it may be empty. The balanced core of $E$ may also be characterized as $\cap\{aE : |a| \geq 1\}$ since, for $A \subset X$, $x \in aA$ for all $|a| \geq 1$ iff $ax \in A$ for all $|a| \leq 1$. From the latter characterization of balanced core as an intersection, it is clear that if $E$ is closed then so is its balanced core. Furthermore, as is straightforward to verify, if $E$ is absorbent, so is $B(E)$.

**Example 4.1.6** BALANCED CORE

(a) The balanced core of $[1,2] \subset \mathbf{R}$ is empty.

(b) The balanced core of $[-1,2] \subset \mathbf{R}$ is $[-1,1]$.

(c) The balanced core of $[-1,2] \subset \mathbf{C}$ is $\{0\}$. ■

As intersections of absorbent sets may fail to be absorbent [Example 4.1.3(f)], a notion of absorbent hull, analogous to balanced hull, cannot be defined.

**Theorem 4.1.7** FINITE INTERSECTIONS OF ABSORBENT SETS Finite intersections of absorbent sets are absorbent.

**Proof.** If $E_1, \ldots, E_n$ are absorbent and $x$ is any vector, there exist positive numbers $r_1, \ldots, r_n$ such that $x \in aE_i$ for scalars $a$ such that $|a| \leq r_i$ for $i = 1, \ldots, n$. For $r = \min_i r_i$, then, $x \in a[\cap_{i=1}^n E_i]$ for $|a| \leq r$. ■

**Theorem 4.1.8** LINEAR IMAGES OF ABSORBENT AND BALANCED SETS Let $X$ and $Y$ be linear spaces over $\mathbf{F}$ and let $A : X \to Y$ be linear.

(a) If $E \subset X$ is balanced then $A(E)$ is balanced.

(b) If $E \subset X$ is absorbent and $A$ is surjective then $A(E)$ is absorbent.

(c) Inverse images under $A$ of absorbent or balanced subsets of $Y$ are absorbent or balanced, respectively.

**Proof.** We use the notation of the statement. (a) If $E$ is balanced then $C(0,1)E \subset E$, so $C(0,1)A(E) = A(C(0,1)E) \subset A(E)$.

(b) Suppose that $E$ is absorbent. If $A$ is surjective then for any $y \in Y$ there is an $x \in X$ such that $y = Ax$. Since $E$ is absorbent, there is an $r > 0$ such that $C(0,r)x \subset E$, which implies that $C(0,r)y \subset A(E)$.

(c) Let $E$ be a balanced subset of $Y$. Then $C(0,1)E \subset E$ and therefore $A^{-1}(C(0,1)E) = C(0,1)A^{-1}(E) \subset A^{-1}(E)$.

Let $E$ be an absorbent subset of $Y$ and let $x \in X$. Since $E$ is absorbent, there is an $r > 0$ such that $C(0,r)Ax \subset E$. Hence $C(0,r)x \subset A^{-1}(E)$ and $A^{-1}(E)$ is absorbent. ■

It follows immediately from Theorem 4.1.8 that if $E$ is balanced or absorbent then nonzero scalar multiples of $E$ are also balanced or absorbent, respectively. It is straightforward to verify that if $S$ and $T$ are balanced then so is their sum $S + T$. Hence linear combinations $aS + bT$ of balanced sets are balanced for any scalars $a$ and $b$. As long as $a$ and $b$ are not both 0, linear combinations $aS + bT$ of absorbent sets $S$ and $T$ are absorbent.

Two useful observations about balanced sets are:

**Theorem 4.1.9** If $E$ is balanced then:

(a) $aE = |a|E$ for any scalar $a$;

(b) if $|a| \leq |b|$ then $aE \subset bE$.

The notion of absorbency may be generalized as follows. If $A$ and $B$ are subsets of the linear space $X$, we say that $A$ *absorbs* $B$ if $A$ can be sufficiently inflated to contain $B$. More precisely, $A$ *absorbs* $B$ if there is an $r > 0$ such that for all scalars $a$ such that $|a| \geq r$, $aA \supset B$; equivalently, $A$ absorbs $B$ iff for some positive $r$, $C(0,r)B \subset A$.

## 4.2 CONVEXITY—ALGEBRAIC

A subset $E$ of the plane $\mathbf{R}^2$ is *convex* if the line segment $[x,y]$ connecting any two of its points, $x$ and $y$, lies wholly within $E$. Thus, the interiors of figures which are "bowed outward," such as circles and triangles, are convex, whereas those with dents (the interior of the letter "B," for example) or holes, such as an annulus, are not. After introducing a suitable generalization of line segment connecting two points in an arbitrary vector space, we define convex set in general. A reason for doing this is that the most widely studied kind of TVS is the "locally convex" TVS, a kind for which there is a base of convex sets at 0. We take them up in Chap. 5 and then really do not consider anything else from then on.

**Definition 4.2.1** CONVEXITY Let $x$ and $y$ be points in a linear space $X$. The set $L(x.y) = \{ax + (1-a)y : a \in \mathbf{R}\}$ is called the *line through $x$ and $y$*. The set $[x,y] = \{ax + (1-a)y : 0 \le a \le 1\} = \{ax + by : a + b = 1, a \ge 0, b \ge 0\}$ is called the *line segment joining* $x$ *and* $y$; sets such as $[x,y) = [x,y] \setminus \{y\}$, etc. are also called *intervals.* A subset $S$ of $X$ is called *convex* if it contains the line segment joining any two points in $S$. Equivalently, $S$ is *convex* if $aS + (1-a)S \subset S$ for all $a \in [0,1]$. Note that for any $a \in [0,1]$, $S \subset aS + (1-a)S$, so the inclusion of the preceding statement is actually an equality. ■

In $\mathbf{R}$, $\mathbf{R}^2$, and $\mathbf{R}^3$, "line" and "line segment" have their usual geometric sense so sets such as the interior of a hexagon are convex while the interior of a crescent is not. (What are the only convex subsets of $\mathbf{R}$?) Translates of linear subspaces are clearly convex. If $f$ is a linear functional on a vector space $X$ and $a$ is a scalar then $f^{-1}(a) = \{x \in X : f(x) = a\}$ is convex; if $a$ is positive, $\{x \in X : |f(x)| < a\}$ and $\{x \in X : |f(x)| \le a\}$ are each convex.

If $(X,p)$ is a seminormed linear space, the triangle inequality shows that any closed or open sphere is convex. As convexity is invariant under translation [Theorem 4.2.4], we show that a sphere centered at 0 is convex to prove this last contention. If $x, y \in C(0,r) = \{x \in X : p(x) \le r\}$ and $a \in [0,1]$ then

$$p(ax + (1-a)y) \le ap(x) + (1-a)p(y) \le ar + (1-a)r = r$$

The defining requirement of convexity of a set $S$ is that $aS + bS \subset S$ for all positive numbers, $a$ and $b$, which add up to 1.

Linear combinations $\sum_{i=1}^{n} a_i x_i$ in which the $a$'s are nonnegative and $\sum_{i=1}^{n} a_i = 1$ are called *convex combinations* of the $x$'s. In this terminology, a convex set is one that contains all its convex combinations.

Unions of convex sets are not convex generally (consider two disjoint convex sets in the plane) but intersections obviously are. The latter fact enables us to consider a smallest convex set containing a given set.

**Definition 4.2.2** CONVEX HULL The *convex hull* (*convex envelope*) $E_c$ of a subset $E$ of a linear space is the intersection of all convex sets which contain $E$. ■

We characterize the convex hull of a set $E$ as the collection of all convex combinations of elements from $E$ in Theorem 4.2.3; we also get a form for the convex hull of a family of convex sets.

**Theorem 4.2.3** CONVEX HULL (a) The convex hull $E_c$ of a set $E$ is the collection $S$ of all convex combinations $\sum_{i=1}^{n} a_i x_i$, $n \in \mathbf{N}$, where the $x$'s are from $E$.

(b) FAMILIES OF CONVEX SETS Let $\{K_i : i \in I\}$ be a family of convex sets. The convex hull of $\cup_{i \in I} K_i$ is the set $S$ of all convex combinations $\sum_{i=1}^{n} a_i x_i$, $n \in \mathbf{N}$ where each $x_i \in K_i$. Thus, for a finite family $\{K_i : i = 1, 2, \ldots, k\}$ of convex sets, the convex hull of $\cup_{i=1}^{k} K_i$ is the set $S$ of all convex combinations $\sum_{i=1}^{k} a_i x_i$ where each $x_i \in K_i$.

**Proof.** We use the notation of the statement. (a) To see that $S$ is convex, suppose that $\sum_{i=1}^{n} a_i x_i$ and $\sum_{i=1}^{n} b_i x_i$ belong to $S$ (we may assume that each sum involves the same $x$'s by using some zero coefficients). If $a$ and $b$ are nonnegative scalars such that $a+b=1$ then

$$a\sum_{i=1}^{n} a_i x_i + b\sum_{i=1}^{n} b_i x_i \in S$$

because $\sum_{i=1}^{n} aa_i + \sum_{i=1}^{n} bb_i = a+b = 1$. Therefore, $S \supset E_c$. Clearly $E_c \supset S$.

(b) Let $S$ be as in the statement of (b). Since $S$ is clearly a subset of $K_c$, we need only prove that $S$ is convex. To this end, let $n \in \mathbf{N}$ and let $x = \sum_{i=1}^{n} a_i x_i$ and $y = \sum_{i=1}^{n} b_i y_i$ be points of $S$. By using some zero coefficients if necessary, we may use the same $n$ for $x$ and $y$. Let $a$ and $b$ be positive numbers such that $a+b=1$. Consider

$$z = a\sum_{i=1}^{n} a_i x_i + b\sum_{i=1}^{n} b_i y_i$$

but rewrite it as

$$z = \sum_{i=1}^{n} (a_i a + b_i b)\left[\frac{a_i a}{a_i a + b_i b} x_i + \frac{b_i b}{a_i a + b_i b} y_i\right]$$

The $K$'s are convex, so for each $i$, $\left[\frac{a_i a}{a_i a+b_i b} x_i + \frac{b_i b}{a_i a+b_i b} y_i\right] \in K_i$. It only remains to note that the terms $a_i a + b_i b$ are nonnegative and add up to 1 to prove that $z \in S$. ■

It follows from our next result that scalar multiples and translates of convex sets are convex. The proofs of both parts follow directly from the definition of convexity and are omitted.

**Theorem 4.2.4** LINEAR COMBINATIONS AND IMAGES OF CONVEX SETS
(a) If $K_1$ and $K_2$ are convex then so is $aK_1 + bK_2$ for any scalars $a$ and $b$.

(b) Let $X$ and $Y$ be vector spaces and let $A : X \to Y$ be a linear map. If $K$ is a convex subset of $X$, so is $A(K)$. If $K$ is a convex subset of $Y$, so is $A^{-1}(K)$.

Our next result is a sort of distributive law for convex sets and *positive* scalars.

**Theorem 4.2.5** If $a$ and $b$ are positive scalars and $K$ is convex then $(a+b)K = aK + bK$.

**Proof.** We use the notation of the statement. For $a, b > 0$ and any set $K$, $(a+b)K \subset aK + bK$. For $y \in aK + bK$, then $y = ax + bw$ for some

$x, w \in K$. For convex $K$, $ax/(a+b) + bw/(a+b) = y/(a+b) \in K$ and $y \in (a+b)K$. ■

Hence, if $K$ is convex and $a_1, \ldots, a_n$ are positive numbers whose sum is 1 then $\sum_{i=1}^{n} a_i S = S$.

The convex hull $K = \{x_0, x_1, \ldots, x_n\}_c$ of a finite set of points is called a *polytope*. If $x_1 - x_0, \ldots, x_n - x_0$ are linearly independent then $K$ is called an *n-simplex* with *vertices* $x_0, \ldots, x_n$. A 0-simplex is a point, a 1-simplex, a line segment, a 2-simplex, a triangle, and so on. A point $x$ in an $n$-simplex can be written in a unique way as a convex combination $x = \sum_{i=1}^{n} a_i x_i$ of the vertices, the numbers $a_0, \ldots, a_n$ being called the *barycentric coordinates of* $x$. We return to this theme in Chap. 9 when we consider the Krein-Milman theorem and its progeny.

**Example 4.2.6** Convex Hulls in Finite-Dimensional Spaces Let $K$ be a subset of $\mathbf{F}^n$. If $\mathbf{F} = \mathbf{C}$, the convex hull $K_c$ of $K$ consists of convex combinations of $2n+1$ elements of $K$, i.e.,

$$K_c = \left\{ \sum_{i=1}^{2n+1} a_i x_i : 0 \le a_i \le 1, \ \sum_{i=1}^{2n+1} a_i = 1, \ x_i \in K \right\}$$

If $\mathbf{F} = \mathbf{R}$, $K_c$ consists of convex combinations of $n+1$ elements of $K$. We prove it for the complex case by treating $\mathbf{C}^n$ as a real vector space. To prove it for $\mathbf{R}^n$ merely replace the $2n+1$ in the argument below by $n+1$. View $\mathbf{C}^n$ as a real vector space and let $K$ be a subset of $\mathbf{C}^n$. For each $m \in \mathbf{N}$, let $L_m = \{\sum_{i=1}^{m} a_i x_i : x_i \in K, \ 0 \le a_i \le 1, \ \sum_{i=1}^{m} a_i = 1\}$. The convex hull $K_c$ of $K$ is given by $\cup_{m \in \mathbf{N}} L_m$. We show now that $K_c = L_{2n+1}$. Clearly, $L_m \subset L_{m+1}$ for every $m$. We show that $L_m \subset L_{m-1}$ for $m > 2n+1$, so that the $L_m$ are equal from $m = 2n+1$ on. For $m > 2n+1$, let $x = \sum_{i=1}^{m} a_i x_i \in L_m$. We may assume that none of the $a_i$ is 0 or $x$ would already belong to $L_{m-1}$. Since the dimension of $\mathbf{C}^n$ is $2n$ and $m-1 > 2n$, the $m-1$ vectors $x_i - x_m$, $1 \le i \le m-1$, are linearly dependent. Therefore there are real numbers $b_1, \ldots, b_{m-1}$, not all 0, such that $\sum_{i=1}^{m-1} b_i (x_i - x_m) = 0$. Letting $b_m = -\left(\sum_{i=1}^{m-1} b_i\right)$ then $\sum_{i=1}^{m} b_i x_i = 0$ and $\sum_{i=1}^{m} b_i = 0$.

Let $|b_k|/a_k = \max\{|b_i|/a_i : 1 \le i \le m, \ a_i \ne 0\}$ and note that $|b_k|/a_k$ must be positive. Let $c_i = a_i - b_i a_k / b_k$ for $1 \le i \le m$. Each $c_i$ is non-negative by the following argument: If, for some $i$, $a_i - b_i a_k / b_k < 0$ then $0 < a_i < b_i a_k / b_k \le |b_i| a_k / |b_k|$, i.e., $|b_i|/a_i > |b_k|/a_k$ which contradicts the maximality of $|b_k|/a_k$. Note also that $c_k = 0$. Since $\sum_{i=1}^{m} b_i = 0$, it follows that

$$\sum_{i=1}^{m} c_i = \sum_{i \ne k} c_i = \sum_{i=1}^{m} (a_i - b_i a_k / b_k) = \sum_{i=1}^{m} a_i - (a_k / b_k) \sum_{i=1}^{m} b_i = \sum_{i=1}^{m} a_i = 1$$

Since $\sum_{i=1}^{m} b_i x_i = 0$,

$$\sum_{i=1}^{m} c_i x_i = \sum_{i \neq k} c_i x_i = \sum_{i=1}^{m} (a_i - b_i a_k / b_k)\, x_i$$
$$= \sum_{i=1}^{m} a_i x_i - (a_k/b_k) \sum_{i=1}^{m} b_i x_i = \sum_{i=1}^{m} a_i x_i = x$$

Since $\sum_{i \neq k} c_i = 1$ and $x = \sum_{i \neq k} c_i x_i$, it follows that $x \in L_{m-1}$. ■

**Definition 4.2.7** DISK A subset $D$ of a linear space $X$ is said to be a *disk* or to be *absolutely convex* or *disked* if it is convex and balanced. ■

Open or closed spheres in any seminormed space are disks and so is any linear subspace. If $f$ is a linear functional and $a$ is a positive number, $\{x \in X : |f(x)| \leq a\}$ is a disk, too.

**Theorem 4.2.8** DISK CONDITION A subset $D$ of a linear space is a disk iff for all scalars $a$ and $b$ such that $|a| + |b| \leq 1$, $aD + bD \subset D$; equivalently if $|c| \geq |a| + |b|$ then $aD + bD \subset cD$.

**Proof.** We use the notation of the statement. If the condition holds, $D$ is obviously balanced (just choose $b = 0$). As for $D$'s convexity, if $a$ and $b$ are nonnegative scalars which add up to 1 then $aD + bD \subset D$ so $D$ is convex.

Conversely, suppose that $D$ is convex and balanced, that $x, y \in D$ and that $|a| + |b| \leq 1$. If $a = 0$ or $b = 0$, the condition is satisfied because $D$ is balanced, so suppose that neither is 0. Since $D$ is balanced, $(a/|a|)\, x$ and $(b/|b|)\, y$ each belong to $D$. Since $D$ is convex,

$$z = \frac{|a|}{|a| + |b|} \frac{a}{|a|} x + \frac{|b|}{|a| + |b|} \frac{b}{|b|} y \in D \qquad (*)$$

Since $D$ is balanced and $|a| + |b| \leq 1$, we conclude that $(|a| + |b|)\, z = ax + by \in D$. ■

By a simple modification of the argument surrounding $(*)$ [like the induction used in Example 4.2.6], we obtain:

**Theorem 4.2.9** If $D$ is absolutely convex then for any $n \in \mathbf{N}$ and scalars $a_1, a_2, \ldots, a_n$ such that $\sum_{i=1}^{n} |a_i| \leq 1$, it follows that $\sum_{i=1}^{n} a_i D \subset D$; equivalently, if $|c| \geq \sum_{i=1}^{n} |a_i|$ then $\sum_{i=1}^{n} a_i D \subset cD$.

A stronger sort of factorization theorem than that of Theorem 4.2.5 prevails for disks $D$: namely, for any scalars $a_1, \ldots, a_n$, $\sum_{k=1}^{n} a_k D = \left(\sum_{k=1}^{n} |a_k|\right) D$. Intersections of disks are clearly disks, so we may consider a *disked* or *absolute convex hull* of a set, namely:

**Definition 4.2.10** DISKED HULL The *disked* (*absolute convex, balanced convex*) *hull* $E_{bc}$ of a subset $E$ of a linear space is the intersection of all disks containing $E$. ■

Some motivation for the notation $E_{bc}$ is provided by the following result.

**Theorem 4.2.11** DISKED HULL For any subset $E$ of a linear space, $E_{bc} = (E_b)_c$, the convex hull of the balanced hull of $E$. [$(E_b)_c$ may not be the same as $(E_c)_b$ because $(E_c)_b$ may not be convex; see Example 4.2.12.] Moreover, $E_{bc}$ consists of the set $S$ of all linear combinations $\sum a_k x_k$ of elements $x_k$ from $E$ and scalars $a_k$ for which $\sum |a_k| \leq 1$.

**Proof.** We use the notation of the statement. It is straightforward to verify that $(E_b)_c$ is balanced. Hence $(E_b)_c$ is absolutely convex and therefore $E_{bc} \subset (E_b)_c$. Using Theorem 4.1.5 on balanced hulls and then Theorem 4.2.3(a) on convex hulls, it is easy to see that typical members of $(E_b)_c$ are of the form $\sum a_k x_k$ described in the statement. Since $E_{bc}$ is absolutely convex, all such elements are in $E_{bc}$ [Theorem 4.2.9]; hence $(E_b)_c \subset E_{bc}$. ■

**Example 4.2.12** $(E_c)_b$ NEED NOT BE CONVEX In $\mathbf{R}^2$ let $E$ be the convex hull of $(0,0), (0,1)$, and $(1,1)$. Its balanced hull $E_b$ is not convex since the line connecting $(-1,-1)$ and $(0,1)$ falls outside $E_b$. ■

**Remark 4.2.13** There is an extensive literature about convexity. Its systematic investigation was begun by H. Minkowski and H. Brunn and helped integrate geometric ideas into analysis. It is especially well developed in $\mathbf{R}^n$ and there are many variations—projective convexity, spherical convexity, hyperconvexity, and quasi-convexity—to name just four. We discuss uniform and strict convexity in normed spaces in Chapter 16. You consider convexity in any partially ordered set $(X, <)$ by defining a subset $A$ to be *convex* if whenever $x, y \in A$ and $x < w < y$ then $w \in A$. In a metric space $(X, d)$, a subset $S$ is said to be *metrically convex* if for each pair of distinct points $x, y \in S$, there is a point $w \in S$ such that $d(x, y) = d(x, w) + d(w, y)$. Various notions of convexity have also been considered in topological groups, and there is a "complex" convexity, called *pseudo-convexity*, which plays an important role in the theory of functions of several complex variables. A survey of these and other notions of convexity, with an extensive bibliography, can be found in Klee [1963]. Another excellent survey is Bonnesen and Fenchel [1934]. Some introductory books on convexity are Valentine [1964], Yaglom and Boltyanski [1961], van Tiel [1984] and Varberg and Roberts [1973].

Following somewhat Grünbaum and Klee [1967], we list a number of convex-type notions in Table 4.2.1; everything takes place in a linear space (over $\mathbf{R}$ or $\mathbf{C}$, as usual).

| Type of set | Together with $x$ and $y$ the set also contains | Whenever $a, b \in \mathbf{F}$ and: |
|---|---|---|
| Subspace | $ax + by$ | no other condition |
| Affine set | $ax + by$ | $a + b = 1$, $a, b \in \mathbf{R}$ |
| Cone | $ax + by$ | $a \geq 0, b > 0$ |
| Midpoint convex | $(1/2)(x + y)$ | no other condition |
| Convex | $ax + by$ | $a + b = 1, a, b \geq 0$ |
| Absolutely convex | $ax + by$ | $\|a\| + \|b\| \leq 1$ |
| $p$-convex $(0 < p \leq 1)$ | $ax + by$ | $a, b \geq 0, a^p + b^p = 1$ |
| Absolutely $p$-convex $(0 < p \leq 1)$ | $ax + by$ | $\|a\|^p + \|b\|^p \leq 1$ |

**Table 4.2.1 Variations on Convexity**

In order to investigate topological vector spaces over ultravalued fields $(K, |\cdot|)$, a notion of convexity is desirable. One such notion is called *$K$-convexity:* If $X$ is a linear space over the ultravalued field $K$ and $D \subset X$ then $D$ is said to be *absolutely $K$-convex* if whenever $a, b \in C(0,1) \subset K$ then $aD + bD \subset D$. Translates $x + D$ of absolutely $K$-convex sets are called *$K$-convex*. Some further discussion of $K$-convexity may be found in Exercises 4.202–4.204 as well as Narici et al. [1971], Springer [1965], van Tiel [1965], Prolla [1982], and von Rooij [1978].

## 4.3 BASIC PROPERTIES

In this section we point out some of the most elementary properties of topological vector spaces. Some of them only depend on the additive topological group structure, so we need only quote our earlier results. As was the case for topological groups, topological vector spaces are very localized structures: Overall behavior is determined by what happens at 0 and we devote attention to the properties of the neighborhoods of 0 in Theorem 4.3.6. Conversely, in Sec. 4.5, we investigate conditions under which a collection of sets will be a neighborhood base at 0 for a vector topology. Ways of manufacturing new topological vector spaces from given ones—products and quotients—are discussed in Sec. 4.7.

We denote the filter of neighborhoods of $x$ by $V(x)$. Some properties of topological vector spaces that have already been proved are:

**Theorem 4.3.1** BASIC PROPERTIES Let $X$ be a TVS over any topological field $E$. Then:

(a) Let $x_0 \in X$. Translation maps, $x \mapsto x_0 + x$, inversion maps $x \mapsto -x$, and multiplier maps $x \mapsto ax$ ($a \in \mathbf{F}$, $a \neq 0$) are linear homeomorphisms of $X$ onto $X$. For any $x$, $a \mapsto ax$ is a continuous map of $\mathbf{F}$ into $X$.

(b) $X$ is a homogeneous topological space [Definition 2.1.5].

(c) If $\mathcal{B}$ is a neighborhood base at 0 in $X$ then $x + \mathcal{B}$ is a neighborhood base at $x$ for any $x \in X$.

(d) Every TVS is a regular topological space—for any neighborhood $V$ of 0, there is a neighborhood $U$ of 0 such that $\operatorname{cl} U \subset V$. $X$ is a $T_0$ topological space iff $X$ is Hausdorff, and $X$ is Hausdorff iff $\cap V(0) = \{0\}$.

**Proof.** We use the notation of the statement. (a) The statement for translations and inverses follows from Theorem 2.1.3. That $x \mapsto ax$ is bijective follows directly from the fact that $a \neq 0$. The continuity of the maps $x \mapsto ax$ and $a \mapsto ax$ follows from the continuity of scalar multiplication. For example, to see that $a \mapsto ax$ $(x \neq 0)$ is continuous, suppose $U$ is a neighborhood of $ax$. By the continuity of scalar multiplication, there is some $r > 0$ and neighborhood $V$ of $x$ such that $C(a, r) x \subset C(a, r) V \subset U$.

(b) Theorem 2.1.6.

(c) Theorem 2.1.6.

(d) Theorem 2.2.1 and Theorem 2.2.2. ■

An interesting consequence of Theorem 4.3.1(a) is that

**Theorem 4.3.2** CLOSURE For any subset $A$ of a TVS $X$ and any $x \in X$, $\operatorname{cl}(x + A) = x + \operatorname{cl} A$; for any scalar $a$, $\operatorname{cl}(aA) = a \operatorname{cl} A$.

By Theorem 2.2.3, the corresponding result for topological groups, it follows that:

**Theorem 4.3.3** In a TVS over any topological field, if $A$ is a nonempty open set and $B$ is any subset then $A + B$ is open.

We observed in Sec. 2.1 that the discrete topology is a group topology. It is not a vector topology because every topological vector space (over **R** or **C**) is connected [Theorem 4.3.4]. TVS over other topological fields may be disconnected, however [Example 4.3.5(c)].

**Theorem 4.3.4** CONNECTEDNESS OF TVS Every $X$ is connected.

**Proof.** Let $X$ be a TVS. Recall that a collection of connected sets with a point in common has a connected union. Note that $X = \cup_{x \in X} \mathbf{F}x$, where $\mathbf{F}$ denotes **R** or **C**. Next observe that each $\mathbf{F}x$ is connected, since it is the continuous image of a connected set and 0 belongs to every $\mathbf{F}x$. ■

It is now easy to give examples of topologized linear spaces $X$ that are not topological vector spaces.

**Example 4.3.5** Let $X$ be a nontrivial vector space.

(a) The discrete topology is not a vector topology. On the other hand, any linear space over any topological field with the trivial topology $\{X, \varnothing\}$ is a topological linear space.

(b) Since $X \neq \{0\}$, $X$ has an infinite number of elements. The cofinite topology (a set is open iff its complement is finite) is not a vector topology for $X$ because it is not regular as it must be by Theorem 4.3.1(d): the cofinite topology is not regular because open sets are infinite and the closure of any infinite subset is $X$.

(c) The connectedness of any TVS establishes a profound difference between real and complex TVS and TVS over arbitrary valued fields. In the proof of Theorem 4.3.4, it is the connectivity of **R** and **C** that yields the connectedness of TVS. Ultravalued fields $(K, |\cdot|)$ are totally disconnected (the biggest connected set to which any point $x$ belongs is just $\{x\}$). In particular, if $X$ is a vector space over a trivially valued field then the discrete topology is a vector topology for $X$. If $X$ is an ultranormed space, it is Hausdorff and topologically zero-dimensional—hence totally disconnected.

(d) If $X$ and $Y$ are complex linear spaces and $\mathcal{L}(X,Y)$ denotes the vector space of all linear maps of $X$ into $Y$ then $\mathcal{L}(X,Y)$ with finite topology [Example 2.3.3] is not a TVS since the finite topology is totally disconnected [Example 2.3.3]. ■

**Theorem 4.3.6** NEIGHBORHOODS OF 0 In a TVS $X$ with filter $V(0)$ of neighborhoods of 0:

(a) any $V \in V(0)$ contains a $U \in V(0)$ such that $U + U \subset V$;

(b) all neighborhoods of 0 are absorbent;

(c) if $a \neq 0$ and $V \in V(0)$ then $aV \in V(0)$;

(d) each neighborhood of 0 contains a balanced neighborhood of 0; and

(e) there is a neighborhood base at 0 consisting of closed, balanced neighborhoods.

**Proof.** We use the notation of the statement. (a) Use Theorem 2.1.10, the corresponding statement for topological groups.

(b) Let $U$ be a neighborhood of 0, and let $x \in X$. Since $0x = 0$ then, by the continuity of scalar multiplication, there exist neighborhoods $C(0,r) \subset \mathbf{F}$ $(r > 0)$ of 0 and $V$ of $x$ such that $C(0,r)V \subset U$. Thus $C(0,r)x \subset U$ and the absorbency of $U$ follows.

(c) Use Theorem 4.3.1(a).

(d) Consider the continuous map $f : \mathbf{F} \times X \to X$, $(a,x) \mapsto ax$. Given a neighborhood $U$ of 0 in $X$, then $f^{-1}(U)$ is a neighborhood of $(0,0)$ in $\mathbf{F} \times X$. Consequently, $f^{-1}(U)$ must contain a set of the form $C(0,r) \times V$ for some positive $r$ and $V \in V(0)$, i.e., $f(C(0,r) \times V) = C(0,r)V \subset U$. As $C(0,r)V = \cup_{a \in C(0,r)} aV$, $C(0,r)V$ is a neighborhood of 0 in $X$. For $b \in C(0,r)$ and a scalar $|a| \leq 1$, $ab \in C(0,r)$ which implies that $C(0,r)V$ is balanced.

(e) First, note that the closure of a balanced set is balanced. To prove (e), let $W$ be a neighborhood of 0 in $X$. By the regularity of TVS [Theorem 4.3.1(d)], $W$ contains a closed neighborhood $V$ of 0. By (d) of this theorem, $V$ contains a balanced neighborhood $U$ of 0. Now $\operatorname{cl} U$ is a closed and balanced neighborhood of 0 which is contained in $W$. ■

## 4.4 CONVEXITY—TOPOLOGICAL

In this section we consider questions like: If $K$ is a convex subset of a TVS, what about its closure? Generally, we begin to probe the interactions between the algebraic and topological structures in a topological vector space. We also define here the most important type of topological vector space, the *locally convex* topological vector space.

**Theorem 4.4.1** Closure and Interior In a topological vector space $X$ over $\mathbf{F} = \mathbf{R}$ or $\mathbf{C}$ (or any topological field):

(a) the sum of a compact set and a closed set is closed (sums of closed sets are not always closed as Example 2.2.4 shows);

(b) for any subsets $A$ and $B$ of $X$, $\operatorname{cl} A + \operatorname{cl} B \subset \operatorname{cl}(A+B)$;

(c) the closure of a subspace is a subspace;

(d) if $\mathcal{B}$ is a base for the filter of neighborhoods of 0 and $E$ is any subset of $X$ then $\operatorname{cl} E = \cap_{B\in\mathcal{B}}(E+B)$;

(e) for any subsets $A$ and $B$ of $X$, if $\operatorname{int} B \neq \varnothing$ then $\operatorname{int} A + \operatorname{int} B \subset A + \operatorname{int} B \subset \operatorname{int}(A+B)$.

**Proof.** We use the notation of the statement. (a) and (b) follow from the corresponding statements for topological groups, namely Theorems 2.2.6 and 2.2.7.

(c) By Theorem 2.4.3, if $M$ is a subspace of $X$, its closure $\operatorname{cl} M$ is an additive topological group. To show that $\operatorname{cl} M$ is closed under scalar multiplication, consider the continuous map $f : \mathbf{F} \times X \to X$, $(a, x) \mapsto ax$. By continuity, $f(\operatorname{cl}(\mathbf{F} \times M)) \subset \operatorname{cl} f(\mathbf{F} \times M)$ so

$$\begin{aligned} f(\operatorname{cl}(\mathbf{F} \times M)) &= f(\operatorname{cl}\mathbf{F} \times \operatorname{cl} M) = f(\mathbf{F} \times \operatorname{cl} M) \\ &= \mathbf{F}\operatorname{cl} M \subset \operatorname{cl} f(\mathbf{F}\times M) = \operatorname{cl} M \end{aligned}$$

(d) Suppose $x \in A = \cap_{B\in\mathcal{B}}(E+B)$. Let $B$ be a balanced neighborhood of 0 and choose $B' \in \mathcal{B}$ such that $B' \subset B$. Since $x \in E + B'$, there exists $e \in E$ such that $x \in e + B'$. Hence $e \in x - B' \subset x - B = x + B$. Thus every neighborhood of $x$ contains a point of $E$, and it follows that $x \in \operatorname{cl} E$, i.e., $A \subset \operatorname{cl} E$. Conversely, suppose that $x \in \operatorname{cl} E$; choose $B \in \mathcal{B}$ and a balanced subneighborhood $B'$ of 0 of $B$. Then $(x + B') \cap E \neq \varnothing$, which implies that $x \in E - B' = E + B' \subset E + B$. Hence $\operatorname{cl} E \subset A$. (For more on closure, see Exercise 4.107.)

(e) It suffices to prove that $A + \operatorname{int} B \subset \operatorname{int}(A+B)$. By Theorem 4.3.3 $A + \operatorname{int} B$ is open and it is clearly contained in $A + B$. Since $\operatorname{int}(A+B)$ is a maximal open subset of $A + B$, it follows that $A + \operatorname{int} B \subset \operatorname{int}(A+B)$. ■

Next, we consider the durability of convexity and balancedness with respect to closures, etc.

**Theorem 4.4.2** CONVEXITY, ETC. If $E$ is a subset a of a TVS $X$ then:

(a) if $E$ is balanced and $0 \in \operatorname{int} E$ [i.e., if $E \in V(0)$] then the interior of $E$ is balanced;

(b) if $E$ is an open neighborhood of 0 then the balanced hull of $E$ is open;

(c) if $E$ is balanced, so is $\operatorname{cl} E$;

(d) the interior of an absorbent set is not generally absorbent; any superset of an absorbent set is absorbent; hence if $E$ is absorbent then so is $\operatorname{cl} E$;

(e) if $E$ is open then so is the convex hull of $E$;

(f) if $E$ is convex then $\operatorname{int} E$ is convex;

(g) if $E$ is convex, so is $\operatorname{cl} E$.

**Proof.** We use the notation of the statement. (a) Since $E$ is balanced and $0 \in \operatorname{int} E$, it suffices to show that $a(\operatorname{int} E) \subset \operatorname{int} E$ for scalars $a$ such that $0 < |a| \leq 1$. In this case, $a(\operatorname{int} E) \subset aE \subset E$ since $E$ is balanced. Since $a \neq 0$, $a(\operatorname{int} E)$ is an open subset of $E$ and is therefore contained in $\operatorname{int} E$.

(b) For any $a \neq 0$, $aE$ is open; hence $E_b = \cup_{0<|a|\leq 1} aE$ is open.

(c) Suppose $E$ is balanced and consider a scalar $a$ such that $0 < |a| \leq 1$. For such $a$, by Theorem 4.3.1(a), $a \operatorname{cl} E = \operatorname{cl}(aE) \subset \operatorname{cl} E$, so $\operatorname{cl} E$ is balanced.

(d) The interior$\{(a,b) \in \mathbf{R}^2 : |b| > a^2\}$ of the absorbent set $E = \{(a,b) \in \mathbf{R}^2 : |b| > a^2\} \cup [-1,1]$ cannot absorb any point of $\mathbf{R}$.

(e) For any finite set of nonzero scalars $a_1, \ldots, a_n$, the set $S = \sum_{i=1}^n a_i E$ is open. If the $a$'s are positive and add up to 1, then $S \subset E_c$. Since any $x \in E_c$ must belong to some such $S$, each $x$ is an interior point of $E_c$. It follows that $E_c$ is open.

(f) Suppose that $E$ is convex and that $a$ and $b$ are positive and add up to 1. By Theorem 4.4.1(e) and $E$'s convexity

$$a\operatorname{int}E + b\operatorname{int}E = \operatorname{int}aE + \operatorname{int}bE \subset \operatorname{int}(aE + bE) = \operatorname{int}E$$

(g) For $0 < a < 1$, using Theorem 4.4.1(b),

$$a \operatorname{cl} E + (1-a)\operatorname{cl} E = \operatorname{cl}(aE) + \operatorname{cl}(1-a)E \subset \operatorname{cl}(aE + (1-a)E) = \operatorname{cl} E \quad \blacksquare$$

As the intersection of closed convex sets is closed and convex, we can define the *closed convex hull* $K(E)$ of a subset $E$ of a TVS $X$ to be the intersection of the closed convex supersets of $E$, $X$ itself being one such set.

**Theorem 4.4.3** CLOSED CONVEX HULL For any subset $E$ of a TVS $X$, the closed convex hull $K(E)$ of $E$ is given by $\operatorname{cl}(E_c) = (\operatorname{cl} E)_c$. We denote it by $\operatorname{cl} E_c$.

**Proof.** We use the notation of the statement. Let $K(E)$ denote closed convex hull of $E$. By Theorem 4.4.2(g), $\operatorname{cl}(E_c)$ is a closed convex superset of $E$; therefore $K(E) \subset \operatorname{cl}(E_c)$. Since $E \subset K(E)$ and $K(E)$ is convex, $E_c \subset K(E)$; since $K(E)$ is closed, $\operatorname{cl}(E_c) \subset K(E)$. As to the equality $\operatorname{cl}(E_c) = \operatorname{cl}(E)_c$, $(\operatorname{cl} E)_c$ is a closed convex set that contains $E$, hence $\operatorname{cl}(E_c) \subset$

$(\mathrm{cl}\, E)_c$. Conversely, since $E \subset E_c$, $\mathrm{cl}\, E \subset \mathrm{cl}\,(E_c)$. Since $\mathrm{cl}\,(E_c)$ is convex, $(\mathrm{cl}\, E)_c \subset \mathrm{cl}\,(E_c)$. ■

The convex hull of a compact set is generally not compact as shown in Example 4.8.8. On the positive side, we have the following two results.

**Theorem 4.4.4** CONVEX HULLS OF CONVEX COMPACT SETS In any TVS $X$ the convex hull of a finite union of convex compact sets is compact.

**Proof.** Let $K_1, \ldots, K_n$ be convex compact subsets of the TVS $X$.. Since each $K_i$ is convex, the convex hull of $\cup_{i=1}^n K_i$ is the collection of all points of the form $\sum_{i=1}^n a_i x_i$, where each $a_i$ is nonnegative, $\sum_{i=1}^n a_i = 1$, and $x_i \in K_i$ [Theorem 4.2.3(b)]. Now consider the set

$$S = \left\{ (a_1, \ldots, a_n) \in \ell_\infty^{\mathbf{R}}(n) : a_i \geq 0,\ \sum_{i=1}^n a_i = 1 \right\}$$

$S$ is obviously bounded in $\ell_\infty^{\mathbf{R}}(n)$ and since convergence of a sequence in $\ell_\infty^{\mathbf{R}}(n)$ is equivalent to componentwise convergence, $S$ is closed as well. Hence $S$ is compact. Since each $K_i$ is compact, it follows that $\Pi_{i=1}^n K_i$ is compact. The map

$$S \times \Pi_{i=1}^n K_i \to X,\ \left((a_i), (x_i)\right) \mapsto \sum_{i=1}^n a_i x_i$$

is continuous since it is the composite of continuous maps. The range of this map, however, is the convex hull of $\cup_{i=1}^n K_i$; in other words, the convex hull of $\cup_{i=1}^n K_i$ is the continuous image of a compact set. ■

**Definition 4.4.5** LOCAL CONVEXITY A topological vector space $X$ is said to be *locally convex* if there is a neighborhood base at 0 of convex sets. We say simply that "$X$ is locally convex" or that "$X$ is a locally convex space (LCS)." The topology for such a space is called a *locally convex topology.* If $X$ is Hausdorff and locally convex, we abbreviate this as "$X$ is an LCHS." ■

The first to consider LCS was von Neumann [1935, p. 4] who called them *convex spaces*; the term "locally convex space" was first used by Tihonov [1935, p. 768] which was adopted by Dieudonné [1942, p. 110, 1981b, p. 238] who required them to be Hausdorff as well.

Since balls, closed or open, determined by seminorms are convex [Theorem 5.2.1(d)], any normed linear space is locally convex. There is an intimate connection between locally convex spaces and seminorms: A linear topology is locally convex iff it is generated by a family of seminorms [Theorem 5.5.2]. It is routine to verify that any subspace of a locally convex space is locally convex in its relative topology. The same is true of appropriately topologized quotient spaces of locally convex spaces as we discuss in Sec. 4.7. We exhibit a nonlocally convex space in Example 4.6.2.

**Theorem 4.4.6** Base at 0 in LCS If $X$ is locally convex then there are neighborhood bases at 0 consisting of:

(a) disks as well as

(b) closed disks and open disks.

**Proof.** (a) If $V$ is any neighborhood of 0, $V$ contains a convex neighborhood $B$ of 0. By Theorem 4.3.6(d), $B$ must contain a balanced neighborhood $U$ of 0. Since $B$ is convex, $B$ contains the convex hull $U_c$ of $U$. Since $U$ is balanced, so is its convex hull, as is easy to verify. Thus $V$ contains the disked neighborhood $U_c$.

(b) Let $\mathcal{B}$ denote a base of disked neighborhoods of 0. Each $B \in \mathcal{B}$ contains a closed neighborhood $U$ of 0 by Theorem 4.3.1(d) and $U$ must contain some $B' \in \mathcal{B}$. The closure of $B'$ is a disk [Theorem 4.4.2(c,g)] and yields the result about closed disked neighborhoods of 0. The collection $\{\operatorname{int} B : B \in \mathcal{B}\}$ is a base of open disked neighborhoods at 0 [Theorem 4.4.2(a,f)]. ■

## 4.5 GENERATING VECTOR TOPOLOGIES

In Sec. 3.2 we examined some conditions which, if satisfied by a filter $\mathcal{B}$ of subsets of a group $X$, generate a group topology for $X$ in which $\mathcal{B}$ is the filter of neighborhoods of 0. We consider similar results in this section for TVS.

**Theorem 4.5.1** Base at 0 for Vector Topology Let $X$ be a linear space over $\mathbf{F} = \mathbf{R}$ or $\mathbf{C}$. Consider a filterbase $\mathcal{B}$ of subsets of $X$ for which:

$(FB_1)$ Each $B \in \mathcal{B}$ is balanced and absorbent.

$(FB_2)$ For each $B \in \mathcal{B}$ there is a $U \in \mathcal{B}$ such that $U + U \subset B$.

Then $\mathcal{B}$ is a neighborhood base at 0 for a vector topology for $X$.

**Proof.** We use the notation of the statement. Since each $B \in \mathcal{B}$ is balanced, each $B$ is symmetric ($B = -B$). Thus, the basis condition $(FB)$ of Theorem 2.3.1 is satisfied, and $\mathcal{B}$ is a neighborhood base at 0 for a group topology for $X$. Let $V(0)$ denote the filter generated by $\mathcal{B}$ so that the family of neighborhoods of any $x$ is given by $x + V(0)$. To prove continuity of scalar multiplication, we show first that

$$\text{Given any } B \in \mathcal{B} \text{ and } a \neq 0, \text{ there exists } V \in \mathcal{B} \text{ such that } aV \subset B \tag{1}$$

*Proof of (1):* Give any $B \in \mathcal{B}$, there exists $V \in \mathcal{B}$ such that $2V = \{2v : v \in V\} \subset V + V \subset B$ by $(FB_2)$. Hence, for any positive integer $n$, there is a $V \in \mathcal{B}$ such that $2^n V \subset B$. For $a \neq 0$, choose $n \in \mathbf{N}$ such that $|a| < 2^n$. Since $V \in \mathcal{B}$, $V$ is balanced and it follows that $aV \subset 2^n V \subset B$ and (1) is verified.

For $a_0 \in \mathbf{F}$, $x_0 \in X$, and $B \in \mathcal{B}$, consider the basic neighborhood $a_0 x_0 + B$ of $a_0 x_0$. To prove continuity of scalar multiplication, we must exhibit $r > 0$ and $V \in \mathcal{B}$ such that $C(a_0, r)(x_0 + V) \subset a_0 x_0 + B$.

By $(FB_2)$, there exists $W \in \mathcal{B}$ such that $W + W + W + 0 \subset B$. Suppose that $a_0 = 0$. Since $W$ is absorbent, there is a positive number $r$ such that

$C(0,r)\,x_0 \subset W \subset B$. Since $W$ is balanced, there is some positive $r \le 1$ such that

$$C(0,r)(x_0 + W) = C(0,r)\,x_0 + C(0,r)\,W \subset W + W \subset B = 0x_0 + B$$

Now suppose that $a_0 \neq 0$ and let $W$ be as above. By (1), there is a $W' \in \mathcal{B}$ such that $a_0 W' \subset W$. Since $\mathcal{B}$ is a filterbase, there is $V \in \mathcal{B}$ such that $V \subset W \cap W'$; since $V$ is absorbent, there is a positive $r \le 1$ such that $C(0,r)\,x_0 \subset V$. Now if $a \in C(a_0, r)$ and $x \in x_0 + V$ then the following three conditions hold: Since $C(0,r)\,x_0 \subset V$

$$(a - a_0)\,x_0 \in V \subset W \tag{2}$$

Since $V$ is balanced,

$$(a - a_0)(x - x_0) \in C(0,r)\,V \subset V \subset W \tag{3}$$

and

$$a_0 (x - x_0) \in a_0 V \subset a_0 W' \subset W \tag{4}$$

Combining (2)–(4), we have

$$(a - a_0)\,x_0 + (a - a_0)(x - x_0) + a_0 (x - x_0) = ax - a_0 x_0 \subset W + W + W \subset B$$

so that $ax \in a_0 x_0 + B$; in other words, $C(a_0, r)(x_0 + V) \subset a_0 x_0 + B$. ■

Next we look at basis conditions for a locally convex topology. Another way, using seminorms, is discussed in Chap. 5.

**Theorem 4.5.2** BASE FOR LOCALLY CONVEX TOPOLOGY Let $\mathcal{B}$ be a filterbase of subsets of the linear space $X$. In order for $\mathcal{B}$ to be a neighborhood base at 0 for a locally convex topology, it suffices that $\mathcal{B}$ satisfy the following two conditions:

$(LC_1)$ Each $B \in \mathcal{B}$ is an absorbent disk.

$(LC_2)$ For each $B \in \mathcal{B}$ there is an $a \in (0, 1/2]$ such that $aB \in \mathcal{B}$.

**Proof.** We use the notation of the statement. Clearly if $\mathcal{B}$ is a base for a vector topology, that topology is locally convex. Comparing the hypothesis of this theorem with that of Theorem 4.5.1, we see that we only have to show that $(FB_2)$ is satisfied. To that end, let $B \in \mathcal{B}$. By $(LC_2)$ there is an $a \in (0, 1/2]$ such that $aB \in \mathcal{B}$. Since $aB$ is convex, $aB + aB = 2aB$ [Theorem 4.2.5]; since $B$ is balanced and $2a \le 1$, $2aB \subset B$. ■

If the absorbent disks in $\mathcal{B}$ of the preceding theorem do not satisfy $(LC_2)$ then the collection $\{aB : a > 0,\ B \in \mathcal{B}\}$ is a base at 0 for a locally convex topology for which the sets $\{aB : a \in (0, 1/2],\ B \in \mathcal{B}\}$ are a base at 0.

If the filterbase $\mathcal{B}$ of Theorem 4.5.1 or 4.5.2 also satisfies:

$(OB)$ For any $B \in \mathcal{B}$ and any $x \in B$ there is a $V \in \mathcal{B}$ such that $x + V \subset B$ then $\mathcal{B}$ is an *open* base at 0.

In Theorem 4.5.3(a) we focus attention on conditions on a collection $\mathcal{S}$ of subsets that ensure that $\mathcal{S}$ is a subbase at 0 for a vector topology. In Theorem 4.5.3(b) we return to locally convex topologies; we do not quite get a subbase (hence the quotation marks: "subbase"), because we have to consider *positive multiples* of finite intersections.

**Theorem 4.5.3** SUBBASES; LOCALLY CONVEX "SUBBASES" Let $\mathcal{S}$ be a nonempty collection of subsets of a linear space $X$.

(a) If $\mathcal{S}$ satisfies $(FB_1)$ and $(FB_2)$ of Theorem 4.5.1, it is a neighborhood subbase at 0 for a vector topology for $X$.

(b) If $\mathcal{S}$ consists of absorbent disks then a neighborhood base at 0 for a locally convex topology for $X$ is given by the collection $\mathcal{B}$ of positive multiples of finite intersections of sets from $\mathcal{S}$.

**Proof.** We use the notation of the statement. (a) If $\mathcal{S}$ satisfies $(FB_1)$ and $(FB_2)$ then so does the collection $\mathcal{B}$ of finite intersections of sets from $\mathcal{S}$ and the result follows from Theorem 4.5.1.

(b) If $\mathcal{S}$ consists of absorbent disks then so does the collection $\mathcal{B}$ of positive multiples of finite intersections of sets from $\mathcal{S}$. Moreover, $\mathcal{B}$ is a filterbase for, given $0 < a < b$, and elements $a \cap_{i=1}^{n} S_i$ and $b \cap_{i=1}^{k} S_i'$ of $\mathcal{B}$ then [Theorem 4.1.9]

$$a \bigcap_{i,j} (S_j \cap S_i') \subset \left( a \bigcap_{j=1}^{n} S_j \right) \bigcap \left( b \bigcap_{i=1}^{k} S_i' \right)$$

Clearly, for each $a \in [0, 1/2)$ and $B \in \mathcal{B}$, $aB \in \mathcal{B}$. The result now follows from Theorem 4.5.2. ■

The topologies obtained in Theorem 4.5.3 are the weakest vector topologies for which each $S \in \mathcal{S}$ is a neighborhood of 0.

**Example 4.5.4** SEMINORM TOPOLOGIES Let $X$ be a linear space.

(a) ONE SEMINORM Let $p$ be a seminorm on $X$. Then $\overline{V}_p = \{x \in X : p(x) \leq 1\}$ is an absorbent disk and so is each $r\overline{V}_p : r > 0$, as is easy to verify. Thus, condition $(LC_1)$ of the basis theorem [4.5.2] for locally convex topologies. Clearly $\mathcal{B}_p = \{r\overline{V}_p : r > 0\}$ also satisfies condition $(LC_2)$ of Theorem 4.5.2. Hence, seminormed—in particular normed—spaces are LCS.

(b) FAMILY OF SEMINORMS Let $P$ be a family of seminorms on $X$. Since $\mathcal{S} = \{\overline{V}_p : p \in P\}$ consists of absorbent disks, the collection of positive multiples of finite intersections of sets from $\mathcal{S}$ is a base at 0 for a locally convex topology $\mathcal{T}_P$ for $X$ [Theorem 4.5.3(b)]. It is called the topology *determined* or *generated* by $P$; it is clearly the weakest topology with respect to which each $p \in P$ is continuous.

(c) FINEST LOCALLY CONVEX TOPOLOGY If $P$ is the family of all seminorms on $X$, the topology $\mathcal{T}_{lc}$ determined by $P$ is called the *finest locally convex topology.* ■

## 4.6 A NON-LOCALLY CONVEX SPACE

We need some preliminary results to exhibit a space that is not locally convex in Example 4.6.2. For $p > 0$, by a "$p$th-power summable" $\mathbf{F}$-valued function $x$ on $(-\infty, \infty)$, we mean that the Lebesgue integral $\int_{\mathbf{R}} |x(t)|^p \, dt$ is finite. Let $L_p(-\infty, \infty)$ denote the set of complex-valued, $p$th-power summable functions $x$ defined on $\mathbf{R}$ made into a vector space with pointwise operations. For $0 < p \leq 1$, we topologize $L_p(-\infty, \infty)$ by means of the map

$$\|x\|^* = \int_{-\infty}^{\infty} |x(t)|^p \, dt \quad (x \in L_p(-\infty, \infty))$$

which fails to be a seminorm only in that $\|ax\|^* \neq |a| \|x\|^*$ $(a \in \mathbf{F}, x \in X)$. The topology it generates is not locally convex because no $\|\cdot\|^*$-sphere is convex; indeed, as we show in Example 4.6.2, the only convex neighborhood of 0 is $L_p(-\infty, \infty)$. We show next that $\|\cdot\|^*$ is a *p-seminorm,* i.e., that $\|\cdot\|^*$ is a seminorm in all respects save for absolute homogeneity: instead of $\|ax\|^* = |a| \|x\|^*$ for $a \in \mathbf{F}$ we have $\|ax\|^* = |a|^p \|x\|^*$. (The terminology $p$-seminorm is only used for $0 < p \leq 1$.) In regard to $\|\cdot\|^*$ being a $p$-seminorm, the only thing not obviously satisfied is the triangle inequality which follows from Theorem 4.6.1(b) below.

**Theorem 4.6.1** (a) If $0 < r \leq s$ and $a_1, \ldots, a_n$ are nonnegative reals then $\left(\sum_{i=1}^n a_i^s\right)^{1/s} \leq \left(\sum_{i=1}^n a_i^r\right)^{1/r}$.

(b) If $0 < t \leq 1$, and $a_1, \ldots, a_n$ are nonnegative reals then $\left(\sum_{i=1}^n a_i\right)^t \leq \sum_{i=1}^n a_i^t$.

**Proof.** We use the notation of the statement. (a) Let $d = \sum_{i=1}^n a_i^r$ and note that the result is clear when $d = 0$. Suppose $d \neq 0$. Since $a_i^r/d \leq 1$ for $i = 1, 2, \ldots, n$ and $s/r \geq 1$,

$$\frac{\left(\sum_{i=1}^n a_i^s\right)^{1/s}}{d^{1/r}} = \left(\sum_{i=1}^n \frac{a_i^s}{d^{s/r}}\right)^{1/s} = \left[\sum_{i=1}^n \left(\frac{a_i^r}{d}\right)^{s/r}\right]^{1/s} \leq \left(\sum_{i=1}^n \frac{a_i^r}{d}\right)^{1/s}$$

Since

$$\sum_{i=1}^n \frac{a_i^r}{d} = \frac{1}{d} \sum_{i=1}^n a_i^r = 1$$

it follows that $\left(\sum_{i=1}^n a_i^s\right)^{1/s} \leq d^{1/r} = \left(\sum_{i=1}^n a_i^r\right)^{1/r}$.

(b) In the inequality of (a), let $s = 1$ and $r = t$. This yields $\sum_{i=1}^{n} a_i \leq \left(\sum_{i=1}^{n} a_i^t\right)^{1/t}$ or $\left(\sum_{i=1}^{n} a_i\right)^t \leq \sum_{i=1}^{n} a_i^t$. ■

Since, for $p > 0$, $f(t) = t^p$ is an increasing function of $t \geq 0$ and, for all $a, b \in \mathbf{F}$, $|a+b| \leq |a| + |b|$, it follows that for $0 < p \leq 1$, $|a+b|^p \leq (|a| + |b|)^p \leq |a|^p + |b|^p$ [Theorem 4.6.1(b)]. This implies that $\|\cdot\|^*$ satisfies the triangle inequality. To see that the filterbase $\mathcal{B}$ of balls $C(0,r) = \{x \in L_p(-\infty,\infty) : \|x\|^* \leq r\}$, $r > 0$, is a base at 0 for a vector topology, we show that the conditions $(FB_1)$ and $(FB_2)$ of Theorem 4.5.1 are satisfied. As to $(FB_1)$, each $C(0,r)$ is clearly absorbent and balanced. The induced pseudometric $d(x,y) = \|x - y\|^*$ is clearly translation invariant so it generates a group topology by the discussion after Definition 2.7.2; the continuity of addition means that $\mathcal{B}$ satisfies $(FB_2)$. $\left(L_p(-\infty,\infty), \|\cdot\|^*\right)$ is therefore a TVS.

**Example 4.6.2** $\left(L_p(-\infty,\infty), \|\cdot\|^*\right)$, $0 < p < 1$, IS NOT LOCALLY CONVEX

We specialize to the case $p = 1/2$, although the result is valid for any $p \in (0,1)$. We contend that the only convex neighborhood of 0 in $L_{1/2}(-\infty,\infty)$ is $L_{1/2}(-\infty,\infty)$, in other words, that no proper neighborhood of 0 is convex. Specifically, we show that the convex hull of any $C(0,r)$ is $L_{1/2}(-\infty,\infty)$. For the sake of this argument, we denote $C(0,r)$ by just $C_r$. We show that

$$C_r \subset \left(C_{r/\sqrt{2}}\right)_c \qquad \text{for any } r > 0 \qquad (*)$$

To show this, let $r > 0$ be given, consider $x \in L_{1/2}(-\infty,\infty)$ such that $\|x\|^* = r$, and consider the function

$$w(t) = \int_{-\infty}^{t} |x(t)|^p \, dt$$

Using some obvious abbreviations for limits, we observe that $w(-\infty) = 0$ and $w(\infty) = r$. Since $w$ must be continuous because of the way it is defined, there must be some $t' \in \mathbf{R}$ such that $w(t') = r/2$. Now consider the $p$th-power summable functions $y$ and $z$ defined as

$$y(t) = \begin{cases} x(t) & \text{for } t \leq t' \\ 0 & \text{for } t > t' \end{cases} \qquad z(t) = \begin{cases} 0 & \text{for } t < t' \\ x(t) & \text{for } t \geq t' \end{cases}$$

Clearly, $x = y + z$ and $\|y\|^* = \|z\|^* = r/2$; hence

$$x = (1/2)(2y + 2z) \qquad \text{and} \qquad \|2y\|^* = 2^{1/2}\|y\|^* = \|2z\|^* = r/\sqrt{2}$$

This last line shows that $2y$ and $2z$ belong to $C_{r/\sqrt{2}}$; hence $x \in \left(C_{r/\sqrt{2}}\right)_c$ which proves $(*)$ for $x$ such that $\|x\|^* = r$. If $x \in C_r$ and $\|x\|^* = b < r$ then, by the argument just given, $x \in \left(C_{b/\sqrt{2}}\right)_c \subset \left(C_{r/\sqrt{2}}\right)_c$ which establishes $(*)$.

By replacing $r$ by $\sqrt{2}r$ in $(*)$, it is evident that $C_{r\sqrt{2}} \subset (C_r)_c$. Doing this again, we see that $C_{2r} \subset \left(C_{\sqrt{2}r}\right) \subset (C_r)_c$, and so on, so that for any $n \in \mathbf{N}$, $C_{2^n r} \subset (C_r)_c$. Hence $L_{1/2}(-\infty,\infty) = \cup_{n\in\mathbf{N}} C_{2^n r} \subset (C_r)_c$.

To complete the proof, suppose that $U$ is a convex neighborhood of 0 in $L_{1/2}(-\infty,\infty)$. Since $\{C_r : r > 0\}$ is a base at 0, $U$ must contain some $C_r$. Since $U$ is convex, it must also contain the convex hull $(C_r)_c$, which implies that $L_{1/2}(-\infty,\infty) \subset U$; in other words, there are no proper convex neighborhoods of 0. ■

## 4.7 PRODUCTS AND QUOTIENTS

As we did with topological groups in Example 2.4.2, we approach products of TVS through the portal of inverse image topologies.

**Example 4.7.1** INVERSE IMAGE TOPOLOGY Let $X$ be a TVS and let $\{X_s : s \in S\}$ be a family of TVS such that for each $s \in S$ there is a linear map $A_s : X \to X_s$. The *inverse image topology* $\mathcal{T}$ for $X$ *determined by the family* $\{A_s : s \in S\}$ is the coarsest topology with respect to which each $A_s$ is continuous [Def. 2.4.1]. For each $s \in S$, let $\mathcal{V}_s(0)$ denote the filterbase of balanced neighborhoods of 0 in $X_s$. A base $\mathcal{V}_{\mathcal{T}}(0)$ at 0 for the inverse image topology $\mathcal{T}$ for $X$ determined by the family $\{A_s : s \in S\}$ is given by intersections of the form $\cap_{k\in K} A_k^{-1}(V_k)$ where $K$ is a finite subset of $S$ and $V_k \in \mathcal{V}_k(0)$ for each $k \in K$. As noted in Example 2.4.5, it is a group topology. Therefore $(\mathrm{FB}_2)$ of Theorem 4.5.1 (i.e., $U + U \subset V$) is satisfied. As to $(\mathrm{FB}_1)$ of Theorem 4.5.1, the sets $\cap_{k\in K} A_k^{-1}(V_k)$, where $K$ is a finite subset of $S$, are clearly balanced and absorbent by Theorem 4.1.7(c). Thus $\mathcal{T}$ is a vector topology. We also say that $\mathcal{T}$ is the *projective limit topology* of $\{X_s : s \in S\}$ with respect to $\{A_s : s \in S\}$. If each $X_s$ is a LCS, each possesses a neighborhood base of disks; it follows that the inverse image topology is locally convex by Theorem 4.2.4(b).

Suppose that $X$ is a vector space and $Y$ is a TVS with filter $\mathcal{V}_Y(0)$ of neighborhoods of 0. Let $A : X \to Y$ be a surjective linear map and let $\mathcal{T}_A$ be the inverse image topology induced by $A$ on $X$ with neighborhood filter$\{A^{-1}(V) : V \in \mathcal{V}_Y(0)\}$ at 0. Then $A : (X, \mathcal{T}_A) \to Y$ is an open map by the following argument. Suppose $A$ is any map, linear or not, and let $U$ be an open subset of $X$. As such, there is an open subset $V$ in $Y$ such that $U = A^{-1}(V)$; since $A$ is onto, $A(U) = A\left(A^{-1}(V)\right) = V$. ■

Let $X = \Pi_{s\in S} X_s$ denote the Cartesian product of a family $\{X_s : s \in S\}$ of TVS endowed with the product topology. For elements $x = (x_s)$ and $y = (y_s)$ of $X$ and any scalar $a$, define $x + y = (x_s + y_s)$ and $ax = (ax_s)$. These operations clearly make $X$ into a vector space. Since projections are linear maps and the product topology is clearly the inverse image topology determined by them, the product topology is a vector topology. The basic neighborhoods of 0 are of the form:

$$\Pi_{s\in S}V_s \text{ where each } V_s \text{ is a balanced neighborhood of 0 in } X_s$$
$$\text{and } V_s = X_s \text{ for all but a finite number of indices } s$$

Whenever we speak of a product of TVS, it is always with respect to the pointwise operations and the product topology. As is generally true for products, $\Pi_{s\in S}X_s$ is Hausdorff iff each $X_s$ is.

Suppose that $M$ is a subspace of the TVS $X$. The relative topology on $M$ is the inverse image topology determined by the injection $x \mapsto x$ from $M$ into $X$ so it is a vector topology for$M$.

Another way to look at the inverse image topology for $X$ determined by the family $\{A_s : X \to X_s : s \in S\}$ of linear maps is as the the inverse image topology $\mathcal{T}$ induced on $X$ by the single linear map

$$\begin{array}{rcl} A : X & \to & \prod_{s\in S} X_s \\ x & \mapsto & (A_s x)_{s\in S} \end{array}$$

**Example 4.7.2** DIRECT IMAGE AND QUOTIENT TOPOLOGY As first discussed in Sec. 2.5, if $X$ is a topological space, $Y$ any set, and $f : X \to Y$ any map, the finest topology $\mathcal{T}_f$ for $Y$ with respect to which $f$ is continuous is the *direct image* (*final, identification*) topology for $T$ *determined* or *induced by* $f$. A set $U \subset Y$ is $\mathcal{T}_f$-open iff $f^{-1}(U)$ is open in $X$. If $X$ is a topological group, $Y$ a group, and $f$ a homomorphism then the direct image topology is a group topology for $Y$ [Theorem 2.5.1]. A neighborhood base for $\mathcal{T}_f$ at 0 is given by the collection $\mathcal{B}$ of all symmetric subsets $B$ of $Y$ such that $f^{-1}(B)$ is a neighborhood of 0 in $X$.

Similar considerations obtain for TVS. As noted in Sec. 2.5, the direct image topology induced by a non-surjective map is disconnected. Hence, for direct image topologies, we restrict consideration to surjective linear maps $A$ of a TVS $X$ onto a vector space $Y$. We show next that the direct image topology $\mathcal{T}_A$ determined by $A$ on $Y$ is a vector topology by showing that conditions $(FB_1)$ and $(FB_2)$ of the basis theorem [4.5.1] are satisfied. A neighborhood base at 0 for $\mathcal{T}_A$ is given by the filterbase $\mathcal{B}$ of all balanced subsets $B$ of $Y$ such that $A^{-1}(B)$ is a neighborhood of 0 in $X$. For any $B \in \mathcal{B}$, $A(B)$ is absorbent [Theorem 4.1.8(b)] because $A$ is surjective so $\mathcal{B}$ satisfies $(FB_1)$. Since $A^{-1}(B)$ is a neighborhood of 0, there is a balanced neighborhood $U$ of 0 in $X$ such that $U + U \subset A^{-1}(B)$. $A(U)$ is balanced by Theorem 4.1.8(a). Since $U \subset A^{-1}(A(U))$, $A^{-1}(A(U))$ is a neighborhood of 0 in $X$. Since $A(U) + A(U) = A(U + U) \subset B$, $\mathcal{B}$ satisfies $(FB_2)$.

**Quotient topology.** If $M$ is a subspace of TVS $X$ then the *quotient topology* on $X/M$ is the direct image topology determined by the canonical (linear) map $k : X \to X/M$, $x \mapsto \dot{x} = x + M$. Note that for any $x \in X$, $k^{-1}(\dot{x}) = x + M$ and that the open subsets of $X/M$ are of the form $k(U)$ where $U$ is an open subset of $X$. If $\mathcal{B}$ is the filterbase of balanced neighborhoods of 0 in $X$ then $k(\mathcal{B})$ is a neighborhood base at 0 for the quotient topology. Whenever we speak of a quotient space of a TVS, it carries the quotient topology. Some other facts about quotient topologies follow in Theorem 4.7.3. ■

Similar to what we did for quotients of pseudometrizable groups in Example 2.7.5, if $M$ is a subspace of a seminormed space $(X, p)$, the canonical seminorm $\dot{p}$ on $X/M$ is defined by taking $\dot{p}(x+M) = \inf\{p(x+m) : m \in M\}$. This seminorm induces the quotient topology by an argument similar to the one for groups in Example 2.7.5.

**Theorem 4.7.3** QUOTIENT TOPOLOGY Let $M$ be a subspace of the topological vector space $X$ and let $k : X \to X/M$ denote the canonical map $x \mapsto x + M = \dot{x}$. Then:

(a) $k$ is a continuous, open linear map;
(b) $X/M$ is Hausdorff iff $M$ is closed;
(c) $k$ is not generally a closed map;
(d) if $X$ is locally convex, so is $X/M$.

**Proof.** We use the notation of the statement. (a) Clear from the discussion in Sec. 2.5 and Example 4.7.2.

(b) Use the corresponding statement for topological groups, Theorem 2.5.4.

(c) View the complex numbers $\mathbf{C}$ in its usual topology as a real TVS and consider the subset $S = \{1+i, 2+(1/2)i, \ldots, n+(1/n)i, \ldots\}$. Clearly $S$ is closed. Consider the subspace $\mathbf{R}$ and associate with each coset $a+bi+\mathbf{R}$ the unique real number $b$. Let $B(a+bi, r)$ denote the open disk of radius $r>0$ about $a+bi$. Letting $k$ stand for the canonical map from $\mathbf{C}$ onto $\mathbf{C}/\mathbf{R}$ then $k(B(a+bi,r)) = (b-r, b+r)\,i+\mathbf{R}$, so a base for the quotient topology, reflected onto $\mathbf{R}$, consists of the open intervals $(b-r, b+r)$ $, r>0,\ b \in \mathbf{R}$. The canonical image of $S$ under this identification is $\{1, 1/2, \ldots, 1/n, \ldots\}$ which is not a closed subset of $\mathbf{R}$. [See Theorem 2.5.3(b) for the analogous statement on topological groups.]

(d) If $\dot{U}$ is a neighborhood of $\dot{0}$ in $X/M$ then $k^{-1}(\dot{U})$ is a neighborhood of 0. Since $X$ is locally convex, $k^{-1}(\dot{U})$ contains some convex neighborhood $V$ of 0. Hence $k(V)$ is a convex neighborhood of 0 contained in $\dot{U}$. ■

In any TVS $X$, $N = \operatorname{cl}\{0\}$ is a closed linear subspace and $X/N$ is a HTVS by Theorem 4.7.3(b). If $X$ and $Y$ are topological vector spaces over any topological field, a linear map $A : X \to Y$ may be dissected in a way similar to the way in which homomorphisms between topological groups were deconstructed in Sec. 2.5. The main facts are contained in Theorem 4.7.4.

**Theorem 4.7.4** DECOMPOSITION OF A LINEAR MAP $X$ and $Y$ are topological vector spaces over a topological field and $A : X \to Y$ is linear. $N = A^{-1}(0)$ and $k$ denotes the canonical map $x \mapsto \dot{x} = x+N$; $\dot{A} : X/N \to A(X)$, $\dot{x} \mapsto Ax$, is called the *canonical isomorphism* associated with $A$; $I : A(X) \to Y$, $Ax \mapsto Ax$, is called the *canonical injection* of $A(X)$ into $Y$.

(a) $I$ is a linear homeomorphism of $A(X)$ into $Y$ and $\dot{A}$ is a vector space isomorphism of $X/N$ onto $A(X)$;

(b) $A$ is continuous iff $\dot{A}$ is continuous; and

(c) $A$ is relatively open iff $\dot{A}$ is open. [Note that if we had dispensed with $I$ and viewed $\dot{A}$ as mapping $X/N$ into $Y$, this would be phrased as: $A$ maps open subsets of $X$ onto relatively open subsets of $A(X)$ iff $\dot{A}$ maps open subsets of $X/N$ onto relatively open subsets of $A(X)$.]

**Proof.** We use the notation of the statement. (a) Clear.

(b,c) Use Theorem 2.5.5(b,c), respectively, the analogous statements for topological groups. ■

Combining (b) and (c), it follows that $\dot{A}$ is a linear homeomorphism iff $A$ is open and continuous.

# 4.8 METRIZABILITY AND COMPLETION

We discuss pseudometrizability and completion of a TVS here. Since we have already discussed these topics for topological groups in Secs. 2.7 and 3.7, not much remains to be done. There is a little surprise in regard to pseudometrizability of TVS in that the topology of a pseudometrizable TVS may always be obtained from an $F$-seminorm [Definition 4.8.1 and Theorem 4.8.3]. With completion available, we return to a question mentioned in Sec. 4.4: When is the convex hull of a compact set compact? Example 4.8.8 shows that the convex hull of a compact set need not be compact. Conditions under which it is are given in Theorem 4.8.9. We return to the topics of pseudometrizability and completion in the context of locally convex spaces in Secs. 5.6 and 5.11.

Let us recall some things. A pseudometric $d$ on an additive topological group $X$ is called *(translation-)invariant* if $d(x,z) = d(x+y, z+y)$ for all $x, y$, and $z$ in $X$. The *value* $|\cdot| = d(\cdot, 0)$ associated with $d$ has properties like ordinary absolute value: It is nonnegative, satisfies the triangle inequality and satisfies $|-x| = |x|$ for every $x \in X$ [Def. 2.7.2]. There is a 1-1 correspondence between *invariant* pseudometrics and values (see after Definition 2.7.2); the balls $C(0,r) = \{x \in X : |x| \leq r\}$, $r > 0$, form a base at 0 for a group topology for a group $X$.

Does an invariant pseudometric on a TVS $X$ induce a *linear* topology? As the trivial metric $d$ is decidedly invariant, inducing the discrete topology, and TVS are connected [Theorem 4.3.4], the answer is no. For the associated trivial value $|\cdot| = d(\cdot, 0)$, if $x \neq 0$ then $|x/n| = 1$ for every $n \in \mathbf{N}$, so $x/n \nrightarrow 0$ and scalar multiplication is discontinuous. The triangle inequality makes for continuity of addition, but additional conditions must be imposed to guarantee continuity of scalar multiplication.

**Definition 4.8.1** F-SEMINORMS An *F-seminorm* is a map $p$ of a vector space $X$ into the nonnegative reals such that for each $a \in \mathbf{F}$ and $x$ and $y$ in $X$:

(a) $p(ax) \leq p(x)$ for $|a| \leq 1$;
(b) $p(x/n) \to 0$; and
(c) $p(x+y) \leq p(x) + p(y)$.

If $p(x)$ is 0 only when $x = 0$ then $p$ is an *F-norm*. ■

Although invariant pseudometrics do not necessarily determine vector topologies, $F$-seminorms do.

**Theorem 4.8.2** $F$-seminorms determine pseudometrizable vector topologies.

**Proof.** Let $p$ be an $F$-seminorm on a vector space $X$. We show that $p$ is a value on $X$. By Def. 4.8.1(a), $p(-x) \leq p(x)$ for every $x \in X$. Consequently $p(x) \leq p(-x)$ as well and $p(x) = p(-x)$. Clearly $p(0) \geq 0$ and for $n \in \mathbf{N}$, $p(0) \leq p(x/n) + p(-x/n) = 2p(x/n)$ so $(1/2)\,p(0) \leq p(x/n) \to 0$; it follows that $p(0) = 0$. As the triangle inequality follows from Def. 4.8.1(c), this shows that $p$ is a value on $X$. As such, the invariant pseudometric $d(x,y) = p(x-y)$ determines a group topology $\mathcal{T}$ on $X$ (see before Theorem 2.7.3) and therefore satisfies $(FB_2)$ of the basis theorem [4.5.1]. Since $p(x/n) \to 0$, each closed ball $C(0,r) = \{x \in X : p(x) \leq r\}$, $r > 0$, is absorbent; since $p(ax) \leq p(x)$ for $|a| \leq 1$, $C(0,r)$ is balanced. Therefore $\mathcal{T}$ satisfies $(FB_1)$ of the basis theorem [4.5.1] and $\{C(0,1/n) : n \in \mathbf{N}\}$ is a base at 0 for a vector topology. Since $\mathcal{T}$ has a denumerable base at 0, it is pseudometrizable by Theorem 2.7.3. ■

We show next that a TVS $X$ is pseudometrizable iff its topology is generated by an $F$-seminorm.

**Theorem 4.8.3** PSEUDOMETRIZABILITY AND F-SEMINORMS A TVS $X$ is pseudometrizable iff it has a countable base at 0; in this case its topology is generated by an $F$-seminorm. The pseudometrizable group $X$ is metrizable iff the topology is Hausdorff, in which case its topology is generated by an $F$-norm.

**Proof.** The pseudometrizability assertion follows immediately from Theorem 2.7.3. Now suppose that a TVS $X$ is pseudometrizable so that it has a countable base $(U_n)$ of balanced neighborhoods of 0. By Theorem 2.7.3 we also know that there is an invariant pseudometric $d$ whose associated value $p(\cdot) = d(\cdot,0)$ determines the topology on $X$. We now show that, by a slight modification of the proof of Theorem 2.7.3, there exists an $F$-seminorm which generates the topology on $X$. Since the $U_n$ are balanced, the function $f$ of Theorem 2.7.3 is such that $f(tx,0) \leq f(x,0)$ for $|t| \leq 1$. This means that the invariant pseudometric $d$ of Theorem 2.7.3 satisfies $d(tx,0) \leq d(x,0)$ for $|t| \leq 1$, and therefore that $p(\cdot) = d(\cdot,0)$ satisfies (a) of Definition 4.8.1. Since the $U_n$ are absorbent, $p$ satisfies (b) of the definition; $p$ satisfies the triangle inequality because $d$ does. The remaining parts are clear. ■

The results of Theorem 2.7.4 and Example 2.7.5 can now be adapted to yield:

**Theorem 4.8.4** QUOTIENTS Let $M$ be a subspace of the TVS $X$. If $X$ is pseudometrizable then so is $X/M$. If $p$ is any $F$-seminorm on $X$ then the quotient $F$-seminorm $\dot{p}(\dot{x}) = \inf\{p(x+m) : m \in M\}$ determines the quotient topology; if $p$ is an $F$-norm and $M$ is closed then $X/M$ is metrizable.

A TVS is *complete* if it is complete as a topological group [Definition 3.1.2]. A complete metrizable *locally convex* space is called a *Fréchet space*. Fréchet was the first to use the term "Banach space." Banach repaid this favor by coining the term "Fréchet space" for complete metrizable TVS; Banach did not require local convexity, which was later appended by Bourbaki. Whenever you encounter "Fréchet space," check to see if local convexity is included as both meanings are in use.

**Definition 4.8.5** COMPLETION A complete TVS $\widehat{X}$ which contains a linearly homeomorphic image of a TVS $X$ as a dense subspace is called a *completion* of $X$. ■

Given a pseudometrizable TVS $(X, d)$ ($d$ invariant), let $\widehat{X}$ denote the collection of Cauchy sequences from $X$. Given $(x_n), (y_n) \in \widehat{X}$ and a scalar $a$, define $a(x_n) = (ax_n)$ and $(x_n) + (y_n) = (x_n + y_n)$. With respect to these definitions, $\widehat{X}$ is a vector space; moreover, $\widehat{d}\,[(x_n), (y_n)] = \lim_n d(x_n, y_n)$ defines an invariant pseudometric on $\widehat{X}$ with respect to which $\widehat{X}$ is complete. The map $f : X \to \widehat{X},\ x \mapsto (x)$, where $(x)$ denotes the sequence consisting solely of $x$'s, is an isometric linear isomorphism and $f(X)$ is dense in $\widehat{X}$, i.e., $\left(\widehat{X}, \widehat{d}\right)$ is a completion of $(X, d)$. (See Theorem 3.7.1, the corresponding result for a pseudometrizable topological group.) Let $\widehat{V}(0)$ denote the neighborhoods of 0 in $\widehat{X}$ and let $N$ denote the subspace $\mathrm{cl}\{0\} = \cap \widehat{V}(0)$. Then $\widehat{X}/N$ is a completion of $X$ as a metric TVS.

With minor modifications to Theorem 3.7.2 on completion of a topological group to cover scalar multiplication we have:

**Theorem 4.8.6** COMPLETION Any TVS $X$ possesses a completion.

It follows from Theorem 3.7.4, on completeness of quotients of a topological group, that:

**Theorem 4.8.7** COMPLETENESS OF QUOTIENT If $X$ is a complete pseudometrizable TVS and $M$ is a closed subspace then $X/M$ is complete.

With completion available, we can now return to a point raised in Sec. 4.4 about compactness of the convex hull of a compact set.

**Example 4.8.8** COMPACT SET WITH NONCOMPACT CONVEX HULL As noted in Theorem 3.4.2, a compact subset of a topological group is complete. To show that there is a compact set with noncompact convex hull, it therefore suffices to exhibit a compact set $K$ whose convex hull $K_c$ is not complete. If $K_c$ were complete then $K_c = \operatorname{cl} K_c$, so it suffices to exhibit a compact set $K$ whose closed convex hull $\operatorname{cl} K_c$ is not complete.

View the vector space $\varphi$ of "finite sequences" (sequences which are almost always 0) as a subspace of the Hilbert space $\ell_2$. For each $n \in \mathbf{N}$, let $x_n$ be the sequence which has $1/n$ as its $n$th entry and 0's everywhere else. Clearly, $x_n \to 0$ in $\ell_2$ so the terms of the sequence together with 0 form a compact subset $K$ of $\varphi$. With $s_n = \sum_{j=1}^{n} 2^{-j} x_j$, a direct computation shows that $s_n$ converges to the sequence $(1/j2^j)$ in $\ell_2$. With $a_n = \sum_{j=1}^{n} 2^{-j}$ then, for every $n$, $s_n/a_n = (\sum_{j=1}^{n} 2^{-j} x_j)/(1/\sum_{j=1}^{n} 2^{-j})$ belongs to the convex hull $K_c$ of $K$. Since $s_n/a_n \to (\lim_n s_n)/1 = (1/j2^j)$ has *no* 0 entries, it is certainly not a vector of $\varphi$, hence not of $\operatorname{cl}_\varphi K_c$ either. It follows that $K_c$ is not compact. ■

As $\varphi$ is incomplete (the sequence $x_k = (1, 1/2, \ldots, 1/k, 0, 0, , \ldots)$ is Cauchy but has no limit in $\varphi$), maybe some sort of completeness condition provides a framework in which the convex hull of a compact set is compact. Indeed, Mazur [1930] proved that the closed convex hull of a compact subset of a Banach space is compact. Lindenstrauss and Tzafriri [1977, p. 30] show that a closed subset of a Banach space is compact iff it is contained in the closed convex hull of a null sequence (the "if" part being straightforward). In Theorem 4.8.9 we show that the convex hull of a totally bounded subset of a locally convex space is totally bounded; hence, if $X$ is a complete LCS then the closed convex hull of a compact set is compact.

**Theorem 4.8.9** HULLS OF COMPACT SETS (a) BALANCED HULL In any TVS $X$ the balanced hull $K_b$ of a totally bounded or compact set $K$ is totally bounded or compact, respectively.

(b) CONVEX HULL IN LCS If $K$ is totally bounded subset of an LCS $X$ then so is its convex hull $K_c$ and therefore (by (a)) its disked hull $K_{bc}$.

(c) COMPLETENESS If $K$ is a compact subset of an LCS then its convex hull $K_c$ and disked hull $K_{bc}$ are compact iff $K_c$ and $K_{bc}$ are complete, respectively. Thus, if $X$ is complete, $\operatorname{cl} K_c$ and $\operatorname{cl} K_{bc}$ are compact.

**Proof.** (a) Let $\widehat{X}$ be a completion of the TVS $X$. If $K \subset X$ is totally bounded, so is its closure $\operatorname{cl}_{\widehat{X}} K$ in $\widehat{X}$ by Theorem 3.3.6; $\operatorname{cl}_{\widehat{X}} K$ is therefore compact by Theorem 3.4.1.

The closed unit disk $D$ of $\mathbf{F}$ is compact; hence $D \times \operatorname{cl}_{\widehat{X}} K$ is compact in the product topology on $\mathbf{F} \times \widehat{X}$. By the continuity of scalar multiplication, the algebraic product $D\,\mathrm{cl}_{\widehat{X}} K$ is therefore compact. Since the balanced hull $K_b = DK \subset D(\mathrm{cl}_{\widehat{X}} K)$, $K_b$ is totally bounded.

Clearly, if $K$ is compact, so is $DK = K_b$.

(b) Suppose that $K$ is a totally bounded subset of the LCS $X$. Given a neighborhood $U$ of 0, choose a disked neighborhood $V$ of 0 such that $V+V \subset U$. Since $K$ is totally bounded, there exist $x_1, \ldots, x_n$ in $K$ such that $K \subset \cup_{i=1}^n (x_i + V) \subset \{x_1, \ldots, x_n\}_c + V$. As this last set is convex [Theorem 4.2.4(a)], $K_c \subset \{x_1, \ldots, x_n\}_c + V$. Since $\{x_1, \ldots, x_n\}_c$ is compact [Theorem 4.4.4], there exist $y_1, \ldots, y_k \in \{x_1, \ldots, x_n\}_c$ such that $\{x_1, \ldots, x_n\}_c \subset \cup_{i=1}^k (y_i + V)$. Thus

$$K_c \subset [\cup_{i=1}^k (y_i + V)] + V = \cup_{i=1}^k (y_i + V + V) \subset \cup_{i=1}^k (y_i + U)$$

(c) By Theorem 3.4.2, completeness is equivalent to compactness for totally bounded sets. Since compactness implies total boundedness, it follows from (b) that if $K$ is compact then $K_b$ and $K_{bc}$ are totally bounded. ■

## 4.9 TOPOLOGICAL COMPLEMENTS

Two subspaces $M$ and $N$ of a vector space $X$ are algebraic complements if $M \cap N = \{0\}$ and $X = M + N$. Under these circumstances each vector $x$ in $X$ has a representation of the form $m+n$ for a unique $m$ in $N$ and a unique $n$ in $N$; we write $X = M \oplus N$ and say $X$ is the *algebraic direct sum* of $M$ and $N$ and that $M$ and $N$ are an *algebraic direct sum decomposition of* $X$ and that $M$ and $N$ are *algebraic complements* of each other. The space $X$ has been split into two parts and recovered by "adding" them back together again. In this section we investigate analogous considerations for TVS $X$—reconstitute $X$ topologically as well as algebraically when $M$ and $N$ carry their subspace topologies and $M \times N$ the product topology. We have to consider the question: If $X = M \oplus N$, under what circumstances is the product topology on $M \times N$ equal to the original on $X$? In other words, when is the map $S$ ("$S$" for "sum")

$$\begin{array}{rccc} S: & M \times N & \to & M + N \\ & (m, n) & \mapsto & m + n \end{array} \qquad (*)$$

a homeomorphism? $S$ is always linear, bijective, and continuous (by continuity of addition), so the question reduces to: When is $S$ an open map?, i.e., when does $m + n \to 0$ imply that $m \to 0$ and $n \to 0$?

When $S$ is a homeomorphism, we say that $M$ and $N$ are *topological complements* (or *supplements*) and $M$ is (*topologically*) *complemented*, that $X$ is the *topological direct sum* of $M$ and $N$, and write $X = M \oplus N$ (top). The common usage is to say that $M$ is complemented, rather than *topologically* complemented. Note that "is a topological complement of" is a symmetric relation.

Any subspace $M$ of a vector space $X$ has an algebraic complement $N$: Extend a Hamel basis $B$ for $M$ to a basis $B_X$ for $X$; then the linear span $N$

of $B_X \setminus B$ is an algebraic complement of $M$. As we show in Example 4.9.7, not every subspace has a topological complement. For one thing, to have a topological complement a subspace of an HTVS must be closed [Theorem 4.9.3] but even a closed subspace of a Banach space can be "uncomplemented." We show in Theorem 7.3.5 that any finite-dimensional (or codimensional) subspace of an LCHS has a topological complement.

**Definition 4.9.1** PROJECTION Let $X$ be a vector space. A linear map $P : X \to X$ such that $P^2 = P \circ P = P$ (i.e., such that $P$ is idempotent) is called a *projection.* ■

The range $P(X) = \{x \in X : Px = x\}$ of a projection $P$ and its null space are algebraically complementary: Any $x \in X$ can be uniquely represented as $x = Px + (x - Px)$. Conversely, if $M$ and $N$ are algebraic complements in $X$, the map sending $x = m + n \in M + N$ into $m$ is a projection called the *projection on* $M$ *along* (or *parallel to*) $N$.

**Theorem 4.9.2** TOPOLOGICAL COMPLEMENT Let $M$ be a linear subspace of a TVS $X$. An algebraic complement $N$ of $M$ is a topological complement of $M$ iff either of the following conditions holds:

(a) PROJECTIONS The projection $P_M$ on $M$ along $N$ is continuous. Hence $M$ is complemented in $X$ iff there is a continuous projection of $X$ onto $M$.

(b) QUOTIENTS The canonical isomorphism $n \mapsto n + M$ of $N$ onto $X/M$ is bicontinuous. Hence any topological complement of $M$ is linearly homeomorphic to $X/M$.

**Proof.** We use the notation of the statement. (a) Let $M$ have a topological complement $N$ and let $P$ be the projection on $M$ along $N$. Suppose that the net $m_s + n_s \to 0$, $m_s \in M$, $n_s \in N$; then, since the map $S$ of $(*)$ is a homeomorphism, $m_s \to 0$ and $n_s \to 0$. Hence $m_s + n_s \to 0$ implies that $m_s = P(m_s + n_s) \to 0$ and $P_M$ is continuous.

Conversely, suppose that $P$ is a continuous projection with range $M$. It is elementary to show that the null space $N$ of $P$ is an algebraic complement of $M$. To prove that $N$ is a topological complement, suppose that $m_s + n_s \to 0$, where $(m_s)$ and $(n_s)$ are nets from $M$ and $N$, respectively. By the continuity of $P$, $m_s = P(m_s + n_s) \to 0$ and $n_s = (m_s + n_s) - P(m_s + n_s) \to 0$ and it follows that $N$ is a topological complement of $M$.

(b) If $N$ is an algebraic complement of $M$ then we may decompose the projection $P$ on $N$ along $M$ as follows:

$$\begin{array}{rcccc} P: & X = M \oplus N & \xrightarrow{k} & X/M & \xrightarrow{h} & N \\ & (m,n) & \mapsto & n + M & \mapsto & n \end{array}$$

($X/M$ carries its quotient topology, and $N$ its subspace topology.) $N$ is a topological complement of $M$ iff $P$ is continuous by (a). $P = h \circ k$ is continuous, however, iff $h$ is continuous. Since $h$ is open ($n_s \to 0 \Rightarrow n_s + M \to 0 + M$), the continuity of $h$ is equivalent to the bicontinuity of the canonical isomorphism $h^{-1}$. ■

As an immediate consequence of Theorem 4.9.2(a), we show next that a complemented subspace of an HTVS is closed.

**Theorem 4.9.3** COMPLEMENTED IMPLIES CLOSED A topologically complemented subspace $M$ of a Hausdorff TVS is closed.

**Proof.** Let $N$ be a topological complement of $M$ of a Hausdorff TVS and let $P_N$ be the (continuous) projection on $N$ along $M$ so that $M = P_N^{-1}(0)$. ■

An infinite-dimensional Banach space $X$ for which there exist infinite-dimensional subspaces $M$ and $N$ such that $X = M \oplus N$ (top) is called *decomposable.* Gowers and Maurey [1993, pp. 862–863] invented a Banach space $X_G$ to negatively settle the *unconditional basic sequence problem*, namely: Does every infinite-dimensional Banach space contain an infinite-dimensional closed subspace with an unconditional Schauder basis [Definition 15.8.1]? Gowers and Maurey quote W. Johnson [*ibid.* p. 852] as having observed that not only is $X_G$ not decomposable, it does not even have a decomposable subspace; in other words, it is *hereditarily indecomposable.* Two other properties of this exotic specimen $X_G$ are (1) every complemented subspace of $X_G$ is of finite dimension or codimension and (2) it is reflexive.

A subspace $M$ of a vector space $X$ is *maximal* if there exists no proper subspace $N$ of $X$ (i.e., $N \neq X$) which contains $M$ properly. If $X$ is a TVS then $\operatorname{cl} M$ is a subspace, so a maximal subspace of a TVS must be closed or dense in $X$. It follows immediately from Theorem 1.5.1(b,c) that:

**Theorem 4.9.4** CLOSED MAXIMAL SUBSPACES AND CONTINUOUS LINEAR FUNCTIONALS Let $X$ be a TVS over $\mathbf{F} = \mathbf{R}$ or $\mathbf{C}$.

(a) $M$ is a closed maximal subspace of $X$ iff $M$ is the null space of a nontrivial continuous linear functional on $X$.

(b) $H$ is a closed hyperplane in $X$ iff $H = \{x \in X : f(x) = a\}$ for some nontrivial continuous linear functional $f$ on $X$ and $a \in \mathbf{F}$.

We now show that a closed maximal subspace of a TVS always has a topological complement.

**Theorem 4.9.5** CLOSED MAXIMAL SUBSPACES ARE COMPLEMENTED If $M$ is a closed maximal subspace of a TVS $X$ then any algebraic complement $N$ of $M$ is a topological complement.

**Proof.** Let $M$ be a closed maximal subspace of a TVS $X$. Since $M$ is maximal, $\operatorname{codim} M = 1$ [Theorem 1.5.1(a)]; hence if $N$ is an algebraic complement of $M$, $\dim N = 1$. Therefore there must be some $x \neq 0$ such that $N = \mathbf{F}x$. Consequently $X = M \oplus \mathbf{F}x$ and each vector $y$ has a unique representation in the form $y = m + tx$, $m \in M$, $t \in \mathbf{F}$. To show that $\mathbf{F}x$ is a topological complement of $M$, we use the criterion of Theorem 4.9.2(a): We show that the projection $P$ on $\mathbf{F}x$ along $M$, $tx + m \mapsto tx$, is continuous.

To this end note that $P^{-1}(0) = M$. As $M$ is closed and $x \notin M$, there exists a balanced neighborhood $U$ of 0 in $X$ such that $(x+U) \cap M = \varnothing$. We claim that if $m + tx \in U$ then $|t| < 1$. If $|t| \geq 1$ then, as $U$ as balanced, $(1/t)(m+tx) = m/t + x \in U$. Hence $x - (m/t + x) = -m/t \in (x+U) \cap M$, which contradicts $(x+U) \cap M = \varnothing$. Hence if $0 < r < 1$ and $m + tx \in rU$ then $|t| < r$.

To establish the continuity of $P$ at 0, suppose that the net $m_s + t_s x \to 0$. As such, for any $0 < r < 1$, $m_s + t_s x \in rU$ eventually. Therefore $|t_s| < r$ eventually. In other words, $t_s \to 0$, which implies that $t_s x = P(m_s + t_s x) \to 0$ and proves the continuity of $P$. ■

**Example 4.9.6** COMPLEMENTS

(a) HILBERT SPACE Any closed subspace $M$ of a Hilbert space $(X, \langle\cdot,\cdot\rangle)$ is complemented by its orthogonal complement $M^{\perp} = \{n \in X : \langle m, n\rangle = 0$ for all $m \in M\}$ [Bachman and Narici 2000, p. 172]. Do any other Banach spaces have the property that every closed subspace is complemented? The question:

> If every closed subspace $M$ of a Banach space $X$ is complemented, is $X$ linearly homeomorphic to a Hilbert space?

is known as *the complemented subspaces problem.* Lindenstrauss and Tzafriri [1971] answered it affirmatively. To put it another way, any Banach space $X$ has uncomplemented closed subspaces unless $X$ is linearly homeomorphic to a Hilbert space.

(b) NULL SPACE OF A LINEAR FUNCTIONAL Let $X$ be a TVS and let $f : X \to \mathbf{F}$ be a nontrivial continuous linear functional. Since its null space is maximal [Theorem 1.5.1(b)], it has a topological complement by Theorem 4.9.5.

(c) For any subset $A \subset [0,1]$ of positive Lebesgue measure with characteristic function $k_A$, the map $P : L_1[0,1] \to L_1[0,1]$, $x \mapsto xk_A$, is a continuous linear projection. For $x \in L_1[0,1]$, let $\operatorname{coz} x = \{t \in [0,1] : x(t) \neq 0\}$. Then $P(X) = \{x \in L_1[0,1] : \operatorname{coz} x \subset A\}$ has $N = P^{-1}(0) = \{x \in L_1[0,1] : \operatorname{coz} x \subset$ $\complement A$ as a topological complement.

(d) $C[0,1]$, $L_p[0,1]$ and $\ell_p$, $1 \leq p \leq \infty$, $p \neq 2$, have closed uncomplemented subspaces [Köthe 1983, pp. 430–431].

(e) FINITE-DIMENSIONAL SUBSPACES OF LCHS Any finite-dimensional or codimensional subspace $M$ of a locally convex Hausdorff space is complemented [Theorem 7.3.5]. But see (f) below.

(f) No finite-dimensional subspace of the non-locally convex space $L_p[0,1]$, $0 < p < 1$, [Example 4.6.2] has a topological complement [Köthe 1983, p. 158].

The first closed uncomplemented subspace was invented by Banach and Mazur [1933, footnote, p. 111]. Murray [1937] showed that $\ell_p$, $1 < p \neq 2$, has uncomplemented subspaces. Bennett et al. 1977 showed that for $1 < p < 2$, $L_p[0,1]$ has an uncomplemented subspace that is linearly homeomorphic to a Hilbert space. Phillips [1940] discovered that $c_0$ is uncomplemented in $\ell_\infty^{\mathbf{R}}$, a fact also discovered by Sobczyk [1940]. We present Whitley's [1966] simplified version of Phillips's proof in Example 4.9.7.

**Example 4.9.7** $c_0$ UNCOMPLEMENTED IN $\ell_\infty^{\mathbf{R}}$ Let $(\ell_\infty^{\mathbf{R}}, \|\cdot\|_\infty)$ be the Banach space of bounded sequences of real numbers and $c_0$ the closed subspace of null sequences. To show that $c_0$ is uncomplemented, we show the criterion of Theorem 4.9.2(a) to be violated, that there is no continuous projection of $\ell_\infty^{\mathbf{R}}$ onto $c_0$. What moves the proof is the following ingenious observation:

(a) Any denumerable set $I$ has an uncountable family $\{U_s : s \in S\}$ of infinite subsets, each of whose pairwise intersections is finite.

*Proof* (a): We may suppose that $I$ is the set of rationals in $(0,1)$. Let $S$ denote the irrationals in $(0,1)$ and for each $s$ in $S$ let $U_s$ consist of the elements of a sequence of rationals in $(0,1)$ which converges to $s$.

(b) $\ell_\infty^{\mathbf{R}}/c_0$ in its quotient topology is a normed space. The canonical norm of the coset $x + c_0$ is given by $\|x + c_0\| = \inf\{\|x+y\|_\infty : y \in c_0\}$ [Theorem 5.11.2].

(c) If $h$ is a continuous linear functional on a normed linear space $(X, \|\cdot\|)$ then $h$ is bounded on the unit ball $U = \{x \in X : \|x\| \leq 1\}$ for otherwise there exists a sequence $(x_n)$ from $U$ such that $|h(x_n)| > n$ for each $n \in \mathbf{N}$. As $h(x_n/n) \not\to 0$, $h$ is discontinuous.

A collection $T$ of linear functionals on a vector space $X$ is called *total* if the only vector on which each member of $T$ vanishes is 0. The coefficient functionals $f_j : \ell_\infty^{\mathbf{R}} \to \mathbf{F}$, $(a_n) \mapsto a_j$, $(j \in \mathbf{N})$ for example, are a total set of continuous linear functionals on $\ell_\infty^{\mathbf{R}}$. Moreover, if $X$ is linearly isomorphic to $Y$, under an isomorphism $A : X \to Y$ say, and $T$ is a total set of linear functionals on $X$ then $\{f \circ A^{-1} : f \in T\}$ is a total set of linear functionals for $Y$. If $E$ is a continuous projection of $\ell_\infty^{\mathbf{R}}$ onto $c_0$ then the closed subspace $M = E^{-1}(0)$ is a topological complement of $c_0$. Since the continuous dual of $\ell_\infty^{\mathbf{R}}$—all continuous linear functionals on $\ell_\infty^{\mathbf{R}}$—has $\{f_j : j \in \mathbf{N}\}$ as a countable total subset, so does $M$ and therefore so does $\ell_\infty^{\mathbf{R}}/c_0$ since $\ell_\infty^{\mathbf{R}}/c_0$ is linearly homeomorphic to $M$ [Theorem 4.9.2(b)]. We now seek to contradict this fact about $\ell_\infty^{\mathbf{R}}/c_0$.

Let $\{U_s : s \in S\}$ be an uncountable family of infinite subsets of $\mathbf{N}$ with finite pairwise intersections as in (a). For each $s \in S$ let $k_s$ be the characteristic function of $U_s$; note that each $k_s \in \ell_\infty^{\mathbf{R}}$. Let $\dot{k}_s$ denote the coset $k_s + c_0$. For a nontrivial continuous linear functional $f$ on $\ell_\infty^{\mathbf{R}}/c_0$ and $n \in \mathbf{N}$, let $B(n) = \{\dot{k}_s : \left|f\left(\dot{k}_s\right)\right| \geq 1/n, s \in S\}$. We show that

$$\{\dot{k}_s : f(\dot{k}_s) \neq 0, s \in S\} = \bigcup_{n \in \mathbf{N}} B(n)$$

is countable by showing that each $B(n)$ is finite. To this end, suppose that $\dot{g}_1, \ldots, \dot{g}_m \in B(n)$ and let $b_i = \operatorname{sgn} f(\dot{g}_i) = f(\dot{g}_i)/|f(\dot{g}_i|$, $1 \leq i \leq m$. We claim that $y = \sum_{i=1}^m b_i \dot{g}_i$ has norm 1 by the following argument. If $g_1$ and $g_2$ are the characteristic functions of $U_{s_1}$ and $U_{s_2}$ $(s_1, s_2 \in S)$, respectively, then the only values that $g_1 + g_2$ can assume are $0, 1$, and 2, and it may be 2 only finitely many times since $U_{s_1} \cap U_{s_2}$ is finite. Let $v = -k_{U_{s_1} \cap U_{s_2}}$. Clearly $v \in c_0$ and

$$\|g_1 + g_2 + v\|_\infty = 1 \geq \|\dot{g}_1 + \dot{g}_2\|$$

For any $w \in c_0$, there is a positive integer $p_w$ such that $\sup_{j \geq p_w} |w(j)| < 1$; hence, since $U_{s_1}$ and $U_{s_1}$ are infinite, $\sup_{j \geq p_w} |g_1(j) + g_2(j) + w(j)| \geq 1$. Hence $\|g_1 + g_2 + w\|_\infty \geq 1$ which implies that $\|\dot{g}_1 + \dot{g}_2\| \geq 1$. It follows that $\|\dot{g}_1 + \dot{g}_2\| = 1$; a similar argument shows that $y = \sum_{i=1}^m b_i \dot{g}_i$ also has norm 1.

Since $f$ is continuous, it is bounded by some positive $K$ on the unit ball $U$ of $\ell_\infty^{\mathbf{R}}/c_0$ by (c); hence $|f(y)| \leq K$. Thus, since $\dot{g}_1, \ldots, \dot{g}_m \in B(n)$,

$$K \geq |f(y)| = \left| \sum_{i=1}^m [\operatorname{sgn} f(\dot{g}_i)]\, f(\dot{g}_i) \right| = \sum_{i=1}^m |f(\dot{g}_i)| \geq m/n$$

Therefore $m \leq Kn$ and the finiteness of $B(n)$ is established.

Now let $\{f_i : i \in \mathbf{N}\}$ be a countable collection of continuous linear functionals on $\ell_\infty^{\mathbf{R}}/c_0$. By the previous argument, any particular $f_i$ can be nonzero on only countably many $\dot{k}_s$. Therefore, since $S$ is uncountable, there exists some $s \in S$ such that $f_i(\dot{k}_s) = 0$ for each $i$. Thus $\{f_i : i \in \mathbf{N}\}$ is not a total set of continuous linear functionals on $\ell_\infty^{\mathbf{R}}/c_0$ and there can be no countable total set of continuous linear functionals on $\ell_\infty^{\mathbf{R}}/c_0$. The proof is now complete. ■

If we restrict consideration to **separable** Banach superspaces $X$ of $c_0$ then there is a continuous projection $P$ of norm less than or equal to 2 of $X$ onto $c_0$. The proof is widely available: some sources are Köthe 1979, p. 21, Diestel 1984, p. 71, or Lindenstrauss and Tzafriri 1977, p. 105.

If $X$ and $Y$ are sets and each is in 1-1 correspondence with a subset of the other then $X$ is in 1-1 correspondence with $Y$. This result is known by some combination of the names of Cantor, Schroeder and Bernstein. For TVS $X$ and $Y$, if $X$ is linearly homeomorphic to a subspace of $Y$ and vice-versa, must $X$ and $Y$ be linearly homeomorphic? By considering $X = C[0,1]$ and $Y = C[0,1] \oplus \ell_1$, Banach and Mazur [1933, p. 106] showed that this was false. In proving that $\ell_\infty$ and $L_\infty[a,b]$ are linearly homeomorphic, Pelczynski [1958], displaying a little sample of his genius, developed his *decomposition method* of Theorem 4.9.8. This led to consideration of the *Schroeder-Bernstein Problem* for Banach spaces that we discuss after the theorem. We denote linear homeomorphism by $\cong$.

**Theorem 4.9.8** THE DECOMPOSITION METHOD Let $X$ and $Y$ be TVS such that $X \cong X \oplus X$ and $Y \cong Y \oplus Y$; in addition, suppose that $Y$ contains a complemented copy of $X$ and $X$ contains a complemented copy of $Y$. Then $X \cong Y$.

**Proof.** We use the notation of the statement. The proof is quite simple. Since $X$ and $Y$ contain a complemented copy of $Y$ and $X$, respectively, there exist subspaces $X_1$ and $Y_1$ such that $X \cong Y \oplus X_1$ and $Y \cong X \oplus Y_1$. Hence

$$X \cong (Y \oplus Y) \oplus X_1 \cong Y \oplus (Y \oplus X_1) \cong Y \oplus X$$

Likewise

$$Y \cong X \oplus Y_1 \cong (X \oplus X) \oplus Y_1 \cong X \oplus (X \oplus Y_1) \cong X \oplus Y \quad \blacksquare$$

Among its varied uses, the decomposition method can be used to show that every infinite dimensional complemented subspace of $X = c_0$ or $\ell_p$, $1 \leq p < \infty$, is linearly homeomorphic to $X$ [Lindenstrauss and Tzafriri 1977, p. 54]. The proof relies on the fact that each space is linearly homeomorphic to a complemented subspace of the other. That fact and other arguments using it, ultimately led to reframing Banach's problem as:

> **The Schroeder-Bernstein Problem.** If $X$ and $Y$ are Banach spaces and each is linearly homeomorphic to a *complemented* subspace of the other, is $X \cong Y$?

Gowers [1996] solved the Schroeder-Bernstein negatively by exhibiting a Banach space $X$ such that $X \cong X \oplus X \oplus X$ but $X \ncong X \oplus X$. Hence $X$ is complemented in $Y = X \oplus X$ and $Y$ is complemented in $X \oplus X \oplus X \cong X$, but $X \ncong Y$.

## 4.10 FINITE-DIMENSIONAL AND LOCALLY COMPACT SPACES

There are two points to this section. One is that there is only one $n$-dimensional Hausdorff TVS for each $n$ in $\mathbf{N}$ up to linear homeomorphism [Theorem 4.10.2]. The other [Theorem 4.10.4] is that local compactness is an overwhelming hypothesis on a TVS: it is equivalent to finite-dimensionality. We begin with two lemmas.

**Theorem 4.10.1** ONE-DIMENSIONAL SPACES If $X$ is a one-dimensional Hausdorff TVS over $\mathbf{F}$ then for any nonzero vector $x$, the map $g : \mathbf{F} \to X$, $a \mapsto ax$, is a linear homeomorphism of $\mathbf{F}$ onto $X$.

**Proof.** We use the notation of the statement. Clearly $g(a) = ax$ $(a \in \mathbf{F})$ establishes a surjective linear isomorphism between $X$ and $\mathbf{F}x$; continuity of $g$ follows from the continuity of scalar multiplication. It remains to show that $g$ is open. To this end let $r$ be positive and $C(0, r) = \{a \in \mathbf{F} : |a| \leq r\}$. We show that $g$ is open by showing that there is a neighborhood $V$ of 0 in $X$ such that if $y = bx \in V$ then $|b| \leq r$; in other words, that that the relatively open neighborhood $\mathbf{F}x \cap V \subset C(0,r)x = g(C(0,r))$. Let $a \in C(0,r)$, $a \neq 0$. Since $X$ is Hausdorff and $ax \neq 0$, there is a balanced neighborhood $V$ of 0 in $X$ such that $ax \notin V$. Now choose $y = bx \in \mathbf{F}x \cap V$ for some scalar $b$. If $|b| > r \geq |a|$ then $|a/b| < 1$. Since $V$ is balanced and $y \in V$, $(a/b)\,y = a(bx)/b = ax \in V$ which contradicts $ax \notin V$. We conclude that $|b| \leq r$ and therefore that $\mathbf{F}x \cap V \subset C(0,r)x$. ■

**Theorem 4.10.2** Finite-Dimensional Spaces For any $n \in \mathbf{N}$, if $X$ is an $n$-dimensional Hausdorff TVS over $\mathbf{F}$ then every surjective linear isomorphism $A : \ell_\infty(n) \to X$ is bicontinuous.

**Proof.** Let $X$ be an $n$-dimensional Hausdorff TVS over $\mathbf{F}$ and let $A : \ell_\infty(n) \to X$ be a surjective linear isomorphism. We proceed by induction on the dimension $n$. The case $n = 1$ has already been covered in Theorem 4.10.1. Suppose that $\dim X = n$ and that the theorem holds for dimensions $\leq n-1$. Let $(e_i)$ denote the standard basis for $\mathbf{F}^n$, i.e., $e_i$ is the $n$-tuple whose entries are all 0 but the $i$th which is 1. Since $A$ is a linear isomorphism, $x_i = Ae_i$, $1 \leq i \leq n$, is a basis for $X$. Let $M$ be the linear span of $x_1, \ldots, x_{n-1}$ in $X$. Endowed with its relative topology, $M$ is a Hausdorff TVS. By the induction hypothesis, the linear isomorphism $(a_1, \ldots, a_{n-1}) \mapsto \sum_{k=1}^{n-1} a_k x_k$ is bicontinuous. Since $\ell_\infty(n-1)$ is complete [Theorem 3.5.4], $M = A(\ell_\infty(n-1))$ is complete by Theorems 3.5.6 and 3.5.1. Since $X$ is Hausdorff, $M$ is closed by Theorem 3.1.9. Since $\dim X/M = 1$, $M$ is a maximal subspace of $X$. $N = \mathbf{F}x_n$ is clearly an algebraic complement of $M$ and is therefore also a topological complement of $M$ by Theorem 4.9.5; hence the linear isomorphism of $M \times N$ onto $X$

$$\begin{array}{ccc} M \times N & \to & X \\ \left(\sum_{k=1}^{n-1} a_k x_k, a_n x_n\right) & \mapsto & \sum_{k=1}^{n} a_k x_k \end{array}$$

is bicontinuous. By the induction hypothesis, the maps $a \mapsto ax_n$ and $(a_1, \ldots, a_{n-1}) \mapsto \Sigma_{k=1}^{n-1} a_k x_k$ are bicontinuous, so the linear isomorphism

$$\begin{array}{ccc} \ell_\infty(n-1) \times \ell_\infty(1) & \to & M \times N \\ ((a_1, \ldots, a_{n-1}), a_n) & \mapsto & \left(\sum_{k=1}^{n-1} (a_k x_k, a_n x_n)\right) \end{array}$$

is also bicontinuous. Combined with the previous observations, it follows that the map $((a_1, \ldots, a_{n-1}), a_n) \mapsto \Sigma_{k=1}^{n} a_k x_k$ is a bicontinuous linear isomorphism. To complete the proof it only remains to observe that the map

$$\begin{array}{ccc} \ell_\infty(n) & \to & \ell_\infty(n-1) \times \mathbf{F} \\ (a_1, \ldots, a_n) & \mapsto & ((a_1, \ldots, a_{n-1}), a_n) \end{array}$$

is a bicontinuous linear isomorphism of $\ell_\infty(n)$ onto $\ell_\infty(n-1) \times \mathbf{F}$. ■

As a consequence of Theorem 4.10.2, finite-dimensional spaces have the following properties.

**Theorem 4.10.3** FINITE-DIMENSIONAL PROPERTIES Let $X$ and $Y$ be TVS over $\mathbf{F}$; let $X$ be $n$-dimensional with basis $\{x_1, x_2, \ldots, x_n\}$. Then:

(a) If $X$ and $Y$ are Hausdorff and $\dim X = \dim Y = n$ then every linear isomorphism of $X$ onto $Y$ is bicontinuous.

(b) UNIQUENESS OF TOPOLOGY If $\mathcal{T}_1$ and $\mathcal{T}_2$ are Hausdorff vector topologies for $X$ then $\mathcal{T}_1 = \mathcal{T}_2$. In particular, all norms are equivalent on a finite-dimensional HTVS. [See also the discussion after Theorem 14.5.1.]

(c) COMPLETENESS If $X$ is Hausdorff then the topology for $X$ is generated by a norm which makes $X$ a Banach space and $X$ is linearly homeomorphic to $\ell_\infty(n)$.

(d) FINITE-DIMENSIONAL IMPLIES CLOSED If $X$ is a subspace of a Hausdorff TVS $Y$ then $X$ is closed.

(e) If $M \subset Y$ is a closed subspace and $N \subset Y$ a finite-dimensional subspace then $M + N$ is closed.

(f) FINITE-DIMENSIONAL DOMAIN If $A : X \to Y$ is a linear map and $X$ and $Y$ are Hausdorff then $A$ is continuous. In particular, any linear functional defined on $X$ is continuous. (We show in Corollary 5.6.3 that a normed linear space $X$ is finite-dimensional iff every linear functional on $X$ is continuous.)

(g) FINITE-DIMENSIONAL RANGE If $A : Y \to X$ is a continuous surjective linear map and $X$ is Hausdorff then $A$ is an open map. Furthermore, $X$ is linearly homeomorphic to $Y/A^{-1}(0)$. In other words, any linear surjection with finite-dimensional Hausdorff range must be open. In particular, nontrivial linear functionals must be open.

**Proof.** We use the notation of the statement. (a) Let $A$ be a linear isomorphism of $X$ onto $Y$. Consider the surjective linear isomorphisms $B : X \to \ell_\infty(n)$, $\sum_{i=1}^n a_i x_i \mapsto (a_1, a_2, \ldots, a_n)$ and $C : \ell_\infty(n) \to Y$, $(a_1, a_2, \ldots, a_n) \mapsto \sum_{i=1}^n a_i A x_i$. $B$ and $C$ are bicontinuous by Theorem 4.10.2; so therefore is $A = C \circ B$.

(b) This was already proved in Theorem 1.7.2 but, by (a), the identity map of $(X, \mathcal{T}_1)$ onto $(X, \mathcal{T}_2)$ is a linear homeomorphism.

(c) Norm $X$ by taking $\|\Sigma a_i x_i\|_\infty = \max_i |a_i|$, $(a_i \in \mathbf{F})$. Since $X$ is Hausdorff, the norm topology must coincide with the original topology by (b). The surjective linear homeomorphism $B : X \to \ell_\infty(n)$ of (a) is a unimorphism by Theorem 3.5.1. Since $\ell_\infty(n)$ is complete [Theorem 3.5.4], $X$ is complete by Theorem 3.5.6.

(d) By (c), $X$ is a complete Hausdorff TVS in its relative topology; it is therefore closed by Theorem 3.1.9(b).

(e) Since $Y$ is not necessarily Hausdorff, $N$ need not be closed. Since $M$ is closed, $Y/M$ is Hausdorff by Theorem 4.7.3(b). Let $k : Y \to Y/M$ be the canonical map $y \mapsto y + M$. Since $k$ is linear, $k(N)$ is a finite-dimensional subspace of the Hausdorff space $Y/M$, hence closed in $Y/M$ by (d). Since $k$ is continuous, $k^{-1}(k(N)) = N + M$ is closed in $Y$.

(f) Since $A$ is linear, $A(X)$ is finite-dimensional, in fact, a finite-dimensional Hausdorff TVS in its relative topology. Let $N = A^{-1}(0)$ and decompose $A$

into $\dot{A} \circ k$ as in Theorem 4.7.4:

$$\begin{array}{ccccccc} A: & X & \xrightarrow{k} & X/N & \xrightarrow{\dot{A}} & A(X) \\ & x & \mapsto & x+N & \mapsto & Ax \end{array}$$

Since $N$ is finite-dimensional, $N$ is closed by (d), so $X/N$ is Hausdorff and therefore the isomorphism $\dot{A}$ is bicontinuous by (a). Since $k$ is continuous, the continuity of $A$ follows.

(g) Since $X$ is Hausdorff, $\{0\}$ is closed, so $N = A^{-1}(0)$ is closed in $Y$ and $Y/N$ is Hausdorff. $X$ is linearly isomorphic to $Y/N$ under the map $\dot{A}$ sending $x + N$ into $Ax$. By (a), $\dot{A}$ is bicontinuous and the openness of $A$ follows from Theorem 4.7.4(c). ■

Just as homeomorphic topological spaces may differ in significant ways—completeness, for example, is not generally preserved by homeomorphism, nor are other properties related to the uniform structure—linearly homeomorphic TVS may differ in consequential ways. Since a linear homeomorphism must be a unimorphism [Definition 3.5.5 and Theorem 3.5.1], properties related to the uniform structure *are* preserved by linear homeomorphisms, but the property of "1-injectivity" of real normed linear spaces is not. In particular $\ell_\infty^{\mathbf{R}}(n)$ is 1-injective but its linear homeomorph $\ell_2^{\mathbf{R}}(n)$ is not (see Example 10.5.4(a,b)).

**Theorem 4.10.4** LOCALLY COMPACT ⇔ FINITE-DIMENSIONAL A Hausdorff TVS $X$ is locally compact iff $X$ is finite-dimensional.

**Proof.** Let $X$ be a Hausdorff TVS of dimension $n \in \mathbf{N}$. Since a base at 0 for $\ell_\infty(n)$ is given by finite products of closed, bounded intervals sets, $\ell_\infty(n)$ is locally compact. Since $X$ is linearly homeomorphic to $\ell_\infty(n)$ by Theorem 4.10.3(c), $X$ is locally compact. To prove the converse we use the fact that if a subspace $M$ contains a neighborhood $V$ of 0 then, since $V$ is absorbent, $M = X$.

Assume that the Hausdorff TVS $X$ is locally compact and let $V$ be a balanced compact neighborhood of 0. We show that the sets $2^{-n}V$, $n \in \mathbf{N}$, form a neighborhood base at 0. Let $B$ be any neighborhood of 0 and choose a balanced neighborhood $U$ of 0 such that $U + U \subset B$. Since $V$ is compact, it is totally bounded by Theorem 3.4.2; hence there exists a finite subset $S$ of $X$ such that $V \subset S + U$. Since $U$ is absorbent, there exists $a \geq 1$ such that $S \subset tU$ for $|t| \geq a$. Since $U$ is balanced, for any $|t| \geq 1$, $U \subset tU$ so, for $|t| \geq a$,

$$V \subset S + U \subset tU + U \subset tU + tU \subset tB$$

and it follows that the sets $2^{-n}V$, $n \in \mathbf{N}$, form a base at 0.

Since $V$ is totally bounded, there is a finite subset $D = \{x_1, x_2, \ldots, x_n\} \subset X$ such that $V \subset D + (1/2)V$. We show that $\dim X \leq n$. The linear span $M$ of $D$ is closed by Theorem 4.10.3(d). Since $V \subset M + (1/2)V$ and $tM = M$ for all nonzero scalars $t$, $(1/2)V \subset M + (1/4)V$. Hence, multiplying by 2,

$$V \subset M + (1/2)V \subset M + (M + (1/4V) = M + (1/4)V$$

By induction, $V \subset \cap_{n\in\mathbf{N}}(M + 2^{-n}V)$. This last set is the closure of $M$ by Theorem 4.4.1(d) which is just $M$. Thus, $M$ contains a neighborhood of 0 and therefore $M = X$. ■

**Theorem 4.10.5** CONVEX HULLS Let $X$ be an $n$-dimensional space TVS. If $K \subset X$ is compact, then its convex hull $K_c$ is compact.

**Proof.** We use the notation of the statement. By Theorem 4.10.3(c), we may assume that $X = \ell_\infty(n)$. As shown in Example 4.2.6, for $\mathbf{F} = \mathbf{R}$ or $\mathbf{C}$, $K_c$ consists of all convex combinations of the form $\sum_{i=1}^{n+1} a_i x_i$ or $\sum_{i=1}^{2n+1} a_i x_i$, $0 \le a_i \le 1$, $\sum a_i = 1$, $x_i \in K$, respectively. By the argument used to show that $(\cup_{i=1}^{n} K_i)_c$ is compact in Theorem 4.4.4, $K_c$ is seen to be compact. ■

# 4.11 EXAMPLES

**Example 4.11.1** TOPOLOGY OF UNIFORM CONVERGENCE INCOMPATIBLE Let $F(T, \mathbf{C}, u)$ be the linear space of complex-valued functions on the topological space $T$ with pointwise operations and the topology of uniform convergence [Example 2.6.5]. As shown there, the topology $\mathcal{T}_u$ of uniform convergence is a group topology for $F(T, \mathbf{C})$. It is not generally a linear topology, however. In particular, if $T = \mathbf{R}$ and $x = t^2$ (or any other unbounded function) then $(1/n)\,t^2$ does not converge to 0 with respect to $\mathcal{T}_u$; i.e., does not converge to 0 uniformly in the usual function-theoretic sense. Thus, scalar multiplication, $(a, x) \mapsto ax$, is not continuous. ■

We consider a particular inverse image topology [Example 4.7.1] next.

**Example 4.11.2** SUPREMUM TOPOLOGY Let $X$ be a linear space and let $\{\mathcal{T}_j : j \in J\}$ be a family of vector topologies for $X$. By the *supremum topology* $\mathcal{T}$ for $X$, we mean the topology generated by the sets $\cup_{j\in J}\mathcal{T}_j$ (as a subbase) and we write $\mathcal{T} = \sup\{\mathcal{T}_j : j \in J\}$. Let $X_j$ be $X$ topologized by $\mathcal{T}_j$ and let $I_j : X \to X_j$, $x \mapsto x$, denote the canonical injection of $X$ into $X_j$. $\mathcal{T}$ is readily identified as the weakest topology for $X$ with respect to which each $I_j$ is continuous, i.e., the inverse image topology for $X$ determined by $\{I_j : j \in J\}$. As such, $\mathcal{T}$ is a vector topology [4.7.1] with neighborhood base at 0 for $\mathcal{T}$ given by sets of the form $\cap_{k\in K} V_k$ where $K$ is a finite subset of $J$ and each $V_k$ is a $\mathcal{T}_k$-neighborhood of 0. If each $\mathcal{T}_j$ is locally convex then so is $\mathcal{T}$. ■

**Example 4.11.3** DIRECT IMAGE TOPOLOGIES FOR FAMILIES OF MAPS (a) Let $X$ be a vector space and let $\{X_j : j \in J\}$ be a family of TVS such that for each $j \in J$ there is a surjective linear map $A_j : X_j \to X$. The trivial topology for $X$ is a vector topology with respect to which each $A_j$ is continuous so the set $Q$ of such topologies is not empty. By the result of Example 4.11.2, $\mathcal{T} = \sup Q$ is a vector topology for $X$. A neighborhood base at 0 for $\mathcal{T}$ is given by finite intersections $\cap_{i=1}^{n} V_i$, $n \in \mathbf{N}$, where each $V_i$ is a neighborhood of 0 in a topology for $X$ with respect to which each $A_i$ is continuous. Since, for each $j \in J$,

$$A_j^{-1}\left(\bigcap_{i=1}^n V_i\right) = \bigcap_{i=1}^n A_j^{-1}(V_i)$$

is a neighborhood of 0 in $X_j$, each $A_j$ is $\mathcal{T}$-continuous. Therefore $\mathcal{T} \in Q$ and it follows that $\mathcal{T}$ is the finest topology for $X$ with respect to which each $A_j$ is continuous. $\mathcal{T}$ is called the *direct image (final) topology* for $X$ *determined by the family* $\{A_j : j \in J\}$.

(b) DIRECT IMAGE LOCALLY CONVEX TOPOLOGY (cf. Examples 5.6.4 and 5.11.7) Let $X$, $\{X_j : j \in J\}$ and $\{A_j : j \in J\}$ be as in (a). Consider the collection $Q_c$ of all *locally convex* topologies for $X$ with respect to which each $A_j$ is continuous. The statements above retain their validity. The significant difference is that $\mathcal{T} = \sup Q_c$ is a locally convex topology for $X$ by the result of Example 4.11.2. It is the finest locally convex topology for $X$ with respect to which each $A_j$ is continuous and $\mathcal{T}$ is called the *direct image* (or *final*) *locally convex topology* or *inductive limit topology* determined by the family $(A_j)$, a topic we return to in Chapter 12. An important special kind of inductive limit topology is discussed in Sec. 12.1. If $M$ is a subspace of the LCS $X$ then the quotient topology is locally convex [Theorem 4.7.3(d)]; hence the quotient topology coincides with the direct image locally convex topology determined by the canonical map $k$.

BASE AT 0 FOR $\mathcal{T}$ A neighborhood base at 0 for $\mathcal{T}$ is given by the filterbase $\mathcal{D}$ of all absorbent disks $D$ of $X$ such that $A_j^{-1}(D)$ is a neighborhood of 0 in $X_j$ for each $j \in J$. To see that this is so, note first that $\mathcal{D}$ is a neighborhood base at 0 in $X$ for a locally convex topology $\mathcal{T}'$ since $\mathcal{D}$ satisfies the conditions of the basis theorem for locally convex topologies [Theorem 4.5.2]. Every $A_j$ is certainly $\mathcal{T}'$-continuous, so $\mathcal{T}' \in Q_c$ and it follows that $\mathcal{T}' \subset \mathcal{T}$.

Conversely, as noted in (a), a neighborhood base at 0 for $\mathcal{T}$ is given by sets of the form $V = \cap_{i=1}^n V_i$, $n \in \mathbf{N}$, where each $V_i$ is a disked neighborhood of 0 in a topology $\mathcal{T}_i$ for $X$ with respect to which $A_j^{-1}(V_i)$ is a neighborhood of 0 in $X_j$ for each $j \in J$. Since each such $V$ belongs to $\mathcal{D}$, $\mathcal{T} \subset \mathcal{T}'$. The gauges $p_D$ of the sets $D \in \mathcal{D}$ are seminorms. The weakest topology for $X$ with respect to which each $p_D$ is continuous is $\mathcal{T}$.

(c) FINEST LOCALLY CONVEX TOPOLOGY Let $P$ denote the collection of all seminorms on a vector space $X$ and for each $p \in P$ let $X_p$ denote the seminormed space $(X, p)$. For each $p \in P$ consider the injection $A_p : X_p \to X$, $x \mapsto x$. The direct image topology for $X$ determined by $\{A_p : p \in P\}$ is the finest locally convex topology $\mathcal{T}_{lc}$ of Example 4.5.4(c). ■

**Example 4.11.4** DIRECT SUMS TVS If $\{X_s : s \in S\}$ is a family of vector spaces, the (*external*) *direct sum* $\oplus_{s \in S} X_s$ is that topological vector subspace of the product $\Pi_{s \in S} X_s$ consisting of elements $(x_s)$ which are almost always 0. If $S$ is finite, the direct sum and the product coincide. $X \oplus X$, $X \oplus X \oplus X$, etc. are called the (*Cartesian*) *square*, *cube*, etc. of $X$. The *canonical injections* are the maps.

$$\begin{array}{rccc} I_s: & X_s & \to & \oplus_{s\in S} X_s \\ & x & \mapsto & (y_t) \end{array} \quad \text{where } y_t = \begin{cases} 0, & t \neq s \\ x, & t = s \end{cases}$$

If each $X_s$ is locally convex then the direct sum endowed with the direct image locally convex topology $\mathcal{T}_d$ determined by the canonical injections $I_s$ [Example 4.11.3(b)] is called the *locally convex direct sum* of the $X_s$. If we identify each $X_s$ with its canonical image $I_s(X_s)$ in $\oplus_{s\in S} X_s$ then a base at 0 for $\mathcal{T}_d$ is given by the set of all absorbent disks $V$ in $\oplus_{s\in S} X_s$ such that $V \cap X_s$ is a neighborhood of 0 in $X_s$ for each $s$ in $S$. ■

## 4.12 EXERCISES

4.101 PRODUCTS OF CONVEX SETS Let $\{X_s : s \in S\}$ be a family of vector spaces. If, for each $s \in S$, $K_s \subset X_s$ is convex, show that $\Pi_{s\in S} K_s$ is convex.

4.102 CONTINUITY Let $Y$ be a TVS, let $\{X_s : s \in S\}$ be a family of TVS, and let $X$ be a vector space.

(a) PRODUCTS Let $pr_s$ denote the projection of $\Pi_{s\in S} X_s$ onto $X_s$. Show that a function $f : Y \to \Pi_{s\in S} X_s$ is continuous iff each composite map $pr_s \circ f$ is continuous for each index $s$.

(b) INVERSE IMAGE TOPOLOGY For each $s$ in $S$, let $A_s : X \to X_s$ be a linear map. Topologize $X$ with the inverse image topology induced by $\{A_s : s \in S\}$, the weakest topology on $X$ that makes each $A_s$ continuous [Example 4.7.1]. If $A : Y \to X$ is linear, show that $A$ is continuous iff $A_s \circ A$ is continuous for each $s$ in $S$.

(c) DIRECT IMAGE TOPOLOGY Let $\{A_s : s \in S\}$ be a family of linear maps of $X_s$ into $X$. Topologize $X$ with the direct image locally convex topology induced by the $\{A_s : s \in S\}$, the strongest locally convex topology with respect to which each $A_s$ is continuous [Example 4.11.3(b)]. Show that a linear map $A : X \to Y$ is continuous iff $A \circ A_s$ is continuous for each $s$ in $S$.

(d) QUOTIENT Let $M$ be a subspace of the LCS $X$, let $k$ be the canonical map $x \mapsto x + M = \dot{x}$ of $X$ onto $X/M$ and let $Y$ be any TVS. If $A : X/M \to Y$ is linear then $A$ is continuous iff the composite map $A \circ k : X \to Y$ is continuous. $A$ is open iff $A \circ k$ is open.

4.103 BALANCED, CONVEX, AND ABSORBENT SETS $E$ is a subset of a vector space $X$.

(a) If $E$ is balanced then $E$ is absorbent iff for each $x$ in $X$, there is a positive number $r$ such that $x \in rE$.

(b) If $E$ is a disk and the linear span of $E$ is $X$ then $E$ is absorbent.

(c) If $E$ is convex then for any $a, b \geq 0$ for which $a+b>0$, $aE+bE=(a+b)E$.

(d) A convex set $E$ is balanced iff $aE \subset E$ for all scalars $a$ of magnitude 1.

(e) Give an example of a set which is balanced but not convex.

(f) Give an example of a convex set whose balanced hull is not convex.

(g) Show that a convex subset $E$ of a TVS must be connected.

(h) Let $X$ be a normed space and let $x, y \in X$. Show that for all $z \in [x, y]$, $\|x-y\| = \|x-z\| + \|z-y\|$.

4.104 LINEAR FUNCTIONALS AND CONVEX SETS Let $f$ be a linear functional on the real vector space $X$.

(a) If $f(x) \leq 1$ for each $x$ in a subset $E$ of $X$ then $f$ is also $\leq 1$ on the convex hull $E_c$ of $E$.

(b) If $E$ is a convex subset of $X$ and $a \in \mathbf{R} \setminus f(E)$ then either $f(x) > a$ for each $x$ in $E$ or $f(x) < a$ for each $x$ in $E$.

4.105 CONVEX HULL Let $E$ and $G$ be subsets of a vector space $X$. Then:

(a) $(aE)_c = aE_c$ for any scalar $a$.

(b) $(E+G)_c = E_c + G_c$.

(c) Give an example of a closed set whose convex hull is not closed.

4.106 INTERIORS Let $X$ be a TVS.

(a) If $E \subset X$ has nonempty interior, show that $E-E$ is a neighborhood of 0.

(b) The only linear subspace of $X$ which has an interior point is $X$.

(c) In reference to Theorem 4.4.1(e), give a specific TVS $X$ and subsets $A$ and $B$ for which the inclusions $\operatorname{int} A + \operatorname{int} B \subset A + \operatorname{int} B \subset \operatorname{int}(A+B)$ are proper.

(d) If $B$ is a disk with nonempty interior, show that $0 \in \operatorname{int} B$.

(e) Give an example of a closed balanced set with nonempty interior for which 0 is not an interior point.

4.107 CLOSURE Let $X$ be a TVS.

(a) In reference to Theorem 4.4.1(d), give an example of a TVS $X$ and a subset $E$ such that $\operatorname{cl} E \neq \cap \{U : E \subset U,\ U \text{ open}\}$.

(b) If $A$ and $B$ are subsets of $X$ then $\operatorname{cl}[\operatorname{cl} A + \operatorname{cl} B] = \operatorname{cl}(A+B)$.

(c) For any subset $A$ of $X$ and any neighborhood $V$ of 0, $\operatorname{cl} A \subset A + V$.

4.108 CLOSED CONVEX HULL Let $E$ be a subset of a TVS $X$. Recall [Theorem 4.4.3] that the closed convex hull $\operatorname{cl} E_c$ of $E$ is the intersection of the closed convex sets which contain $E$. Show that:

(a) If $\operatorname{cl} E_c$ is compact then $\operatorname{cl}(E+G)_c = \operatorname{cl} E_c + \operatorname{cl} G_c$ for any subset $G$ of $X$.

(b) If $\operatorname{cl} E_c$ and $\operatorname{cl} G_c$ are compact then $\operatorname{cl}(E \cup G)_c = (\operatorname{cl} E_c \cup \operatorname{cl} G_c)_c$.

4.109 CLOSED DISKED HULL The *closed disked hull* of a subset $E$ of a TVS $X$ is the intersection of all closed disks that contain $E$. Show that the closed disked hull of $E$ is given by $\operatorname{cl}(E_{bc})$.

4.110 CASE WHEN THE CANONICAL MAP IS CLOSED Generally [Theorem 4.7.3] the canonical map $k$, $x \mapsto x + M = \dot{x}$, of a TVS $X$ into a quotient space $X/M$ is not closed. Show that if $M = \operatorname{cl}\{0\}$ then $k$ is closed.

4.111 ALGEBRAIC CLOSURE AND INTERIOR Let $A$ be a subset of a vector space $X$. The *algebraic interior* $\operatorname{aint} A$ of $A$ consists of those $x \in A$ such that for every line $L$ through $x$, $L \cap A$ contains a line segment $(y, z)$ to which $x$ belongs. A point $x \in X$ is *linearly accessible* from $A$ if there exists $y \in A$ such that $[y, x) \subset A$. The *algebraic closure* $\operatorname{acl} A$ of $A$ consists of the union of $A$ and those $x \in X$ that are linearly accessible from $A$. If $K \subset X$ is convex, show that:

(a) $\operatorname{aint} K$ and $\operatorname{acl} K$ are convex.

(b) If $x \in \operatorname{aint} K$ and $y \in \operatorname{acl} K$ then $[x, y) \subset \operatorname{aint} K$.

(c) GAUGE If $0 \in \operatorname{aint} K$, we define the *gauge* or *Minkowski functional* $p_K(x) = \inf\{a \geq 0 : x \in aK\}$ of $K$ for each $x \in X$. Show that $p_K$ is a positive sublinear functional, i.e., is nonnegative, positive homogeneous [$p_K(ax) \geq 0$ for any $x \in X$ and any $a \geq 0$] and subadditive [$p_K(x+y) \leq p_K(x) + p_K(y)$ for any $x, y \in X$].

(d) Let $p_K$ be as in (c). Show that

$$\operatorname{aint}(\operatorname{aint} K) = \operatorname{aint} K = \operatorname{aint}(\operatorname{acl} K) = \{x \in X : p_K(x) < 1\}$$

(e) GAUGE OF ABSORBENT CONVEX SET Let $D$ be an absorbent convex set. Define its gauge to be the function $p_D(x) = \inf\{a \geq 0 : x \in aD\}$ for each $x \in X$. Show that $p_D$ is a positive sublinear functional.

4.112 TIETZE TOPOLOGY Let $\{X_t : t \in T\}$ be an infinite family of nontrivial HTVS over the topological field $E$ and let $X$ be the linear space $\Pi_{t \in T} X_t$. The *Tietze* or *box topology* $\mathcal{T}$ for $X$ is that topology which has as a base at 0 sets of the form $\Pi_{t \in T} V_t$ where each $V_t$ is a neighborhood of 0 in $X_t$.

(a) Show that $\mathcal{T}$ is compatible with the additive group structure of $X$.

(b) Unless $E$ carries the discrete topology, show that $\mathcal{T}$ is not a vector topology.

(c) Let $M$ be the the subspace consisting of those elements $(x_t) \in \Pi_{t \in T} X_t$ for which $x_t = 0$ for almost all indices $t$. Show that $\mathcal{T}$ induces a vector topology $\mathcal{T}'$ on $M$. If each $X_t$ is locally convex then so is $M$.

(d) With $\mathcal{T}'$ as in (c), show that if each $X_t$ is complete then $(M, \mathcal{T}')$ is complete.

4.113 LINEAR COMPACTNESS Call a translate $x + M$ $(x \in X)$ of a subspace $M$ of a vector space $X$ a *linear variety*. A TVS $X$ is *linearly compact* if every collection $\{H_s : s \in S\}$ of closed linear varieties has the finite intersection property (Sec. 10.3). The most elementary properties of linear compactness follow.

(a) CLOSED SUBSPACE A closed subspace of a linearly compact space is linearly compact.

(b) CONTINUOUS LINEAR IMAGE The continuous linear image of a linearly compact space is linearly compact.

(c) QUOTIENT SPACE The quotient of a linearly compact space by any subspace is linearly compact.

(d) "TIHONOV THEOREM" The Cartesian product of linearly compact spaces is linearly compact.

4.114 BALANCED SUBSETS OF $\mathbf{F}$ Let $U$ and $D$ denote the open and closed unit disks, respectively, in $\mathbf{F}$ (= $\mathbf{R}$ or $\mathbf{C}$). Show that the only balanced subsets of $\mathbf{F}$ are the trivial ones $\{0\}$ and $\mathbf{F}$ and positive multiples of $U$ or $D$.

4.115 DIRECT SUMS If $\{X_s : s \in S\}$ is a family of TVS, their topological direct sum $\oplus_{s \in S} X_s$ is the subspace of the product $\Pi_{s \in S} X_s$ consisting of elements $(x_s)$ which are almost always 0.

(a) Show that $\oplus_{s \in S} X_s$ is dense in $\Pi_{s \in S} X_s$.

(b) DIRECT VERSUS TOPOLOGICAL DIRECT If $M_1, \ldots, M_n$ are subspaces of a TVS $X$ and $X = \oplus_{i=1}^n M_i$ then the canonical bijection $\Pi_{i=1}^n M_i \to X$, $(x_i) \mapsto \sum_{i=1}^n x_j$, is continuous but not necessarily a homeomorphism (counterexample?), so $X$ is not necessarily the *topological* direct sum of the $M_i$. $X$ *is* the topological direct sum of the $M_i$ iff for each $i$, the map $X \to M_i$, $\sum_{i=1}^n x_j \mapsto x_i$, is continuous.

4.116 LOCALLY COMPACT SPACES Let $X$ be a TVS over a nontrivially ultravalued field $(K, |\cdot|)$. If $(K, |\cdot|)$ is complete then $X$ is locally compact iff $K$ is locally compact and, for some positive integer $n$, $X$ is linearly homeomorphic to $(K^n, \|\cdot\|_\infty)$ where $\|(a_1, \ldots, a_n)\|_\infty = \max_i |a_i|$ [cf. Exercise 5.206].

4.117 FINEST VECTOR TOPOLOGY Let $X$ be a vector space.

(a) Show that there is a finest vector topology $\mathcal{T}_f$ on $X$.

(b) For any TVS $Y$ and any linear map $A : (X, \mathcal{T}_f) \to Y$, show that $A$ is continuous.

(c) If the dimension of $X$ is uncountable, show that $\mathcal{T}_f$ is not locally convex.

4.201 ONE-DIMENSIONAL SUBSPACES Let $X$ be a Hausdorff TVS over a nontrivially valued field $K$.

(a) Show that if $X$ is one-dimensional over $K$ then $X$ is linearly homeomorphic to the one-dimensional TVS $K$.

(b) Show that the result of (a) fails if $K$ carries the trivial valuation.

4.202 $K$-CONVEXITY Let $X$ be a TVS over an ultravalued field $(K, |\cdot|)$; let $U = \{a \in K : |a| \leq 1\}$. A subset $E$ of $X$ is *absolutely $K$-convex* or a *$K$-disk* if $VE + VE \subset E$. Translates $x + E$ $(x \in X)$ of absolutely $K$-convex sets $E$ are called *$K$-convex.*

(a) If $0 \in E$ and $E$ is $K$-convex, show that $E$ is absolutely $K$-convex.

(b) $E$ is $K$-convex iff $ax + by + cz \in E$ for all $x, y$, and $z$ in $E$ and all $a, b, c \in U$ such that $a + b + c = 1$.

(c) If the characteristic of $K$ is not 2 then $E$ is $K$-convex iff for each $a$ in $U$, $ax + (1 - a)y \in E$ for all $x$ and $y$ in $E$.

(d) The *$K$-convex hull* of a subset $E$ of $X$ is defined in formally the same way as for real or complex vector spaces [Definition 4.2.2]. It comprises all linear combinations of the form $x + \Sigma a_i x_i$, where each $x_i \in E$, $x$ is a fixed vector in $E$, and the $a_i$ are from $U$.

(e) The *absolute $K$-convex hull* $E_{bc}$ of a subset $E$ of $X$ is also defined in formally the same way as for real and complex spaces [Definition 4.2.10]. $E_{bc}$ consists of all linear combinations of vectors in $E$ with scalars from $U$. If $E$ is open, so is $E_{bc}$.

(f) If $E$ is absolutely $K$-convex then $\operatorname{int} E = E$ or $\operatorname{int} E = \varnothing$.

4.203 LOCALLY K-CONVEX SPACES A TVS $X$ over an ultravalued field $(K, |\cdot|)$ is *locally $K$-convex* if it has a base of $K$-convex neighborhoods of 0. $X$ is said to be a *non-Archimedean TVS* if it has a base of balanced neighborhoods $B$ of 0 such that $B + B \subset B$.

(a) If $X$ is non-Archimedean then $X$ is zero-dimensional in the sense that there is a base of clopen sets for the topology.

(b) $X$ is locally $K$-convex iff $X$ is non-Archimedean.

4.204 C-COMPACTNESS Let $X$ be a TVS over a nontrivially ultravalued field $(K, |\cdot|)$. A filterbase $\mathcal{B}$ of $X$ of $K$-convex sets is called a *$K$-convex filterbase*. $X$ is said to be *c-compact* if every $K$-convex filterbase has an adherence point. The notion of spherical completeness [Exercise 3.205] is pivotal in discussing the Hahn-Banach theorem for non-Archimedean normed spaces $X$ over $K$ [Exercise 10.201] and is equivalent to $c$-compactness.

(a) Any compact set is $c$-compact and any closed subset of a $c$-compact set is $c$-compact.

(b) A maximal convex filterbase $\mathcal{B}$ on a $K$-convex subset $E$ of $X$ converges to $x \in E$ iff $x$ is an adherence point of $\mathcal{B}$.

(c) If $E$ is $K$-convex and $c$-compact then $E$ is closed.

(d) If $X$ is $c$-compact then $X$ is linearly compact [see Exercise 4.113].

(e) The Cartesian product of $c$-compact spaces is $c$-compact.

(f) $K$ is spherically complete iff $K$ is $c$-compact.

(g) If $X$ is locally $K$-convex then the absolute $K$-convex hull of a finite union of $c$-compact $K$-disks is $c$-compact.

## Hints

4.103 (g) Consider any two distinct points $x, y \in E$ and consider the map $t \mapsto tx + (1-t)y$ of $[0,1]$ into $A$.

(h) For $y \neq x$, let $u = (1/\|y-x\|)(y-x)$ and $z(t) = x + tu$ for $0 \le t \le \|y-x\|$. Show that $\|z-x\| = t$ and $\|y-z\| = \|y-x\| - t$.

4.105 (c) Let $E = \{(s,t) \in \mathbf{R}^2 : t = \pm(1/s),\ s > 0\}$.

4.106 (c) Let $X = \mathbf{R}$, let $A$ be the rationals, and let $B$ be the irrationals together with 0.

4.107 (a) Let $X = \mathbf{R}$ and let $E = \{1, 1/2, 1/3, \ldots\}$.

4.108 (a) Use the fact that the sum of a compact and a closed set is closed [Theorem 4.4.1(a)]. (b) Consider the map $F : \operatorname{cl} E_c \times \operatorname{cl} G_c \times [0,1] \to (\operatorname{cl} E_c \cup \operatorname{cl} G_c)_c$ which sends $(x,y,t)$ into $tx + (1-t)y$. Show that the range of $F$ is compact and convex.

4.110 If $E$ is a closed subset of $X$ and $k(x) \notin k(E)$, choose a neighborhood $U$ of 0 such that $(x + U + U) \cap E = \varnothing$ and use the fact that $U \supset M$.

4.111 (b) Consider two cases: $y \in C$ and $y \in \operatorname{acl} C \setminus C$.

4.114 Use Theorem 4.1.9 to show that if $B$ is a balanced proper subset of $\mathbf{F}$ then $B$ must be bounded. Let $\sup\{|a| : a \in B\} = b > 0$ and show that $bU \subset B \subset bD$. If $bU \neq B$ then there is some $a$ in $B$ such that $|a| = b$; this implies that $B = bD$.

4.115 (b) See Example 4.9.7.

4.117 (a) Consider the collection of all vector topologies for $X$ and use the result of Example 4.11.2. (b) Consider the inverse image topology induced by $A$ on $X$. (c) Let $\{x_s : s \in S\}$, $S$ uncountable, be a Hamel basis for $X$. Show that $D = \left\{x \in X : x = \sum_{s\in S} a_s x_s, \sum_{s\in S} |a_s|^{1/2} \leq 1\right\}$ is a neighborhood of 0 in the finest vector topology that does not contain an absorbent disk.

# Chapter 5

# Locally Convex Spaces and Seminorms



We call the weakest topology that makes each member of a family $P$ of seminorms on a vector space continuous a "seminorm topology," a topic we first discussed in Example 4.5.4. The local convexity of a seminorm topology stems from the fact that the open ball $V_p = \{x \in X : p(x) < 1\}$ determined by a seminorm $p$ is convex, a consequence of the fact that $p$ satisfies the triangle inequality. Indeed, a topology is locally convex iff it is generated by a family of seminorms. Two prominent seminorms are:

> Let $C(\mathbf{R}, \mathbf{R})$ be the linear space of all continuous maps of $\mathbf{R}$ into $\mathbf{R}$. For each $n \in \mathbf{N}$, let $p_n(x) = \sup |x([-n, n])|$, $x \in C(\mathbf{R}, \mathbf{R})$. Each $p_n$ is a seminorm and the topology determined by $P = \{p_n : n \in \mathbf{N}\}$ on $C(\mathbf{R},\mathbf{R})$ is called the *compact-open* topology.

Let $X^*$ denote the algebraic dual—the linear space of all linear functionals on the vector space $X$. For any $f \in X^*$, the map defined by $p_f(x) = |f(x)|$ $(x \in X)$ is a seminorm.

Every locally convex topology is a seminorm topology because each convex neighborhood of 0 produces a seminorm by way of the following mechanism: If $E$ an absorbent set in a vector space $X$ then each vector $x$ will belong to $aE$ for scalars $a$ of sufficiently large magnitude. By choosing the "smallest" $a \geq 0$ for which this happens, we define a function $p_E$ on $X$ called the *gauge* or *Minkowski functional* of $E$. If $E$ is a disk as well, $p_E$ is a seminorm.

Aside from looking at seminorm topologies in general, we examine a few concrete cases in some detail such as $C(T, \mathbf{F}, c)$, the space of $\mathbf{F}$-valued continuous functions on the topological space $T$ with the compact-open topology in Sec. 5.8. We determine conditions on $T$ that make $C(T, \mathbf{F}, c)$ metrizable and complete and also prove Ascoli's venerable theorem [Theorem 5.10.4] characterizing the compact subsets of $C(T, \mathbf{F}, c)$ as essentially the equicontinuous ones.

An ordering $\leq$ on a real vector space $X$ such that, for $x, y, z \in X$ and $a \in \mathbf{F}$, $x \leq y$ implies $x + z \leq y + z$, and $a \geq 0$ and $x \geq 0$ imply that $ax \geq 0$ is called a *linear* ordering and $X$ an *ordered* vector space. We dip briefly into the large subject of ordered vector spaces in Sec. 5.12. An ordered vector space is *Dedekind* or *order complete* if every majorized subset has a least upper bound. For completely regular spaces $T$, we prove that the pointwise ordered space $C(T, \mathbf{R})$ of continuous real-valued functions is order complete iff $T$ is "extremally disconnected" in the sense that disjoint open subsets of $T$ have disjoint closures. Extremal disconnectedness makes a prominent appearance in Theorem 10.7.5 where a question related to the vector-valued Hahn-Banach theorem is resolved.

## 5.1 SEMINORMS

A *seminorm* is a map $p : X \to \mathbf{R}^+ \cup \{0\}$ of a real or complex vector space $X$ over $\mathbf{F}$ which is *absolutely homogeneous*, $p(ax) = |a|\, p(x)$ for all $a \in \mathbf{F}$ and $x \in X$, and *subadditive*, $p(x + y) \leq p(x) + p(y)$ for all $x, y \in X$. If $p(x)$ can be 0 only when $x = 0$ then $p$ is a *norm*. An immediate consequence of the subadditivity of a seminorm is the *second triangle inequality:*

$$|p(x) - p(y)| \leq p(x - y)$$

If $p$ is a seminorm then $p(x - y)$ is an invariant pseudometric.

Another evident property of seminorms is:

**Theorem 5.1.1** SUMS AND MAXIMA If $a_1, \ldots, a_n$ are nonnegative scalars and $p_1, \ldots, p_n$ are seminorms then $\max_i a_i p_i$ and $\sum_{i=1}^n a_i p_i$ are seminorms.

Seminorms arise naturally in many ways in analysis, through integrals, evaluation at a point, and suprema of functions on sets. Also any norm is a seminorm, so each of the norms of Chap. 1 is an example of a seminorm. Here are some others.

**Example 5.1.2** SEMINORMS (a) If $A$ is a linear map of a vector space $X$ over $\mathbf{F}$ into a seminormed space $(Y, p)$ then $p \circ A$ is a seminorm on $X$. In particular, if $f$ is a linear functional on $X$ then $x \mapsto |f(x)|$ is a seminorm on $X$. (When is it a norm?)

(b) If $X$ is a linear space of integrable $\mathbf{F}$-valued functions on some set $T$ then $p(x) = \left|\int_T x\right|$ is a seminorm on $X$,

(c) If $X$ is a vector space of $\mathbf{F}$-valued functions on a set $T$ then, for any $t_0 \in T$, the map $x \to |x(t_0)|$ is a seminorm on $X$.

(d) Let $C(T, \mathbf{F})$ be the linear space of all continuous maps of the topological space $T$ into $\mathbf{F}$ and let $K$ be a compact subset of $T$. Then $p_K(x) = \sup |x(K)|$ $(x \in C(T, \mathbf{F}))$ is a seminorm. ■

## 5.2 CONTINUITY OF SEMINORMS

**Notation.** If $p$ is a seminorm on a vector space $X$ then the *open unit ball* or *sphere* determined by $p$ is the set $V_p = \{x \in X : p(x) < 1\}$; the *closed unit ball* is $\overline{V}_p = \{x \in X : p(x) \leq 1\}$. ■

Some properties of the balls $V_p$ and $\overline{V}_p$ follow.

**Theorem 5.2.1** PROPERTIES OF $V_p$ Let $p$ be a seminorm on a vector space $X$. Then:

(a) if $q$ is a seminorm on $X$ then $p \leq q$ [i.e., $p(x) \leq q(x)$ for every $x$] iff $V_q \subset V_p$;

(b) for any positive $r$, $rV_p = \{x \in X : p(x) < r\} = V_{(1/r)p}$;

(c) for any $x \in X$, $x + V_p = \{y \in X : p(x - y) < 1\}$;

(d) $V_p$ is an absorbent disk;

(e) if $X$ is a TVS and $p$ is continuous then $\operatorname{cl} V_p = \overline{V}_p$.

**Proof.** Note that the analogs of (a)–(d) hold for $\overline{V}_p$. As (a)–(d) require only routine verification, we prove only (e). To that end, suppose $p \neq 0$ is a continuous seminorm. For $x \in \operatorname{cl} V_p$, consider a net of points $x_s \in V_p$ that converges to $x$. Then $p(x_s) \to p(x)$ and $p(x_s) < 1$ for each $s$ imply that $p(x) \leq 1$. Hence $p(\operatorname{cl} V_p) \subset \operatorname{cl} p(V_p) = \operatorname{cl}[0, 1) = [0, 1]$. Hence, $\operatorname{cl} V_p \subset \overline{V}_p$. Conversely, suppose that $x \in \overline{V}_p$. If $p(x) < 1$ then $x \in V_p$ so suppose $p(x) = 1$. Let $t_n = 1 - 1/n$ so that $t_n x \in V_p$ for every $n$. Since $t_n x \to x$, it follows that $x \in \operatorname{cl} V_p$. ■

An immediate consequence of (a) is that $p = q$ iff $V_q = V_p$. In regard to (e), note that if $(T, d)$ is a metric space and $x$ a point of $T$ then the closure of the open ball $B(x,r) = \{y \in T : d(x,y) < r\}$, $r > 0$, is not necessarily equal to the closed ball $\{y \in T : d(x,y) \leq r\}$. If $d$ is the trivial metric, for example (the distance between distinct points is 1, otherwise 0) then $\operatorname{cl} B(x,1)$ is $\{x\} \neq \{y \in T : d(x,y) \leq 1\} = T$.

**Theorem 5.2.2** CONTINUITY OF SEMINORMS If $p$ is a seminorm on the TVS $X$ with filter $V(0)$ of neighborhoods of 0 then the following are equivalent:

(a) $p$ is uniformly continuous.
(b) $V_p$ is an open set.
(c) $\overline{V}_p$ is a neighborhood of 0.
(d) $p$ is continuous at 0.
(e) There is a continuous seminorm $q$ such that $p \leq q$.

**Proof.** In the notation of the statement, the chain of implications from (a) to (c) is clear. To see that (c) implies (d), suppose that $\overline{V}_p \in V(0)$. Since, for any nonzero $a$, the map $x \mapsto ax$ is a homeomorphism of $X$ onto $X$, it follows that $r\overline{V}_p$ is a neighborhood of 0 for any positive $r$. Consequently, if the net $x_s \to 0$ in $X$, $x_s \in rV_p$ eventually or, equivalently, $p(x_s) \leq r$ eventually. In other words, $p(x_s) \to 0$.

(d) $\Rightarrow$ (a): Continuity of $p$ at 0 means that for any $r > 0$ there is a neighborhood $V$ of 0 such that $p(V) \subset [0,r)$. By Theorem 4.3.6(a,d) there is a balanced neighborhood $U$ of 0 such that $U - U \subset V$. For $x, y \in U$ then $x - y \in V$ so $p(x-y) < r$. The second triangle inequality yields the uniform continuity of $p$.

To capture the equivalence of (d) and (e), note that if $p$ is continuous at 0 then (e) holds trivially: $p \leq p$. Conversely, note that for any two seminorms $p$ and $q$, $p \leq q$ iff $\overline{V}_q \subset \overline{V}_p$ [Theorem 5.2.1(a)] Thus, if $q$ is continuous, $\overline{V}_q$ is a neighborhood of 0 and therefore so is $\overline{V}_p$. The desired result now follows from (c). ∎

If $(X,p)$ is a seminormed space then certainly a positive multiple $rp$, $r > 0$, of $p$ is a continuous seminorm. These are essentially the only ones by the following argument: If $q$ is a continuous seminorm on $(X,p)$ then for some $r > 0$, $q\left(r\overline{V}_p\right) \leq 1$, i.e., $\{x \in X : p(x) \leq r\} = r\overline{V}_p = \overline{V}_{(1/r)p} \subset \overline{V}_q$ which is equivalent to $q(x) \leq (1/r)\,p(x)$ for every $x \in X$ [Theorem 5.2.1(a,b)]. In other words:

**Theorem 5.2.3** CONTINUOUS SEMINORMS ON A SEMINORMED SPACE For a seminorm $q$ to be continuous on a seminormed space $(X,p)$ it is necessary and sufficient for $q$ to be dominated by a positive multiple of $p$.

As an immediate application, we have the following two simple but useful results.

**Example 5.2.4** (a) CONTINUITY OF COEFFICIENT FUNCTIONALS ON $\ell_p$ For $1 \leq p < \infty$ the *coefficient functionals* $q_k : \ell_\infty \to \mathbf{F}$, $(a_n) \mapsto a_k$, $(k \in \mathbf{N})$ are continuous linear functionals. They are clearly linear and (therefore) $p_k = |q_k|$ is a seminorm. Moreover $p_k$ is continuous iff $q_k$ is. To prove continuity of $p_k$, suppose that $x = (a_n) \in \ell_p$. For any $k \in \mathbf{N}$, $|a_k|^p \leq \sum_{j\in\mathbf{N}} |a_j|^p = \left(\|x\|_p\right)^p$. Hence $|a_k| = p_k(x) \leq \|x\|_p$, i.e., $p_k(\cdot) \leq \|\cdot\|_p$ and the continuity of $p_k$ follows from Theorem 5.2.3. Finally, note that $p_k = |q_k|$ is continuous iff $q_k$ is continuous.

(b) CONTINUITY OF EVALUATION MAPS ON $(\ell_\infty(T), \|\cdot\|_\infty)$ For any set $T$ and any $t \in T$, the map $t'(x) = x(t)$ $(x \in \ell_\infty(T))$ is called the *evaluation map determined by* $t$. Clearly the seminorm $p_t(\cdot) = |t'(\cdot)|$ is continuous iff $t'$ is. Since $p_t \leq \|\cdot\|_\infty$, the continuity of $p_t$ follows from Theorem 5.2.3. ■

## 5.3 GAUGES

We noted in Sec. 5.2 that if $p$ is a seminorm then $V_p = \{x : p(x) < 1\}$ and $\overline{V}_p = \{x : p(x) \leq 1\}$ are absorbent disks. There is a closer connection between absorbent disks and seminorms. If $D$ is an absorbent disk, there is a seminorm $p$ such that $V_p \subset D \subset \overline{V}_p$, as we show in the next two results.

**Theorem 5.3.1** GAUGES If $D$ is an absorbent disk in the linear space $X$ then the *gauge* (or *Minkowski functional*) of $D$, defined at each $x$ in $X$ by

$$p_D(x) = \inf\{r > 0 : x \in rD\}$$

is a seminorm on $X$. The French term for "gauge" is *jauge*; in German it is *Strahldistanz* or *Distanzfunktion*. An alternative way to compute $p_D(x)$ is obviously given by

$$p_D(x) = \frac{1}{\sup\{c > 0 : cx \in D\}}$$

**Proof.** Let $D$ be an absorbent disk. Since $D$ is absorbent, $\{r > 0 : x \in rD\}$ is nonempty. Since $D$ is convex then $aD + bD = (a+b)D$ for any positive $a$ and $b$ [Theorem 4.2.5]. Thus, if $x \in aD$ and $y \in bD$ then $x + y \in (a+b)D$. Consequently, $a + b \geq p_D(x+y)$. Since $a$ and $b$ are arbitrary, it follows that $p_D(x+y) \leq p_D(x) + p_D(y)$.

As to the absolute homogeneity of $p_D$, consider a nonzero scalar $b$. For $a > 0$, $bx \in aD$ if and only if $x \in (a/b)D$. Since $|a/b| = a/|b|$ and $D$ is balanced then, by Theorem 4.1.9, $(a/b)D = (a/|b|)D$. Hence

$$\begin{aligned} p_D(bx) &= \inf\left\{a > 0 : x \in \frac{a}{|b|}D\right\} = |b| \inf\left\{\frac{a}{|b|} > 0 : x \in \frac{a}{|b|}D\right\} \\ &= |b| \inf\{a > 0 : x \in aD\} = |b| p_D(x) \quad ■ \end{aligned}$$

**Theorem 5.3.2** UNIT BALL OF A GAUGE If $D$ is an absorbent disk in the TVS $X$ then $\operatorname{int} D \subset V_{p_D} \subset D \subset \overline{V}_{p_D} \subset \operatorname{cl} D$.

**Proof.** Let $D$ be an absorbent disk in the TVS $X$. If $x \in \operatorname{int} D$ then $D$ is a neighborhood of $x$. Since $(1+1/n)x \to x$, $(1+1/n)x \in \operatorname{int} D$ eventually; since $1/(1+1/n) < 1$, this implies that $p_D(x) < 1$ and $\operatorname{int} D \subset V_{p_D}$. For $x \in V_{p_D}$, there exists $a \in [0,1)$ such that $x \in aD$. Since $D$ is balanced, $aD \subset D$ and therefore $V_{p_D} \subset D$. Clearly, $x \in D$ implies that $p_D(x) \leq 1$ so $D \subset \overline{V}_{p_D}$.

To prove that $\overline{V}_{p_D} \subset \operatorname{cl} D$, first consider $x \in X$ such that $p_D(x) < 1$. There must exist $a \in [0,1)$ such that $x \in aD \subset D \subset \operatorname{cl} D$. If $p_D(x) = 1$ then for any $0 < a < 1$, $ax \in V_{p_D} \subset D$. By the continuity of scalar multiplication, for any neighborhood $V$ of $x$, there is some open ball $B(1,r) \subset \mathbf{F}$, $0 < r < 1$, such that $B(1,r)x \subset V$. It follows that $V$ contains points of $D$. Hence $x \in \operatorname{cl} D$. ■

The criteria for continuity of seminorms of Theorem 5.2.2 apply, of course, to gauges. For gauges, we have:

**Theorem 5.3.3** CONTINUITY OF GAUGES Let $p_D$ denote the gauge of an absorbent disk $D$ of the TVS $X$. Then:

(a) if $D$ is open then $D = V_{p_D}$;

(b) $p_D$ is continuous iff $D$ is a neighborhood of 0; and

(c) if $p_D$ is continuous then $V_{p_D} = \operatorname{int} D$ and $\overline{V}_{p_D} = \operatorname{cl} D$.

**Proof.** Let $D$ be an absorbent disk in the TVS $X$. (a) If $D$ is open then $D = \operatorname{int} D$ and the fact that $D = V_{p_D}$ follows immediately from Theorem 5.3.2.

(b) If $p_D$ is continuous, it is clear that $V_{p_D}$ is an open neighborhood of 0. Conversely, suppose that $D$ is a neighborhood of 0. Since $D \subset \overline{V}_{p_D}$ [Theorem 5.3.2], $\overline{V}_{p_D}$ is a neighborhood of 0 and the continuity of $p$ follows from Theorem 5.2.2(c).

(c) If $p_D$ is continuous then $V_{p_D}$ is open. Since $V_{p_D} \subset D$ [Theorem 5.3.2], $V_{p_D} \subset \operatorname{int} D$ and equality follows from Theorem 5.3.2. Likewise, if $p_D$ is continuous then $\overline{V}_{p_D}$ is closed. By Theorem 5.3.2, $D \subset \overline{V}_{p_D} \subset \operatorname{cl} D$ so $\operatorname{cl} D \subset \overline{V}_{p_D}$. ■

## 5.4 SUBLINEAR FUNCTIONALS

Banach invented the notion of *sublinear functional* when he proved his version of the Hahn-Banach theorem.

**Definition 5.4.1** SUBLINEAR FUNCTIONAL A map $p : X \to \mathbf{R}^+ \cup \{0\}$ defined on a vector space $X$ over $\mathbf{F}$ such that, for all $x, y \in X$,

(a) (subadditive) $p(x+y) \leq p(x) + p(y)$ and

(b) (positive homogeneous) for $r \geq 0$, $p(rx) = rp(x)$

is called a *sublinear functional.* If $p(x) \geq 0$ for all $x \in X$ then $p$ is called a *positive* sublinear functional.

Any linear map is sublinear so the map $t \mapsto -t$ on $\mathbf{R}$ is sublinear. Hence sublinear functionals can be negative. For any sublinear functional $p$, the positive homogeneity implies that $p(0) = 0$ and therefore also that $0 \leq p(x) + p(-x)$.

Seminorms are clearly sublinear and the next result indicates a way to generate a seminorm from a sublinear functional on a real vector space.

**Theorem 5.4.2** SUBLINEAR TO SEMINORM Let $p$ be a sublinear functional on a real vector space $X$. For $x \in X$, define $q(x) = \max[p(x), p(-x)]$. Then $q$ is a seminorm called the *associated seminorm.*

**Proof.** We use the notation of the statement. Since one of $p(x)$ and $p(-x)$ must be nonnegative, it follows that $q(x) \geq 0$ for every $x \in X$. By definition, $q(0) = p(0) = 0$. If $a > 0$ then $p(ax) = ap(x)$ and $p(-ax) = ap(-x)$; hence $q(ax) = \max[p(ax), p(-ax)] = a\max[p(x), p(-x)] = aq(x)$. Clearly $q(-x) = q(x)$ so for $a < 0$, $q(ax) = q(-ax) = (-a)q(x)$; therefore $q(ax) = |a|q(x)$. For $x, y \in X$, $q(x+y) = p(x+y)$ or $p(-x-y)$; as each is $\leq \max[p(x), p(-x)] + \max[p(y), p(-y)] = q(x) + q(y)$, the subadditivity of $q$ follows. ■

# 5.5 SEMINORM TOPOLOGIES

As noted in Example 4.5.4(b), a seminorm topology determined by a family $P$ of seminorms is locally convex. As a further note on them we have:

**Theorem 5.5.1** SEMINORM TOPOLOGIES Let $P$ be a family of seminorms on the vector space $X$. Let $V_P(0)$ denote the filter of neighborhoods of 0 for the seminorm topology $\mathcal{T}_P$. Then:

(a) $\mathcal{T}_P$ is Hausdorff iff for each nonzero $x$ in $X$ there is a $p \in P$ such that $p(x) \neq 0$; and

(b) an open base for $V_P(0)$ is given by positive multiples of finite intersections of open balls $V_p = \{x \in X : p(x) < 1\}$, $p \in P$. Since each $V_p$ is convex, it follows that $\mathcal{T}_P$ is a locally convex topology.

**Proof.** We use the notation of the statement. (a) If $\mathcal{T}_P$ is Hausdorff then for any vector $x \neq 0$, there is a basic neighborhood $r \cap_{j=1}^{n} V_{p_j}$ of 0 to which $x$ does not belong. Therefore, $p_j(x) \neq 0$ for some $j$, $1 \leq j \leq n$. Conversely, if $p(x) \neq 0$ for some $p$ in $P$ then $x \notin p(x)V_p$. Therefore $\mathcal{T}_P$ is Hausdorff by Theorem 2.2.2(d).

(b) Suppose that $p_1, \ldots, p_n \in P$ and $x \in r[\cap_{j=1}^{n} V_{p_j}]$, $r > 0$. For $a > 0$ such that $a + \max_j p_j(x) < r$ then $x + a[\cap_{j=1}^{n} V_{p_j}] \subset r[\cap_{j=1}^{n} V_{p_j}]$. Hence the open base condition (OB) mentioned after Theorem 4.5.2 is satisfied. ■

Having seen that families of seminorms generate locally convex topologies, we show next that this is the only way—that any locally convex topology is determined by a family of seminorms.

To this end let $X$ be a locally convex space, let $\mathcal{V}(0)$ denote the neighborhood filter at 0 in $X$, and let $P$ denote the collection of all continuous seminorms on $X$. Certainly, $P$ is not empty, for there must be disked neighborhoods of 0 and the gauges of such sets populate $P$. Let $\mathcal{V}_P(0)$ denote the neighborhood filter at 0 in the seminorm topology determined by $P$. Since each $p \in P$ is continuous, $\mathcal{V}_P(0) \subset \mathcal{V}(0)$ . Conversely, if $V \in \mathcal{V}(0)$, there must be some open disk $U$ such that $U \subset V$. With $p_U$ denoting the gauge of $U$, $V_{p_U} \subset U \subset V$ [Theorem 5.3.2]; since $V_{p_U} \in \mathcal{V}_P(0)$, $V \in \mathcal{V}_P(0)$ and $\mathcal{V}_P(0) = \mathcal{V}(0)$. We summarize this now for future reference.

**Theorem 5.5.2** SEMINORMS AND LOCALLY CONVEX SPACES A TVS $X$ is locally convex iff its topology is generated by a family of (perforce continuous) seminorms. In particular, the topology is determined by the gauges $p_U$ of all open disks $U$ in $X$.

Since Cartesian products of convex sets are convex, the product of LCS is a LCS. It is easy to verify the following assertion about a family of seminorms that generates the product topology.

**Theorem 5.5.3** SEMINORMS AND PRODUCTS OF LCS Let $X = \Pi_{s \in S} X_s$ where each $X_s$ is a LCS. For each $s \in S$, let $P_s$ denote a family of seminorms that determines the topology on $X_s$ and let $pr_s : \Pi_{t \in S} X_t \to X_s$, $(x_t) \mapsto x_s$, denote the projection of $\Pi_{t \in S} X_t$ onto $X_s$. Then the seminorms $\{p_s \circ pr_s : p_s \in P_s,\ s \in S\}$ determine the product topology.

**Definition 5.5.4** BASE OF CONTINUOUS SEMINORMS Let $P$ be the class of all continuous seminorms on the TVS $X$. A subset $Q$ of $P$ is said to be a *base of continuous seminorms* if, for any $p \in P$, there is a $q \in Q$ and a constant $r > 0$ such that $p \leq rq$. ■

If $Q$ is a base of continuous seminorms then sets of the form $aV_q$ where $a > 0$ and $q \in Q$ are a neighborhood base at 0 as is easy to show. It follows from Theorem 5.2.3 that the singleton $\{p\}$ is a base for the continuous seminorms on a seminormed space $(X, p)$. If $X$ carries a topology $\mathcal{T}_P$ generated by a family $P$ of seminorms and $Q$ is a base of continuous seminorms then the topology generated by $Q$ is a base for $\mathcal{T}_P$, as follows from the fact that $p \leq rq$ $(r > 0)$ iff $(1/r)V_q = V_{rq} \subset V_p$.

By Theorem 5.1.1, it follows that if $p_1, \ldots, p_n$ are seminorms then so is $p = \max_j p_j$ and $V_p = \cap_{j=1}^n V_{p_j}$. This motivates us to consider the following notion.

**Definition 5.5.5** SATURATED FAMILIES A family $P$ of seminorms is *saturated* if $\max_j p_j \in P$ for any $p_1, \ldots, p_n \in P$. ■

For saturated families $P$ of seminorms, a typical basic neighborhood of 0 is of the form $rV_p$ where $r > 0$ and $p \in P$—no intersections needed.

**Example 5.5.6** COMPACT-OPEN TOPOLOGY Let $C(\mathbf{R},\mathbf{R})$ denote the linear space of all continuous maps of $\mathbf{R}$ into $\mathbf{R}$. For any compact subset $K$ of $\mathbf{R}$, let $p_K(x) = \sup|x(K)|$, $x \in C(\mathbf{R},\mathbf{R})$. Each $p_K$ is a seminorm on $C(\mathbf{R},\mathbf{R})$ and the seminorm topology determined by $\{p_K : K \text{ compact} \subset T\}$ is called the *compact-open* topology, a topic first mentioned in Example 2.6.9 and one we return to in Sec. 5.8. $C(\mathbf{R},\mathbf{R},c)$ denotes $C(\mathbf{R},\mathbf{R})$ equipped with the compact-open topology. Since $\mathbf{R}$ is *hemicompact* in the sense that it has a countable family of compact subsets, namely, $K_n = [-n,n]$, $n \in \mathbf{N}$, such that any compact subset of $\mathbf{R}$ is contained in one of them, it is easy to see that the gauges $p_{K_n}$, $n \in \mathbf{N}$, are a base of continuous seminorms in $C(\mathbf{R},\mathbf{R},c)$; they are obviously saturated as well. ■

We discuss weak and weak-$*$ topologies at length in Chap. 8.

**Example 5.5.7** WEAK TOPOLOGIES $\sigma(X,X')$ and $\sigma(X',X)$ Let $X'$ denote the continuous dual of $X$, the linear space of all continuous linear functionals on the TVS $X$.

(a) WEAK TOPOLOGY $\sigma(X,X')$ Consider the family $P$ of seminorms $p_f(\cdot) = |f(\cdot)|$, $f \in X'$. The topology $\sigma(X,X')$ generated by $P$ on $X$ is called the *weak* or *weakened* (*affaiblié*) topology. This locally convex topology is clearly weaker than the original topology on $X$.

(b) WEAK-$*$ TOPOLOGY $\sigma(X',X)$ The topology $\sigma(X',X)$ generated by $\{p_x : x \in X\}$ on $X'$ where $p_x(f) = |f(x)|$, $f \in X'$, is called the *weak-$*$* (read: weak-star) *topology*. ■

## 5.6 METRIZABILITY OF LCS

Suppose that $X$ is a LCS whose topology is determined by a *countable* family $\{p_n\}$ of seminorms. There is no loss of generality in assuming that the $p_n$ are increasing (i.e., $p_1 \le p_2 \le \cdots$) for we may replace each $p_n$ by $q_n = \max\{p_j : 1 \le j \le n\}$ and still get the same the topology. Since $(p_n)$ is increasing, the balls $V_{p_n} = \{x \in X : p_n(x) < 1\}$ are decreasing [Theorem 5.2.1(a)]. Moreover, as $\{(1/j)V_{p_n} : (j,n) \in \mathbf{N}\times\mathbf{N}\}$ is a countable base at 0, $X$ is pseudometrizable by Theorem 4.8.3. In Theorem 5.6.1 we exhibit an $F$-seminorm [Definition 4.8.1] that generates the topology.

**Theorem 5.6.1** PSEUDOMETRIZABLE LCS An LCS $X$ is pseudometrizable iff its topology is generated by an increasing sequence $(p_n)$ of continuous seminorms. In this case the topology is generated by the $F$-seminorm $p$ defined by $p(x) = \sum_{n\in\mathbf{N}} 2^{-n}[p_n(x)/(1+p_n(x))]$ $(x \in X)$. We mention that if $X$ is metrizable, $p$ is an $F$-norm.

**Proof.** The assertion about pseudometrizability of an LCS $X$ being implied by the presence of an increasing sequence $(p_n)$ of seminorms follows from the discussion above. Conversely, if $X$ is pseudometrizable, it has a countable base

$\{V_n : n \in \mathbf{N}\}$ of neighborhoods of 0 [Theorem 4.8.3] which we may assume to be decreasing. By the local convexity, each $V_n$ contains a disked neighborhood $U_n$ of 0 and we can assume that the $U_n$ are decreasing as well. Consequently the gauges $p_n$ of the $U_n$ are an increasing family of seminorms which generate the topology. We show now that the $p$ of the statement is an $F$-seminorm. The series by which $p$ is defined is seen to converge by comparison with $\Sigma_{n\in\mathbf{N}}2^{-n}$ and obviously $p \geq 0$. The function $h(t) = t/(1+t)$ is increasing for $t \geq 0$ and, for $|t| \leq 1$ and $x \in X$, $p_n(tx) = |t|\, p_n(x) \leq p_n(x)$. Therefore, for each $n \in \mathbf{N}$, $p_n(tx)/[1 + p_n(tx)] \leq p_n(x)/[1 + p_n(x)]$; it follows that, $p(tx) \leq p(x)$ for $|t| \leq 1$. To show that $p$ satisfies the triangle inequality, we show that, for every $n \in \mathbf{N}$ and $x, y \in X$,

$$\frac{p_n(x+y)}{1+p_n(x+y)} \leq \frac{p_n(x)+p_n(y)}{1+p_n(x)+p_n(y)} \leq \frac{p_n(x)}{1+p_n(x)} + \frac{p_n(y)}{1+p_n(y)} \tag{1}$$

To accomplish this, we prove the following lemma: For $a, b, c \geq 0$ such that $c \leq a + b$,

$$\frac{c}{1+c} \leq \frac{a}{1+a} + \frac{b}{1+b} \tag{2}$$

If $c = 0$ or $a + b = 0$, (2) is clearly satisfied so we assume that each of $a, b, c$ is nonzero, in which case $1/c \geq 1/(a+b)$. Therefore $1 + \frac{1}{c} \geq 1 + \frac{1}{a+b}$ which implies that

$$\frac{1}{1+\frac{1}{c}} \leq \frac{1}{1+\frac{1}{a+b}}$$

or

$$\frac{c}{1+c} \leq \frac{a+b}{1+a+b} = \frac{a}{1+a+b} + \frac{b}{1+a+b} \leq \frac{a}{1+a} + \frac{b}{1+b}$$

Since each $p_n$ satisfies the triangle inequality, it only remains to take $c = p_n(x+y)$, $a = p_n(x)$ and $b = p_n(y)$ to establish (1) from which the triangle inequality for $p$ follows.

To see that $p(x/n) \to 0$, let $x \in X$ and $r > 0$ be given. Choose $k \in \mathbf{N}$ such that $\Sigma_{n>k}2^{-n} < r/2$. By the continuity of each $p_n$ and $h$ at 0, there exists $a > 0$ such that

$$\frac{p_n(x/a)}{2^n[1+p_n(x/a)]} < \frac{r}{2k} \quad \text{for } n = 1, \ldots, k$$

It follows that $p(x/a) < r$ and that $p$ is an $F$-seminorm.

To see that the topologies determined by $p$ and $\{p_n : n \in \mathbf{N}\}$ coincide, let $r > 0$ be given and choose $k$ such that $\sum_{n\geq k+1} 2^{-n} < r/2$. Since $s/(1+s) \to 0$ as $s \to 0$, we may choose $t > 0$ such that $tr/(1+tr) < r/2$. For any $x \in X$ such that $p_k(x) < rt$, since the $p_n$ are increasing,

$$\sum_{j=1}^{k} 2^{-j} \frac{p_j(x)}{1+p_j(x)} < \sum_{j=1}^{k} 2^{-j} \frac{rt}{1+rt} < \frac{r}{2} \tag{$*$}$$

Since $\Sigma_{n>k}2^{-n} < r/2$ and $p_n(x)/[1+p_n(x)] < 1$ for all $n$, it follows from $(*)$ that $p(x) < r$; therefore $rtV_{p_k} \subset rV_p$ and the topology determined by $p$ is seen to be weaker than that determined by $p_k$, hence weaker than that determined by $\{p_n : n \in \mathbf{N}\}$.

Conversely, consider the basic neighborhood $rV_{p_n}$ of 0, $r > 0$, $n \in \mathbf{N}$, in the original topology. Consider $t > 0$ such that $t < r/[2^n(1+r)]$. For $x \in X$ such that $p(x) < t$ then, by the way $p$ is defined,

$$\frac{p_n(x)}{2^n(1+p_n(x))} \leq p(x) < t < \frac{r}{2^n(1+r)}$$

Since $s/(1+s)$ is increasing, this implies that $p_n(x) < r$ and therefore that $tV_p \subset rV_{p_n}$. Hence the topology determined by $\{p_n : n \in \mathbf{N}\}$ is coarser than that determined by $p$ and the two are seen to be equal. ■

We use the following result to show that a certain space is *not* metrizable in Example 5.6.4.

**Theorem 5.6.2** Discontinuous Linear Functionals Discontinuous linear functionals exist on any infinite-dimensional pseudometrizable TVS $X$. Therefore the continuous dual $X'$ is a proper subset of the algebraic dual $X^*$. If all linear functionals are continuous—if $X' = X^*$—then $X$ is finite-dimensional.

**Proof.** Let $\{x_n\}$ be a denumerable linearly independent subset of the infinite-dimensional pseudometrizable TVS $X$ and let $(U_n)$ be a decreasing sequence of neighborhoods of 0 which are a base at 0. Choose positive numbers $t_n$ for each $n \in \mathbf{N}$ such that $t_n x_n \in U_n$. Clearly, $t_n x_n \to 0$. Extend $\{x_n\}$ to a Hamel base $B$. A discontinuous linear functional $f$ is defined on $X$ by taking $f(x_n) = 1/t_n$ for each $n \in \mathbf{N}$ and 0 on the elements of the set difference $B \setminus \{x_n\}$. ■

**Corollary 5.6.3** A normed linear space $X$ is finite-dimensional iff every linear functional on $X$ is continuous.

**Proof.** If the normed space $X$ is finite-dimensional, the continuity follows from Theorem 4.10.3(f). By Theorem 5.6.2 , if all linear functionals are continuous on $X$ then $X$ is finite-dimensional. ■

**Example 5.6.4** Finest Locally Convex Topology Let $X$ be any linear space endowed with the finest locally convex topology $\mathcal{T}_{lc}$ of Examples 4.5.4(c) and 4.11.3(c). Because of the correspondence between gauges of absorbent disks and seminorms, $\mathcal{T}_{lc}$ may also be described as the topology having the family of all absorbent disks as a base at 0. If $D$ is the unit disk in the scalar field $\mathbf{F}$ and $f$ is any linear functional on $X$ then $f^{-1}(D)$ is an absorbent disk in $X$, hence a $\mathcal{T}_{lc}$-neighborhood of 0. Therefore every linear functional on $X$ is continuous. Since there are no discontinuous linear functionals on $(X, \mathcal{T}_{lc})$, $\mathcal{T}_{lc}$ is *not pseudometrizable on any infinite-dimensional space* by Theorem 5.6.2. ■

## 5.7 CONTINUITY OF LINEAR MAPS

From Theorems 2.1.8 and 3.5.1 on topological groups, it is clear that:

**Theorem 5.7.1** CONTINUITY OF LINEAR MAPS If $X$ and $Y$ are topological vector spaces over the same topological field and $A : X \to Y$ is a linear map then:

(a) $A$ is continuous iff it is continuous at one point of $X$.

(b) If $A$ is continuous then it is uniformly continuous.

We use the following convergence result for nets in LCS to characterize continuity of certain linear maps,

**Theorem 5.7.2** CONVERGENCE IN LCS Let $P$ be a base of continuous seminorms [Definition 5.5.4] in the LCS $X$. For each net $(x_s)$ of points of $X$:

(a) $x_s \to x$ iff $p(x_s - x) \to 0$ for each $p$ in $P$;

(b) if $(x_s)$ is a Cauchy net then so is $(p(x_s))$ for each $p \in P$.

**Proof.** We use the notation of the statement. (a) Clearly, if $x_s \to x$ then $x_s - x \to 0$ and $p(x_s - x) \to p(0) = 0$ for any continuous $p \in P$. Conversely, suppose that $p(x_s - x) \to 0$ for each $p$ in $P$. Since $X$ is locally convex, it suffices to show that $x_s - x$ eventually belongs to any open disk $D$. The gauge $p_D$ of $D$ is a continuous seminorm [Theorem 5.3.3(b)]. Since $P$ is a base, for some $r > 0$ and $p \in P$, $p_D \leq rp$. By hypothesis, $rp(x_s - x) < 1$ eventually, so $x_s - x \in V_{P_D}$ eventually. Since $D$ is open, $V_{P_D} = D$ by Theorem 5.3.3(a).

(b) Any continuous seminorm is uniformly continuous by Theorem 5.2.2(a) and the uniformly continuous image of a Cauchy net is Cauchy by Theorem 3.5.2. ■

**Theorem 5.7.3** CONTINUITY OF LINEAR MAPS The linear map $A$ mapping the TVS $X$ into the LCS $Y$ is continuous iff for each continuous seminorm $q$ on $Y$ there is a continuous seminorm $p$ on $X$ such that $q \circ A \leq p$. In particular, for a TVS $X$, a linear functional $f : X \to \mathbf{F}$ is continuous iff there is a continuous seminorm $p$ on $X$ such that $|f| \leq p$. Note that if $X$ is real, $f : X \to \mathbf{R}$ a linear functional and $p$ a seminorm, then $f \leq p$ iff $|f| \leq p$.

**Proof.** We use the notation of the statement. If $A$ is continuous and $q$ is a continuous seminorm then $q \circ A$ is a continuous seminorm on $X$ which satisfies $q \circ A \leq q \circ A$.

If $q$ is a seminorm on $Y$, $q \circ A$ is a seminorm on $X$. If the condition holds then, for each continuous seminorm $q$ on $Y$, there is a continuous seminorm

$p$ on $X$ which dominates $q \circ A$. Therefore, $q \circ A$ is continuous by Theorem 5.2.2(e). To prove that $A$ is continuous suppose that $x_s \to 0$ in $X$. If so, then $p(x_s) \to 0$ which implies that $q(Ax_s) \to 0$. Since $q$ is arbitrary, $Ax_s \to 0$ by Theorem 5.7.2(a). ■

We generalize these connections between continuity of linear maps and domination by seminorms in Theorem 5.7.4. We say that a subset $B$ of a seminormed space $(X,p)$ is *bounded* if $p(B)$ is a bounded set of scalars. If $(X,p)$ and $(Y,q)$ are seminormed spaces and $g : X \to Y$ maps bounded sets into bounded sets, we say that $g$ is *locally bounded.* We deviate from this convention for linear maps on seminormed spaces for historical reasons and often say "bounded" linear map instead of *locally* bounded linear map.

**Theorem 5.7.4** CONTINUITY BETWEEN SEMINORMED SPACES If $(X,p)$ and $(Y,q)$ are seminormed spaces and $A : X \to Y$ is linear then the following statements are equivalent:

(a) $A$ is continuous.

(b) $A$ is bounded on $\overline{V}_p$.

(c) $A$ is locally bounded.

**Proof.** We use the notation of the statement. (a) $\Leftrightarrow$ (b) Since $q$ is a seminorm on $Y$, $q \circ A$ is a seminorm on $X$. Since $q$ is continuous, $q \circ A$ is continuous iff $A$ is. By Theorem 5.2.3, $q \circ A$ is continuous iff there is some positive constant $r$ such that $q \circ A \leq rp$, i.e., continuity of $A$ is equivalent to the statement that for any $x \in \overline{V}_p$, $(q \circ A)(x) \leq r$.

(c) $\Leftrightarrow$ (b) Since $\overline{V}_p$ is bounded, (c) implies (b). To show that (b) implies (c), let $r > 0$ be such that $(q \circ A)(\overline{V}_p) \subset [0, r]$ and let $B$ be a bounded subset of $X$. As such, there is some positive $s$ such that $B \subset s\overline{V}_p$. Hence $A(B) \subset sA(\overline{V}_p)$ and $(q \circ A)(B) \subset [0, sr]$. ■

We show in Theorem 6.5.1 that if there is a neighborhood $V$ of 0 in a TVS $X$ such that $f(V)$ is a bounded subset of $\mathbf{F}$ then $f$ is continuous.

**Example 5.7.5** THE BIDUAL Let $X$ be a normed space with closed unit ball $U = \{x \in X : \|x\| \leq 1\}$ and $X'$ its dual, normed by the *canonical norm*: For $f \in X'$, $\|f\| = \sup\{|f(U)|\}$. The dual $X''$ of $X'$ is called the *bidual* of $X$ and is normed in the same way: For $F \in X''$, $\|F\| = \sup\{|F(f)| : f \in X', \|f\| \leq 1\}$. Each $x \in X$ determines a linear functional $Jx$ on $X'$ by taking $Jx(f) = f(x)$ $(f \in X')$. To show that $Jx$ is continuous, we show it to be bounded on the unit ball of $X'$ : If $f \in X'$, $\|f\| \leq 1$, then $|Jx(f)| = |f(x)| \leq \|x\|$. The map $J : X \to X''$ is generally a linear isometry [Theorem 8.4.5]. If $J$ is surjective, we say that $X$ is *reflexive*. ■

We show in Theorem 5.7.7 that a linear functional is continuous iff its null space is closed. First, we need a lemma.

**Lemma 5.7.6** Let $f$ be a nontrivial linear functional on a vector space $X$ over $\mathbf{F} = \mathbf{R}$ or $\mathbf{C}$ with null space $N(f) = f^{-1}(0)$. If $f(x) = 1$ and $U \subset X$ is balanced then $(x + U) \cap N(f) = \varnothing$ iff $|f(u)| < 1$ for all $u$ in $U$.

**Proof.** We use the notation of the statement. Suppose $f(x) = 1$. Clearly, if $|f(u)| < 1$ for each $u$ in $U$ then for any $u \in U$, $f(x + u) = f(x) + f(u) = 1 + f(u) \neq 0$; hence $(x + U) \cap N(f) = \varnothing$. Conversely, if $|f(u)| \geq 1$ for some $u \in U$ then $-u/f(u) \in U$ and $x - u/f(u) \in (x + U) \cap N(f)$. ■

**Theorem 5.7.7** CONTINUOUS IFF CLOSED NULL SPACE Let $f$ be a linear functional on a TVS $X$ over $\mathbf{F} = \mathbf{R}$ or $\mathbf{C}$. Then $f$ is continuous iff (a) or (b) holds:

(a) $N(f)$ is closed;
(b) $\operatorname{Re} f$ is continuous;
(c) If $f$ is continuous and nontrivial then $f$ is an open map;
(d) If $X$ is a normed space and $f$ is continuous then $\|f\| = \|\operatorname{Re} f\|$.

**Proof.** We use the notation of the statement. (a) If $f$ is continuous then clearly $N(f)$ is closed. Suppose conversely that $N(f)$ is closed and $f(z) = 1$. Since $N(f)$ is closed, there is a balanced neighborhood $U$ of 0 such that $(z + U) \cap N(f) = \varnothing$. By Lemma 5.7.6, therefore, $U \subset \{x \in X : |f(x)| < 1\}$ and $\{x \in X : |f(x)| < 1\}$ is a neighborhood of 0. Since $f$ is linear, for each $r > 0$, $\{x \in X : |f(x)| < r\} = r\{x \in X : |f(x)| < 1\}$ and it follows that $f$ is continuous at 0.

(b) We may write $f(x) = \operatorname{Re} f(x) - i \operatorname{Re} f(ix)$ by Theorem 1.4.2. The equivalence of the continuity of $f$ and $\operatorname{Re} f$ is now apparent.

(c) This follows from Theorem 4.10.3(g).

(d) Suppose $f$ is continuous. Clearly $|\operatorname{Re} f(x)| \leq |f(x)|$ for every $x \in X$ which implies that $\|\operatorname{Re} f\| \leq \|f\|$. For any unit vector $x \in X$, there exists a unit scalar $a \in \mathbf{C}$ such that $af(x) = f(ax) = |f(x)|$. Since $\|ax\| = 1$, $|f(x)| = f(ax) = \operatorname{Re} f(ax) = |\operatorname{Re} f(ax)| \leq \|\operatorname{Re} f\|$ which implies that $\|f\| \leq \|\operatorname{Re} f\|$. ■

## 5.8 THE COMPACT-OPEN TOPOLOGY

**Notation.** Throughout the section $T$ denotes a topological space and $C(T, \mathbf{F})$ the linear space of continuous maps of $T$ into $\mathbf{F} = \mathbf{R}$ or $\mathbf{C}$. ■

We consider $C(T, \mathbf{F})$ endowed with the (locally convex) compact- and point-open topologies. We characterize when $C(T, \mathbf{F})$ with the compact-open topology is metrizable ($T$ hemicompact) and complete ($T$ a $k_R$-space).

Let $\mathcal{K}$ denote the class of compact subsets of $T$. For $K \in \mathcal{K}$, $p_K(x) = \sup |x(K)|$, $x \in C(T, \mathbf{F})$, is a seminorm. (For singletons $K = \{t\}$, we write $p_t$ instead of $p_{\{t\}}$.) The topology $\mathcal{T}_c$ generated by $\{p_K : K \in \mathcal{K}\}$ is called the *compact-open* topology, topology of *uniform convergence on compact sets*, or topology of *compact convergence*. Since $\{p_K\}$ is evidently a saturated family of seminorms, a base at 0 for $\mathcal{T}_c$ is given by positive multiples of open balls $V_{p_K} = \{x \in C(T, \mathbf{F}) : p_K(x) < 1\}$, $K \in \mathcal{K}$. We have already encountered the compact-open topology in Examples 2.6.9 and 5.5.6.

Since $\mathcal{T}_c$ is generated by a family of seminorms, it is a locally convex topology. For a nonzero $x \in C(T, \mathbf{F})$, there exists $t \in T$ such that $x(t) \neq 0$. Hence, the continuous seminorm $p_t(x) = |x(t)| \neq 0$ and it follows that $\mathcal{T}_c$ is Hausdorff [Theorem 5.5.1(a)]. We denote $C(T, \mathbf{F})$ equipped with $\mathcal{T}_c$ by $C(T, \mathbf{F}, c)$.

If $T$ is compact, the compact-open topology simplifies considerably: $\{p_T\}$ is a base of continuous seminorms and $C(T, \mathbf{F}, c)$ is a normed space. When $T$ is compact, we refer to $p_T$ as the *uniform norm* or *sup norm* and denote it by $\|\cdot\|_\infty$.

We can also view the compact-open topology as an inverse image topology. For each $K \in \mathcal{K}$, the restriction of $p_K$ to $C(K, \mathbf{F})$ is a norm and we suppose $C(K, \mathbf{F})$ to carry this norm topology. Now consider the family of linear maps

$$A_K : C(T, \mathbf{F}) \to C(K, \mathbf{F}), \; x \mapsto x|_K \qquad (*)$$

where $x|_K$ denotes the restriction of $x$ to $K$. Certainly, a base at 0 for the norm topology on $C(K, \mathbf{F})$ is given by the collection of sets of the form $B_{Kr} = \{y \in C(K, \mathbf{F}) : \sup |y(K)| < r\}$, $r > 0$, so [Example 4.7.1] a subbase at 0 for the weakest topology on $C(T, \mathbf{F})$ with respect to which each $A_K$ is continuous is given by sets of the form $A_K^{-1}(B_{Kr}) = rV_{p_K}$ $(K \in \mathcal{K}, \|\cdot\|_\infty$ $r > 0)$. Consequently, this inverse image topology and the compact-open topology coincide.

As follows from Theorem 5.7.2(a), convergence in the compact-open topology is the usual function-theoretic notion of uniform convergence on compact sets:

**Example 5.8.1** UNIFORM CONVERGENCE ON COMPACT SETS A net $(x_s)_{s \in S}$ in $C(T, \mathbf{F}, c)$ converges to $x$ iff for each compact subset $K$ of $T$ and $r > 0$, there exists an index $s_0 \in S$ such that $s \geq s_0$ implies $|x_s(t) - x(t)| < r$ for all $t$ in $K$.

The following criterion for convergence of sequences obtains.

**Theorem 5.8.2** CONTINUOUS CONVERGENCE When $T$ is first countable, a sequence $(x_n)$ in $C(T, \mathbf{F}, c)$ converges to $x \in C(T, \mathbf{F}, c)$ iff $x_n(t_n) \to x(t)$ for each $t \in T$ and sequence $t_n \to t$. This latter type of convergence is called *continuous convergence* so on first countable spaces $T$, continuous convergence (of continuous functions) is equivalent to uniform convergence on all compact subsets. ■

**Proof.** We use the notation above. Suppose that $x_n \to x$ in the compact-open topology and $t_n \to t$. Since $\{t_n\} \cup \{t\} = K$ is compact, $x_n \to x$ uniformly on $K$. The continuity of $x$ and the inequality

$$|x_n(t_n) - x(t)| \leq |x_n(t_n) - x(t_n)| + |x(t_n) - x(t)| \quad (n \in \mathbf{N})$$

imply that $x_n(t_n) \to x(t)$. Note that no use was made here of the first countability of $T$, so we can say that uniform convergence on compact sets always implies continuous convergence. Conversely, suppose that $x_n \to x$ continuously but for some compact set $K$, $p_K(x_n - x) \nrightarrow 0$. Then for some $r > 0$ there exist integers $n_1 < n_2 < \cdots$ and points $t_k \in K$ such that $|x_{n_k}(t_k) - x(t_k)| \geq r$ for every $k$. Since $K$ is compact and first countable, we may assume that $t_k$ converges to some $t \in K$ (otherwise, use a convergent subsequence). Since $x_n \to x$ continuously, it follows that $x_{n_k}(t_k) \to x(t)$. Since $x$ is continuous, we have

$$|x_{n_k}(t_k) - x(t_k)| \leq |x_{n_k}(t_k) - x(t)| + |x(t) - x(t_k)| \to 0$$

which is a contradiction. ■

We noted in Example 5.2.4(b) that for $t \in T$, the *evaluation map* $t'(x) \mapsto x(t)$ associated with $t$ on $\ell_\infty(T)$, is continuous; since $|t'(\cdot)|$ is a seminorm, it follows from Theorem 5.2.1 that $t'$ is uniformly continuous. Essentially the same argument proves that:

**Theorem 5.8.3** EVALUATION MAPS UNIFORMLY CONTINUOUS ON $C(T, \mathbf{F}, c)$.

The notion of hemicompactness completely characterizes metrizability of $C(T, \mathbf{F}, c)$.

**Definition 5.8.4** HEMICOMPACT A topological space $T$ is *hemicompact* if it has a countable family $(K_n)$ of compact subsets such that each compact set in $T$ is contained in some $K_n$. ■

Compact spaces are hemicompact as are locally compact $\sigma$-compact spaces. The space $\mathbf{Q}$ of rationals in its Euclidean topology is not hemicompact. For a little more about hemicompactness in general, see Exercise 5.201.

**Theorem 5.8.5** METRIZABILITY OF $C(T, \mathbf{F}, c)$ For completely regular Hausdorff spaces $T$, $C(T, \mathbf{F}, c)$ is metrizable iff $T$ is hemicompact.

**Proof.** We use the notation of the statement. As observed earlier, the compact-open topology is generally Hausdorff. Suppose that $T$ is hemicompact and let $(K_n)$ be a countable family of compact sets such that each compact set in $T$ is contained in some $K_n$. If the compact set $K$ is contained in $K_n$ then $p_K \leq p_{K_n}$ and the seminorms $\{p_{K_n} : n \in \mathbf{N}\}$ are therefore a base of continuous seminorms. Since the compact-open topology is generated by a countable family of seminorms, it is metrizable by Theorem 5.6.1.

Conversely, if $C(T, \mathbf{F}, c)$ is metrizable, the compact-open topology $\mathcal{T}_c$ must have a countable base at 0. Hence there must be positive numbers $a_n$ and compact sets $K_n$ such that $\{a_n V_{p_{K_n}} : n \in \mathbf{N}\}$ is a base at 0. Consequently, for any compact set $K$, there is some $n \in \mathbf{N}$ such that $a_n V_{p_{K_n}} \subset V_{p_K}$. If $K \not\subset K_n$, there is some point $t$ in $K \setminus K_n$. Since $T$ is a completely regular Hausdorff space, there exists $x \in C(T, \mathbf{F})$ such that $x(t) = |a_n|$ and $x(K_n) = \{0\}$. Hence $x \in a_n V_{p_{K_n}}$ but $x \notin V_{p_K}$ which is contradictory. Thus $K \subset K_n$ and the proof is complete. ∎

**Definition 5.8.6** K$_R$-SPACES A completely regular Hausdorff space $T$ is a *$k_R$-space* if a scalar-valued function $x$ whose restriction to each compact subset of $T$ is continuous must be continuous on $T$. ∎

First countable completely regular Hausdorff spaces are $k_R$-spaces (Exercise 5.202).

**Theorem 5.8.7** COMPLETENESS OF $C(T, \mathbf{F}, c)$ For a completely regular Hausdorff space $T$, $C(T, \mathbf{F}, c)$ is complete iff $T$ is a $k_R$-space.

**Proof.** Let $T$ be a completely regular Hausdorff space and suppose that $T$ is a $k_R$-space. Note that if $T$ is compact then $C(T, \mathbf{F}, c)$ is complete since (1) a Cauchy net from $C(T, \mathbf{F}, c)$ is a "uniform" Cauchy net in the usual function-theoretic sense whose pointwise limit is therefore a uniform limit, and (2) a uniform limit of continuous functions is continuous. For noncompact $T$, suppose that $(x_s)_{s \in S}$ is a Cauchy net in $C(T, \mathbf{F}, c)$. By Theorem 5.7.2, it follows that $(p_K(x_s))$ is a Cauchy net for each compact subset $K$ of $T$. Hence the net $(x_s|_K)$ of restrictions of $x_s$ to $K$ is a Cauchy net in the complete space $C(K, \mathbf{F}, c)$. Let $x_K \in C(K, \mathbf{F}, c)$ be its limit. Since each $t$ in $T$ belongs to some compact set, we can define the function $x(t) = x_K(t)$, where $K$ is some compact set in $T$ to which $t$ belongs. To see that $x$ is well defined, suppose that $t$ belongs to the compact set $H$ as well as $K$. Since, for any $s \in S$,

$$|x_K(t) - x_H(t)| \le |x_K(t) - x_s(t)| + |x_s(t) - x_H(t)|$$

it follows that $x$ is well defined. Since $x$ is continuous on compact subsets of the $k_R$-space $T$, $x$ is continuous on $T$. Since, for any compact set $K$, $p_K(x - x_s) \to 0$, $x = \lim x_s$ [Theorem 5.7.2(a)].

Now assume that $C(T, \mathbf{F}, c)$ is complete and let $x$ be a bounded map of $T$ into $\mathbf{F}$ which is continuous on compact subsets of $T$. We have to show that $x$ is continuous on $T$. Let $\beta T$ be the Stone–Cech compactification of $T$ [Example 8.3.3]. Since $\beta T$ is a normal topological space, for any compact subset $K$ of $T$, $x|_K$ can be extended to a continuous function $y_K$ defined on $\beta T$ in a sup-preserving way—$\sup |y_K(\beta T)| = \sup |x(K)|$—by the Tietze-Urysohn extension theorem. Let $x_K$ denote the restriction of $y_K$ to $T$. The class $\mathcal{K}$ of compact subsets of $T$ is a directed set with respect to the ordering: for $K, H \in \mathcal{K}$, $K \le H$ iff $K \subset H$. To show that the net $(x_K)_{K \in \mathcal{K}}$ in $C(T, \mathbf{F}, c)$ is

Cauchy, let $H$ be a compact subset of $T$, $a > 0$, and consider the neighborhood $aV_{p_H}$ of 0. For any compact subsets $L$ and $K$ of $T$ which contain $H$ (i.e., $L, K \geq H$) then

$$p_H(x_L - x_K) = p_H(x_L|_H - x_K|_H) = p_H(x_H - x_H) = 0$$

Hence $x_L - x_K \in aV_{p_H}$ and $(x_K)$ is seen to be a Cauchy net in $C(T, \mathbf{F}, c)$. Moreover, since $x_K(t) = x(t)$ for any compact set $K \supset \{t\}$, $x_K(t) \to x(t)$ at each $t$ in $T$ ($x_K$ is "pointwise convergent" to $x$). As $C(T, \mathbf{F}, c)$ is complete, the Cauchy net $(x_K)$ has a limit $y$ in $C(T, \mathbf{F}, c)$. We conclude that $x = y$ and therefore that $x \in C(T, \mathbf{F}, c)$.

Now suppose that $x$ is an unbounded real-valued function on $T$ with continuous restriction to each compact subset $K$ of $T$ and consider the following truncations of $x$: For each positive integer $n$

$$x_n(t) = \begin{cases} x(t) & \text{if } |x(t)| \leq n \\ n & \text{if } x(t) > n \\ -n & \text{if } x(t) < -n \end{cases}$$

As the maximum or minimum of continuous maps, each $x_n$ is continuous and so is the restriction of $x_n$ to any compact set $K$. Hence, by the previous argument for bounded functions, each $x_n \in C(T, \mathbf{F})$ and $(x_n)$ obviously converges pointwise to $x$ on $T$. For any compact set $K$, the continuous function $x|_K$ is bounded and therefore equal to $x_n|_K$ for sufficiently large $n$. Thus $x_n \to x$ uniformly on each compact set. It follows that $(x_n)$ is a Cauchy sequence on $C(T, \mathbf{F}, c)$. By the completeness of $C(T, \mathbf{F}, c)$, it follows that $x_n \to y \in C(T, \mathbf{F}, c)$. As $(x_n)$ must also converge to $y$ pointwise, it follows that $x = y$ and proves that $x$ is continuous. If $x$ is an unbounded complex-valued function with continuous restriction to each compact set, the argument just used applies to the real and imaginary parts of $x$. ■

## 5.9 THE POINT-OPEN TOPOLOGY

**Notation.** Throughout the section $T$ denotes a topological space and $C(T, \mathbf{F})$ the linear space of continuous maps of $T$ into $\mathbf{F} = \mathbf{R}$ or $\mathbf{C}$. ■

We first discussed the point-open topology in Example 2.6.8.

**Example 5.9.1** POINT-OPEN TOPOLOGY Let $C(T, \mathbf{F})$ denote the linear space of continuous maps of the topological space $T$ into $\mathbf{F} = \mathbf{R}$ or $\mathbf{C}$. For each $t$ in $T$, $p_t(x) = |x(t)|$, $x \in C(T, \mathbf{F})$, is a seminorm on $C(T, \mathbf{F})$. The topology $\mathcal{T}_p$ generated by $\{p_t : t \in T\}$ is called the *point-open* topology or topology of *pointwise convergence*. We denote $C(T, \mathbf{F})$ endowed with the point-open topology by $C(T, \mathbf{F}, p)$. ■

Since singletons are compact, the point-open topology is clearly weaker than the compact-open topology; it is a locally convex topology since it is generated by seminorms, it is Hausdorff because $x \neq 0$ implies $p_t(x) \neq 0$ for some $t \in T$. Since $V_{p_t} \cap V_{p_s} = V_{p_{(s,t)}}$ $(s, t \in T)$, it is clear that finite subsets of $T$ also "generate" the point-open topology. For this reason it is also called the topology of *uniform convergence on finite sets.*

Evidently:

**Theorem 5.9.2** POINTWISE CONVERGENCE A net $(x_s)$ in $C(T, \mathbf{F}, p)$ converges to $x$ in $C(T, \mathbf{F}, p)$ iff for each $t$ in $T$, $x_s(t) \to x(t)$.

The point-open topology can also be described as an inverse image topology.

**Theorem 5.9.3** POINT-OPEN TOPOLOGY AS INVERSE IMAGE TOPOLOGY The point-open topology $\mathcal{T}_p$ is is the inverse image topology [Example 4.7.1; cf. Example 2.4.2] induced by the linear map

$$\begin{array}{rccc} I: & C(T, \mathbf{F}, p) & \to & \prod_{t \in T} F_t = \mathbf{F}^T \\ & x & \mapsto & (x(t))_{t \in T} \end{array}$$

Indeed, $I$ is a linear homeomorphism.

**Proof.** We use the notation above and compare subbasic neighborhoods in the two topologies. A $\mathcal{T}_p$-subbasic neighborhood of 0 in $C(T, \mathbf{F}, p)$ is given by

$$rV_{p_t} = \{x \in C(T, \mathbf{F}) : |x(t)| < r\} \quad (t \in T, \ r > 0)$$

A typical product subbase neighborhood of 0 [Example 2.4.5] is of the form $V = \Pi_{s \in T} V_s$, where $V_s = \mathbf{F}$ for all $s$ except $s = t$ and $V_t = \{b \in \mathbf{F} : |b| < r\}$, $r > 0$. Thus, a subbasic inverse image neighborhood of 0 induced by $I$ is of the form

$$I^{-1}(V_t) = \{x \in C(T, \mathbf{F}) : |x(t)| < r\} = rV_{p_t}$$

$I$ is clearly 1-1 and open. ■

# 5.10 ASCOLI'S THEOREM

**Notation.** Throughout the section $T$ denotes a topological space and $C(T, \mathbf{F})$ the linear space of continuous maps of $T$ into $\mathbf{F} = \mathbf{R}$ or $\mathbf{C}$. ■

After introducing the notion of equicontinuous sets of continuous functions, we prove Ascoli's (pronounced AS-coli) theorem [Theorem 5.10.4] about compactness of sets $H$ of continuous functions in the compact-open topology; it says essentially that $H \subset C(T, \mathbf{F}, c)$ is compact iff it is equicontinuous. The connection between some condition like compactness and equicontinuity of a family of linear maps is one we return to in Secs. 8.6 and 11.9 (on the Banach-Steinhaus theorem).

**Definition 5.10.1** EQUICONTINUITY Suppose that $T$ is a topological space, $X$ a topological group, and $F(T,X)$ the set of all functions mapping $T$ into $X$. To say that $H \subset F(T,X)$ is *equicontinuous at* $t \in T$ means that given any neighborhood $V$ of 0 in $X$, there is some neighborhood $U$ of $t$ such that $x(U) \subset x(t)+V$ for each $x \in H$. To say that $H$ is *equicontinuous* means that $H$ is equicontinuous at each $t \in T$. ■

It immediately follows that: (1) $H$ equicontinuous at $t$ implies that each $x$ in $H$ is continuous at $t$; and (2) any finite subset of $C(T,\mathbf{F})$ is equicontinuous. If $T = \mathbf{R}$ and $H = \{x \in C(T,\mathbf{F}) : x$ is differentiable on the open interval $(a,b)$ and there exists $M > 0$ such that $|x'(t)| \leq M$ for all $x \in H$ and $t \in (a,b)$ then $H$ is equicontinuous on $(a,b)$ because for each $x \in H$, by the mean value theorem,

$$|x(t) - x(t')| \leq M\,|t - t'| \qquad (t, t' \in (a,b))$$

Thus, given any $r > 0$, we may take $U$ to be $(t - r/2M, t + r/2M) \cap (a,b)$ for any $t \in (a,b)$ to make $|x(t) - x(t')| < r$ for any $t \in (a,b)$.

We obtain the close relationship between equicontinuity and compactness in $C(T,\mathbf{F},c)$ in Theorem 5.10.4. First, we need the preliminary results Lemmas 5.10.2 and 5.10.3.

**Lemma 5.10.2** INDUCED TOPOLOGIES ON EQUICONTINUOUS SUBSETS Let $T$ be a topological space. The same topology is induced on equicontinuous subsets $H$ of $C(T,\mathbf{F})$ by the point-open topology $\mathcal{T}_p$ and the compact-open topology $\mathcal{T}_c$: with $H \cap \mathcal{T} = \{H \cap G : G \in \mathcal{T}\}$, we show that $H \cap \mathcal{T}_c = H \cap \mathcal{T}_p$.

**Proof.** Let $H \subset C(T,\mathbf{F})$ be equicontinuous. Generally, $\mathcal{T}_p \subset \mathcal{T}_c$, so we need only show that $\mathcal{T}_c \cap H \subset \mathcal{T}_p \cap H$. To this end, let $x_0 \in H$. A $\mathcal{T}_c \cap H$-subbasic neighborhood of $x_0$ is of the form

$$U_c = \{x \in H : p_K(x - x_0) < r\}$$

where $r > 0$ and $K$ is a compact subset of $T$. We now find $r' > 0$ and $t_1, \ldots, t_n \in T$ such that the $\mathcal{T}_p \cap H$-neighborhood of $x_0$

$$U_p = \{x \in H : p_{\{t_1,\ldots,t_n\}}(x - x_0) < r'\} \subset U_c$$

To this end, let $K$ and $r > 0$ be given. Since $H$ is equicontinuous, there exist neighborhoods $V_t$ of $t$ for each $t \in T$ such that $|x(s) - x(t)| < r/4$ for each $s \in V_t$ and each $x \in H$. A finite number $V_{t_1}, \ldots, V_{t_n}$ of the $V_t$ cover the compact set $K$. Now consider $x \in H$ such that $p_{\{t_1,\ldots,t_n\}}(x - x_0) < r/2\}$. For $t \in K$, there is an $i$, $1 \leq i \leq n$, such that $t \in V_{t_i}$. Thus

$$\begin{aligned} |x(t) - x_0(t)| &\leq |x(t) - x(t_i)| + |x(t_i) - x_0(t_i)| + |x_0(t_i) - x_0(t)| \\ &< r/4 + r/2 + r/4 = r \end{aligned}$$

Since $t \in K$ is arbitrary, it follows that $x \in U_c$. ■

**Lemma 5.10.3** POINTWISE CLOSURES OF EQUICONTINUOUS SETS Let $H$ be an equicontinuous subset of the linear space $F(T, \mathbf{F})$ of all functions mapping the topological space $T$ into $\mathbf{F}$. Then the pointwise closure (i.e., the $\mathcal{T}_p$-closure) $\mathrm{cl}_p\, H$ of $H$ in $F(T, \mathbf{F}, p)$ is an equicontinuous subset of $C(T, \mathbf{F})$.

**Proof.** We use the notation of the statement. Let $H \subset F(T, \mathbf{F})$ be equicontinuous. Given $r > 0$ and $t \in T$, there is a neighborhood $V$ of $t$ such that

$$|x(s) - x(t)| < r \tag{**}$$

for each $x \in H$ and $s \in V$. We show that $(**)$ holds for each $y \in \mathrm{cl}_p\, H$ as well. To this end let $y \in \mathrm{cl}_p\, H$ and let $s \in V$. Since $y \in \mathrm{cl}_p\, H$ then, given $r' > 0$, there is an $x \in H$ such that $p_{\{s,t\}}(x - y) < r'$. Hence

$$|y(s) - y(t)| \leq |y(s) - x(s)| + |x(s) - x(t)| + |x(t) - y(t)| \leq r' + r + r'$$

It follows that $\mathrm{cl}_p\, H$ is equicontinuous. ■

As an immediate consequence of Lemmas 5.10.2 and 5.10.3, the $\mathcal{T}_c$-closure of an equicontinuous subset of $F(T, \mathbf{F})$ is equicontinuous in $C(T, \mathbf{F}, c)$. In our next result we obtain some results connecting pointwise boundedness of equicontinuous sets and compactness.

**Theorem 5.10.4** EQUICONTINUITY AND COMPACTNESS (Ascoli) Let $T$ be a topological space. If $H$ is an equicontinuous subset of $C(T, \mathbf{F}, c)$ such that $t'(H) = \{x(t) : x \in H\}$ is a bounded subset of $\mathbf{F}$ for each $t \in T$ then $H$ is relatively compact. Conversely, if $T$ is locally compact and $H$ is a relatively compact subset of $C(T, \mathbf{F}, c)$ then $H$ is equicontinuous and $t'(H)$ is a bounded subset of $\mathbf{F}$ for each $t \in T$.

**Proof.** We use the notation of the statement. Suppose first that $H$ is equicontinuous and that $t'(H)$ is bounded for each $t \in T$. As such, each $t'(H)$ is relatively compact by the Heine-Borel theorem. Let $\mathrm{cl}_p\, H$ and $\mathrm{cl}_c\, H$ denote the $\mathcal{T}_p$- and $\mathcal{T}_c$-closures of $H$ in $C(T, \mathbf{F})$, respectively [ Lemmas 5.10.2 and 5.10.3]. By Lemma 5.10.3, the pointwise closure $\mathrm{cl}_p\, H$ of $H$ computed in $F(T, \mathbf{F}, p)$ is an equicontinuous subset of $C(T, \mathbf{F})$. For each $t \in T$, let $F_t = \mathbf{F}$ and consider the homeomorphism of Theorem 5.9.3:

$$\begin{array}{rccc} I: & C(T, \mathbf{F}, p) & \to & \prod_{t \in T} F_t = F(T, \mathbf{F}, p) \\ & x & \mapsto & (x(t))_{t \in T} \end{array}$$

To see that $\mathrm{cl}_p\, I(H)$ is compact in $F(T, \mathbf{F})$, consider

$$\mathrm{cl}_p I(H) = I(\mathrm{cl}_p H) = \{(x(t))_{t \in T} : x \in \mathrm{cl}_p H\} = \prod_{t \in T} t'(\mathrm{cl}_p H) \subset \prod_{t \in T} \mathrm{cl}_{\mathbf{F}} t'(\mathrm{cl}_p H)$$

where $\mathrm{cl}_{\mathbf{F}}$ denotes closure computed in $\mathbf{F}$. It follows from Tihonov theorem's on products of compact sets that $\mathrm{cl}_p\, I(H)$ is compact and therefore $\mathrm{cl}_p\, H$ is

$\mathcal{T}_p$-compact. Since $H$ is equicontinuous, $\operatorname{cl}_p H = \operatorname{cl}_c H$ by Lemma 5.10.2 and $\operatorname{cl}_c H$ is $\mathcal{T}_c$-compact.

Conversely, suppose that $T$ is locally compact and that $H \subset C(T, \mathbf{F}, c)$ is relatively compact. For any $t \in T$, the evaluation map $t'$ is continuous on $C(T, \mathbf{F}, c)$ by Theorem 5.8.3; hence each $t'(\operatorname{cl}_c H)$ is compact. It follows that $t'(H)$ is relatively compact—hence bounded—for any $t \in T$.

To show equicontinuity of $H$, consider $t \in T$ and $r > 0$. Since $T$ is locally compact, we may choose a compact neighborhood $K$ of $t$. Let $V_{p_K} = \{x \in C(T, \mathbf{F}) : p_K(x) < 1\}$. Since $H$ is relatively $\mathcal{T}_c$-compact, there exists $n \in \mathbf{N}$ and $x_i \in \operatorname{cl}_c H$, $i = 1, \ldots, n$, such that $\{x_i + (r/3)V_{p_K} : i = 1, \ldots, n\}$ cover $H$. Thus for each $s \in K$ and $x \in H$, there is some $x_i$, $1 \leq i \leq n$, such that $|x(s) - x_i(s)| < r/3$. Since each $x_i$ is continuous, there exist neighborhoods $V_i$ $1 \leq i \leq n$, of $t$ such that

$$|x_i(s) - x_i(t)| < r/3 \qquad \text{for each } s \in V_i$$

Now for any $x \in H$, any $s \in U = K \cap V_1 \cap \cdots \cap V_n$, we have, for some $i$ whose choice depends on $x$,

$$|x(s) - x(t)| < |x(s) - x_i(s)| + |x_i(s) - x_i(t)| + |x_i(t) - x(t)| < r/3 + r/3 + r/3$$

Hence $x(U) \subset x(t) + B(0, r)$ and the equicontinuity of $H$ is established. ■

## 5.11 PRODUCTS AND QUOTIENTS

This section has two principal purposes: (1) To describe the families of seminorms that generate the topologies of products and quotients of LCS and (2) to discuss completions of LCS.

If $X$ is locally convex then its topology is determined by the gauges $p_U$ (seminorms) of open disks $U$ of $X$ [Theorem 5.5.2]. A subspace $M$ of $X$ is locally convex and its topology is generated by $\{p|_M : p \in P\}$. $X/M$ in its quotient topology is locally convex as well [Theorem 4.7.3(d)]. We identify the seminorms that generate the quotient topology next.

**Theorem 5.11.1** QUOTIENT TOPOLOGY Let $\mathcal{D}$ denote the family of open disks in the locally convex space $X$. Let $M$ be a subspace of $X$ and let $k : X \to X/M$ denote the canonical map $x \mapsto \dot{x} = x + M$. The open disks of $X/M$ are $\{k(D) : D \in \mathcal{D}\}$. For $D \in \mathcal{D}$, let $\dot{D} = k(D)$ with gauge $p_{\dot{D}}(\dot{x}) = \inf \left\{ r > 0 : \dot{x} \in r\dot{D} \right\}$, $x \in X$. The quotient topology is generated by $\{p_{\dot{D}} : D \in \mathcal{D}\}$ and for any $x \in X$ and $D \in \mathcal{D}$,

$$p_{\dot{D}}(\dot{x}) = \inf p_D(x + M) \tag{1}$$

where $p_D(x) = \inf \{r > 0 : x \in rD\}$, the gauge of $D$.

**Proof.** We use the notation of the statement. By Theorem 4.7.3(a), for any open disk $D$ of $X$, $k(D) = \dot{D}$ is an open disk in $X/M$. Conversely, if $W$ is an open disk in $X/M$, $k^{-1}(W) = D$ is an open disk in $X$ and $W = k(D) = \dot{D}$. Consequently the gauges $\{p_{\dot{D}} : D \in \mathcal{D}\}$ generate the quotient topology of $X/M$. For any $x \in X$ and any open disk $D$ of $X$,

$$p_{\dot{D}}(\dot{x}) = \inf\left\{r > 0 : \dot{x} \in r\dot{D}\right\}$$

For $r > 0$ and $x \in X$, $\dot{x} \in r\dot{D}$ iff there is some $m \in M$ such that $x + m \in rD$. Hence, $r \geq p_D(x+m) \geq \inf p_D(x+M)$. Since $r$ is arbitrary, it follows that $p_{\dot{D}}(\dot{x}) \geq \inf p_D(x+M)$. To reverse the inequality, let $m \in M$ and suppose that $r > 0$ is such that $x + m \in rD$. It follows that $r \geq p_{\dot{D}}(\dot{x})$ and, since $r$ is arbitrary, that $p_D(x+m) \geq p_{\dot{D}}(\dot{x})$. Formula (1) now follows from the arbitrariness of $m$. ■

More generally, the following, easily verified result obtains.

**Theorem 5.11.2** SEMINORMS IN QUOTIENT TOPOLOGIES If $P$ is a base of continuous seminorms for the locally convex space $X$ and $M$ is a subspace of $X$ then a base $\dot{P}$ of continuous seminorms for $X/M$ is given by seminorms of the form $\dot{p}$ $(p \in P)$ where $\dot{p}(\dot{x}) = \inf p(x+M)$ $(x \in X)$.

It follows that if $M$ is a subspace of a seminormed space $(X,p)$ then the quotient topology on $X/M$ is determined by $\dot{p}$. In particular, if $M$ is a closed subspace of a normed space $X$, we always norm $X/M$ by taking $\|x+M\| = \inf\{\|x+M\| : m \in M\}$.

To illustrate how the seminorms $\dot{p}$ mentioned in Theorem 5.11.2 behave, consider $\ell_p^{\mathbf{R}}(2)$. For the sake of this discussion, we denote $\|\cdot\|_2$ by $p$. Let $M = \{0\} \times \mathbf{R}$. We may identify $X/M$ with $\mathbf{R} \times \{0\}$. Having done so, $\dot{p}(a,0) = \inf\{p(a,b) : b \in \mathbf{R}\} = \inf\{(a^2+b^2)^{1/2} : b \in \mathbf{R}\} = |a|$.

If $p$ is a seminorm on a vector space $X$ then clearly the null space (or kernel) $N_p = p^{-1}(0)$ of $p$ is a subspace of $X$. Moreover, for any $y \in N_p$, $p(x) \leq p(x+y) \leq p(x)$, so $\dot{p}(x+N_p) = p(x)$ for any $x \in X$. This implies that $\dot{p}$ is actually a norm on $X/N_p : \dot{p}(x+N_p) = 0 \iff p(x) = 0 \iff x \in N_p$. We call $\dot{p}$ the *factor norm* on $X/N_p$.

**Theorem 5.11.3** COMPLETION OF SEMINORMED SPACE A seminormed space $(X,p)$ has a completion $\left(\widehat{X}, \widehat{p}\right)$ as a seminormed space. If $p$ is a norm then $(X,p)$ has a completion as a Banach space.

**Proof.** A seminorm $p$ on a vector space $X$ determines an invariant pseudometric $d$ when we take $d(x,y) = p(x-y)$ $(x,y \in X)$. Consequently, by Theorem 3.7.1, a seminormed space $(X,p)$ possesses a completion $\left(\widehat{X}, \widehat{d}\right)$ as a complete pseudometrizable group where $\widehat{d}$ extends $d$. $\widehat{X}$ consists of Cauchy sequences $(x_n)$ from $X$ and is a topological group with respect to pointwise addition. With respect to scalar multiplication defined as $a(x_n) = (ax_n)$ $(a \in \mathbf{F})$, $\widehat{X}$ becomes a TVS. We can extend $p$ to $\widehat{X}$ by taking $\widehat{p}((x_n))$ to be the limit of the Cauchy sequence $p(x_n)$ [Theorem 5.7.2(b)] and $\widehat{p}$ is a seminorm on $\widehat{X}$. For details on the completion of a normed space, see Bachman and Narici 2000, pp. 118–121. ■

We make use of some of these notions now in obtaining the characterization of locally convex Hausdorff spaces of Theorem 5.11.4. We identify them as a subspace of a space of continuous functions in Theorem 8.4.8(b).

**Theorem 5.11.4** LOCALLY CONVEX SPACES Let $X$ be a locally convex space whose topology is generated by a family $P$ of seminorms.

(a) $X$ is linearly homeomorphic to a subspace of a product of seminormed spaces.

(b) If $X$ is Hausdorff then $X$ is is linearly homeomorphic to a subspace of a product of Banach spaces.

**Proof.** We use the notation of the statement. (a) For each $p \in P$, let $X_p$ denote the seminormed space $(X, p)$ and $A_p$ the continuous linear map $x \mapsto x$ from $X$ onto $X_p$. Now consider the injective map

$$\begin{array}{rcl} A : X & \to & \prod_{p \in P} X_p \\ x & \mapsto & (A_p x)_{p \in P} \end{array} \tag{2}$$

By considering subbasic neighborhoods $rV_q \times \prod_{p \neq q} X_p$ of 0 in the product, we see that the the inverse image topology induced by $A$ [Example 4.7.1] is just $X$'s original topology. Consequently, $A$ is a continuous open map of $X$ onto $A(X)$ [Example 4.7.1]. That $A$ is a linear isomorphism is clear.

(b) Suppose that $X$ is Hausdorff. For $p \in P$, let $N_p$ denote the closed subspace $p^{-1}(0)$ and, for any $x \in X$, $\dot{p}(\dot{x}) = \inf p(x + N_p)$. Let $X_p$ denote the normed space $(X/N_p, \dot{p})$. For each $p \in P$, let $Y_p$ denote a completion of $X_p$ as a Banach space [Theorem 5.11.3]. Let $k_p$ denote the canonical map $x \mapsto \dot{x}$ from $X$ onto $X_p$ and let $V_p = \{x \in X : p(x) < 1\}$ denote the open unit ball determined by $p$ and note that $V_p = k_p^{-1}(V_{\dot{p}})$. Since a base $\mathcal{B}$ at 0 for the topology for $X$ is given by finite intersections of the unit balls $V_p$, $\mathcal{B}$ is a base at 0 for the inverse image topology for $X$ determined by the maps $k_p$ $(p \in P)$ or, equivalently [Example 4.7.1] as the inverse image topology for $X$ determined by the map

$$\begin{array}{rcl} A : X & \to & A(X) \subset \prod_{p \in P} Y_p \\ x & \mapsto & (k_p(x)) \end{array} \tag{3}$$

That $A$ is 1-1 follows immediately from the fact that $X$ is Hausdorff ($x \neq 0 \Rightarrow$ there exists $p \in P$ such that $p(x) \neq 0 \Rightarrow x \notin N_p$). The openness of $A$ follows from the fact that $A$ is onto $A(X)$ [Example 4.7.1]. ■

The converse of each part of Theorem 5.11.4 is apparent. It is worth noting that an infinite product of normed spaces is never normable [Theorem 6.4.5].

**Theorem 5.11.5** COMPLETION OF LCS Every LCS $X$ possesses a completion which is a LCS.

**Proof.** Let $X$ be a LCS and let $P$ be the collection of continuous seminorms on $X$. Each seminormed space $(X, p)$, $p \in P$, has a completion $Y_p$ which is also a seminormed space by Theorem 5.11.3. By Theorem 3.5.4, the product $\Pi_{p \in P} Y_p$ is complete. It is locally convex by Theorem 5.5.3. As in Theorem 5.11.4(a), the map $A : X \to Y_p$, $x \mapsto (x_p)$, where $x_p = x$ for each $p \in P$ is a linear homeomorphism. The closure of $A(X)$ in $\Pi_{p \in P} Y_p$ is therefore the desired completion of $X$. ■

Let $(X_s)_{s \in S}$ be a family of LCS. For each $s \in S$, let $P_s$ be the family of continuous seminorms on $X_s$. As we have already observed [Theorem 5.5.3], the product $\Pi_{s \in S} X_s$ is locally convex. With $\mathrm{pr}_s$ denoting the projection of $\Pi_{s \in S} X_s$ onto $X_s$, the topology is generated by the seminorms $\{p \circ \mathrm{pr}_s : s \in S,\ p \in P_s\}$. This result is generalized in Example 5.11.6.

**Example 5.11.6** LOCALLY CONVEX INVERSE IMAGE TOPOLOGY AND SEMINORMS Consider a linear space $X$, a family $\{X_s : s \in S\}$ of locally convex spaces and a family of linear maps $A_s : X \to X_s$. Let $\mathcal{T}$ denote the inverse image topology determined by $\{A_s : s \in S\}$, the weakest (necessarily locally convex) topology with respect to which each $A_s$ is continuous [Example 4.7.1]. Letting $P_s$ be the family of continuous seminorms on $X_s$ then for each index $s$, we may take $\mathcal{B}_s$ to be the filterbase $\{V_p : p \in P_s\}$ as a base at 0 in $X_s$. The subbasic neighborhoods of the inverse image topology are $U_s = A_s^{-1}(V_p) = (p \circ A_s)^{-1}([0,1))$. It follows that $\mathcal{T}$ is determined by the seminorms $p \circ A_s$ where $s \in S$ and $p \in P_s$. ■

**Example 5.11.7** LOCALLY CONVEX DIRECT IMAGE TOPOLOGY AND SEMINORMS Let $\{(X_s, \mathcal{T}_s) : s \in S\}$ be a family of locally convex subspaces of the vector space $X$ and let $I_s : X_s \to X$ denote the canonical injection $x \mapsto x$ for each index $s$. Let $P$ denote the collection of seminorms $p$ on $X$ with respect to which the restriction of $p$ to each $X_s$ is $\mathcal{T}_s$-continuous. As pointed out in Example 4.11.3(b), the direct image locally convex topology $\mathcal{T}_c$ for $X$ determined by the family $\{I_s : s \in S\}$ has a neighborhood base at 0 given by the filterbase $\mathcal{D}$ of all absorbent disks $D \subset X$ such that $I_s^{-1}(D) = D \cap X_s$ is a neighborhood of 0 in $X_s$ for each $s \in S$. It follows that $\mathcal{T}_c$ is generated by $P$. ■

## 5.12 ORDERED VECTOR SPACES

**All vector spaces in this section are real.**

The most important spaces of functional analysis are spaces of functions. When they consist only of real-valued functions, a natural ordering is available: namely, $x \leq y$ iff $x(t) \leq y(t)$ for each $t$ in the domain. It would be profligate not to avail ourselves of the information that the ordering has to offer, just as we help ourselves to complex function theory when dealing with complex spaces. We begin a brief introduction to the large subject of ordered vector spaces by promulgating the criteria that link the order to the linear structure.

An ordered vector space $X$ is *order complete* if every majorized subset has a least upper bound. The deepest result of the section characterizes those topological spaces $T$ for which the pointwise ordered space $C(T, \mathbf{R})$ of continuous functions is order complete. Roughly, it says that $C(T, \mathbf{R})$ is order complete iff $T$ is "extremally disconnected."

A *preorder* $\leq$ on a set $T$ is a reflexive ($t \leq t$ for every $t \in T$), transitive ($r \leq s$ and $s \leq t \Rightarrow r \leq t$ for all $r, s, t \in T$) binary relation, while an *order* is an antisymmetric ($r \leq s$ and $s \leq r \Rightarrow r = s$ $r, s \in T$) preorder. If $(T, \leq)$ is preordered, $S \subset T$ is *bounded above* by $t \in T$ if $s \leq t$ for each $s \in S$; we also say that $t$ is an *upper bound* for $S$. Analogous conventions apply to *lower bounded* and *bounded below*. $S \subset T$ is *bounded* if it is bounded above and below. An upper bound $t$ for $S$ is the *least upper bound* or *supremum* for $S$ if, for any upper bound $u$ for $S$, $t \leq u$; we write $t = \sup S$. Analogous conventions apply to *greatest lower bound* or *infimum* $t$ of $S$, denoted $t = \inf S$. For two-point sets $\{x, y\}$, we write and $x \vee y = \sup(x, y)$ and $x \wedge y = \inf(x, y)$. A preorder or order on a real vector space $X$ is said to be a *vector preorder* or *order*, respectively, or to be *compatible* with the linear structure if the order is preserved by translation and multiplication by positive numbers: For all $x, y$, and $z$ in $X$:

(a) $x \leq y$ implies $x + z \leq y + z$; and

(b) $x \leq y$ implies $ax \leq ay$ for any $a > 0$.

A vector space $X$ with a compatible preorder $\leq$ is called a *preordered vector space*. Sets such as $[x, y] = \{w \in X : x \leq w \leq y\}$ and $(x, y) = \{w \in X : x < w < y\}$, etc., are called *order intervals.* A subset $S$ which is contained in an order interval is called *order bounded.* It is easy to verify that order intervals are convex and that, for $x \geq 0$, $[-x, x]$ is balanced.

In Example 5.12.1, we discuss some examples which involve the so-called pointwise order.

**Example 5.12.1** Pointwise Ordering

(a) Let $T$ be any set and $X$ any linear space of real-valued functions on $T$. For $x, y \in X$, define $x \leq y$ iff $x(t) \leq y(t)$ for each $t$ in $T$. This is called the *pointwise order* for $X$.

(b) Particular spaces that are usually ordered in this manner include the spaces $\ell_\infty(T, \mathbf{R})$ of bounded real-valued maps on the set $T$ and $C(T, \mathbf{R})$ of continuous maps of the topological space $T$ into $\mathbf{R}$. For the discrete space $T = \{1, 2, \ldots, n\}$, $\mathbf{R}^n$ is a special case of or $C(T, \mathbf{R})$. As other special cases we have $\ell_\infty^{\mathbf{R}}$ and the subspace $c_0$ of null real sequences.

(c) Let $L_\infty(\mathbf{R}, \mathbf{R})$ denote the space of almost-everywhere-bounded real-valued Lebesgue measurable functions mapping $\mathbf{R}$ into $\mathbf{R}$. A vector preordering is defined by taking, for $x, y \in L_\infty(\mathbf{R}, \mathbf{R})$, $x \le y$ iff $x(t) \le y(t)$ almost everywhere; this is easily generalized to more general spaces of measurable real-valued functions with respect to a positive measure. We refer to this preorder as the *almost everywhere pointwise preorder.* If we identify functions that are equal a.e., $\le$ is an order. ■

**Theorem 5.12.2** BASIC PROPERTIES In a preordered vector space $(X, \le)$, for each $x, y$ in $X$:

(a) $x \ge 0$ and $y \ge 0$ imply $x + y \ge 0$;
(b) $x \le y$ iff $-y \le -x$;
(c) $x \le y$ and $a < 0$ imply that $ax \ge ay$.

**Proof.** We use the notation of the statement. (a) $x + y \ge 0 + y \ge 0 + 0$.
(b) $x \le y \Rightarrow 0 \le y - x \Rightarrow -y \le -x$.
(c) For $x \le y$ and $a < 0$, $(-a)(y - x) \ge 0$. ■

The set $P = \{x : x \ge 0\}$ of *positive* elements has the following three properties:

(1) $P + P \subset P$;

(2) $aP \subset P$ for $a \ge 0$; and

(3) $P \cap -P = \{0\}$ (where $-P = \{-x : x \in P\}$) iff $\le$ is an order.

[To prove (3), note that if $P \cap -P = \{0\}$, $x \le y$ and $y \le x$ then $y - x \in P \cap -P$. Conversely, if $\le$ is an order and $x \in P \cap -P$ then $x \ge 0$ and $x \le 0$; thus $x = 0$.]

Generally, when a subset $C$ of a vector space has properties (1) and (2), it is called a *cone* (with *vertex* 0). (Many authors call such a thing a *convex* cone or a *wedge*.) In accordance with this terminology, we refer to the set $P$ of positive elements of an ordered vector space as the *positive cone*. If a cone $P$ has been specified in a real vector space, a compatible preordering that has $P$ as its set of positive elements is produced by taking $y \ge x$ to mean that $y - x \in P$, as is trivial to verify. Put another way, on any real vector space $X$, there is a 1-1 correspondence between vector preorders and cones. Because of this, one commonly sees expressions such as "let $X$ be ordered by the positive cone $P$." A case in which this is the most natural way to do things appears in the following example.

**Example 5.12.3** DUAL CONE AND DUAL ORDERING If $(X, \le)$ is an ordered vector space, we say that a linear functional $f$ on $X$ is *positive* if $x \ge 0$ implies that $f(x) \geq 0$ for each $x \in X$. (Integrals with respect to positive measures are positive linear functionals if the space of measurable functions carries the pointwise or pointwise almost everywhere ordering.) By the linearity of $f$, this is equivalent to $f$ being *order preserving*, i.e., $x \leq y$ implies $f(x) \leq f(y)$. If $X^*$ denotes the linear space of all linear functionals on $X$ then the set $P^*$ of all positive linear functionals on $X$ is a cone in $X^*$ called the *dual cone*. The preorder induced by it is called the *dual preorder* for $X^*$. ■

The real numbers are "Archimedean ordered" in the sense that for any positive number $a$, $na \to \infty$ or, equivalently, $1/n \to 0$. The generalization of this is contained in the following definition.

**Definition 5.12.4** ARCHIMEDEAN ORDERS; ORDER UNIT A preordered vector space $(X, \leq)$ is *Archimedean preordered* if the only elements $x \in X$ for which $\{nx : n \in \mathbf{N}\}$ is bounded above are $x \leq 0$. If, as happens with $\mathbf{R}$, there is a positive element $u$ in the preordered space $X$ such that each vector $x \in [-ru, nu] = r[-u, u]$ for some $r > 0$—in other words, $[-u, u]$ is absorbent—then $u$ is called an *order unit.* ■

**Example 5.12.5** ARCHIMEDEAN ORDERS

(a) Each of the pointwise ordered spaces of Example 5.12.1 are Archimedean ordered.

(b) LEXICOGRAPHIC ORDER The lexicographic order for $\mathbf{R}^2$ is that induced by the positive cone $P = \{(a, b) : a > 0 \text{ or } a = 0 \text{ and } b \geq 0\}$. It is not Archimedean since $r(0, 1) \leq (1, 1)$ for each positive $r$. ■

**Example 5.12.6** ORDER UNITS (a) ORDER UNITS FOR $\ell_\infty(T, \mathbf{R})$ The function $u$ which is identically 1 is an order unit for the pointwise ordered space $\ell_\infty(T, \mathbf{R})$ [Example 5.12.1(b)]; indeed, any positive constant function is an order unit. Likewise, if $T$ is compact then $C(T, \mathbf{R})$ has positive constants as order units.

(b) ORDER UNITS FOR $\mathbf{R}^T$ Consider the pointwise ordered set $\mathbf{R}^T$ of all functions from the set $T$ into $\mathbf{R}$. If $T$ is finite, consisting of $n$ elements say then the $n$-tuple $u = (1, 1, \ldots, 1)$ is an order unit. If $T = \mathbf{N}$ then consider a positive element $x = (a_n)$ of $\mathbf{R}^\mathbf{N}$ such that $a_n \to \infty$. As $e^{a_n}/a_n \to \infty$, there cannot be a positive integer $k$ such that $e^{a_n} \leq ka_n$ for sufficiently large $n$ and $x$ cannot be an order unit. Generally, $\mathbf{R}^T$ has an order unit iff $T$ is finite.

(c) NO ORDER UNIT FOR $C(\mathbf{R}, \mathbf{R})$ OR $L_p[0, 1]$ for $1 \leq p < \infty$ By arguments similar to that of (b), $C(\mathbf{R}, \mathbf{R})$ has no order unit and neither do the spaces $L_p[0, 1]$ for $1 \leq p < \infty$. ■

**Theorem 5.12.7** ARCHIMEDEAN ORDER A preordered vector space $(X, \leq)$ with an order unit $u$ is Archimedean preordered iff $nx \leq u$ for each $n \in \mathbf{N}$ implies that $x \leq 0$.

**Proof.** We use the notation of the statement. The condition is clearly necessary. Conversely, suppose that the condition holds and that for some $x, y \in X$, $nx \leq y$ for each $n$ in $\mathbf{N}$. Since $u$ is an order unit, there is some positive integer $k$ such that $y \leq ku$; thus $(n/k)x \leq u$ for all $n \in \mathbf{N}$. By letting $n = mk$, $m \in \mathbf{N}$, it follows that $mx \leq u$ for all $m \in \mathbf{N}$. The condition implies that $x \leq 0$. ■

As shown in our next two results, order units generate sublinear functionals and, in the case of Archimedean ordered spaces, norms.

**Theorem 5.12.8** A SUBLINEAR FUNCTIONAL On any preordered (real) vector space $X$ with order unit $u$, the map $p(x) = \inf\{t \in \mathbf{R} : x \leq tu\}$ $(x \in X)$ is a sublinear functional.

**Proof.** We use the notation of the statement. As to the positive homogeneity of $p$, consider $p(0x) = p(0) = \inf\{t \in \mathbf{R} : 0 \leq tu\} = 0$. For $x \in X$, let $P(x) = \{t \in \mathbf{R} : x \leq tu\}$ so that $p(x) = \inf P(x)$. It is easy to see that, for $a > 0$, $P(ax) = aP(x)$ which implies that $p(ax) = ap(x)$. Likewise, it is easy to verify that, for any $x, y \in X$, $P(x) + P(y) \subset P(x+y)$ which yields the subadditivity by standard properties of infima. ■

**Theorem 5.12.9** ORDER UNIT NORM On any Archimedean ordered (real) vector space $X$ with order unit $u$, let $U$ denote the absorbent disk $[-u, u]$. Then the gauge $p_U$ of $U$ $[p_U(x) = \inf\{r > 0 : x \in r[-u, u]\},\ x \in X]$ is a norm called the *order unit norm*; furthermore $p_U(u) = 1$ and the closed ball $\overline{V}_{p_U}$ determined by $p_U$ is the order interval $[-u, u]$.

**Proof.** We use the notation of the statement. As a gauge, $p_U$ is a seminorm [Theorem 5.3.1]. We now use the Archimedean nature of the ordering of $X$ to show that $-p_U(x)u \leq x \leq p_U(x)u$ for any $x \in X$, from which it follows that $p_U(x) = 0$ implies that $x = 0$, i.e., that $p_U$ is a norm. For any $n \in \mathbf{N}$ and $x \in X$, $x \leq [p_U(x) + 1/n]u$ so $nx \leq np_U(x)u + u$ or $n(x - p_U(x)u) \leq u$. Since the ordering is Archimedean, $x - p_U(x)u \leq 0$ or $x \leq p_U(x)\,u$. Replacing $x$ by $-x$ and using the fact that $p_U$ is a seminorm, this implies that $-x \leq p_U(x)\,u$ from which the result follows. Clearly $p_U(u) \leq 1$. If $p_U(u) = r < 1$ then $u \leq ru$ which implies that $u \leq 0$, a contradiction. The assertion $\overline{V}_{p_U} = [-u, u]$ is straightforward to verify. ■

**Example 5.12.10** ORDER UNIT NORM FOR $\ell_\infty(T, \mathbf{R})$ Let $T$ be any set and $X = \ell_\infty(T, \mathbf{R})$ be the pointwise Archimedean ordered linear space of all bounded real-valued functions on $T$ (or any linear subspace thereof) containing the order unit $u$, $u(t) \equiv 1$, normed by

$$\|x\|_\infty = \sup|x(T)| = \max(\sup x(T), -\inf x(T)),\ x \in X$$

Let $p_U$ be the gauge of $U = [-u, u]$. Since $\|x\|_\infty \leq 1$ iff $-1 \leq x(t) \leq 1$ for each $t \in T$ and this is equivalent to $-u \leq x \leq u$, $\|x\|_\infty \leq 1$ iff $p_U(x) \leq 1$. Hence

$p_U$ and $\|\cdot\|_\infty$ determine the same unit balls which means that $\|\cdot\|_\infty = p_U$ [Theorem 5.2.1(a)]. We can also recover $\|\cdot\|_\infty$ from the order by means of the sublinear functional $p(x) = \inf\{t \in \mathbf{R} : x \leq tu\}$ of Theorem 5.12.8: $\|x\|_\infty = \max(p(x), p(-x))$. ■

Thus, the closed balls of $C(T, \mathbf{R}, \|\cdot\|_\infty)$ ($T$ compact), $(c_0, \|\cdot\|_\infty)$, $\ell_\infty^{\mathbf{R}}(n)$, $\ell_\infty^{\mathbf{R}}$, and $L_\infty(\mathbf{R}, \mathbf{R})$ are order intervals. Note, however, that while every closed ball is an interval in an Archimedean ordered space with order unit, not every order interval is a closed ball. In $\ell_\infty^{\mathbf{R}}(2)$ with pointwise ordering, for example, with $w = (0, 1)$ and order unit $u = (1, 1)$, the order interval $[w, u] = [0, 1] \times \{1\}$, is not a closed $\|\cdot\|_\infty$-ball. [Generally, an order interval $[x, y]$ is a closed ball with respect to $\|\cdot\|_\infty$ iff $y - x = ru$, for some $r > 0$.]

Many of the ordered pointwise spaces $X$ have the additional property that $\sup(x, y) \in X$ for each $x, y \in X$. In particular, if $X = \ell_\infty(T, \mathbf{R})$ or $C_b(T, \mathbf{R})$, bounded and bounded continuous functions, respectively, on the set $T$ or the topological space $T$ then $X$ has this property. We define supremum and infimum in ordered vector spaces in formally the same way as for subsets of $\mathbf{R}$. As we show in Theorem 5.12.12, if $\sup(x, y) \in X$ exists for $x, y \in X$, so does $\inf(x, y)$.

**Definition 5.12.11** VECTOR LATTICE A preordered linear space $X$ which contains the supremum of each pair of its elements is called a *(vector) lattice* or *Riesz space*. ■

Other synonyms for vector lattices are *k-lineal spaces* in the Russian literature and *semiordered linear spaces* in the terminology of Nakano and his school.

$C(T, \mathbf{R})$ is a vector lattice (pointwise suprema) but, as the sup of differentiable functions need not be differentiable, the subspace $D([0, 1], \mathbf{R})$ of differentiable functions on $[0, 1]$ is not; $\ell_\infty(T, \mathbf{R})$ is a vector lattice as well.

**Theorem 5.12.12** SUPS AND INFS Let $(X, \leq)$ be a preordered vector space. For any $x, y \in X$:

(a) $x \geq y$ iff $x = \sup(x, y)$ and iff $y = \inf(x, y)$;

(b) if $\sup(x, y)$ exists then $\inf(-x, -y)$ exists; moreover $\inf(-x, -y) = -\sup(x, y)$;

(c) $\sup(x, y) \in X$ iff $\inf(x, y) \in X$ and if either of them exists then, for all $w \in X$, $\sup(x+w, y+w) = \sup(x, y)+w$ and $\inf(x+w, y+w) = \inf(x, y)+w$; moreover $x + y = \inf(x, y) + \sup(x, y)$. If $\sup(x, 0)$ exists for each $x$ in $X$ then $X$ is a vector lattice.

**Proof.** We use the notation of the statement. (a) Obvious.

(b) If $z = \sup(x, y)$ then $-z = \inf(-x, -y)$ as is straightforward to verify.

(c) If $\sup(x, y) = z$ then $z+w$ is an upper bound for $\{x+w, y+w\}$ for any $w$ in $X$. If $v$ majorizes $\{x+w, y+w\}$ then $v-w$ majorizes $\{x, y\}$ so $v-w \geq z$ or $v \geq z + w$. Thus $z + w = \sup(x, y) + w = \sup(x + w, y + w)$. An identical sort of argument proves the dual statement about infima. Now suppose that $\sup(x,y)$ exists and let $w = -x - y$ in the previous equality. This yields $\sup(-x,-y) = \sup(x,y) - x - y$. Since $\sup(-x,-y) = -\inf(x,y)$ by (b) we get that $\inf(x,y) = x + y - \sup(x,y)$, the desired result. Finally, if $\sup(x,0)$ exists for each $x \in X$ then $\sup(x-y,0)$ exists and $\sup(x-y,0)+y = \sup(x,y)$. ■

**Definition 5.12.13** LATTICE ISOMORPHISM A linear map $A : X \to Y$ between vector lattices $X$ and $Y$ is called a *lattice homomorphism* or *Riesz homomorphism* if $A$ preserves suprema and infima of two-point sets. An injective lattice homomorphism is a *lattice isomorphism.* ■

We omit the simple proof of the following result.

**Theorem 5.12.14** LATTICE HOMOMORPHISM A linear map $A : X \to Y$ between vector lattices $X$ and $Y$ is a *lattice homomorphism* iff any of the following conditions hold:

(a) $A$ is order preserving;
(b) $A$ is *positive* in the sense that $x \geq 0 \Rightarrow Ax \geq 0$;
(c) $A$ preserves suprema of two-point sets;
(d) $A$ preserves infima of two-point sets.

As is well known, Dedekind's way to complete the rationals was to supply suprema for all bounded sets. With that in mind, we now make the following definition.

**Definition 5.12.15** ORDER COMPLETENESS An ordered (real) vector space $X$ is *order complete* or *Dedekind complete* (or to be a *complete vector lattice* or *complete Riesz space*) if every nonempty majorized set has a supremum. ■

As observed in Theorem 5.12.12(b), if $B$ is a two-point subset of an ordered vector space which possesses a supremum then $\sup(-B) = -\inf B$ which implies that $\inf(-B) = -\sup B$. Hence each nonempty minorized subset of an ordered space has an infimum iff each majorized subset has a supremum.

**Example 5.12.16** ORDER COMPLETENESS (a) The vector lattice $C([0,2],\mathbf{R})$ of real-valued continuous functions on $[0,2]$ with pointwise ordering is not order complete: The set $x_n(t) = t^n$, for $0 \leq t < 1$ and $x_n(t) = 1$ for $1 \leq t \leq 2$, $n \in \mathbf{N}$, for example, is decreasing, so its pointwise limit is its infimum; yet the infimum is discontinuous.

(b) The pointwise ordered vector space $\mathbf{R}^{\mathbf{N}}$ of all real sequences is order complete—the supremum of any upper-bounded set being just the pointwise supremum. The pointwise ordered linear space $\mathbf{R}^T$ of all real functions on the set $T$ is order complete by essentially the same argument as for $\mathbf{R}^{\mathbf{N}}$.

(c) $\ell_\infty(T, \mathbf{R})$ is order complete. The subspace $\varphi$ of $\mathbf{R}^{\mathbf{N}}$ of sequences $(t_n)$ such that almost all $t_n = 0$, is not order complete: The bounded sequence of sequences $x_n \in \varphi$ where $x_n$ has 1's in the first $n$ places and 0's thereafter has no supremum in $\varphi$. ■

Our next goal is to characterize order-complete spaces of continuous functions $C(T, \mathbf{R})$. To do so, we need the notion of an *extremally disconnected* space, a topological space in which each open set has open closure, as, for example, any discrete space. For an alternative characterization of extremal disconnectedness:

**Theorem 5.12.17** EXTREMAL DISCONNECTEDNESS A topological space $T$ is extremally disconnected iff for all pairs of disjoint open subsets $A$ and $B$, $\operatorname{cl} A \cap \operatorname{cl} B = \varnothing$.

**Proof.** We use the notation of the statement. Suppose that $T$ is extremally disconnected and that $A$ and $B$ are disjoint open subsets. Then $B$ is contained in the closed set $\complement A$; hence $\operatorname{cl} B \subset \complement A$, so $\complement \operatorname{cl} B \supset A$. Since $\operatorname{cl} B$ is open, this implies that the closed set $\complement \operatorname{cl} B \supset \operatorname{cl} A$ and therefore that $\operatorname{cl} A \cap \operatorname{cl} B = \varnothing$.

Conversely, suppose that the condition holds and let $G$ be open. $G$ and $\complement \operatorname{cl} G$ are then disjoint open sets; therefore $\operatorname{cl} G \cap \operatorname{cl}\left(\complement \operatorname{cl} G\right) = \varnothing$ as well. Hence $\operatorname{cl}\left(\complement \operatorname{cl} G\right) \subset \complement \operatorname{cl} G$, which implies that $\complement (\operatorname{cl} G)$ is closed. ■

Since a normal Hausdorff space $T$ is extremally disconnected iff its Stone–Cech compactification $\beta T$ is extremally disconnected [Engelking 1977, p. 453], it follows that $\beta\mathbf{N}$, $\mathbf{N}$ discrete, is extremally disconnected. Cartesian products of extremally disconnected spaces need not be extremally disconnected [Engelking 1977, p. 465] and $\beta\mathbf{N} \backslash \mathbf{N}$ is not extremally disconnected [Engelking 1977, p. 453], so the property is not hereditary either. Stone [1937a] implicitly introduced the notion of extremal disconnectedness and proved [1973b; 1940; 1941; 1949, Th. 17; cf. Kakutani 1941, Th. 12, and Semadeni 1971, p. 457ff.] that every complete boolean algebra is isomorphic as a Boolean algebra to the Boolean algebra of all clopen (= closed and open) subsets of an extremally disconnected compact Hausdorff space.

**Theorem 5.12.18** ORDER COMPLETENESS OF $C(T, \mathbf{R})$ If $T$ is extremally disconnected then the space $C(T, \mathbf{R})$ of continuous functions mapping $T$ into $\mathbf{R}$ with pointwise order is order complete. If $T$ is completely regular and $C(T, \mathbf{R})$ is order complete then $T$ is extremally disconnected.

**Proof.** We use the notation of the statement. First we assume that $T$ is extremally disconnected and show that $C(T, \mathbf{R})$ is order complete. If $E \subset C(T, \mathbf{R})$ is bounded below by $y$ then $E - y = \{x - y : x \in E\}$ is bounded below by 0. If $E - y$ has $w$ as an infimum then $\inf E = y + w$ [cf. Theorem 5.12.12(c)]. Hence it suffices to assume that for each $x \in E$, $x(t) \geq 0$ for all $t$ in $T$ and show that $E$ has an infimum.

For each $x \in E$ and positive $r$, let $G_{xr} = \{t \in T : x(t) < r\}$. Clearly, $G_{xr}$ is open and therefore $G_r = \cup_{x \in E} G_{xr}$ is open. Note too that $\{G_r : r > 0\}$ is increasing in the sense that $p < q$ implies that $G_p \subset G_q$ and that $T = \cup_{r>0} G_r$. Since $T$ is extremally disconnected, each $\operatorname{cl} G_r$ is open. Since $T = \cup_{r>0} \operatorname{cl} G_r$, we may define $y(t) = \inf\{r > 0 : t \in \operatorname{cl} G_r\}$ for each $t$ in $T$. We now show that $y$ is continuous and next that $y = \inf E$.

To show that $y$ is continuous, we show, for $0 \leq a < b$, that $y^{-1}((a, b))$ is open. Since $y^{-1}((a, b)) = y^{-1}([0, b)) \setminus y^{-1}([0, a])$, it suffices to show that $y^{-1}([0, b))$ is open and that $y^{-1}([0, a])$ is closed. Thus, since

$$y^{-1}([0, b)) = \bigcup_{r<b} \operatorname{cl} G_r \qquad \text{and} \qquad y^{-1}[(0, a]) = \bigcup_{r \leq a} \operatorname{cl} G_r = \operatorname{cl} G_a$$

$y$ is seen to be continuous. To see that $y$ is a lower bound for $E$, consider any $t$ in $T$. Either $y(t) = 0$ or there is some $r$ such that $0 < r < y(t)$. The latter condition implies that $t \notin \operatorname{cl} G_r$, i.e., that $x(t) \geq r$ for each $x$ in $E$. It follows that $x(t) \geq y(t)$ for every $x \in E$ or that $y$ minorizes $E$.

Finally, let $w$ be any lower bound for $E$ and let $t \in T$. For any $r > 0$, if $s \in G_r$, there exists $x$ in $E$ such that $x(s) < r$; hence $w(s) \leq x\,(s) < r$. Since $w$ is continuous, it follows $w(t) \leq r$ for any $t \in \operatorname{cl} G_r$. In other words, $w(t)$ is a lower bound for the set $\{r > 0 : t \in \operatorname{cl} G_r\}$. As $y(t) = \inf\{r > 0 : t \in \operatorname{cl} G_r\}$, $w(t) \leq y(t)$. Since $t$ is arbitrary, $w \leq y$.

Now suppose that $T$ is completely regular and that $C(T, \mathbf{R})$ is order complete. Let $G$ be any proper nonempty open subset of $T$ and let $E = \{x \in C(T, \mathbf{R}) : x(T) \subset [0, 1] \text{ and } x(\complement G) = \{0\}\}$. The function $u$, $u(t) \equiv 1$, is an upper bound for $E$, so $E$ has a supremum, $y$ say. By the complete regularity of $T$, for each $t \in G$ there is some $x \in E$ such that $x(t) = 1$. Hence $y$ is identically equal to 1 on $G$. Since $y$ is continuous, $y$ is identically 1 on $\operatorname{cl} G$ as well. For $s \notin \operatorname{cl} G$, there exists $z \in C(T, [0, 1])$ such that $z(s) = 0$ and $z = 1$ on $\operatorname{cl} G$. Hence $z$ is an upper bound for $E$ and therefore $z \geq y \geq 0$. This, however, implies that $y(s) = 0$, from which we conclude that $y$ is identically 0 on $\complement \operatorname{cl} G$. Thus, the continuous function $y$ is the characteristic function of $\operatorname{cl} G$; hence $y^{-1}\,(1/2, 3/2) = \operatorname{cl} G$ is open. ■

We note that the result above implies that $\mathbf{R}^n$ is order complete with respect to the pointwise ordering since $\mathbf{R}^n = C(\{1, 2, \ldots, n\}, \mathbf{R})$ with $\{1, , \ldots, n\}$ endowed with the discrete topology.

In any vector lattice, $|x| = \sup(x, -x)$ is called the *absolute value* of $x$; note that the range of $|\cdot|$ is in $X$, not $\mathbf{R}$, however. For all $x, y \in \mathbf{R}$:

(a) $|x| \geq 0$ and $= 0$ iff $x = 0$;

(b) $|ax| = |a|\,|x|$ for all $a \in \mathbf{R}$;

(c) $|x + y| \leq |x| + |y|$; and

(d) $||x| - |y|| \leq |x - y|$.

If $X$ has a norm as well and $|x| \leq |y|$ implies that $\|x\| \leq \|y\|$ [as happens with $C(T, \mathbf{R}, \|\cdot\|_\infty)$, $T$ compact, for example] then the norm is called a *lattice norm* and $X$ a *normed lattice*. If, in addition, $X$ is a Banach space then $X$ is called a *Banach lattice*. The linear space $L_\infty(\mathbf{R}, \mathbf{R})$ of all essentially bounded Lebesgue-measurable real-valued functions on $\mathbf{R}$ with ess sup as norm is also a Banach lattice when it carries the almost everywhere pointwise ordering [Example 5.12.1(c)]. $C(T, \mathbf{R}, \|\cdot\|_\infty)$, $T$ compact Hausdorff, and $L_\infty(\mathbf{R}, \mathbf{R})$ are each Banach lattices with the additional property

$$\|\sup(x, y)\| = \sup(\|x\|, \|y\|) \qquad \text{for } x, y \geq 0 \tag{M}$$

Generally, a Banach lattice which satisfies (M) is called an *M-space* or *abstract M-space* (*AM-space*), $C(T, \mathbf{R})$ and $L_\infty(\mathbf{R}, \mathbf{R})$, being concrete $M$-spaces.

If the Banach lattice $X$ is *additive on the* (positive) *cone*,

$$\|x + y\| = \|x\| + \|y\| \qquad \text{for } x, y \geq 0 \tag{L}$$

then $X$ is called an (*abstract*) *L-space*. $L_1(\mathbf{R}, \mathbf{R})$, for example, the linear space of all Lebesgue-measurable maps $x$ of $\mathbf{R}$ into $\mathbf{R}$ such that $\int_{\mathbf{R}} |x(t)|\, dt = \|x\|_1 < \infty$, with almost everywhere pointwise order, is an example of such a space.

A unit $u$ of an abstract $M$-space $X$ is a positive element such that $\|u\| = 1$ and $\|x\| \leq 1$ implies $x \leq u$; equivalently, a unit is the largest element of the closed unit ball. Kakutani [1941, p. 998, Th. 2] showed that if $X$ is an abstract $M$-space with a unit then there is a compact Hausdorff space $T$ and a surjective linear isometry $A : X \to C(T, \mathbf{R}, \|\cdot\|_\infty)$ such that $A(\sup(x, y)) = \sup(Ax, Ay)$ for all $x, y \in X$. Kakutani [1941, p. 1021, Th. 15] also showed that the continuous dual of an abstract $M$ space in the dual order [Example 5.12.3] is an abstract $L$-space, and vice versa (cf. Kelley and Namioka 1963, p. 239). For further discussion of Kakutani's results, see Schaefer [1971, p. 246ff.]; cf. also Peressini [1967, p. 113ff.], Semadeni [1971, p. 457ff], and Asimow and Ellis [1980, p. 67].

Finally, we consider a link between a vector ordering and a topology. Suppose that $X$ is a vector lattice. A subset $E$ of $X$ is *order convex* if for all $x, y$ in $E$, the order interval $[x, y] \subset E$.

**Definition 5.12.19** TOPOLOGICAL VECTOR LATTICE A vector lattice $X$ which is also a TVS (real, of course) is called a *topological vector lattice* if

(a) the cone of positive elements is closed; and

(b) there exists a base of order convex neighborhoods of 0. ■

The Banach lattices mentioned above are topological vector lattices [Peressini 1967, p. 61ff.]; other references to ordered vector spaces and topological vector lattices include Schaefer [1971, p. 235ff], Cristescu [1977, p. 125ff.], Nachbin [1965, p. 83ff.], and Jameson [1970, p. 78ff.]. An excellent general reference on Riesz spaces is Luxemburg and Zaanen 1971.

## 5.13 EXERCISES

5.101 For $x = \{t_1, \ldots, t_n) \in \mathbf{C}^n$, let $p_j(x) = |t_j|$, $j = 1, 2, \ldots, n$. Show that the topology determined by the family $(p_j)$ of seminorms is the usual Euclidean topology [cf. Exercise 5.203(e)].

5.102 SEMINORMS

(a) Let $p$ and $q$ be seminorms on $X$. Show that if $p(x) \leq 1$ implies $q(x) \leq 1$ for every $x \in X$ then $q \leq p$.

(b) If $p_1, \ldots, p_n$ are seminorms and $a$ and $b$ are positive numbers such that $\max_j p_j(x) < a$ implies $q(x) < b$ for every $x \in X$ then $q \leq (b/a)\sum_{j=1}^n p_j$.

(c) SEPARATION AND DOMINATION Let $f$ be a nontrivial linear functional on a real vector space $X$, let $H = f^{-1}(1)$ and let $p$ be a seminorm on $X$. With $V_p = \{x \in X : p(x) < 1\}$, show that $f \leq p$ iff $f^{-1}(1) \cap V_p = \varnothing$.

5.103 Let $X$ be a vector space. If $D$ is an absorbent disk in $X$ then there is an absorbent disk $G$ such that $G + G \subset D$.

5.104 SEMINORMS AND CLOSURE Let $X$ be an LCS whose topology is generated by a family $P$ of seminorms.

(a) For any subset $E$ of $X$, show that $x \in \operatorname{cl} E$ iff for any positive $r$ and $p_1, \ldots, p_n \in P$ there exists $y \in E$ such that $\sum_{j=1}^n p_j(x - y) < r$.

(b) $\operatorname{cl}\{0\} = \cap_{p \in P} p^{-1}(0)$.

5.105 GAUGES Let $A$ and $B$ be absorbent disks in the TVS $X$ with gauges $p_A$ and $p_B$, respectively. Then:

(a) If $t \neq 0$, $p_{tA} = (1/|t|)p_A$.

(b) If $A \subset B$ then $p_A \geq p_B$.

(c) If $p$ is a seminorm then $p$ is the gauge of every absorbent disk $D$ for which $\{x \in X : p(x) < 1\} \subset D \subset \{x \in X : p(x) \leq 1\}$. Give an example of an absorbent disk $D$ for which strict inclusions prevail on both sides.

(d) The gauge of the absorbent disk $D$ is a norm iff $D$ contains no subspace of positive dimension.

5.106 FINEST LOCALLY CONVEX TOPOLOGY Let the vector space $X$ carry the finest locally convex topology $\mathcal{T}_{lc}$ [Example 5.6.4], the topology generated by the collection of all seminorms on $X$. Show that:

(a) $\mathcal{T}_{lc}$ is Hausdorff.

(b) If $M$ is a subspace of $(X, \mathcal{T}_{lc})$ then $M$ is closed.

(c) Any linear map of $(X, \mathcal{T}_{lc})$ into any LCS is continuous.

5.107 CONTINUITY OF MULTILINEAR MAPS [cf. Theorem 5.7.3] Let $A$ be a multilinear map (i.e., linear in each argument when the others are held fixed) mapping a product $\Pi_{i=1}^n X_i$ of TVS $X_1, \ldots, X_n$ into a LCS $Y$ whose topology is determined by a family $Q$ of seminorms.

(a) $A$ is continuous iff for each seminorm $q$ in $Q$ there are seminorms $p_1, \ldots, p_n$ on $X_1, \ldots, X_n$, respectively, and $r > 0$ such that for all $(x_1, \ldots, x_n) \in \Pi_{i=1}^n X_i$, $q(A(x_1, \ldots, x_n)) \leq r p_1(x_1) \cdots p_n(x_n)$.

(b) $A$ is continuous iff for each seminorm $q$ in $Q$ there is some neighborhood of 0 in $\Pi_{i=1}^n X_i$ on which $q \circ A$ is bounded.

5.108 CONES Let $E$ be a subset of a real vector space $X$.

(a) If $E$ is a nonempty cone (i.e., $E + E \subset E$ and $rE \subset E$ for $r \geq 0$ as in Sec. 5.12) then the linear span $[E]$ of $E$ is the set $E - E = \{x - y : x, y \in E\}$.

(b) If $E$ is a cone then $E \cap (-E)$ is a subspace of $X$ and is the largest subspace contained in $E$.

(c) If $A$ is a linear map of $X$ into the real vector space $Y$ and $E$ is a cone in $X$ then so is $A(E)$; if $E$ is a cone in $Y$ then so is $A^{-1}(E)$.

(d) The cone $K(E)$ *generated* by $E$ is defined to be the intersection of all cones which contain $E$ (such as $X$, for example, or the linear span of $E$). $K(E)$ consists of all finite sums of the form $\sum_i a_i x_i$ where each $x_i \in E$ and each $a_i$ is a positive number.

(e) If $\{E_s : s \in S\}$ is a family of cones then the cone generated by their union consists of all finite sums of the form $\sum a_s x_s$, where $x_s \in E_s$ and each $a_s$ is positive.

5.109 EXTENDING SEMINORMS Let $p$ be a seminorm on a subspace $M$ of a TVS $X$. If $q$ is a seminorm on $X$ such that $q|_M \geq p$ then there is a seminorm $r$ on $X$ which reduces to $p$ on $M$ and is such that $r \leq q$.

5.110 COMPLETENESS

(a) Let $\mathcal{S}$ be a covering of the topological space $T$ and let $F_{\mathcal{S}}(T, \mathbf{F})$ denote the linear space of all functions mapping $T$ into the scalar field $\mathbf{F}$ with $\mathcal{S}$-topology $\mathcal{T}_{\mathcal{S}}$ [Example 2.6.2 and Sec. 3.2]. Show that $\mathcal{T}_{\mathcal{S}}$ is locally convex, Hausdorff, and that $F_{\mathcal{S}}(T, \mathbf{F})$ is complete.

(b) If $X$ is a vector space and $X^*$ its algebraic dual, the linear space of all linear functionals on $X$ then $X^*$ is a linear subspace of $F(X, \mathbf{F})$. Let $F(X, \mathbf{F}, p)$ denote $F(X, \mathbf{F})$ equipped with the point-open topology of Example 5.9.1. Show that $X^*$ is a complete subspace of $F(X, \mathbf{F}, p)$.

5.111 $C(\mathbf{R},\mathbf{R},c)$ METRIZABLE Let $C(\mathbf{R},\mathbf{R},c)$ be the LCS of continuous maps of $\mathbf{R}$ into $\mathbf{R}$ with the compact-open topology introduced in Example 5.5.6 and Sec. 5.8. Show that $C(\mathbf{R},\mathbf{R},c)$ is metrizable and complete.

5.112 METRIZABILITY OF $C(T,\mathbf{F},p)$ Let $C(T,\mathbf{F},p)$ be the linear space of continuous maps of the completely regular Hausdorff space $T$ into $\mathbf{F}$ with the point-open topology as in Example 5.9.1. Show that $C(T,\mathbf{F},p)$ is metrizable iff $T$ is countable.

5.201 HEMICOMPACTNESS A topological space $T$ is *hemicompact* if it has a countable family $(K_n)$ of compact subsets such that any compact subset of $T$ is contained in one of them. $\mathbf{R}$, for example, is hemicompact since any compact subset is contained in $K_n = [-n,n]$ for some $n \in \mathbf{N}$.

(a) HEMI- VERSUS $\sigma$-COMPACTNESS A topological space $T$ is *$\sigma$-compact* if there is a countable family $(K_n)$ of compact subsets which cover $T$. $\mathbf{R}$ in its usual topology, for example, is $\sigma$-compact since $\mathbf{R} = \bigcup_{n\in\mathbf{Z}}[n,n+1]$. Show that any hemicompact space is $\sigma$-compact and that the converse holds if $T$ is locally compact.

(b) Any locally compact second countable space is hemicompact.

(c) If $T$ is first countable and hemicompact then $T$ is locally compact.

(d) Give an example of a $\sigma$-compact space which is not hemicompact.

5.202 $k_R$-SPACES A *$k_R$-space* is a completely regular Hausdorff space $T$ in which the continuity of an $\mathbf{F}$-valued function is implied by its continuity on compact subsets of $T$ [Definition 5.8.6].

(a) On any first countable completely regular Hausdorff space, show that sequential continuity implies continuity.

(b) If sequential continuity implies continuity on the completely regular Hausdorff space $T$ then $T$ is a $k_R$-space.

(c) $k$-SPACES A completely regular Hausdorff space $T$ is a *$k$-space* if a subset $G$ of $T$ is open iff its intersection $G \cap K$ with each compact subset $K$ of $T$ is open in $K$. If $T$ is a $k$-space then $T$ is a $k_R$-space.

(d) Any locally compact or first countable completely regular Hausdorff space is a $k$-space.

5.203 FAMILIES OF SEMINORMS

(a) Let $S$ be a set and for each $s \in S$ let $P_s$ be a family of seminorms on a vector space $X$. Let $\mathcal{T}_s$ denote the topology generated by $P_s$ on $X$. Let $\mathcal{T}$ be the the topology generated by $P = \cup_{s \in S} P_s$. Show that $\mathcal{T} = \sup_{s \in S} \mathcal{T}_s$ [cf. Example 4.11.2].

(b) Let $P$ and $Q$ be families of seminorms on the vector space $X$ generating the topologies $\mathcal{T}_P$ and $\mathcal{T}_Q$, respectively. If $P \subset Q$ then $\mathcal{T}_P \subset \mathcal{T}_Q$.

(c) If $P$ and $Q$, as in (b), determine the same topology then $P$ is said to be *equivalent* to $Q$. Show that $P$ is equivalent to $Q$ iff for each $p \in P$ there are $q_1, \ldots, q_n$ ($n \in \mathbf{N}$) in $Q$ and $r > 0$ such that $p \leq r \max_j q_j$, and vice-versa for each $q$ in $Q$.

(d) If $P$ is a countable family of seminorms, show that there is an equivalent family $(q_n)$ of seminorms such that $q_1 \leq q_2 \leq \cdots$.

(e) The topology generated by a finite collection $p_1, \ldots, p_n$ of seminorms is the same as that generated by $\sum_{j=1}^{n} p_j$ and $\max_j p_j$ [cf. Exercise 5.101].

(f) DIRECTED FAMILY A family $P$ of seminorms is *directed* if for each $p, q \in P$, there is an $r \in P$ such that $\max(p, q) \leq r$. (This is sometimes called "filtrant.") Thus, "directed" is somewhat weaker than "saturated" [Definition 5.5.5]. Show that "base of continuous seminorms" in Theorem 5.11.2 may be replaced by "directed system of continuous seminorms."

5.204 ULTRASEMINORMS Let $X$ be a vector space over an ultravalued field $(K, |\cdot|)$. A seminorm $p$ on $X$ is defined in formally the same way as for real or complex spaces. If $p$ also satisfies $p(x+y) \leq \max[p(x), p(y)]$ for all $x$ and $y$ in $X$ then $p$ is said to be an *ultraseminorm* or a *non-Archimedean seminorm*.

(a) If $p$ is an ultraseminorm on $X$ then $p(x) \neq p(y) \Rightarrow p(x+y) = \max[p(x), p(y)]$ for any $x, y \in X$.

(b) If $p$ is an ultraseminorm then $V_p = \{x \in X : p(x) < 1\}$ is an absorbent $K$-disk [Exercise 4.202].

(c) If $X$ is a TVS over $K$ and $D$ is an absorbent $K$-disk in $X$ then the gauge of $D$, $p_D(x) = \inf\{|a| : x \in aD\}$, $x \in X$, is an ultraseminorm. Moreover [cf. Theorem 5.3.2]:

(d) $\{x \in X : p_D(x) < 1\} \subset D \subset \{x \in X : p_D(x) \leq 1\}$; and

(e) if $D$ has nonempty interior then $p_D$ is continuous (note the result of Exercise 4.202(f)).

5.205 LOCALLY $K$-CONVEX SPACES Let $X$ be a vector space over a nontrivially ultravalued field $(K, |\cdot|)$.

(a) ULTRASEMINORMS If $P$ is a family of ultraseminorms [Exercise 5.204] on $X$ then the (inverse image) topology generated by $P$ on $X$ is locally $K$*-convex* [Exercise 4.203]. Conversely, if $X$ is locally $K$-convex, the gauges (same formal definition) of any collection of disked neighborhoods of 0 which are a base at 0 are a collection of continuous ultraseminorms which generate the topology on $X$.

(b) CONTINUITY Let $X$ and $Y$ be locally $K$-convex spaces over $K$. The linear map $A : X \to Y$ is continuous iff for each continuous ultraseminorm $q$ on $Y$ there is a continuous ultraseminorm $p$ on $X$ and positive number $c$ such that $q \circ A \leq cp$.

(c) PSEUDOMETRIZABILITY If $X$ is locally $K$-convex then the following conditions are equivalent.

i. There is an ultrapseudometric $d$ [i.e., $d(x,y) \leq \max((d(x,z), d(z,y))$ for all $x, y, z \in X$] which generates the topology on $X$.

ii. There is a countable base for the topology at 0.

iii. There exists an increasing sequence $(p_n)$ of ultraseminorms which generates the topology on $X$.

If the locally $K$-convex space $X$ satisfies any of these conditions and is complete then $X$ is called a *non-Archimedean Fréchet space.*

5.206 LOCALLY C-COMPACT SPACES Let $X$ be a TVS over a nontrivially ultravalued field $(K, |\cdot|)$.

(a) [Cf. Exercise 4.116.] If $X$ is an ultranormed space [Exercise 5.204] and is *locally c-compact* in the sense that there exists a $c$-compact [Exercise 4.204] $K$-disked [Exercise 4.202] neighborhood of 0 then the underlying field $K$ is $c$-compact and $X$ is finite-dimensional.

(b) If $X$ is locally $K$-convex [Exercise 4.203] and locally $c$-compact then $X$ is $c$-compact.

## Hints to Exercises

5.109 Consider the sets $B_p = \{x \in M : p(x) \leq 1\}$ and $\overline{V}_q = \{x \in X : q(x) \leq 1\}$. Let $B = \{ax + (1-a)y : x \in B_p,\ y \in V_q\}$ and let $r$ be the gauge of $B$.

5.111 (a) Use Theorems 2.6.4 and 3.2.2.

5.201 (a) If $T$ is locally compact and $\sigma$-compact then there are compact sets $\{K_n : n \in \mathbf{N}\}$ such that $T = \cup_{n \in \mathbf{N}} K_n$. Show that there exists a relatively compact open set $U_1$ which contains $K_1$. Then show that there is a relatively compact open set $U_2$ which contains $\operatorname{cl} U_1 \cup K_2$. Produce a cover $(U_n)$ of $T$ of relatively compact open sets such that $\operatorname{cl} U_{n-1} \subset U_n$ for all $n \geq 2$. Show that any compact set $K$ is covered by a finite number of the $\operatorname{cl} U_n$.

(c) Let $\{K_n : n \in \mathbf{N}\}$ be an increasing cover of compact subsets of $T$ such that each compact subset of $T$ is contained in one of them. Choose a point $t \in T$ and let $(V_n)$ be a decreasing base of open neighborhoods of $t$. If no $\operatorname{cl} V_n$ is compact then $V_n$ is not a subset of $K_n$ for any $n$. For $t_n \in V_n \backslash K_n$, $t_n \to t$, so $\{t_n\} \cup \{t\}$ is compact. Therefore there is some $m$ such that $\{t_n\} \cup \{t\} \subset K_m$.

(d) The first countable space $\mathbf{Q}$ of rationals is not locally compact. Now use (c).

5.202 (b) If $t_n \to t$ then $\{t_n\} \cup \{t\}$ is compact.

(c) If $x : T \to \mathbf{F}$ is continuous on the compact subsets $K$ of $T$ then for each open subset $G$ of $\mathbf{F}, x^{-1}(G) \cap K = x|_K^{-1}(G)$ is open in $K$.

For further facts about $k$- and $k_R$-spaces, see Beckenstein et al. [1977, pp. 63–73, 116–124].

# Chapter 6

# Bounded Sets



The idea of bounded set in a TVS is a natural extension of what it is in a normed space:

> A subset $B$ of a normed space is bounded if it is contained in all sufficiently large multiples $kU$ of the unit ball $U$.
>
> A subset $B$ of a TVS is bounded if it is contained in all sufficiently large multiples $kV$ of any neighborhood $V$ of 0.

Boundedness has interesting connections with metrizability and continuity. For example: A TVS is seminormable iff it has a bounded convex neighborhood of 0 [Theorem 6.2.1]. Although normed spaces have "small" neighborhoods of 0, more general TVS do not: If a TVS has a bounded neighborhood of 0, it must be pseudometrizable. Unlike the situation for normed spaces, there are many metric TVS which do not have any bounded neighborhood of 0 [Examples 6.1.7 and 6.2.2].

A continuous map is often described as one which transforms small changes into small changes, despite the fact that the continuous map $t \mapsto 1/t$ transforms the interval $(0, 1)$ into $(1, \infty)$. A continuous *linear* map, however, of one TVS into another is "bounded" in the sense that it transforms bounded sets into bounded sets.

Must a bounded linear map be continuous? Generally, "no" [Example 13.2.8(c)], but it must be in a wide variety of special cases. For linear maps between normed spaces, for example, local boundedness implies continuity. More generally, if the domain is pseudometrizable and the range any TVS, local boundedness implies continuity [Theorem 6.5.2]. If the range is an LCS then the class of domains for which local boundedness implies continuity is the "bornological" spaces, a topic we take up in detail in Chap. 13.

## 6.1 BOUNDED SETS

A subset $B$ of a normed space $X$ is bounded if it is contained in a sufficiently large ball $C(0,r) = \{x \in X : \|x\| \leq r\}$ of radius $r > 0$ about 0. In other words, $B$ is bounded if it is contained in a sufficiently large multiple of the unit ball. If $A$ and $B$ are subsets of a TVS, we say that $A$ *absorbs* $B$ if $B \subset aA$ for all scalars $a$ of sufficiently large magnitude. Thus, a subset $B$ of a normed space $X$ is bounded if it is absorbed by the unit ball.

**Definition 6.1.1** BOUNDED SETS A subset $B$ of a TVS is said to be *bounded* (*beschränkt, borné*) if $B$ is contained in all sufficiently large multiples of any neighborhood $V$ of 0; thus, given $V$, there must be some $k > 0$ such that whenever $|a| \geq k$, $B \subset aV$. Equivalently, for all scalars $b$ of sufficiently small magnitude, $bB \subset V$. ■

For $B$ to be bounded, it suffices that it be absorbed by each neighborhood of a basic system of neighborhoods of 0. As there is a basis of balanced neighborhoods of 0 in any TVS, we can say that $B$ is bounded if for every balanced neighborhood $V$ of 0, there exists $a \in \mathbf{F}$ such that $B \subset aV$. In a LCS whose topology is generated by a family $P$ of seminorms, $B$ is bounded iff $p(B)$ is bounded for each $p$ in $P$ [Theorem 6.1.5]. The finer the topology, the fewer bounded sets there will be; if a set is bounded in one topology, it remains so in all coarser topologies. Finite sets are obviously bounded.

Is there a base of bounded neighborhoods at 0 in the general TVS? More basically: When is a neighborhood of 0 bounded? As it turns out (see Sec. 6.2), the existence of so much as one bounded neighborhood of 0 implies that the space is pseudometrizable. There can only be a bounded convex neighborhood of 0 in a seminormed space [Sec. 6.2].

A type of set frequently classified as "small" is the compact set. Our next result shows that compact sets are bounded. Recall that a subset $E$ of a topological group $X$ is totally bounded if, for all neighborhoods $V$ of 0, there exists a finite number of elements $x_1, \ldots, x_n$ from $X$ such that the sets $x_1 + V, \ldots, x_n + V$ cover $E$ [Definition 3.3.1]. As observed before Theorem 3.4.1, relatively compact sets are totally bounded and complete totally bounded sets are relatively compact [Theorem 3.4.1].

**Theorem 6.1.2** TOTALLY BOUNDED IMPLIES BOUNDED In any TVS $X$ totally bounded sets are bounded.

**Proof.** Let $B$ be totally bounded and let $V$ be a neighborhood of 0. Choose a balanced neighborhood $U$ of 0 such that $U + U \subset V$. Since $B$ is totally bounded, there is a finite subset $S$ of $X$ such that $B \subset S + U$; since $U$ is absorbent and balanced there is an $a \geq 1$ such that $S \subset aU$. Thus, by Theorem 4.1.9, for $|b| \geq a$, $B \subset bU + U \subset bU + bU \subset bV$, and it follows that $B$ is bounded. ■

In the very important case of weak topologies, the converse holds: weakly bounded implies weakly totally bounded [Theorem 8.2.8].

If $(x_n)$ is a Cauchy sequence in a TVS then $\{x_n : n \in \mathbf{N}\}$ is Cauchy bounded [Definition 3.3.4], hence totally bounded by Theorem 3.3.5. Therefore any Cauchy sequence (strictly speaking, the set of points which comprise the sequence) is bounded.

We mentioned above that boundedness and norm boundedness are identical notions in a normed space. In metrizable TVS this coincidence does not take place.

**Example 6.1.3** BOUNDEDNESS STRONGER THAN METRIC BOUNDEDNESS In a pseudometric space $(S, d)$, a set $B \subset S$ is *metrically bounded* or *d-bounded* if $\sup\{d(x, y) : (x, y) \in B \times B\} = d < \infty$. In this case $B$ is said to be of *finite diameter* and $d$ is called the *diameter* (or *d-diameter*) of $B$.

(a) METRICALLY BOUNDED $\not\Rightarrow$ BOUNDED In his thesis, Fréchet [1906, pp. 39–40] invented the following remarkable invariant metric for the linear space $(s)$ of all real (it works for complex as well) sequences with pointwise operations. For $x = (a_n)$, $y = (b_n) \in (s)$,

$$d(x, y) = \sum_{n \in \mathbf{N}} 2^{-n} \frac{|a_n - b_n|}{1 + |a_n - b_n|}$$

It follows as in the proof of Theorem 5.6.1 that, with each $p_n = p$, $d(x, 0) = p(x)$ is an $F$-norm and so induces a vector topology on $(s)$. Obviously $(s)$ is $d$-bounded by 1. It is not bounded, however. The proper neighborhoods of 0 are of the form $C(0, r)$, $0 < r < 1$, and we can choose $a > 0$ such that $a/(1 + a)$ is arbitrarily close to 1; by so doing, the sequence $(a)$ is such that $\sum_{n \in \mathbf{N}} 2^{-n} |a| / (1 + |a|) > r$ so $(s)$ is not contained in $C(0, r)$.

(b) BOUNDED IMPLIES METRICALLY BOUNDED If $(X, d)$ is a pseudometrizable TVS, there is an $F$-seminorm $p$ which generates the topology on $X$ by Theorem 4.8.3. By the triangle inequality, $p(kx) \leq kp(x)$ for any positive integer $k$. Hence for any ball about the origin, $C(0, r) = \{x \in X : p(x) \leq r\}$, $r > 0$, it follows that $kC(0, r) \subset C(0, kr)$. If $B \subset X$ is bounded then, for some $k \in \mathbf{N}$, $B \subset kC(0, 1) \subset C(0, k)$. Hence, for any $x$,$y \in B$, $p(x - y) \leq p(x) + p(y) \leq 2k$. Since we may assume that $p(x - y) = d(x, y)$, $B$ is metrically bounded. ■

In metrizable LCS, the notions of bounded and metrically bounded coincide [Theorem 8.8.7].

Komolgorov [1934, pp. 29–30] defined a subset $B$ of a TVS $X$ to be bounded (*beschränkt*) if, for any sequence $(a_n)$ of real numbers, $a_n \to 0$, and any sequence $(x_n)$ of vectors, $a_n x_n \to 0$, the same definition that had been introduced by Mazur and Orlicz [1933, p. 152] for metrizable TVS. Komolgorov immediately put the notion to use in proving that a TVS is seminormable iff it has a bounded convex neighborhood of 0 [Theorem 6.2.1(b)]. The version of boundedness we use, introduced by von Neumann [1935], is equivalent to Komolgorov's as we show in Theorem 6.1.4. For a detailed history of the "bounded set," see Przeworska-Rolewicz and Rolewicz [2005].

**Theorem 6.1.4** BOUNDEDNESS AND SEQUENCES If $B$ is a subset of a TVS $X$ then:

(a) $B$ is bounded iff for any sequence $(x_n)$ from $B$ and any sequence $(a_n)$ of scalars which tends to 0, $a_n x_n \to 0$;

(b) for $B$ to be bounded, it suffices that all denumerable subsets be bounded.

**Proof.** Let $X$ be a TVS, suppose that $B$ is bounded and let $V$ be a neighborhood of 0. Since $B$ is bounded, there is some $k > 0$ such that $aB \subset V$ for all scalars $a$ such that $|a| \leq k$. Hence, if $a_n \to 0$, $a_n B \subset V$ eventually; it follows that, for any sequence $(x_n)$ from $B$, $a_n x_n \in V$ eventually.

Conversely, if $B$ is not bounded, there must be some balanced neighborhood $U$ of 0 such that, for each $n \in \mathbf{N}$, there is some $x_n \in B$ such that $x_n \notin nU$. Thus $(1/n)x_n \nrightarrow 0$. and (a) follows. As the set $\{x_n : n \in \mathbf{N}\}$ is an unbounded denumerable subset of $B$, (b) follows. ■

If $p$ is a seminorm on $X$ and $V_p = \{x \in X : p(x) < 1\}$ then $B \subset aV_p$ iff $p(B) \subset [0, |a|)$. As a result of this observation, it is clear that we may characterize bounded subsets of locally convex spaces as follows.

**Theorem 6.1.5** BOUNDEDNESS IN LOCALLY CONVEX SPACES In a locally convex space $X$ whose topology is generated by a family of seminorms $P$, a subset $B$ is bounded iff $p(B)$ is bounded for each $p$ in $P$. By Theorem 6.1.4 then $B$ is bounded iff for any denumerable subset $D$ of $B$, $p(D)$ is bounded for each $p$ in $P$.

As an immediate application of Theorem 6.1.5, let $C(T, \mathbf{F}, c)$ denote the TVS of continuous scalar-valued functions on the topological space $T$ with compact-open topology $\mathcal{T}_c$ (Sec. 5.8), i.e., with topology defined by the seminorms $p_K(x) = \sup |x(K)|$, where $K$ is a compact subset of $T$. For a subset of $B \subset C(T, \mathbf{F}, c)$ to be $\mathcal{T}_c$-bounded, we must have $p_K(B) < \infty$ for each compact set $K$. Stated another way, for $B \subset C(T, \mathbf{F}, c)$ to be bounded, $B$ must be uniformly bounded on each compact subset of $T$.

As mentioned in Example 5.5.7, the weakened topology $\sigma(X, X')$ on a TVS $X$ is that generated by the seminorms $x \to |f(x)|$, $f \in X'$. Hence $B \subset X$ is $\sigma(X, X')$-bounded [i.e., weakly bounded] iff $|f(B)|$, is bounded for each $f \in X'$.

Generally subspaces of a TVS are not bounded because they contain lines $\mathbf{R}x$. Our next result characterizes bounded subspaces. Note that among its consequences is the fact that no nontrivial subspace of a Hausdorff TVS can be bounded.

**Theorem 6.1.6** BOUNDED SUBSPACES A subspace $M$ of a TVS $X$ is bounded iff $M \subset \text{cl}\{0\} = \cap V(0)$.

**Proof.** If $B \subset \text{cl}\{0\} = \cap V(0)$ then $B$ is a subset of any neighborhood $V$ of 0; therefore any subset of $\text{cl}\{0\}$ is bounded.

Conversely, suppose that $M$ is a bounded subspace and that $x$ is an element of $M$ which is not in $\text{cl}\{0\}$. Since $nx \in M$ for any positive integer $n$ and $M$ is bounded and $1/n \to 0$, Theorem 6.1.4 implies that $\lim(1/n)(nx) = 0$. It follows that $x \in \text{cl}\{0\}$ which is contradictory. (Since $X$ is not necessarily Hausdorff, sequences may have more than one limit.) ■

In Example 6.1.7 we exhibit three TVS which have unbounded basic neighborhoods of 0. The first two are metrizable spaces. In each case we demonstrate unboundedness of a set $V$ by showing that $V$ contains a nontrivial linear subspace.

**Example 6.1.7** UNBOUNDED NEIGHBORHOODS OF 0 (a) Let $C(\mathbf{R}, \mathbf{R}, c)$ be the LCHS of continuous maps of $\mathbf{R}$ into $\mathbf{R}$ with compact-open topology [Example 5.5.6]. A saturated family of seminorms [Definition 5.5.5] for $C(\mathbf{R}, \mathbf{R}, c)$ is given by $p_n(x) = \sup |x[-n, n]|$ $(x \in C(\mathbf{R}, \mathbf{R}, c)$, $n \in \mathbf{N})$. For any positive integer $n$, $V_{p_n} = \{x \in C(\mathbf{R}, \mathbf{R}, c) : p_n(x) \leq 1\}$ contains the nontrivial subspace $M_n = \{x \in C(\mathbf{R}, \mathbf{R}, c) : x([-n, n]) = 0\}$. It follows from Theorem 6.1.6 that *no* $V_{p_n}$ is bounded and therefore that no neighborhood of 0 in $C(\mathbf{R}, \mathbf{R}, c)$ is bounded.

(b) Let $(s)$ be the metric TVS of all sequences of complex numbers with the metric of Example 6.1.3(a). Given a basic neighborhood of 0, $B(0, r) = \{x \in (s) : d(x, 0) < r\}$, $0, < r < 1$, choose $n \in \mathbf{N}$ such that $2^{-n} < r$. Let $e_n$ be the sequence with zeros everywhere but the $n$th position where there is a 1. For any complex number $c$, $d(ce_n, 0) = 2^{-n}[|c| / (1 + |c|)] \leq 2^{-n} < r$. It follows that the subspace $\mathbf{C}e_n \subset B(0, r)$; therefore $B(0, r)$ is unbounded.

(c) WEAK NEIGHBORHOODS OF 0 UNBOUNDED Let $X$ be a TVS whose continuous dual $X'$, the linear space of all continuous linear forms on $X$, is infinite-dimensional. The weak topology $\sigma(X, X')$ for $X$ [Example 5.5.7(a)] has basic neighborhoods of 0 of the form $V(0, f_1, \ldots, f_n, r) = \{x \in X : |f_i(x)| < r,\ i = 1, \ldots, n\}$, for $f_1, \ldots, f_n \in X'$, $n \in \mathbf{N}$ and $r > 0$. Let $N = \cap_{i=1}^n f_i^{-1}(0)$. Since $X'$ is infinite-dimensional there exists $g \in X'$ linearly independent of $f_1, \ldots, f_n$; consequently, there exists $x \in N$ such that $g(x) = 1$ [Theorem 1.4.1(f)] so $x \neq 0$. Therefore $\mathbf{R}x \subset V(0, f_1, \ldots, f_n, 1)$. Could $\mathbf{R}x$ be bounded? If so then $\mathbf{R}x \subset \cap V(0)$ [Theorem 6.1.6]. Since $g(2x) = 2$, however, $\mathbf{R}x \not\subset V(0, g, 1)$. The weak topology is Hausdorff—$\text{cl}\{0\} = \{0\}$—iff $X'$ is total in the sense that for every nonzero $x \in X$, there exists $f \in X'$ such that $f(x) \neq 0$ [Theorem 5.5.1(a)], a condition satisfied by every LCHS [Theorem 7.7.7(b)]. ■

## 6.2 METRIZABILITY

This section has to do with the consequences of there being a bounded neighborhood of 0 in a TVS $X$. If there is a bounded neighborhood of 0 then $X$ is pseudometrizable; $X$ is seminormable iff it possesses a bounded *convex* neighborhood of 0. We have already established a result like this [Theorem 4.10.4]: A Hausdorff TVS $X$ has a compact neighborhood of 0 iff it is finite-dimensional and therefore normable by Theorem 4.10.3(c).

**Theorem 6.2.1** METRIZABILITY AND NORMABILITY Let $X$ be a TVS. Then:

(a) If $X$ has a bounded neighborhood of 0 then $X$ is pseudometrizable; the converse is false [Example 6.2.2].

(b) If $X$ has a bounded convex neighborhood of 0 then $X$ is seminormable.

If $X$ is a Hausdorff space then the "pseudo" and the "semi" may be omitted in (a) and (b), respectively.

**Proof.** Let $X$ be a TVS. (a) If $B \subset X$ is a bounded neighborhood of 0 and $V$ any neighborhood of 0 then $B \subset kV$ for some $k$ in $\mathbf{N}$. Hence $\{(1/n)B : n \in \mathbf{N}\}$ is a base at 0 in $X$ and $X$ is pseudometrizable by Theorem 4.8.3.

(b) Suppose now that $B$ is a bounded convex neighborhood of 0. Since $B$ must contain a balanced neighborhood of 0 [Theorem 4.3.6(d)] and since the convex hull of a balanced set is balanced, we may suppose that $B$ is a disk. We now show that the gauge $p_B$ of $B$ determines the topology on $X$.

Since $B$ is bounded, if $V$ is any neighborhood of 0, there is some positive $r$ such that $rB \subset V$. Hence, letting $V_{p_B}$ denote the open unit ball determined by $p_B$, $rV_{p_B} \subset rB \subset V$ [Theorem 5.3.2]. The positive multiples of $V_{p_B}$ are therefore a base at 0 for the topology on $X$.

The last statement follows directly from Theorem 4.8.3. ■

If $(X, p)$ is a seminormed space then positive multiples $rV_p$ of the open unit ball $V_p$ determined by $p$ are a base at 0. In particular, since any neighborhood $V$ of 0 must contain some $rV_p$, i.e., $V_p \subset (1/r)V$, $V_p$ is a bounded convex neighborhood of 0. In other words, the existence of a bounded neighborhood of 0 of Theorem 6.2.1(b) is necessary as well as sufficient for seminormability.

Unlike the situation for normability, a TVS can be pseudometrizable without having a bounded neighborhood of 0, as Examples 6.2.2(a)–(d) show.

**Example 6.2.2** METRIZABLE $\nRightarrow$ BOUNDED NEIGHBORHOOD OF 0

We consider four metrizable TVS below that are not normable:

(a) $C(\mathbf{R}, \mathbf{R}, c)$ The LCHS $C(\mathbf{R}, \mathbf{R}, c)$ of continuous maps with compact-open topology has no bounded neighborhood of 0 [Example 6.1.7(a)], hence no bounded convex neighborhood of 0 and is therefore not normable. Since $\mathbf{R}$ is hemicompact [Example 5.5.6], $C(\mathbf{R}, \mathbf{R}, c)$ *is* metrizable by Theorem 5.8.5.

(b) The metric TVS $(s)$ of all sequences of complex numbers with the metric of Example 6.1.3 does not have a bounded neighborhood of 0 either, as shown in Example 6.1.7(b). Therefore, $(s)$ is not normable.

(c) The space $L_p(-\infty, \infty)$, $0 < p < 1$, of Example 4.6.2 is a pseudometric TVS whose only convex neighborhood of 0 is the whole space. Since $L_p(-\infty, \infty)$ is not bounded [Theorem 6.1.6], it is not seminormable.

(d) Let $D[0, 2\pi]$ be the LCHS of infinitely differentiable real-valued functions on $[0, 2\pi]$ with topology generated by the increasing sequence of seminorms $p_n$, $n = 0, 1, \ldots$, where $p_n(x) = \sup\{\left|x^{(j)}([0, 2\pi])\right| : 0 \leq j \leq n\}$, $x \in D[0, 2\pi]$. $D[0, 2\pi]$ is metrizable since its topology is defined by a countable family of seminorms [Theorem 5.6.1]. As $(p_n)$ is a saturated family of seminorms, to see that $D[0, 2\pi]$ has no bounded neighborhood of 0, it suffices to show that none of the unit balls $V_{p_n}$ is bounded. As $p_1(\{\sin nt : n \in \mathbf{N}\}) = \mathbf{N}$, $\{\sin nt : n \in \mathbf{N}\}$ is unbounded by Theorem 6.1.5. As $\{\sin nt : n \in \mathbf{N}\} \subset V_{p_0}$, it follows that $V_{p_0}$ is unbounded. Similar considerations show that none of the $V_{p_n}$, $n = 0, 1, \ldots$, are bounded. ■

## 6.3 STABILITY OF BOUNDED SETS

In this section we examine the stability of boundedness and total boundedness in regard to the formation of Cartesian products and balanced and convex hulls. In regard to convex hulls we see that in LCS, bounded sets have bounded convex hulls. The analogous result for totally bounded sets has already been proved in Theorem 4.8.9.

**Theorem 6.3.1** STABILITY OF BOUNDED SETS In any TVS $X$ with neighborhood filter $\mathcal{V}(0)$ of 0:

(a) subsets of bounded sets (hence intersections) are bounded;

(b) finite unions of bounded sets are bounded (arbitrary unions obviously can be unbounded);

(c) finite sums and scalar multiples of bounded sets are bounded;

(d) the "quotient" of a bounded set is bounded in a quotient topology;

(e) PRODUCTS if $\{X_s : s \in S\}$ are TVS and $B_s$ is a bounded subset of $X_s$ for each $s \in S$ then $\Pi_{s \in S} B_s$ is a bounded subset of $\Pi_{s \in S} X_s$.

**Proof.** Parts (a) and (b) are clear, as is the boundedness of scalar multiples of bounded sets.

(c) To prove that $B_1 + \cdots + B_n$ is bounded if each $B_i$ is, let $U \in \mathcal{V}(0)$ and choose $V \in \mathcal{V}(0)$ such that $V + V + \cdots + V$ ($n$ terms) $\subset U$. Since each $B_i$ is bounded, there is an $r > 0$ such that

$$B_1 + \cdots + B_n \subset rV + \cdots + rV = r(V + \cdots + V) \subset rU$$

(d,e) As noted in Example 4.7.2, with $k$ denoting the canonical map of $X$ into the quotient, if $\mathcal{B}$ is the filterbase of balanced neighborhoods of 0 in $X$ then $k(\mathcal{B})$ is a neighborhood base at 0 for the quotient topology. Hence, if $B \subset rV$ then $k(B) \subset rk(V)$. To prove (e), it suffices to show that $\Pi_{s\in S}B_s$ is absorbed by each basic neighborhood $\Pi_{s\in S}V_s$ of 0 in $\Pi_{s\in S}X_s$ where the neighborhoods $V_s = X_s$ for each $s \in S$ except $s = s_1, \ldots, s_n$. As there is some $r > 0$ such that $B_{s_i} \subset rV_{s_i}$ $(i = 1, \ldots, n)$, it follows that $\Pi_{s\in S}B_s \subset r\Pi_{s\in S}V_s$. ■

**Theorem 6.3.2** Closure and Balanced Hull In any TVS $X$ over $\mathbf{F}$ with neighborhood filter $V(0)$ of 0:

(a) the closure (and interior) of a bounded set is bounded;
(b) the balanced hull of a bounded set is bounded;
(c) the closure of a totally bounded set is totally bounded;
(d) the balanced hull of a totally bounded set is totally bounded.

**Proof.** Let $B$ be a bounded subset of the TVS $X$. By Theorem 4.3.6, there is a base of closed neighborhoods $V$ of 0 so that $B \subset rV \Rightarrow \operatorname{cl} B \subset rV$ $(r > 0)$ and (a) follows. The proof of (b) is quite similar, only here we use the fact that there is a base of balanced neighborhoods of 0. We have already shown (c) to hold in any topological group in Theorem 3.3.6. As for (d), let $E$ be totally bounded, let $V$ be a neighborhood of 0, and choose a balanced neighborhood $U$ of 0 such that $U + U \subset V$. Let $D$ denote the closed unit disk in $\mathbf{F}$. Since $E$ is totally bounded, there is a finite subset $S \subset X$ such that $E \subset S + U$. Hence the balanced hull $E_b = DE \subset D(S + U) \subset DS + U$. Since $S$ is finite, $DS$ is compact and can therefore be covered by $S' + U$ for some finite subset $S'$ of $X$. It follows that $E_b \subset S' + U + U \subset S' + V$. ■

The following results are for locally convex spaces.

**Theorem 6.3.3** Convex Hulls in LCS In a locally convex space, the convex hull $B_c$ of a bounded set $B$ is bounded and this is not true in general.

**Proof.** Let $B$ be bounded and suppose that $U$ is a convex neighborhood of 0. Let $r > 0$ be such that $B \subset rU$. The desired result now follows from the fact that $B_c \subset (rU)_c = rU$.

As for the ill-fated general statement, consider the $p$-seminormed TVS $L_p(-\infty, \infty)$, $0 < p < 1$, of Example 4.6.2. Since $\|\cdot\|^*$ is a $p$-seminorm and boundedness is equivalent to pseudometric boundedness in any $p$-seminormed space, it follows that that the closed $\|\cdot\|^*$-balls

$$C(0, r) = \{x \in L_p(-\infty, \infty) : \|x\|^* \le r\},\ r > 0,$$

are bounded. As shown in Example 4.6.2, the convex hull of any $C(0, r)$ is $L_p(-\infty, \infty)$ which is unbounded. ■

A base $\mathcal{B}$ for a neighborhood filter $V(x)$ is a collection of neighborhoods of $x$ such that each $V \in V(x)$ contains some $B \in \mathcal{B}$ as a subset. A notion of "base" for the system of bounded sets in a TVS is defined next.

**Definition 6.3.4** BASE OF BOUNDED SETS In a TVS $X$ a collection $\mathcal{B}$ of bounded sets is a *base* (*fundamental system*) of bounded sets if for any bounded subset $E$ of $X$ there is a $B \in \mathcal{B}$ such that $E \subset B$. ■

The closed intervals $[-n, n]$ ($n \in \mathbf{N}$) are a base of bounded sets in $\mathbf{R}$, as are the closed balls $C(0, n)$ about 0 of radius $n \in \mathbf{N}$ in any seminormed space.

**Theorem 6.3.5** In any LCS the closed bounded disks are a base of bounded sets.

**Proof.** Let $X$ be an LCS, suppose that $B \subset X$ is bounded and let $C$ be the closure of the disked hull $B_{bc}$ of $B$. Since $B$ is bounded, $B_b$ is bounded by Theorem 6.3.2(b). Since $X$ is locally convex, $B_{bc}$ is bounded by Theorem 6.3.3; therefore, so is $\operatorname{cl} B_{bc} = C$, by Theorem 6.3.2(a). ■

## 6.4 CONTINUITY

For TVS $X$ and $Y$, we say that $f : X \to Y$ is *locally bounded* if $f$ maps bounded sets into bounded sets. Having seen that local boundedness and continuity of a linear map between seminormed spaces are equivalent [Theorem 5.7.4], we investigate this connection in more general situations here and in the next section. (We take up the subject in detail again in Chap. 13.) What emerges is that continuity of a linear map always implies local boundedness [Theorem 6.4.1] while the converse holds in a wide variety of special cases. An example of a locally bounded discontinuous linear map is given in Example 13.2.8(c).

**Theorem 6.4.1** CONTINUOUS "HOMOGENEOUS" IMAGE OF BOUNDED SET Let $X$ and $Y$ be TVS, and suppose that $f : X \to Y$ is continuous and such that $f(ax) = a^r f(x)$ for some $r > 0$ for all $a > 0$ and any $x \in X$. If $B \subset X$ is bounded then $f(B)$ is bounded. It follows that continuous linear maps are locally bounded.

**Proof.** With notation as above, let $V_X(0)$ and $V_Y(0)$ denote the neighborhood filters of 0 in $X$ and $Y$, respectively. Since $f(ax) = a^r f(x)$ for some $r > 0$, for all $a > 0$ and any $x \in X$, it follows that $f(0) = 0$. Since $f$ is continuous, for any $U \in V_Y(0)$, $V = f^{-1}(U) \in V_X(0)$. Since $B$ is bounded in $X$, there is some positive number $a$ such that $B \subset aV$. Hence $f(B) \subset a^r f(V) \subset a^r U$. ■

It follows from Theorem 6.4.1 that if a subset $B$ of an arbitrary product $\Pi_{s \in S} X_s$ of TVS $\{X_s : s \in S\}$ is bounded then so are all its projections (continuous linear maps). Conversely, if all projections $\operatorname{pr}_s(B)$ of $B \subset \Pi_{s \in S} X_s$

are bounded, so is $B$ since $B \subset \Pi_{s\in S}\,\mathrm{pr}_s\,(B)$ and $\Pi_{s\in S}\,\mathrm{pr}_s\,(B)$ is bounded by Theorem 6.3.1(e). We summarize this now for future reference.

**Theorem 6.4.2** BOUNDEDNESS AND PRODUCTS A subset of a product of TVS is bounded iff all its projections are.

**Definition 6.4.3** MULTILINEAR MAP Given finitely many vector spaces $X_1$, $X_2, \ldots, X_n$ and $Y$, a map $A : \Pi_{i=1}^n X_i \to Y$ is *multilinear* if it is linear in each argument while the others are held fixed. ■

The following statement is an immediate consequence of Theorems 6.4.1 and 6.4.2.

**Theorem 6.4.4** CONTINUOUS MULTILINEAR MAPS LOCALLY BOUNDED An image of a finite product of bounded sets under a continuous multilinear map is bounded. Equivalently, we could say: A continuous multilinear image of a bounded set is bounded.

It is interesting to compare our next result with the embedding results of locally convex spaces in products of seminormed spaces of Theorem 5.11.4.

**Theorem 6.4.5** PRODUCT OF NORMED SPACES NOT NORMABLE Let $\{X_s : s \in S\}$ be an infinite family of nontrivial seminormed spaces. The product $\Pi_{s\in S}X$ is not seminormable.

**Proof.** With notation as above, if $\Pi_{s\in S}X$ were seminormable, it would have a bounded convex neighborhood of 0. Hence it would have a bounded basic neighborhood of 0 in the product topology. We now show that no basic neighborhood of 0 is bounded.

To this end, let $\Pi_{s\in S}V_s$ be a basic neighborhood of $0 : V_s = X_s$ for all $s$ except $s = s_1, \ldots, s_n$. For any $s \notin \{s_1, \ldots, s_n\}$, $\mathrm{pr}_s(\Pi_{s\in S}V_s) = X_s$, which is not bounded. Consequently, $\Pi_{s\in S}V_s$ is not bounded by Theorem 6.4.2. ■

The next result characterizes boundedness in a weak topology generated by a family of linear maps.

**Theorem 6.4.6** BOUNDED SETS IN AN INVERSE IMAGE TOPOLOGY Let $\{X_s : s \in S\}$ be a family of TVS. For each $s \in S$, let $A_s : X \to X_s$ be a linear map from the vector space $X$ into $X_s$ and let $X$ carry the inverse image topology generated by the family $\{A_s : s \in S\}$ [Example 4.7.1]. Then $B \subset X$ is bounded iff $A_s(B)$ is bounded for each $s \in S$.

**Proof.** With notation as above, since the inverse image topology for $X$ is determined by the map

$$\begin{array}{rcl} A : X & \to & \prod_{s\in S} X_s = Y \\ x & \mapsto & (A_s x)_{s\in S} \end{array}$$

$A$ is continuous. Therefore, by Theorem 6.4.1, if $B \subset X$ is bounded then $A(B)$ is bounded. Hence the projection $\mathrm{pr}_s\, A(B) = A_s(B)$ is bounded for each $s$ in $S$ [Theorem 6.4.2]. Conversely, if each $A_s(B)$ is bounded, so is $\Pi_{s \in S} A_s(B)$ by Theorem 6.4.2 and $A(B) \subset \Pi_{s \in S} A_s(B)$. Since $A(B)$ is bounded, $B$ is bounded by the way the inverse image topology is defined. ■

Using Theorem 6.4.6, we have another [see after Theorem 6.1.5] way of showing that a subset $B$ of $C(T, \mathbf{F}, c)$ is bounded iff, for each $x \in B$, $x(K)$ is uniformly bounded on each compact subset $K$ of $T$. By Theorem 6.4.6, each $A_K(B) = \{x|_K : b \in B\}$ is a bounded subset of $C(K, \mathbf{F}, c)$ [see (*) in Sec. 5.8]. For $A_K(B)$ to be bounded in $C(K, \mathbf{F}, c)$ it is necessary and sufficient that for each $t \in K$ there exist $k_K > 0$ such that for all $x \in B$, $|x(t)| \leq k_K$.

## 6.5 WHEN LOCALLY BOUNDED IMPLIES CONTINUOUS

We have seen that continuous linear maps are locally bounded [Theorem 6.4.1]. The converse is false [Example 8.8.8] but there are many important special cases in which local boundedness implies continuity—normed spaces, for example, or more generally bornological spaces [Chap. 13]. We develop a criterion for continuity in Theorem 6.5.1 and then show that locally bounded maps with pseudometrizable domains must be continuous.

**Theorem 6.5.1** BOUNDED ON NEIGHBORHOOD OF 0 If a linear map $A : X \to Y$ takes a neighborhood $V$ of 0 in the TVS $X$ into a bounded subset of the TVS $Y$ then $A$ is continuous.

**Proof.** With notation as above, since $A(V)$ is bounded, then for any neighborhood $U$ of 0 in $Y$, there is some positive number $r$ such that $rA(V) = A(rV) \subset U$. The continuity of $A$ follows from the fact that $rV$ is a neighborhood of 0 in $X$. ■

Recall that a TVS is pseudometrizable iff it has a countable base of neighborhoods of 0 [Theorem 4.8.3].

**Theorem 6.5.2** BOUNDED ON A PSEUDOMETRIZABLE DOMAIN Suppose that $A$ is a linear map taking a pseudometrizable TVS $X$ into a TVS $Y$. Then if $A$ is locally bounded, it is continuous.

**Proof.** With notation as above, let $(U_n)$ be a countable neighborhood base at 0 in $X$. By taking some intersections, it is easy to see that we may assume that $U_1 \supset U_2 \supset \cdots$. If $A$ is not continuous, there is a balanced neighborhood $V$ of 0 in $Y$ such that $A^{-1}(V)$ is not a neighborhood of 0. Thus, for every $n$, $(1/n)U_n \not\subset A^{-1}(V)$, so there is a sequence of elements $u_n \in (1/n)U_n$ such that for each $n$, $Au_n \notin V$. Since $(U_n)$ is decreasing, it follows that $nu_n \to 0$.

Hence $\{nu_n : n \in \mathbf{N}\}$ is relatively compact and therefore bounded. Since $A$ is a locally bounded map, $\{nAu_n : n \in \mathbf{N}\}$ is bounded; consequently, there is some $a > 0$ such that $\{nAu_n : n \in \mathbf{N}\} \subset aV$. Since $V$ is balanced, for $n > a$, $Au_n \in (a/n)V \subset V$ which contradicts the way in which the $u_n$ were chosen. ■

**Example 6.5.3** UNIFORM NORM FOR $L(X,Y)$ Let $X$ and $Y$ be normed spaces and $L(X,Y)$ be the linear space of all continuous linear maps $A : X \to Y$. (a) Each $A \in L(X,Y)$ is bounded on the unit ball $\{x \in X : \|x\| \leq 1\}$ [Example 5.7.4(a)] and it is straightforward to verify that $\|A\| = \sup\{\|Ax\| : \|x\| \leq 1\}$ is a norm for $L(X,Y)$ called the *uniform norm,* the topology it induces, the *uniform operator topology.* Some other ways of computing $\|A\|$ are $\|A\| = \sup\{\|Ax\| : \|x\| = 1\} = \inf\{k : \|Ax\| \leq k\|x\| \text{ for every } x\}$. Since the uniform limit of continuous linear maps is linear and continuous, the normed space $L(X,Y)$ is complete iff $Y$ is. For another approach to the uniform norm topology, see Example 11.2.4(a). In the normed space context. we assume $L(X,Y)$ to carry the uniform norm unless we indicate otherwise.

(b) In the special case when $Y = \mathbf{F}$ we write $X'$ instead of $L(X,\mathbf{F})$ with the uniform norm. We refer to the uniform norm in this case as the *canonical norm for* $X'$; we assume $X'$ to carry the canonical norm unless we specifically indicate otherwise. Since $\mathbf{F}$ is complete, $X'$ is a Banach space even if $X$ is not. We refer to $X''$ in its norm topology as the *bidual* of $X$.

## 6.6 LIOUVILLE'S THEOREM

The classical Liouville theorem states that a bounded entire function must be constant. After introducing suitable analogs of "entire" and "bounded" we prove a version of Liouville's theorem for entire vector-valued functions $f : \mathbf{C} \to X$ where $X$ is a TVS with sufficiently many continuous linear functionals. A striking application of this version of Liouville's theorem is given in Example 7.7.9 where we prove a Cauchy integral theorem for vector-valued functions. The Gelfand-Mazur theorem is another, namely, that a complex Banach division algebra is linearly isometric to $\mathbf{C}$ [Bachman and Narici 2000, p. 321; Beckenstein et al., 1977, p. 212].

**Definition 6.6.1** BOUNDED MAPS Let $X$ be a TVS and $f$ a map from a subset $G \subset \mathbf{C}$ into $X$; $f$ is called *bounded* if $f(G)$ is a bounded subset of $X$. ■

**Definition 6.6.2** ANALYTICITY Let $G$ be an open subset of the complex plane and let $X$ be a TVS. The map $x : G \to X$ is *analytic* (or *holomorphic*) in $G$ if the limit

$$\lim_{s \to t} \frac{x(s) - x(t)}{s - t} = x'(t)$$

exists for each $t \in G$. If $x$ is analytic throughout $\mathbf{C}$ then $x$ is called *entire*. ■

**Theorem 6.6.3** If $f$ is a continuous linear functional on the TVS $X$ and $x : G \to X$ is analytic in $G$ then the composite map $f \circ x : G \to \mathbf{C}$ is analytic in $G$.

**Proof.** We use the notation of the statement. Since $f$ is continuous and linear,

$$\lim_{s \to t} \frac{(f \circ x)(s) - (f \circ x)(t)}{s - t} = f\left( \lim_{s \to t} \frac{x(s) - x(t)}{s - t} \right) = f(x'(t))$$

exists at each $t_0 \in G$. ■

In the paragraph preceding Definition 6.6.1, the expression "sufficiently many" continuous linear functionals was used. We specify what "sufficiently many" is in the following definition.

**Definition 6.6.4** TOTAL SUBSET A subset $S$ of linear functionals on a vector space $X$ is called *total* if $f(x) = 0$ for each $f \in S$ implies that $x = 0$. ■

In Theorem 6.6.5 we hypothesize that the space $X'$ of all continuous linear functionals on a TVS $X$ is total. For any LCHS $X$, $X'$ is total [Theorem 7.7.7(b)].

**Theorem 6.6.5** LIOUVILLE'S THEOREM Let $X$ be a TVS with a total dual. If $x : \mathbf{C} \to X$ is entire and bounded then $x$ is constant.

**Proof.** We use the notation of the statement. Since the continuous homogeneous image of a bounded set is bounded [Theorem 6.4.1], if $x$ is bounded then $f \circ x$ is bounded and entire for any $f \in X'$ [Theorem 6.6.3]. Thus the standard Liouville theorem applies to each of the entire functions $f \circ x$ and we conclude that $f \circ x$ is constant for each $f \in X'$. Hence for any $a, b \in \mathbf{C}$, $f(x(a)) = f(x(b))$, i.e., $f(x(a) - x(b)) = 0$ for every $f$ in a total set of linear functionals. We conclude that $x(a) = x(b)$. Since $a$ and $b$ are arbitrary, the constancy of $x$ is proved. ■

## 6.7 BORNOLOGIES

A general notion of boundedness for subsets of a set $T$ is introduced briefly here. The basic idea is to abstract the properties of bounded sets of Theorem 6.3.1(a,b).

**Definition 6.7.1** BORNOLOGY A *bornology* (or *boundedness*) on a set $T$ is a collection $\mathcal{B}$ of subsets of $T$ which covers $T$ and which is closed with respect to the formation of subsets and finite unions. The elements $B \in \mathcal{B}$ are called *bounded sets*. $(T, \mathcal{B})$ is called a *bounded structure*. ■

If $X$ is a TVS, its collection of bounded sets constitutes a bornology called the *usual bornology* or *natural boundedness.* A similar convention about "natural" and "usual" prevails in any metric space. On any set $T$, the collection $\mathcal{P}$ of all subsets as well as $\mathcal{F} = \{\varnothing\}$ are bornologies. $\mathcal{P}$ is called the *discrete bornology* for $T$ and $\mathcal{F}$ the *trivial.*

**Example 6.7.2** COMPACT BORNOLOGY If $T$ is a topological space, the family of relatively compact sets is a bornology on $T$ called the *compact bornology.* ■

**Example 6.7.3** BOUNDEDNESS IN TOPOLOGICAL RINGS If $S$ is a subset of a commutative topological ring $X$, we say that $S$ is *bounded* if for each neighborhood $V$ of 0 there exists a neighborhood $W$ of 0 such that $SW \subset V$. If, for example, $K$ is a valued field, this notion of boundedness coincides with the metric boundedness induced by the valuation. Kaplansky [1947] (cf. Narici et al. 1971, pp. 10–18) used this notion of boundedness to obtain necessary and sufficient conditions for a topological field $K$ to admit a valuation. Calling $a \in K$ *nilpotent* if neither $(a^n)$ nor $(a^{-n})$ approaches 0, $K$ admits a valuation preserving the topology iff the set of nilpotent elements is open and $B \subset K$ being bounded away from 0 (i.e., disjoint from a neighborhood of 0) implies that $\{b^{-1} : b \in B\}$ is bounded. ■

In a bounded structure $(T, \mathcal{B})$, if $T$ is not bounded, the collection $\complement\mathcal{B} = \{\complement B : B \in \mathcal{B}\}$ of complements of bounded sets forms a filter, called the *filter at infinity*, with empty intersection.

**Definition 6.7.4** BASES AND SUBBASES A *base* $\mathcal{B}$ for a bornology $\mathcal{B}'$ is a collection of bounded sets such that each $B' \in \mathcal{B}'$ is a subset of some $B \in \mathcal{B}$. A *subbase* $\mathcal{S}$ for $\mathcal{B}'$ is a collection of bounded sets such that the collection of finite unions of sets of $\mathcal{S}$ is base for $\mathcal{B}'$ [Definition 6.3.4]. ■

Thus, the collection of disks $\{C(0,n) : n \in \mathbf{N}\}$ about the origin in $\mathbf{C}$ is a base for the usual bornology of $\mathbf{C}$. If $(S, \mathcal{B})$ and $(S, \mathcal{B}')$ are bounded structures then $\mathcal{B} \times \mathcal{B}' = \{B \times B' : B \in \mathcal{B},\}$ is a base for a bornology for $S \times T$. As intersections of bornologies are clearly bornologies, given any collection $\mathcal{S}$ of subsets of a set $T$, there is a smallest bornology $[\mathcal{S}]$ containing $\mathcal{S}$ called the *bornology generated (determined) by* $\mathcal{S}$. If $\mathcal{S}$ covers $T$, $[\mathcal{S}]$ consists of subsets of finite unions of sets from $\mathcal{S}$.

**Definition 6.7.5** WEAKER AND STRONGER A bornology $\mathcal{A}$ is *weaker* (*coarser*) than a boundedness $\mathcal{B}$—or $\mathcal{B}$ is *stronger* (*finer*) than $\mathcal{A}$—on a set $T$ if $\mathcal{A} \subset \mathcal{B}$.* ■

*Many authors (e.g., Hogbe-Nlend 1971 and Waelbroeck 1971) use the reverse convention and say that $\mathcal{A}$ is weaker than $\mathcal{B}$ if $\mathcal{A} \supset \mathcal{B}$. Our usage is consistent with that of Hu [1949].

Clearly, the trivial bornology on a set is the weakest possible, the discrete, the strongest.

If $(S, \mathcal{A})$ and $(T, \mathcal{B})$ are bounded structures and $f$ a map from $S$ into $T$ which takes bounded sets into bounded sets, $f$ is called *locally bounded.* If $f$ is a bijection and $f$ and $f^{-1}$ are locally bounded, $f$ is a *bornomorphism.*

**Example 6.7.6** INVERSE AND DIRECT IMAGE BORNOLOGIES Consider $f : S \to T$. If $(T, \mathcal{B})$ is a bounded structure, the *inverse image (initial) bornology* $\mathcal{B}_i$ induced by $f$ on $S$ is that which has $f^{-1}(\mathcal{B})$ as a base. $\mathcal{B}_i$ is the strongest bornology for $S$ with respect to which $f$ is locally bounded. If $(S, \mathcal{B})$ is a bounded structure, $T$ any set, and $f$ is surjective then the *direct image (final) bornology* $\mathcal{B}_f$ *determined by* $f$ on $T$ is the bornology $[f(\mathcal{B})]$ generated by $f(\mathcal{B})$. $\mathcal{B}_f$ is the weakest bornology for $T$ with respect to which $f$ is locally bounded. ■

Evidently, intersections and unions of bornologies are bornologies.

**Theorem 6.7.7** SUP AND INF Let $(\mathcal{B}_i)_{i \in I}$ be a family of bornologies on a set $T$. The supremum $\vee_{i \in I} \mathcal{B}_i$ and infimum $\wedge_{i \in I} \mathcal{B}_i$ of $(\mathcal{B}_i)_{i \in I}$ are the bornologies $[\cup_{i \in I} \mathcal{B}_i]$ and $[\cap_{i \in I} \mathcal{B}_i]$ determined by $\cup_{i \in I} \mathcal{B}_i$ and $\cap_{i \in I} \mathcal{B}_i$, respectively.

(a) $\wedge_{i \in I} \mathcal{B}_i = \cap_{i \in I} \mathcal{B}_i$; $\wedge_{i \in I} \mathcal{B}_i$ consists of all sets of the form $\cap_{i \in I} B_i$, where $B_i \in \mathcal{B}_i$ for each $i \in I$.

(b) $\vee_{i \in I} \mathcal{B}_i$ consists of all subsets of $T$ of the form $\cup_{i \in I} B_i$, where $B_i \in \mathcal{B}_i$ and $B_i = \varnothing$ in all but finitely many cases.

**Proof.** Clear. ■

The next result is similar in appearance to the corresponding results for topologies [Examples 4.7.1 and 4.11.3].

**Theorem 6.7.8** INVERSE AND DIRECT IMAGE BORNOLOGY (a) consider a set $S$, a family of bounded structures $(T_i, \mathcal{B}_i)_{i \in I}$ and maps $f_i : S \to T_i$ $(i \in I)$. The *Inverse image (initial) bornology* $\mathcal{B}$ for $S$ determined by the maps $(f_i)$ is the strongest bornology for $S$ with respect to which each $f_i$ is locally bounded and is given by $\cap_{i \in I} [f_i^{-1}(\mathcal{B}_i)]$.

(b) Consider a set $T$, a family $(S_i, \mathcal{B}_i)_{i \in I}$ of bounded structures and surjective maps $f_i : S_i \to T$. The *direct image (final) bornology* $\mathcal{B}$ for $T$ determined by the maps $(f_i)$ is the weakest bornology for $T$ with respect to which each $f_i$ is locally bounded. If $\mathcal{A}_i$ denotes the bornology generated by $f_i(\mathcal{B}_i)$ then $\mathcal{B} = \vee_{i \in I} \mathcal{A}_i$.

**Proof.** We use the notation of the statement and prove only part (a). By Example 6.7.6, $[f_i^{-1}(\mathcal{B})]$ is the strongest bornology for $S$ with respect to which $f_i$ is locally bounded. Since, for each $i$, $\cap_{j \in I} [f_j^{-1}(\mathcal{B}_j)] \subset [f_i^{-1}(\mathcal{B}_i)]$, it follows that $\cap_{i \in I} [f_i^{-1}(\mathcal{B}_i)] \subset \mathcal{B}$. Since $\mathcal{B}$ makes each $f_i$ locally bounded, however, $\mathcal{B}$ is weaker than each $[f_i^{-1}(\mathcal{B}_i)]$. ■

With these results, we can consider some natural bornologies for subsets and products of bounded structures $(T, \mathcal{B})$. For $S \subset T$, we "bornologize" $S$ with the initial bornology induced by the injection $t \mapsto t$ of $S$ into $T$, the finest bornology for $S$ such that $t \mapsto t$ is locally bounded. We call this bornology $S \cap \mathcal{B}$ for $S$ the *subspace bornology.* For products $\Pi_{i \in I} T_i = T$ of bounded structures $(T_i, \mathcal{B}_i)_{i \in I}$, the *product bornology* for $T$ is the initial bornology induced by the projections $(\text{pr}_i)_{i \in I}$. This, the strongest bornology for $T$ with respect to which each projection is locally bounded, has the sets $\Pi_{i \in I} B_i$, $B_i \in \mathcal{B}_i$, as a base.

We can now introduce bornologies to vector spaces.

**Definition 6.7.9** VECTOR BORNOLOGY A bornology on a vector space $X$ is a *vector bornology* (is *compatible with the linear structure* of $X$) if the maps

$$\begin{array}{ccc} \mathbf{F} \times X \to X & \text{and} & X \times X \to X \\ (a, x) \mapsto ax & & (x, y) \mapsto x + y \end{array}$$

are each locally bounded when $\mathbf{F}$ carries its natural bornology and $\mathbf{F} \times X$ and $X \times X$ carry their product bornologies. A *convex bornology* is a vector bornology which has a base of convex sets. ■

If $X$ is a TVS and $\mathcal{B}$ consists of the bounded (in the sense of Definition 6.1.1) subsets of $X$ then $\mathcal{B}$ is a vector bornology, for vector addition and scalar multiplication are each easily seen to be locally bounded. Moreover, in LCS the closed absolutely convex bounded sets form a base for $\mathcal{B}$ [Theorem 6.3.5].

An equivalent description of vector bornology follows.

**Theorem 6.7.10** VECTOR BORNOLOGY The bornology $\mathcal{B}$ is a vector bornology on the linear space $X$ iff finite sums and balanced hulls of bounded sets are bounded.

**Proof.** Let $C(0,1) = \{a \in \mathbf{F} : |a| \leq 1\}$. If $\mathcal{B}$ is a vector bornology the linear space $X$ and $B \in \mathcal{B}$, its balanced hull $B_b = C(0,1)B$ is bounded because the map $(a, x) \mapsto ax$ $(a \in \mathbf{F}, x \in X)$ is locally bounded. The finite sum condition follows by induction. Conversely, suppose that the condition holds. We need only prove that $(a, x) \mapsto ax$ is locally bounded. It suffices to show that $C(0, r)B$, $r > 0$, is bounded when $B$ is bounded and balanced. Since $B$ is balanced then for $n \geq r > 0$, $(n \in \mathbf{N})$

$$C(0, r)B \subset nB \subset B + B + \cdots + B \quad (n \text{ summands})$$

As a finite sum of bounded sets, $B + B + \cdots + B$ is bounded. ■

We conclude the section with two more examples.

**Example 6.7.11** BORNOLOGY OF EQUICONTINUITY Let $C(T, \mathbf{F})$ denote the linear space of continuous scalar-valued functions on the topological space $T$.

The collection $\mathcal{B}$ of equicontinuous subsets of $C(T,\mathbf{F})$ forms a vector bornology for $C(T,\mathbf{F})$. If $C(T,\mathbf{F})$ is endowed with the compact-open topology then, by Ascoli's theorem [5.10.4], $\mathcal{B}$ is the compact boundedness of $C(T,\mathbf{F},c)$ first mentioned in Example 6.7.2 ■

**Example 6.7.12** LCS OF LOCALLY BOUNDED FUNCTIONS Let $(T,\mathcal{B})$ be a bounded structure and let $LB(T,\mathbf{F})$ be the linear space of locally bounded maps of $T$ into the scalar field $\mathbf{F}$. The topology $\mathcal{T}_b$ for $LB(T,\mathbf{F})$ determined by the seminorms $p_B$, $B \in \mathcal{B}$, where $p_B(x) = \sup|x(B)|$ for $x \in LB(T,\mathbf{F})$ is called the *topology of uniform convergence on bounded sets.* We remark that the LCS $LB(T,\mathbf{F})$ is complete. If a subset $E$ of $LB(T,\mathbf{F})$ is bounded then each $p_B$ must be bounded on $E$. In other words, for each $B \in \mathcal{B}$, the set $\cup_{x\in E}x(B)$ is a bounded subset of $\mathbf{F}$. In this case we say that $E$ is *equibounded.* An important special case occurs when $T$ is a TVS $X$, and the bornology $\mathcal{B}$ consists of the $\sigma(X,X')$-bounded subsets of $X$ or some subcollection thereof, such as the finite subsets of $X$ or the $\sigma(X,X')$-compact disks and we restrict consideration to the subspace $X'$ of $LB(X,\mathbf{F})$ (see Secs. 8.5 and 11.2). ■

## 6.8 EXERCISES

6.101 FINER TOPOLOGIES If $\mathcal{T}'$ is a finer vector topology than the vector topology $\mathcal{T}$ on the vector space $X$ then each $\mathcal{T}'$-bounded set is $\mathcal{T}$-bounded. In particular, if $(X,\mathcal{T})$ is a TVS, every $\mathcal{T}$-bounded set is $\sigma(X,X')$-bounded where $\sigma(X,X')$ is as in Example 5.5.7(a). Theorem 8.8.7 shows that the converse is true too in LCHS.

6.102 SUBSPACES A subset $B$ of a subspace $M$ of a TVS $X$ is bounded in $M$ iff it is bounded in $X$.

6.103 Let $(X,d)$ be a metrizable TVS and let $|x| = d(x,0)$. Show that:

(a) for each $n \in \mathbf{N}$, $\frac{1}{n}|x| \leq \left|\frac{x}{n}\right|$;

(b) if $B$ is a bounded subset of $X$ then $\sup|B| < \infty$.

6.104 BOUNDEDNESS VERSUS METRIC BOUNDEDNESS For each real number $t$, define $p(t) = |t|/(1+|t|)$. Show that $p$ induces a compatible metric topology on $\mathbf{R}$ by showing that $p$ is an $F$-seminorm [Definition 4.8.1]. For the metric $d(s,t) = p(s-t)$ $(s,t \in \mathbf{R})$, show that $(\mathbf{R},d)$ is bounded but the TVS $(\mathbf{R},p)$ is not.

6.105 Consider the sequence space $\ell_p$, $0 < p < 1$, of sequences of scalars $x = (a_n)$ such that $q(x) = \sum|a_n|^p < \infty$. Show that $q$ induces a vector topology on $X$.

(a) Show that boundedness and metric boundedness coincide in $\ell_p$.

(b) Show that the closed unit ball determined by $q$ is bounded but that its convex hull is not.

6.106 Show that a LCS $X$ is seminormable iff it has a nonempty bounded open set.

6.107 Let $(X, \mathcal{T})$ be a TVS with a bounded neighborhood of 0. Show that the supremum of the locally convex topologies weaker than $\mathcal{T}$ [Example 4.11.2] can be defined by a single seminorm.

6.108 Let $B$ be a subset of an LCS $X$. If $(a_n)$ is a sequence of nonzero scalars such that, for every sequence $(x_n)$ from $B$, $(a_n x_n)$ is bounded then $B$ is bounded.

6.109 QUASI-COMPLETENESS If each closed bounded subset of a TVS $X$ is complete then $X$ is said to be *quasi-complete.*

(a) Show that complete implies quasi-complete and that quasi-complete implies sequentially complete.

(b) The result of Theorem 5.8.7 says that for a completely regular Hausdorff space $T$, the space $C(T, \mathbf{F}, c)$ of continuous scalar-valued maps on $T$ with compact-open topology is complete iff $T$ is a $k_R$-space. Show that $C(T, \mathbf{F}, c)$ is quasi-complete iff $T$ is a $k_R$-space.

(c) Products of quasi-complete spaces are quasi-complete.

The converses of the implications in (a) are generally false; in this connection see Jarchow [1981, p. 227, Prop. 1; p. 71, 3.6G].

6.110 CONTINUITY OF LINEAR FUNCTIONALS Let $f$ be a nontrivial linear form on a TVS $X$ and let $N = f^{-1}(0)$ be its null space. The following statements are then equivalent.

(a) $f$ is continuous;

(b) $N$ is closed;

(c) $N$ is not dense in $X$;

(d) for some neighborhood $U$ of 0 in $X$, $f(U)$ is bounded.

6.111 If $U$ is a subset of the pseudometrizable TVS $X$ and $U$ absorbs each null sequence [i.e., each $(x_n)$ such that $x_n \to 0$] then $U$ is a neighborhood of 0 in $X$.

6.112 BORNIVOROUS SETS A subset of a TVS which absorbs each bounded set is said to be *bornivorous* or to be a *bornivore.* Neighborhoods of 0 are bornivorous but there can be other bornivorous sets as well.

(a) If $X$ is a pseudometrizable TVS, show that each bornivore is a neighborhood of 0.

(b) Let $A$ be linear map of a TVS $X$ into a TVS $Y$. Show that $A$ is locally bounded iff $A^{-1}$ takes bornivorous sets into bornivorous sets.

6.113 SEQUENTIAL DESCRIPTION A linear map $A$ of a seminormed space $(X, p)$ into a LCS $Y$ is locally bounded iff $A$ maps null sequences into bounded sequences [cf. Exercise 13.201(d)].

6.114 SEQUENTIAL CONTINUITY IMPLIES LOCAL BOUNDEDNESS A sequentially continuous linear map of one TVS into another is locally bounded.

6.115 TOTALLY BOUNDED SETS

(a) Show that the continuous linear image of a totally bounded set is totally bounded.

(b) In a LCS, show that a linear combination of totally bounded sets is totally bounded.

6.116 BOUNDED SETS IN FUNCTION SPACES Let $X$ be a linear subspace of the linear space $F(T, Y)$ of all functions mapping the set $T$ into the TVS $Y$. As in Sec. 2.6, for a subset $S$ of $T$ and a neighborhood $V$ of 0 in $Y$, let $W(S, V) = \{x \in X : x(S) \subset V\}$.

(a) Show that $W(S, V)$ is absorbent iff $x(S)$ is bounded for each $x$ in $X$ [see Theorem 11.2.1].

(b) If $\mathcal{S}$ is a collection of subsets of $T$ then the sets $W(S, V)$, where $S \in \mathcal{S}$ and $V$ is a neighborhood of 0 in $Y$, form a subbase at 0 for a vector topology iff $x(S)$ is bounded for each $x$ in $X$ and $S$ in $\mathcal{S}$ [cf. Theorem 11.2.2].

6.117 DIRECT SUMS

(a) Let $\{X_s : s \in S\}$ be a family of Hausdorff TVS over a nondiscrete topological field $K$ and let $X$ be the direct sum $\oplus X_s$ as in Example 4.11.4; endow $X$ with the relative box topology of Exercise 4.112. Show that a subset $B$ of $X$ is bounded iff $B$ is contained in a product $\Pi\{X_s : s \in H\} = \{(x_s) : s \notin H \Rightarrow x_s = 0\}$ where $H \subset S$ is finite and the projection $\text{pr}_s B$ onto $X_s$ is bounded for each $s \in H$.

(b) Let $(X_s)$ be a family of LCS and let $X$ be their locally convex direct sum [Example 4.11.4]. Show that a subset of $X$ is bounded iff the conditions of (a) are met.

6.118 SEMINORMS FROM DISKS Let $D$ be a disk in a vector space $X$ and let $X_D$ denote the linear span $[D]$ of $D$ in $X$. View $X_D$ as a seminormed space where the seminorm is the gauge $p_D$ of $D$ in $X_D$. (Verify that $D$ is absorbent in $X_D$.)

(a) If $X$ is a TVS and $D$ is a bounded disk, show that the topology $X_D$ receives from $p_D$ is finer than that which it inherits as a subspace of $X$.

(b) With everything as in (a), show that if the topology on $X$ is Hausdorff then $p_D$ is a norm.

6.119 BOUNDED SETS IN TVS OVER VALUED FIELDS Let $X$ be a TVS over a nontrivially valued field $K$. As in the real or complex case, a subset $B$ of $X$ is called *bounded* if it is absorbed by each neighborhood $V$ of 0 in $X$: $B \subset aV$ for sufficiently large $|a|$.

(a) Show that if $X$ has a bounded neighborhood of 0 then $X$ is pseudo-metrizable.

(b) Suppose that the valuation on $K$ is non-Archimedean. Then the topology on $X$ is generated by a non-Archimedean seminorm [Exercise 5.204] iff $X$ has an absolutely $K$-convex [Exercise 4.202] bounded neighborhood of 0.

6.120 C-COMPACT DOES NOT IMPLY BOUNDED A $c$-compact [Exercise 4.204] subset of a locally $K$-convex space [Exercise 4.203] is not necessarily bounded.

6.201 MACKEY'S COUNTABILITY CONDITION Let $(B_n)$ be a sequence of bounded subsets of the metrizable LCS $X$. Show that there exists a bounded set $B$ and a sequence $(a_n)$ of positive scalars such that $B_n \subset a_n B$ for each $n \in \mathbf{N}$.

6.202 P-NORMS A subset $E$ of a vector space $X$ is called *$p$-convex* if $ax+by \in E$ for all $x$ and $y$ in $E$ and all scalars $a, b \geq 0$ such that $a^p + b^p = 1$ for some $1 \geq p > 0$; $E$ is called *absolutely $p$-convex* or a *$p$-disk* if $ax + by \in E$ whenever $|a|^p + |b|^p \leq 1$. A *$p$-seminorm*, $1 \geq p > 0$, $q$ on a vector space $X$ is a map that only fails to be a seminorm because it satisfies $q(ax) = |a|^p q(x)$ $(a \in \mathbf{F}, x \in X)$ instead of being absolutely homogeneous. An example of a $p$-seminorm that is not a norm is given in Example 4.6.2. Show that a TVS $X$ is $p$-seminormable, $0 < p \leq 1$, iff it has a bounded $p$-convex neighborhood of 0.

6.203 BOUNDED STRUCTURES Let $T$ be a topological space and let $\mathcal{B}$ be a bornology for $T$. The bornology $\operatorname{cl}\mathcal{B}$ generated by the sets $\operatorname{cl} B$, $B \in \mathcal{B}$, is called the *closure* of $\mathcal{B}$; similarly, $\operatorname{int}\mathcal{B} = \{\operatorname{int} B : B \in \mathcal{B}\}$ determines a bornology called the *interior* of $\mathcal{B}$.

(a) Show that $\operatorname{int}\mathcal{B} \subset \mathcal{B} \subset \operatorname{cl}\mathcal{B}$.

(b) If $\mathcal{A}$ and $\mathcal{B}$ are bornologies and $\mathcal{A} \supset \mathcal{B}$ then $\operatorname{cl}\mathcal{A} \supset \operatorname{cl}\mathcal{B}$ and $\operatorname{int}\mathcal{A} \supset \operatorname{int}\mathcal{B}$. $\mathcal{B}$ is called *closed* if $\mathcal{B} = \operatorname{cl}\mathcal{B}$, *open* if $\mathcal{B} = \operatorname{int}\mathcal{B}$, *proper* if it is both open and closed.

(c) Show that (1) $\mathcal{B}$ is closed iff (2) the closed subsets of $\mathcal{B}$ generate $\mathcal{B}$ which is equivalent to (3) the closure of each bounded set is bounded. $T$ is called *locally bounded* if each point of $T$ possesses a bounded neighborhood.

(d) Every compact subset of a locally bounded space is bounded.

(e) The compact bornology [Example 6.7.2] of a topological space $S$ makes $S$ locally bounded iff $S$ is locally compact.

(f) Show that the natural boundedness of any pseudometrizable space is proper and has a countable base. $(T, \mathcal{B})$ is called *boundedly pseudometrizable* if there is a pseudometric $d$ for $T$ which determines the topology on $T$ and for which the $d$-bounded sets are $\mathcal{B}$.

(g) METRIZATION THEOREM $(T, \mathcal{B})$ is boundedly pseudometrizable iff the following three conditions are satisfied:

i. The topological space $T$ is pseudometrizable;

ii. $T$ is locally bounded; and

iii. $\mathcal{B}$ is proper and has a countable base.

## Hints to Exercises

6.104 To see that $\mathbf{R}$ is not a bounded subset of the TVS $(\mathbf{R}, p)$, let $x_n = n$, and $t_n = 1/n$, $n \in \mathbf{N}$, and use Theorem 6.1.4.

6.108 If $B$ is not bounded and $(a_n)$ is a sequence of nonzero scalars, there exists a continuous seminorm $p$ and a sequence $(x_n)$ from $B$ such that $p(x_n) \geq n/|a_n|$.

6.109 (b) See proof of Theorem 5.8.7.

6.111 Let $(V_n)$ be a decreasing neighborhood base at 0. If $U$ is not a neighborhood of 0 then for every $n$, there exists $x_n$ in $(1/n)V_n \cap \complement U$; $nx_n \to 0$ but $nx_n \notin nU$, so $U$ does not absorb $\{nx_n : n \in \mathbf{N}\}$.

6.112 (a) Use Exercise 6.111.

6.113 Use Theorem 6.1.4.

6.114 Use Exercise 6.113.

6.119 (b) Use the fact that $\text{cl}\{0\}$ is the intersection of the neighborhoods of 0.

6.201 Let $(p_n)$ be a countable family of seminorms which determines the topology on $X$. For each $n \in \mathbf{N}$, there exists a sequence $(a_{n_j})$ of positive numbers such that $p_j(x) \leq a_{n_j}$ for each $x \in B_n$ and $j \in \mathbf{N}$. Define $a_n = \max(a_{1_n}, a_{2_n}, \ldots, a_{n_n})$ and let $B = \{x \in X : p_n(x) \leq a_n$ for every $n \in \mathbf{N}\}$.

6.202 If $(X, q)$ is a $p$-seminormed space then $U = \{x \in X : q(x) \leq 1\}$ is a bounded $p$-convex set. Conversely, if $V$ is a bounded $p$-convex neighborhood of 0 in a TVS $X$ then $V$ contains a balanced neighborhood $W$ of 0. Let $U$ be the *absolute $p$-convex hull* of $W$, the set of all linear combinations of the form $\sum a_i x_i$, where the $x_i \in W$ and $\sum |a_i|^p \leq 1$. Show that $U \subset V$ and that $r(x) = \inf\{a^p : a \geq 0,\ x \in aU\}$ is a $p$-seminorm that generates the topology.

# Chapter 7

# Hahn–Banach Theorems



The Hahn–Banach theorem has the power that only existence theorems can have. It has rightly been called the crown jewel of functional analysis. Its deep and diverse applications include proof of the existence of Green's and Neumann's functions, a Cauchy integral theorem for Banach algebras, proof of the existence of finitely additive measures defined on all subsets of **R**, the basis for the duality theory of LCHS and many other things. We prove various versions of the theorem and provide some of its history. We trace the evolution of the theorem and also discuss the role of Eduard Helly in its development. We argue that the theorem should be called the *Helly*-Hahn–Banach theorem.

## 7.1 WHAT IS IT?

Two principal versions of the Hahn–Banach theorem are as a continuous extension theorem (analytic form) and as a separation theorem (geometric form) about inserting a hyperplane between open convex sets.

ANALYTIC FORMS

**Dominated extension.** Let $f$ be a linear functional defined on a subspace $M$ of a real vector space (no norm) $X$, $p$ a sublinear functional defined on $X$ and $f \le p$ on $M$ ("dominated" by $p$), $f$ can be extended to a linear functional $F$ defined on $X$ with $F \le p$:

$$\begin{array}{llccl} F: & X & & & F \le p \\ & \mid & \searrow & & \\ f: & M & \longrightarrow & \mathbf{R} & f \le p \end{array}$$

For complex spaces, we mainly need some absolute values: If $X$ is complex, and $p$ a *seminorm* such that $|f| \le p$ on $M$ then $|F| \le p$ [Theorem 7.3.2].

**Continuous extension.** If $X$ is normed space over $\mathbf{F} = \mathbf{R}$ or $\mathbf{C}$ and $f : M \longrightarrow \mathbf{F}$ is a continuous linear functional then there exists a continuous linear functional $F$ extending $f$ defined on all of $X$ [Corollary 7.3.3]; in particular, there is a continuous extension $F$ such that $\|F\| = \|f\|$ [Theorem 7.4.1].

GEOMETRIC FORM

If a ball $B$ lies to one side of a line $L$ in $\mathbf{R}^3$ then there is a plane containing $L$ that lies to one side of $B$. The plane is not unique unless the line is tangent to $B$. The geometric form [Theorem 7.7.3] of the Hahn–Banach theorem generalizes this idea: Let $M$ be a linear subspace of a real or complex topological vector space $X$. If the linear variety $x + M$ $(x \in X)$ does not meet the open convex set $G$ then there exists a closed hyperplane $H$ containing $x + M$ that does not meet $G$. Mazur [1933] deduced the geometric form from the analytic form but he made no mention of the converse possibility. In a 1941 article, Dieudonné [1981b] refers to the geometric form as *the* Hahn–Banach theorem, so he was apparently aware of the equivalence of the two. It is first called the *geometric form* by Bourbaki.

The analytic form is a cousin of Tietze's theorem that a bounded continuous $f : K \to [a, b]$ defined on a closed subset $K$ of a normal space $T$ possesses a continuous extension $F : T \to [a, b]$. The geometric form resembles Urysohn's lemma about separating disjoint closed subsets of a normal space by a continuous function that assumes different values on the sets.

There are many other versions of the theorem—for vector lattices, modules, boolean algebras, bilinear functionals, groups, semigroups and more. It has many applications not only outside functional analysis but outside mathematics. Feinberg and Lavine [1983], for example, develop thermodynamics using the Hahn–Banach theorem, Neumann and Velasco [1994] apply Hahn–Banach type theorems to develop feasibility results on the existence of flows and potentials and Delbaen and Schachermayer [1994] use it to develop a fundamental theorem of asset pricing.

## 7.2 THE OBVIOUS SOLUTION

Suppose that $X$ is a vector space—no norm—over $\mathbf{F} = \mathbf{R}$ or $\mathbf{C}$ and the linear functional $f$ maps a subspace $M \subset X$ into $\mathbf{F}$. An easy way to linearly extend $f$ to $X$ is to take an algebraic complement $N$ of $M$, consider the projection $P_M$ on $M$ along $N$ and take $F = f \circ P_M$. In effect, take $F$ to be 0 outside $M$. (A nice application of this result is the proof of the existence of additive maps of $\mathbf{R}$ into $\mathbf{R}$ which are not of the form $x \mapsto tx$ for some $t \in \mathbf{R}$ [Exercise 7.102]). Will this technique work for *continuous* linear functionals $f$ defined on a closed subspace $M$ (extend $f$ by continuity to $\operatorname{cl} M$ if $M$ is not closed) of a TVS $X$? If $P_M$ is continuous then $F = f \circ P_M$ is a continuous linear extension of $f$. Generally, however, we cannot rely on this method because $P_M$ is continuous if and only if $M$ has a *topological* complement [Theorem 4.9.2(a)] and uncomplemented subspaces are common—$c_0$, for example, is an uncomplemented subspace of $\ell_\infty$ [Example 4.9.7] so there is no continuous projection of $\ell_\infty$ onto $c_0$. $C[0,1]$, $L_p[0,1]$ and $\ell_p$, $1 \leq p \leq \infty$, $p \neq 2$, have closed uncomplemented subspaces [Köthe 1983, pp. 430–431] and no finite-dimensional subspace of $L_p[0,1]$, $0 < p < 1$, [Example 4.6.2] has a topological complement [Köthe 1983, p. 158]. In fact, any Banach space $X$ has uncomplemented closed subspaces unless $X$ is linearly homeomorphic to a Hilbert space [Lindenstrauss and Tzafriri 1971]. Some instances in which a subspace $M$ of a locally convex Hausdorff space is complemented are $M$ finite-dimensional or -codimensional, or $M$ a closed subspace of a Hilbert space $(X, \langle \cdot, \cdot \rangle)$, in which case its orthogonal complement $M^\perp = \{x \in X : \langle m, x \rangle = 0 \text{ for all } m \in M\}$ is a topological complement. We say a little more about the Hilbert space situation in Example 16.4.5, this being a case in which $f \circ P_M$ is the only continuous linear extension of $f$ with the same norm.

## 7.3 DOMINATED EXTENSIONS

Recall that a sublinear functional is a map $p$ of a vector space $X$ into $\mathbf{R}$ which is subadditive ($p(x+y) \leq p(x) + p(y)$ for all $x, y \in X$) and positive homogeneous ($p(ax) = ap(x)$ for all $x \in X$ and $a \geq 0$).

The first Hahn–Banach theorem [Helly 1921] showed that a continuous linear functional defined on a subspace of a certain type of normed space has an extension of the same norm. Helly proved the theorem for certain normed subspaces of $\mathbf{C}^{\mathbf{N}}$, the general normed space not having been defined yet. Hahn [1927, p. 217] defined a general Banach space and used Helly's technique—the one-dimensional extension—to prove the norm-preserving version for general

Banach spaces. Banach [1929, Part I, p. 213], unaware of Hahn's result, generalized the result from the norm-preserving version to the dominated extension one that we prove in Theorem 7.3.2(a). The key inductive step is the extension of $f$ from $M$ to $M \oplus \mathbf{R}x$ for $x \notin M$ [Theorem 7.3.1]. We prove the dominated version in quite a different way in Sec. 7.6 after identifying *minimal* sublinear functionals as linear. Banach's and Hahn's versions were more general than Helly's in three ways: (1) They were valid in a general Banach space; (2) Banach circumvented the norm and proved a dominated version for sublinear functionals; and (3) each eliminated a separability requirement on a type of dual space that Helly used. They paid a stiff price, however: While Helly treated real and complex spaces, Hahn and Banach only dealt with the real case.

Helly's proof used mathematical induction. Hahn and Banach's generalization used *transfinite* induction. We use transfinite induction in the form of Zorn's lemma after establishing the existence of one-dimensional extensions in Theorem 7.3.1. The decisive inequality (7.1) of the proof was established for $C[a,b]$ by Helly [1912, pp. 273–274].

**Theorem 7.3.1** DOMINATED ONE-DIMENSIONAL EXTENSIONS Let $M$ be a subspace of a real vector space $X$, $x \notin M$, $p$ a sublinear functional defined on $X$ and let $f$ be a linear functional defined on $M$ with $f \le p$ on $M$. Then there exists a linear functional $F$ defined on $M \oplus \mathbf{R}x$ such that $F \le p$.

**Proof.** With notation as in the statement, since $f \le p$ on $M$, then for arbitrary $m, n \in M$,

$$\begin{aligned} f(m-n) = f(m) - f(n) &\le p(m + x - x - n) \\ &\le p(m+x) + p(-x-n) \end{aligned}$$

It follows that

$$-p(-x-n) - f(n) \le p(m+x) - f(m)$$

which implies that

$$a = \sup_{n \in M} [-p(-x-n) - f(n)] \le \inf_{m \in M} [p(x+m) - f(m)] = b$$

Hence, for any $c \in [a,b]$ and $m \in M$,

$$-p(-x-m) - f(m) \le c \le p(x+m) - f(m) \tag{7.1}$$

For $d \in \mathbf{R}$, define $F$ on $m + dx \in M \oplus \mathbf{R}x$ as $f(m) + dc$. Clearly $F$ is a linear extension of $f$. It remains to show that $F \le p$. If $d = 0$, this follows from the hypothesis. For $d > 0$, in the right half of Ineq. (7.1), replace $m$ by $m/d$ to get

$$c \le p(m/d + x) - f(m/d)$$

Since $p$ is sublinear, we can multiply both sides by $d$ to get

$$F(m+dx) = f(m) + cd \leq p(m+dx)$$

that is, that $F \leq p$. Now suppose that $d < 0$ and replace $m$ by $m/d$ in the left side of Ineq. (7.1) to get

$$-p(-x - m/d) - f(m/d) \leq c$$

Multiplying by $-d$, we see that $F(m+dx) = f(m) + cd \leq p(m+dx)$. ∎

The extension $F$ of the theorem will be unique iff we have equalities in (7.1), a theme (uniqueness) we return to in Theorem 16.4.2.

To establish the general form of the Hahn–Banach theorem, we first recall:

> Zorn's Lemma Let $(T, \leq)$ be an ordered set (Sec. 5.12). A subset $S$ of $T$ is *totally (linearly) ordered* if for all $s, t \in S$, $s \leq t$ or $t \leq s$. An upper bound $s' \in T$ for $S$ is such that $s \leq s'$ for all $s \in S$. A *maximal element* $s'$ of $S$ is an element of $S$ such that, for all $s \in S$, $s' \leq s$ implies $s = s'$. $(T, \leq)$ is *inductively ordered* if any totally ordered subset of $T$ has an upper bound in $T$. Zorn's lemma says that every nonempty inductively ordered set has a maximal element.

Note that upper bounds for $S \subset T$ need not belong to $S$ but maximal elements do.

The dominated extension theorem below enables us to prove continuous extension theorems [Corollary 7.3.3] because a linear functional $f$ on a TVS $X$ is continuous iff there is a continuous seminorm $p$ on $X$ such that $|f| \leq p$ [Theorem 5.7.3]. The key to obtaining the complex version from the real one—Theorem 7.3.2(b)—is the 1-1 correspondence between real and complex linear functionals of Theorem 1.4.2, namely,

| complex linear functional | | real linear functional |
|---|---|---|
| $f(x) = \operatorname{Re} f(x) - i \operatorname{Re} f(ix)$ | $\leftrightarrow$ | $\operatorname{Re} f(x)$ |

**Theorem 7.3.2** Dominated Extensions Let $M$ be a subspace of a vector space $X$ over $\mathbf{F} = \mathbf{R}$ or $\mathbf{C}$, let $p$ be a sublinear functional defined on $X$ and let $f$ be a linear functional on $M$.

(a) Real If $\mathbf{F} = \mathbf{R}$ and $f \leq p$ on $M$ then there exists a linear functional $F$ defined on $X$ which extends $f$ such that $F \leq p$. If $p$ is a seminorm then $|F| \leq p$.

(b) Complex If $\mathbf{F} = \mathbf{C}$, $p$ is a seminorm and $|f| \leq p$ on $M$ then there is a linear functional $F$ on $X$ which extends $f$ and such that $|F| \leq p$.

**Proof.** We use the notation of the statement. (a) Let $f$ be a linear functional defined on $M$ majorized by a sublinear functional $p$ on $M$. Consider the set

$\mathcal{M}$ of pairs $(N, g_N)$ of linear functionals $g_N$ defined on a subspace $N \supset M$ which extend $f$ with $g_N \leq p$. $\mathcal{M}$ is not empty by Theorem 7.3.1. Order $\mathcal{M}$ by taking $(N_1, g_{N_1}) \leq (N_2, g_{N_2})$ if $N_1 \subset N_2$ and $g_{N_2}|_{N_1} = g_{N_1}$. We show that $\mathcal{M}$ is inductively ordered by showing that if $\mathcal{N}$ is a totally ordered subset of $\mathcal{M}$, an upper bound for $\mathcal{N}$ is given by $(N\hat{}, g\hat{})$ where $N\hat{} = \cup_{(N,g_N)\in\mathcal{N}} N$ and $g\hat{}\,(x) = g_N(x)$ if $x \in N$. As to the definition of $g\hat{}$, note that if $x \in N$ and $x \in N'$ then, since $g_{N'}$ extends $g_N$ or vice-versa, $g\hat{}$ is well-defined. It is easy to verify that $N\hat{}$ is a subspace and that $g\hat{}$ is linear and extends $f$. Next, we show that $(N\hat{}, g\hat{})$ is an upper bound for $\mathcal{N}$. For any vector $x \in N\hat{}$, $N$ such that $(N, g_N) \in \mathcal{N}$ and $x \in N$, $g\hat{}\,(x) = g_N(x) \leq p(x)$ so $g\hat{} \leq p$. Hence $(N\hat{}, g\hat{}) \in \mathcal{M}$ and $(N, g_N) \leq (N\hat{}, g\hat{})$, i.e., $(N\hat{}, g\hat{})$ is an upper bound for $\mathcal{M}$. $\mathcal{M}$ is therefore totally ordered and so has a maximal element $(N, F)$. If $N \neq X$, choose $z \notin N$. By Theorem 7.3.1, we may extend $F$ to a linear map $F\hat{} : N + \mathbf{F}z \to \mathbf{R}$ with $F\hat{} \leq p$ and thereby contradict the maximality of $F$. We conclude that $N = X$. If $p$ is a seminorm then $F \leq p$ implies that $F(-x) \leq p(-x) = p(x)$ $(x \in X)$ and therefore that $|F| \leq p$.

(b) Now suppose that $p$ is a seminorm on $X$ and $|f| \leq p$ on $M$. View $X$ as a real vector space. Since, by Theorem 1.4.2, $|f(m)| = |\operatorname{Re} f(m) - i \operatorname{Re} f(im)| \leq p(m)$ $(m \in M)$. Since $\operatorname{Re} f(m) \leq |f(m)|$, it follows that $\operatorname{Re} f \leq p$ on $M$. By (a) there is a real linear functional $G$ which extends $\operatorname{Re} f$ and such that $G \leq p$. Let $F(x) = G(x) - iG(ix)$, $x \in X$, a complex linear functional by Theorem 1.4.2. Therefore $F = f$ on $M$. To see that $|F| \leq p$, suppose $x \in X$ and $F(x) = re^{it} \neq 0$, $r > 0$, so that $F(e^{-it}x) = r$, a purely real quantity. Thus $F(e^{-it}x) = G(e^{-it}x)$, so $|F(x)| = F(e^{-it}x) = G(e^{-it}x) \leq p(e^{-it}x) = p(x)$. ■

**Corollary 7.3.3** CONTINUOUS EXTENSIONS ON LCS Let $M$ be a subspace of a LCS $X$ over $\mathbf{F} = \mathbf{R}$ or $\mathbf{C}$ and let $f : M \to \mathbf{F}$ be a continuous linear functional. Then $f$ has a continuous linear extension $F \in X'$.

**Proof.** We use the notation of the statement. As a subspace of a LCS, $M$ is locally convex in its relative topology. Moreover, if $P$ is the family of continuous seminorms on $X$ then $P|_M = \{p|_M : p \in P\}$ generates the relative topology on $M$. Thus, if $f$ is a continuous linear form on $M$, there exists $p \in P$ such that $|f| \leq p$ on $M$ [Theorem 5.7.3]. By Theorem 7.3.2 there exists a linear extension $F$ of $f$ such that $|F| \leq p$. Continuity of $F$ follows from Theorem 5.7.3. ■

Thus, local convexity is sufficient for the *Hahn–Banach extension property* (HBEP), the ability to extend a continuous linear functional from any subspace to the whole space. Although local convexity is not necessary for HBEP, it helps [Theorem 7.7.8]. We discuss the linkage between local convexity and HBEP in Sec. 8.14.

What about uniqueness of the extension? In the proof of Theorem 7.3.1 we chose $c \in [a, b]$ in Inequality (7.1); a different choice would produce a different extension. The extension is unique more often than you might think, however—on subspaces of any Hilbert space or $\ell_p$, $1 < p < \infty$, for example, to mention two. We explore the subject of uniqueness in Sec. 16.4.

As an immediate consequence of the dominated version [7.3.2], we have Theorem 7.3.4. In it, $p$ could be a norm in which case if $w \neq 0$ then $p(w) \neq 0$; hence, if every continuous linear functional vanishes on $w$ then $w = 0$.

**Theorem 7.3.4** Let $p$ be a sublinear functional on a real vector space $X$. For any $w \in X$, there exists a linear functional $F$ on $X$ such that:
(a) $F(w) = p(w)$;
(b) $-p(-x) \leq F(x) \leq p(x)$ for all $x \in X$;
(c) If $p$ is a seminorm then $|F| \leq p$.
In either case, if $p$ is continuous then $F$ is continuous.

**Proof.** We use the notation of the statement. (a,b) Let $M = \mathbf{R}w$ and consider $f : M \to \mathbf{R}$, $aw \mapsto ap(w)$ $(a \in \mathbf{R})$. Clearly $f$ is a linear functional; we show that $f \leq p$ on $M$ by showing that $f(aw) = p(aw)$ for $a \geq 0$ and $f(aw) \leq p(aw)$ for $a < 0$. For $a \geq 0$, $f(aw) = ap(w) = p(aw)$. Note that $0 = p(w - w) \leq p(w) + p(-w)$; hence $p(w) \geq -p(-w)$ and for $a < 0$, $ap(w) \leq -ap(-w) = p(aw)$. Thus, $f(aw) = ap(w) \leq p(aw)$. By Theorem 7.3.2(a), there exists a linear functional $F$ on $X$ extending $f$ such that $F \leq p$ everywhere. Since $-F(x) = F(-x) \leq p(-x)$, it follows that $-F(-x) = F(x) \geq -p(-x)$. As to (c), if $p$ is a seminorm, $p(-x) = p(x)$ and the result follows from (b). The assertion about continuity follows by proving continuity at 0. ■

In Sec. 4.9 we discussed topological complements and showed, in particular, that the subspace $c_0$ (null sequences) of $(\ell_\infty, \|\cdot\|_\infty)$ (bounded sequences) did not have a topological complement. We can now prove an affirmative result about complements.

**Theorem 7.3.5** FINITE-DIMENSIONAL IMPLIES COMPLEMENTED Let $X$ be a LCHS over $\mathbf{R}$ over $\mathbf{C}$ and let $M$ be a finite-dimensional (or codimensional) subspace. Then $M$ has a topological complement.

**Proof.** We use the notation of the statement. By Theorem 4.9.2(a), it suffices to show that there is a continuous projection $P$ from $X$ onto $M$. Let $\{x_1, \ldots, x_n\}$ be a basis for $M$. For any scalars $\{a_1, a_2, \ldots, a_n\}$ and each $1 \leq i \leq n$, let $f_i(\sum a_j x_j) = a_i$. Since $M$ is finite-dimensional, the $f_i$ are continuous [Theorem 4.10.3(f)] and therefore may be extended to continuous linear functionals $F_i$ $(1 \leq i \leq n)$ on $X$ by Corollary 7.3.3. For each $1 \leq i \leq n$, the map $x \mapsto F_i(x)x_i$ is continuous; hence so is the map $P : X \to M$, $x \mapsto \sum_i F_i(x)x_i$. Since, for each $i$, $P(x_i) = f_i(x_i)\,x_i = x_i$, it follows that $P \circ P = P$. ■

## 7.4 CONSEQUENCES

We deduce the norm-preserving version of the Hahn–Banach here and a number of important consequences of it. We prove in Theorem 7.4.3 that there are always continuous extensions of greater norm. In Theorem 10.1.2 we show that linear *maps,* rather than functionals, can be extended for certain codomains. We assume as always that the dual $X'$ of a normed space $X$ carries the canonical norm of Example 5.7.5.

**Theorem 7.4.1** NORM-PRESERVING HAHN–BANACH THEOREM Let $X$ be a normed space over **R** or **C**. Then if $f$ is a continuous linear functional defined on a linear subspace $M$ of $X$, there exists a continuous linear extension $F$ of $f$ such that $\|F\| = \|f\|$.

**Proof.** We use the notation of the statement. Let the linear functional $f$ be continuous on $M$. As such, $|f(\cdot)| \leq \|f\| \|\cdot\|$ on $M$ by the comment after Theorem 5.7.4. Theorem 7.3.2 yields a continuous linear extension $F$ of $f$ such that $|F(\cdot)| \leq \|f\| \|\cdot\|$ on $X$. This immediately implies that $\|F\| \leq \|f\|$, whereas the fact that the unit ball of $M$ is a subset of the unit ball of $X$ implies that

$$\begin{aligned} \|f\| &= \sup\{|f(m)| : \|m\| = 1,\ m \in M\} \\ &= \sup\{|F(m)| : \|m\| = 1,\ m \in M\} \\ &\leq \sup\{|F(x)| : \|x\| = 1,\ x \in X\} = \|F\| \quad \blacksquare \end{aligned}$$

Some immediate implications of Theorem 7.4.1 are collected in Theorem 7.4.2.

**Theorem 7.4.2** CONSEQUENCES Let $(X, \|\cdot\|)$ be a normed space over $\mathbf{F} = \mathbf{R}$ or **C** with normed dual space $X'$; let $SU' = \{f \in X' : \|f\| = 1\}$.

(a) For any $x \in X$, there is a continuous linear functional $f \in SU'$ such that $f(x) = \|x\|$ Therefore $x = 0$ iff every $f \in X'$ vanishes on $x$.

(b) For any $x \in X$, $\|x\| = \sup\{|f(x)| : f \in SU'\}$.

(c) Suppose $M$ is a subspace of $X$ and $w \notin \operatorname{cl} M$. Then $d(w, M) = \inf\{\|w - m\| : m \in M\} = d > 0$ and there exists $f \in SU'$ that vanishes on $M$ and $f(w) = d$. Hence $w \in \operatorname{cl} M$ iff every continuous linear functional that vanishes on $M$ vanishes on $w$.

(d) Suppose $M$ is a subspace of $X$ and $w \notin \operatorname{cl} M$. Then there exists $g \in X'$ that vanishes on $M$, $g(w) = 1$ and $\|g\| = 1/d$.

(e) Given a finite set $\{x_1, x_2, \ldots, x_n\}$ of linearly independent vectors, there exists $\{f_1, f_2, \ldots, f_n\} \subset X'$ such that $f_i(x_j) = \delta_{ij}$ for all $i, j \in \{1, 2, \ldots, n\}$.

**Proof.** We use the notation of the statement. (a) The result is clear if $x = 0$ so assume $x \neq 0$. Let $M = \mathbf{F}x$ and define $g(ax) = a\|x\|$ for each $a \in \mathbf{F}$. As the unit vectors of $M$ are $\{ax : |a| = 1/\|x\|\}$, it follows that $\|g\| = 1$. Now use Theorem 7.4.1 to linearly extend $g$ to $f \in X'$ with $\|g\| = \|f\|$.

(b) For any $x \in X$, $\sup\{|f(x)| : f \in SU'\} \leq \|x\|$. The result now follows from (a).

(c) With notation as in the statement, for $m + aw \in N = M \oplus \mathbf{F}w$, define $f(m + aw) = ad$. Clearly $f$ is linear, vanishes on $M$, and $f(w) = d$. For any $a \in \mathbf{F}$ and any $m \in M$, $|a| d \leq \|aw - am\|$; indeed,

$$|a| d \leq \|aw + m\| \quad (a \in \mathbf{F},\ m \in M) \tag{$*$}$$

To see that $f$ is continuous, suppose $(m_n)$ and $(a_n)$ are sequences from $M$ and $\mathbf{F}$, respectively, such that $m_n + a_n w \to 0$. By $(*)$, for every $n$, $|a_n| d \leq \|a_n w + m_n\|$; hence $a_n \to 0$. Therefore $f(m_n + a_n w) = a_n d \to 0$ and $f$ is continuous. To see that $\|f\| = 1$, consider $x = m + aw \in N$, $a \neq 0$. Since $f(w - m) = d$,

$$\|x\| = \|m + aw\| = \left\| a\left(w - \left(-\frac{m}{a}\right)\right)\right\| \geq |a| d = |f(x)| \tag{$**$}$$

Hence $\|f\| \leq 1$. Conversely, given $r > 0$, choose $m \in M$ such that $\|w - m\| < d + r$ and let $y = \frac{w-m}{\|w-m\|} \in N$. Then

$$|f(y)| = \frac{|f(w - m)|}{\|w - m\|} > \frac{d}{d + r}$$

Since $\|y\| = 1$, it follows that $\|f\| \geq 1$ and therefore that $\|f\| = 1$. Now use Theorem 7.4.1 to extend $f$ to $X$.

(d) For $f$ as in (c), let $g = (1/d) f$.

(e) Let $M$ denote the linear span of $\{x_1, x_2, \ldots, x_n\}$. For each $i = 1, 2, \ldots, n$, there exist linear functionals $f_i$ defined on $M$ such that $f_i(x_j) = \delta_{ij}$. As $M$ is finite-dimensional, each $f_i$ is continuous [Theorem 4.10.3(f)]. It only remains to extend each $f_i$ to $X$ by Theorem 7.4.1. ■

We show next that there are extensions of continuous linear functionals of strictly greater norm on any subspace of any normed space.

**Theorem 7.4.3** Extensions of Greater Norm Always Exist If $f$ is a continuous linear functional defined on the closed proper subspace $M$ of the normed space $X$ over $\mathbf{F} = \mathbf{R}$ or $\mathbf{C}$ then there are continuous linear extensions $F$ of $f$ with $\|F\| > \|f\|$.

**Proof.** Let $f, M$ and $X$ be as above. Choose a unit vector $u \notin M$, let $d$ be the positive distance from $u$ to $M$ and choose a scalar $b > \|f\|$. Define $g$ on $M \oplus \mathbf{F}u$ by taking $g(m + au) = f(m) + ab$ for $a \in \mathbf{F}$ and $m \in M$. Clearly $g$ is linear. As to its continuity, suppose $(m_n)$ and $(a_n)$ are sequences from $M$ and $\mathbf{F}$, respectively, such that $m_n + a_n u \to 0$. By $(*)$ in the proof of Theorem 7.4.2(c), $|a_n| d \leq \|m_n + a_n u\|$ for every $n$ and therefore $a_n \to 0$; this implies that $m_n \to 0$. Therefore $g(m_n + a_n u) = f(m_n) + a_n b \to 0$ and $g$ is continuous. Since $g(u) = b > \|f\|$, it follows that $\|g\| > \|f\|$. Finally, extend $g$ to $F \in X'$ with $\|g\| = \|F\|$ by Theorem 7.4.1. ■

### 7.4.1 The Dual of $C\,[0,1]$

$C([0,1],\mathbf{C},\|\cdot\|_\infty)$ or just $C\,[0,1]$ denotes the Banach space of continuous complex-valued functions on the closed interval $[0,1]$. We show that its continuous dual $C[0,1]'$ "is" the space $BV[0,1]$ of complex-valued functions of bounded variation on $[0,1]$ with seminorm $v\,(x)$, the variation of the function $x$ [Example 1.7.9]. By $\ell_\infty\,[0,1]$, we mean the Banach space of all bounded complex-valued functions on $[0,1]$ of Example 1.7.5. In Theorem 7.4.4 we show that any $f \in C[0,1]'$ can be represented as the Riemann-Stieltjes integral $f(\cdot) = \int_0^1 (\cdot)dg(t)$ for some $g \in BV[0,1]$. The $g$ is not unique, however: If $g$ serves to "represent" $f$ in this manner, so does $g+c$ for any scalar $c$. And this is not the only way in which "representative" $g$'s may differ. There is a linear isometry between $C\,[0,1]'$ and a certain subspace of $BV[0,1]$, however; the details may be found in Bachman and Narici [2000, p. 221ff.].

**Theorem 7.4.4** $C[0,1]'$ "=" $BV[0,1]$ Let $BV[0,1]$ be as above. For any $f \in C[0,1]'$, there is a function $g \in BV[0,1]$ such that $f(\cdot) = \int_0^1 (\cdot)dg(t)$ (Riemann-Stieltjes integral) and $\|f\| = v\,(g)$, the variation of $g$, where $\|f\| = \sup\{|f\,(x)| : \|x\|_\infty \leq 1, x \in C\,[0,1]\}$.

**Proof.** Let $f \in C[0,1]'$ and let $k_t$ denote the characteristic function of $[0,t]$. View $C\,[0,1]$ as a subspace of $\ell_\infty\,[0,1]$ and extend $f$ to a linear functional $F$ defined on $\ell_\infty\,[0,1]$ with $\|F\| = \|f\|$ [Theorem 7.4.1]. If $g$ "represents" $f$ as in the statement of the theorem then for each $0 \leq t \leq 1$, $F(k_t) = \int_0^1 k_t(s)dg(s) = g(t) - g(0)$. Moreover, since adding a constant to $g$ does not change the value of the integral, we may assume that $g(0) = 0$; thus, a necessary condition for $g$ to satisfy is that $F(k_t) = g(t)$ for $0 \leq t \leq 1$. We now prove that $g(t) = F(k_t)$ is the desired function.

Let $0 = t_0 < t_1 < \cdots < t_n = 1$ be a partition of $[0,1]$. For $k = 1,\ldots,n$, let $s_k$ and $r_k > 0$ be such that

$$g(t_k) - g(t_{k-1}) = r_k e^{is_k} \qquad \text{and} \qquad a_k = \begin{cases} e^{-is_k} & \text{if } g(t_k) \neq g(t_{k-1}) \\ 0 & \text{if } g(t_k) = g(t_{k-1}) \end{cases}$$

Consider the step function $x(t) = a_1$ for $t \in [0,t_1]$ and $x(t) = a_j$ for $t \in (t_{j-1},t_j]$, $j = 2,\ldots,n$. Clearly, $x = \sum_{j=1}^n a_j(k_{t_j} - k_{t_{j-1}})$ and it follows that $F(x) = \sum_{j=1}^n a_j[F(k_{t_j}) - F(k_{t_{j-1}})] = \sum_{j=1}^n a_j[g(t_j) - g(t_{j-1})] = \sum_{j=1}^n r_j = \sum_{j=1}^n |g(t_j) - g(t_{j-1})|$. As $\|x\|_\infty \leq 1$, it follows that $|F(x)| = \sum_{j=1}^n |g(t_j) - g(t_{j-1})| \leq \|F\| = \|f\|$. Hence, $g$ is of bounded variation and $v\,(g) \leq \|f\|$.

Now let $x \in C\,[0,1]$ and let $z = \sum_{j=1}^n x(t_{j-1})[k_{t_j} - k_{t_{j-1}}]$. By the linearity of $F$, $F(z) = \sum_{j=1}^n x(t_{j-1})[g(t_j) - g(t_{j-1})]$. By the uniform continuity of $x$, given $r > 0$, for partitions $0 = t_0 < t_1 < \cdots < t_n = 1$ of sufficiently small gauge, $\|z - x\|_\infty < r$—in other words, as the gauge goes to 0, $z \to x$ while $F(z) \to \int_0^1 x(t)dg(t)$. Thus, since $F$ is continuous and $x \in C\,[0,1]$, $F(x) = f(x) = \int_0^1 x(t)dg(t)$. Finally, since $|f(x)| = \left|\int_0^1 x(t)dg(t)\right| \leq \|x\|_\infty v\,(g)$ for any $x \in C[0,1]$, it follows that $\|f\| \leq v(g)$. Combining this with the observation that $v(g) \leq \|f\|$ at the end of the preceding paragraph, it follows that $\|f\| = v(g)$. ■

## 7.5 THE MAZUR–ORLICZ THEOREM

The Mazur–Orlicz [1953] theorem is equivalent to the Hahn–Banach theorem. We present Ptak's [1956] simplified proof in Theorem 7.5.1. A word on notation: We use $v$ to denote the vector-valued function in the theorem and $r$ for the real-valued function. We state it for real vector spaces $X$ but $X$ could be complex; in that case the linear functional $F$ would be a real linear functional.

**Theorem 7.5.1** MAZUR–ORLICZ THEOREM Let $X$ be a real vector space, $p$ a sublinear functional and $T$ any set. For any two maps $v : T \to X$ and $r : T \to \mathbf{R}$:

(a) there exists a linear functional $F$ such that $F \leq p$ and $r \leq F \circ v$ iff

(b) for any finite subset $\{a_1, a_2, \ldots, a_n\}$ of nonnegative scalars and any $\{t_1, t_2, \ldots, t_n\} \subset T$, $\sum_{i=1}^n a_i r(t_i) \leq p\left(\sum_{i=1}^n a_i v(t_i)\right)$.

$$\begin{array}{llll} X & & & \sum_{i=1}^n a_i r(t_i) \leq p\left(\sum_{i=1}^n a_i v(t_i)\right) \\ \uparrow v & \overset{F}{\searrow} & & F \leq p \text{ on } X \\ T & \overset{r}{\longrightarrow} & \mathbf{R} & r \leq F \circ v \text{ on } T \end{array}$$

**Remark 7.5.2** Before proving the theorem, we note that it yields the Hahn–Banach theorem. To see that, let $T$ be a subspace $M$ of $X$, let $v$ be the injection $I : M \to X$, $m \mapsto m$, and let $r$ be a linear functional $f$ defined on $M$. For $n = 1$, $a_1 = 1$ and $t_1 = m \in M$, $f(m) \leq p(m)$ by (b), in other words, $f \leq p$ on $M$. By (a), there exists a linear functional $F$ on $X$ that satisfies $r \leq F \circ v$ on $M$, i.e., $f \leq F$ on $M$.

$$\begin{array}{llll} X & & & F \leq p \text{ on } X \\ \uparrow I & \overset{F}{\searrow} & & F|_M = f \\ M & \overset{f}{\longrightarrow} & \mathbf{R} & f \leq p \text{ on } M \end{array}$$

Since $f$ and $F$ are linear, $f \leq F$ on $M$ implies that $f(-m) \leq F(-m)$ for all $m \in M$ and therefore that $F = f$ on $M$.

**Proof of Theorem 7.5.1.** We use the notation of the statement. (a) $\Rightarrow$ (b). With notation as in the statement, suppose that $F \leq p$ and $r \leq F \circ v$. Let $\{a_1, a_2, \ldots, a_n\}$ be nonnegative scalars and let $\{t_1, t_2, \ldots, t_n\}$ be points in $T$. Since $r \leq F \circ v$, $a_i r(t_i) \leq a_i F(v(t_i))$ for each $i$; since $F \leq p$ is linear,

$$\sum_{i=1}^n a_i r(t_i) \leq \sum_{i=1}^n a_i F(v(t_i)) = F\left(\sum_{i=1}^n a_i v(t_i)\right) \leq p\left(\sum_{i=1}^n a_i (v(t_i))\right)$$

(b) $\Rightarrow$ (a). Given $p, v$ and $r$ as in the statement of the theorem, we construct a linear functional $F$ such that $F \leq p$ and $r \leq F \circ v$. Given $n \in \mathbf{N}$ and $x \in X$, consider the set

$$Q_n = \left\{ p\left(x + \sum_{i=1}^{n} a_i v(t_i)\right) - \sum_{i=1}^{n} a_i r(t_i) \right\}$$

as $\{a_1, a_2, \ldots, a_n\}$ ranges through all $n$-tuples of nonnegative scalars and $\{t_1, t_2, \ldots, t_n\} \subset T$. We show next that $q(x) = \inf_{n \in \mathbf{N}} Q_n$ is a sublinear functional. First, we show that $q$ is well-defined by showing that $-p(-x)$ is a lower bound for $Q_n$ for any $n$. By hypothesis, for any choice of nonnegative scalars $a_j$ and points $t_j$ of $T$,

$$\begin{aligned} \textstyle\sum_{i=1}^{n} a_i r(t_i) &\leq p\left(\textstyle\sum_{i=1}^{n} a_i v(t_i)\right) \\ &= p\left(x + \textstyle\sum_{i=1}^{n} a_i v(t_i)) - x\right) \\ &\leq p\left(x + \textstyle\sum_{i=1}^{n} a_i v(t_i)\right) + p(-x) \end{aligned}$$

which implies that

$$-p(-x) \leq p\left(x + \sum_{i=1}^{n} a_i v(t_i)\right) - \sum_{i=1}^{n} a_i r(t_i)$$

*Subadditive:* Suppose $x, y \in X$, $n \in \mathbf{N}$, $\{a'_1, a'_2, \ldots, a'_n\}$ and $\{a_1, a_2, \ldots, a_n\}$ are nonnegative scalars and $\{t'_1, t'_2, \ldots, t'_n\}, \{t_1, t_2, \ldots, t_n\} \subset T$. Then

$$\begin{aligned} &q(x+y) = \inf_{n \in \mathbf{N}} \left\{p\left(x + y + \textstyle\sum_{i=1}^{n} a_i v(t_i)\right) - \textstyle\sum_{i=1}^{n} a_i r(t_i)\right\} \\ &\quad \leq p\left[x + y + \textstyle\sum_{i=1}^{n} a_i v(t_i) + \textstyle\sum_{i=1}^{n} a'_i v(t'_i)\right] - \textstyle\sum_{i=1}^{n} a_i r(t_i) - \textstyle\sum_{i=1}^{n} a'_i r(t'_i) \\ &\quad \leq p\left[x + \textstyle\sum_{i=1}^{n} a_i v(t_i)\right] - \textstyle\sum_{i=1}^{n} a_i r(t_i) + p\left[y + \textstyle\sum_{i=1}^{n} a'_i v(t'_i)\right] - \textstyle\sum_{i=1}^{n} a'_i r(t'_i) \end{aligned}$$

which implies that $q(x+y) \leq q(x) + q(y)$.

*Positive homogeneous:* For $a \geq 0$ and $x \in X$,

$$\begin{aligned} q(ax) &= \inf_n \left[ p\left(ax + \sum_{i=1}^{n} a_i v(t_i)\right) - \sum_{i=1}^{n} a_i r(t_i) \right] \\ &= \inf_n \left[ p\left(ax + \sum_{i=1}^{n} a a_i v(t_i)\right) - \sum_{i=1}^{n} a a_i r(t_i) \right] \\ &= a \inf_n \left[ p\left(x + \sum_{i=1}^{n} a_i v(t_i)\right) - \sum_{i=1}^{n} a_i r(t_i) \right] = a q(x) \end{aligned}$$

$F \leq p:$ Clearly $q \leq p$ and by Theorem 7.3.4 there exists a linear functional $F \leq q$.

$r \leq F \circ v:$ By the definition of $q$, for any $t \in T$,

$$q(-v(t)) \leq p(-v(t) + v(t)) - r(t) = -r(t)$$

Since $F \leq q$, $F(-v(t)) \leq q(-v(t)) \leq -r(t)$ and therefore $F(v(t)) \geq r(t)$. ■

## 7.6 MINIMAL SUBLINEAR FUNCTIONALS

We denote the class of sublinear functionals on a vector space $X$ by $X^{\#}$. $X^{\#}$ is not a linear space but it is closed under the pointwise operations of addition and multiplication by numbers $a \geq 0$. We order $X^{\#}$ by taking, for $p, q \in X\#$, $p \leq q$ to mean that $p(x) \leq q(x)$ for every $x$ in $X$.

The goal of the section is a different proof of the dominated extension theorem, Theorem 7.3.2(a). We prove it by establishing two things: (1) The linear functionals on a real vector space $X$ are the minimal sublinear functionals, i.e., the minimal elements of $X^{\#}$ [Theorem 7.6.3]; and (2) "beneath" each sublinear functional, there is a linear one [Theorem 7.6.4]. Theorem 7.6.1 is an elementary criterion for linearity of a sublinear functional.

**Theorem 7.6.1** LINEAR IFF $p(x)+p(-x) \leq 0$ If $p$ is a sublinear functional on a real vector space $X$ then the following are equivalent:
(a) $p$ is a linear functional;
(b) for every $x \in X$, $p(x) + p(-x) = 0$;
(c) for every $x \in X$, $p(x) + p(-x) \leq 0$.

**Proof.** We use the notation of the statement. (a) $\Rightarrow$ (b) If $p$ is linear then $0 = p(0) = p(x - x) = p(x) + p(-x)$. Clearly, (b) $\Rightarrow$ (c). (c) $\Rightarrow$ (a): If (c) holds then $0 = p(0) = p(x - x) \leq p(x) + p(-x) \leq 0$, so $p(x) + p(-x) = 0$; hence $p(-x) = -p(x)$ and it follows that, for any scalar $a$, $p(ax) = ap(x)$. For all $x, y \in X$, $p(x) = p(x + y - y) \leq p(x + y) + p(-y) = p(x + y) - p(y)$ which implies that $p(x) + p(y) = p(x + y)$. ■

The following result has many important applications; in particular, it creates a sublinear functional beneath a given sublinear functional.

**Theorem 7.6.2** AUXILIARY FUNCTIONALS If $p$ is a sublinear functional on a real vector space $X$ then the *auxiliary functional* $q(x) = \inf\{p(x+tw) - tp(w) : t \geq 0, w \in X\}$ $(x \in X)$ is a sublinear functional such that $q \leq p$.

**Proof.** Given $p$ in $X^{\#}$, $x, w \in X$ and $t \geq 0$, $p(tw) \leq p(x + tw) + p(-x)$. Thus $-p(-x) \leq p(x + tw) - p(tw)$ and we can define the *auxiliary functional* (associated with $p$)

$$q(x) = \inf\{p(x + tw) - tp(w) : t \geq 0,\ w \in X\} \qquad (*)$$

By letting $t = 0$, it is clear that $q \leq p$. We now show that $q$ is sublinear. If $a = 0$ then $p(ax + tw) - tp(w) = p(tw) - tp(w) = 0$ for all $t \geq 0$ and $w \in X$ so $q(0x) = q(0) = 0$. For $a > 0$,

$$\begin{aligned} q(ax) &= \inf\{p(ax + tw) - tp(w) : t \geq 0, w \in X\} \\ &= \inf\{ap(x + (t/a)w) - a(t/a)p(w) : t \geq 0, w \in X\} \\ &= a \inf\{p(x + (t/a)w) - (t/a)p(w) : t \geq 0, w \in X\} = aq(x) \end{aligned}$$

Subadditivity: For $x, y \in X$ and $s, t \geq 0$ but not both 0, let $w = (1/(s+t))(sx+ty)$ so that $sx + ty = (s+t)w$. Then

$$\begin{aligned} q(x+y) &\leq p(x+y+(s+t)w) - (s+t)p(w) \\ &\leq p(x+sw+y+tw) - sp(w) - tp(w) \\ &\leq p(x+sw) - sp(w) + p(y+tw) - tp(w) \end{aligned}$$

which implies that $q(x+y) \leq q(x) + q(y)$. ■

**Theorem 7.6.3** SUBLINEAR LINEAR IFF MINIMAL A sublinear functional $p$ on a real vector space $X$ is a minimal element of $X^{\#}$ iff $p$ is linear.

**Proof.** Suppose that $p, q \in X^{\#}$, $q \leq p$, $p$ is linear and $x \in X$. Since $q$ is sublinear, $0 = q(x-x) \leq q(x) + q(-x)$ so $-q(-x) \leq q(x)$. Since $q(-x) \leq p(-x) = -p(x)$, it follows that $p(x) \leq -q(-x) \leq q(x)$, i.e., $p \leq q$. Therefore $q = p$ and $p$ is a minimal element of $X^{\#}$.

Conversely, suppose that $p$ is a minimal element of $X^{\#}$ and let $q$ be the auxiliary sublinear functional of Theorem 7.6.2. Since $q \leq p$, the minimality of $p$ implies that $q = p$. If we let $t = 1$ and $x = -w$ then, by the definition of $q$ in $(*)$, $p(-w) = q(-w) \leq p(-w+w) - p(w) = -p(w)$. Hence $p(-w) + p(w) \leq 0$ for any $w \in X$ and the linearity of $p$ follows from Theorem 7.6.1(a,c). ■

**Theorem 7.6.4** LINEAR BELOW SUBLINEAR For any sublinear functional $p$ on a real vector space $X$ there is a linear functional $f$ on $X$ such that $f \leq p$.

**Proof.** Let $p \in X^{\#}$. Clearly, the class $L = \{q \in X^{\#} : q \leq p\}$ is nonempty. We now show that $L$ is inductively ordered so that it will have a minimal element $f$ by Zorn's lemma; $f$ must be linear by Theorem 7.6.3.

Let $P$ be a totally ordered subset of $L$. We claim that for any $x \in X$, $\{q(x) : q \in P\}$ is bounded below. If not, then for each $n \in \mathbf{N}$ there exists $p_n \in P$ such that $p_n(x) \leq -n$. Since $P$ is totally ordered, $q_n = \min(p_1, \ldots, p_n) \in P$ for each $n \in \mathbf{N}$. Thus $(q_n)$ is a decreasing sequence of sublinear functionals with the property that $q_n(x) \leq -n$ for every $n$. Hence

$$0 = q_n(x-x) \leq q_n(x) + q_n(-x) \leq -n + q_n(-x)$$

Hence $n \leq q_n(-x) \leq q_1(-x)$ for every $n$. As this contradicts the fact that $q_1$ is real-valued, $p^*(x) = \inf\{q(x) : q \in P\}$ is a real number for every $x \in X$. For any $q \in P$, $p^* \leq q \leq p$, i.e., $p^* \leq p$. Hence it only remains to show that $p^*$ is sublinear to prove that $p^*$ is a lower bound for $P$.

By standard properties of infima, $p^*(0) = 0$ and $p^*(tx) = tp^*(x)$ for all $t \geq 0$. Let $r, q \in P$. Since $P$ is totally ordered, we may assume that $r \leq q$; hence, for any $x, y \in X$, $p^*(x+y) \leq r(x+y) \leq r(x) + q(y)$ and, similarly, $p^*(x+y) \leq q(x) + r(y)$. Therefore $p^*(x+y) \leq \inf_{q \in P}[q(x) + r(y)] = p^*(x) + r(y) \leq p^*(x) + p^*(y)$. ■

We can now reprove the dominated version of the Hahn–Banach theorem. It suffices to establish the real case, the complex version following exactly as in Theorem 7.3.2(b).

**Theorem 7.6.5** DOMINATED EXTENSION THEOREM Let $X$ be a real vector space, $p$ a sublinear functional on $X$, $M$ a subspace of $X$ and $f : M \to \mathbf{R}$ a linear functional on $M$ such that $f \leq p$. Then there is a linear functional $F$ on $X$ which extends $f$ and such that $F \leq p$.

**Proof.** With notation as above, for any $m \in M$, $f(-m) \leq p(-m)$. For $x \in X$, therefore $-p(-x) + f(-m) \leq -p(-x) + p(-m)$. Thus,

$$-p(-x) \leq -p(-x) + p(-m) - f(-m) \tag{1}$$

Since $p(-m) \leq p(x-m)+p(-x)$, $-p(-x)+p(-m) \leq p(x-m)$ and (1) yields $-p(-x) \leq p(x-m) + f(m)$. The map

$$q(x) = \inf\{p(x-m) + f(m) : m \in M\} \quad (x \in X)$$

is therefore real-valued. Two things are immediate: $q \leq f$ on $M$ (let $m = x$) and $q \leq p$ on $X$ (let $m = 0$). We now show that $q$ is sublinear.

For each $m$ in $M$, $f(-m) \leq p(-m)$; hence $f(m) + p(-m) \geq 0$ for every $m$ in $M$ so $q(0) \geq 0$. By letting $m = 0$ in $p(0 - m) + f(m)$, it follows that $0 \in \{f(m) + p(-m) : m \in M\}$, so $q(0) \leq 0$ and therefore that $q(0) = 0$.

For $a > 0$, $q(ax) = \inf\{p(ax - m) + f(m) : m \in M\} = \inf\{ap(x - m/a) + af(m/a) : m \in M\} = aq(x)$. Given $x, y \in X$ and $r > 0$, there exist $m, n \in M$ such that

$$p(x-m) + f(m) < q(x) + r/2 \text{ and } p(y-n) + f(n) < q(y) + r/2$$

Therefore,

$$\begin{aligned} q(x) + q(y) &> p(x-m) + p(y-n) + f(m+n) - r \\ &\geq p(x+y-(m+n)) + f(m+n) - r \\ &\geq q(x+y) - r \end{aligned}$$

and the subadditivity of $q$ follows. By Theorem 7.6.4 there exists a linear functional $F \leq q \leq p$ on $X$. Since $F \leq q \leq f$ on $M$, it follows that $F = f$ on $M$ since $f$ is a minimal element of $M^\#$ [Theorem 7.6.3]. ■

## 7.7 GEOMETRIC FORM

The geometric form of the Hahn–Banach theorem [Theorem 7.7.3] generalizes the notion that if a line $L$ does not pierce a sphere $B$ in $\mathbf{R}^3$ then $L$ is contained in a plane disjoint from $B$. In the geometric form, we replace "line" by "linear variety," "ball" by "open convex set," and "plane" by "hyperplane." This version is equivalent to the Hahn–Banach extension theorem [Theorem 7.3.2].

Before proving the geometric version, we establish a connection between domination by a sublinear functional and separation [Theorem 7.7.1] and that any open convex set $G$ in a TVS is expressible in the form $\{x : p(x-w) < 1\}$ for some sublinear functional $p$ [Theorem 7.7.2].

Let $p$ be a sublinear functional on the real or complex vector space $X$ and let $V_p = \{x \in X : p(x) < 1\}$. It is easy to verify [cf. Theorem 5.2.1] that:

(1) $V_p$ is convex;
(2) $rV_p = \{x \in X : p(x) < r\}$ for any $r > 0$;
(3) for $w \in X$, $w + V_p = \{x \in X : p(x-w) < 1\}$.

If a sublinear functional $p$ is such that $p(x) \geq 0$ for every $x \in X$ then $p$ is called *positive*. In regard to Theorem 7.7.1(b), note that a linear functional $f$ on a TVS is continuous iff there is a continuous seminorm $p$ such that $|f| \leq p$ [Theorem 5.7.3].

**Theorem 7.7.1** Let $f$ be a nontrivial linear functional on a real vector space $X$, let $p$ be a positive sublinear functional and let $V_p = \{x \in X : p(x) < 1\}$. Then

(a) SEPARATION AND DOMINATION $f \leq p$ iff $f^{-1}(1) \cap V_p = \varnothing$.

(b) CONTINUITY AND DOMINATION If $X$ is a TVS, $p$ is continuous and $f \leq p$ then $f$ is continuous.

**Proof.** We use the notation of the statement. (a) Suppose that $f \leq p$ and let $H = \{x \in X : f(x) = 1\} = f^{-1}(1)$. Then if $x \in V_p$, $f(x) < 1$, so $x \notin H$, i.e., $f^{-1}(1) \cap V_p = \varnothing$. Conversely, suppose there exists $x \neq 0$ such that $f(x) > p(x) \geq 0$ and let $a = 1/f(x)$. Then $p(ax) < f(ax) = 1$ and $ax \in f^{-1}(1) \cap V_p$.

(b) Suppose $p$ is continuous and $f \leq p$. Since $V_p$ is an open convex neighborhood of 0, it contains a disked subneighborhood $D$ of 0; let $q$ be the gauge of $D$. Since $D \subset V_p$, $p \leq q$. Since $f \leq p$, for every $x \in X$, $f(-x) \leq p(-x) \leq q(-x) = q(x)$, i.e., $|f| \leq q$. Since $D$ is a neighborhood of 0, $q$ is continuous [Theorem 5.3.3(b)]; hence, $f$ is continuous by Theorem 5.7.3. ■

In Secs. 5.1–5.3 we noted some connections between seminorms and absorbent disks. If $p$ is seminorm then $V_p = \{x : p(x) < 1\}$ and $\overline{V}_p = \{x : p(x) \leq 1\}$ are absorbent disks. If $D$ is an absorbent disk in a TVS then its gauge

$$p_D(x) = \inf\{a > 0 : x \in aD) \qquad (x \in X) \tag{$*$}$$

is a seminorm and $V_{p_D} \subset D \subset \overline{V}_{p_D}$. Also, a seminorm $p$ on a TVS is continuous iff $V_p$ is open [Theorem 5.2.2].

If we take an absorbent convex set $D$ instead of an absorbent disk—i.e., if we drop "balanced"—and consider the gauge function $p_D$ of $(*)$, the only difference is that $p_D$ is positive homogeneous instead of absolutely homogeneous; in other words, the gauge $p_D$ of an absorbent convex set is a positive sublinear functional.

One difference between positive sublinear functionals and seminorms is the loss of the second triangle inequality $|p(x) - p(y)| \leq p(x - y)$. It is replaced by the weaker condition of Theorem 7.7.2(a) below.

**Theorem 7.7.2** CONTINUITY OF SUBLINEAR FUNCTIONALS If $X$ is a vector space over $\mathbf{R}$ or $\mathbf{C}$ and $p$ is a sublinear functional on $X$ then:

(a) $p(x) - p(y) \leq p(x - y)$ for any $x, y \in X$.

If $X$ is a TVS then

(b) $p$ is uniformly continuous iff $p$ is continuous at 0;

(c) if $p$ is positive then $p$ is continuous iff $V_p = \{x \in X : p(x) < 1\}$ is open;

(d) if $D$ is an open convex neighborhood of 0 and $p_D$ is its gauge (as in Eq. $(*)$ above) then $D = V_{p_D}$ and $p_D$ is continuous;

(e) the open convex subsets of $X$ are those of the form $w + V_p = \{x \in X : p(x - w) < 1\}$ for some $w \in X$ and continuous positive sublinear functional $p$.

**Proof.** Let $p$ be a sublinear functional on $X$. (a) For any $x, y \in X$, $p(x) \leq p(x - y) + p(y)$.

(b) We prove only the sufficiency. Suppose $p$ is continuous at 0 and $x_s \to x$. Then $x_s - x \to 0$ and $x - x_s \to 0$. Thus, given $r > 0$, for sufficiently large $s$, $p(x - x_s) < r$ and $p(x_s - x) < r$. By (a) it follows that $|p(x) - p(x_s)| < r$.

(c) Again we prove only sufficiency. Suppose $V_p$ is open and $x_s \to 0$. For any $r > 0$, $rV_p = \{x \in X : p(x) < r\}$ is open so $x_s \in rV_p$ eventually, i.e., $p(x_s) < r$ eventually. Thus $p(x_s) \to 0$ and $p$ is continuous at 0; $p$ is therefore continuous by (b).

(d) Suppose $D$ is an open convex neighborhood of 0 and $p_D$ is its gauge. Let $x \in D$. Since $(1 + 1/n)x \to x$, $x \in (1 + 1/n)^{-1}D$ eventually. Therefore $p_D(x) < 1$ and $D \subset V_{p_D}$. Conversely, if $p_D(x) < 1$, there exists $b > 1$ such that $bx \in D$. Since $D$ is convex and $0 \in D$, the line segment $[0, bx] = \{(1 - t)0 + tbx : 0 \leq t \leq 1\} \subset D$, which implies that $x \in D$. The continuity of $p_D$ follows from (c).

(e) If $G$ is an open convex set and $w \in G$ then $D = G - w$ is an open convex neighborhood of 0. Let $p_D$ be the gauge of $D$ as in Eq. $(*)$. By (d), $G - w = D = V_{p_D}$. ■

The result below was essentially first proved by Mazur [1933] for normed spaces. Bourbaki subsequently called it the *geometric form* of the Hahn–Banach theorem.

**Theorem 7.7.3** GEOMETRIC FORM Let $G$ be an open convex subset of a TVS $X$ over $\mathbf{F} = \mathbf{R}$ or $\mathbf{C}$ and let $M$ be a linear variety of $X$. If $M \cap G = \varnothing$ then there is a closed hyperplane $H \supset M$ such that $H \cap G = \varnothing$.

**Proof.** With notation as above, first note that we may translate things so that $M$ is a subspace and prove the existence of a maximal subspace $H \supset M$ which does not meet $G$. Second, it suffices to prove the theorem in the case $\mathbf{F} = \mathbf{R}$. If that has been done and $\mathbf{F} = \mathbf{C}$, view $X$ as a real vector space and $M$ as an $\mathbf{R}$-subspace. The result for the real case then yields a maximal $\mathbf{R}$-subspace $H \supset M$ which does not meet $G$. Then $iH \supset iM = M$ and $H \cap iH$ is a maximal complex subspace by Theorem 1.5.2(a) which does not meet $G$.

Suppose that $\mathbf{F} = \mathbf{R}$ and that $M$ is a subspace of $X$. Since $G$ is an an open convex set, there exists a continuous positive sublinear functional $p$ such that $G = w + V_p = \{x \in X : p(x - w) < 1\}$ for some $w \in X$ [Theorem 7.7.2(e)]. Since $G \cap M = \varnothing$, $p(m - w) \geq 1$ for each $m \in M$. Define $f$ on $M + \mathbf{R}w$ by taking $f(m + aw) = -a$ $(m \in M,\ a \in \mathbf{R})$. We show next that $f \leq p$ on $M \oplus \mathbf{R}w$. For $a \geq 0$, $f(m + aw) = -a \leq 0 \leq p(m + aw)$. For $a < 0$, $f(m+aw) = -a \leq -ap(-m/a - w)$ [since $p(y - w) \geq 1$ for $y \in M$] $= p(m+aw)$. Thus $f \leq p$ on $M \oplus \mathbf{R}w$. By the Hahn–Banach extension theorem [7.3.2(a)], $f$ has a linear extension $F$ such that $F \leq p$ everywhere on $X$. By Theorem 7.7.1(b), $F$ is continuous. Since $f$ vanishes on $M$, the null space $N(F)$ of $F$ is a closed maximal subspace of $X$ which contains $M$ [Theorem 4.9.4(a)]. To see that $N(F) \cap G = \varnothing$, we show that if $x \in N(F)$ then $x \notin G$. If $F(x) = 0$ then, since $F \leq p$, $F(x-w) = F(x) - F(w) = -f(w) = 1 \leq p(x-w)$, so $x \notin G$. ■

It is common to describe separation of sets in topological spaces by means of real-valued continuous functions—as in completely regular spaces, for example. Theorems 7.7.4 and 7.7.5 effect separations of this type as do all the results of Sec. 7.8.

**Theorem 7.7.4** SEPARATING OPEN CONVEX SET AND SUBSPACE Let $M$ be a subspace of a TVS $X$ over $\mathbf{F} = \mathbf{R}$ or $\mathbf{C}$ and $G$ a nonempty open convex subset which does not meet $M$. Then there is a continuous linear functional $f$ on $X$ such that $f = 0$ on $M$ and

(a) $f > 0$ on $G$ if $\mathbf{F} = \mathbf{R}$, or

(b) $\operatorname{Re} f > 0$ on $G$ if $\mathbf{F} = \mathbf{C}$.

**Proof.** We adhere to the notation of the statement. (a) $\mathbf{F} = \mathbf{R}$: We know that there is a closed maximal subspace $H \supset M$ which does not meet $G$ by Theorem 7.7.3. We may assume that $H$ is the null space $N(f)$ of some continuous linear functional $f$ by Theorem 4.9.4(a). Can $f$ change sign on $G$? If there are vectors $x$ and $y$ in $G$ such that $f(x) = a > 0$ and $f(y) = b < 0$, let $c = -a/b$ or $-b/a$, whichever makes $c \geq 1$. Now, with $c = -a/b$, $x/(1+c) + cy/(1+c) = w \in G$ and $f(w) = 0$ which contradicts the fact that $H \cap G = \varnothing$. It only remains to choose $f$ or $-f$.

(b) $\mathbf{F} = \mathbf{C}$: View $X$ as a real vector space. By (a) there is a real linear functional $g$ such that $g > 0$ on $G$ and $g = 0$ on $M$. Let $f$ be the continuous linear functional defined by taking $f(x) = g(x) - ig(ix)$ at each $x \in X$ and let

$H = g^{-1}(0)$. By Theorem 1.5.2(a), $N(f) = H \cap iH \supset M$ (since $iM = M$). ■

**Theorem 7.7.5** POINT AND CLOSED SUBSPACE Let $X$ be a LCS over **R** or **C**, $M$ a closed subspace of $X$ and $x \notin M$. Then there is a continuous linear functional $f$ on $X$ which vanishes on $M$ but not on $x$.

**Proof.** We use the notation of the statement. Since $X$ is locally convex and $x$ is not in the closed set $M$, there is an open convex neighborhood $G$ of $x$ which does not meet $M$. By the previous result, there is a continuous linear functional $f$ which vanishes on $M$ but not on $G$, hence not on $x$. ■

**Theorem 7.7.6** ABUNDANCE OF CONTINUOUS LINEAR FUNCTIONALS Let $X$ be a LCS over **R** or **C** and $x \notin \mathrm{cl}\{0\}$. Then there is a continuous linear functional $f$ such that $f(x) \neq 0$.

**Proof.** Since $\mathrm{cl}\{0\}$ is a closed subspace we need only apply the preceding result. ■

As a consequence of Theorem 7.7.6, if $X$ is a LCHS and $x \notin \{0\} = \mathrm{cl}\{0\}$, there exists a continuous linear functional $f$ which does not vanish on $x$—the dual $X'$ of a LCHS $X$ is total, in other words. This and a few other corollaries like it are listed in Theorem 7.7.7 below.

**Theorem 7.7.7** CONTINUOUS FUNCTIONALS ON LCHS Let $X$ be an LCHS and $X'$ its continuous dual. Then:

(a) $X'$ separates the points of $X$ in the sense that if $x \neq y$ $(x, y \in X)$, there is $f \in X'$ such that $f(x) \neq f(y)$; if $x$ and $y$ are linearly independent, there exists $f \in X'$ such that $f(x) = 0$ and $f(y) = 1$.

(b) If $f(x) = 0$ for all $f \in X'$ then $x = 0$.

(c) If $M$ is a subspace of $X$ and $w$ a vector such that, for any $f \in X'$, $f(M) = 0$ implies $f(w) = 0$ then $w \in \mathrm{cl}\, M$.

**Theorem 7.7.8** NONTRIVIAL DUAL AND CONVEX NEIGHBORHOOD OF 0 A TVS $X$ has a nontrivial dual iff there is a proper convex neighborhood of 0.

**Proof.** Let $V$ be a proper convex neighborhood of 0 in then TVS $X$ and let $p_V$ be its gauge [Eq. $(*)$ above]. For all $r > 0$ and $x \in rV$, $p_V(x) \leq r$ so $p_V$ is continuous at 0 and therefore continuous by Theorem 7.7.2(b). Since $V$ is proper, there exists $w \notin V$ so that $p_V(w) > 0$. By Theorem 7.3.4(a) there is a continuous linear functional $f$ on $X$ such that $f(w) = p_V(w)$. Conversely, if $f$ is a nontrivial continuous linear functional $f$ on $X$, $V = \{x \in X : |f(x)| \leq 1\}$ is a proper convex neighborhood of 0. ■

Since the space $\left(L_p(-\infty, \infty), \|\cdot\|^*\right)$ of Example 4.6.2 has no proper convex neighborhood of 0, It follows that $\left(L_p(-\infty, \infty), \|\cdot\|^*\right)$ has a trivial dual.

**Example 7.7.9** CAUCHY INTEGRAL THEOREM If $x$ maps a region $D$ of $\mathbf{C}$ into a TVS $X$, we have a notion of analyticity for $x$ [Definition 6.6.2] and proved a Liouville-type theorem [Theorem 6.6.5] for TVS $X$ with total duals. By Theorem 7.7.7(b), LCHS possess total duals, thus validating the Liouville theorem for them. We can now prove: If $x$ maps a region $D$ of $\mathbf{C}$ bounded by a rectifiable Jordan arc $C$ into a Banach space $(X, \|\cdot\|)$ and $x$ is analytic in $D$ and continuous on $C$ then $\int_C x(s)ds = 0$ (line integral limit in the norm $\|\cdot\|$).

**Proof.** Let $f$ be a continuous linear functional on the TVS $X$. With $y = \int_C x(s)\, ds$, since $f$ is linear and continuous, $f(y) = \int_C f(x(s))\, ds$. Since $x$ is analytic in $D$ and continuous on $C$, so is the complex-valued function $f(x(\cdot))$. Therefore $f(y) = 0$ by the ordinary Cauchy integral theorem. Since $f$ is an arbitrary continuous linear functional, $y = 0$ by Theorem 7.7.7(b). ■

## 7.8 SEPARATION OF CONVEX SETS

In Theorem 7.7.4 we showed that an open convex set $G$ and a subspace $M$ could be separated by a continuous linear functional $f$ in the sense that $f = 0$ on $M$ but is positive on $G$. Thus

$$\sup f(M) < \inf f(G) \tag{s}$$

Separating convex sets in this manner is the theme of this section. We consider three kinds:

**Definition 7.8.1** SEPARATED BY A HYPERPLANE Let $A$ and $B$ be disjoint convex subsets of a vector space $X$ and let $f$ be a real nontrivial linear functional on $X$. For $c \in \mathbf{R}$ and $H = f^{-1}(c)$ we say that $A$ and $B$ are:

(a) *separated by the hyperplane* $H$ if for all $a \in A$ and $b \in B$, $f(a) \leq c \leq f(b)$, i.e., $\sup f(A) \leq \inf f(B)$; equivalently, $A$ and $B$ are separated by a hyperplane iff $\{0\}$ is separated from the algebraic difference $B - A$, a convex set, by a hyperplane;

(b) *strictly separated by the hyperplane* $H$ if for all $a \in A$ and $b \in B$, $f(a) < c < f(b)$; equivalently, $\{0\}$ is strictly separated from $B - A$ by a hyperplane;

(c) *strongly separated by the hyperplane* $H$ if there is some $r > 0$ such that for all $a \in A$ and $b \in B$, $f(a) \leq c - r < c + r \leq f(b)$; equivalently, $\{0\}$ is strongly separated from $B - A$ by a hyperplane, or $\sup f(A) < \inf f(B)$].

Let us interpret these things geometrically now. With notation as in Definition 7.8.1, the hyperplane $H = f^{-1}(c)$ splits $X$ into a lower, a middle and an upper convex part:

$$L = \{x \in X : f(x) \leq c\}, \ H, \quad \text{and} \quad U = \{x \in X : f(x) \geq c\}$$

$L$ and $U$ are called the *half spaces* determined by $H$. When defined by strict inequalities, $L$ and $U$ are called the *strict half spaces* determined by $H$. If a set $E$ is wholly in $L$ or $U$, we say that $E$ *lies to one side of* $H$. With respect to these conventions, $A$ and $B$ are

(a) separated if they lie in the distinct half spaces determined by $H$; and

(b) strictly separated if they lie in the strict half spaces determined by $H$.

The geometric interpretation of strong separation is that $A$ and $B$ are on opposite sides of hyperplanes [namely, $f^{-1}(c-r)$ and $f^{-1}(c+r)$] "parallel" to $H$, lying a positive distance to either side of it. In the first case, $A$ and $B$ separated, $A$ and $B$ can actually meet, while if they are strictly separated, they cannot—although possibly $\sup f(A) = \inf f(B)$. They are strongly separated iff

$$\sup f(A) < \inf f(B) \tag{ss}$$

As we show next, a convex set does not meet a hyperplane iff it lies strictly to one side of it.

**Theorem 7.8.2** Convex Sets and Half Spaces Let $X$ be a real vector space. A convex subset $G$ of $X$ is strictly to one side of a hyperplane $H$ iff $G \cap H = \varnothing$. If $H = \{x \in X : f(x) = c\}$ for some linear functional $f$ on $X$ and real number $c$, this means that $G \cap H = \varnothing$ implies that $f(G) \subset (-\infty, c)$ or $f(G) \subset (c, \infty)$.

**Proof.** We use the notation of the statement. The necessity of the condition is obvious. Conversely, suppose that $H = f^{-1}(c)$ for some linear functional $f$ on the real vector space $X$ and real number $c$. If the convex set $G \subset X$ is not strictly on one side of $H$, there exist $x, y \in G$ such that $f(x) = a \le c \le b = f(y)$. Since $c \in [a, b]$, there exist $\alpha, \beta \ge 0$, $\alpha + \beta = 1$, such that $\alpha a + \beta b = c$. Since $c = f(\alpha x + \beta y)$ and $G$ is convex, $\alpha x + \beta y \in G \cap H$. ■

So far everything has been purely algebraic. Our main interest, of course, centers on what happens in *topological* vector spaces, in situations where $f$ is a *continuous* linear functional ($H$ is a *closed* hyperplane) and the disjoint convex sets $A$ and $B$ have additional topological properties such as openness [Theorem 7.8.4] or one compact and the other closed [Theorem 7.8.6]. As our first entrance into these waters, we note the following fact.

**Theorem 7.8.3** Half Spaces Let $X$ be a real TVS and let $H = \{x \in X : f(x) = c\}$ for some nontrivial continuous linear functional $f$ on $X$ and real number $c$. If $G$ has nonempty interior and lies to one side of $H$ then $\operatorname{cl} G$ is also on one side of $H$ and $\operatorname{int} G$ is strictly to one side of $H$.

**Proof.** Suppose $f$ is a nontrivial continuous linear functional on the real TVS $X$, that $G \subset X$ has nonempty interior and $f(G) \subset (-\infty, c]$ for some $c \in \mathbf{R}$. Then $f(\operatorname{cl} G) \subset \operatorname{cl} f(G) \subset (-\infty, c]$. Since $f$ is nontrivial, $f$ is an open map [Theorem 5.7.7(c)]. Thus, since $f(G) \subset (-\infty, c]$, $\operatorname{int} f(G) = f(\operatorname{int} G) \subset (-\infty, c)$. ■

Having established some preliminary results, we can now prove our first separation theorems for convex sets. In Theorems 7.8.4 and 7.8.5, the complex variant follows trivially from the real version.

**Theorem 7.8.4** SEPARATING OPEN CONVEX SETS Let $A$ and $B$ nonempty disjoint convex subsets of a TVS $X$ over $\mathbf{F} = \mathbf{R}$ or $\mathbf{C}$.

(a) If $\mathbf{F} = \mathbf{R}$ and $A$ is open then $A$ and $B$ are separated by a closed hyperplane; if $B$ is open as well then $A$ and $B$ are strictly separated by a closed hyperplane.

(b) If $\mathbf{F} = \mathbf{C}$ and $A$ is open, there is a continuous linear functional $g$ on $X$ and real number $c$ such that $\operatorname{Re} g > c$ on $A$ and $\operatorname{Re} g \le c$ on $B$; if $B$ is open too then $\operatorname{Re} g < c$ on $B$.

**Proof.** We use the notation of the statement. (a) The algebraic difference $A - B = \cup_{x \in B}(A - x)$ is open, convex, and does not meet $\{0\}$. Therefore, by Theorem 7.7.4(a), there is a continuous linear functional $f$ on $X$ which is positive on $A - B$. Hence $f(x) > f(y)$ for any $x \in A$ and $y \in B$ and $f(A)$ is bounded below. Let $a = \inf f(A)$. Then for any $x \in A$ and $y \in B$, $f(x) \ge a \ge f(y)$. Since $f$ is nontrivial, $f$ is an open map [Theorem 5.7.7(c)]. Since $A$ is open and $f(A) \subset [a, \infty)$ [Theorem 7.8.3], $\operatorname{int} f(A) = f(\operatorname{int} A) = f(A) \subset (a, \infty)$. If $B$ is open as well then $f(B) \subset (-\infty, a)$, by the same argument.

(b) View $X$ as a real TVS, let $f$ be as in (a), and let $g$ be the complex linear functional, $g(x) = f(x) - if(ix)$ [Theorem 1.4.2] whose real part is $f$. ■

**Theorem 7.8.5** STRONG SEPARATION Let $A$ and $B$ be disjoint nonempty convex subsets of a real or complex TVS $X$.

(a) $A$ and $B$ are strongly separated by a closed hyperplane iff there is an open convex neighborhood $V$ of 0 such that $(A+V) \cap B = \varnothing$ [or, equivalently, $V \cap (B - A) = \varnothing$].

(b) If $X$ is a LCS then $A$ and $B$ are strongly separated by a closed hyperplane iff $0 \notin \operatorname{cl}(B - A)$.

**Proof.** We use the notation of the statement. (a) Let $V$ be an open convex neighborhood of 0 such that $(A+V) \cap B = \varnothing$. Since $A+V$ is open and convex [Theorems 2.2.3 and 4.2.4] and disjoint from $B$, there is a closed hyperplane $f^{-1}(c)$ for some continuous real linear functional $f$ and $c \in \mathbf{R}$ which separates $(A+V)$ and $B$ by Theorem 7.8.4(a). Since $f$ is nontrivial, it is an open map by Theorem 5.7.7(c). Hence there is some positive number $r$ such that the open interval $(-r, r) \subset f(V)$ and we may choose $v \in V$ such that $f(v) < 0$. Since $f^{-1}(c)$ separates $(A+V)$ and $B$, it follows that, for all $a \in A$ and $b \in B$, $f(a) + f(v) \ge f(b)$ (using $-f$, if necessary). Hence $\inf f(A) > \sup f(B)$.

Conversely, suppose that $A$ and $B$ are strongly separated by a closed hyperplane $f^{-1}(c)$ as above so that there is some $r > 0$ such that for all $a \in A$ and $b \in B$

$$f(b) \leq c - r < c + r \leq f(a) \qquad (*)$$

Let $V = f^{-1}(-r, r)$. Clearly, $V$ is an open convex neighborhood of 0 and if $(A + V) \cap B \neq \varnothing$, there are elements $a \in A$, $b \in B$, and $v \in V$ such that $f(b) = f(a) + f(v)$. This implies that $|f(b) - f(a)| = |f(v)| < r$, which contradicts $(*)$. We conclude that $(A + V) \cap B = \varnothing$.

(b) If $A$ and $B$ are strongly separated then, by (a), there is a convex neighborhood $V$ of 0 such that $V \cap (B - A) = \varnothing$, so $0 \notin \text{cl}(B - A)$. Conversely, if $0 \notin \text{cl}(B - A)$, there is a convex neighborhood $V$ of 0 such that $V \cap (B - A) = \varnothing$, i.e., $(A + V) \cap B = \varnothing$ and the strong separation follows from (a). ■

**Theorem 7.8.6** CLOSED AND COMPACT CONVEX SETS Let $A$ and $B$ be nonempty disjoint convex subsets of the LCS $X$ over $\mathbf{F} = \mathbf{R}$ or $\mathbf{C}$.

If $\mathbf{F} = \mathbf{R}$ then:

(a) If $A$ is closed and $B$ is compact then they are strongly separated by a closed hyperplane, i.e., there exists a continuous linear functional $f$ on $X$ such that $\sup f(A) < \inf f(B)$.

(b) If $A$ is closed and $w \notin A$, there is a continuous linear functional $f$ on $X$ such that $f(w) > \sup f(A)$. [With $g = -f$, note that $g(w) < \inf g(A)$.]

If $\mathbf{F} = \mathbf{C}$ then:

(c) If $A$ is closed and $B$ is compact then there is continuous linear functional $g$ on $X$ and real number $c$ such that $\text{Re}\, g < c$ on $A$ and $\text{Re}\, g > c$ on $B$.

(d) If $A$ is closed and $w \notin A$ then there is a continuous linear functional $g$ on $X$ such that $\text{Re}\, g(w) > \sup(\text{Re}\, g)(A)$.

**Proof.** We use the notation of the statement. Once (a) is established, (b) follows by taking $B = \{w\}$; (c) and (d) follow from (a) and (b) by viewing $X$ as a real space, taking $f$ as in (a) and (b), respectively, and letting $g$ be the complex linear functional whose real part is $f$. Therefore we prove only (a).

Suppose that $A$ is closed and $B$ is compact. The compact set $B$ is contained in the open set $\complement A$ so, by Theorem 2.2.5 and the local convexity of $X$, there is a convex neighborhood $V$ of 0 such that $B + V \subset \complement A$. The desired result now follows from Theorem 7.8.5(a). ■

In $\mathbf{R}^2$ a square is the intersection of four closed half planes, closed half spaces actually; triangles are the intersection of three closed half planes. A circle is the intersection of the closed half planes determined by its tangents. The generalization of these notions to general convex subsets of locally convex spaces is contained in Theorem 7.8.7(b) below. As we point out in the proof, Theorem 7.8.7(a,b) provides geometric reformulations of Theorems 7.7.5 and Theorem 7.8.6(b), respectively. These descriptions are internal in that they do not rely on functions on the space.

**Theorem 7.8.7** STRUCTURE OF CONVEX SETS AND HYPERPLANES Let $X$ be a LCS over $\mathbf{F} = \mathbf{R}$ or $\mathbf{C}$. Then:

(a) any closed linear variety $M \subset X$ is the intersection of all closed hyperplanes which contain it [cf. Theorem 7.7.5];

(b) viewing $X$ as a real space, any nonempty closed convex set $K$ is the intersection of all closed half spaces that contain it [cf. Theorem 7.8.6(b)].

**Proof.** We use the notation of the statement. (a) Since we can translate things, there is no loss of generality in assuming that $M$ is a subspace. Let $L$ denote the intersection of all closed maximal subspaces $H$ that contain $M$. Clearly $M \subset L$. To reverse the inclusion, suppose that $x \notin M$. By Theorem 7.7.5 there is a continuous linear functional $f$ on $X$ which is 1 on $x$ and 0 on $M$. Thus $M \subset f^{-1}(0)$ and $x \notin f^{-1}(0)$; therefore, $x \notin L$. It follows that $L \subset M$.

We remark that the result of (a) implies Theorem 7.7.5 about separating a point $x$ from a closed subspace $M$ by a continuous linear functional: If (a) holds and $x \notin M$, a closed subspace, then there is a closed maximal subspace $H = f^{-1}(0) \supset M$ such that $x \notin H$ for some continuous linear functional $f$ [Theorem 4.9.4(a)]. Consequently $f$ vanishes on $M$ but $f(x) \neq 0$.

(b) Let $L$ be the intersection of the closed half spaces that contain $K$. Clearly $K \subset L$. If $x \notin K$, by Theorem 7.8.6(b), there is a continuous linear functional $f$ on $X$ such that $a = f(x) > \sup f(K) = b$. Thus $K$ is a subset of the half space $f^{-1}(-\infty, b]$ but $x \notin f^{-1}(-\infty, b]$. Therefore $x \notin L$.

Conversely, if Theorem 7.8.7(b) holds then Theorem 7.8.6(b) (on separating a point $x$ from a closed convex set $K$ by a continuous linear functional) follows for if $x \notin K$, $x$ is not in some half space $S \supset K$. If $S = \{x \in X : f(x) \leq c\}$ then $\sup f(K) \leq c < f(x)$. ■

There are many other, stronger, results like Theorem 7.8.7. For example Mazur [1933] proved that in a Banach space, every closed bounded convex set can be expressed as the intersection of a family of closed balls iff its norm is Fréchet differentiable. For more results like this, see Giles [1982, Chap. 4] and Granero et al. 2004.

The result below is a cousin of Theorem 7.8.6 (separating closed and compact convex sets) which we put to use in proving a theorem of Helly's [7.10.2] on solving systems of linear equations with a constraint.

**Theorem 7.8.8** POINTS AND DISKED NEIGHBORHOODS Let $X$ be a LCS over $\mathbf{F} = \mathbf{R}$ or $\mathbf{C}$, $V$ a balanced, convex neighborhood of 0. For any $x \notin V$, there is a continuous linear functional $f$ on $X$ such that $\sup |f(V)| \leq |f(x)|$.

**Proof.** We use the notation of the statement. Let $V$ be a disk such that $x \notin V$ and let $p$ be the gauge of $V$ so that $p(x) \geq 1$ and $p \leq 1$ on $V$. On the linear space $\mathbf{F}x$ define $g(cx) = cp(x)$ for $c \in \mathbf{F}$. Since $|g| = p$ on $\mathbf{F}x$, $g$ may be extended to a continuous linear functional $f$ on $X$ such that $|f| \leq p$

everywhere by Theorem 7.3.2(b). Hence $\sup|f(V)| \leq \sup p(V) = 1 \leq p(x) = f(x)$. ■

### 7.8.1 Smoothness

The notion of *supporting hyperplane* generalizes the notion of tangent line or plane.

**Definition 7.8.9** SUPPORT POINTS AND SMOOTHNESS (a) Let $K$ be a convex subset of a real linear space $X$, $f$ a linear functional on $X$ and $H = f^{-1}(c)$ $(c \in \mathbf{R})$. We say that $H$ *supports* (*is a supporting hyperplane [hyperplan d'appui] for*) $K$ if $H$ meets $K$ and $K$ lies to one side of $H$. If $H$ supports $K$ and $u \in K \cap H$, $u$ is called a *support point of* $K$. If there is a unique supporting hyperplane $H$ at $u \in K$ we say that $u$ is a *smooth point* of $K$; we call $H$ the *tangent hyperplane.*

(b) SMOOTH NORMED SPACE If at each point $x$ of the unit ball $U(X)$ of a normed space $X$ there is a unique closed supporting hyperplane to $U(X)$ at $x$, we say that $X$ is *smooth*. Smoothness at a point $x$ is equivalent to Gateaux differentiability of the norm at $x$ [Köthe 1983, p. 350]. ■

**Example 7.8.10** SMOOTH POINTS (a) Let $u = (1,1) \in U\left(\ell_\infty^{\mathbf{R}}(2)\right) = U$, consider the two orthogonal projections $\mathrm{pr}_1$ and $\mathrm{pr}_2$ onto $\mathbf{R}\times\{0\}$ and $\{0\}\times\mathbf{R}$, respectively, and the two supporting hyperplanes $\mathrm{pr}_1^{-1}(1) = \{1\} \times \mathbf{R}$ and $\mathrm{pr}_2^{-1}(1) = \mathbf{R} \times \{1\}$ to $U$ at $u$. As these are distinct hyperplanes, $u$ is not a smooth point of $U$.

(b) Let $u = (0,1) \in U\left(\ell_2^{\mathbf{R}}(2)\right) = U$. Let $\mathrm{pr}_2$ denote the orthogonal projection onto $\{0\} \times \mathbf{R}$. Then $\mathrm{pr}_2^{-1}(1) = \mathbf{R} \times \{1\}$ supports $U$ at $u$. Since hyperplanes in $\mathbf{R}^2$ may be identified with lines in $\mathbf{R}^2$ and any other line through $u$ would pierce the circle, $\mathrm{pr}_2^{-1}(1)$ is the unique supporting hyperplane to $U$ at $u$ and $u$ is a smooth point of $U$.

(c) Consider $X = C([0,3], \mathbf{F}, \|\cdot\|_\infty)$, let $U = U(X)$, and consider the constant function $u(t) \equiv 1$ on $[0,3]$. Let $f$ and $g$ denote the evaluation maps $1'$ and $2'$. Clearly $u \in U \subset \{x \in X : f(x) \leq 1\}$ and $u \in U \subset \{x \in X : g(x) \leq 1\}$. Consequently $f^{-1}(1)$ and $g^{-1}(1)$ support $U$ at $u$. As there exist $x, y \in C([0,3], \mathbf{F}, \|\cdot\|_\infty)$ such that $x(1) = 1$, $x(2) = 0$ and $y(1) = 0$, $y(2) = 1$, $f \neq g$, and these are distinct hyperplanes. Therefore $u$ is not a smooth point of $U$. ■

In Sec. 10.2 we discuss the possibility of replacing the field $\mathbf{F}$ by a normed space $Y$ and consider extending continuous linear maps $A : M \to Y$ from a subspace $M$ of a normed space $X$ to $X$. Normed spaces $Y$ for which norm-preserving extensions exist for all $A$, $M$ and $X$ are called *1-injective*. As we discuss after Theorem 10.8.2, a real smooth finite-dimensional space $Y$ is not 1-injective.

Given a unit vector $u \in X$, there exists $f \in SU(X') = \{f \in X' : \|f\| = 1\}$ such that $f(u) = \|u\| = 1$ by Theorem 7.4.2(a). Since $f(U) \subset [-1,1]$,

$U \subset \{x \in X : f(x) \leq 1\}$ so $f^{-1}(1) = H$ supports $U$ at $u$. Smoothness at $u$ is characterized by there being a unique $f$ of this type.

**Theorem 7.8.11** SMOOTHNESS CRITERION If a real normed space $X$ is smooth at a unit vector $u$ then there exists a unique $f \in SU(X') = \{f \in X' : \|f\| = 1\}$ such that $H = f^{-1}(1)$ supports the unit ball $U$ at $u$.

**Proof.** Suppose the real normed space $X$ is smooth at a unit vector $u$. As noted above, there exists $f \in SU(X')$ such that $f(u) = 1$ and $f^{-1}(1)$ supports $U$ at $u$. If $g \in SU(X')$ is such that $g(u) = 1$ and $g^{-1}(1)$ supports $U$ at $u$ then, by the smoothness, $g^{-1}(1) = f^{-1}(1)$ which implies that $f = g$. ■

## 7.9 ORIGIN OF THE THEOREM

The Hahn–Banach theorem arose from attempts to solve infinite systems of linear equations such as the moment and Fourier series problems below. The key to the solvability is determining "compatibility" of the system of equations. For example, the system $x + y = 2$ and $x + y = 4$ cannot be solved because it requires contradictory things and so are "incompatible." The first attempts to determine compatibility for infinite systems of linear equations extended known determinant and row-reduction techniques. It was classical analysis—almost solve the problem in a finite situation, then take a limit. A fatal defect of these approaches was the need for the (very rare) convergence of infinite products.

- **The moment problem.** If all moments $f_n(x) = \int_0^1 t^n x(t)\, dt = c_n$ $(n \in \mathbf{N})$ of a function $x$ are known, find $x$.

- **The Fourier series problem.** If all Fourier cosine coefficients

$$c_n = \frac{1}{\pi} \int_{-\pi}^{\pi} (\cos nt)\, x(t)\, dt = \quad (n \in \mathbf{N})$$

of a function $x$ are known, find $x$.

Riesz and Helly obtained solutions to problems like these in important special cases such as $L_p[0,1]$ and $C[a,b]$. In effect, they discovered that solvability was equivalent to the continuity of a certain linear functional.

Consider more general versions of the problems above: Let $X$ be a normed space with dual $X'$, let $S$ be a set, and let $\{c_s : s \in S\}$ be a collection of scalars.

(V) **The vector problem.** Let $\{f_s : s \in S\}$ be a collection of locally bounded linear functionals on $X$. Find $x \in X$ such that $f_s(x) = c_s$ for every $s$.

and its dual:

(F) **The functional problem.** Let $\{x_s : s \in S\}$ be a collection of vectors from $X$. Find $f \in X'$ such that $f(x_s) = c_s$ for every $s$.

As we discuss after Theorem 7.10.1, if $X$ is "reflexive" then solving (F) also solves (V).

Motivated by Hilbert's work on $L_2[0,1]$, Riesz [1910] invented the spaces $L_p[0,1]$, $1 < p < \infty$ (he didn't consider the $\ell_p$ spaces until 1913). Instead of the moment and Fourier series problems *per se* [1910, 1911], he considered the vector problem (**LP**) below. In doing so, he inadvertently proved a special case of the Hahn–Banach theorem.

(LP) Let $S$ be a set. For $p > 1$ and $1/p + 1/q = 1$, given $y_s \in L_q[a,b]$ [equivalently, consider the functionals $f_s$ of Eq. (7.2)] and scalars $\{c_s : s \in S\}$, find $x \in L_p[a,b]$ such that

$$f_s(x) = \int_a^b x(t)y_s(t)dt = c_s \text{ for each } s \in S \tag{7.2}$$

For there to be a solution $x$, he showed that the following necessary and sufficient connection between the $y$'s and the $c$'s had to prevail: There exists $K > 0$ such that for any finite subset $J$ of $S$ and scalars $\{a_s : s \in J\}$,

$$\left|\sum_{s\in J} a_s c_s\right| \le K\left(\int_a^b \left|\sum_{s\in J} a_s y_s\right|^q\right)^{1/q} = K\left\|\sum_{s\in J} a_s y_s\right\|_q \tag{$*$}$$

Condition $(*)$ implies that if $\sum_{s\in J} a_s y_s = 0$ for a finite set of scalars $\{a_s : s \in J\}$ then $\sum_{s\in J} a_s c_s = 0$. Thus, if we consider the linear functional $g$ on the linear span $M = [y_s : s \in S]$ of the $y$'s in $L_q[a,b]$ defined by taking $g(y_s) = c_s$ $(s \in S)$, $g$ is well-defined. Not only that, for any $y \in M$, $|g(y)| \le K\|y\|_q$ so $g$ is continuous on $M$. If there is an $x \in L_p[a,b]$ which solves (LP) then $g$ has a continuous extension $G$ to $L_q[a,b]$, namely, for any $y \in L_q[a,b]$,

$$G(y) = \int x(t)y(t)\,dt$$

Thus, Riesz showed that:

- (LP) is solvable if and only if a certain linear functional $g$ defined on a subspace of $L_q[a,b]$ is continuous.

- If the system is solvable then $g$ can be extended to a continuous linear functional defined on all of $L_q[a,b]$.

Helly [1912] also solved a functional problem for $C[a,b]$ and developed an early Hahn–Banach theorem. He continued to work on such problems for his *Habilitationsschrift.* He published the results in 1921 in an article Dieudonné [1981, p. 130] subsequently called "a landmark in the history of functional analysis." As Helly says at the beginning of the article, the conditions for solving infinite systems of linear equations had been given by Schmidt [1908] and Riesz [1913] "in the case that the coefficients and solutions satisfy certain inequalities." His aim, he said, was to show that the conditions could be interpreted geometrically. Some high points of Helly's 1921 article are:

**General normed sequence space.** He abandoned special cases and considered a general subset $X \subset \mathbf{C}^{\mathbf{N}}$ equipped with a norm $D$ as we define it today; he did not require that $X$ be a vector space, nor did he use the word *norm* or the notation $\|\cdot\|$.

**Dual space.** Helly took as the "dual space" of $X$ the set $X^d$ of all complex sequences $u = (u_n)$ such that $\sum_{n\in\mathbf{N}} x_n u_n < \infty$ for all sequences $(x_n) \in X$. He did not give $X^d$ a name. He did call a certain seminorm $\Delta$ (see below) on $X^d$ a *polare Abstandsfunktion,* so he may have been thinking of *polare Raum,* the name Hahn later used for the dual in a more general setting. $X^d$ *is* a vector space regardless of what $X$ is. If $X = c$ or $c_0$ then $X^d = \ell_1$; if $X = \ell_1$ then $X^d = \ell_\infty$ but if $X = \ell_\infty$, the $X^d$ you get is a proper subset of what we call the dual of $X$ today. Nowadays, such pairs $(X, X^d)$, subject to *absolute* convergence of $\sum x_n u_n$, are called *Köthe sequence spaces* and *$\alpha$-duals,* respectively.

**Seminorm for the dual.** For $x = (x_n) \in X$ and $u = (u_n) \in X^d$, Helly defines an analog of an inner product $\langle \cdot, \cdot \rangle$ on $X \times X^d$: for $x \in X$ and $u \in X^d$, $\langle x, u \rangle = \sum_{n\in\mathbf{N}} x_n u_n$. (If $X$ is a vector space, $\langle \cdot, \cdot \rangle$ is a bilinear form and $(X, X^d)$ a dual pair.) Using an idea of Minkowski's, he defines the *polare Abstandsfunktion* $\Delta$ for $X^d$ as

$$\Delta(u) = \sup\{|\langle x, u\rangle| : D(x) = 1\}$$

He observes that $\Delta$ is generally a seminorm, not a norm. He notes that $D$ and $\Delta$ satisfy a Cauchy-Schwarz-type inequality, namely,

$$|\langle x, u\rangle| \le D(x)\,\Delta(u)$$

**The problem.** Helly sought to solve the following vector problem:

- Given sequences $f_n = (f_{nj})$ from $X^d \subset \mathbf{C}^{\mathbf{N}}$ and a sequence $(c_n) \in \mathbf{C}^{\mathbf{N}}$, find $x = (x_j) \in X$ such that

$$\langle x, f_n \rangle = \sum_{j\in\mathbf{N}} x_j f_{nj} = c_n \text{ for each } n \in \mathbf{N}$$

His method of attack is quite original: He doesn't seek the $x$ right away. Apparently with a belief in reflexivity (when he began his investigation) he seeks:

(1) a continuous linear functional $h \in X^{dd} = \left(X^d\right)^d$ such that $h\left(f_n\right) = c_n$ for each $n$ then

(2) $x \in X$ such that $h\left(f_n\right) = f_n\left(x\right)$ for every $n$.

He discovered that the $x \in X$ corresponding to $h$ did not always exist, thus showing that some spaces are not reflexive.

**The first Hahn–Banach theorem.** In order to establish (1), Helly extended a locally bounded linear functional $f$ from a subspace $M$ to the whole space. *Assuming that $X^d$ is separable* so he could use induction, the key step was the one-dimensional extension: For $x \notin M$, find a linear functional $F$ such that, for $\mathbf{F} = \mathbf{R}$ or $\mathbf{C}$,

$$\begin{array}{llcll} F: & M \oplus \mathbf{F}x & & & |F| \leq k\,\|\cdot\| \\ & \quad | & \searrow & & \\ f: & \quad M & \longrightarrow & \mathbf{F} & |f| \leq k\,\|\cdot\| \text{ (for some } k) \end{array}$$

Hahn [1927] and Banach [1929] also employed the technique of the one-dimensional extension to prove what we call the Hahn–Banach theorem today. As they used transfinite induction rather than ordinary induction, they generalized Helly's result not only to the general normed space but also eliminated the separability requirement on the dual. Each acknowledged Helly's work. Their gain in generality was offset by the fact that their proof required the choice of a number between two others and so only applied to real spaces whereas Helly used his intersection theorem [10.3.5(a)], a result that applied to the real and complex cases simultaneously.

Riesz [1910b] solved the functional problem in some special cases. Helly [1912] did so in more general situations. In one of the first important applications of the Hahn–Banach theorem, Banach [1932, pp. 55–57, Theorems 4 and 5] solved the general functional problem. In the proof of sufficiency in Theorem 7.9.1 below, he used condition (**) to create a continuous linear functional on a subspace which he then continuously extended to the whole space. Theorem 7.9.1 is often referred to as *Helly's theorem*. We prove a slightly more general version in Theorem 7.10.1.

**Theorem 7.9.1** Let $X$ be a real normed space, let $\{x_s\}$ and $\{c_s\}$, $s \in S$, be sets of vectors and scalars, respectively. Then there is a continuous linear functional $f$ on $X$ such that $f\left(x_s\right) = c_s$ for each $s \in S$ if and only if there exists $K > 0$ such that for all finite subsets $J$ of $S$ and scalars $\{a_j : j \in J\}$

$$\sum_{j \in J} \left|a_j c_j\right| \leq K \left\| \sum_{j \in J} a_j x_j \right\| \qquad (**)$$

## 7.10 FUNCTIONAL PROBLEM SOLVED

Given a vector $x \neq 0$ from a LCHS $X$ over $\mathbf{F} = \mathbf{R}$ or $\mathbf{C}$ and a scalar $c$, there exists a continuous linear functional $f$ such that $f(x) = c$ [Theorem 7.4.2(a)]. If we consider the analogous question for two vectors $x$ and $y$ and two scalars $c$ and $d$, we meet the situations $\{x, y\}$ linearly dependent and $\{x, y\}$ linearly independent separately. If $\{x, y\}$ is linearly dependent, if $y = 2x$ say, then we must also have $d = 2c$. If $x$ and $y$ are linearly independent, the problem is easy to solve: Let $g$ be defined on $\mathbf{F}x$ by taking $g(ax) = ad$ for $a \in \mathbf{F}$; let $h(by) = bc$ on $\mathbf{F}y$ for each $b \in \mathbf{F}$; for any $z = ax + by \in [x, y] = \mathbf{F}x + \mathbf{F}y$, let $k(z) = g(ax) + h(by) = ad + bc$. Since its domain is finite-dimensional, $k$ is continuous [Corollary 5.6.3]. Now let $f$ be the continuous linear extension of $k$ to $X$. These ideas easily extend to any finite collection of vectors $x_1, \ldots, x_n$ and scalars $c_1, \ldots, c_n$: If the vectors are linearly independent, there is always a continuous linear functional on $X$ which maps $x_j$ into $c_j$ for $j = 1, \ldots, n$; if the vectors are linearly dependent then the problem is solvable iff the scalars are "compatible": For normed spaces, the condition of the next theorem guarantees compatibility. Before proceeding, recall that a linear functional $f$ on a normed space $X$ is continuous [Theorem 5.7.4(a,b)] iff $f$ is bounded on the unit ball $U = \{x \in X : \|x\| \leq 1\}$; $\sup\{|f(x)| : \|x\| \leq 1\} = \|f\|$ is a norm on the dual space $X'$ of all continuous linear functionals on $X$.

Theorem 7.10.1 shows that to solve a certain system of linear equations, it is necessary and sufficient that a continuity-type condition be satisfied.

**Theorem 7.10.1** THE FUNCTIONAL PROBLEM Let $X$ be a normed space over $\mathbf{F} = \mathbf{R}$ or $\mathbf{C}$, let $\{x_s : s \in S\}$ and $\{c_s : s \in S\}$ be sets of vectors and scalars, respectively. Then there is a continuous linear functional $f$ on $X$ such that $f(x_s) = c_s$ for each $s \in S$ iff there exists $K > 0$ such that

$$\left|\sum_{s\in S} a_s c_s\right| \leq K \left\|\sum_{s\in S} a_s x_s\right\| \tag{1}$$

for any choice of scalars $\{a_s : s \in S\}$ for which $a_s = 0$ for all but finitely many $s \in S$ ("almost all" the $a_s = 0$).

**Proof.** We use the notation of the statement. NECESSITY. Let $f$ be a continuous linear functional on $X$ such that $f(x_s) = c_s$ for each $s \in S$ and let $\{a_s : s \in S\}$ be a collection of scalars, almost all of which are 0. Then $\|f\|$ functions as the $K$ of condition (1):

$$\left|\sum_{s\in S} a_s c_s\right| = \left|\sum_{s\in S} a_s f(x_s)\right| \leq \|f\| \left\|\sum_{s\in S} a_s x_s\right\|$$

SUFFICIENCY. Suppose that (1) holds and let $M$ be the subspace spanned by $\{x_s\}$. For $x \in M$, there exists a finite subset $J$ of $S$ and scalars $\{a_s : s \in J\}$

such that $x = \sum_{s \in J} a_s x_s$; define $g$ on $M$ by taking $g(x) = \sum_{s \in J} a_s c_s$. Is $g$ well-defined? If $\{x_s : s \in S\}$ is linearly independent, the representation $x = \sum_{s \in J} a_s x_s$ is unique and there is no problem but suppose that $\sum_{s \in J} a_s x_s = x = \sum_{s \in J} b_s x_s$. (By using some zero coefficients, we may assume that $J$ is a finite subset of $S$ to which the indices of the scalars $a_s$ and $b_s$ belong.) By the condition,

$$\left|\sum_{s \in J} a_s c_s - \sum_{s \in J} b_s c_s\right| = \left|\sum_{s \in J} (a_s - b_s) c_s\right| \leq K \left\|\sum_{s \in J} (a_s - b_s) x_s\right\| = 0$$

and $g$ is seen to be well-defined. Clearly, $g$ is linear on $M$ and for any $x = \sum_{s \in J} a_s x_s \in M$,

$$|g(x)| = \left|g\left(\sum_{s \in J} a_s x_s\right)\right| = \left|\sum_{s \in J} a_s c_s\right| \leq K \left\|\sum_{s \in J} a_s x_s\right\| = K \|x\|$$

so $g$ is continuous $M$. As such, by Theorem 7.4.1, $g$ may be extended from $M$ to a continuous linear functional $f$ on $X$. ■

In reflexive spaces [Example 5.7.5], solving the functional problem solves the vector problem. Suppose $\{f_s : s \in S\} \subset X'$ and $\{c_s : s \in S\} \subset \mathbf{F}$. If $X$ is reflexive, the preceding result applied to $X'$ implies that there is an $f \in X''$—hence a corresponding $x \in X$ by reflexivity—such that $f(f_s) = (Jx)(f_s) = f_s(x) = c_s$ for each $s \in S$ if there is some $K > 0$ such that $\left|\sum_{s \in S} a_s c_s\right| \leq K \left\|\sum_{s \in S} a_s f_s\right\|$ for all families $(a_s)$ of scalars which are almost all 0.

Banach used the Hahn–Banach theorem to prove Theorem 7.10.1 but Theorem 7.10.1 implies the Hahn–Banach theorem: Assuming that Theorem 7.10.1 holds, let $\{x_s\}$ be the vectors of a subspace $M$, let $f$ be a continuous linear functional on $M$; for each $s \in S$, let $c_s = f(x_s)$. Since $f$ is continuous, (1) is satisfied and $f$ possesses a continuous extension to $X$.

For normed spaces which are not necessarily reflexive, we have the following result on consistency of a finite set of linear equations subject to the constraint $\|x\| \leq K + r$.

**Theorem 7.10.2** Finite Systems with a Constraint Let $X$ be a normed space over $\mathbf{F} = \mathbf{R}$ or $\mathbf{C}$, $K > 0$, $f_1, \ldots, f_n \in X'$, and $c_1, \ldots, c_n \in \mathbf{F}$ ($n \in \mathbf{N}$). Then for $r > 0$, there exists $x = x(r) \in X$, $\|x\| \leq K + r$, such that $f_i(x) = c_i$ $(1 \leq i \leq n)$ iff for all $n$-tuples $(a_i) \in \mathbf{F}^n$,

$$\left|\sum_{i=1}^{n} a_i c_i\right| \leq K \left\|\sum_{i=1}^{n} a_i f_i\right\| \tag{2}$$

**Proof.** We use the notation of the statement. Necessity. Let $r > 0$ be given, suppose $\|x\| \leq K + r$, $f_i \in X'$ such that $f_i(x) = c_i$ $(1 \leq i \leq n)$ and let

$a_1, \ldots, a_n \in \mathbf{F}$. With $f = \sum_{i=1}^n a_i f_i$,

$$\begin{aligned} |f(x)| &= \left|\sum_{i=1}^n a_i f_i(x)\right| = \left|\sum_{i=1}^n a_i c_i\right| \\ &\leq \|f\| \, \|x\| = \left\|\sum_{i=1}^n a_i f_i\right\| \|x\| \leq \left\|\sum_{i=1}^n a_i f_i\right\| (K + r) \end{aligned}$$

Since $r$ is arbitrary, (2) follows.

SUFFICIENCY. Suppose that the condition holds and that the $f_i$ $(1 \leq i \leq n)$ are linearly independent. Consider the linear map $A : X \to \ell_2(n)$, $x \mapsto (f_1(x), \ldots, f_n(x))$. Since the $f_i$ are linearly independent, $A$ is surjective by the following argument: For $f_k$ to be a linear combination of $f_i$, $1 \leq i \leq n$, $i \neq k$, it is necessary and sufficient that $\cap_{i \neq k} N(f_i) \subset N(f_k)$ [Theorem 1.4.1(f)]. Consequently, there exist $x_i \in X$, $1 \leq i \leq n$, such that $f_i(x_k) = \delta_{ik}$, $1 \leq i, k \leq n$; thus $Ax_k = e_k$, $(1 \leq k \leq n)$, the $k$th standard basis for $\ell_2(n)$ with $k$th entry 1 and the rest 0. That $A$ is onto is now evident. Let $\mathrm{pr}_i$ denote the $i$th projection of $\ell_2(n)$ onto $\mathbf{F}$. Note that $\mathrm{pr}_i \circ A = f_i$ for each $1 \leq i \leq n$. As each $f_i$ and $\mathrm{pr}_i$ are continuous, $A$ is continuous. Since $\ell_2(n)$ is a Hausdorff space, the openness of $A$ follows from Theorem 4.10.3(g).

Let $D = \{x \in X : \|x\| \leq K + r\}$. Since $A$ is an open map, $A(D)$ is a disked neighborhood of 0 in $\ell_2(n)$. We want to find $x \in D$ such that $f_i(x) = c_i$, $1 \leq i \leq n$, i.e., $x \in D$ such that $Ax = (c_1, \ldots, c_n)$. If no such $x$ exists then $(c_1, \ldots, c_n) \notin A(D)$. Since $A(D)$ is a disked neighborhood of 0, there exists a continuous linear functional $g$ on $\ell_2(n)$ such that $\sup |g(A(D))| \leq g(c_1, \ldots, c_n)$ [Theorem 7.8.8]. Letting $g(e_k) = b_k$, $1 \leq k \leq n$, for any $(a_i) \in \ell_2(n)$, $g((a_i)) = \sum_{i=1}^n a_i b_i$. Let $h = \sum_{i=1}^n b_i f_i$ and note that $\sup\{|h(x)| : x \in C(0, K + r) = (K + r) C(0, 1)\} = (K + r) \|h\|$, a fact that we use on the second line below:

$$\begin{aligned} |g(c_1, \ldots, c_n)| &= \left|\sum_{i=1}^n c_i b_i\right| \geq \sup_{x \in D} |g(Ax)| = \sup_{x \in D} \left|\sum_{i=1}^n b_i f_i(x)\right| \\ &= (K + r) \left\|\sum_{i=1}^n b_i f_i\right\| > K \left\|\sum_{i=1}^n b_i f_i\right\| \end{aligned}$$

As $|\sum_{i=1}^n c_i b_i| > K \|\sum_{i=1}^n b_i f_i\|$ this proves the contrapositive of the desired result when $\{f_i\}$ is linearly independent.

Now suppose that $\{f_i\}$ is linearly dependent and, renumbering if necessary, let $\{f_1, \ldots, f_k\}$ be a maximal linearly independent subset. By the result above, there is an $x \in X$, $\|x\| \leq K + r$, such that $f_i(x) = c_i$ for $1 \leq i \leq k$. We show next that (2) guarantees that $f_m(x) = c_m$ for $m > k$ as well. For $m > k$ choose $a_i$ such that $f_m = \sum_{i=1}^k a_i f_i$. Letting $a_m = -1$ and $a_i = 0$, for

$i > k$, we have by (2),

$$\left| \sum_{i=1}^{k} a_i c_i - c_m \right| \leq K \left\| \sum_{i=1}^{k} a_i f_i - f_m \right\| = 0$$

Therefore $f_m(x) = \sum_{i=1}^{k} a_i f_i(x) = \sum_{i=1}^{k} a_i c_i = c_m$. ■

## 7.11 THE AXIOM OF CHOICE

As the axiom of choice, we take the ability to form a set consisting of one element from each of a family of nonempty sets. By teasing out a maximal element $F$ from the dominating extensions of a linear functional $f$, the standard proof of the Hahn–Banach theorem (HB), Theorem 7.3.2, uses the axiom of choice (AC) in its Zorn's lemma formulation, so AC $\Rightarrow$ HB.

Does HB imply AC? as Tihonov's theorem on products of compact sets does? Can we call it "the analyst's form of AC?" In a word: "No." The details are as follows.

It follows from Zorn's lemma that every filter of sets is contained in a filter that is not properly contained in any other filter, an *ultrafilter*. This property is known as the *ultrafilter theorem* (UT); so

$$\text{AC} \implies (\text{UT})$$

Halpern [1964] proved that UT $\not\Rightarrow$ AC. Łoś and Ryll-Nardzewski [1951] and Luxemburg [1962, 1967a,b] proved that UT $\Rightarrow$ HB. Pincus [1972, 1974] proved that HB $\not\Rightarrow$ UT. We therefore have the following irreversible hierarchy:

$$\text{AC} \implies \text{UT} \implies \text{HB}$$

The "prime ideal theorem for Boolean algebras" asserts that there is a function $F$ defined on the class of all Boolean algebras $B$ such that $F(B)$ is a prime ideal of $B$ for each $B$. Using techniques from non-standard analysis, Luxemburg [1962] showed that the prime ideal theorem implies the Hahn–Banach theorem and conjectured that the prime ideal and Hahn–Banach theorems might be equivalent. Halpern [1964], however, proved that the prime ideal theorem is strictly weaker than AC. Luxemburg [1967b] showed that a modified form of the Hahn–Banach theorem is valid if and only if every Boolean algebra admits a nontrivial measure. The modification consists of allowing the extended linear functional on the real Banach space $X$ to take values in a "reduced power of the reals" (as used in nonstandard analysis) rather than $\mathbf{R}$; the modified version is also equivalent to the unit ball of the dual of the normed space $X$ being convex-compact in the weak-$*$ topology [Example 5.5.7(b)], i.e., that every family of weak-*-closed convex sets that satisfies the finite intersection condition has nonempty intersection. Luxemburg and Väth [2001] proved that the assertion that any Banach space has at least one nontrivial locally bounded linear functional implies the Hahn–Banach theorem.

### 7.11.1 Avoiding the Axiom of Choice

Various people have proved weaker versions of the Hahn–Banach theorem that do not rely on the axiom of choice.

►Axiom of Dependent Choices (ADC): Let $R$ be a relation on a set $T$ such that $R_t = \{s \in T : tRs\} \neq \varnothing$ for each $t \in T$. Then given $t_0 \in T$, there exists a sequence $(t_n)$ of elements of $T$ such that $t_0 R t_1, t_1 R t_2, \ldots$.

ADC is stronger than the countable axiom of choice (the ability to form a set consisting of one element from each of a denumerable family of nonempty sets) but weaker than AC. Garnir, de Wilde and Schmets use only ADC to prove the following version of the Hahn–Banach theorem for *separable* spaces.

**Theorem 7.11.1** [Garnir et al. 1968, p. 183] Let $X$ be an LCHS over $\mathbf{R}$ or $\mathbf{C}$ whose topology is generated by a family $P$ of seminorms for which the seminormed space $(X, p)$ is separable for each $p \in P$. If $f$ is a linear functional defined on a subspace $M$ of $X$ and $|f| \leq kp$ for some $k > 0$ and $p \in P$ (i.e., $f$ is continuous on $M$) then there is a linear extension $F$ of $f$ such that $|F| \leq kp$ on $X$.

►Ishihara [1989] proved another "constructive" version.

►Mulvey and Pelletier [1991]. *Locales* generalize the lattice of open sets of a space without reference to the points of the space. Mulvey and Pelletier avoid dependence on AC and use locales to prove a version of the Hahn–Banach theorem in any Grothendieck topos. Coquand and Cederquist [2000] give a direct constructive proof of the Hahn–Banach theorem for normed spaces in a topos.

►Cederquist, Coquand and Negri [1998] present a constructive proof of the theorem in the setting of formal topology; they have a good bibliography about other such "pointfree" arguments.

►Coquand [2006] has a constructive proof of the geometric form.

►Dodu and Morillon [1999] add a little and take a little. They suppose that the Banach space $X$ satisfies the stronger completeness requirement that Cauchy *nets* converge. They then prove the Hahn–Banach theorem for uniformly convex Banach spaces whose norm is Gateaux differentiable without AC. Still assuming that the Banach space $X$ satisfies the stronger completeness requirement, Albius and Morillon [2001] show that to have the Hahn–Banach theorem, it suffices to have a strengthened differentiability condition, *uniform smoothness*, namely, the uniform convergence of $(\|x+h\| + \|x-h\| - 2\|x\| \,/\, \|h\|)$ as $h \to 0$ for all $x$ on the surface of the unit ball of $X$.

Pietsch [2007, pp. 579–588] has a thorough discussion of the axiom of choice in functional analysis.

## 7.12 NOTES ON THE HAHN–BANACH THEOREM

For the moment let us take as the Hahn–Banach theorem the version of Theorem 7.3.2(a), that a linear functional $f$ defined on a subspace $M$ of a real linear space $X$ and majorized on $M$ by a sublinear functional $p$ can be extended to a linear functional $F$ on $X$ which is majorized by $p$.

Some consequences of the Hahn–Banach theorem that we have not proved are:

1. The existence of Green's and Neumann's functions [Garabedian and Schiffman 1954].

2. Generalized limits [Banach 1932, pp. 33–34; Berberian 1974, pp. 117–122]. On the Banach space $\ell_\infty^{\mathbf{R}}$ of all bounded sequences of real numbers, there is a linear functional $L$ which essentially makes nonconvergent sequences have limits. The properties of $L$ are, for $x = (a_k) \in \ell_\infty$,

   (a) if $x_j = (0, 0, \ldots, 0, a_{j+1}, a_{j+2}, \ldots)$ then $L(x_j) = L(x)$ for each $j \in \mathbf{N}$;

   (b) $\inf_k a_k \leq L(x) \leq \sup_k a_k$;

   (c) $\liminf_k a_k \leq L(x) \leq \limsup_k a_k$;

   (d) $L(x) = \lim_k a_k$ whenever $x = (a_k)$ is convergent.

   For another interesting approach to "Banach limits," see Larsen [1973, p. 264f].

3. The solution of the "easy" problem of measure (Banach 1932, pp. 33–34; Bachman and Narici 2000, pp. 188–194). There is a finitely additive (hence the "easy") translation-invariant set function $m$ such that $m([0, 1]) = 1$ defined on *all* subsets of $\mathbf{R}$.

In addition to these, there are applications (to name just a random few) to control theory [Leigh 1980, p. 88ff], convex programming [Balakrishnan 1981], game theory [König 1982], and a development of thermodynamics based on it [Feinberg and Lavine 1983].

The complex version Theorem 7.3.2(b) was obtained by F. J. Murray for $L_p$ spaces but in a perfectly general way which enabled Bohnenblust and Sobczyk [1938] to obtain it for general normed spaces; they were the first to denominate it the Hahn–Banach theorem. The complex version was also obtained independently by Soukhomlinov [1938]. Banach, who only treated real spaces in his 1932 book might have treated complex spaces if he had had the complex Hahn–Banach theorem. Mazur [1933] proved the geometric form Theorem 7.7.3 in normed spaces. It is first called the "geometric form" by Bourbaki. In a 1941 article, Dieudonné [1981b, pp. 231–234] refers to the geometric form as *the* Hahn–Banach theorem so he was apparently aware of the equivalence of the two. Apropos of what geometric separation can mean, we mention James's result (see Holmes 1975, p. 161) that a real Banach space is reflexive iff each pair of disjoint closed convex subsets, one of which is bounded, can be strictly separated by a hyperplane.

The approach to the Hahn–Banach theorem by way of minimal sublinear functionals in Sec. 7.6 was developed in a series of papers by König [1968, 1970, 1972a, 1978, 1982], Fuchsteiner and König [1978], and Simons [1970a, b; 1975]. The principal result of this approach is the Hahn–Banach sandwich theorem [Exercise 7.105(c)]. A more general version of the sandwich theorem [Exercise 7.202] is the following:

**Theorem 7.12.1** [Fuchsteiner and König 1978; König 1982] Let $S$ be any subset of the real linear space $X$, $p$ a sublinear functional on $X$, and $f$ any map of $S$ into $\mathbf{R}$ with $f \leq p|_S$. If there exist $a, b > 0$ such that

$$\inf_{s\in S}[p(s - ax - by) - f(s) + af(x) + bf(y)] \leq 0 \quad \text{for all } x, y \in S \qquad (*)$$

then there is a linear functional $F$ on $X$ such that $F \leq p$ and $f \leq F|S$.

A number of applications of this version to general minimax theorems such as the following appear in König [1982].

**Theorem 7.12.2** Let $T$ be a compact Hausdorff space and let *USC(T)* denote the class of upper semicontinuous maps $x$ [i.e., $x$ such that $x^{-1}[c, \infty)$ is closed for each real $c$] of $T$ into $\mathbf{R} \cup \{-\infty\}$. *USC(T)* is closed under pointwise addition and multiplication by nonnegative scalars. Let $S$ be a nonempty subset of $USC(T)$ and suppose that $a, b, c, d > 0$, $a + b = 1 = c + d$, are such that: $(L)$ for any $x, y \in S$ and $r > 0$, there exists $z \in S$ such that $z \leq ax + by + r$ and $(R)$ for any $s, t \in T$ and $r > 0$ there exists $u \in T$ such that $cx(s) + dx(t) \leq x(u) + r$. Then, with $\max x = \sup x(T)$ and $\inf_{x\in S} x$ denoting the pointwise infimum, there exists $t \in T$ such that $\inf_{x\in S} \max x = \inf_{x\in S} x(t)$; in other words (which is where the name "minimax" comes from), $\inf_{x\in S} \max x = \max \inf_{x\in S} x$.

König gets many consequences from this and theorems like it. He shows, for example, that the Silov boundary is the closure of the Choquet boundary [Sec. 9.5], proves the existence of Jensen measures, and deduces the Farkas lemma, namely: If $X$ is a real vector space and $f_1, \ldots, f_n$ and $f$ are linear functionals on $X$ such that for any $x \in X$, $f_1(x), \ldots, f_n(x) \leq 0$ implies $f(x) \leq 0$ then there are numbers $t_1, \ldots, t_n \geq 0$ such that $f = \sum t_i f_i$. (The converse is obvious.) An even more general theorem of this type is that of Rodé 1978; see also König's [1980] simplified version of Rodé's proof.

Blatter and Seever [1975] consider the general question of when one class of functions can be sandwiched between two others. If $T$ is any set and $A$,

$B$, and $C$ are families of maps of $T$ into $\mathbf{R}$, they say that $C$ *interposes* $A$ *and* $B$, $A \leq C \leq B$, if for any $a \in A$ and $b \in B$ there is a $c \in C$ such that $a \leq c \leq b$ (pointwise). If, for instance, $T$ is a commutative semigroup then the class $C$ of additive functions on $T$ interposes $A$ and $B$, where $A$ denotes the superadditive functions $[x(s+t) \geq x(s) + x(t)]$ and $B$ the subadditive functions [Kaufman 1966a].

Andenaes [1970] proved the following variant of the Hahn–Banach theorem, which has a number of interesting consequences.

**Theorem 7.12.3** Let $M$ be a subspace of a real vector space $X$, $p$ a sublinear functional on $X$, $f$ a linear functional on $M$, $f \leq p|M$, and let $S$ be any subset of $X$. Then there exists a linear functional $F$ on $X$ which extends $f$ and which is majorized by $p$ which is (pointwise) maximal on $S$ in the class of such linear functionals, i.e., if $G$ is a linear extension of $f$ which is majorized by $p$ and $G \geq F$ on $S$ then $G|S = F|S$ (see Exercise 7.201).

Using this theorem, Andenaes gets an efficient proof of a version of the Choquet-Bishop-de Leeuw theorem, the existence of extreme measures, and part of the Choquet-Meyer characterization of simplexes.

For information on other ordered Hahn–Banach theorems as well as on the Hahn–Banach theorem generally, some expository articles with good bibliographies are Buskes 1993, Narici and Beckenstein 1997, Narici 2007, Herrero 2003, Rodríguez-Salinas 1971 and Horvath 1973.

A normed algebra $X$ is an algebra over $\mathbf{F} = \mathbf{R}$ or $\mathbf{C}$ equipped with a submultiplicative norm $[\|xy\| \leq \|x\| \, \|y\|]$. A homomorphism $h : X \to \mathbf{F}$ is a linear multiplicative map. If $M$ is a subalgebra of $X$, $h : M \to \mathbf{F}$ a homomorphism which is majorized by a sublinear functional $p$, is there a homomorphic extension $H$ of $h$ which is also majorized by $p$? As we now show, even if $p$ is multiplicative, there may not be such an extension. For example, view $\mathbf{R}$ as a subalgebra of $\mathbf{C}$ with the usual absolute value and consider the identity homomorphism $I : \mathbf{R} \to \mathbf{R}$, $x \mapsto x$. Let $p$ be absolute value. If $I^*$ were a homomorphic extension of $I$ to $\mathbf{C}$ then $I^*(i)^2 = I^*(i^2) = I^*(-1) = I(-1) = -1$ which is not satisfiable since $I^*(i)$ is a real number. Some positive results about Hahn–Banach-type theorems for homomorphisms on normed algebras may be found in Grilliot [1967] (cf. Sec. 9.5).

**Theorem 7.12.4** Let $T$ be any topological space and let $C_b(T, \mathbf{R}, \|\cdot\|_\infty)$ be the Banach algebra of all bounded continuous functions of $T$ into $\mathbf{R}$. Let $M$ be any subalgebra of $C_b(T, \mathbf{R}, \|\cdot\|_\infty)$. Then any continuous homomorphism $h : M \to \mathbf{R}$ can be extended to a continuous homomorphism $H : C_b(T, \mathbf{R}, \|\cdot\|_\infty) \to \mathbf{R}$ of the same norm.

There is, of course, the following standard sufficient condition about extendibility of homomorphisms (see Naimark 1959, p. 212f. or Zelazko 1973, p. 61ff.) for which we need to define the *Silov boundary* of a commutative Banach algebra with identity $X$. A closed subset $F$ of the space $\mathcal{M}$ of maximal ideals of $X$ with the Gelfand topology is called a *determining* or *maximizing set* if each of the functions $\hat{x} : \mathcal{M} \to \mathbf{C}$, $M \mapsto x(M)$, achieves its maximum absolute value there. A minimal (with respect to set inclusion) maximizing set is called the (the uniqueness is justified by the theorem below) *Silov boundary* and:

**Theorem 7.12.5** Every commutative complex Banach algebra $X$ with identity has a unique minimal maximizing set. If $Y$ is a commutative Banach subalgebra of $X$ with identity then each continuous complex homomorphism of $Y$ whose kernel is in the Silov boundary of $Y$ may be extended to a continuous homomorphism on $X$.

There may, however, be other continuous homomorphisms on subalgebras than those with kernel in the Silov boundary which permit continuous extension to the whole space.

## 7.13 HELLY

Eduard Helly was born in Vienna in 1884 and got his PhD from the University of Vienna in 1907. (A reproduction of the first page of his *handwritten* dissertation appears on p. 130 of Butzer et al. 1980; that article and another by Butzer et al. [1984] are excellent sources for information about Helly and his work.) By means different from and simpler than Riesz [1911], Helly [1912] also solved a moment problem, proved special cases of the Hahn–Banach theorem and the Banach-Steinhaus theorem [1912, p. 268] for linear functionals on $C[a,b]$ as well as [1912, p. 267] an early version of the Alaoglu theorem [8.4.1] for $C[a,b]$; for Helly's version, see Theorem 8.4.2.

Helly volunteered for the Austrian Army in 1914 and went to the Russian front in 1915. He was wounded by a bullet through the lungs in September 1915, a wound that ultimately caused his death. He spent almost the next five years as a prisoner of war in a camp near Tobolsk, Siberia. He endured eastern Siberia's frigidity along with a Hungarian university student named Tibor Radó (1895–1965). Helly tutored Radó in the camp and imbued him with a taste for mathematical research. The Great War ended but peace did not come to Russia. The White Russian forces contended with the Red armies. Other players on the scene were a Czech army of some 50,000 escaped prisoners who joined the White Russians. Japan saw an opportunity to pick up some Russian territory and sent troops; so did the Americans, the British and others. In the midst of this chaos, there was no repatriation of POWs. Radó escaped from the camp in 1919 and went north! With the help of some Eskimos, he traversed thousands of kilometers on his way west and reached Hungary in 1920. He abandoned civil engineering and switched to mathematics at the University of Szeged. His teachers included Frigyes Riesz and Alfred Haar. His most famous work is his solution to the Plateau problem

[1930a-c; see also Gray and Micallef 2008] concerning bounding contours for minimal surfaces.

As of the summer of 1920, Helly was still a POW in Tobolsk but by going east to Japan, then to the Middle East and Egypt, Helly got back to Vienna in November 1920. In order to be a professor in the Austrian system, it was necessary to write a post-doctoral thesis called a *Habilitationsschrift.* As with a doctoral dissertation, it is reviewed by and defended before an academic committee. It is necessary to attain the Habilitation (the qualification) to be a *Privatdozent*, one who may supervise doctoral students. Helly had presented talks about what became his 1921 paper to the Wiener Mathematischen Gesellschaft before the war. He successfully presented his Habilitation thesis to the faculty of the University of Vienna in 1921 and then applied for a professorship there. Largely as a result of Hahn's opposition, he did not get one. Helly's wife, Dr. Elise Bloch, also a mathematician, attributed Hahn's opposition to two sources: Helly was (1) Jewish and (2) too old, Helly being 37 at the time. Helly did become a Privatdozent in August 1921, a position that paid nothing. To support himself, he went to work for a bank. As a Privatdozent, he supervised three doctoral students and taught practically every semester from 1921 until 1938. The bank failed in 1929 and he got a job in 1930 in the actuarial department of the insurance company Lebensversicherungs-Gesellschaft Phönix where his co-workers included the mathematicians Eugene Lukacs (who had taken courses with Helly and Hahn at the University of Vienna) and Z. W. Birnbaum. In a 1979 letter [Butzer et al. 1980, p. 139], Birnbaum said:

> Helly was a delightful man, cheerful in the face of adversities, with a gentle sense of humor. There were three mathematicians in the Phönix office who were my immediate superiors. One of them had the title "Prokurist" while Helly, to my knowledge, did not get that high. Whenever a non-routine question came up, the difference between Helly and the other two became apparent: Helly gave the problem a mathematical formulation and obtained a solution which could be used over and over again in similar cases; the other two worked the problem numerically in each case, by trial and error, grinding it out on their hand-operated Odhner desk calculators. Incidentally, even the manner in which he handled his desk calculator was ingenious, devising shortcuts and step-saving routines.

On March 13, 1938, the day after the *Anschluss Österreichs,* the political union of Germany and Austria, Jews were ordered to appear in evening dress and scrub the streets. Their stores and apartments were pillaged. In May 1938 the Nazis promulgated the Nuremberg racial laws. These excluded Jews from most professions, barred them from attending universities and forced them to wear a yellow badge. All Jewish women had to take the name Sarah as part of their name, all Jewish men the name Israel. All Jewish bank accounts were frozen and all licenses held by Jews—even driver's licenses—were revoked. As a result, Helly was fired by the Phönix and could no longer teach at the university. It was still possible, indeed encouraged, for Jews to emigrate after paying an emigration tax, the *Reichsfluchtsteuer* and 130,000 did. Among the 30,000 who came to the United States were Helly, his wife, and their eight-year old son Walter Sigmund; they emigrated to Brooklyn in 1938. (Birnbaum and Lukacs also emigrated to the United States.) There were so many qualified *émigrés* that Helly was unable to secure a university position, even though he had letters of recommendation from Einstein, Oswald Veblen and Hermann Weyl. He survived by tutoring high school students. Eventually, he found employment at some junior colleges in New Jersey. In 1943, upon the recommendation of Karl Menger, among others, he was offered the position of visiting lecturer at Illinois Institute of Technology in Chicago. This turn of good fortune did him no good, however. His second heart attack, a remnant of his WWI wound, killed him on November 28, 1943. Gödel, a thesis student of Hahn's, summed it up in a note to Walter in the funeral book: "Now all is well but . . . Papa dies." Walter got a PhD in physics from the Massachusetts Institute of Technology and later became Professor of Operations Research at the Polytechnic Institute of Brooklyn, our *alma mater*, now called NYU-Polytechnic Institute. He is noted for "$(p,q)$-Helly cliques" and also his 1975 book *Urban Systems Models.* We were at "Poly" as students and teachers but never met him. A friend of ours, Maurice Figueres, took a course with him there in 1987 and was frequently driven back to Manhattan by him after class. He said to Maurice, "My dad was a *real* mathematician."

The Hahn–Banach theorem should be called the *Helly*-Hahn–Banach theorem but, as with so many other misnamed results, this will never happen, a final piece of bad luck for Eduard Helly.

## 7.14 EXERCISES

**Notation.** $X$ and $Y$ stand for vector spaces over $\mathbf{R}$ or $\mathbf{C}$ unless otherwise noted. $X'$ denotes the linear space of all continuous linear functionals on a TVS $X$. $X^*$ denotes the linear space of *all* linear functionals on $X$, the algebraic dual of $X$.

7.101 Consider the subspace $M = \mathbf{R}\times\{0\} \subset \ell_2^{\mathbf{R}}(2)$ and the linear functional $f(a,0) = a$ defined on $M$. Let $y \in \ell_2^{\mathbf{R}}(2)$ be a unit vector of angle $\beta \neq 0, \pi$ with the $x$-axis. Show that the subspace $N = \mathbf{R}y\times\{0\}$ is a topological complement of $M$. Show that, for any such $N$, the projection $P_M$ on $M$ along $N$ is a continuous extension of $f$ and $f \circ P_M$ is a continuous extension of $f$ of norm $|\csc\beta|$.

7.102 Let $X$ and $Y$ be vector spaces over the same field $K$.

(a) Let $M$ be a subspace of $X$ and let $A : M \to Y$ be a linear map. Show that there is a linear map $\hat{A} : X \to Y$ which extends $A$.

(b) Prove that there exist additive maps $f : \mathbf{R} \to \mathbf{R}$ (i.e., $f(x+y) = f(x)+f(y)$) other than those of the form $f(x) = tx$ for some $t \in \mathbf{R}$. (These latter ones are the only continuous ones, however.)

7.103 SUBLINEAR FUNCTIONALS Let $X$ be a real vector space and let $X^{\#}$ denote the class of sublinear functionals on $X$.

(a) MAX AND SUP If $p_1, \ldots, p_n \in X^{\#}$ then $p(x) = \max_j p_j(x)$, $x \in X$, is a sublinear functional. If $P$ is any subset of $X^{\#}$ such that for each $x \in X$, $q(x) = \sup_{p \in P} P(x) < \infty$ then $q \in X^{\#}$.

(b) SPECIAL CASE Show that the algebraic dual $(\mathbf{R}^n)^*$ of $\mathbf{R}^n$ is linearly isomorphic to $\mathbf{R}^n$. In particular, if for $x = (a_1, \ldots, a_n) \in \mathbf{R}^n$ and $f \in (\mathbf{R}^n)^*$, $f(x) = \sum_{i=1}^{n} t_i a_i$, consider the map $f \mapsto (t_1, \ldots, t_n)$. Show that $f((t_1, \ldots, t_n)) \le \|(t_1, \ldots, t_n)\|_2$ iff $\|(t_1, \ldots, t_n)\|_2 \le 1$. With $\|(a_1, \ldots, a_n)\|_\infty = \max(a_1, \ldots, a_n)$, show that $f((t_1, \ldots, t_n)) \le \|(t_1, \ldots, t_n)\|_\infty$ iff $t_1, \ldots, t_n \ge 0$ and $\sum_i t_i = 1$.

7.104 CONVEX AND CONCAVE FUNCTIONS Let $f$ be an extended real-valued map $f$ defined on a real vector space $X$ or a convex subset thereof; $f$ is called *convex* if $f(ax+by) \le af(x) + bf(y)$ for all $a, b > 0$ such that $a + b = 1$ and all $x$ and $y$ in the domain of $f$. If $x \ne y$ implies that $f(ax+by) < af(x) + bf(y)$, we say that $f$ is *strictly convex.* Geometrically, this means that the line connecting any two points of the graph of $f$ is above the curve. (Unfortunately, such functions are usually called "concave upward" in calculus courses.) Real linear functionals and norms are convex functions. With $a$ and $b$ as above, if $f(ax+by) \ge af(x)+bf(y)$ then $f$ is called *concave.* If $f$ is convex and concave—i.e., if equality holds—then $f$ is called *affine.*

(a) Let $f$ be a real function defined on the open interval $(c, d)$. Show that if $f$ is twice differentiable then $f$ is convex iff $f'' \ge 0$ on $(c, d)$ and strictly convex if $f'' > 0$.

(b) A normed space $X$ is called *strictly convex* if for all distinct unit vectors $x$ and $y$, $\|x+y\| < \|x\| + \|y\|$ or, equivalently, if $\left\|\frac{x+y}{2}\right\| < 1$. Show that $f(x) = (1/2)\|x\|^2$ is convex and that $X$ is strictly convex iff $f$ is strictly convex.

(c) If $g$ is a linear functional on $\mathbf{R}^n$ then show that $x \mapsto g(x) + c$ is an affine map for any real number $c$.

A certain duality exists between convex and concave functions which stems from (d).

(d) $f$ is convex iff $-f$ is concave.

(e) EPIGRAPH The *epigraph of* $f : K \to \mathbf{R}$ is the set $\operatorname{epi} f = \{(x, r) \in K \times \mathbf{R} : f(x) \le r\}$. Show that $f$ is convex iff $\operatorname{epi} f$ is a convex set. If $X$ is a TVS and $K$ is closed, show that $f$ is lower semicontinuous [i.e.$\{x \in K : f(x) \le r\}$ is closed for each $r \in \mathbf{R}$] iff $\operatorname{epi} f$ is closed.

(f) If $\{f_s : s \in S\}$ is a family of convex functions defined on $K$ then $f(x) = \inf_{s \in S} f_s(x)$ $(x \in K)$ is a convex function.

7.105 HAHN–BANACH SANDWICH THEOREM Part (c) generalizes Theorem 7.3.2; it is further generalized in Exercise 7.202.

Let $S$ be any subset of the real vector space $X$, let $p$ be a sublinear functional on $X$, and let $f : S \to \mathbf{R}$ be any map with $f \leq p$ on $S$.

(a) Show that the auxiliary functional $g(x) = \inf\{p(x+ts) - tf(s) : s \in S,\ t \geq 0\} = \inf\{p(x+ts) - tf(s) : s \in S,\ t > 0\}$, $x \in X$, is real-valued and $\leq p$.

(b) Let $g$ and $S$ be as in (a). If $h$ is a linear functional on $X$ then $h \leq g$ iff $f \leq h|S$ and $h \leq p$. Moreover, if $S$ is convex and $f$ concave then $g$ is sublinear.

(c) SANDWICH THEOREM If $S$ is convex and $f$ concave then there is a linear functional $F$ on $X$ such that $F \leq p$ and $f \leq F$ on $S$.

Let $X^{\#}$ and $X^{*}$ denote the sublinear and linear functionals, respectively, on $X$. The result of (d) is sometimes called the *convex version* of the Hahn–Banach theorem.

(d) If $S$ is a nonempty convex subset of $X$ and $p \in X^{\#}$, there exists $F \in X^{*}$ such that $F \leq p$ and $\inf F(S) = \inf p(S)$.

(e) For $p \in X^{\#}$ and any $x \in X$, $\{f(x) : f \in X^{*},\ f \leq p\} = [-p(-x), p(x)]$.

7.106 SEPARATION THEOREMS Things take place in a vector space $X$.

(a) Let $D$ be an absorbent convex set in $X$ and let $p_D$ be its gauge as defined in Eq. $(*)$ of Sec. 7.7. For a real linear functional $f$ on $X$, show that $f \leq p_D$ iff $f \leq 1$ on $D$. If $D$ is an absorbent disk then $|f| \leq p_D$ iff $|f| \leq 1$ on $D$.

(b) Let $D$ be an absorbent disk and let $f$ be a real linear functional defined on a subspace $M$ of $X$ such that $|f| \leq 1$ on $D \cap M$. Then there is a linear extension $F$ defined on $X$ of $f$ such that $|F| \leq 1$ on $D$.

(c) For any subset $A$ of $X$, let $K(A) = \{x \in A : A - x \text{ is absorbent}\}$. Let $A$ and $B$ be nonempty convex sets such that $K(A) \neq \varnothing$. Then $A$ and $B$ are separated by a hyperplane iff $B \cap K(A) = \varnothing$.

(d) Two convex sets $A$ and $B$ are strongly separated by a hyperplane iff there is an absorbent convex set $U$ such that $(A+U) \cap B = \varnothing$.

7.201 MAXIMAL EXTENSIONS Let $f$ be a linear functional on a subspace $M$ of a real linear space $X$ which is dominated by a sublinear functional $p$ and let $S$ be any subset of $X$. A linear extension $F$ of $f$, $F \leq p$, is $(S,p)$-*maximal* if whenever $G$ is another linear extension of $f$, $G \leq p$, such that $G|_S \geq F|_S$ then $G|_S = F|_S$.

(a) There exist $(S, p)$-maximal extensions.

(b) Let $p'(x) = \inf\{f(x) + p(x - m) : m \in M\}$. If $S = \{w\}$ $(w \in X)$ then for any $(S, p)$-maximal extension $F$ of $f$, $F(w) = p'(w)$.

(c) UNIQUENESS Let $p'$ be as in (b). If $S$ is a cone [Sec. 5.12] containing $M$ then the $(S, p)$-maximal extensions of $f$ coincide on the subspace $M(S)$ spanned by $M$ and $S$ iff $p'$ is additive on $S$.

7.202 GENERALIZED HAHN–BANACH SANDWICH THEOREM [Fuchsteiner and König 1978; König 1982]: Part (a) below is a lemma to the main theorem in (b) and its corollary (c). The decisive greater generality of (c) over the sandwich theorem of Exercise 7.105(c) stems from the weakening of the statement "for all $a, b > 0$, $a + b = 1$" to merely "there exist $a, b > 0$." $X$ denotes a real linear space.

(a) Let $\mathbf{R}^- = \mathbf{R} \cup \{-\infty\}$, let $K$ be a nonempty cone in $X$, and let $g : K \to \mathbf{R}^-$ be such that for each $x \in K$ the map $t \mapsto g(tx)$ is upper semicontinuous on $(0, \infty)$ [i.e., $g^{-1}[c, \infty)$ is closed for each $c \in (0, \infty)$]. Suppose there exist $a, b > 0$ such that

$$g(ax + by) \le ag(x) + bg(y) \text{ for all } x, y \in K \qquad (*)$$

Define $h : K \to \mathbf{R}^-$ to be $h(x) = \inf_{t>0} t^{-1} g(tx)$ for each $x \in K$. Then $h$ is sublinear. Moreover, if $a + b \ne 1$ then

$$h(x) = \lim_n \frac{g[(a+b)^n x]}{(a+b)^n} \qquad \text{for all } x \in K$$

where the limit exists since $g((a+b)x) \le (a+b)g(x)$ for each $x \in K$ by $(*)$.

(b) GENERALIZED SANDWICH THEOREM Let $p$ be a sublinear functional on the real vector space $X$, let $W$ be a nonempty subset of $X$, and let $f : W \to \mathbf{R}$ be any map which is $\le p|_W$.
If there exist $a, b > 0$ such that, for all $u, v \in W$,

$$\inf_{w \in W} [p(w - au - bv) - f(w) + af(u) + bf(v)] \le 0 \qquad (**)$$

then there is a linear functional $F$ on $X$ such that $F \le p$ and $f \le F|_W$.

(c) Let $p, W$, and $f$ be as in (b). If there exist $a, b > 0$ such that $u, v \in W \implies au + bv \in W$ and $f(au + bv) \ge af(u) + bf(v)$ then there is a linear functional $F$ on $X$ such that $F \le p$ and $f \le F|_W$.

(d) GENERALIZED CONVEX VERSION Let $p$ be a sublinear functional on $X$ and let $W$ be a nonempty subset of $X$. If there exist $a, b > 0$ such that $a + b = 1$ and $\inf_{w \in W}(pw - au - bv) \le 0$ for all $u, v \in W$ then there is a linear functional $F$ on $X$ such that $F \le p$ and $\inf F(W) = \inf p(W)$.

(e) Let $p$ be a sublinear functional on $X$ and let $W$ be a nonempty subset of $X$. Suppose that there are $a, b > 0$ such that

$$\inf_{w \in W} p(w - au - bv) \leq 0 \text{ for all } u, v \in W$$

If $p|_W \geq 0$ then there is a linear functional $F$ on $X$ such that $F \leq p$ and $F|_W \geq 0$.

## Hints to Exercises

7.102 (b) View $\mathbf{R}$ as a vector space over the rationals $\mathbf{Q}$. Use (a) to extend the identity map $I : \mathbf{Q} \to \mathbf{R}$, $x \mapsto x$. If $x$ and $y$ are linearly independent over $\mathbf{Q}$ and $I^*$ is an extension of $I$ for which $I^*x = 0$ and $I^*y = 1$, can $I^*(\cdot) = t(\cdot)$ for any $t \in \mathbf{R}$?

7.105 (d) In the case when $p(S)$ is bounded below, let $k = \inf p(S)$ and define $f : S \to \mathbf{R}$ to be the constant $k$. Now use (c).

(e) For any $f \in X^*$, $f \leq p$ implies that $-p(-x) \leq f(x) \leq p(x)$ for every $x \in X$. To get the reverse inclusion, use (a) with $S = \{x\}$ and $\{-x\}$, respectively. This yields $f, g \in X^*$ such that $f(x) = p(x)$ and $g(-x) = p(-x)$.

7.107 (a) Let $P = \{x \in X : x \geq 0\}$. Show that $M + P = M - P = M + P - P$. Deduce that even though $P$ is not generally a subspace, $N = M + P$ is, and also, for any $n \in N$, there exist $m, m' \in M$ such that $m \leq n \leq m'$. The strategy is to extend $f$ positively to $N$ and then take *any* extension (as in Exercise 7.102(b), for example) to $X$. For any $n \in N$, let $p(n) = \inf\{f(u) : n \leq u \in M\}$. Show that $p$ is sublinear and that $p = f$ on $M$. Now use the dominated extension theorem [Theorem 7.3.2(a)], to get an extension $F$ of $f$ to $N$ with $F \leq p$. Clearly, $x \leq 0$ implies $p(x) \leq 0$ and therefore that $F(x) \leq 0$, so that $F$ is positive.

(b) Show that any subspace containing an order unit is cofinal.

(c) If $F$ is a positive extension of $f$, let $U = \{x \in X : F(x) \leq 1\}$. Conversely, let $p$ be the gauge of $U - P$ (as defined in Sec. 7.7), and use Exercise 7.106(d) and the Hahn–Banach extension theorem [7.3.2(a)].

7.201 (a) You can assume without loss of generality that $S$ is a cone since a linear functional $F$ is $(S, p)$-maximal iff $F$ is $(K(S), p)$-maximal where $K(S)$ denotes the cone generated by $S$ [Exercises 5.108(d)]. If $S \subset M$, the result follows from the dominated extension theorem [7.3.2(a)]. Therefore, suppose that $S \setminus M \neq \varnothing$ and let $M(S)$ be the subspace spanned by $M$ and $S$. Extend $f$ to $M(S)$. Consider the collection of pairs $(f', M')$ where $f'$ is a linear functional on a subspace $M'$ of $M(S)$. Order the pairs by extension: $(f', M') \preceq (f'', M'')$ iff $M' \subset M''$ and $f''|_{M'} = f'$ and let $\mathcal{A}$ consist of pairs $(f', M')$ such that

(a) (1) $(f, M) \preceq (f', M')$,

(2) $f' \leq p|_{M'}$,

(3) $f'$ is maximal on $M' \cap S$ (i.e., $f'$ is $(M' \cap S,\ p|_{M'})$-maximal, and

(4) $M'$ is the linear span of $M$ and $M' \cap S$.

Show that $(f, M)$ satisfies these conditions (which is routine) and that any totally ordered subset $\{(f_t, M_t) : t \in T\}$ of $\mathcal{A}$ has an upper bound in $\mathcal{A}$, namely, $(f_T, M_T)$, where $M_T = \cup_{t \in T} M_t$ and $f_T(x) = f_t(x)$ where $t$ is such that $x \in M_t$. Conclude that $\mathcal{A}$ has a maximal element $(f_0, M_0)$ with respect to $\preceq$. Suppose now that $M(S) \setminus M_0 \neq \varnothing$ and let $x_1 \in M(S) \setminus M_0$. Let $M_1$ be the subspace spanned by $M$ and $x_1$. Since $M_0$ satisfies (4), $M_1$ must also. Define a linear functional $f_1$ on $M_1$ by taking $f_1(x + tx_1) = f_0(x) + tf_1(x_1)$ for $x \in M_0$ and $t \in \mathbf{R}$ where $f_1(x_1) = \inf\{f_0(x) + p(x_1 - x) : x \in M_0\}$. Now $(f_0, M_0) \preceq (f_1, M_1)$ and $f_1 \leq p|_{M_1}$. Let $g_1$ be a linear functional on $M_1$ such that $g_{1M} = f$, $g_1 \leq p|_{M_1}$, and $g_1|_{M_1 \cap S} \geq f_1|_{M_1 \cap S}$. Since $f_0$ is maximal on $M_0 \cap S$, it follows that $g_1|_{M_0 \cap S} = f_0|_{M_0 \cap S}$. Hence $g_1$ and $f_0$ coincide on the linear span of $M$ and $M_0 \cap S$; by (4), $g_1|_{M_0} = f_0$. We have $g_1(x_1) \geq f_1(x_1)$ but if $g_1(x_1) > f_1(x_1)$ then, for some $x \in M_0$, $g_1(x_1) > f_0(x) + p_1(x_1 - x)$, i.e., $g_1(x_1 - x) > p(x_1 - x)$, contradicting $g_1 \leq p|_{M_1}$. Thus $g_1(x_1) = f_1(x_1)$, so $g_1 = f_1$. In particular, $f_1$ is maximal on $M_1 \cap S$, so $(f_1, M_1) \in \mathcal{A}$, and $(f_1, M_0) \prec (f_1, M_1)$ which contradicts the choice of $(f_0, M_0)$. Therefore, $M_0 = M(S)$. If $M(S) \neq X$, use the dominated extension theorem [7.3.2(a)] to extend $f_0$ to a linear functional $F$ on $X$ such that $F \leq p$. $F$ is then the desired $(S, p)$-maximal extension.

(b) For any linear extension $F$ of $f$, $F \leq p$, it is also true that $F \leq p'$. Using a construction like the one used to define $f_1$ in part (a), there is a linear functional $F_0$ extending $f$, $F_0 \leq p$, such that $F_0(w) = p'(w)$. The conclusion now follows from the definition of $(S, p)$-maximality.

(c) Assume that all $(S, p)$-maximal extensions of $f$ coincide on $M(S)$. By (b) there exists, for each $x \in S$, an $(S, p)$-maximal extension $F_x$ such that $F_x(x) = p'(x)$. This implies that $p'(x + y) = F_{x+y}(x + y) = F_{x+y}(x) + F_{x+y}(y) = F_x(x) + F_y(y) = p'(x) + p'(y)$. Conversely, suppose that $p'$ is additive on $S$. From the definition of $p'$, $p'|_M = f$, $p' \leq p$, and $p'$ is a sublinear functional. Since $S$ is a cone and $M \subset S$, $M(S) = S - S$. Define $f'$ on $M(S)$ by $f'(x - y) = p'(x) - p'(y)$ for $x, y \in S$; $f'$ is well-defined and $f'|_M = f$. Also, $f'(x - y) = p'(x) - p'(y) \leq p'(x - y) \leq p(x - y)$ for all $x, y \in S$. Thus $f' \leq p|_{M(S)}$. An extension $F$ of $f'$, $F \leq p$, is $(S, p)$-maximal since $F|_S = p'|_S$. For any linear extension $G$ of $f$, $G \leq p$, we have $G|_S \leq p'_S$ so if $G$ is $(S, p)$-maximal, $G|_S = F|_S$.

7.202 (a) (i) Define $H : K \to \mathbf{R}^-$ to be $H(x) = \inf_{n\in\mathbf{Z}} g((a+b)^n x)/(a+b)^n = \lim_n g((a+b)^n x)/(a+b)^n$ for each $x \in K$. It follows that $H(ax+by) \leq aH(x) + bH(y)$ for all $x, y \in K$ and $H((a+b)x) = (a+b)H(x)$ for each $x \in K$. Hence for each $x \in K$, the map $t \mapsto H(tx)$ is upper semicontinuous on $(0, \infty)$.

(ii) Show that $H$ is convex on $K$ as follows. Let $M \subset [0,1]$ consist of those $t \in [0,1]$ such that $H((1-t)x + ty) \leq (1-t)H(x) + tH(y)$ for all $x, y \in K$ [with the convention $0 \times (-\infty) = 0$]. Then

1. $0, 1 \in M$,
2. $s, t \in M \implies (as+bt)/(a+b) \in M$; from now on let $d = (as+bt)/(a+b)$.
   In fact for $x, y \in K$ we have

$$
\begin{aligned}
&H((1-d)x + dy) = H[\left(\tfrac{a(1-s)+b(1-t)}{a+b}\right)x + \tfrac{as+bt}{a+b}y] \\
&= [\tfrac{1}{a+b}]H[a((1-s)x+sy) + b[(1-t)x+ty]]) \\
&\leq [\tfrac{a}{a+b}]H((1-s)x+sy) + [\tfrac{b}{a+b}]H((1-t)x+ty) \\
&\leq [\tfrac{a}{a+b}][(1-s)H(x) + sH(y)] + [\tfrac{b}{a+b}][(1-t)H(x)+tH(y)] \\
&= (1-d)H(x) + dH(y)
\end{aligned}
$$

From (1) and (2) it follows that

3. $\operatorname{cl} M = [0,1]$, and
4. since $M$ is closed, $M = [0,1]$. To see that $M$ is closed, note that the set difference $[0,1] \setminus M$ consists of $t \in (0,1)$ such that there exist $x, y \in K$ with $H((1-t)x + ty) > (1-t)H(x) + tH(y)$ or $H(x+y) > (1-t)H(x/(1-t)) + tH(y/t)$ which must be open by the upper semicontinuity of $H$ mentioned in (i).

(iii) For $x \in K$ we have

$$
\begin{aligned}
\inf_{t>0} t^{-1}H(tx) &= \inf_{t>0}\inf_{n\in\mathbf{Z}} g((a+b)^n tx)/(a+b)^n t \\
&= \inf_{t>0} t^{-1} g(tx) = h(x)
\end{aligned}
$$

Next, show that $h$ is subadditive and therefore sublinear on $K$. In particular, for $x, y \in K$ and $s, t > 0$, we have, by (ii),

i. $h(x+y) \leq (s+t)H\left((s+t)^{-1}(x+y)\right)$
   $= (s+t)H\left(\frac{s}{s+t}\frac{x}{s} + \frac{t}{s+t}\frac{y}{t}\right) \leq sH(x/s) + tH(y/t)$.

Hence $h(x+y) \leq h(x) + h(y)$.

(iv) Assume now that $a+b \neq 1$. We claim that $H(tx) = tH(x)$ for each $x \in K$ and $t > 0$ so that $H = h$ by the following argument. For $p \in \mathbf{Z}$ with $0 < t < (a+b)^p$ and $n \in \mathbf{Z}$ we have

i. $H\left(\left[\left((a+b)^p - t\right)(a+b)^n + t\right]x\right)$
$= (a+b)^p H\left(\left(1 - \frac{t}{(a+b)^p}\right)(a+b)^n x + \frac{t}{(a+b)^p} x\right)$
$\leq (a+b)^p \left\{\left[(1 - \frac{t}{(a+b)^p}\right](a+b)^n H(x) + \left[\frac{t}{(a+b)^p}\right] H(x)\right\}$
$= \left[(a+b)^p - t\right](a+b)^n H(x) + tH(x)$

The map $t \mapsto H(tx)$ is convex on $(0,\infty)$ and therefore is always $-\infty$ or always finite-valued and hence continuous. We can assume the latter case. Show that $H(tx) \leq tH(x)$ for all $x \in K$ and $t > 0$ to establish the equality.

(b) Consider the auxiliary functional $h(x) = \inf\{p(x+tw) - tf(w) : w \in W,\ t > 0\}$ discussed in Exercise 7.105(a). By Theorem 7.6.4 it suffices to prove that $h$ is sublinear.

To do this, introduce the functional $g$ defined at each $x \in X$ by $g(x) = \inf_{w \in W}[p(x+w) - f(w)]$; $g$ is real-valued and $g(ax+by) \leq ag(x) + bg(y)$ for all $x, y \in X$. Moreover, for $x \in X$, $s, t > 0$, and $w \in W$ we have $[p(tx+w) - f(w)] - [p(sx+w) - f(w)] \leq p((t-s)x) \leq n((t-s)x) = |t-s|\, n(x)$, where $n(x) = \max[p(x), p(-x)]$ is a seminorm [Theorem 5.4.2]. Hence $|g(tx) - g(sx)| \leq |t-s|\, n(x)$ for all $x \in X$ and $s, t > 0$. Thus, for fixed $x \in X$, the function $t \mapsto g(tx)$ is continuous on $(0,\infty)$. Finally, show that $h(x) = \inf_{t>0}(1/t)\, g(x)$ and apply the result of (a).

(c) Let $w = au + bv$ in (**) of (b).

(d) See the hint to Exercise 7.105(d).

(e) Let $f = 0$.

# Chapter 8

# Duality



The linear space $X'$ of continuous linear functionals on a TVS $X$ is called the dual space of $X$. "Duality theory" comprises mainly two things:

1. *Special Duality Theory:* identifying the duals of particular spaces—the dual of $\ell_2$ "is" $\ell_2$ for example, or that $\ell_p' = \ell_q$.

2. *General Duality Theory:* using things about $X'$ to discover things about $X$.

This chapter is mostly about general duality theory.

Suppose $X$ and $X'$ are just vector spaces, untopologized, and that $\langle \cdot, \cdot \rangle$ is a bilinear functional on $X \times X'$, in which case we say that $X$ and $X'$ are *paired spaces.* Any $x' \in X'$ determines a linear functional on $X$, namely $x \mapsto \langle x, x' \rangle$. We say that a vector topology $\mathcal{J}$ on $X$ is a *topology of the pair* $(X, X')$ if $X'$ is the continuous dual $(X, \mathcal{J})'$ of $(X, \mathcal{J})$ [Definition 8.7.1]. It happens that some things depend only on the pair. The bounded subsets of a locally convex space $X$, for example, are the same for any topology of the pair $(X, X')$ [Theorem 8.8.7]. Thus, the specific topology on $X$ can be bypassed to some extent when investigating the bounded sets of $X$. Sometimes the topology on $X$ can be replaced by another more manageable topology of the pair $(X, X')$.

If the topology on $X$ is weak enough, then $X'$ can be very small—too small to be "sensitive" to events in $X$. If $X$ carries the trivial topology, for example, then $X' = \{0\}$. If $X$ is an LCHS, then a consequence of the Hahn–Banach theorem [7.7.7] guarantees that $X'$ distinguishes the points of $X$ [i.e., if $x \neq y$, then there exists $f \in X'$ such that $f(x) \neq f(y)$]. Some of the strongest results require this hypothesis; the invariance of bounded sets mentioned above does, for example.

One of the major results of duality theory, the Mackey–Arens theorem [8.7.4], establishes the extent to which a topology of a pair $(X, X')$ can vary: A locally convex topology $\mathcal{J}$ of the pair must fall between the weak topology $\sigma(X, X')$ [Example 8.5.3] and the "Mackey topology $\tau(X, X')$" [Example 8.5.4].

There is a symmetry to many of the results of the chapter that stems from the fact that each vector in $X$ may be viewed as a linear functional on $X'$ by way of the map $f \mapsto f(x)$ on $X'$. There are times when we say things like "by the dual form of" a certain result. Still, there is an *asymmetry* that should be mentioned. If $X$ is a normed space, then there is a "natural" topology for its dual $X'$, its canonical norm topology [Example 6.5.3(b)]. As the dual of a normed space is always complete, here is an instance in which $X$ and $X'$ cannot be treated equally. If $X$ is a TVS, then there is no "natural" topology for $X'$. Instead, there are various, reasonable, possibilities. Which to choose? One choice is not consistently made. Rather, different topologies are employed in different circumstances. For reflexivity, for example, it is customary to endow $X'$ with the "strong topology" $\beta(X', X)$ [Example 8.5.5].

The origins of duality theory date from the time when duals of certain spaces were considered and identified for the first time. Some confusion undoubtedly was caused initially by the fact that the duals of $\ell_2$ and $L_2[0, 2\pi]$ were $\ell_2$ and $L_2[0, 2\pi]$, respectively. Riesz's investigation of the spaces $\ell_p'$, $p \geq 1$, made it clear that this was generally untrue but suggested a different false trail. The dual $\ell_p'$ of $\ell_p$ for $p \geq 1$ $(1/p + 1/q = 1)$ "is" $\ell_q$. Hence $\ell_p'' = \ell_q' = \ell_p$, thus suggesting that all spaces were "reflexive," i.e., that $X'' = X$ generally. Subsequently, of course, it was realized that this was not the case either.

## 8.1 PAIRED SPACES

A *bilinear functional* $\langle \cdot, \cdot \rangle : X \times Y \to \mathbf{F}$, $(x, y) \mapsto \langle x, y \rangle$, is a map which is linear in either argument when the other is fixed. We usually omit explicit reference to the bilinear functional $\langle \cdot, \cdot \rangle$ and just refer to $(X, Y)$ as a *pairing* or *pair* or say that $X$ and $Y$ are *paired spaces.* The order is unimportant: We also refer to $(Y, X)$ as a pair with respect to $\langle y, x \rangle' = \langle x, y \rangle$, i.e., the same bilinear functional. If for each nonzero $x \in X$ there exists $y \in Y$ such that $\langle x, y \rangle \neq 0$ then $Y$ is said to *distinguish points* of $X$; the analogous meaning is attached to $X$ *distinguishes points of* $Y$. If each vector space distinguishes the points of the other, then we call $(X, Y)$ a *dual pair.*

If $X$ is a linear space and $X^*$ its algebraic dual—the vector space of all linear functionals on $X$—then the *natural pairing* of $X$ and $X^*$ is that arising from the *natural* or *canonical* bilinear functional on $X \times X^*$: $(x, x^*) \mapsto \langle x, x^* \rangle = x^*(x)$. The expression "natural map" is also used to indicate this bilinear functional acting on the product of $X$ and any subspace of $X^*$. If $X$ is a TVS, then its *topological* (*continuous*) *dual* $X'$, the space of all *continuous* linear functionals on $X$, is a conspicuous subspace of $X^*$ to consider natural pairings of $X$ with. If $X$ and $Y$ are paired spaces and $y \in Y$, then the map $y^*$

$$y^* : X \to \mathbf{F}, \quad x \mapsto \langle x, y \rangle$$

is obviously a linear functional on $X$; that is, $y^* \in X^*$. The map

$$D : Y \to X^*, \ y \mapsto y^* \qquad (*)$$

is generally not 1-1: It is possible that $y^* = w^*$ even though $y \neq w$. For it to be injective, it is clearly necessary and sufficient that

$$\langle x, y \rangle = 0 \text{ for each } x \in X \Longrightarrow y = 0$$

or, equivalently,

$$y \neq 0 \Rightarrow \text{there is some } x \in X \text{ such that } \langle x, y \rangle \neq 0$$

i.e., when $X$ *distinguishes points* of $Y$; in this case, we will usually not distinguish between $y$ and the linear functional $y^*$ and refer to $y$ itself as a member of $X^*$.

**Example 8.1.1** PAIRS Let $X$ be a vector space.

(a) Under the natural pairing $(X, X^*)$ is a dual pair.

(b) If $M$ is a subspace of $X^*$, then $(X, M)$ is a dual pair with respect to the natural pairing iff $M$ is a total subspace [Definition 6.6.4].

(c) If $X$ is a TVS and $f \neq 0$ ($f \in X'$) then there exists $x \in X$ such that $f(x) \neq 0$, i.e., $X$ distinguishes points of $X'$.

(d) If $X$ is an LCHS, then $X'$ distinguishes points of $X$ by the Hahn–Banach theorem [7.7.7(b)] so $(X, X')$ is a dual pair.

(e) For $X = \mathbf{R}^n$, for $x = (a_i) \in \mathbf{R}^n$ and $y = (b_i) \in \mathbf{R}^n$, $\langle x, y \rangle = \sum_{i=1}^n a_i b_i$ establishes a pairing for $(X, X)$.

(f) If $(X, \langle \cdot, \cdot \rangle)$ is an inner product space, then $(X, X)$ does not form a paired system with respect to $\langle \cdot, \cdot \rangle$ since the inner product is sesquilinear, not bilinear: $\langle x, ay \rangle = \overline{a} \langle x, y \rangle$ instead of $a \langle x, y \rangle$ $(a \in \mathbf{C})$. ■

In many of the results to follow, having a *dual* pair is not necessary: In some, validity remains intact for any pairing; in others, only something like "$Y$ distinguishes points of $X$" is needed. We have not made a consistent attempt in the sequel to always provide minimal hypotheses, however.

Early terms for *dual* include *polare Raum* [Hahn 1927, p. 219], *espace conjugué* and *transponierte Raum* [both due to Schauder 1930, p. 184], the former also used by Banach [1932, p. 188], adjoint space [Alaoglu 1940, p. 252]; *dual* is due to Bourbaki [1938, p. 1702].

## 8.2 WEAK TOPOLOGIES

In an inner product space $(X, \langle \cdot, \cdot \rangle)$ to say that $x_n \to x$ "weakly" means that $\langle x_n, y \rangle \to \langle x, y \rangle$ for all $y \in X$. Riesz and Hilbert made extensive and effective use of weak convergence from the early 1900s on. As a result of the invention of the Lebesgue integral there came many new types of convergence. There was nothing sacrosanct about norm convergence for the pioneers; if anything, weak convergence was the preferred form. Banach [1929 Part II, p. 231] introduced weak convergence of vectors in normed spaces—namely $x_n \to x$ *weakly* if $f(x_n) \to f(x)$ for all $f \in X'$—and the analogous notion for convergence of a sequence $(f_n)$ from $X'$, what we call weak-$*$ convergence today. We first considered weak topologies on a TVS $X$ in Example 5.5.7.

**Definition 8.2.1** WEAK TOPOLOGIES If $X$ and $Y$ are paired vector spaces, each over the real or complex field $\mathbf{F}$, then the map $p_y(\cdot) = |\langle \cdot, y \rangle|$ determines a seminorm on $X$ for each $y \in Y$. The weakest topology $\sigma(X, Y)$ for $X$ for which the seminorms $\{p_y : y \in Y\}$ are continuous is called the *weak topology* (*topologie faible, schwache Topologie*) on $X$ for the pair $(X, Y)$; the space of the second argument determines the topology on the first. An analogous meaning is attached to $\sigma(Y, X)$. ■

The following simple result is useful.

**Theorem 8.2.2** PAIRS For a pair $(X, Y)$, the following are equivalent:

(a) $X$ distinguishes points of $Y$;

(b) the map $D : Y \to X^*, y \mapsto y^*$, of $(*)$ of the preceding section is injective;

(c) $\sigma(Y, X)$ is Hausdorff.

**Proof.** We prove only (a) $\Rightarrow$ (c). If $X$ distinguishes the points of $Y$, then for each nonzero $y \in Y$, there is some $x \in X$ such that $0 \neq |\langle x, y\rangle| = p_x(y)$. Hence $\sigma(Y, X)$ is Hausdorff by Theorem 5.5.1(a). ■

**Remark 8.2.3** For any TVS $X$, $X$ distinguishes the points of $X'$ [Example 8.1.1(c)] so $\sigma(X', X)$ is always Hausdorff.

A neighborhood base at 0 for $\sigma(X, Y)$ is given by finite intersections of sets $\{x \in X : |\langle x, y\rangle| < r\}$, $r > 0$, $y \in Y$, i.e., by sets such as

$$V(0, y_1, \ldots, y_n, r) = \{x \in X : |\langle x, y_j\rangle| < r, 1 \leq j \leq n\} \qquad (*)$$

where $n \in \mathbf{N}$ and $y_1, \ldots, y_n \in Y$. A typical neighborhood of $z \in X$, namely $z + V(0, y_1, \ldots, y_n, r)$ is $V(z, y_1, \ldots, y_n, r) = \{x \in X : |\langle z - x, y_j\rangle| < r, 1 \leq j \leq n\}$.

**Definition 8.2.4** WEAKENED TOPOLOGIES If $(X, \mathcal{J})$ is a TVS and $X'$ its dual, view $X$ and $X'$ as paired by the natural pairing. The weak topology $\sigma(X, X')$ induced on $X$ by $X'$ is sometimes called the *weakened topology* (*topologie affaiblié*). $X'$ endowed with $\sigma(X', X)$ is called the *weak dual* of $X$. ■

For a TVS $X$, a base for $\sigma(X, X')$ at 0 is given by the collection of $\mathcal{J}$-open neighborhoods $V(0, f_1, \ldots, f_n, r)$ of $(*)$ $(f_1, \ldots, f_n \in X', r > 0)$, $\sigma(X, X')$ is seen to be coarser than the original topology on $X$. Hence convergence implies weakened convergence. Although we tend to think of neighborhoods as "small," weak neighborhoods of 0 are generally of considerable girth. As to their topological obesity, weak neighborhoods of 0 in infinite-dimensional LCHS are unbounded [Example 6.1.7(c)]. As to their algebraic breadth, for infinite-dimensional $X$, for any $f \in X'$, $V(0, f, 1)$ contains $f^{-1}(0)$, a subspace of codimension 1! The weak topology $\sigma(X', X)$ is called the *weak-$*$* (read: weak star) topology for $X'$.

In many cases the dual $X'$ of a TVS $X$ can be completely characterized such as those listed in Table 1.7.1. If $X$ is a finite-dimensional HTVS, then the dual of $X$ is essentially just $X$. If $(X, \langle \cdot, \cdot \rangle)$ is a Hilbert space and $f$ is a continuous linear functional on $X$, then there is a unique $y \in X$ such that $f = \langle \cdot, y\rangle$ [Example 15.7.1]. The map $f \mapsto y$ is an additive bijection of $X'$ onto $X$ but it is not an isomorphism since $af \to \bar{a}y$, $\bar{a}$ denoting the complex conjugate of $a$.

**Example 8.2.5** WEAKENED CONVERGENCE If $X$ is a TVS, it is easy to characterize "weakened" convergence of a net $(x_s)$: $x_s \to x$ in $\sigma(X, X')$ iff for each $x' \in X'$, $x'(x_s) \to x'(x)$. This follows from the facts that (1) the seminorms $\{|\langle \cdot, x'\rangle| : x' \in X'\}$ generate $\sigma(X, X')$ and (2) [Theorem 5.7.2(a)] a net $(w_s)$ in an LCS converges to $w$ iff $p(w_s - w) \to 0$ for each seminorm $p$ from a family of seminorms which generates the topology on $X$.

The weak topology with its comparatively huge neighborhoods is so coarse that it makes convergence much easier. If $(X, \langle \cdot, \cdot \rangle)$ is a Hilbert space, for example, then all continuous linear functionals on $X$ are of the form $x \mapsto \langle x, y \rangle$ $(y \in X)$ [Example 15.7.1]. If $(x_n)$ is a sequence of orthonormal vectors, then $(x_n)$ converges weakly to 0 but does not converge to 0 in the norm topology: By Bessel's inequality—$\sum_{n \in N} |(x_n, y)| \leq \|y\|^2$ for any $y \in X$—it follows that $|(x_n, y)| \to 0$ for any $y \in X$; since $\|x_n\| = 1$ for every $n$, $x_n \nrightarrow 0$ in the norm topology. Thus, in infinite-dimensional Hilbert spaces, the weakened topology is strictly coarser than the original topology. ■

As follows from the definitions, if $(X, Y)$ is a pair a net $(x_s)$ in $X$, is as Cauchy net with respect to $\sigma(X, Y)$—a *weak* or $\sigma(X, Y)$-*Cauchy net*—if $\langle x_s, y \rangle$ is a Cauchy net for each $y \in Y$. $X$ is *weakly* or $\sigma(X, Y)$-*complete* if every weak Cauchy sequence converges.

**Example 8.2.6** COMPLETE $\not\Rightarrow$WEAKLY COMPLETE As we show in Example 15.7.3(a), the dual of the Banach space $(c_0, \|\cdot\|_\infty)$ of null sequences is $\ell_1$ in that any $f \in c_0'$ is of the form $f(x) = \sum_{n \in \mathbf{N}} a_n b_n$ $(x = (a_n) \in c_0)$ for some $(b_n) \in \ell_1$. Consider the sequence $x_n = (1, 1, 1, 0, 0, \ldots)$ which is 0 after the $n$th entry. For $f \in X'$, for some $(b_n) \in \ell_1$, $f(x_n) = \sum_{j=1}^{n} b_j$; hence $(x_n)$ is a weak Cauchy sequence without a weak limit. ■

**Nomenclature.** If $X$ and $Y$ are paired spaces and $K \subset X$ is compact with respect to $\sigma(X, Y)$, we say "$K$ is $\sigma(X, Y)$-compact" or the less precise "$K$ is weakly compact." Similarly, if $f$ is a scalar-valued function on $X$ which is continuous when $X$ carries $\sigma(X, Y)$, we say that "$f$ is weakly continuous," or the safer "$f$ is $\sigma(X, Y)$-continuous." We denote the $\sigma(X, Y)$-closure of a set $S$ by "$\text{cl}_{\sigma(X,Y)} S$." ■

**Example 8.2.7** WEAK TOPOLOGY AS INVERSE IMAGE TOPOLOGY Let $(X, Y)$ be a pair and consider the canonical map $y \mapsto y^*$ of $Y$ into the algebraic dual $X^*$ of $X$ : for $y \in Y$, $y^*(x) = \langle x, y \rangle$ $(x \in X)$. Then $\sigma(X, Y)$ is the inverse image topology [Example 4.7.1] determined by the linear maps $\{y^* : y \in Y\}$. It is straightforward to verify that this is the same as the inverse image topology determined by the linear map

$$\begin{aligned} A: \quad & X \to \prod_{y \in Y} \mathbf{F}_y \quad \text{where } \mathbf{F}_y = \mathbf{F} \text{ for each } y \in Y \\ & x \mapsto (\langle x, y \rangle)_{y \in Y} \end{aligned} \tag{$*$}$$

Note that $A$ is 1-1 iff $Y$ distinguishes points of $X$; in this case $A$ is a relatively open map [Example 4.7.1] and $(X, \sigma(X, Y))$ is linearly homeomorphic to $A(X)$. ■

Certain properties of $\sigma(X, Y)$ are immediate because it is a seminorm topology. For example: a subset $B \subset X$ is weakly bounded iff $\langle B, y \rangle$ is a bounded set of scalars for each $y \in Y$ [Theorem 6.1.5]. We can say more about weak boundedness, however, namely: Weak boundedness and weak *total* boundedness are the same thing as we show in Theorem 8.2.8.

**Theorem 8.2.8** WEAKLY BOUNDED = WEAKLY TOTALLY BOUNDED If $(X,Y)$ is a pair and $Y$ distinguishes points of $X$, then $B \subset X$ is $\sigma(X,Y)$-bounded iff $B$ is $\sigma(X,Y)$-totally bounded.

**Proof.** Since totally bounded sets are generally bounded [Theorem 6.1.2], we need only prove that weakly bounded implies weakly totally bounded. To this end, suppose that $B$ is $\sigma(X,Y)$-bounded. According to Theorem 6.1.5, $B$ is $\sigma(X,Y)$-bounded iff $\langle B,y\rangle$ is a bounded set of scalars for each $y \in Y$. By the Heine-Borel theorem, however, a set of scalars is bounded iff it is relatively compact. Thus, by the Tihonov product theorem, $\Pi_{y\in Y} \operatorname{cl}\langle B,y\rangle$ is compact. With $A$ as in $(*)$ above, this implies that $A(B) \subset \Pi\langle B,y\rangle$ is relatively compact, hence totally bounded. Since $Y$ distinguishes points of $X$, $A$ is 1-1 and it follows that $B = A^{-1}(A(B))$. Since $A$ is a linear homeomorphism of $(X,\sigma(X,Y))$ onto $A(X)$ [Example 8.2.7], the linear map $A^{-1}$ is uniformly continuous [Theorem 3.5.1]. Since the uniformly continuous image of a totally bounded set is totally bounded [Theorem 3.5.3], $B$ is $\sigma(X,Y)$-totally bounded. ■

Our next result shows that the pointwise limit of linear functionals is a linear functional.

**Theorem 8.2.9** $X^*$ IS $\sigma(X^*,X)$-COMPLETE If $X$ is any vector space and $X^*$ its algebraic dual, then $(X^*,\sigma(X^*,X))$ is complete.

**Proof.** We use the notation of the statement. If $(f_s)$ is a Cauchy net in $(X^*,\sigma(X^*,X))$ then, for any $x \in X$, $(f_s(x))$ is a Cauchy net of scalars by Theorem 5.7.2(b). Define the function $f$ at each $x \in X$ to be $\lim f_s(x)$. The linearity of $f$ follows immediately from the linearity of each $f_s$ and the continuity of scalar operations. ■

We noted in Theorem 3.4.1 that a closed, totally bounded subset of a complete topological group is compact. Hence, Theorems 8.2.8 and 8.2.9 show that weak algebraic duals $(X^*,\sigma(X^*,X))$ resemble finite-dimensional spaces in that their closed and bounded subsets are compact. Generally, LCHS with this property are called *semi-Montel spaces* [Exercise 12.203].

What are the $\sigma(X,Y)$-continuous linear functionals on $X$? Certainly, each $y \in Y$ is such a functional, but are there others? As we show next, there are not.

**Theorem 8.2.10** WEAK REPRESENTATION THEOREM For any pair $(X,Y)$, the set $Y$ comprises all $\sigma(X,Y)$-continuous linear functionals on $X$:

(a) If $g$ is a weakly continuous linear functional on $X$, then there exists $y \in Y$ such that for each $x \in X$, $g(x) = \langle x,y\rangle$; $y$ is unique iff $X$ distinguishes points of $Y$.

(b) Letting $X^{\perp} = \{y \in Y : \langle X, y \rangle = \{0\}\}$ then, whether $X$ distinguishes points of $Y$ or not, $(X, \sigma(X,Y))'$ may be identified with $Y/X^{\perp}$.

**Proof.** We use the notation of the statement. For a seminorm $q$ on $X$, $V_q = \{x \in X : q(x) < 1\}$. As noted after Def. 5.5.5, if $p_1, \ldots, p_n$ are seminorms, then $\max_j p_j = p$ is a seminorm, $V_p = \cap_{j=1}^n V_{p_j}$ and $p^{-1}(0) = \cap_{j=1}^n p_j^{-1}(0)$. Continuity of a linear functional $f$ on a TVS $X$ is equivalent to $|f| \leq p$ for some continuous seminorm $p$ on $X$ [Theorem 5.7.3]. This means that for a $\sigma(X,Y)$-continuous linear functional $g$, there exist $y_1, \ldots, y_n \in Y$ such that for each $x \in X$, $|g(x)| \leq \max_j |\langle x, y_j \rangle|$. For $1 \leq j \leq n$, let $p_{y_j}(\cdot) = |\langle \cdot, y_j \rangle|$, let $p = \max_j p_{y_j}$ and identify $y_j \in Y$ with the linear functional $\langle \cdot, y_j \rangle$. Since $p^{-1}(0) = \cap_{j=1}^n p_{y_j}^{-1}(0) = \cap_{j=1}^n y_j^{-1}(0) \subset g^{-1}(0)$, it follows that $g$ may be expressed [Theorem 1.4.1(f)] as a linear combination $g = \sum a_i y_i = y$ $(a_i \in \mathbf{F})$. The uniqueness assertion and (b) are clear. ■

The smaller a family of functions, the coarser the initial topology it generates. The following consequence of Theorem 8.2.10 is in this spirit.

**Theorem 8.2.11** PROPER SUBSPACES, COARSER TOPOLOGIES If $(X,Y)$ is a pair and $M$ is a proper subspace of $Y$ such that $(X,M)$ is a dual pair with respect to the restriction of $\langle \cdot, \cdot \rangle$ to $X \times M$, then $\sigma(X,M)$ is strictly coarser than $\sigma(X,Y)$.

**Proof.** With notation as above, if $y \in Y$ but $y \notin M$, then the map $x \mapsto \langle x, y \rangle$ is $\sigma(X,Y)$-continuous but not $\sigma(X,M)$-continuous by Theorem 8.2.10. ■

## 8.3 POLARS

In Sec. 2.6 we topologized groups of functions. We considered the additive group $F(T,X)$ of all functions mapping the set $T$ into the commutative topological group $X$. For any collection $\mathcal{S}$ of subsets of $T$, there is a group topology $\mathcal{T}_{\mathcal{S}}$ for $F(T,X)$ called the $\mathcal{S}$-topology formed by taking as a subbase at 0 sets of the form

$$W(S,V) = \{f \in F(T,X) : f(S) \subset V\}, \ S \in \mathcal{S}, \ V \text{ a neighborhood of 0 in } X$$

functions which are "small" on the sets of $\mathcal{S}$. If the range $X$ is $\mathbf{F} = \mathbf{R}$ or $\mathbf{C}$, the subbasic neighborhoods of 0 are denoted

$$W(S,r) = \{f \in F(T,\mathbf{F}) : \sup |f(S)| < r\}, \ r > 0$$

If $(X,Y)$ is a pair, we can view $X$ as a subspace of $F(Y,\mathbf{F})$. The $\mathcal{S}$-topology for $X$ determined by the collection $\mathcal{S}$ of one-point subsets of $Y$ is $\sigma(X,Y)$.

We begin our approach to polar topologies by singling out sets of functions in $Y$ which are small on certain subsets of $X$ and developing some of their properties. For each statement about $Y$, there is, of course, a corresponding dual statement about $X$.

**Definition 8.3.1** POLARS Let $(X, Y)$ be a pair. If $E \subset X$ then the *polar* $E^{\circ}$ of $E$ is $\{y \in Y : \sup |\langle E, y\rangle| \leq 1\}$. If $E \subset Y$, then the polar of $E$ is given by $E^{\circ} = \{x \in X : \sup |\langle x, E\rangle| \leq 1\}$. ■

**Remark 8.3.2** REAL VS. ABSOLUTE POLAR Some authors use the term *absolute* polar for what we call polar and use "polar" of $E \subset X$ for $\{y \in Y : \sup \operatorname{Re} \langle E, y\rangle \leq 1\}$. Clearly, $E^{\circ} \subset E^{r}$ but if $E$ is balanced, then $E^{r} = E^{\circ}$ by the following argument: If $\operatorname{Re} \langle x, y\rangle \leq 1$ for all $x$ in the balanced set $E$ then $\operatorname{Re} \langle ax, y\rangle \leq 1$ for all $|a| \leq 1$—in particular for $a = e^{i\theta}$ for all $\theta$—which implies that $|\langle x, y\rangle| \leq 1$. For more information on real polars, see Exercise 8.104. ■

**Example 8.3.3** POLARS OF UNIT BALLS $U(X)^{\circ} = U(X')$ AND $U(X')^{\circ} = U(X)$ Let $X$ be a normed space with solid unit ball $U(X)$; let $SU(X) = \{x \in X : \|x\| = 1\}$. $X'$ denotes the continuous dual of $X$ equipped with its canonical norm topology—$\|f\| = \sup |f(U(X))| = \sup |f(SU(X))|$ for $f \in X'$. $U(X')$ and $SU(X')$ denote the unit ball and surface of the unit ball of $X'$, respectively. Then $U(X)^{\circ} = \{f \in X' : \sup |\langle U(X), f\rangle| \leq 1\} = \{f \in X' : \|f\| \leq 1\} = U(X')$. As in the proof of Theorem 7.4.2(b), $\|x\| = \sup |\langle x, U(X')\rangle|$ so

$$U(X')^{\circ} = \{x \in X : \sup |\langle x, U(X')\rangle| \leq 1\} = \{x \in X : \|x\| \leq 1\} = U(X). \quad ■$$

**Example 8.3.4** POLAR OF A SUBSPACE If $(X, Y)$ is a pair and $M$ is a subspace of $X$, then $M^{\circ} = \{y \in Y : \langle M, y\rangle = 0\}$ for if $|\langle m, y\rangle| \leq 1$ for each $m \in M$, then for any $r > 0$, $|\langle m/r, y\rangle| \leq 1$, i.e., $|\langle m, y\rangle| \leq r$. ■

In dealing with $\mathcal{S}$-topologies, we considered functions that are arbitrarily small on certain sets; in singling out polars, we only consider functions that are "$\leq 1$" on certain sets—not very small. Should we consider "$r$-polars"

$$E^{r} = \{y \in Y : \sup |\langle E, y\rangle| \leq r\} \qquad r > 0$$

instead? Since we are going to consider positive multiples of polars and

$$|\langle x, y\rangle| \leq r \Leftrightarrow |\langle x, y/r\rangle| \leq 1$$

$y \in E^{r}$ iff $y/r \in E^{\circ}$. Thus, $rE^{\circ} = E^{r}$ so it is unnecessary to deal with the sets $E^{r}$.

We use the subbase Theorem 4.5.3(b) to define polar topologies in Sec. 8.5. Therefore, we want absorbent sets. The following result characterizes sets with absorbent polars.

**Theorem 8.3.5** WEAKLY BOUNDED IFF ABSORBENT POLAR If $(X, Y)$ is a pair, then $B \subset X$ is $\sigma(X, Y)$-bounded iff $B^{\circ}$ is an absorbent subset of $Y$.

**Proof.** We use the notation of the statement. By Theorem 6.1.5 , $B \subset X$ is $\sigma(X, Y)$-bounded iff $\langle B, y \rangle$ is a bounded set of scalars for each $y \in Y$. For any $y \in Y$, $\langle B, y \rangle$ is bounded iff $|\langle B, ay \rangle| \leq 1$ for sufficiently small values of $|a|$. In other words, $\langle B, y \rangle$ is bounded iff $ay \in B^\circ$ for sufficient small values of $|a|$—i.e., iff $B^\circ$ absorbs $y$. ■

The elementary properties of polars are summarized in Theorem 8.3.6.

**Theorem 8.3.6** POLAR FACTS Let $(X, Y)$ be a pair and let $A$ and $B$ be subsets of $X$. Then:
(a) $A^\circ$ is a $\sigma(Y, X)$-closed disk;
(b) if $A \subset B$, then $A^\circ \supset B^\circ$;
(c) for $a \neq 0$, $(aA)^\circ = a^{-1} A^\circ = |a|^{-1} A^\circ$;
(d) $A \subset A^{\circ\circ} = (A^\circ)^\circ$, and $A^\circ = A^{\circ\circ\circ}$; $A^{\circ\circ}$ is called the *bipolar* of $A$.

**Proof.** We use the notation of the statement. (a) To prove that $A^\circ$ is absolutely convex, let $a$ and $b$ be scalars such that $|a| + |b| \leq 1$, and let $y, z \in A^\circ$. For any $x \in A$

$$|\langle x, ay + bz \rangle| \leq |a| \, |\langle x, y \rangle| + |b| \, |\langle x, z \rangle| \leq |a| + |b| \leq 1$$

To see that $A^\circ$ is $\sigma(Y, X)$-closed, note that for any $x$ in $A$ the set $\{y \in Y : |\langle x, y \rangle| \leq 1\}$ is $\sigma(Y, X)$-closed since $\sigma(Y, X)$ is the inverse image topology determined by the maps $\{\langle x, \cdot \rangle : x \in X\}$. It only remains to observe that

$$A^\circ = \bigcap_{x \in A} \{y \in Y : |\langle x, y \rangle| \leq 1\}$$

(b) and (c) follow directly from the definition of polar.
(d) The inclusion $A \subset A^{\circ\circ}$ follows directly from the definition. Thus $A^\circ \subset (A^\circ)^{\circ\circ} = A^{\circ\circ\circ}$. By (b), however, $A \subset A^{\circ\circ}$ implies that $A^\circ \supset A^{\circ\circ\circ}$. ■

The results of Theorem 8.3.7 show that a set can be swollen in certain ways without affecting its polar. This simplifies certain things about polar topologies [Theorem 8.5.2].

**Theorem 8.3.7** POLARS OF HULLS AND CLOSURES Let $(X, Y)$ be a pair and let $S$ be a subset of $X$. Then:
(a) $S^\circ = (S_b)^\circ$;
(b) $S^\circ = (S_c)^\circ$;
(c) $S^\circ = \left(\mathrm{cl}_{\sigma(X,Y)} S\right)^\circ$;
(d) $S^\circ = \left(\mathrm{cl}_{\sigma(X,Y)} (S_{bc})\right)^\circ$.

**Proof.** We use the notation of the statement. All parts follow from Theorem 8.3.6 and their proofs are virtually identical, so we prove only (d) here. To see this, note that $S \subset \mathrm{cl}_{\sigma(X,Y)} S_{bc} \subset S^{\circ\circ}$ [Theorem 8.3.6(a)]. Hence $S^\circ \supset (\mathrm{cl}_{\sigma(X,Y)} S_{bc})^\circ \supset S^{\circ\circ\circ} = S^\circ$. ■

The preceding results about polars follow directly from definitions. The following central characterization of bipolars relies on the Hahn–Banach theorem.

**Theorem 8.3.8** THE BIPOLAR THEOREM Let $(X, Y)$ be a pair and let $S$ be a subset of $X$. Then $S^{\circ\circ} = \mathrm{cl}_{\sigma(X,Y)} S_{bc}$.

**Proof.** We use the notation of the statement. Theorem 8.3.6(a,b) imply that $S \subset \mathrm{cl}_{\sigma(X,Y)} S_{bc} \subset S^{\circ\circ}$. By Theorem 7.8.6(b), if $w \notin \mathrm{cl}_{\sigma(X,Y)} S_{bc}$, there exists a $\sigma(X, Y)$-continuous real linear functional $f$ on $X$ such that $\sup f\left(\mathrm{cl}_{\sigma(X,Y)} S_{bc}\right) = a < f(w)$. Since $\mathrm{cl}_{\sigma(X,Y)} S_{bc}$ is balanced, $0 \in \mathrm{cl}_{\sigma(X,Y)} S_{bc}$ and therefore $0 = f(0) < a$. Hence, we may replace $f$ by $g = f/a$ to get $\sup g\left(\mathrm{cl}_{\sigma(X,Y)} S_{bc}\right) = 1 < g(w)$. Now consider the complex linear functional $h(x) = g(x) - ig(ix)$ $(x \in X)$. By Theorem 8.2.10(a), there is some $y \in Y$ such that $h(x) = \langle x, y\rangle$ for all $x \in X$. Since $\mathrm{cl}_{\sigma(X,Y)} S_{bc}$ is balanced and $\mathrm{Re}\langle x, y\rangle = g(x) \leq 1$ for all $x \in \mathrm{cl}_{\sigma(X,Y)} S_{bc}$, it follows that $|\langle x, y\rangle| \leq 1$ for all $x \in \mathrm{cl}_{\sigma(X,Y)} S_{bc}$ [Remark 8.3.2]; hence $y \in S^{\circ}$. Since $|\langle w, y\rangle| \geq |g(w)| > 1$, it follows that $w \notin (S^{\circ})^{\circ}$, whence $S^{\circ\circ} \subset \mathrm{cl}_{\sigma(X,Y)} S_{bc}$. ■

We characterize polars of unions and intersections next. Theorem 8.3.9(a), on unions, has actually been proved already in a more general setting in Theorem 2.6.1(e).

**Theorem 8.3.9** UNIONS AND INTERSECTIONS Let $(X, Y)$ be a pair, let $\{S_i : i \in I\}$ be a collection of subsets of $X$. Then:

(a) $(\cup_{i\in I} S_i)^{\circ} = \cap_{i\in I} S_i{}^{\circ}$;

(b) If each $S_i$ is a $\sigma(X, Y)$-closed disk, then $(\cap_{i\in I} S_i)^{\circ} = \mathrm{cl}_{\sigma(X,Y)} (\cup_{i\in I} S_i{}^{\circ})_{bc}$.

**Proof.** Part (a) follows from the observation that

$$\begin{aligned} y \in (\cup_{i\in I} S_i)^{\circ} &\Leftrightarrow \sup|\langle \cup_{i\in I} S_i, y\rangle| \leq 1 \\ &\Leftrightarrow \text{for each } i \in I,\ \sup|\langle S_i, y\rangle| \leq 1 \end{aligned}$$

As for (b), note that the bipolar theorem [8.3.8] implies that $S_i = S_i{}^{\circ\circ} = \mathrm{cl}_{\sigma(X,Y)} (S_i)_{bc}$ for each $i \in I$. By (a), $\cap_{i\in I} S_i = \cap_{i\in I} S_i{}^{\circ\circ} = (\cup_{i\in I} S_i{}^{\circ})^{\circ}$, which implies that $(\cap_{i\in I} S_i)^{\circ} = (\cup_{i\in I} S_i{}^{\circ})^{\circ\circ} = \mathrm{cl}_{\sigma(X,Y)} (\cup_{i\in I} S_i{}^{\circ})_{bc}$. ■

## 8.4 ALAOGLU

In this section we prove Alaoglu's theorem, that neighborhoods of 0 in a TVS have weak-∗ compact polars. This very important result—maybe *the* most important fact about the weak-∗ topology—echoes throughout functional analysis. To illustrate its potency, we put it to immediate use in showing: (1) that an LCHS is embedded in a space of continuous functions [Theorem 8.4.8] and (2) to prove the existence of the Stone-Cech compactification $\beta T$ of a completely regular Hausdorff space $T$ [Example 8.4.10]. We call it the Alaoglu theorem because it is commonly done; Banach–Alaoglu [Banach proved a version [1932, p. 123] for separable normed spaces] and Bourbaki–Alaoglu are also in common use. Noting that *weak-$*$-compactness theorem* would be more appropriate, Pietsch [2007, p. 77] lists 12 mathematicians who can lay claim to it or an important predecessor.

**Theorem 8.4.1** ALAOGLU THEOREM If $U$ is a neighborhood of 0 in the TVS $X$ over $\mathbf{F}$ then its polar $U^\circ$ is $\sigma(X', X)$-compact.

**Proof.** Let $U$ be a neighborhood of 0 in the TVS $X$ and let $X^*$ denote the algebraic dual of $X$. Since $U$ is absorbent and $U \subset U^{\circ\circ}$, it follows that $U^{\circ\circ}$ is absorbent. This implies that $U^{\circ\circ\circ} = U^\circ$ is $\sigma(X', X)$-bounded by Theorem 8.3.5. Since $X$ distinguishes points of $X'$, boundedness and total boundedness coalesce in $\sigma(X', X)$ [Theorem 8.2.8] and $U^\circ$ is therefore $\sigma(X', X)$-totally bounded. Since $\sigma(X^*, X) \cap X' = \sigma(X', X)$, it follows that $U^\circ$ is $\sigma(X^*, X)$-totally bounded as well. Since $X^*$ is $\sigma(X^*, X)$-complete [Theorem 8.2.9], if $U^\circ$ can be shown to be $\sigma(X^*, X)$-closed, its $\sigma(X', X)$-compactness will follow from Theorem 3.4.1. To see that $U^\circ$ is $\sigma(X^*, X)$-closed, consider the polar computed in the algebraic dual

$$U^d = \{y \in X^* : \sup |\langle U, y\rangle| \leq 1\}$$

If $y \in U^d$, then $y$ is bounded on a neighborhood 0 in $X$. Therefore it is continuous [Theorem 6.5.1] and $U^d = U^\circ$. $U^d$ is $\sigma(X^*, X)$-closed by Theorem 8.3.6(a). ■

Let $U$ be the unit ball in an infinite-dimensional normed space $X$. By Example 8.3.3, $U^\circ = U'$, the unit ball of $X'$, which is $\sigma(X', X)$-compact by the Alaoglu theorem [8.4.1]; $U^\circ$ is certainly not compact in the norm topology for $X'$, however, because that would mean that $X'$ is finite-dimensional [Theorem 4.10.4]. Note too that the Alaoglu theorem provides something of a Heine-Borel property for the weak-$*$ topology, namely that closed norm-bounded subsets of $X'$ are weak-$*$ compact.

Helly [1912, p. 267] proved the following early version of the Alaoglu theorem for $C[a, b]$.

**Theorem 8.4.2** Given a norm-bounded infinite subset $B \subset \{f \in C[a, b]' : \|f\| \leq r\}$, $r > 0$, of $C[a, b]'$, there exists a sequence $(f_n)$ from $B$ and $f \in C[a, b]'$, $\|f\| \leq r$, such that $f_n \to f$ pointwise on $C[a, b]$.

In other words, the unit ball of $C[a, b]'$ is relatively countably $\sigma(X', X)$-compact. In the same paper [p. 283; cf. Natanson 1961, p. 222] he got something similar, what is now called the *Helly selection* [or *choice*] *theorem*. Note that $BV[a, b]$ is essentially $C[a, b]'$ [Theorem 7.4.4].

**Theorem 8.4.3** HELLY SELECTION THEOREM Let $BV[a, b]$ denote the set of functions of bounded variation on $[a, b]$. Given an infinite bounded subset

$B \subset BV[a,b]$, $\sup_{x \in B} |x|[a,b] \le a$, $a > 0$, say, such that the total variation $v(x) \le b$, $b > 0$, for each $x \in B$, then there exists $x \in BV[a,b]$ and a sequence $(x_n)$ from $B$ such that $x_n \to x$ pointwise on $[a,b]$.

Banach [1932, p. 123, Th. 3] showed that, for any *separable* Banach space $X$, the unit ball of $X'$ is sequentially $\sigma(X', X)$-compact. (Sequential compactness is the only kind that Banach considers in his book.)

Hahn [1927, p. 219] introduced the canonical embedding of $X$ in $X^{**}$ of the next definition.

**Definition 8.4.4** CANONICAL EMBEDDING IN $X^{**}$ Let $X$ be a vector space. Given any $x \in X$, $\langle x, \cdot \rangle \in X^{**}$. We call the map $J : X \to X^{**}$, $x \mapsto \langle x, \cdot \rangle$, the *canonical embedding* of $X$ in $X^{**}$. ■

$J$ is clearly linear. Since $X^*$ distinguishes the points of $X$, $Jx = 0$ implies that $Jx(x^*) = \langle x, x^* \rangle = 0$ for all $x^* \in X^*$ and therefore that $x = 0$; hence $J$ is 1-1. When $J$ is 1-1, we usually identify $X$ and $J(X)$. If $X$ is a normed space, it follows immediately from Theorem 7.4.2(a,b) that:

**Theorem 8.4.5** $J$ IS AN ISOMETRY If $X$ is a normed linear space, the canonical embedding $J$ is a linear isometry.

This enables us to show that finite-dimensional normed spaces are reflexive:

**Theorem 8.4.6** FINITE-DIMENSIONAL SPACES For a normed space $X$ the following are equivalent: For $n \in \mathbf{N}$

(a) $\dim X = n$;

(b) $\dim X' = n$;

(c) $\dim X'' = n$.

Since $\dim X'' = n = \dim X$, it follows from the proof of (c) $\Rightarrow$ (a) that $J$ is onto, i.e., that $X$ is reflexive.

**Proof.** We use the notation of the statement. (a) $\Rightarrow$ (b) Let $\{x_1, \ldots, x_n\}$ be a basis for the normed space $X$. For $i, j = 1, 2, \ldots, n$, define the linear functional $f_i(x_j) = \delta_{ij}$. The $f_i$ are continuous because $X$ is finite-dimensional [Theorem 4.10.3(f)]. To see that the $f_i$ are a basis for $X'$—the *dual* basis—consider $f \in X'$ and let $a_i = f(x_i)$ so that $f = \sum_{i=1}^n a_i f_i$. The $f_i$ are linearly independent because if $\sum_{i=1}^n b_i f_i = 0$ then, for all $j$, $0 = \sum_{i=1}^n b_i f_i(x_j) = b_j$. The same argument shows that (b) $\Rightarrow$ (c). As to (c) $\Rightarrow$ (a), suppose $\dim X'' = n$. Since $J$ is a linear isometry, $\dim X = \dim J(X) \le \dim X'' = n$. Hence, $X$ is finite-dimensional, and therefore, by (a)$\Rightarrow$ (c), $\dim X = \dim X'' = n$. ■

How big is the normed space $X$ in its bidual $X''$? As Goldstine's [1938] theorem below shows, it is $\sigma(X'', X')$-dense in $X''$.

**Theorem 8.4.7** DENSITY OF $X$ IN $X''$ If $X$ is a normed space then $X$ is $\sigma(X'', X')$-dense in $X''$.

**Proof.** Let $U$, $U'$ and $U''$ denote the unit balls of the normed space $X$, $X'$ and $X''$, respectively. It suffices to show that $U$ is $\sigma(X'', X')$-dense in $U''$. By Theorem 8.4.5, we may view $X$ as a subset of $X''$. For the dual pair $(X'', X')$, it follows from the bipolar theorem [8.3.8] that $\text{cl}_{\sigma(X'',X')} U = U^{\circ\circ}$. By Example 8.3.3, $U^\circ = U'$ and $U^{\circ\circ} = (U')^\circ = U''$. ■

The Alaoglu theorem points to a necessary condition for a normed space to be *reflexive* in the sense that $J(X) = X''$: If $X$ is to be the (normed) dual of a normed space $X'$, then its unit ball $U(X)$ must be $\sigma(X, X')$-compact. Banach [1932, p. 189, Th. 13] took the first step in proving that this condition was also sufficient—he showed that if the unit ball of a *separable* Banach space is *sequentially* $\sigma(X, X')$-compact, then $X$ is reflexive. We prove in Theorem 15.5.1 that a Banach space $X$ is reflexive iff $U(X)$ is $\sigma(X, X')$-compact. (We show in Theorem 15.4.8 that, for a $\sigma(X, X')$-closed subset $K$ of a normed space $X$, the notions of $\sigma(X, X')$-compactness, $\sigma(X, X')$-sequential compactness and $\sigma(X, X')$-countable compactness coincide.)

The direct ancestor of the representation theorem, Theorem 8.4.8(a), is a theorem of Banach and Mazur [Banach 1932, p. 185] identifying a separable real Banach space as a closed subspace of $C([0,1], \mathbf{R}, \|\cdot\|_\infty)$. Their argument is quite different from the one given here, however, using the fact that every compact metric space is the continuous image of Cantor's ternary set. The space $C(T, \mathbf{F}, c)$ of continuous functions on a topological space $T$ with compact-open topology is discussed in Sec. 5.8.

**Theorem 8.4.8** REPRESENTATION AS SUBSPACE OF $C(T, \mathbf{F}, \|\cdot\|_\infty)$

(a) NORMED SPACE Each normed space $X$ is linearly isometric to a subspace of $C(T, \mathbf{F}, \|\cdot\|_\infty)$ for some compact set $T$. Since every normed space is contained in some $C(T, \mathbf{F}, \|\cdot\|_\infty)$, the latter spaces are called *universal* for normed spaces. (Every *separable* Banach space is linearly isometric to a subspace of $C([0,1], \mathbf{F}, \|\cdot\|_\infty)$.)

(b) LCHS If $X$ is an LCHS then there is a locally compact topological space $T$ such that $X$ is linearly homeomorphic to a subspace of $C(T, \mathbf{F}, c)$.

**Proof.** (a) Let $X$ be a normed space and consider the solid unit ball $U'$ of $X'$. Since $U' = U^\circ$, the polar of the unit ball $U$ of $X$ [Example 8.3.3], $U'$ is $\sigma(X', X)$-compact by the Alaoglu theorem [8.4.1]. With respect to $(X, X')$ in the natural pairing, consider a slightly modified canonical embedding, the linear isomorphism

$$K : X \to C(U', \mathbf{F}, \|\cdot\|_\infty), \; x \mapsto \langle x, \cdot \rangle$$

By Theorem 7.4.2(b),

$$\|x\| = \sup\{|\langle x, x' \rangle| : x' \in U'\} = \|Kx\|_\infty \qquad (*)$$

Hence $K$ is a linear isometry. Since $K$ maps Cauchy sequences into Cauchy sequences, if $X$ is a Banach space, then it is linearly isometric to a *closed* subspace of $C(U', \mathbf{F}, c)$.

(b) Since any LCHS $X$ is linearly homeomorphic to a subspace of a product $\Pi_{s\in S}X_s$ of Banach spaces $(X_s, \|\cdot\|_s)$ [Theorem 5.11.4(b)], it suffices to demonstrate the theorem for such products. By identifying the Banach space $X_s$ with its canonical image in $\Pi_{s\in S}X_s$, we may assume that the spaces $\{X_s : s \in S\}$ are disjoint. For each $s \in S$, let $U_s'$ be the solid unit ball in the normed space $X_s'$. Each $U_s'$ is compact in its relative $\sigma(X_s', X_s)$-topology $\mathcal{T}_s$. The sets $\cup\{\mathcal{T}_s : s \in S\}$ form a base of open sets for a topology on $T = \cup_{s\in S}U_s'$. Since each $U_s'$ is compact, it follows that $T$ is locally compact. If $E$ is a subset of $T$ which intersects infinitely many of the disjoint sets $U_s'$ then $E$ cannot be compact. Consequently, any compact subset $K$ of $T$ must be contained in finitely many of the $U_s'$ and therefore the compact-open topology on $C(T, \mathbf{F}, c)$ is determined by the seminorms $p_K = \sup|(\cdot)(K)|$ where $K$ is a finite union of the $U_s'$. (The collection of such $p_K$ forms a *base* of seminorms in the sense of Definition 5.5.4.)

Next, consider the map $A : \Pi_{s\in S}X_s \to C(T, \mathbf{F}, c)$ defined as follows: Let $x = (x_s) \in \Pi_{s\in S}X_s$ and $t \in \cup_{s\in S}U_s'$. Since the sets $U_s'$ are disjoint, there is a unique $s \in S$ such that $t = u_s \in U_s$ and we define $Ax(t) = \langle x_s, u_s\rangle$. Since each $x_s$ is $\mathcal{T}_s$-continuous on $U_s'$ and a convergent net in $T$ must eventually belong to some $U_s'$, $Ax$ is continuous, i.e., $Ax \in C(T, \mathbf{F})$. $A$ is clearly a linear isomorphism. For a finite subset $\{s_1, s_2, \ldots, s_n\}$ of $S$, $K = \cup_{i=1}^n U_{s_i}'$ and $x \in C(T, \mathbf{F}, c)$,

$$\begin{aligned} p_K(x) &= \sup|x|\left(U_{s_1}' \cup \cdots \cup U_{s_n}'\right) = \sup(\sup|x|\left(U_{s_1}'\right), \ldots, \sup|x|\left(U_{s_n}'\right)) \\ &= \sup(\|x_{s_1}\|_{s_1}, \ldots, \|x_{s_n}\|_{s_n}) \end{aligned}$$

Hence, for any $r > 0$, $p_K(x) < r$ iff $\sup(\|x_{s_1}\|_{s_1}, \ldots, \|x_{s_n}\|_{s_n}) < r$ so

$$A\left(\left\{x \in \Pi_{s\in S}X_s : \sup(\|x_{s_1}\|_{s_1}, \ldots, \|x_{s_n}\|_{s_n}) < r\right\}\right) = rV_{p_K}$$

and $A$ is a linear homeomorphism. ■

**Definition 8.4.9** COMPACTIFICATION A *compactification* $S$ of a topological space $T$ is a compact space which contains a dense homeomorphic copy of $T$. ■

We now prove that a completely regular Hausdorff space $T$ possesses a compactification $\beta T$ with the property that each bounded continuous map of $T$ into $\mathbf{R}$ has a continuous extension to $\beta T$. The construction depends on Tihonov's theorem on compactness of products by way of the Alaoglu theorem [8.4.1].

**Example 8.4.10** STONE–CECH COMPACTIFICATION A completely regular Hausdorff space $T$ possesses a compactification $\beta T$ with the property that each bounded continuous map $x$ of $T$ into $\mathbf{F}$ has a continuous extension $\beta x$ to $\beta T$. Since $T$ is dense in $\beta T$, the bounds on $\beta x$ are the same as for $x$. It is customary to call $\beta T$ the *Stone–Cech compactification* of $T$.

Consider the Banach space $X = C_b(T, \mathbf{F}, \|\cdot\|_\infty)$ of all bounded $\mathbf{F}$-valued continuous functions on $T$. Let $U'$ denote the solid unit ball of of the normed dual $X'$ of $X$ and $SU(X') = \{f \in X' : \|f\| = 1\}$.

(a) $T' \subset SU(X')$ For each $t \in T$, the evaluation map $t'$ on $C_b(T, \mathbf{F}, \|\cdot\|_\infty)$ is a continuous linear functional [Theorem 5.8.3] on $X$. Since $T$ is completely regular, for any $t \in T$, $\|t'\| = \sup\{|t'(x)| : \|x\|_\infty = 1\} = 1$.

$$\begin{array}{c} T' \subset SU(C_b(T, \mathbf{F}, \|\cdot\|_\infty))' \\ | \\ C_b(T, \mathbf{F}, \|\cdot\|_\infty) \\ | \\ T \end{array}$$

(b) The map $f : T \to (U', \sigma(X', X))$, $t \mapsto t'$, is a homeomorphism.

Clearly, $f$ is 1-1 and if a net $t_s \to t$ in $T$, then, for any $x \in C_b(T, \mathbf{F}, \|\cdot\|_\infty)$, $x(t_s) \to x(t)$; hence $t'_s(x) \to t'(x)$ for every $x$ in $X$, which means that $t'_s \to t'$ in $\sigma(X', X)$; $f$, therefore, is continuous. To see that $f$ is a homeomorphism, we show that $f$ maps closed subsets $K$ of $T$ into closed sets. To that end, let $f(K) = K' = \{t' : t \in K\}$. Since $T$ is a completely regular Hausdorff space, if $t \notin K$, then there exists $x \in C_b(T, \mathbf{R})$ such that $x(t) = 0$ and $x(K) = \{2\}$. Hence $\{x\}^\circ$ is a neighborhood of $t'$ which is disjoint from $K'$; therefore $t' \notin \mathrm{cl}_{\sigma(X',X)} K'$, and it follows that $K' = f(K)$ is a closed subset of $T'$.

(c) We now identify $T$ and $T'$. Since $U'$ is $\sigma(X', X)$-compact by the Alaoglu theorem [Example 8.3.3 and Theorem 8.4.1], the closure $\beta T$ of $T'$ in $U'$ is a compactification of $T$.

(d) To see that each $x \in C_b(T, \mathbf{F}, \|\cdot\|_\infty)$ has a continuous extension $\beta x$ to $\beta T$, consider the canonical embedding $J : X \to X''$, $x \mapsto \langle x, \cdot\rangle$. Clearly, $Jx$ is $\sigma(X', X)$-continuous on $X'$; so, therefore, is its restriction $\beta x$ to $\beta T$ and this is the desired extension of $x$ to $\beta T$. ∎

For $T = \mathbf{N}$ with the discrete topology, $C_b(T, \mathbf{F}, \|\cdot\|_\infty)$ is just $\ell_\infty$ and $\mathbf{N}$ is homeomorphic to the evaluation maps $\{n' : n \in \mathbf{N}\}$ so $\beta\mathbf{N} = \mathrm{cl}_{\sigma(\ell'_\infty, \ell_\infty)} \mathbf{N}$. That is about as concrete as we can get about $\beta\mathbf{N}$.

To avoid the pathological situation of Hausdorff spaces whose only continuous functions were constant, Tihonov [1930, p. 545] introduced "complete regularity" and constructed a Hausdorff compactification for a completely regular space. (This is the same article in which he proved the "Tihonov theorem.") His technique was elaborated by Cech [1937, p. 831] who introduced the notation $\beta T$ for it. Stone [1937b, pp. 461–463, 476], using very different methods, also constructed $\beta T$. Nowadays Tihonov's name is never associated with $\beta T$ even though his was the first work on the subject and was cited by both Stone and Cech.

# 8.5 POLAR TOPOLOGIES

Suppose $T$ is a set, $X$ a topological group and $\mathcal{S}$ a collection of subsets of $T$. The (group) $\mathcal{S}$-topology $\mathcal{T}_{\mathcal{S}}$ on the group $F(T,X)$ of maps of $T$ into $X$ has as subbasic neighborhoods of 0 the sets [Sec. 2.6]

$$W(S,V) = \{x \in F(T,X) : x(S) \subset V\}, S \subset T, V \text{ a neighborhood of 0 in } X.$$

We consider some special $\mathcal{S}$-topologies in this section. Given a pair $(X,Y)$, we view $X$ as a collection of functions (namely, $x^* = \langle x, \cdot \rangle, x \in X$) on $Y$, more exactly we view $X$ as a subgroup of $F(Y, \mathbf{F})$. We then consider a collection $\mathcal{S}$ of subsets $S$ of $Y$ and topologize $X$ by means of the sets $W(S,1) = S^\circ = \{x \in X : \sup |\langle x, S\rangle| \leq 1\}$.

**Notation.** $(X,Y)$ denotes a pair for the remainder of the section.

**Polar topologies.** If $\mathcal{S}$ is a collection of $\sigma(Y,X)$-bounded subsets of $Y$, $\mathcal{S}^\circ = \{S^\circ : S \in \mathcal{S}\}$ determines a topology $\mathcal{T}_{\mathcal{S}}$ for $X$. By considering $\sigma(Y,X)$-bounded sets $S$ of $Y$, we guarantee that each $S^\circ$ is absorbent in $X$ [Theorem 8.3.5]. By Theorem 8.3.6(a), each $S^\circ$ is a disk. Hence, by the subbase theorem [4.5.3(b)], the collection of positive multiples of finite intersections of sets from $\mathcal{S}^\circ$ is a base at 0 for a locally convex topology $\mathcal{T}_{\mathcal{S}}$ for $Y$ that we call the *polar topology determined by* $\mathcal{S}$. The complication of having to consider finite intersections is frequently unnecessary in practice. In many cases (Examples 8.5.3 to 8.5.5, for example), $\mathcal{S}$ is "directed" in the sense that for any $A, B \in \mathcal{S}$, there exists $C \in \mathcal{S}$ such that $A \cup B \subset C$; therefore $(A \cup B)^\circ = A^\circ \cap B^\circ \supset C^\circ$; hence positive multiples of the sets $S^\circ$ themselves form a base at 0 for $\mathcal{T}_{\mathcal{S}}$.

We show in Theorem 8.6.6 that every locally convex topology is a polar topology. Since polar topologies $\mathcal{T}_{\mathcal{S}}$ are locally convex, we know that they are generated by seminorms. What seminorms? We show next that $\mathcal{T}_{\mathcal{S}}$ is generated by the gauges $p_{S^\circ}$ of the sets $S^\circ$ $(S \in \mathcal{S})$.

**Polar topologies as seminorm topologies.** Consider a collection $\mathcal{S}$ of $\sigma(Y,X)$-bounded subsets of $Y$. Then $\langle x, S\rangle$ is a bounded set of scalars for each $S \in \mathcal{S}$ and $x \in X$ [Theorem 6.1.5]. Thus, we can consider the function $p_S : X \to \mathbf{R}$ $(S \in \mathcal{S})$ defined by

$$p_S(x) = \sup |\langle x, S\rangle| \quad (x \in X)$$

It is routine to verify that each such $p_S$ is a seminorm on $X$. If $p_S(x) = r > 0$, then $x/r \in S^\circ$ or $x \in rS^\circ$. If $t < r$, then $x/t \notin S^\circ$. In other words, $r = p_{S^\circ}(x) = \inf\{a > 0 : x \in aS^\circ\}$. The polar topology determined by $\mathcal{S}$ has as a subbase at 0 positive multiples of the sets

$$S^\circ = \{x \in X : p_S(x) = \sup |\langle x, S\rangle| \leq 1\} = \overline{V}_{p_{S^\circ}} \quad \blacksquare$$

Since polar topologies are $\mathcal{S}$-topologies, several things follow immediately. For example, by Theorem 2.6.4, an $\mathcal{S}$-topology is Hausdorff if $\mathcal{S}$ covers $Y$. For polar topologies in dual pairs:

**Theorem 8.5.1** HAUSDORFF POLAR TOPOLOGIES Let $(X, Y)$ be a dual pair. A collection $\mathcal{S}$ of $\sigma(X, Y)$-bounded subsets of $X$ determines a Hausdorff polar topology $\mathcal{T}_{\mathcal{S}}$ on $Y$ iff the linear span $M$ of $\cup\mathcal{S}$ is $\sigma(X, Y)$-dense in $X$.

**Proof.** We use the notation of the statement. To demonstrate sufficiency, we consider an element $y \in Y$ and suppose that $p_S(y) = \sup|\langle S, y\rangle| = 0$ for every $S$ in $\mathcal{S}$. Since $y$ vanishes on each $S \in \mathcal{S}$, then $y$ vanishes on the linear span $M$ of $\cup\mathcal{S}$ as well. Since $y$ is $\sigma(X, Y)$-continuous, it must also vanish on $\mathrm{cl}_{\sigma(X,Y)} M = X$. Since $(X, Y)$ is a *dual* pair, this means that $y$ must be 0. By Theorem 5.5.1(a), it follows that $\mathcal{T}_{\mathcal{S}}$ is Hausdorff.

Conversely, suppose that $\mathrm{cl}_{\sigma(X,Y)} M \neq X$. By the Hahn–Banach theorem [7.7.5] there is a nonzero $\sigma(X, Y)$-continuous linear functional $y$ on $X$ such that $y\left(\mathrm{cl}_{\sigma(X,Y)} M\right) = \{0\}$. By the weak representation theorem [8.2.10] we may assume that $y \in Y$. Thus $y \neq 0$ and $p_S(y) = 0$ for each $S \in \mathcal{S}$; hence $\mathcal{T}_{\mathcal{S}}$ is not Hausdorff [Theorem 5.5.1(a)]. ■

Generally, if $\mathcal{S}$ is replaced by the collection $\mathcal{S}'$ of all subsets of finite unions of sets in $\mathcal{S}$, the $\mathcal{S}$-topology is unaffected [Theorem 2.6.3]. For polar topologies we have the following stronger result.

**Theorem 8.5.2** MODIFICATIONS OF $\mathcal{S}$ If $(X, Y)$ is a pair and $\mathcal{S}$ a collection of $\sigma(X, Y)$-bounded subsets of $X$ then the polar topology determined by $\mathcal{S}$ on $Y$ is not altered if $\mathcal{S}$ is replaced by any of the following collections of [$\sigma(X, Y)$-bounded] subsets of $X$:

(a) subsets of finite unions of sets in $\mathcal{S}$;

(b) the sets $aS$, where $a$ is a scalar and $S \in \mathcal{S}$;

(c) balanced hulls $S_b$ or convex hulls $S_c$ of sets $S \in \mathcal{S}$;

(d) $\sigma(X, Y)$-closures of sets in $\mathcal{S}$;

(e) $\sigma(X, Y)$-closures of the balanced convex hulls of sets in $\mathcal{S}$.

**Proof.** $X, Y$ and $\mathcal{S}$ are as above and $\mathcal{V}_{\mathcal{S}}(0)$ denotes the filter of neighborhoods of 0 in the polar topology on $Y$ determined by $\mathcal{S}$.

(a) If $\mathcal{S}_1$ is the collection of subsets of finite unions of sets of $\mathcal{S}$, then the polar topology $\mathcal{T}_{\mathcal{S}_1}$ is finer than $\mathcal{T}_{\mathcal{S}}$. Conversely, suppose that $S$ is a subset of $\cup_{j=1}^n S_j$, where $S_j \in S$, for $j = 1, \ldots, n$. Then $\left(\cup_{j=1}^n S_j\right)^\circ = \cap_{j=1}^n S_j{}^\circ \subset S^\circ$ by Theorem 8.3.9(a), from which it follows that $S^\circ$ is a $\mathcal{T}_{\mathcal{S}}$-neighborhood of 0.

In (b)–(e) $S$ represent an element of $\mathcal{S}$.

(b) If $a \neq 0$, then $(aS)^\circ = |a|^{-1} S^\circ$ and $|a|^{-1} S^\circ \in \mathcal{V}_{\mathcal{S}}(0)$. If $a = 0$, then $(aS)^\circ = Y$. In either case $(aS)^\circ \in \mathcal{V}_{\mathcal{S}}(0)$.

(c) By Theorem 8.3.7(a,b), $S^\circ = (S_b)^\circ = (S_c)^\circ \in \mathcal{V}_{\mathcal{S}}(0)$.

(d) By Theorem 8.3.7(c), $S^\circ = \left(\mathrm{cl}_{\sigma(X,Y)} S\right)^\circ \in \mathcal{V}_{\mathcal{S}}(0)$.

(e) By Theorem 8.3.7(d), $S^\circ = \left(\mathrm{cl}_{\sigma(X,Y)} S_{bc}\right)^\circ \in \mathcal{V}_{\mathcal{S}}(0)$. ■

The following polar topologies are discussed in Examples 8.5.3–8.5.5. For a pair $(X, Y)$, $\mathcal{S}$ denotes a collection of of $\sigma(Y, X)$-bounded subsets of $Y$.

| $\mathcal{S}$ | Topology | Name |
|---|---|---|
| finite subsets | $\sigma(X,Y)$ | weak |
| $\sigma(Y,X)$-compact disks | $\tau(X,Y)$ | Mackey |
| $\sigma(Y,X)$-bounded sets | $\beta(X,Y)$ | strong |
| equicontinuous subsets of $Y$ | $\varepsilon(X,Y)$ | see Theorem 8.6.6 |
| Polar Topologies | | |

**Example 8.5.3** WEAK TOPOLOGY $\sigma(X,Y)$ If $\mathcal{S}$ denotes the collection of one-point subsets of $Y$, then $\mathcal{T}_{\mathcal{S}} = \sigma(X,Y)$. Thus $\sigma(X,Y)$ is a *topology of pointwise convergence* [Example 2.6.8]. In view of Theorem 8.5.2, $\mathcal{S}$ may be expanded to the collection of $\sigma(Y,X)$-closed absolutely convex hulls of finite subsets of $Y$—all sets of the form $\{\sum_i a_i y_i : \sum_i |a_i| \leq 1\}$ for finite collections $\{a_i\}$ of scalars and vectors $\{y_i\} \subset Y$—without affecting $\mathcal{T}_{\mathcal{S}}$. The basic $\sigma(X,Y)$-neighborhoods of 0 are of the form $V(0, y_1, \ldots, y_n, r) = \{x \in X : |\langle x, y_i\rangle| \leq r \text{ for } i = 1, 2, \ldots, n\}$ where $r > 0$ and $y_1, \ldots, y_n \in Y$; the basic $\sigma(X,Y)$-neighborhoods of $x_0 \in X$ are of the form

$$\begin{aligned} V(x_0, y_1, \ldots, y_n, r) &= x_0 + V(0, y_1, \ldots, y_n, r) \\ &= x_0 + r\{y_1, \ldots, y_n\}^\circ \\ &= \{x \in X : |\langle x - x_0, y_i\rangle| \leq r \text{ for } i = 1, 2, \ldots, n\} \end{aligned}$$ ■

**Example 8.5.4** MACKEY TOPOLOGY $\tau(X,Y)$ As the class of $\sigma(Y,X)$-bounded sets, consider the collection $\mathcal{S}$ of all $\sigma(Y,X)$-compact disks of $Y$. The polar topology $\tau(X,Y)$ determined by $\mathcal{S}$ on $X$ is called the *Mackey topology.* Note that Theorem 8.5.2 notwithstanding, $\tau(X,Y)$ is not generated by the class of $\sigma(Y,X)$-compact *subsets* of $Y$ since the balanced convex hull of a $\sigma(Y,X)$-compact set need not be $\sigma(Y,X)$-compact [Example 4.8.8]. ■

As we already know, each $y \in Y$ determines a continuous linear functional on $(X, \sigma(X,Y))$ and there are no others by the representation theorem, Theorem 8.2.10. In other words, the dual of $(X, \sigma(X,Y))$ is $Y$. It turns out that $(X, \tau(X,Y))' = Y$, too, even though $\tau(X,Y)$ is generally a finer topology than $\sigma(X,Y)$. We prove in Theorem 8.7.4 that $\tau(X,Y)$ is the finest locally convex topology for $X$ which leaves $Y$ as the continuous dual of $X$. Many prominent LCS $X$ carry $\tau(X,X')$: For example, every pseudometrizable LCS $X$ carries $\tau(X,X')$ [Example 8.8.10]; so does every infrabarreled space $X$ [Theorem 11.11.6] and every bornological space [Theorem 13.2.10(d)].

**Example 8.5.5** STRONG TOPOLOGY $\beta(X,Y)$ The strongest possible polar topology for $X$ is that determined by the class of *all* $\sigma(Y,X)$-bounded subsets of $Y$. This topology, $\beta(X,Y)$—"$\beta$" as in "b"ounded—the topology of uniform convergence on weakly bounded subsets of $Y$, is called the *strong* (*strongest polar*) topology. Generally, since positive multiples and finite unions of bounded sets are bounded, the sets $\{B^\circ : B \text{ is } \sigma(Y,X)\text{-bounded}\}$ form a base (not just a subbase) at 0 for $\beta(X,Y)$. Since a weakly compact set must be weakly bounded, the Mackey topology $\tau(X,Y)$ of Example 8.5.4 is coarser than $\beta(X,Y)$. Given a TVS $X$, $(X', \beta(X',X))$ is referred to as the *strong dual* of $X$ and $\beta(X',X)$-bounded subsets of $X'$ are called *strongly bounded.*

If $(X,\mathcal{T})$ is a normed or barreled space then $\mathcal{T} = \beta(X,X')$ [Example 8.8.9 and Theorem 11.3.6]. When we consider reflexivity for TVS $X$, we suppose $X'$ and $X''$ to be equipped with $\beta(X',X)$ and $\beta(X'',X')$, respectively [Sec. 15.1]. Finally, note that $\beta(X,Y)$ may be viewed in the context of the topologies discussed in Example 6.7.12 as a topology of uniform convergence on weakly bounded sets. ■

**Example 8.5.6** REAL AND COMPLEX WEAK TOPOLOGIES If $X$ is a complex TVS, it can also be viewed as a real one. The point of this example is to show that the weakened topology on $X$ is the same, regardless of which perspective is adopted. Let the "real" and "complex" duals of $X$ be denoted $X'(\mathbf{R})$ and $X'(\mathbf{C})$, respectively. For $f \in X'(\mathbf{C})$, it follows from Theorem 1.4.2 that $f(x) = r(x) - ir(ix)$, where $r$ denotes the real linear functional $\text{Re} f$. Since $f$ is continuous, so is $r$ by Theorem 5.7.7(b). Thus $r(\cdot)$ and $r(i(\cdot))$ are continuous, i.e., they belong to $X'(\mathbf{R})$. The equality of $\sigma(X, X'(\mathbf{C}))$ and $\sigma(X, X'(\mathbf{R}))$ follows from the observation that $(1/\sqrt{2})\{r, r(i\cdot)\}^\circ \subset \{f\}^\circ \subset \{r, r(i(\cdot))\}^\circ$. ■

## 8.6 EQUICONTINUITY

Ascoli's theorem [5.10.4] shows a close connection between equicontinuity of a set of continuous functions and compactness in the compact-open topology. A similar sort of consanguinity exists between equicontinuous [Definition 5.10.1] subsets of $X'$ and $\sigma(X',X)$-compactness [Theorem 8.6.5]. The principal result of the section is Theorem 8.6.6: Every locally convex topology for a vector space $X$ is a polar topology generated by the equicontinuous subsets of $X'$.

**Notation** If $H$ is a collection of maps and $U$ some subset of their common domain, then $H(U)$ denotes the set $\cup_{h\in H} h(U)$. If $p$ and $q$ are seminorms on $X$ and $p(x) \le q(x)$ for each $x \in X$, then we write $p \le q$. $L(X,Y)$ denotes the linear space of all continuous linear maps of the TVS $X$ into the TVS $Y$. ■

We considered equicontinuity for maps on a topological space taking values in a topological group in Definition 5.10.1. When we consider collections of linear maps, it simplifies as follows. A collection $H$ of linear maps from the TVS $X$ into the TVS $Y$ is *equicontinuous* if for each neighborhood $V$ of 0 in $Y$, there is a neighborhood $U$ of 0 in $X$ such that $H(U) \subset V$. In other words,

$H$ is equicontinuous if it is equicontinuous at 0—$h(x+U) = h(x) + h(U) \subset h(x) + V$ for any $x \in X$ and $h \in H$. Hence $H$ is equicontinuous at every point of $X$. Moreover, since $U$ depends only on $V$ and not on $x$, an expression such as "uniformly equicontinuous" might be more appropriate to describe equicontinuity of linear maps. Linear or not, equicontinuity of a family implies that each member of the family is continuous.

Let $X$ be a TVS and $F(X, \mathbf{F})$ and $C(X, \mathbf{F})$ the spaces of functions and continuous functions, respectively, mapping $X$ into $\mathbf{F}$. By Lemma 5.10.2, the same topology is induced on equicontinuous subsets of $X' \subset C(X, \mathbf{F})$ by the topology $\sigma(X', X)$ of pointwise convergence on $X'$ and the topology of uniform convergence on compact subsets of $X$. By Lemma 5.10.3 it follows that the pointwise closure of an equicontinuous subset $H$ of $X'$ in $F(X, \mathbf{F})$ is an equicontinuous subset of $C(X, \mathbf{F})$. By an argument similar to that used to prove Theorem 8.2.9, the pointwise closure of $H$ in $C(X, \mathbf{F})$ actually consists of *linear* maps, i.e., is an equicontinuous subset of $X'$. We single this out for future reference.

**Theorem 8.6.1** Closures of Equicontinuous Sets If $X$ is a TVS and $H$ is an equicontinuous subset of its dual $X'$, then the $\sigma(X', X)$-closure of $H$ is also an equicontinuous subset of $X'$.

**Theorem 8.6.2** Disked Hull of Equicontinuous Set Let $L(X, Y)$ denote the linear space of all continuous linear maps of the TVS $X$ into the LCS $Y$. Then the balanced convex hull $H_{bc}$ of an equicontinuous subset $H$ of $L(X, Y)$ is equicontinuous. Since subsets of equicontinuous sets are equicontinuous, it follows that if $H$ is equicontinuous, then so are $H_b$ and $H_c$.

**Proof.** Let $H, X$ and $Y$ be as above. Let $V$ be a balanced convex neighborhood of 0 in $Y$ and choose a neighborhood $U$ of 0 in $X$ such that $H(U) \subset V$. Then

$$\begin{aligned} H_{bc}(U) &= \{\textstyle\sum_{\text{finite}} a_i h_i(u) : u \in U,\ h_i \in H,\ \sum_{\text{finite}} |a_i| \leq 1,\ a_i \in \mathbf{F}\} \\ &\subset H(U)_{bc} = \{\textstyle\sum_{\text{finite}} a_i h_i(u_i) : u_i \in U, h_i \in H,\ \sum_{\text{finite}} |a_i| \leq 1\} \\ &\subset V_{bc} = V \end{aligned}$$

and it follows that $H_{bc}$ is equicontinuous. ■

In Theorem 5.7.3 we showed that for a linear map $A$ taking a TVS $X$ into an LCS $Y$ to be continuous, it is necessary and sufficient that for each continuous seminorm $q$ on $Y$, there be a continuous seminorm $p$ on $X$ such that $q \circ A \leq p$. The following result yields a similar equicontinuity criterion for a family of linear maps.

**Theorem 8.6.3** Seminorm Criterion for Equicontinuity Let $H$ be a family of linear maps mapping the LCS $X$ into the LCS $Y$. Then $H$ is equicontinuous iff for each continuous seminorm $q$ on $Y$, there is a continuous seminorm $p$ on $X$ such that $q \circ h \leq p$ for each $h \in H$.

**Proof.** In the notation above, suppose that the condition holds. To prove equicontinuity, it suffices to consider neighborhoods of 0 in $Y$ of the form $V_q = \{y \in Y : q(y) < 1\}$ where $q$ is a continuous seminorm on $Y$. By the condition, there exists a continuous seminorm $p$ on $X$ such that for each $h \in H$, $q \circ h \leq p$—$p(x) < 1 \Rightarrow h(x) \in V_q$. Hence $H(V_p) \subset V_q$ where $V_p = \{x \in X : p(x) < 1\}$.

Conversely, suppose that $H$ is an equicontinuous subset of $L(X,Y)$. If $q$ is a continuous seminorm on $Y$, then the equicontinuity of $H$ implies that there must be some continuous seminorm $p$ on $X$ such that $H\left(\overline{V}_p\right) \subset V_q$. If $x \in \overline{V}_p$ and $p(x) = 0$, then for all $a > 0$, $p(ax) = 0$, so $(q \circ h)(ax) < 1$ for each $h \in H$; the latter condition implies that $(q \circ h)(x) = 0 = p(x)$. If $p(x) \neq 0$, then $p(x/p(x)) = 1$, so

$$q(h(x/p(x))) = (1/p(x))(q \circ h)(x) \leq 1$$

or $(q \circ h)(x) \leq p(x)$. ■

Theorem 8.6.4 demonstrates that the equicontinuous subsets of $X'$ are essentially just the polars of neighborhoods of 0 in $X$.

**Theorem 8.6.4** EQUICONTINUITY IN $X'$ Let $X$ be a TVS and let $X'$ be its dual. A subset $H$ of $X'$ is equicontinuous iff (a) or (b) hold.

(a) $H$ is contained in the polar of some neighborhood of 0 in $X$.

(b) $H^\circ$ is a neighborhood of 0 in $X$.

**Proof.** Let $X$ be a TVS. (a) If $H \subset X'$ is equicontinuous, then there is some neighborhood $V$ of 0 in $X$ such that $H(V) \subset D$, the closed unit disk of $\mathbf{F}$ which implies that $H \subset V^\circ$. Conversely, suppose $H \subset V^\circ$ for some neighborhood $V$ of 0 in $X$. For $r > 0$, then $|h(rv)| \leq r$ for all $h \in H$ and $v \in V$, i.e., $H(rV) \subset rD$.

(b) $\Leftrightarrow$ (a) Let $V$ be a neighborhood of 0 in $X$. If $H \subset V^\circ$ then $H^\circ \supset V^{\circ\circ} \supset V$ and $H^\circ$ is a neighborhood of 0 in $X$. Conversely, if $H^\circ$ is a neighborhood of 0 in $X$, then $H \subset H^{\circ\circ}$. ■

Ascoli's theorem [5.10.4] shows a close connection between compactness in the compact-open topology and equicontinuity. The following result bears some resemblance; see also Theorem 11.2.6.

**Theorem 8.6.5** EQUICONTINUITY $\Rightarrow$ RELATIVE $\sigma(X',X)$-COMPACTNESS If $X$ is a TVS and $H$ is an equicontinuous subset of $X'$, then $H$ is relatively $\sigma(X',X)$-compact.

**Proof.** Let $X$ and $H$ be as above. Since $H$ is equicontinuous, $H$ is contained in the polar of some neighborhood $V$ of 0 in $X$ by Theorem 8.6.4(a). By the Alaoglu theorem [8.4.1], $V^\circ$ is $\sigma(X',X)$-compact. Since $\sigma(X',X)$ is Hausdorff [Remark 8.2.3], $\text{cl}_{\sigma(X',X)} H$ is $\sigma(X',X)$-compact. ■

As a consequence of Theorem 8.6.5, an equicontinuous set of linear functionals is weak-$*$ bounded; we denote by $\varepsilon(X, X')$ the polar topology determined by the equicontinuous subsets of $X'$. We put this to immediate use in Theorem 8.6.6, a central result in the theory of LCS.

**Theorem 8.6.6** Every Locally Convex Topology Is Polar If $\mathcal{T}$ is a locally convex topology for the linear space $X$ then $\mathcal{T} = \varepsilon(X, X')$, the polar topology determined by the equicontinuous subsets of $X'$.

**Proof.** Let $(X, \mathcal{T})$ be an LCS. Consider the neighborhood base $\mathcal{B}$ at 0 for $(X, \mathcal{T})$ of closed disks. By the weak representation theorem [8.2.10], it follows that the dual of $(X, \sigma(X, X'))$ is $X'$. Hence the closed half spaces of $(X, \sigma(X, X'))$ are the same as those of $(X, \mathcal{T})$. Since any $B \in \mathcal{B}$ is the intersection of all closed half spaces that contain it [Theorem 7.8.7(a)], $B$ is $\sigma(X, X')$-closed. By the bipolar theorem [8.3.8], $B = B^{\circ\circ}$ for each $B \in \mathcal{B}$. If $H \subset X'$ is equicontinuous, then $H^\circ$ is a $\mathcal{T}$-neighborhood of 0 by Theorem 8.6.4(b); consequently $\varepsilon(X, X') \subset \mathcal{T}$. On the other hand, $B^\circ$ is equicontinuous for any $B \in \mathcal{B}$ [Theorem 8.6.4(a)]; therefore $B = B^{\circ\circ} \in \varepsilon(X, X')$ and $\mathcal{T} \subset \varepsilon(X, X')$. ■

## 8.7 TOPOLOGIES OF PAIRS

By weakening a topology on a given space, its space of continuous functions generally diminishes. Yet if $(X, \mathcal{T})$ is an LCS and $\mathcal{T}$ is weakened to $\sigma(X, X')$, the class $X'$ of continuous linear functionals on $X$ remains the same [Theorem 8.2.10]. Similarly, the stronger the topology on a given set, the more continuous functions there are. And just as $\mathcal{T}$ may be weakened somewhat without altering $X'$, $\mathcal{T}$ may also be strengthened without changing $X'$: indeed, $\mathcal{T}$ may be strengthened up to $\tau(X, X')$, the Mackey topology, or topology of uniform convergence on the $\sigma(X', X)$-compact disks of $X'$ without affecting $X'$ [Theorem 8.7.4]. This profound result concerning the allowable variability in $\mathcal{T}$ without affecting $X'$ is called the Mackey–Arens theorem (after Mackey [1946] and Arens [1947]) and is the main result of this section. It is central in duality theory and its consequences reverberate throughout the sequel.

For paired spaces $X$ and $Y$, we usually we identify the element $y \in Y$ with the linear functional $\langle \cdot, y \rangle = y^*$. Since $X$ distinguishes points of $Y$, the correspondence $y \mapsto y^*$ is injective [Theorem 8.2.2(b)]. In this sense we can make the following definition.

**Definition 8.7.1** Topology of the Pair Let $(X, Y)$ be a pair in which $X$ distinguishes points of $Y$. A locally convex topology $\mathcal{T}$ for $X$ is a *topology of the pair* (is *compatible with the pairing*) if $Y = (X, \mathcal{T})'$. ■

The weak topology $\sigma(X, Y)$ for a pair $(X, Y)$ in which $X$ distinguishes points of $Y$ is a topology of the pair as shown by the weak representation theorem [8.2.10]. If the normed space $X$ is not reflexive then the norm topology on $X'$ is not a topology of the pair $(X', X)$.

The following result says that the continuous linear functionals on a TVS are precisely the linear functionals that are bounded on neighborhoods of 0.

**Theorem 8.7.2** DUAL OF A TVS Let $X$ be a TVS, let $X^*$ be its algebraic dual and consider the pair $(X, X^*)$. If $\mathcal{B}$ is a base at 0 in the TVS $X$, then the dual $X' = \cup_{B \in \mathcal{B}} B^\circ$ (polars computed in $X^*$).

**Proof.** Let $\mathcal{B}$ be a base at 0 in the TVS $X$. Any $f \in \cup_{B \in \mathcal{B}} B^\circ$ is bounded on a neighborhood of 0, hence continuous by Theorem 6.5.1. Conversely, if $f \in X'$, then $|f|$ must be $\leq 1$ on some $B \in \mathcal{B}$ by continuity. ■

**Theorem 8.7.3** TOPOLOGIES OF THE PAIR (Mackey–Arens I) Let $(X, Y)$ be a pair in which $X$ distinguishes points of $Y$. A locally convex topology $\mathcal{T}$ on $X$ is a topology of the pair iff $\mathcal{T}$ is a polar topology determined by a collection $\mathcal{S}$ of $\sigma(Y, X)$-compact disks of $Y$ which cover $Y$.

**Proof.** If $\mathcal{T}$ is a topology of the pair $(X, Y)$ on $X$ then $Y = (X, \mathcal{T})'$. Consequently $\mathcal{T} = \varepsilon(X, Y)$, the polar topology determined by the class $\mathcal{E}$ of equicontinuous subsets of $Y$ [Theorem 8.6.6]. Since singletons are equicontinuous, $\mathcal{E}$ covers $Y$. As follows from Theorem 8.5.2, $\varepsilon(X, Y)$ is unchanged if $\mathcal{E}$ is replaced by the collection $\mathcal{S}$ of $\sigma(Y, X)$-closures of disked hulls of sets in $\mathcal{E}$. Such sets are disks by Theorem 4.4.2(c,g). The disked hull of an equicontinuous set is equicontinuous [Theorem 8.6.2] and the closure of an equicontinuous set is equicontinuous [Theorem 8.6.1], so the sets of $\mathcal{S}$ are $\sigma(Y, X)$-compact by Theorem 8.6.5.

Conversely, suppose that $\mathcal{S}'$ is a cover of $Y$ consisting of $\sigma(Y, X)$-compact disks. If $\mathcal{S}'$ is replaced by the collection $\mathcal{S}$ of positive multiples of disked hulls of finite unions of sets in $\mathcal{S}'$, the polar topology $\mathcal{T}$ induced by it is unchanged. Since the disked hull of a finite union of *convex* compact sets is compact [Theorem 4.4.4], the sets in $\mathcal{S}$ are $\sigma(Y, X)$-compact disks. Thus, $\mathcal{S}^\circ = \{S^\circ : S \in \mathcal{S}\}$ is a base at 0 for $\mathcal{T}$. Let $S^{\circ *}$ denote the polar of $S^\circ$ computed in $X^*$. By Theorem 8.7.2, the dual of $(X, \mathcal{T})' = \cup \mathcal{S}^{\circ *}$. Since each $S \in \mathcal{S}$ may be viewed as a subset of $X^*$, it follows from the bipolar theorem [8.3.8] that $S^{\circ *} = \mathrm{cl}_{\sigma(X^*, X)} S$ for each $S \in \mathcal{S}$. Since each $S \in \mathcal{S}$ is $\sigma(Y, X)$-compact and $\sigma(Y, X)$ is the topology induced by $\sigma(X^*, X)$ on $Y$, $S$ is $\sigma(X^*, X)$-compact, therefore $\sigma(X^*, X)$-closed. Hence $S^{\circ *} = \mathrm{cl}_{\sigma(X^*, X)} S = S$ and therefore $(X, \mathcal{T})' = \cup \mathcal{S}^{\circ *} = \cup \mathcal{S}$. Since $\mathcal{S}$ covers $Y$, $(X, \mathcal{T})' = \cup \mathcal{S} = Y$. ■

Theorem 8.7.4 is a direct consequence of Theorem 8.7.3 and is also called the Mackey–Arens theorem.

**Theorem 8.7.4** BOUNDS ON TOPOLOGIES OF THE PAIR (Mackey–Arens II) Let $(X, Y)$ be a pair in which $X$ distinguishes points of $Y$. A locally convex topology $\mathcal{T}$ is a topology of the pair $(X, Y)$ iff $\sigma(X, Y) \subset \mathcal{T} \subset \tau(X, Y)$.

**Proof.** With notation as above, if $\mathcal{T}$ is a topology of the pair $(X,Y)$ then $\mathcal{T}$ is a polar topology determined by *some* collection of $\sigma(Y,X)$-compact disks of $Y$ by Theorem 8.7.3. Hence $\mathcal{T}$ is coarser than the Mackey topology $\tau(X,Y)$—the polar topology determined by *all* $\sigma(Y,X)$-compact disks of $Y$. On the other hand, we know that $\sigma(X,Y)$ is a topology of the dual pair by Theorem 8.2.10(a) and that it is the coarsest topology on $X$ with respect to which each member of $Y$ is continuous. Hence $\sigma(X,Y) \subset \mathcal{T}$. Conversely, if $\sigma(X,Y) \subset \mathcal{T} \subset \tau(X,Y)$, it follows from Theorems 8.2.10(a) and 8.7.3 that $Y = (X, \sigma(X,Y))' \subset (X,\mathcal{T})' \subset (X, \tau(X,Y))' = Y$. ■

As another application of Theorem 8.7.3, we obtain a characterization of the dual of the space $C(T,\mathbf{F},c)$ of continuous $\mathbf{F}$-valued functions on the topological space $T$ with compact-open topology [Example 5.5.6]. For $t \in T$ consider the associated evaluation map (continuous linear functional) first mentioned in Sec. 5.8

$$t' : C(T,\mathbf{F},c) \to \mathbf{F}, \ x \mapsto x(t)$$

We characterize $C(T,\mathbf{F},c)'$ in terms of evaluation maps in Theorem 8.7.5 from which it follows that when $T$ is compact, $(T')_{bc}$ is $\sigma(X^*,X)$-dense in $C(T,\mathbf{F},c)'$.

**Theorem 8.7.5** $C(T,\mathbf{F},c)'$ AND EVALUATION MAPS Let $T$ be a topological space, $X = C(T,\mathbf{F},c)$, $\mathcal{K}$ the compact subsets $K$ of $T$, $K' = \{t' : t \in K\}$ and $X^*$ the algebraic dual of $X$. Then $C(T,\mathbf{F},c)'$ is the linear span

$$\left[\cup_{K\in\mathcal{K}} \operatorname*{cl}_{\sigma(X^*,X)} (K')_{bc}\right] = H \text{ of } \cup_{K\in\mathcal{K}} \operatorname*{cl}_{\sigma(X^*,X)} (K')_{bc}$$

If $T$ is compact, then this simplifies to

$$C(T,\mathbf{F},c)' = \left[\mathrm{cl}_{\sigma(X^*,X)} (T')_{bc}\right]$$

**Proof.** We use the notation of the statement. Since $t' \in H$ for each $t \in T$, $(X,H)$ is a dual pair. To show that $H = C(T,\mathbf{F},c)'$, we show that the compact-open topology is the polar topology generated by a collection of $\sigma(H,X)$-compact disks of $H$ [Theorem 8.7.3], namely the sets $\{\mathrm{cl}_{\sigma(X^*,X)} (K')_{bc} : K \in \mathcal{K}\}$. To see that each $K^* = \mathrm{cl}_{\sigma(X^*,X)} (K')_{bc}$ is $\sigma(H,X)$-compact, recall three things:

(1) $(X^*, \sigma(X^*,X))$ is complete [Theorem 8.2.9].

(2) A closed and totally bounded subset of a complete topological group is compact [Theorem 3.4.1].

(3) A set is $\sigma(X^*,X)$-totally bounded iff it is $\sigma(X^*,X)$-bounded [Theorem 8.2.8].

As $\sigma(X^*,X)$-compactness implies $\sigma(H,X)$-compactness, it suffices to show that each $K^*$ $(K \in \mathcal{K})$ is $\sigma(X^*,X)$-bounded. To do this, we use the criterion of Theorem 6.1.5 and show that $\sup|\langle x, K^*\rangle| < \infty$ for each $x$ in $X$. To this end, let $t_1, \ldots, t_n \in K$ and $a_1, \ldots, a_n \in \mathbf{F}$ be such that $\sum_{i=1}^n |a_i| \le 1$. Then $\sum_{i=1}^n a_i t_i' \in (K')_{bc}$ and for any $x \in X$,

$$\begin{aligned} \left| \left\langle x, \sum_{i=1}^n a_i t_i' \right\rangle \right| &= \left| \sum_{i=1}^n a_i t_i' (x) \right| = \left| \sum_{i=1}^n a_i x (t_i) \right| \\ &\le \left( \sum_{i=1}^n |a_i| \right) p_K (x) \le p_K (x) = \sup |x (K)| \end{aligned}$$

By Theorem 6.3.2(a), it follows that $K^*$ is $\sigma (X^*, X)$-bounded. Finally, we show that

$$(K^*)^\circ = \overline{V}_{p_K} = \{x \in X : p_K (x) \le 1\}$$

By the argument above, $|f (x)| \le p_K (x)$ for each $x \in X$ and each $f \in K^*$; hence $\overline{V}_{p_K} \subset (K^*)^\circ$. To obtain the reverse inclusion, consider any $x \in (K^*)^\circ$ and any $t \in K$. Then $|x (t)| \le 1$ and it follows that $p_K (x) = \sup |x (K)| \le 1$. ■

## 8.8 PERMANENCE IN DUALITY

The result of Theorem 7.8.7(b) states that any closed convex set in a real LCS $X$—and any LCS can be viewed as real—is the intersection of all the closed half spaces that contain it, a *half space* being a set of the form $L = \{x : f(x) \le c)\}$ where $f$ is a linear functional and $c$ a real number. A half space $L$ is closed iff $f$ is continuous. Thus, if $\mathcal{T}_1$ and $\mathcal{T}_2$ are topologies of the pair $(X, X')$, since $(X, \mathcal{T}_1)' = (X, \mathcal{T}_2)'$, the class of closed *convex* sets is the same in either topology—we emphasize the *convex* here since because it is not generally true that closed implies weakly closed. We record this for future reference.

**Theorem 8.8.1** PERMANENCE OF CLOSED CONVEX SETS Let $(X, Y)$ be a pair. Then

(a) the class of closed convex subsets of $X$ is the same with respect to any topology of the pair; and

(b) the closure of a convex subset of $X$ is the same in any topology of the pair.

(c) For any topology $\mathcal{T}$ of the pair and any $\mathcal{T}$-closed disk $B$ of $X$, $B = B^{\circ\circ}$ by the bipolar theorem [8.3.8].

The very definition of topology of a pair is one which ensures a certain kind of permanence, namely, of the space of continuous linear functionals. Theorem 8.8.1 is an example of another kind of immutability. Our main result in this section [Theorem 8.8.7] asserts that, for any LCHS $X$, the bounded sets of $X$ are the same in any topology of the dual pair $(X, X')$.

**Definition 8.8.2** BARRELS A *barrel* (*tonneau*) is a closed absorbent disk. ■

Certainly, there is no dearth of barrels in a TVS since the closure of the absolute convex hull of any absorbent set is a barrel; in particular, the closed absolute convex hull of any neighborhood of 0 is a barrel. In LCS there is a base of barrels at 0 [Theorem 4.4.6]. In many cases—any Banach space, for example—every barrel is a neighborhood of 0, such LCS being called *barreled spaces*; their special properties are examined in Chap. 11. Last, we note that if $X$ is an LCS, it is not necessary to distinguish between "barrel" and "weak barrel" by virtue of Theorem 8.8.1. Since dealing with polars forces consideration of weak closures, this observation simplifies (as well as makes possible) many results, such as Theorems 8.8.3 and 8.8.4.

In Theorem 8.6.4(a) we showed that a subset $H$ of the dual $X'$ of a TVS $X$ is equicontinuous iff $H$ is contained in the polar of a neighborhood of 0 in $X$. Since equicontinuous sets are weakly totally bounded [Theorem 8.6.5], they are weakly bounded.

**Theorem 8.8.3** WEAK-$*$ BOUNDEDNESS AND POLARS OF BARRELS Let $X$ be an LCS. A subset $H$ of $X'$ is $\sigma(X',X)$-bounded iff $H \subset B^\circ$ for some barrel $B$ in $X$.

**Proof.** Let $B$ be a barrel in the LCS $X$. By Theorem 8.3.5 to show that $B^\circ$ is $\sigma(X',X)$-bounded, we have to show that $B^{\circ\circ}$ is absorbent. By the bipolar theorem [8.3.8], $B^{\circ\circ} = B$, an absorbent set; therefore $B^\circ$ is $\sigma(X',X)$-bounded and so is any subset $H$ of $B^\circ$. Conversely, suppose that $H$ is $\sigma(X',X)$-bounded. Then $H^\circ$ is a $\sigma(X,X')$-closed disk by Theorem 8.3.6(a); it is absorbent by Theorem 8.3.5 and therefore a barrel in $X$. Since $H^\circ$ is closed and convex, it is a barrel in the original topology on $X$ as well by Theorem 8.8.1(a). $H \subset H^{\circ\circ}$ by Theorem 8.3.6(d). ■

Having established a connection between weak boundedness and polars of barrels, we consider next the connection between barrels and polars of $\sigma(Y,X)$-bounded sets.

**Theorem 8.8.4** BARREL = POLAR OF $\sigma(Y,X)$-BOUNDED SET Let $\mathcal{T}$ be a topology of the pair $(X,Y)$. Then $B$ is a barrel in the LCS $(X,\mathcal{T})$ iff $B$ is the polar of a $\sigma(Y,X)$-bounded subset of $Y$.

**Proof.** Let $\mathcal{T}$ be a topology of the pair $(X,Y)$. If $B$ is a barrel in the LCS $(X,\mathcal{T})$ then, since $B$ is absorbent, $B^\circ$ is $\sigma(Y,X)$-bounded [Theorem 8.3.5]. By the bipolar theorem [8.3.8], $B = B^{\circ\circ}$. Conversely, if $H \subset X$ is $\sigma(Y,X)$-bounded then $H^\circ$ is absorbent by Theorem 8.3.5, a $\sigma(X,Y)$-closed disk by Theorem 8.3.6(a). ■

Our next two results have to do with the absorbent properties of barrels. The more significant one is Theorem 8.8.5, Theorem 8.8.6 being a direct consequence of it. Even though Theorem 8.8.4 is stated for barrels $B$, the convexity of $B$ is not necessary: The result remains true for closed balanced absorbent sets. This is not the case in Theorem 8.8.6, however.

**Theorem 8.8.5** BARRELS ABSORB CONVEX COMPACT SETS If $B$ is a barrel in the TVS $X$, then $B$ absorbs each convex compact subset $K$ of $X$.

**Proof.** With notation as in the statement, suppose that there is some $x \in K$ such that

$$K \cap (x + V) \subset nB \text{ for some neighborhood } V \text{ of } 0 \text{ and } n \in \mathbf{N} \qquad (*)$$

In other words, suppose that $K$ contains a relative neighborhood of $x$ which is absorbed by $B$. Translated to the origin, the condition becomes $(K - x) \cap V \subset nB - x$. We contend that the existence of such an $x, V$, and $n$ suffices to prove the theorem. To see this, we argue as follows: $K$ is compact, so $K - x$ is bounded and there must be some $a \geq 1$ such that $(K - x) \subset aV$. Since $K$ is convex and $0 \in K - x$, it follows that for any $y \in K - x$,

$$(1/a)y = (1/a)y + (1 - 1/a)0 \in K - x$$

i.e., $K - x \subset a(K - x)$. Since $(K - x) \subset aV$, $K - x \subset a(K - x) \cap aV \subset a(nB - x)$ which implies that

$$K \subset a(nB - x) + x = anB + (1 - a)x$$

As $B$ is absorbent, there exists $r > 0$ such that $(1 - a)x \in rB$. Since $B$ is convex, $K \subset anB + rB = (an + r)B$ by Theorem 4.2.5 and $B$ absorbs $K$.

We now show that if $(*)$ is not satisfied, then $B$ is not absorbent. If $K \cap (x + V) \not\subset nB$ for any $n$ and $V$, then for $n = 1$, any open neighborhood $V_0$ of 0 and any $x_0 \in K$, there exists $x_1 \in K \cap (x_0 + V_0) \cap \complement B$. Since $(x_0 + V_0) \cap \complement B$ is open, there is an open neighborhood $V_1$ of 0 such that $x_1 + \operatorname{cl} V_1 \subset (x_0 + V_0) \cap \complement B$. Similarly, there exists $x_2 \in K \cap (x_1 + V_1) \cap \complement(2B)$ and an open neighborhood $V_2$ of 0 such that

$$x_2 + \operatorname{cl} V_2 \subset (x_1 + V_1) \cap \complement(2B)$$

and so on, to yield a decreasing sequence $(K \cap (x_n + \operatorname{cl} V_n)$ of closed nonempty subsets of $K$. Since $K$ is compact, the sequence must have a nonempty intersection. If $y$ is a member of this intersection, however, then $y \notin nB$ for any positive integer $n$—in other words, $B$ is not absorbent, which is contradictory. Hence the condition must be satisfied and the proof is complete. ■

**Theorem 8.8.6** BARRELS ABSORB CONVEX BOUNDED COMPLETE SETS IN LCHS If $B$ is a barrel in the LCHS $X$, then $B$ absorbs each convex bounded complete subset $E$ of $X$.

**Proof.** With notation as above, if $B$ does not absorb $E$, then there is some sequence $(x_n)$ of points of $E$ such that, for each $n$, $x_n \notin n^2 B$. Since $E$ is bounded, $(1/n)x_n \to 0$ by Theorem 6.1.4(a). Since $X$ is Hausdorff, the completeness of $E$ implies that it is closed [Theorem 3.1.9(b)]. Therefore $0 \in E$. Since $0 \in E$ and $E$ is convex $(1/n)x_n \in E$ for every $n$. Thus, $H = \{0, x_1, (1/2)x_2, (1/3)\,x_3, \ldots\}$ is a compact subset of $E$. Since $X$ is locally convex, the convex hull $H_c$ of $H$ is totally bounded by Theorem 4.8.9(b). Since $E$ is closed, $\operatorname{cl} H_c \subset E$. As a closed subset of a complete set, $\operatorname{cl} H_c$ is complete by Theorem 3.1.9(a). As a complete and totally bounded set, it is compact by Theorem 3.4.2. Since $\operatorname{cl} H_c$ is convex and compact, $B$ absorbs $\operatorname{cl} H_c$ by Theorem 8.8.5. Since the points $(1/n)\,x_n$ of $H$ do not belong to $nB$ for any $n$, $B$ does not absorb $\operatorname{cl} H_c$, a contradiction. ■

Next, we prove Mackey's theorem that the bounded sets of an LCHS $X$ are the same in any topology of the dual pair $(X, X')$. Consequently, we need not distinguish between bounded and weakly bounded subsets of an LCHS.

**Theorem 8.8.7** PERMANENCE OF BOUNDED SETS IN DUAL PAIRS For any LCHS $(X, \mathcal{T})$ the bounded sets are the same in any topology of the pair $(X, X')$.

**Proof.** Let $(X, \mathcal{T})$ LCHS. If a set is $\mathcal{T}$-bounded, then it is bounded in any topology which is coarser than $\mathcal{T}$. Since $\sigma(X, X') \subset \mathcal{T} \subset \tau(X, X')$ by the Mackey–Arens theorem [8.7.4], it suffices to show that any $\sigma(X, X')$-bounded subset $B$ of $X$ is $\tau(X, X')$-bounded. To this end, let $V$ be a closed balanced convex $\tau(X, X')$-neighborhood of 0. Since $V$ is a neighborhood of 0, $V^\circ$ is $\sigma(X', X)$-compact by the Alaoglu theorem [8.4.1]. Since $B$ is $\sigma(X, X')$-bounded, $B^\circ$ is $\sigma(X', X)$-barrel in $X'$ by the dual form of Theorem 8.8.4. As such, $B^\circ$ must absorb $V^\circ$ by Theorem 8.8.5. Consequently, $V = V^{\circ\circ}$ absorbs $B^{\circ\circ} \supset B$ and the proof is complete. ■

We showed that discontinuous linear functionals exist on any infinite-dimensional pseudometrizable linear space [Theorem 5.6.2]. The next result provides instances in which a linear map that takes bounded sets into bounded sets is discontinuous.

**Example 8.8.8** LOCALLY BOUNDED DISCONTINUOUS LINEAR MAP We observed in Example 8.2.5 that the weak topology $\sigma(X, X')$ on an infinite-dimensional Hilbert space $X$ is strictly weaker than the norm topology: Any orthonormal sequence $(x_n)$ of vectors converges weakly to 0 but does not converge to 0 in the norm topology. In particular, for the orthonormal sequence $(\cos nt/\sqrt{\pi})$ in the Hilbert space $L_2[-\pi, \pi]$, for any $x \in L_2[-\pi, \pi]$, $\int_{-=\pi}^{\pi} x(t) \cos nt/\sqrt{\pi}\, dt \to 0$, a fact often called the Riemann–Lebesgue lemma [Bachman et al. 2000, p. 104]. Thus, the identity map $I : (X, \sigma(X, X')) \to (X, \|\cdot\|), x \mapsto x$, is discontinuous, something that remains true for any TVS

$X$ where $\sigma(X, X')$ is strictly weaker than the original topology. The identity map takes bounded sets into bounded sets by Theorem 8.8.7. ■

**Example 8.8.9** FOR NORMED $X$, $X'$ CARRIES $\beta(X', X)$ Let $X$ be a normed space and let $X'$ carry its norm topology $\mathcal{T}_n$ [Example 6.5.3(b)]. Let $U(X)$ and $U(X')$ denote the closed unit balls of $X$ and $X'$, respectively. By Example 8.3.3, $U(X') = U(X)^\circ$. The norm topology is a topology of the dual pair $(X, X')$ so the norm-bounded set $U(X)$ is $\sigma(X, X')$-bounded by Theorem 8.8.7. As such, $U(X)^\circ \in \beta(X', X)$ and $\mathcal{T}_n \subset \beta(X', X)$.

Conversely, if $B$ is a $\sigma(X, X')$-bounded subset of $X$, it is norm bounded by Theorem 8.8.7. Therefore, there exists $r > 0$ such that $B \subset rU(X)$. Hence $B^\circ \supset (1/r)U(X)^\circ = (1/r)U(X')$ and $B^\circ$ is seen to be a norm neighborhood of 0; thus $\beta(X', X) \subset \mathcal{T}_n$. A normed space $X$ need not carry $\beta(X, X')$, however [Exercise 11.103]. ■

Another consequence of Theorem 8.8.7 is that if $(X, \mathcal{T})$ is a metrizable LCS, then $\mathcal{T} = \tau(X, X')$ as we demonstrate in Example 8.8.10.

**Example 8.8.10** METRIZABLE LCS $X$ CARRY $\tau(X, X')$ Let $(X, \mathcal{T})$ be a metrizable LCS. By the Mackey–Arens theorem [8.7.4], we know that $\mathcal{T} \subset \tau(X, X')$. To see that each $\tau(X, X')$-neighborhood $U$ of 0 is a $\mathcal{T}$-neighborhood of 0, let $(V_n)$ be a decreasing base of balanced $\mathcal{T}$-neighborhoods of 0 in $X$. If $U$ is not a $\mathcal{T}$-neighborhood of 0, then for each $n \in \mathbf{N}$, there exists $x_n \in (1/n)V_n \setminus U$. Since $nx_n \to 0$, $\{nx_n\}$ is $\mathcal{T}$-bounded. By Theorem 8.8.7, $\{nx_n\}$ is $\tau(X, X')$-bounded as well and therefore should be absorbed by $U$. Hence there should be $k \in \mathbf{N}$ such that $\{nx_n : n \in \mathbf{N}\} \subset kU$ which implies that $kx_k \in kU$ or that $x_k \in U$ which contradicts the way the $x_n$ were chosen.

## 8.9 ORTHOGONALS

The notion of "orthogonal" of this section provides a sort of generalization of the geometric notion of orthogonality. Orthogonals facilitate conversion of certain statements in $X$ to assertions in $X'$ and generally play a role in duality theory. For example [Theorem 8.12.1] for a subspace $M$ of an LCS $X$, $(X/M)'$ is linearly isomorphic to $M^\perp$ and $M'$ is linearly isomorphic to $X'/M^\perp$.

**Notation.** If $X$ and $Y$ are paired spaces and $S$ is a subset of $X$, the linear span of $S$ in $X$ is denoted (as usual) by $[S]$. The $\sigma(X, Y)$-closure of $S$ in $X$ is denoted by $\mathrm{cl}_{\sigma(X,Y)} S$ the first time it appears in a proof and by $\mathrm{cl}_\sigma S$ in subsequent appearances. ■

**Definition 8.9.1** ORTHOGONALS Let $X$ and $Y$ be paired spaces. For $S \subset X$ the set $S^\perp = \{y \in Y : \langle s, y \rangle = 0 \text{ for each } s \in S\}$ is the *orthogonal* (or *annihilator*) of $S$. Analogous conventions hold for subsets of $Y$. ■

It is immediate that $S^\perp$ is a subspace no matter what $S$ is. If $S$ is a sub*space*, however, then $S^\circ = S^\perp$ [Example 8.3.4].

**Theorem 8.9.2** CALCULUS OF ORTHOGONALS Let $X$ and $Y$ be paired linear spaces and let $H$ and $S$ be subsets of $X$. Then:

(a) if $S \subset H$, then $H^\perp \subset S^\perp$;

(b) $S^\perp = [S]^\perp = (\mathrm{cl}_{\sigma(X,Y)}[S])^\perp$;

(c) $S \subset S^{\perp\perp} = (S^\perp)^\perp$ and $S^\perp = S^{\perp\perp\perp}$;

(d) $S^\perp$ is $\sigma(Y,X)$-closed;

(e) $S^{\perp\perp} = \mathrm{cl}_{\sigma(X,Y)}[S]$; thus, if $M$ is a $\sigma(X,Y)$-closed subspace of $X$, $M = M^{\perp\perp}$;

(f) if $(S_i)_{i\in I}$ is a collection of subsets of $X$ then $(\cup_{i\in I} S_i)^\perp = \cap_{i\in I} S_i^\perp$;

(g) if $(M_i)_{i\in I}$ is a family of $\sigma(X,Y)$-closed subspaces of $X$, then

$$(\cap_{i\in I} M_i)^\perp = \operatorname*{cl}_{\sigma(Y,X)} [\cup_{i\in I} M_i^\perp]$$

**Proof.** We use the notation of the statement. (a) Clear.

(b) Since $S \subset [S] \subset \mathrm{cl}_{\sigma(X,Y)}[S]$, it follows that $S^\perp \supset [S]^\perp \supset (\mathrm{cl}_\sigma[S])^\perp$. Using the bilinearity of $\langle\cdot,\cdot\rangle$, it is straightforward to verify that $S^\perp = [S]^\perp$, so it only remains to show that $[S]^\perp \subset (\mathrm{cl}_\sigma[S])^\perp$. To this end suppose that $y \in [S]^\perp$ and $x \in \mathrm{cl}_\sigma[S]$; we have to show that $\langle x,y\rangle = 0$. Choose a net $(x_s)$ from $[S]$ which converges weakly to $x$. Since $\langle x_s,y\rangle = 0$ for every $s$ and $\langle\cdot,y\rangle$ is $\sigma(X,Y)$-continuous, it follows that $\langle x,y\rangle = \lim_s \langle x_s,y\rangle = 0$.

(c) For any $x \in S$ and $y \in S^\perp$, $\langle x,y\rangle = 0$. Hence $S \subset S^{\perp\perp}$. By (a) this implies that $S^{\perp\perp\perp} \subset S^\perp$; equality follows from the fact that $S^\perp \subset (S^\perp)^{\perp\perp}$.

(d) $S^\perp = [S]^\perp = [S]^\circ$ which is $\sigma(Y,X)$-closed by Theorem 8.3.6(a).

(e) By (b) and (c), $S^{\perp\perp} = (\mathrm{cl}_{\sigma(X,Y)}[S])^{\perp\perp} \supset \mathrm{cl}_\sigma[S]$. If this inclusion is proper, there is an $x \in S^{\perp\perp} \setminus \mathrm{cl}_\sigma[S]$. By a consequence of the Hahn–Banach theorem [7.7.5], there exists a $\sigma(X,Y)$-continuous linear functional $f$ on $X$ such that $f(x) = 1$ that vanishes on $\mathrm{cl}_\sigma[S]$. By the weak representation theorem [8.2.10], there exists $y \in Y$ such that $f = \langle\cdot,y\rangle$. Thus $y \in (\mathrm{cl}_\sigma[S])^\perp = S^\perp$ but $y \notin (S^{\perp\perp})^\perp$ which contradicts (c).

(f) Clear.

(g) By (e) $M_i = M_i^{\perp\perp}$ for each $i \in I$, so $\cap_{i\in I} M_i = \cap_{i\in I} M_i^{\perp\perp} = (\cup_{i\in I} M_i^\perp)^\perp$ $= (\mathrm{cl}_{\sigma(Y,X)}[\cup_{i\in I} M_i^\perp])^\perp$. Hence, by (e), $(\cap_{i\in I} M_i)^\perp = (\mathrm{cl}_{\sigma(Y,X)}[\cup_{i\in I} M_i^\perp])^{\perp\perp} = \mathrm{cl}_{\sigma(Y,X)}[\cap_{i\in I} M_i^\perp]$. ■

If $X$ is a normed space and $S \subset X$ then (a) $S^\perp = [S]^\perp$ is $\sigma(X';, X)$-closed by (d) above, hence norm-closed in $X'$ by Mackey's theorem on the closure of a convex set in paired spaces [Theorem 8.8.1(b)]. Similarly, $S^{\perp\perp}$ is norm-closed in $X$.

## 8.10 ADJOINTS

**Notation.** $(X, X')$ and $(Y, Y')$ denote *dual* pairs throughout the section, $A : X \to Y$ is a linear map and $N(A) = A^{-1}(0)$. We regard $(X', X)$ and $(Y', Y)$ as paired spaces as well: if $\langle x, x' \rangle$ is the bilinear functional on $(X, X')$ we take $\langle x', x \rangle = \langle x, x' \rangle$ as the bilinear functional on $(X', X)$. ■

In solving finite systems of linear equations, it is possible to achieve a certain symmetry in phrasing through the use of adjoints; i.e., through the use of the conjugate transpose of the matrix of coefficients. Generalizations include the Fredholm alternative theorem for compact (= completely continuous) maps $A$ mapping a normed space into itself.

Occasionally some fact about the adjoint is easier to obtain than a corresponding fact about the map itself. Particular instances of this appear in Sec. 8.11 concerning continuity of a map versus continuity of its adjoint; in Sec. 8.13, we use the adjoint to obtain some relations between openness of a linear map and weak openness. In addition, the notion of adjoint facilitates the introduction of certain operations to spaces of distributions (duals of certain spaces of functions).

Let $X^*$ and $Y^*$ denote the algebraic duals of $X$ and $Y$, and let $A : X \to Y$ be a linear map. We define the *algebraic adjoint* $A^* : Y^* \to X^*$ as follows: For $y^* \in Y^*$, consider the map $A^*y^* : X \to \mathbf{F}$, $x \mapsto \langle Ax, y^* \rangle$. Evidently $A^*y^* \in X^*$. Thus, the defining equation for $A^*$ is

$$\langle Ax, y^* \rangle = \langle x, A^*y^* \rangle \qquad (x \in X,\ y^* \in Y^*)$$

**Example 8.10.1** MATRICES Let $B = \{e_1, \ldots, e_n\}$ be a basis for $\mathbf{R}^n$. The *dual basis* $\{e_1', \ldots, e_n'\}$ for $(\mathbf{R}^n)^* = \mathbf{R}^n$, is $e_j'(e_i) = \delta_{ij},\ i, j = 1, \ldots, n$. Associate with each linear map $A : \mathbf{R}^n \to \mathbf{R}^n$ the matrix $(a_{ij})$ where $Ae_j = \sum_{i=1}^n a_{ij} e_i$ for each $j$. The matrix associated with $A^*$ with respect to the dual basis is the matrix transpose $(a_{ji})$.

As noted in Sec. 8.1, the continuous duals $X'$ and $Y'$ of TVS $X$ and $Y$ are embedded in $X^*$ and $Y^*$, respectively. Moreover, by the weak representation theorem [8.2.10], $X'$ and $Y'$ are the respective duals of $(X, \sigma(X, X'))$ and $(Y, \sigma(Y, Y'))$. Although we are certainly at liberty to restrict $A^*$ to $Y'$, $A^*|_{Y'} = A'$ need not map $Y'$ into $X'$. A complete answer to the question "When does $A'$ map $Y'$ into $X'$?" is given in Theorem 8.10.3 below.

**Definition 8.10.2** WEAK CONTINUITY Let $(X, X')$ and $(Y, Y')$ denote dual pairs. In saying that linear map $A : X \to Y$ is *weakly continuous,* we mean that $A : (X, \sigma(X, X')) \to (Y, \sigma(Y, Y'))$ is continuous. The analogous meaning is attached to "a linear map $B : Y' \to X'$ is weakly continuous." This latter situation is sometimes also described "$B$ is *weak-$*$ continuous.*" ■

**Theorem 8.10.3** $A'(Y') \subset X' \Leftrightarrow A$ WEAKLY CONTINUOUS Let $(X, X')$ and $(Y, Y')$ denote dual pairs and let $A : X \to Y$ be linear. Then $A'(Y') \subset X'$ iff $A$ is weakly continuous.

**Proof.** Let $(X, X')$ and $(Y, Y')$ denote dual pairs and let $A : X \to Y$ be linear. Suppose $A'(Y') \subset X'$ and that $(x_s)$ is a net from $X$ that is $\sigma(X, X')$-convergent to 0. This means that for any $x' \in X'$, $\langle x_s, x' \rangle \to 0$. In particular, for any $y' \in Y'$, $\langle x_s, A'y' \rangle \to 0$ which is equivalent to $\langle Ax_s, y' \rangle \to 0$; this implies that $(Ax_s)$ is $\sigma(Y, Y')$-convergent to 0, i.e., that $A$ is weakly continuous.

Conversely, suppose that $A$ is weakly continuous. To show that $A'(Y') \subset X'$, suppose that the net $(x_s)$ from $X$ is $\sigma(X, X')$-convergent to 0. Since $A$ is weakly continuous, $(Ax_s)$ is $\sigma(Y, Y')$-convergent to 0. Hence, for any $y' \in Y'$, $(A'y')(x_s) = \langle Ax_s, y' \rangle \to 0$. In other words, $A'y'$ is continuous on $(X, \sigma(X, X'))$. By the weak representation theorem [8.2.10], $A'y' \in X'$. ■

Because of Theorem 8.10.3, **we always assume that the linear map is weakly continuous whenever we speak of adjoints.** It is worth noting here that a continuous linear map $A$ of the LCHS $X$ into the LCHS $Y$ is weakly continuous [Theorem 8.11.3(c)].

**Definition 8.10.4** ADJOINT Let $(X, X')$ and $(Y, Y')$ be dual pairs and let $A : X \to Y$ be a weakly continuous linear map. The restriction $A' : Y' \to X'$ of the algebraic adjoint $A^*$ to $Y'$ is called the *adjoint* (or *transpose*) of $A$. ■

Assuming that the appropriate conditions hold, it is easy to verify that

$$(AB)' = B'A' \text{ and } (A^{-1})' = (A')^{-1} \qquad (*)$$

For example, to see that $(A^{-1})' = (A')^{-1}$, consider $x' \in X'$ and $y \in Y$ and choose $y' \in Y'$ and $x \in X$ such that $A'y' = x'$ and $Ax = y$. Then

$$\begin{aligned} \langle y, (A^{-1})'x' \rangle &= \langle A^{-1}y, x' \rangle = \langle x, x' \rangle = \langle x, A'y' \rangle \\ &= \langle Ax, y' \rangle = \langle y, y' \rangle = \left\langle y, (A')^{-1} x' \right\rangle \end{aligned}$$

**Theorem 8.10.5** $A'' = A$ Let $(X, X')$ and $(Y, Y')$ be dual pairs and let the linear map $A : X \to Y$ be weakly continuous. Then $A' : (Y', \sigma(Y', Y)) \to (X', \sigma(X', X))$ is continuous—i.e., $A'$ is weak-$*$ continuous—and $A'' = (A')' = A$.

**Proof.** Let $(X, X')$ and $(Y, Y')$ be dual pairs and let the linear map $A : X \to Y$ be weakly continuous. To see that $A'$ is weak-$*$ continuous, suppose that $(y'_s)$ is $\sigma(Y', Y)$-convergent to 0. This implies that $\langle Ax, y'_s \rangle \to 0$ for each $x \in X$. Since $\langle x, A'y'_s \rangle \to 0$ for each $x \in X$, $A'y'_s \to 0$ with respect to $\sigma(X', X)$. We define $A''$ by

$$\langle x, A'y' \rangle = \langle A'y', x \rangle = \langle y', A''x \rangle \qquad (x \in X,\ y' \in Y')$$

As $\langle Ax, y' \rangle = \langle x, A'y' \rangle$, the action of $Ax$ on $y'$ is the same as that of $A''x$. Since $(Y, Y')$ is a dual pair, it follows that $A''x = Ax$ and therefore that $A'' = A$. ■

**Theorem 8.10.6** POLARS AND ADJOINTS Let $(X, X')$ and $(Y, Y')$ be dual pairs and let the linear map $A : X \to Y$ be weakly continuous. For any subsets $E \subset X$, $G \subset Y$,

(a) $A(E)^\circ = (A')^{-1}(E^\circ)$;

(b) $A(E) \subset G \Rightarrow A'(G^\circ) \subset E^\circ$;

(c) if $E$ and $G$ are weakly closed disks then $A(E) \subset G$ iff $A'(G^\circ) \subset E^\circ$;

(d) $A'^{-1}(0) = N(A') = A(X)^\perp$;

(e) INJECTIVITY OF ADJOINT $\mathrm{cl}_{\sigma(Y,X)} A(X) = Y$ iff $N(A') = 0$; in other words, $A'$ is 1-1 iff $A(X)$ is weakly dense in $Y$.

**Proof.** We use the notation of the statement. (a) Let $D = \{a \in \mathbf{F} : |a| \leq 1\}$. Then

$$\begin{aligned} y \in A(E)^\circ &\Leftrightarrow |\langle A(E), y \rangle| \subset D \\ &\Leftrightarrow A'y \in E^\circ \\ &\Leftrightarrow y \in (A')^{-1}(E^\circ) \end{aligned}$$

(b) By Theorem 8.3.6(b), $A(E) \subset G \Rightarrow A(E)^\circ \supset G^\circ$. By (a), $A(E)^\circ = (A')^{-1}(E^\circ)$ so $A'(G^\circ) \subset E^\circ$.

(c) Part (b) yields the necessity half of (c). If $E^\circ \supset A'(G^\circ)$, then $E^{\circ\circ} \subset A'(G^\circ)^\circ = (A'')^{-1}(G^{\circ\circ})$ by (a). Since $A'' = A$, and $E^{\circ\circ} = E$, and $G^{\circ\circ} = G$ by the bipolar theorem [8.3.8], the desired result, $E \subset A^{-1}(G)$, follows.

(d) Clearly, $A'y' = 0$ iff $y' \in A(X)^\perp$.

(e) By Theorem 8.9.2(b), $A(X)^\perp = (\mathrm{cl}_{\sigma(Y,X)} A(X))^\perp$. Thus, by (d), if $A(X)$ is weakly dense in $Y$, then $N(A') = 0$. Conversely, if $N(A') = 0$, then $N(A')^\perp = Y = A(X)^{\perp\perp}$ [by (d)]$= \mathrm{cl}_{\sigma(Y,X)} A(X)$ by Theorem 8.9.2(e). ■

## 8.11 ADJOINTS AND CONTINUITY

In this section $A$ denotes a linear map of an LCHS $X$ into an LCHS $Y$. We obtain various relations about and between continuity of $A$ and continuity of $A'$.

Recall from Examples 8.5.4 and 8.5.5 that the Mackey and strong topologies, $\tau(X, X')$ and $\beta(X, X')$, are the polar topologies generated by the $\sigma(X', X)$-compact disks and $\sigma(X', X)$-bounded subsets of $X'$, respectively.

**Definition 8.11.1** STRONG AND MACKEY CONTINUITY Let $(X, X')$ and $(Y, Y')$ be dual pairs. The linear map $A : X \to Y$ is *strongly continuous* if $A : (X, \beta(X, X')) \to (Y, \beta(Y, Y'))$ is continuous, *Mackey continuous* if $A : (X, \tau(X, X')) \to (Y, \tau(Y, Y'))$ is continuous. ■

In our first result, we characterize continuity of $A$ by means of its adjoint $A'$.

**Theorem 8.11.2** CONTINUITY Let $A$ be a weakly continuous linear map of the LCHS $X$ into the LCHS $Y$. Then $A : X \to Y$ is continuous iff $A'$ maps equicontinuous subsets of $Y'$ into equicontinuous subsets of $X'$.

**Proof.** Let $A$, $X$ and $Y$ be as above. Suppose that $A$ is continuous and that $E$ is an equicontinuous subset of $Y'$. To prove that $A'(E)$ is an equicontinuous subset of $X'$, we use the criterion of Theorem 8.6.4(a) and show that $A'(E) \subset V^\circ$ for some neighborhood $V$ of 0 in $X$. Since $E$ is equicontinuous, $E^\circ$ is a neighborhood of 0 in $Y$ by Theorem 8.6.6. Since $A$ is continuous, $V = A^{-1}(E^\circ)$ is a neighborhood of 0 in $X$. By the dual form of Theorem 8.10.6(a), $A'(E)^\circ = A^{-1}(E^\circ) = V$. Thus $V^\circ = A'(E)^{\circ\circ} \supset A'(E)$.

Conversely, suppose that $A'$ preserves equicontinuity. By Theorem 8.6.6, polars of equicontinuous subsets $V \subset Y'$ form a base at 0 in $Y$. By Theorem 8.10.6(a), $A^{-1}(V^\circ) = A'(V)^\circ$. Hence $A^{-1}(V^\circ)$ is a neighborhood of 0 in $X$ by Theorem 8.6.6. ■

The results of Theorem 8.11.3 are of two types: an external type in which some kind of continuity about $A$ implies some kind of continuity about $A'$ and an internal kind, some sort of continuity of $A$ implying another kind about $A$. Results of this type concerning openness of $A$ appear in Sec. 8.13. The result of Theorem 8.11.3(c), that continuity of a linear map implies weak continuity, had been noted by Banach for normed spaces in his book, the false converse by Dunford.

**Theorem 8.11.3** CONTINUITIES Let $A$ be a linear map of the LCHS $X$ into the LCHS $Y$. Then:

(a) if $A$ is weakly continuous, then $A$ is Mackey continuous;
(b) if $A$ is weakly continuous, then $A$ is strongly continuous;
(c) if $A$ is continuous, then $A$ is weakly continuous but not conversely;
(d) if $A$ is continuous, then $A'$ is strongly continuous and Mackey continuous.

**Proof.** Let $A$ be a linear map of the LCHS $X$ into the LCHS $Y$. (a) Suppose that $A$ is weakly continuous and that $V$ is a $\sigma(Y', Y)$-compact disk in $Y'$; as such, $V^\circ$ is a basic $\tau(Y, Y')$-neighborhood of 0 in $Y$. By the weak-$*$ continuity [Theorem 8.10.5] and linearity of $A'$, $A'(V)$ is a $\sigma(X', X)$-compact disk of $X'$. Hence, by Theorems 8.10.5 ($A = A''$) and 8.10.6(a), $A'(V)^\circ = (A'')^{-1}(V^\circ) = A^{-1}(V^\circ)$ is a basic $\tau(X, X')$-neighborhood of 0 in $X$.

(b) Suppose that $A$ is weakly continuous and that $V$ is a $\sigma(Y', Y)$-bounded subset of $Y'$; as such, $V^\circ$ is a basic $\beta(Y, Y')$-neighborhood of 0 in $Y$. Since $A'$ is weak-$*$ continuous [Theorem 8.10.5] and continuity implies local boundedness [Theorem 6.4.1], $A'(V)$ is a $\sigma(X', X)$-bounded subset of $X'$. By Theorem 8.10.6(a), $A'(V)^\circ = A^{-1}(V^\circ)$ and $A^{-1}(V^\circ)$ is a basic $\beta(X, X')$-neighborhood of 0 in $X$.

(c) The weak topology $\sigma(Y, Y')$ is generated by the seminorms $\{q_{y'}(\cdot) = |\langle \cdot, y' \rangle| : y' \in Y'\}$. If $A$ is continuous, then for each $y' \in Y'$, the map $p_{y'}(\cdot) =$

$|\langle A(\cdot), y'\rangle|$ is a continuous seminorm on $X$. It follows from Theorem 5.7.3 that $A$ is weakly continuous. In Example 8.8.8 we mentioned that any sequence $(x_n)$ of orthonormal vectors in an infinite-dimensional Hilbert space $X$ is weakly convergent to 0, but not norm-convergent to 0. Consequently the identity map $I : (X, \sigma(X, X')) \to (X, \|\cdot\|), x \mapsto x$, is weakly continuous but not continuous.

(d) If $A$ is continuous, then $A$ is weakly continuous by (c). Therefore, $A'$ is weak-$*$ continuous by Theorem 8.10.5. Hence, by the dual forms of (a) and (b), $A'$ is strongly and Mackey continuous. ■

Even though weak continuity does not generally imply continuity, it does for linear maps between normed spaces: Normed spaces $X$ carry the strong topology $\beta(X, X')$ [Example 8.8.9] and weak continuity implies strong continuity by Theorem 8.11.3(b). We summarize this below for future reference (see also the corresponding result for barreled spaces, Cor. 11.3.7).

**Corollary 8.11.4** WEAK CONTINUITY = CONTINUITY IN NORMED SPACES. If $X$ and $Y$ are normed spaces, the linear map $A : X \to Y$ is continuous iff it is weakly continuous.

For normed spaces we also have the following result.

**Theorem 8.11.5** $\|A\| = \|A'\|$ IN NORMED SPACES For a continuous linear map $A : X \to Y$ between normed spaces $X$ and $Y$, $\|A\| = \|A'\|$.

**Proof.** We use the notation of the statement. Continuous linear maps and functionals are normed as in Example 6.5.3. For $A \in L(X, Y)$, $y' \in Y'$ and $x \in X$,

$$|(A'y')(x)| = |y'(Ax)| \leq \|y'\| \|Ax\| \leq \|y'\| \|A\| \|x\| \qquad (\blacklozenge)$$

which implies that $\|A'y'\| \leq \|A\| \|y'\|$ and therefore that $\|A'\| \leq \|A\|$. Since $A'' = A$ [Theorem 8.10.5], $\|A\| = \|A''\| \leq \|A'\|$. ■

## 8.12 SUBSPACES AND QUOTIENTS

If $M$ is a subspace of a linear space $X$, then, after making some identifications, $X/M$ is seen to be complementary to $M$ in the sense that $X = M \oplus X/M$. This lies at the base of the duality exhibited in Theorem 8.12.1 below.

All pairings of TVS $X$ are the natural pairings with their continuous duals $X'$. If $M$ is a subspace of a TVS $X$, $X/M$ carries the quotient topology, the direct image topology induced by the canonical map $k : X \to X/M$, $x \mapsto x + M$. The results of Theorem 8.12.1 are purely algebraic.

**Theorem 8.12.1** DUALS OF QUOTIENTS AND SUBSPACES Let $M$ be a subspace of the TVS $X$. Then:

(a) QUOTIENTS $(X/M)'$ is linearly isomorphic to $M^{\perp}$.

(b) SUBSPACES If $X$ is an LCS, then $M'$ is linearly isomorphic to $X'/M^{\perp}$.

**Proof.** Let $M$ be a subspace of the TVS $X$. (a) Given $f \in M^{\perp} = \{f \in X' : f(M) = \{0\}\}$, we define $\dot{f} : X/M \to \mathbf{F}$ by taking $\dot{f}(x+M) = f(x)$. The map $\dot{f}$ is well-defined because $f \in M^{\perp}$; its linearity is clear. The open sets $U \subset X/M$ of the quotient topology are those such that $k^{-1}(U)$ is open in $X$. Since $k$ is continuous and $f = \dot{f} \circ k$, the continuity of $f$ implies the continuity of $\dot{f}$. The map

$$B : M^{\perp} \to (X/M)',\ f \mapsto \dot{f}$$

is linear and injective. It is surjective since, for any $\dot{f} \in (X/M)'$, $\dot{f} \circ k \in M^{\perp}$. It follows that that $M^{\perp}$ and $(X/M)'$ are linearly isomorphic.

(b) Consider the linear map

$$A : X' \to M',\ f \mapsto f|_M$$

where $f|_M$ denotes the restriction of $f$ to $M$. If $X$ is an LCS, then any linear functional which is continuous on $M$ may be extended to a continuous linear functional on $X$ by the Hahn–Banach theorem [Corollary 7.3.3]. That is, if $X$ is an LCS, $A$ is surjective and it follows that $M'$ is linearly isomorphic to $X'/A^{-1}(0)$. It is clear, however, that $A^{-1}(0) = M^{\perp}$. ■

**Theorem 8.12.2** WEAK TOPOLOGIES ON SUBSPACES (a) If $M$ is a subspace of an LCS $X$, then $\sigma(M, M')$ is the relative topology induced by $\sigma(X, X')$ on $M$. Also $\sigma(M, M') = \sigma(M, X'|_M)$, where $X'|M = \{x'|_M : x' \in X'\}$.

(b) If $X$ and $Y$ are paired spaces and $M$ is a subspace of $X$, then $M$ and $Y/M^{\perp}$ are paired spaces and $\sigma(M, Y/M^{\perp})$ is the relative topology induced by $\sigma(X, Y)$ on $M$. If $(X, Y)$ is a dual pair, then so is $(M, Y/_{M^{\perp}})$.

**Proof.** We use the notation of the statement. (a) A base at 0 for $\sigma(M, M')$ is given by polars $H^{\circ}$ of finite subsets $H$ of $M'$ [Example 8.5.3], polars being computed in $M$. Since $X$ is locally convex, any $f \in M'$ may be extended to an $f' \in X'$ [Corollary 7.3.3]. Hence $M' = X'|_M$ and $\sigma(M, M') = \sigma(M, X') = \sigma(M, X'|_M)$. To see that $\sigma(M, M')$ is the relative topology induced by $\sigma(X, X')$ on $M$, note that a relative $\sigma(X, X')$-basic neighborhood of 0 is, for a finite subset $H$ of $X'$,

$$\begin{aligned} M \cap \{m \in X : |\langle m, f\rangle| \le 1,\ f \in H\} &= \{m \in M : |\langle m, f\rangle| \le 1,\ f \in H\} \\ &= \{m \in M : |\langle m, f\rangle| \le 1,\ f \in H|_M\} \end{aligned}$$

which is a basic $\sigma(M, M')$-neighborhood of 0.

(b) Let $X' = (X, \sigma(X, Y))'$. Since $(X, \sigma(X, Y))$ is an LCS, $X'/M^{\perp}$ is linearly isomorphic to $M'$ by Theorem 8.12.1(b) where $M^{\perp}$ is computed in $X'$. By the weak representation theorem [8.2.10(b)], $X'$ may be identified with $Y/X^{\perp}$, $X^{\perp}$ being computed in $Y$. Computing $M^{\perp}$ in $Y/X^{\perp}$ yields $M^{\perp}/X^{\perp}$. Thus $X'/M^{\perp}$ is linearly isomorphic to $(Y/X^{\perp})/(M^{\perp}/X^{\perp})$, which is linearly isomorphic to $Y/M^{\perp}$, $M^{\perp}$ being computed in $Y$. In summary, $M'$ is linearly isomorphic to $Y/M^\perp$ and $M$ and $Y/M^\perp$ are paired spaces with respect to the bilinear functional:

$$\begin{array}{rcl} M \times Y/M^\perp & \to & \mathbf{F} \\ (m, y + M^\perp) & \mapsto & \langle m, y \rangle \end{array}$$

which is well-defined because we are identifying functions $y \in Y$ which vanish on $M$. It is a dual pairing if the original pairing is.

By (a), it follows that $\sigma(M, M') = \sigma(M, Y/M^\perp) = \sigma(M, X') = \sigma(M, Y/X^\perp)$ and that $\sigma(M, Y/X^\perp)$ is the relative topology induced by $\sigma(X, Y/X^\perp)$ on $M$. By the nature of the pairing of $X$ and $Y/X^\perp$, namely that induced by

$$\begin{array}{rcl} X \times Y/X^\perp & \to & \mathbf{F} \\ (x, y + X^\perp) & \mapsto & \langle x, y \rangle \end{array}$$

$\sigma(X, Y/X^\perp)$ is seen to coincide with $\sigma(X, Y)$. Finally, note that the proof of (b) is trivial if $X$ distinguishes points of $Y$. In this case $(X, \sigma(X, Y))' = Y$ and everything fits into the framework of (a). ∎

Given a subspace $M$ of a TVS $X$, here are potentially three topologies for $X/M$: the quotient topology $\mathcal{T}_q$ induced by the original topology on $X$, the weakened topology $\sigma(X/M, (X/M)')$ [when $X/M$ carries $\mathcal{T}_q$], and the quotient topology induced on $X/M$ by $\sigma(X, X')$. Theorem 8.12.3 shows that the latter two topologies are the same.

**Theorem 8.12.3** WEAK TOPOLOGIES ON QUOTIENTS (a) If $M$ is a subspace of a TVS $X$, then the weakened topology $\sigma(X/M, (X/M)')$ for $X/M$ is the quotient topology on $X/M$ induced by $\sigma(X, X')$.

(b) If $X$ and $Y$ are paired spaces and $M$ is a subspace of $X$, then $X/M$ and $M^\perp$ are paired spaces and $\sigma(X/M, M^\perp)$ is the quotient topology on $X/M$ induced by $\sigma(X, Y)$.

**Proof.** We use the notation of the statement. (a) A neighborhood base at 0 for $\sigma(X/M, (X/M)')$ consists of polars in $X/M$ of finite subsets $H$ of $(X/M)'$. Thus a typical weakened basic neighborhood of 0 in $X/M$ is of the form

$$\dot{U} = \left\{ \dot{x} = x + M \in X/M : \left| \left\langle \dot{x}, \dot{f} \right\rangle \right| \leq 1, \ \dot{f} \in H \right\}$$

Associated with each $f \in X'$ is an $\dot{f} \in (X/M)'$ via $f = \dot{f} \circ k$, as in the proof of Theorem 8.12.1(a): $\dot{f}(x + M) = f(x)$. Let

$$\begin{array}{rcl} U & = & k^{-1}(\dot{U}) \\ & = & \{x \in X : |\langle x, f \rangle| \leq 1, f = \dot{f} \circ k, \ \dot{f} \in H\} \\ & = & H^\circ \in \sigma(X, X') \end{array}$$

Since sets such as $U$—polars of finite subsets of $X'$—form a base of balanced neighborhoods of 0 for $\sigma(X, X')$, the family of all such $k(U)$ is a base at 0 for the quotient topology on $(X, \sigma(X, X'))/M$. (See the general discussion of quotient topologies in Example 4.7.2.) It follows that $\sigma(X/M, (X/M)')$ coincides with the quotient topology induced on $X/M$ by $\sigma(X, X')$.

(b) The proof is quite similar to that of Theorem 8.12.2(a). Let $X$ carry $\sigma(X, Y)$. Then, by Theorem 8.12.1(a), identify $(X/M)'$ with $M^{\perp}$. By (a) it follows that $\sigma(X/M, M^{\perp})$ is the quotient topology induced by $\sigma(X, X') = \sigma(X, Y/X^{\perp}) = \sigma(X, Y)$ on $X/M$. The bilinear functional pairing $X/M$ and $M^{\perp}$ is the map $(x + M, y) \mapsto \langle x, y\rangle$; it is a dual pairing if the original pairing is. ■

Equicontinuous subsets of the dual $M'$ of a closed subspace $M$ of an LCHS $X$ and of $(X/M)'$ are completely determined by the equicontinuous subsets of $X'$, as we show next.

**Theorem 8.12.4** EQUICONTINUITY IN SUBSPACES AND QUOTIENTS Let $M$ be a subspace of an LCHS $X$.

(a) SUBSPACES A subset $E$ of the dual $M'$ of $M$ is equicontinuous on $M$ iff there is an equicontinuous subset $E'$ of $X'$ such that $E'|_M = \{x'|_M : x' \in E'\} = E$. Using the identification of $M'$ and $X'/M^{\perp}$ of Theorem 8.12.1(b), we may say that $E \subset M'$ is equicontinuous iff it is the image $k(E')$ under the canonical map $k : X' \to X'/M^{\perp}$ of an equicontinuous subset $E'$ of $X'$.

(b) QUOTIENTS If $M$ is closed [so that $X/M$ is Hausdorff and $(X/M, (X/M')$ form a dual pair], then the equicontinuous subsets of $(X/M)'$ are in 1–1 correspondence with the equicontinuous subsets of of $X'$ which lie in $M^{\perp}$.

**Proof.** We use the notation of the statement.(a) Consider the injection map $I : M \to X$, $m \mapsto m$. Since $I$ is continuous, it is weakly continuous [Theorem 8.11.3(c)] so we can consider its adjoint $I' : X' \to M'$. For $m \in M$ and $m' \in M'$, $\langle Im, m'\rangle = \langle m, m'\rangle = \langle m, I'm'\rangle$; in other words $I'$ is the map $m' \mapsto m'|_M$. Since $I$ is continuous, $I'$ maps the equicontinuous subset $E'$ of $X'$ into the equicontinuous set $E'|_M \subset M'$ [Theorem 8.11.2].

Conversely, if $E$ is an equicontinuous subset of $M'$, then by Theorem 8.6.4(a), there is a closed balanced convex neighborhood $V$ of 0 in $X$ such that $E \subset (M \cap V)^{\circ}$, the polar being computed in $M'$. Let $p_V$ denote the gauge of $V$ in $X$. For any $m \in M$ and $r > 0$ $[m/(p_V(m) + r)] \in V \cap M$. Hence for any $f \in E$, $|f(m/(p_V(m) + r))| \leq 1$. Since $r$ is arbitrary, $|f(m)| \leq p_V(m)$ and $|f| \leq p_V$ on $M$. Therefore $f$ can be extended to $F \in X'$ with $|F| \leq p_V$ on $X$ by the Hahn–Banach theorem [7.3.2(b)]. Hence for $x \in V$, $|F(x)| \leq p_V(x) \leq 1$. It follows that the set $E' = \{F : f \in E\}$ of such extensions is a subset of $V^{\circ}$ and therefore that $E'$ is equicontinuous.

(b) Let $k : X \to X/M$ denote the canonical map and identify $M^{\perp}$ and $(X/M)'$ by Theorem 8.12.1(a). For $x \in X$ and $n' \in M^{\perp}$, $\langle kx, n'\rangle = n'(x) = \langle x, k'n'\rangle$ so the adjoint $k'$ is seen to be the canonical injection $n' \mapsto n'$ of $M^{\perp}$

into $X'$.

$$\begin{array}{ccc} X' & \xleftarrow{k'} & M^{\perp} \cong (X/M)' \\ | & & | \\ X & \xrightarrow{k} & X/M \end{array}$$

Since $k$ is continuous, $k'$ maps equicontinuous subsets of $M^{\perp}$ into equicontinuous subsets of $X'$ [Theorem 8.11.2] which lie in $X'$. Conversely, if $E$ is an equicontinuous subset of $X'$ lying in $M^{\perp}$, there is a neighborhood $V$ of 0 in $X$ such that $E \subset V^{\circ} \cap M^{\perp}$ [Theorem 8.6.4(a)]. Hence $(k')^{-1}(E) = E \subset (k')^{-1}(V^{\circ}) = V^{\circ} \cap M^{\perp}$. It follows that $E$ is an equicontinuous subset of $M^{\perp}$ [Theorem 8.6.4(a)]. ■

## 8.13 OPENNESS OF LINEAR MAPS

If $A$ is a linear map of the TVS $X$ into the TVS $Y$ and $N = A^{-1}(0)$, $A$ can be decomposed as in Diagram 8.13.1. By Theorem 4.7.4, $\dot{A}$ is a linear isomorphism and is continuous iff $A$ is continuous; $A$ maps open subsets of $X$ onto open subsets of $A(X)$ (i.e., $A$ is *relatively open)* iff $\dot{A}$ is relatively open.

$$\begin{array}{cccc} X & \xrightarrow{A} & A(X) & Ax \\ & {}_{k}\searrow & \uparrow \dot{A} & \uparrow \\ & & X/N & x+N \end{array}$$

**Diagram 8.13.1**

In this section we obtain some criteria concerning when $A$ is relatively open and some connections between weak openness and openness. The main tools are the results of the preceding two sections about adjoints and continuity and duals of subspaces and quotients.

**Notation.** $(X, X')$ and $(Y, Y')$ are dual pairs, $A : X \to Y$ is a weakly continuous linear map, $A'$ its adjoint and $N(A) = A^{-1}(0)$; $k$ denotes the canonical map $x \mapsto x + N(A)$ and $\dot{A}$ is as in Diagram 8.13.1. To say that $A$ is relatively open means that $A : (X, \sigma(X, X')) \to (Y, \sigma(Y, Y'))$ is relatively open. ■

As we show next, if a linear map $A$ satisfies an openness condition, its adjoint has a certain closedness property.

**Theorem 8.13.1** WEAK OPENNESS Let $(X, X')$ be $(Y, Y')$ be dual pairs. If $A : (X, \sigma(X, X')) \to (Y, \sigma(Y, Y'))$ is a continuous linear map, then the following conditions are equivalent:

(a) $A$ is relatively open;
(b) $A'(Y')$ is $\sigma(X', X)$-closed in $X'$;
(c) $A'(Y') = N(A)^{\perp}$.

**Proof.** We use the notation of the statement. (a) $\Leftrightarrow$ (b) We prove the equivalence of (a) and (b) for 1-1 maps and then reduce the general case to the 1-1 case through use of the isomorphism $\dot{A}$. Observe that, by the dual form of Theorem 8.10.6(d), $N(A) = A'(Y')^{\perp}$. As $A'(Y')^{\perp\perp} = \text{cl}_{\sigma(X',X)} A'(Y')$, $A$ is 1–1 iff $A'(Y')$ is $\sigma(X', X)$-dense in $X'$. Hence, when $A$ is 1–1, (a) $\Leftrightarrow$ (b) becomes:

$$A \text{ is relatively open iff } A' \text{ is surjective} \qquad (*)$$

Suppose $A$ is injective. If $A$ is relatively open, then $A^{-1}$ is a weakly continuous linear map of $A(X)$ onto $X$. So, for any $x' \in X'$, the map $Ax \mapsto \langle A^{-1}(Ax), x' \rangle$ is a $\sigma(Y', Y)$-continuous linear functional on $A(X)$. As such, by the Hahn–Banach extension theorem [Corollary 7.3.3] and the representation theorem [8.2.10(a)], there must be some linear functional $y' \in (Y, \sigma(Y', Y))' = Y'$ such that for each $(x, x') \in X \times X'$, $\langle Ax, y' \rangle = \langle A^{-1}(Ax), x' \rangle = \langle x, x' \rangle$. It follows that $x' = A'y'$ and $A'$ is seen to be surjective.

Conversely, suppose that $A'$ is surjective and consider the basic $\sigma(X, X')$-neighborhood of 0 in $X$ :

$$V = \left\{ x \in X : \left| \langle x, x_j' \rangle \right| \leq 1, \ j = 1, \ldots, n \right\} \qquad (x_j' \in X', \ n \in \mathbf{N})$$

Since $A'$ is onto, there exist $y_1', \ldots, y_n' \in Y'$ such that $A'y_j' = x_j'$ for each $j$, i.e.,

$$\left| \langle x, x_j' \rangle \right| \leq 1 \text{ iff } \left| \langle x, A'y_j' \rangle \right| = \left| \langle Ax, y_j' \rangle \right| \leq 1$$

With $V' = \left\{ y \in Y : \left| \langle y, y_j' \rangle \right| \leq 1, \ j = 1, \ldots, n \right\}$, it follows that $A(V) = V' \cap A(X)$ and that $A$ is relatively open. We have now shown the equivalence of (a) and (b) for injective maps $A$.

(a) $\Rightarrow$ (c) By Theorem 8.9.2(e), $\text{cl}_{\sigma(X',X)} A'(Y') = A'(Y')^{\perp\perp}$. By the dual form of Theorem 8.10.6(d), $N = N(A) = A'(Y')^{\perp}$. Hence $\text{cl}_{\sigma(X',X)} A'(Y') = A'(Y')^{\perp\perp} = N^{\perp}$. Let $(X/N)'$ denote the dual of $(X, \sigma(X, X'))/N$.

$$\begin{array}{ccc} (X/N)' & \xleftarrow{\dot{A}'} & Y' \\ | & & | \\ X/N & \xrightarrow{\dot{A}} & Y \end{array}$$

By Theorem 4.7.4(c), $A$ is relatively open iff $\dot{A}$ is relatively open. It follows from the preceding argument that if $A$ is relatively open, then $\dot{A}'$ is onto. Now consider the surjective isomorphism $B$ considered in the proof of Theorem 8.12.1(a):

$$B : N^{\perp} \to (X/N)', \ f \mapsto \dot{f}$$

where $\dot{f}(x + N) = f(x)$ for each $x \in X$. Since $A' = B^{-1} \circ \dot{A}'$, then

$$A'(Y') = B^{-1}(\dot{A}(Y')) = B^{-1}((X/N)') = N^{\perp}$$

As $N^{\perp}$ is $\sigma(X, X')$-closed in $X'$, $\text{cl}_{\sigma(X',X)} A'(Y') = A'(Y') = N^{\perp}$. Hence (a) $\Rightarrow$ (c).

(c) $\Rightarrow$ (a) Conversely, if $A'(Y') = N^{\perp}$, then $A'(Y')$ is $\sigma(X', X)$-closed in $X'$. Consequently $A'(Y') = A'(Y')^{\perp\perp} = N^{\perp}$ by Theorem 8.10.6(d). Since $A' = B^{-1} \circ \dot{A}'$, $B^{-1}(\dot{A}'(Y')) = N^{\perp}$. Therefore $\dot{A}'(Y') = B(N^{\perp}) = (X/N)'$; that is, $\dot{A}'$ is onto, whence $\dot{A}$ is relatively open by (∗) which implies that $A$ is relatively open. ■

There is the following duality between a linear map and its adjoint.

**Theorem 8.13.2** SURJECTIVE AND INJECTIVE Let $(X, X')$ and $(Y, Y')$ be dual pairs. The weakly continuous linear map $A : X \to Y$ is surjective iff $A' : (Y', \sigma(Y', Y)) \to (X', \sigma(X', X))$ is injective and relatively open. Consequently $A$ is injective iff $A'$ is surjective and $A$ is bijective iff $A'$ is bijective.

**Proof.** In the notation of the statement, by the dual form of Theorem 8.13.1(a,c), $A'$ is weakly relatively open iff $A(X) = N(A')^{\perp}$. Thus if $A'$ is injective, $A$ is surjective. Conversely, suppose $A$ is onto. Then, by Theorem 8.10.6(d), $N(A') = A(X)^{\perp} = \{0\}$ so $A'$ is injective. Taking orthogonals, $N(A')^{\perp} = Y = A(X)$. Therefore $A$ is weakly relatively open by the dual form of Theorem 8.13.1(c). ■

We have already seen that continuity of $A : X \to Y$ implies weak continuity when $X$ and $Y$ are LCHS [Theorem 8.11.3(c)]. The following result shows that openness implies weak openness.

**Theorem 8.13.3** RELATIVE OPENNESS $\Longrightarrow$ WEAK RELATIVE OPENNESS If $X$ and $Y$ are LCHS and the linear map $A : X \to Y$ is relatively open, then $A$ is weakly relatively open.

**Proof.** Let $X, Y$ and $A$ be as above and $N, k$, and $\dot{A}$ be as in Diagram 8.13.1. As shown by Theorem 8.12.3, the weakened topology $\sigma(X/N, (X/N)')$ [when $X/N$ carries its quotient topology] and the quotient topology induced on $X/N$ by $\sigma(X, X')$ are the same; consequently, $\{k(V) : V \text{ a } \sigma(X, X')\text{-basic neighborhood of } 0 \text{ in } X\}$ is a base at 0 for $\sigma(X/N, (X/N)')$. Since $A$ is relatively open, $\dot{A}^{-1}$ is continuous on $A(X)$—hence weakly continuous by Theorem 8.11.3(c). In other words, $\dot{A}$ is weakly relatively open and the weak relative openness of $A$ follows from Theorem 4.7.4(c). ■

As we show next, in the context of Theorem 8.13.3, if $A$ is continuous, then $A$ is open iff $A$ is weakly open.

**Theorem 8.13.4** RELATIVELY OPEN CONTINUOUS MAPS If $X$ and $Y$ are LCHS and the linear map $A : X \to Y$ is continuous then $A$ is relatively open iff $A$ is weakly relatively open and each equicontinuous subset of $A'(Y')$ is the image of some equicontinuous subset of $Y'$.

**Proof.** Let $X, Y$ and $A$ be as above. *(1) A bijective.* Suppose that $A$ relatively open. Since $A$ is bijective, $A$ is open. Consequently $A$ is weakly open by Theorem 8.13.3 which also implies that $A'$ is bijective. By $(*)$ before Theorem 8.10.5, $(A^{-1})' = (A')^{-1}$. Now suppose that $E \subset X'$ is equicontinuous. Since $A$ is open, $A^{-1}$ is continuous; therefore $(A^{-1})'(E)$ is an equicontinuous subset of $Y'$ [Theorem 8.11.2]. It follows that $(A')^{-1}(E) = (A^{-1})'(E)$ or that $E = A'\left[(A^{-1})'(E)\right]$.

Conversely, suppose that $A$ is weakly relatively open and each equicontinuous subset of $A'(Y')$ is the image of some equicontinuous subset of $Y'$. We show that $A^{-1}$ is continuous by showing that $(A^{-1})'$ maps equicontinuous subsets of $X'$ into equicontinuous subsets of $Y'$ and use Theorem 8.11.2. If $E$ is an equicontinuous subset of $X'$, let $U$ be an equicontinuous subset of $Y'$ such that $A'(U) = E$. Since $(A^{-1})' = (A')^{-1}$,

$$(A^{-1})'(E) = (A')^{-1}(E) = (A')^{-1}(A'(U)) = U$$

and the continuity of $A^{-1}$ follows.

*(2) General case.* Let $N = A^{-1}(0)$, let $k$ be the canonical map of $X$ onto $X/N$, and let $I$ be the injection $y \mapsto y$ of $A(X)$ into $Y$; for any $x \in X$, $\dot{A}(kx) = Ax$.

$$\begin{array}{ccccccc} X' & \overset{k'}{\longleftarrow} & N^{\perp} \cong (X/N)' & \overset{\dot{A}'}{\longleftarrow} & Y'/A(X)^{\perp} \cong A(X)' & \overset{I'}{\longleftarrow} & Y' \\ | & & | & & | & & | \\ X & \overset{k}{\longrightarrow} & X/N & \overset{\dot{A}}{\longrightarrow} & A(X) & \overset{I}{\longrightarrow} & Y \end{array}$$

In the diagram above we have identified $(X/N)'$ with $N^{\perp}$ and $A(X)'$ with $Y'/A(X)^{\perp}$ by Theorem 8.12.1. As in the proof of Theorem 8.12.4(a,b), $I'$ is the canonical map of $Y'$ onto $Y'/A(X)^{\perp}$ and $k'$ is the injection $n' \mapsto n'$ of $N^{\perp}$ into $X'$. By Theorem 4.7.4(c), $A$ is (weakly) relatively open iff $\dot{A}$ is (weakly) open.

Suppose that $A$ is relatively open. Then $\dot{A}$ is open, hence weakly open by Theorem 8.13.3, and therefore $A$ is weakly relatively open. If $E$ is an equicontinuous subset of $N^{\perp}$, then there exists $E' \subset Y'/A(X)^{\perp}$, equicontinuous on $A(X)$, such that $\dot{A}'(E') = E$ by (1) above. $E'$, however, is the image under $k'$ of an equicontinuous subset $U'$ of $Y'$ by Theorem 8.12.4(b). Thus

$$E = \dot{A}'\left(k'(U')\right) = \dot{A}'\left(U'\right)) = A'(U')$$

and the condition on equicontinuous sets is seen to be satisfied.

Conversely, suppose that $A$ is weakly relatively open and each equicontinuous subset of $A'(Y')$ is the image of some equicontinuous subset of $Y'$. To show that $A$ is relatively open, we show that $\dot{A}$ is open. To do this it suffices by (1) to show that $\dot{A}$ is weakly open and that equicontinuous subsets of $N^{\perp} = \dot{A}'(Y'/A(X)^{\perp})$ are images of equicontinuous subsets of $Y'/A(X)^{\perp}$. The weak openness of $\dot{A}$ follows from the weak relative openness of $\dot{A}$ while the equicontinuity condition on $A$ yields the equicontinuity condition on $\dot{A}$. It follows that $\dot{A}$ is open. ■

In Theorem 8.11.2 we proved that a weakly continuous linear map $A$ is continuous iff its adjoint $A'$ maps equicontinuous sets into equicontinuous sets. The following consequence of Theorem 8.13.4 gives a similar characterization of linear homeomorphisms.

**Theorem 8.13.5** LINEAR HOMEOMORPHISMS Let $X$ and $Y$ be LCHS and let $A : X \to Y$ be an injective continuous linear map. $A$ is a linear homeomorphism of $X$ into $Y$ iff each equicontinuous subset of $X'$ is the image of an equicontinuous subset of $Y'$ under $A'$.

**Proof.** Let $X, Y$ and $A$ be as above. If $A$ is a linear homeomorphism, it is relatively open, hence weakly relatively open by Theorem 8.13.3. By the dual form of Theorem 8.13.2, $A'$ is onto. By Theorem 8.13.4, the equicontinuous subsets of $X'$ must be the $A'$-images of equicontinuous subsets of $Y'$.

Conversely, suppose that each equicontinuous subset of $X'$ is the image of an equicontinuous subset of $Y'$ under $A'$. Since one-point sets are equicontinuous, it follows that $A'$ is onto. By Theorem 8.13.2, it follows that $A$ is 1-1 and weakly relatively open. Since the conditions of Theorem 8.13.4 are satisfied, $A$ is relatively open. ■

## 8.14 LOCAL CONVEXITY AND HBEP

By saying that a topological vector space $X$ has the *Hahn–Banach extension property* (HBEP) we mean that any continuous linear functional $f : M \to \mathbf{F}$ on any linear subspace $M$ possesses a continuous extension to $X$. Every locally convex Hausdorff space has HBEP. What about the converse?

In the absence of local convexity, a topological vector space $X$ need not have any nontrivial continuous linear functionals. By Theorem 7.7.8, a TVS $X$ has a nontrivial dual iff there is a proper convex neighborhood of 0. Hence the space $\left(L_p(-\infty,\infty), \|\cdot\|^*\right)$ of Example 4.6.2 has a trivial dual.

A. Shields had observed that given a dual pair $(X, X')$, any topology between the weak $(\sigma(X, X'))$ and the Mackey topologies $(\tau(X, X'))$ has HBEP; he asked if such topologies had to be locally convex. Gregory and Shapiro 1970 showed that if $\sigma(X, X') \neq \tau(X, X')$, there are non-locally convex topologies in between, thereby providing a plethora of non-locally convex topologies with HBEP.

Topological vector spaces can have rich duals and still not have HBEP. For $0 < p < 1$, the non-locally convex spaces $\ell_p$ and the Hardy spaces $H_p$ do not have HBEP but have an abundance of continuous linear functionals such as the evaluation functionals at $n \in \mathbf{N}$ for $\ell_p$ or points $t$ in the open unit disk for $H_p$. Indeed, $\ell_p' = \ell_\infty$ for any $0 < p < 1$ [Kalton et al. 1984].

Kakol 1992 gives an elementary construction for a profusion of vector topologies $\mathcal{T}$ on a fixed infinite-dimensional vector space $X$ such that $(X, \mathcal{T})$ does not have the HBEP even though $X'$ is rich enough to separate the points of $X$.

Let us say that a subspace $M$ of a TVS $X$ has the *separation property* if any $x \notin M$ can be separated from $M$ by a continuous linear functional $f$ in the sense that $f(M) = \{0\}$ and $f(x) = 1$. If it is possible to extend any $f \in M'$ to an element of $X'$, we say that $M$ has the *extension property.* For individual subspaces there is no connection between the separation and extension properties. Duren et al. 1969 showed that there are closed subspaces $M$ of $H^p$, $0 < p < 1$, with the separation property which do not have the extension property and vice-versa. Nevertheless (*ibid.*), for an arbitrary TVS $X$, *every* subspace has the separation property if and only if every subspace has the extension property.

Shapiro 1970 showed that a complete metrizable TVS $X$ *with a Schauder basis* has the HBEP if and only if it is locally convex. Kalton removed the "with a basis" hypothesis. Using the fact that a complete metrizable TVS has HBEP if and only if every closed subspace is weakly closed and developing some basic sequence techniques for complete metrizable TVS, Kalton [1974; Kalton et al. 1984, p. 71] showed that a complete metrizable TVS with HBEP must be locally convex. This is false without metrizability, however— Any vector space $X$ of uncountable dimension with the finest vector topology $\mathcal{T}$ has the Hahn–Banach extension property [Shuchat 1972], but is not:

(1) metrizable [Example 5.6.4] or

(2) locally convex [Exercise 4.117(c)].

## 8.15 EXERCISES

8.101 For paired spaces $(X, Y)$, when is $\sigma(X, Y)$ Hausdorff?

8.102 For any TVS $X$, show that $(X', \sigma(X', X))' = X$.

8.103 How does the Alaoglu theorem [8.4.1] depend on Tihonov's theorem on compactness of products of compact spaces?

8.104 Real Polars Let $X$ and $Y$ be paired spaces. We defined the polar $E^\circ$ of a subset $E$ of $X$ to be those $y$ in $Y$ such that $\sup |\langle E, y\rangle| \le 1$. Many authors call this the *absolute* polar of $E$ and use the following notion instead: The *real* polar $E^r = \{y \in Y : \operatorname{Re}\langle x, y\rangle \le 1\}$.

(a) Generally, $E^\circ \subset E^r$ but if $E$ is balanced, $E^r = E^\circ$.

(b) $E^\circ = (E_b)^r$ where $E_b$ denotes the balanced hull of $E$.

(c) $E^r$ is $\sigma(Y, X)$-closed and convex.

(d) If $(X, Y)$ is a dual pair and $E \subset X$, then $E^{rr} = \operatorname{cl}_{\sigma(X,Y)} (E \cup \{0\})_c$, where $(E \cup \{0\})_c$ denotes the convex hull of $E \cup \{0\}$.

8.105 For any normed space $X$, show that $\sigma(X', X)$-compactness is equivalent to being $\sigma(X', X)$-closed and norm-bounded.

8.106 If $X$ is an infinite-dimensional normed space then the closed unit ball of $X'$ cannot contain any $\sigma(X', X)$-neighborhood of 0.

8.107 METRIZABILITY OF $\beta(X', X)$ Show that, for any LCHS $X$, the strong topology $\beta(X', X)$ is metrizable iff $X$ has a denumerable base of bounded sets [Definition 6.3.4]. Note that this means that although the strong dual of a normed space is a normed space [Example 8.8.9], the strong dual of a metrizable space need not be metrizable since they generally have unbounded neighborhoods of 0. Indeed, if a Fréchet space has a bounded neighborhood of 0, it is normable. Strong duals of Fréchet spaces, *(DF)-spaces*, first investigated by Grothendieck [1954a], are of considerable interest in their own right. Bierstedt and Bonet [2003] have an excellent survey of them.

8.108 TOTAL SETS For any dual pair $(X, Y)$, we say that $E \subset Y$ is *total* with respect to $X$ if $E^{\perp} = \{0\}$.

(a) Show that $E$ is total iff the linear span of $E$ is dense in $(Y, \sigma(Y, X))$.

(b) [Converse to Theorem 8.2.9] The LCS $X$ is $\sigma(X, X')$-complete iff $X = (X')^*$ where $(X')^*$ denotes the algebraic dual of $X'$.

(c) For any vector space $X$, each $\sigma(X, X^*)$-bounded subset of $X$ is finite-dimensional.

(d) For any vector space $X$, show that any linear subspace of $X$ is $\sigma(X, X^*)$-closed.

8.109 For a normed space $X$, let $\mathcal{T}$ denote the norm topology on $X'$ [Example 6.5.3(b)] and let $X''$ denote the continuous dual of $(X', \mathcal{T})$. Show that $\sigma(X', X) \subset \sigma(X', X'') \subset \mathcal{T}$.

8.110 DUAL OF THE COMPLETION Let $(X, \mathcal{T})$ be a Hausdorff TVS and let $(\widehat{X}, \widehat{\mathcal{T}})$ be a completion. Show that:

(a) $X' = (\widehat{X})'$.

(b) If $X$ is an LCHS then the same topology is induced on any equicontinuous subset of $X' = (\widehat{X})'$ by $\sigma(X', X)$ as by $\sigma(X', \widehat{X})$.

(c) If the LCHS $X$ carries $\tau(X, X')$—as any metrizable LCHS does, for example—then $\widehat{\mathcal{T}} = \tau(\widehat{X}, X')$.

8.111 BOUNDED VERSUS STRONGLY BOUNDED Let $\varphi$ denote the linear space of all "finite" sequences $(a_n)$ of scalars such that $a_n = 0$ eventually, and consider $(\varphi, \varphi)$ as a dual pair with the bilinear form $\langle (a_n), (b_n) \rangle = \sum a_n b_n$.

(a) Show that a subset $E$ of $\varphi$ is $\sigma(\varphi, \varphi)$-bounded iff there exists a sequence $(m_n)$ of positive numbers such that for all $(a_n) \in E$, $|a_n| \leq m_n$ for every $n$.

(b) A subset $E$ of $\varphi$ is $\beta(\varphi, \varphi)$-bounded iff $E$ satisfies the condition of (a) and $m_n = 0$ eventually. Thus, show that there are $\sigma(\varphi, \varphi)$-bounded sets that are not $\beta(\varphi, \varphi)$-bounded, weakly bounded sets that are not strongly bounded, in other words.

(c) Show that $\sigma(\varphi, \varphi) = \tau(\varphi, \varphi)$.

8.112 View the space $\varphi$ of all finite sequences as a subspace of $\ell_2$. Identify the continuous dual $\varphi'$ of $\varphi$.

8.113 DUAL OF A DIRECT SUM For TVS $X$ and $Y$ over $\mathbf{F}$, show that the continuous dual of $X \oplus Y$, i.e., $X \times Y$ with the product topology as in Example 4.11.4, is $X' \oplus Y'$. (For infinitely many summands, see Exercises 12.201 and 12.202.)

8.201 SEPARABLE SPACES Let $X$ be a locally convex space. Roughly, there is a certain duality between separability and metrizability.

(a) Let $E$ be a subset of $X$ whose linear span $[E]$ is dense in $X$. Show that $\sigma(X', E)$ induces the $\sigma(X', X)$ topology on any $\sigma(X', X)$-compact subset $K$ of $X'$.

(b) Show that the conclusion of (a) holds if $K$ is equicontinuous.

(c) METRIZABILITY OF EQUICONTINUOUS SETS If $X$ is separable (i.e., has a countable dense subset) and $K$ is an equicontinuous subset of $X'$, then $K$ is metrizable in the relative $\sigma(X', X)$ topology.

(d) SEPARABILITY OF $X'$ If $X$ is a separable metrizable space, then $(X', \sigma(X', X))$ is separable.

(e) SEPARABLE NORMED SPACES A normed space $X$ is separable iff the closed unit ball $U(X')$ of $X'$ is metrizable in its relative $\sigma(X', X)$ topology.

(f) SEPARABLE DUAL If the normed space $X$ has a separable dual $X'$ (in its norm topology), then $X$ is separable.

8.202 DUALITY IN NON-ARCHIMEDEAN SPACES [van Tiel 1965] Let $X$ and $Y$ be locally $K$-convex spaces [Exercise 4.203] over a nontrivially ultra-valued field $(K, |\cdot|)$. So that a Hahn–Banach theorem [Exercise 10.201] holds, we assume that $K$ is spherically complete [Exercise 3.205]. $X'$ and $Y'$ denote the dual spaces of continuous linear functionals (i.e., linear maps into $K$) on $X$ and $Y$, respectively.

(a) POLARS For $E \subset X$, define $E^\circ = \{f \in X' : \sup |f(E)| \leq 1\}$. Some properties that these polars have include:

(1) If $E$ is bounded [Exercise 6.119], then $E^\circ$ is absorbent in $X'$; in fact, $E^\circ$ is a $\sigma(X', X)$-barrel.

(2) ["Alaoglu"; cf. Theorem 8.4.1] If $U$ is a neighborhood of 0 in $X$, then $U^{\circ}$ is $\sigma(X', X)$-$c$-compact, where $c$-compact is as defined in Exercise 4.204.

(3) If $K$ is locally compact and $U$ is a neighborhood of 0 in $X$, then $U^{\circ}$ is $\sigma(X', X)$-compact.

(b) BIPOLARS We compute the bipolar $E^{\circ\circ}$ of $E \subset X$ as $E^{\circ\circ} = \{y \in X : \sup |\langle y, E^{\circ}\rangle| \leq 1\}$. The properties of bipolars include:

(1) If the valuation on $K$ is discrete [Exercise 10.201(e)], then the closure of the absolute-$K$-convex hull $\operatorname{cl} E_{bc} = E^{\circ\circ}$.

(2) If the valuation on $K$ is dense [Exercise 10.201(e)] and $E$ is a closed $K$-disk in $X$, then for all $|a| > 1$, $E^{\circ\circ} \subset aE$.

(c) TOPOLOGIES OF THE DUAL PAIR Dual pairs and polar topologies are defined as in the real or complex case.

(1) The polar topology generated by $\sigma(X', X)$-bounded, $c$-compact [Exercise 4.204] $K$-disks of $X'$ is a compatible topology.

(2) If $K$ is locally compact, then a compatible topology is a polar topology generated by $\sigma(X', X)$-compact $K$-disks of $X'$. By $\tau(X, X')$ and $\tau_c(X, X')$ we mean the polar topologies generated by the polars of $\sigma(X', X)$-compact $K$-disks and $\sigma(X', X)$-$c$-compact and $\sigma(X', X)$-bounded (see Exercise 6.120) $K$-disks of $X'$, respectively.

(3) $\tau_c(X, X')$ is the finest topology of the dual pair $(X, X')$.

(4) If $K$ is locally compact, then $\tau(X, X')$ is the finest topology of the dual pair $(X, X')$.

(5) The bounded sets of $X$ are the same with respect to any topology of the dual pair $(X, X')$.

(6) The $K$-convex $c$-compact sets are the same in all topologies of the dual pair.

## Hints to Exercises

8.105 Shrink the set into the closed unit ball $U(X')$ and use Theorem 8.4.1.

8.106 If it did, $(X', \|\cdot\|)$ would be locally compact.

8.110 (a) A continuous linear functional on $X$ is uniformly continuous [Theorem 3.5.1]. Therefore, it can be extended to $\widehat{X}$ [Theorem 3.6.2].

(b) Let $U$ be a closed disked neighborhood of 0 in $X$. Show that its closure $\operatorname{cl}_{\widehat{X}} U$ in $\widehat{X}$ has the same polar as $U$, each being computed in $X'$. By Theorem 8.6.4, $X$ and $\widehat{X}$ determine the same class of equicontinuous sets in $X'$. The polar $U^{\circ}$ of any neighborhood $U$ of 0 in $X$ is $\sigma(X', X)$-compact, consequently also $\sigma(X', \widehat{X})$-compact. Since $\sigma(X', \widehat{X})$ is finer than $\sigma(X', X)$, the identity map of the compact space $(U^\circ, \sigma(X', \widehat{X}) \cap U^\circ)$ into the Hausdorff space $(U^\circ, \sigma(X', X) \cap U^\circ)$ is a homeomorphism.

(c) Apply the result of (b) to the class of $\sigma(X', X)$-compact disks and $\sigma\left(X', \widehat{X}\right)$-compact disks.

8.113 For $f \in X'$ and $g \in Y'$, define $(f \oplus g)(x \oplus y) = f(x) + g(y)$.

8.201 (a) Note that a continuous injection of a compact space into a Hausdorff space is a homeomorphism as it maps closed sets into closed sets. Consider the identity map of $(K, \sigma(X', X))$ into $(K, \sigma(X', E))$.

(b) If $K$ is equicontinuous, there is some neighborhood $V$ of 0 in $X$ such that $K \subset V^\circ$ by Theorem 8.6.4. $V^\circ$ is $\sigma(X', X)$-compact by Theorem 8.4.1.

(c) If $E$ is a countable dense subset of $X$ then $\sigma(X', X)$ is metrizable by Theorem 5.6.1. Now use (b).

(d) If $(U_n)$ is a countable base of neighborhoods of 0 in $X$ then $X' = \cup_{n \in \mathbf{N}} U_n^\circ$ for if $f \in X'$ then, for some $n \in \mathbf{N}$, $U_n \subset \{f\}^\circ$. Each $U_n^\circ$ is $\sigma(X', X)$-compact [Theorem 8.4.1], therefore metrizable by (c). Any compact metric space $T$ is separable since:

(1) for each $n \in \mathbf{N}$, the open balls $\{B(t, 1/n) : t \in T\}$ have a finite subcover, and

(2) the set of centers of the finite subcovers forms a countable dense set.

(e) If $X$ is separable, use (c). If $U(X')$ is metrizable, there exists a countable base $(E_n^\circ)$ of $\sigma(X', X)$-neighborhoods of 0 in $U(X')$ where each $E_n$ is a finite subset of $X$. Now use Theorem 7.7.7(c).

(f) $X$ is a subspace of the separable space $(X'', \sigma(X'', X'))$.

# Chapter 9

# Krein–Milman and Banach–Stone



If the compact Hausdorff spaces $S$ and $T$ are homeomorphic, the Banach spaces $C(T, \mathbf{F}, \|\cdot\|_\infty)$ and $C(S, \mathbf{F}, \|\cdot\|_\infty)$ are linearly isometric. This is easy. The deeper result is the converse, the Banach–Stone theorem, that linear isometry implies homeomorphism [Theorem 9.6.2]. We prove it using a technique invented by Arens and Kelley. To do that we need a generalization of the notion of vertex of a polygon, something called an *extreme point.* After developing the elementary properties of extreme points, we prove the Krein–Milman theorem which generalizes the notion that if you connect the vertices of a square and fill in the resulting figure, you recover the square. We discuss several variants of the Banach–Stone theorem in Secs. 9.7–9.9.

## 9.1 MIDPOINTS AND SEGMENTS

We recall our conventions about lines and line segments in vector spaces [Definition 4.2.1].

**Definition 9.1.1** LINE SEGMENTS Let $x$ and $y$ be two points in a vector space $X$ over $\mathbf{F}$. Then:

(a) The *closed line segment* (or *interval*) $[x, y]$ with *endpoints* $x$ and $y$ or *line joining* $x$ and $y$ is the set $\{tx + (1-t)\,y : t \in [0,1]\}$ of convex combinations of $x$ and $y$. It is a *proper* line segment if $x \neq y$. If $x = y$, then $[x, y] = \{x\}$.

(b) The *open line segment (interval)* with endpoints $x$ and $y$ is $(x, y) = \{tx + (1-t)y : t \in (0,1)\}$. If $x = y$, $(x, y) = \emptyset$. If $x \neq y$ and $z \in (x, y)$, then we say that $z$ is *between* $x$ and $y$. The point $(1/2)\,(x+y)$ is called the *midpoint* of $[x, y]$ and $(x, y)$ and is between $x$ and $y$ for $x \neq y$.

(c) The *line through $x$ and $y$* is $L\,(x, y) = \{tx + (1-t)\,y : t \in \mathbf{R}\} = y + \mathbf{R}\,(x - y)$. ■

We prove the Banach–Stone theorem in Sec. 9.6 but not the way Banach or Stone proved their versions. Theorem 9.1.2 below, due to Mazur and Ulam [Banach 1932, p. 166], was the key to Banach's proof. Banach called isometries with a fixed point—such as the $A$ in Theorem 9.1.2—*rotations.*

**Theorem 9.1.2** ISOMETRIES AND MIDPOINTS Let $X$ and $Y$ be normed spaces over $\mathbf{F}$ and let $A : X \to Y$ be a surjective isometry such that $A0 = 0$. Then $A$ maps midpoints into midpoints and is $\mathbf{R}$-linear.

**Proof.** With notation as above, for any $x, y \in X$, it is routine to verify that $m = (1/2)\,(x+y)$ is equidistant from $x$ and $y$, namely that $\|m - x\| = \|m - y\| = (1/2)\,\|x - y\|$. There can be other points $w$ equidistant from $x$ and $y$, however. [In $\ell_\infty^{\mathbf{R}}(2)$, for example, with $x = (-1, 0)$ and $y = (1, 0)$, each point $w \in \{0\} \times [-1, 1]$ is equidistant from $x$ and $y$.] We denote the set of such equidistant points by $E_1(x, y) = \{w \in X : \|w - x\| = \|w - y\| = (1/2)\,\|x - y\|\}$ and let $d\,(E_1\,(x, y))$ denote the diameter of $E_1(x, y)$. Now compress $E_1(x, y)$: Consider $E_2(x, y) = \{w \in E_1 : \|w - z\| \leq (1/2)\,d\,(E_1(x, y))$ for each $z \in E_1(x, y)\}$. [In the example just considered, $E_2(x, y) = \{0\} \times [-1/2, 1/2]$.] For $n \geq 2$, let

$$E_n(x,y) = \{w \in E_{n-1}(x,y) : \|w - z\| \leq (1/2)\, d\,(E_{n-1}(x,y)) \ \forall z \in E_{n-1}(x,y)\}$$

If $w, z \in E_2(x,y)$, $\|w - z\| \leq (1/2)\, d\,(E_1(x,y))$. Thus $d\,(E_2(x,y)) \leq (1/2)\, d\,(E_1(x,y))$. Indeed, for each $n \geq 2$,

$$d\,(E_n(x,y)) \leq 1/2^{n-1} d\,(E_1(x,y))$$

Since $E_1\,(x,y) \supset E_2\,(x,y) \supset \cdots$ and $d\,(E_n\,(x,y)) \to 0$, if $\cap_{n\in\mathbf{N}} E_n\,(x,y)$ is not empty, it is a singleton. We now show that $\cap_{n\in\mathbf{N}} E_n(x,y) = \{m\} = \{(1/2)\,(x+y)\}$. First, we need the following result:

For each $n \in \mathbf{N}$, $z \in E_n\,(x,y)$ implies that $\overline{z} = x + y - z \in E_n(x,y)$. We argue by induction. For $n = 1$ and $z \in E_1(x,y)$,

$$\|\overline{z} - x\| = \|y - z\| \text{ and } \|\overline{z} - y\| = \|x - z\|$$

Therefore $\|\overline{z} - x\| = \|\overline{z} - y\| = \|y - z\| = \|x - z\| = \|x - y\|/2$ and $\overline{z} \in E_1(x,y)$. Suppose that $n \geq 2$ and $u \in E_{n-1}(x,y) \Rightarrow \bar{u} = x + y - u \in E_{n-1}(x,y)$. Hence, for $w \in E_{n-1}(x,y)$, $x+y-w \in E_{n-1}$ and for $z \in E_n(x,y)$,

$$\|\overline{z} - w\| = \|x + y - z - w\| = \|x + y - w - z\| \leq (1/2)\, d(E_{n-1}(x,y))$$

Therefore $\overline{z} \in E_n(x,y)$.

We now show that $m \in E_n(x,y)$ for all $n$. Clearly $m \in E_1(x,y)$. Now suppose that $n \geq 2$ and $z \in E_{n-1}(x,y)$. Then

$$2\,\|m - z\| = \|x + y - 2z\| = \|x + y - z - z\| = \|\overline{z} - z\| \leq d(E_{n-1}(x,y))$$

Thus $\|m - z\| \leq (1/2)\, d(E_{n-1}(x,y))$ for all $z \in E_{n-1}(x,y)$; therefore $m = (1/2)\,(x+y) \in E_n(x,y)$. Since $d\,(E_n\,(x,y)) \to 0$, $\{m\} = \cap_{n\in\mathbf{N}} E_n\,(x,y)$. By the same argument, $\cap_{n\in\mathbf{N}} E_n\,(Ax, Ay) = \{(1/2)\,(Ax + Ay)\}$.

We now show that $A\,(E_n(x,y)) = E_n(Ax, Ay)$ for all $n$. Note that $w \in E_1\,(x,y)$ iff $\|w - x\| = \|w - y\| = (1/2)\,\|x - y\|$; since $A$ is an isometry, this is is equivalent to

$$\|Aw - Ax\| = \|Aw - Ay\| = (1/2)\,\|Ax - Ay\|$$

i.e., $Aw \in E_1\,(Ax, Ay)$ and therefore $A\,(E_1(x,y)) = E_1(Ax, Ay)$. Now suppose $n \geq 2$. Since $A$ is a an isometry, $d\,(E_{n-1}\,(Ax, Ay)) = d\,(E_{n-1}\,(x,y))$; hence, since $A$ is surjective,

$$\begin{aligned}
v \in A\,(E_n\,(x,y)) &\Leftrightarrow \exists w \in E_n\,(x,y) \text{ such that } v = Aw \in A\,(E_n\,(x,y)) \\
&\Leftrightarrow \forall z \in E_n\,(x,y) \ \ \|Aw - Az\| \leq (1/2)\, d\,(E_{n-1}\,(Ax, Ay)) \\
&\Leftrightarrow \forall z \in E_n\,(x,y) \ \ \|w - z\| \leq (1/2)\, d\,(E_{n-1}\,(x,y)) \\
&\Leftrightarrow w \in E_n\,(x,y) \Leftrightarrow A\,(E_n(x,y)) = E_n(Ax, Ay)
\end{aligned}$$

Hence

$$A(\cap_{n\in\mathbf{N}}E_n(x,y)) = A(\{m\}) = \cap_{n\in\mathbf{N}}E_n(Ax,Ay) = \{(1/2)(Ax+Ay)\}$$

Thus $Am = (1/2)(Ax+Ay)$. This yields the additivity of $A$ by the following argument:

$$Ax = A\left(\frac{2x+0}{2}\right) = \frac{A(2x)+A0}{2} = \frac{A(2x)}{2} \Rightarrow A(2x) = 2Ax$$

Therefore

$$A(x+y) = A\left(\frac{2x+2y}{2}\right) = \frac{A(2x)+A(2y)}{2} = Ax + Ay$$

It follows easily now that $A(qx) = qAx$ for all $q \in \mathbf{Q}$ and $\mathbf{R}$-linearity follows by continuity. ■

**Theorem 9.1.3** LINEAR FUNCTIONALS MONOTONE ON LINES Any real linear functional is monotone (hence 1-1) or constant on lines and line segments.

**Proof.** Let $f$ be a real linear functional on a vector space and let $x$ and $y$ be distinct vectors. Since $x \neq y$, the map $t \to y + t(x-y)$ establishes a 1-1 correspondence between $\mathbf{R}$ and the line $L(x,y)$ through $x$ and $y$ and we order $L(x,y)$ accordingly: $y + s(x-y) \leq y + t(x-y)$ iff $s \leq t$. Now the restriction $f|_{L(x,y)}$ sends $y+t(x-y)$ into $f(y)+tf(x-y)$. Identifying $\mathbf{R}$ and $L(x,y)$, the derivative of $f$ with respect to $t$ is $f(x-y)$, so $f$ is increasing if $f(x-y) > 0$, constant if $f(x) = f(y)$, and decreasing if $f(x-y) < 0$. ■

**Definition 9.1.4** AFFINE MAP A map $A$ of a vector space $X$ into a vector space $Y$ is said to be *affine* if $A(tx + 1(1-t)y) = tAx + (1-t)Ay$ for all $x, y \in X$ and $t \in [0,1]$, in other words, if $A([x,y]) = [Ax, Ay]$ for all $x, y \in X$. ■

A translate $x \mapsto Bx+v$ of an $\mathbf{R}$-linear map $B: X \to Y$, $v \in Y$, is affine and many authors take this to be the definition of "affine." As this latter notion is equivalent to $A(tx+(1-t)y) = tAx+(1-t)Ay$ for all $t \in \mathbf{R}$ rather than just $t \in [0,1]$, it is a stronger notion. The same proof shows that Theorem 9.1.3 can be strengthened to state that affine functionals are monotone on line segments.

## 9.2 EXTREME POINTS

The corners of a square "characterize" it in a sense: If you connect the dots and fill in the resulting figure, the square is recovered. In other words, the convex hull of the vertices of a square is the square. The same thing happens for any convex polygon: The convex hull of the vertices yields the original figure. This raises some questions:

1. What properties of the vertices permit this? What makes a vertex a vertex?

2. Assuming that some generalization of vertex has been chosen, for what sets $K$ is $K$ the convex hull of its "vertices?"

In answer to Question 1 we take as the intrinsic verticial property that of not being in the interior of any line segment in the set. The term that we ultimately use is "extreme point" [Definition 9.2.1]. Minkowski proved that a compact convex subset of $\mathbf{R}^n$ can always be reconstituted as the convex hull of its extreme points. The Krein–Milman theorem [Theorem 9.4.6] is an infinite-dimensional version of this result: It is the *closure* of the convex hull of the extreme points that reconstitutes the set in this case. Thus, the points of the original set are *approximated* by the points of the convex hull of the extreme points. Extreme points were first considered by Minkowski in $\mathbf{R}^3$ around 1909 [see his 1911 collected works]. Carathéodory [1911, p. 200] and Steinitz [1913/16, Part III, p. 16] generalized it to $\mathbf{R}^n$. Roy [1987] has a nice survey of the development of extreme point.

**Definition 9.2.1** EXTREME POINTS A point $x$ of a convex set $K$ is an *extreme point* of $K$ if $x$ is not between any two distinct points of $K$. The set of extreme points of $K$ is denoted $\operatorname{ext} K$. ■

We omit the easy proof of the following result.

**Theorem 9.2.2** CHARACTERIZING EXTREME POINTS Let $K$ be convex and $x \in K$. Then $x \in \operatorname{ext} K$ iff any of the following conditions hold.
(a) $K \setminus \{x\}$ is convex;
(b) $x$ is not the midpoint of any proper line segment of $K$;
(c) if $y, z \in K$ are such that $x \in [y, z]$ then $x = y = z$;
(d) if $x + y, x - y \in K$, then $y = 0$.

Clearly, $a$ and $b$ are extreme points of $[a, b]$ in $\mathbf{R}$ and the vertices of a square are extreme points of the square; the open interval $(a, b)$ has no extreme points.

If $x \in \operatorname{ext} K$ ($K$ convex), then $x$ can only be written as a trivial convex combination of any two points in $K$. A similar statement holds for convex combinations of more than two points in $K$ : If $x = t_1 x_1 + \cdots + t_n x_n$ for $x_i \in K$, $t_i \in [0, 1]$, $1 \leq i \leq n$, and $\sum_i t_i = 1$, then all the $t$'s must be 0 but one, for suppose $t_1, t_2 \in (0, 1)$. Then $\sum_{i=2}^n [t_i / (1 - t_1)] = 1$ and $w = \left( \sum_{i=2}^n \frac{t_i}{1-t_1} x_i \right) \in K$. Hence

$$\begin{aligned} x &= t_1 x_1 + (1 - t_1) \left( \sum_{i=2}^n \frac{t_i}{1-t_1} x_i \right) \\ &= t_1 x_1 + (1 - t_1) w \end{aligned}$$

which implies that $x \notin \operatorname{ext} K$.

The linear image of an extreme point of a convex set need not be an extreme point of the image. Consider, for example, an equilateral triangle whose base is on the $x$-axis. The projection onto the $x$-axis maps the apex of the triangle into an interior point of the image. As shown in Theorem 9.2.3, however, 1-1 linear maps—hence linear isometries—carry extreme points into extreme points. An important application of this fact propels a relatively simple proof of the Banach–Stone theorem [Theorem 9.6.2].

**Theorem 9.2.3** Injective Affine Maps and Extreme Points Let $X$ and $Y$ be linear spaces, $K$ a convex subset of $X$, and $A$ an affine map of $X$ into $Y$. If $A$ is 1-1, then $\operatorname{ext} A(K) = A(\operatorname{ext} K)$. Hence injective linear maps take extreme points into extreme points.

**Proof.** With notation as above, if $A$ is 1-1, then $Ax \in (Ay, Az)$ implies that $x \in (y, z)$. Thus if $x \in \operatorname{ext} K$, then $Ax$ must be an extreme point of $A(K)$. ■

We give some examples about extreme points in Example 9.2.5 but first we investigate the connection between the purely algebraic notion of extreme point of a convex subset $K$ of a TVS and the topological boundary of $K$. For a subset $A$ of a topological space, the (topological) *boundary* $\operatorname{bdry} A = (\operatorname{cl} A) \cap (\operatorname{cl} \complement A)$ is the set of points that adhere to $A$ and to $\complement A$. It is easy to verify that, for any set $A$, $A = \operatorname{int} A \cup \operatorname{bdry} A$. Comparing boundary points and extreme points, we first note that an extreme point must be *in* the set—a boundary point need not. Extreme points, however, are necessarily boundary points as we show in Theorem 9.2.4; hence open sets have no extreme points.

**Theorem 9.2.4** Ext $K \subset$ Bdry $K$ Let $K$ be a convex subset of a TVS $X$. If $x \in \operatorname{int} K$, then $x$ is not an extreme point of $K$.

**Proof.** Let $K$ be a convex subset of a TVS $X$ and suppose $x \in \operatorname{int} K$. Since the map $g : \mathbf{R} \to X$, $t \mapsto tx$, is continuous, there must be some $r > 0$ such that the segment $g(1 - r, 1 + r) = ((1 - r)x, (1 + r)x) \subset \operatorname{int} K \subset K$. Since $x \in ((1 - r)x, (1 + r)x)$, $x \notin \operatorname{ext} K$. ■

**Notation.** For a normed space $X$, $U(X) = \{x \in X : \|x\| \leq 1\}$ and $SU(X) = \{x \in X : \|x\| = 1\}$. ■.

By Theorem 9.2.4, for any normed space $X$, it follows that $\operatorname{ext} U(X) \subset SU(X)$ and that $\operatorname{bdry} U = SU(X)$. In Examples 9.2.5(a,b) we deal with normed spaces $X$ in which $\operatorname{ext} U(X) = SU(X)$. Spaces like this are called *strictly convex* and have some interesting properties. We study them at some length in Sec. 16.1. One interesting consequence of strict convexity, for example, is the Taylor-Foguel theorem [16.4.8]: It says that norm-preserving extensions of continuous linear functionals defined on any subspace of a normed space $X$ are unique iff $X'$ is strictly convex. **Warning:** In the context of

Banach spaces $X$, you will see the expression "extreme point of $X$." What is meant is extreme point of $U(X)$.

**Example 9.2.5** EXTREME POINTS OF UNIT BALLS

(a) EUCLIDEAN N-SPACE $\operatorname{ext} U\left(\ell_2^{\mathbf{R}}(n)\right) = SU\left(\ell_2^{\mathbf{R}}(n)\right)$ Vertices of convex polygonal figures $E$ in $\ell_2^{\mathbf{R}}(2)$ are their only extreme points as are the points on the circumference of any circle. For circles $C$, $\operatorname{ext} C = \operatorname{bdry} C$. Analogous statements hold in $\ell_2^{\mathbf{R}}(n)$. In particular, every point on the surface of the unit ball $U\left(\ell_2^{\mathbf{R}}(n)\right)$ is an extreme point. More generally, in any inner product space $X$, $\operatorname{ext} U(X) = SU(X)$ [Theorem 16.1.5].

**Proof.** We prove the assertion about balls for $n = 2$. Let $y = (b_1, b_2)$ and $z = (c_1, c_2)$ be unit vectors, let $t \in (0,1)$ and let $x = ty + (1-t)z$. Then

$$\begin{aligned}\|x\|_2^2 &= t^2\left(b_1^2 + b_2^2\right) + 2t(1-t)(b_1c_1 + b_2c_2) + (1-t)^2\left(c_1^2 + c_2^2\right)\\ &= t^2 + 2t(1-t)(b_1c_1 + b_2c_2) + (1-t)^2\end{aligned}$$

This can only be $1 = (t + (1-t))^2$ if $b_1c_1 + b_2c_2 = 1$, in which case $(b_1 - c_1)^2 + (b_2 - c_2)^2 = 0$ which implies that $y = z = x$.

(b) PTH POWER SUMMABLE SEQUENCES $\operatorname{ext} U(L_p) = SU(L_p)$, $1 < p < \infty$ We show in Theorem 16.1.4 that in any normed space $X$, $SU(X) = \operatorname{ext} U(X)$ iff for any nonzero $x$ and $y$, $\|x + y\| = \|x\| + \|y\|$ implies that there exists $a > 0$ such that $y = ax$. For $p > 1$, equality holds in the Minkowski inequality [see Example 1.7.8] iff one vector is a scalar multiple of the other; hence the $L_p$ spaces are such that $\operatorname{ext} U(L_p) = SU(L_p)$ for $p > 1$.

(c) NULL SEQUENCES $\operatorname{ext} U(c_0) = \emptyset$ Let $(c_0, \|\cdot\|_\infty)$ denote the Banach space of all complex null sequences. By Theorem 9.2.4, $\operatorname{ext} U(c_0) \subset \operatorname{bdry} U(c_0) = SU(c_0)$. We show next that no point $x = (t_n) \in SU(c_0)$ is an extreme point. Choose $n \in \mathbf{N}$ such that $|t_n| < \|x\| = 1$. For sufficiently small positive $r$, replace $t_n$ by $t_n - r$ and $t_n + r$ to create vectors $y$ and $z$ in $SU(c_0)$ such that $x = (1/2)(y + z)$.

(d) ORDERED SPACES Let $X$ be a real Archimedean ordered vector space with an order unit $u$ [Definition 5.12.4] as, for example, are the pointwise ordered spaces $\mathbf{R}^n$, $\ell_\infty(T, \mathbf{R})$, $C(T, \mathbf{R})$, and $L_\infty(\mathbf{R})$. An order unit for each of these is the function $u$ that is identically 1. By Theorem 5.12.9, $n(x) = \inf\{r > 0 : -ru \le x \le ru\}$ $(x \in X)$ is a norm—indeed, $n = \|\cdot\|_\infty$ for $\mathbf{R}^n$, $\ell_\infty(T, \mathbf{R})$, $C(T, \mathbf{R})$, and $L_\infty(\mathbf{R})$. The $n$-closed balls are the order intervals $[x, y] = \{w \in X : x \le w \le y\}$ and the unit ball is the order interval $[-u, u]$; $u$ and $-u$ are extreme points of $[-u, u]$ since $u \notin (x, y)$ for any $x, y \in (-u, u)$. These are not usually the only extreme points, however (see (g) below).

(e) CONTINUOUS FUNCTIONS ON COMPACT SETS $\operatorname{ext} U\left(C\left(T, \mathbf{F}, \|\cdot\|_\infty\right)\right) = \{x \in C(T, \mathbf{F}) : |x(t)| \equiv 1\}$ $C\left(T, \mathbf{F}, \|\cdot\|_\infty\right)$ denotes the Banach space of all continuous maps of the compact Hausdorff space $T$ into $\mathbf{F} = \mathbf{R}$ or $\mathbf{C}$. By (d), for $\mathbf{F} = \mathbf{R}$, we know that $u(t) \equiv 1$ is an extreme point of $U\left(C\left(T, \mathbf{R}, \|\cdot\|_\infty\right)\right)$ and so is $-u$. For $\mathbf{F} = \mathbf{C}$, as we prove next, the extreme points of $U\left(C\left(T, \mathbf{F}, \|\cdot\|_\infty\right)\right)$ are those $x$ such that $|x(t)| \equiv 1$; for connected $T$ and $C(T, \mathbf{R}, \|\cdot\|_\infty)$, they are just $x(t) \equiv 1$ and $-x$. Arens and Kelley [1947, p. 503] used this observation to give the simple proof of the Banach–Stone theorem that we use here [Theorem 9.6.2].

By Theorem 9.2.4 we know that if $x \in \operatorname{ext} U(C(T, \mathbf{F}, \|\cdot\|_\infty))$, then $\|x\|_\infty = 1$. Moreover, as in (c), if $|x(s)| < 1$ for some $s$ in $T$, by the complete regularity of $T$, there must be some $y \in C(T, \mathbf{R})$, which vanishes outside a neighborhood of $s$, of sufficiently small norm that $x-y$ and $x+y$ belong to $U(C(T, \mathbf{F}, \|\cdot\|_\infty))$. Hence $x = (1/2)(x+y) + (1/2)(x-y)$ and $x$ is not an extreme point of $U(C(T, \mathbf{F}), \|\cdot\|_\infty)$. On the other hand, if $|x(t)| \equiv 1$ then for each $t \in T$, $x(t)$ is on the circumference of the closed unit disk $D$ of $\mathbf{F}$. In other words, $x(t)$ is an extreme point of $D$ for each $t$ in $T$. Thus, if $x = (1/2)(y+z)$ for $y, z \in U(C(T, \mathbf{F}, \|\cdot\|_\infty))$, then for each $t \in T$, $y(t), z(t) \in D$ and $x(t) = (1/2)[x(t) + y(t)]$ imply that $x(t) = y(t) = z(t)$ [(a) above and Theorem 9.2.2(c)]; therefore $x = y = z$.

Something else is clear now: *The extreme points of* $U(C(T, \mathbf{R}, \|\cdot\|_\infty))$ *are in one-to-one correspondence with the clopen subsets of* $T$. On the one hand, if $K$ is a clopen subset of $T$ then the function that is 1 on $K$ and $-1$ on its complement is an extreme point of $U(C(T, \mathbf{R}, \|\cdot\|_\infty))$ by the result above. On the other hand, if $x \in \operatorname{ext} U(C(T, \mathbf{R}))$ then, since $|x(t)| \equiv 1$, $x^{-1}(\{1\}) = x^{-1}((1/2, 3/2))$ is clopen.

(f) BOUNDED CONTINUOUS FUNCTIONS $\operatorname{ext} U(C_b(T, \mathbf{R}, \|\cdot\|_\infty)) = \{x \in C_b(T, \mathbf{R}) : |x(t)| \equiv 1\}$ $C_b(T, \mathbf{R}, \|\cdot\|_\infty)$ denotes the Banach space of all bounded continuous maps of the completely regular Hausdorff space $T$ into $\mathbf{R}$. Let $\beta T$ be the Stone-Cech compactification of $T$ [Example 8.4.10]. Then $C_b(T, \mathbf{R}, \|\cdot\|_\infty)$ is linearly isometric to $C(\beta T, \mathbf{R}, \|\cdot\|_\infty)$ under the mapping $A : C_b(T, \mathbf{F}, \|\cdot\|_\infty) \to C(\beta T, \mathbf{R}, \|\cdot\|_\infty)$, $x \mapsto \beta x$, the continuous extension of $x$ to $\beta T$ and $A(U(C_b(T, \mathbf{R}, \|\cdot\|_\infty))) = U(C(\beta T, \mathbf{R}, \|\cdot\|_\infty))$. By (e),

$$\operatorname{ext} U(C(\beta T, \mathbf{R}, \|\cdot\|_\infty)) = \{u \in C(\beta T, \mathbf{R}) : |u(t)| = 1 \ \forall\, t \in \beta T\}$$

Since $A$ is an isometry [Theorem 9.2.3],

$$\begin{aligned} \operatorname{ext} U(C_b(T, \mathbf{R}, \|\cdot\|_\infty)) &= A^{-1}(\{u \in C(\beta T, \mathbf{R}) : |u(t)| = 1 \ \forall\, t \in \beta T\}) \\ &= \{u \in C_b(T, \mathbf{R}) : |u(t)| = 1 \ \forall\, t \in T\} \end{aligned}$$

an observation first made by Krein and Milman [1940, pp. 136–137].

(g) BOUNDED SEQUENCES $\operatorname{ext} U(\ell_\infty^{\mathbf{R}}) = \{(t_n) \in \mathbf{R}^{\mathbf{N}} : |t_n| = 1 \ \forall\, n \in \mathbf{N}\}$ Endow $\mathbf{N}$ with the discrete topology so that $C_b(\mathbf{N}, \mathbf{R}, \|\cdot\|_\infty) = \ell_\infty$. The result now follows from (f). It is also easy to see that for each $n \in \mathbf{N}$, $U(\ell_\infty^{\mathbf{R}}(n))$ has $2^n$ extreme points.

(h) SUMMABLE SEQUENCES $\operatorname{ext} U(\ell_1) = \{ae_n : |a| = 1,\ n \in \mathbf{N}\}$ Let $\ell_1$ be the Banach space of all absolutely summable complex sequences $x = (a_n)$ with $\|x\|_1 = \sum_n |a_n|$. The extreme points of $U(\ell_1)$ are unit multiples of the standard basis vectors $e_n$, vectors of the form $ae_n$ where $|a| = 1$ and $e_n$ is the sequence whose $n$th entry is 1 and whose other entries are 0. Similar reasoning shows that, for each $n \in \mathbf{N}$, $U\left(\ell_1^{\mathbf{R}}(n)\right)$ has $2n$ extreme points.

**Proof.** Suppose that $|a| = 1$ and $ae_j = (1/2)(x+y)$ for $x = (s_n)$ and $y = (t_n)$ in $U(\ell_1)$. This immediately implies that $a = (1/2)(s_j + t_j)$ and that $s_n = t_n = 0$ for $n \neq j$. Since $a$ is an extreme point of the solid unit disk $D \subset \mathbf{C}$ (by (a)) and $a = (1/2)(s_j + t_j)$, $(s_j, t_j \in D)$, it follows that $a = s_j = t_j$; hence $x = y = ae_j$. On the other hand, suppose that $x = (a_n) \in SU(\ell_1)$ has two nonzero entries $a_j$ and $a_k$ with $k < j$. Then for some $t \in (0,1)$, $\sum_{n \le k} |a_n| = t$ and $\sum_{n>k} |a_n| = 1 - t$. With $y = (1/t)(a_1, \ldots, a_k, 0, 0, \ldots)$ and $z = (1-t)^{-1}(0, 0, \ldots, 0, a_{k+1}, \ldots)$, it follows that $y \neq z$, $\|y\|_1 = \|z\|_1 = 1$ and $x = ty + (1-t)z$. It follows that $x \notin \operatorname{ext} U(\ell_1)$.

(i) SUMMABLE FUNCTIONS $\operatorname{ext} U(L_1(\mathbf{R})) = \emptyset$ Let $L_1(\mathbf{R})$ be the Banach space of (absolutely) Lebesgue-integrable real- or complex-valued functions $x$ on $\mathbf{R}$ normed by $\|x\|_1 = \int_{\mathbf{R}} |x(t)|\, dt$. By Theorem 9.2.4, to show that $\operatorname{ext} U(L_1(\mathbf{R})) = \emptyset$, it suffices to consider vectors $x$ of norm 1. Choose $r \in \mathbf{R}$ such that $\int_{-\infty}^{r} |x(t)|\, dt = 1/2$ and let $p$ and $q$ denote the characteristic functions of $(-\infty, r)$, and $[r, \infty)$, respectively. With $y = 2xp$ and $z = 2xq$, it follows that $y \neq z$, $\|y\|_1 = \|z\|_1 = 1$, and $x = (1/2)(y+z)$. ■

In any normed space $X$, the unit ball $U(X)$ is convex. Could $\ell_\infty^{\mathbf{R}}(2)$ be linearly isometric to $\ell_p^{\mathbf{R}}(2)$ for some $1 < p < \infty$ under an isometry $A$ say? If so, then $A(\operatorname{ext} U\left(\ell_p^{\mathbf{R}}(2)\right) = \operatorname{ext} U\left(\ell_\infty^{\mathbf{R}}(2)\right)$ [Theorem 9.2.3]. As the latter set contains only four points and the former is infinite, $\ell_\infty^{\mathbf{R}}(2)$ cannot be linearly isometric to $\ell_p^{\mathbf{R}}(2)$.

**Example 9.2.6** FOR WHAT $n$ IS $\ell_1^{\mathbf{R}}(n)$ LINEARLY ISOMETRIC TO $\ell_\infty^{\mathbf{R}}(n)$? For $n = 1$ and 2, $\ell_\infty^{\mathbf{R}}(n)$ is linearly isometric to $\ell_1^{\mathbf{R}}(n)$: For $n = 2$, consider the linear map determined by $e_1 \mapsto (1/2)(e_1 + e_2)$ and $e_2 \mapsto (1/2)(e_1 - e_2)$. For $n > 2$ they are not linearly isometric: By Examples 9.2.5(g,h), $U\left(\ell_1^{\mathbf{R}}(n)\right)$ has $2n$ extreme points while $U\left(\ell_\infty^{\mathbf{R}}(n)\right)$ has $2^n$. It follows from Theorem 9.2.3 that they are not linearly isometric. ■

## 9.3 FACES

Now consider a weaker kind of extremality condition, a generalization of face of a cube rather than vertex.

**Definition 9.3.1** FACE OF A CONVEX SET Let $K$ be a nonempty convex set. A nonempty subset $F$ of $K$ is a *face* (*extremal subset*) *of* $K$ if for any $x$ in $F$, $x$ being between two points of $K$ implies that those points are in $F: x \in (y, z)$ for $y, z$ in $K$ implies $y, z \in F$. ■

**Example 9.3.2** FACES (a) The perimeter of any convex polygon in $\mathbf{R}^2$ is a face of the polygon.

(b) Any convex subset of a vector space is a face of itself.

(c) A singleton $\{x\}$ is a face of a convex set $K$ iff $x \in \operatorname{ext} K$.

(d) "Is a face of" is a transitive relation: For convex sets $F$ and $K$, if $A$ is a face of $F$ and $F$ a face of $K$, then $A$ is a face of $K$. As to this statement, suppose $x \in A$ and $x \in (y, z)$ for $y, z \in K$. Since $F$ is a face of $K$ and $x \in F$, this implies that $y, z \in F$. Since $A$ is a face of $F$, $y, z \in A$ and it follows that $A$ is a face of $K$.

(e) If $F$ is a convex face of the convex set $K$ then $\operatorname{ext} F \subset \operatorname{ext} K$. For if $x \in \operatorname{ext} F$ and $y, z \in K$ are such that $x \in [y, z]$ then $y, z \in F$. Since $x \in \operatorname{ext} F$, $x = y = z$. ■

The notion of face of a convex set $K$ enunciated above is not a good generalization of "edge" of a convex polygon or face of a cube because these are convex and faces need not be (although many authors require a face to be convex). The perimeter of a triangle, for example, is not convex. Two possible ways to improve on the analogy with "edge," for a convex subset $K$ of an LCS $X$, are (1) for any boundary point $x$ of $K$, consider a maximal convex subset of the boundary to which $x$ belongs; or (2) as any point $x$ of $K$ is in some face of $K$ ($K$ itself, for one), consider the intersection $F(x)$ of all faces of $K$ to which $x$ belongs.

**Theorem 9.3.3** BOUNDARIES AND FACES IN LCS Let $K$ be a convex subset of the LCS $X$ with boundary $\operatorname{bdry} K$. If $K \cap \operatorname{bdry} K = F \neq \varnothing$ then $F$ is a face of $K$. Conversely, if $x$ belongs to a proper face $F$ of $K$ (i.e., $F$ is a proper subset of $K$), then $x$ belongs to the boundary of $K$.

**Proof.** Let $K$ be a convex subset of the LCS $X$. Suppose $F = K \cap \operatorname{bdry} K \neq \varnothing$ and $v \in F$. If $v \in \operatorname{ext} K$ then $v$ is not between distinct points of $K$, so suppose $v \notin \operatorname{ext} K$. If $F$ is not a face of $K$ then $v$ can be between two points of $K$, at least one of which is not in the boundary of $K$. In other words, $v$ may be between an interior point $y$ of $K$ and some other point $x$ of $K$. (If $\operatorname{int} K = \emptyset$, then $K = \operatorname{bdry} K$ and theorem is trivial.) Choose $t \in (0, 1)$ such that $v = tx + (1 - t)\, y$. Since $y \in \operatorname{int} K$ and $X$ is locally convex, there must be a disked neighborhood $V$ of 0 such that $y + V \subset \operatorname{int} K$. Therefore, there is a continuous seminorm $p$ (namely, the gauge of $V$) such that $y + \{w \in X : p(w) < 1\} = \{w \in X : p(w - y) < 1\} \subset \operatorname{int} K$. Since $v$ is a boundary point of $K$, for any positive $d$ there must be some $z \notin K$ such that $p(z - v) < d$. Choose $d$ such that $d/(1 - t) < 1$ and let $w = (1 - t)^{-1}(z - tx)$, so $z = tx + (1 - t)\, w$. Since $(z - v) = (1 - t)(w - y)$, $p(z - v) = (1 - t)\, p(w - y) < d$. Then $p(w - y) < d/(1 - t) < 1$ which implies that $w \in \operatorname{int} K$. Since $K$ is convex, this implies that $z = tx + (1 - t)\, w \in K$ which is contradictory.

Conversely, suppose that $x \operatorname{int} K$ is in a proper subset $F$ of $K$. We show that $F$ cannot be a face of $K$. Since $F$ is proper, there is some $w \in K \setminus$

$F$. Since $x \in \operatorname{int} K$, there is a continuous seminorm $p$ on $X$ such that $\{y \in X : p(x-y) < 1\} \subset K$. For any $r > 0$, $p(x - [(1+r)x - rw]) = rp(x-w)$; hence, for sufficiently small $r$, $(1+r)x - rw \in K$. Hence

$$x = \frac{1}{1+r}[(1+r)x - rw] + \frac{r}{1+r}w$$

is between $(1+r)x - rw$ and $w$ and $w \notin F$. Therefore $F$ is not a face of $K$. ■

**Theorem 9.3.4** EXTREMAL VALUES AND FACES Let $K$ be a convex subset of a vector space $X$ and let $f$ be a real linear functional on $X$ which is bounded below on $K$. If $F = \{x \in K : f(x) = \inf f(K)\} \neq \emptyset$ then $F$ is a convex face of $K$. The analogous statement holds for suprema.

**Proof.** We use the notation of the statement. Since the hyperplane $f^{-1}(t)$ is convex for any $t \in \mathbf{R}$, it follows that $F$ is convex. With $a = \inf f(K)$, since the linear image of a convex set is convex, $f(K) = [a,b]$ or $[a,b)$ for some $b \in \mathbf{R}$, or $f(K) = [a,\infty)$. Suppose $x = ty + (1-t)z$ for some $y, z \in K$ and $t \in [0,1]$ and assume that $f(y) \leq f(z)$. Since $a$ is an extreme point of $[a,b]$ or $[a,\infty)$, $f(x) = a = tf(y) + (1-t)f(z)$ implies $f(y) = a = f(z)$, i.e., that $y, z \in F$. ■

In Theorem 9.3.4, if $f$ is only concave [Exercise 7.104] the assertion about infima still holds; if $f$ is convex, then $f^{-1}(\max f|_K) \cap K$ is a face of $K$.

**Theorem 9.3.5** ADJOINTS AND EXTREME POINTS Let $X$ and $Y$ be normed spaces and let their continuous duals $X'$ and $Y'$ carry their norm topologies [Example 6.5.3(b)]. Suppose $A : X \to Y$ is a continuous linear map with adjoint $A'$. If $\|A\| \leq 1$, $f \in \operatorname{ext} U(X')$ and $A'^{-1}(f) \cap U(Y') \neq \varnothing$ then $A'^{-1}(f) \cap U(Y')$ is a convex face of the unit ball $U(Y')$ of $Y'$.

**Proof.** With notation as above, suppose that $f \in \operatorname{ext} U(X')$, $g \in A'^{-1}(f) \cap U(Y')$, and $g \in [h,j]$ where $h, j \in U(Y')$; thus $g = th + (1-t)j$ for some $t \in [0,1]$. Applying $A'$, $A'g = f = tA'h + (1-t)A'j$. By Theorem 8.11.5, $\|A'\| = \|A\|$, so $\|A'k\| \leq \|A\|\,\|k\| \leq \|A\| \leq 1$ for any $k \in U(Y')$; hence $A'(U(Y')) \subset U(X')$. In particular, $A'h$ and $A'j$ must belong to $U(X')$. Since $f \in \operatorname{ext} U(X')$, however, it follows that $f = A'h = A'j$—in other words, that $h, j \in A'^{-1}(f)$. The convexity of $A'^{-1}(f) \cap U(Y')$ is obvious. ■

## 9.4 KREIN–MILMAN THEOREMS

**Notation** $\operatorname{cl} E_c$ stands for the closed convex hull of $E$, i.e., the closure of $E_c$. ■

Any closed half plane of $\ell_2^{\mathbf{R}}(n)$ is a closed convex set with no extreme points. We have already seen [Example 9.2.5(c,i)] that there are bounded closed convex sets that are devoid of extreme points as well, namely, the closed unit balls of $c_0$ and $L_1(\mathbf{R})$. If one thinks of a compact convex set as the generalization of an interval in $\mathbf{R}$, then here is a genus of set that *should* have extreme points. Our first result shows that they do. The idea of the argument in Theorem 9.4.1 below is to go from $K$ to a face $F$ of $K$, then to a face of $F$, and so on, until terminating in an extreme point.

By *the Krein–Milman theorem*, Theorem 9.4.6(b) is most likely what is meant; however, Theorems 9.4.1 and 9.4.3 are also possible. Theorem 9.4.1 is one of those curious hybrids—a topological hypothesis, an algebraic conclusion.

**Theorem 9.4.1** KREIN–MILMAN I: COMPACT CONVEX SETS HAVE EXTREME POINTS If $K$ is a nonempty compact convex subset of an LCHS $X$ then $\operatorname{ext} K \neq \emptyset$.

**Proof.** Let $K$ be as above. Using Zorn's Lemma we get a minimal face $F$ of $K$ and then show that $F$ is a singleton.

Let $f$ be a real continuous linear functional on $X$. Since $K$ is compact, the closed set $F = \{x \in K : f(x) = \min f(K)\}$ is nonempty. Hence $F$ is a convex face of $K$ by Theorem 9.3.4. It follows that the class $\mathcal{F}$ of nonempty closed convex faces of $K$ is nonempty. Order $\mathcal{F}$ by inclusion: $F \leq G$ iff $F \subset G$ for $F, G$ in $\mathcal{F}$. If $\{F_s : s \in S\}$ is a totally ordered subset of $\mathcal{F}$, it must satisfy the finite intersection condition; hence, the compactness of $K$ implies that $E = \cap_{s \in S} F_s$ is nonempty and closed; $E$ is a convex face of $K$ because each $F_s$ is. Thus, $E$ is a lower bound for $\{F_s : s \in S\}$ and Zorn's lemma implies the existence of a minimal element $F \in \mathcal{F}$, i.e., a closed convex face of $K$ which is minimal with respect to set inclusion. If $F$ is a singleton $\{x\}$, then the theorem is proved because, as observed in Example 9.3.2(c), a singleton $\{x\}$ is a face of $K$ iff $x$ is an extreme point of $K$. Suppose that $x$ and $y$ are distinct points of $F$. Since $X$ is an LCHS, there is a real continuous linear functional $g$ on $X$ such that $g(x) \neq g(y)$ by Theorem 7.7.7(a). Since $F$ is compact, $g$ is bounded below on $F$ so, by Theorem 9.3.4, the set $E = \{w \in F : g(w) = \min g(F)\}$ is a closed convex face of $F$, hence also of $K$ [Example 9.3.2(d)]. Since $g(x) \neq g(y)$, only one of $x, y$ can belong to $E$ which means that $E$ is a proper subset of $F$. As this contradicts the minimality of $F$, we conclude that $F$ is a singleton. ■

How important is local convexity for the Krein–Milman theorem? In an arbitrary TVS, a compact convex set need not have any extreme points. Roberts [1976, 1977] (cf. Kalton and Peck 1980) gives an example of a compact convex subset of $L_p[0,1]$, $0 < p < 1$, with no extreme points. If, however, the compact convex subset $K$ of a real HTVS is "strongly" locally convex—for each $x$ in $K$ and each relative neighborhood $U$ of $x$ contained in $K$, there is a relatively open convex neighborhood $V$ of $x$ contained in $U$—then $K$ must have extreme points [Jamison et al., 1976].

Kelley [1950] proved that Tihonov's theorem on the compactness of products of compact spaces implies the axiom of choice. Bell and Fremlin [1972, pp. 168–169] showed that the original Banach space version of Theorem 9.4.1—namely, that the closed unit ball of the dual $X'$ of a normed space $X$ has an extreme point—implies the axiom of choice as well.

Some applications of Theorem 9.4.1 follow in Example 9.4.2. Note the similarity of the results in (c) and (d) to the classical maximum modulus theorem of complex analysis, a theme we return to in Sec. 9.5.

**Example 9.4.2** (a) REFLEXIVITY Let $X$ be a normed space and let $X'$ be its continuous dual. Let $U(X')$ denote the solid unit ball in $X'$. By the Alaoglu theorem [8.4.1], $U(X')$ $(= U(X)^{\circ})$ is $\sigma(X', X)$-compact. Since $U(X')$ is convex, it must have extreme points by Theorem 9.4.1. In other words, the unit ball of the dual of any normed space must have extreme points. As the unit balls of $c_0$ and $L_1(\mathbf{R})$ have no extreme points [Example 9.2.5(c,i)], neither of them can be the dual of a normed space—in particular, they cannot be reflexive.

(b) If $F$ is a closed convex face of a compact convex subset $K$ of an LCHS, then $F$ has an extreme point $x$ by Theorem 9.4.1; $x$ is an extreme point of $K$ by Example 9.3.2(e).

(c) EXTREMA AT EXTREME POINTS If $K$ is a compact convex subset of an LCHS $X$ and $f$ a real continuous linear functional on $X$, then $f$ assumes its extremal values $\max f(K)$ and $\min f(K)$ at extreme points of $K$. Thus, if $f = 0$ on $\operatorname{ext} K$, then $f = 0$ on $K$.

**Proof.** We prove only the statement about minima. Since $K$ is compact, $F = \{x \in K : f(x) = \min f(K)\}$ is nonempty and is therefore a closed convex face of $K$ by Theorem 9.3.4. Consequently, an extreme point of $K$ is in $F$ by (b).

A similar argument establishes the following result which has important consequences in optimization theory.

(d) EXTREMA AT EXTREME POINTS If $K$ is a compact convex subset of an LCHS $X$, then (1) a lower semicontinuous concave function assumes its minimum on $K$ at an extreme point of $K$, and (2) an upper semicontinuous convex function assumes its maximum value at an extreme point of $K$.

(e) LINEAR PROGRAMMING For $(a_1, \ldots, a_n), (t_1, \ldots, t_n) \in \mathbf{R}^n$, a problem such as "minimize $f(t_1, \ldots, t_n) = a_1 t_1 + \cdots + a_n t_n$ subject to the condition that $(t_1, \ldots, t_n)$ satisfy a finite set of linear inequalities (constraints)" is called a "linear programming problem" or an "abstract mathematical program"; $f$ is called "the objective function." The set of $(t_1, \ldots, t_1)$ satisfying all the constraints is called the "set of feasible solutions." As a finite intersection of half spaces, the set of feasible solutions is a convex set. When the set of feasible solutions is closed and bounded, the result of (c) or (d) shows that solutions exist at extreme points of the set of feasible solutions, the "corners" of the set. Moreover, one need only solve the problem on the extreme points to solve the problem. More details about optimization within the framework of functional analysis can be found in Holmes [1972], Krabs [1979], and Balakrishnan [1981]. ■

One version of the Krein–Milman theorem [9.4.6(b)] says that a compact convex set $K$ may be reconstituted as the closed convex hull of its extreme points: $K = \text{cl}(\text{ext}\, K)_c$. Theorem 9.4.3 provides an equivalent description of those subsets $B$ of a convex compact set $K$ for which $\text{cl}\, B_c = K$.

**Theorem 9.4.3** Krein–Milman II: When $K = \text{cl}\, B_c$ Let $B$ be a subset of a compact convex set $K$ in an LCHS $X$. Then:

(a) $\inf f(B) = \inf f(\text{cl}\, B_c)$ for any real continuous linear functional $f$ on $X$ and

(b) $\text{cl}\, B_c = K$ iff for all real continuous linear functionals $f$ on $X$, $\inf f(B) = \min f(K)$.

**Proof.** Let $B$ be a subset of a compact convex set $K$ in an LCHS $X$. (a) Let $f$ be a real continuous linear functional on $X$. Since $K$ is compact, $f(K)$ is bounded below; hence, since $B \subset \text{cl}\, B_c$, $b = \inf f(B) \geq \inf f(\text{cl}\, B_c)$. To reverse the inequality, let $x \in B_c : x = \sum_{i=1}^{n} t_i x_i$ for $x_i \in B$, $t_i \geq 0$ $(1 \leq i \leq n)$, and $\sum_{i=1}^{n} t_i = 1$. Then $f(x) \geq b = \sum_{i=1}^{n} t_i b$, so $\inf f(B_c) \geq b$. Since $f$ is continuous, $f(\text{cl}\, B_c) \subset \text{cl}\, f(B_c)$, so $\inf f(\text{cl}\, B_c) \geq \inf(\text{cl}\, f(B_c)) = \inf f(B_c) \geq b$.

(b) If $K = \text{cl}\, B_c$, then $\min f(K) = \inf f(\text{cl}\, B_c) = \inf f(B)$ by (a). Conversely, suppose that for all real continuous linear functionals $f$ on $X$, $\inf f(B) = \min f(K)$. Clearly, $\text{cl}\, B_c \subset K$. To reverse the inclusion, suppose that $x \in K$. If $x \notin \text{cl}\, B_c$, there is, by Theorem 7.8.6(b), some real-valued $f \in X'$ such that $f(x) < \inf f(\text{cl}\, B_c) = \inf f(B)$ [by (a)] $= \min f(K)$ which is contradictory. ■

The structure of the set of extreme points of a set is unpredictable. They form a $G_\delta$-set in a metrizable LCHS [Exercise 9.107]. If $K$ is a *compact* convex set, then $\text{ext}\, K$ is a Baire space [Definition 11.6.5] in its relative topology [Dixmier 1964, p. 355]. As Example 9.4.4 shows, the set of extreme points of a compact convex set need not even be closed.

**Example 9.4.4** Ext $K$ Need Not Be Closed Consider the line segment $[v, w]$ connecting $v = (0,0,1)$, $w = (0,0,-1)$ in $\ell_2^{\mathbf{R}}(3)$. Let $D$ be the closed disk in the $xy$-plane of radius 1 centered at $(1,0,0)$. Let $K$ be the convex hull of $L \cup D$. The extreme points of the compact set $K$ are $v, w$ and all points on the circumference of $D$ except $(0,0,0)$. ■

**Lemma 9.4.5** Case When $\text{ext}\,\text{cl}\, B_c \subset \text{cl}\, B$ If $B$ is a subset of an LCHS $X$ whose closed convex hull $\text{cl}\, B_c$ is compact, then $\text{ext}\,\text{cl}\, B_c \subset \text{cl}\, B$.

**Proof.** Let $B$ be a subset of the LCHS $X$ such that $\text{cl}\, B_c$ is compact and let $x \in \text{ext}\,\text{cl}\, B_c$. Let $V$ be a closed disked neighborhood of 0. We show that

$(x+V)\cap B\neq\emptyset$. Since $\operatorname{cl}B_c$ is compact, $B$ is totally bounded [Definition 3.3.1] so there exist $x_i\in B$ $(1\le i\le n)$ such that $B\subset\cup_i(x_i+V)$. Each set $\operatorname{cl}[(x_i+V)\cap B]_c=K_i$ $(1\le i\le n)$ is convex and compact [Theorem 4.8.9(b)], so $(\cup_iK_i)_c$ is a compact convex set by Theorem 4.4.4(a). Since $B\subset\cup_iK_i$, it follows that $\operatorname{cl}B_c\subset(\cup_iK_i)_c$. The reverse inclusion holds as well: For each $i$, $K_i\subset\operatorname{cl}B_c$, so $(\cup_iK_i)_c\subset\operatorname{cl}B_c$—hence $\operatorname{cl}B_c=(\cup_iK_i)_c$. Thus any $x\in\operatorname{ext}\operatorname{cl}B_c$ must be a convex combination $\sum_i t_iy_i$, where $y_i\in K_i$ [Theorem 4.2.3(b)], $t_i\ge 0$, $1\le i\le n$, and $\sum_i t_i=1$. As $x$ is an extreme point of $\operatorname{cl}B_c$, this convex combination must be trivial (see discussion after Theorem 9.2.2): for some $j$ we must have $x=y_j\in K_j=\operatorname{cl}[(x_j+V)\cap B]_c$. Since $(x_j+V)\cap B\subset x_j+V$ and $V$ is convex and closed, $\operatorname{cl}[(x_j+V)\cap B]_c\subset x_j+V\subset B+V$. Therefore $x\in B+V$ which implies that $(x+V)\cap B\neq\emptyset$ and proves that $x\in\operatorname{cl}B$. ■

If $X$ is a complete LCHS, it follows from Theorem 4.8.9(c) that if $K\subset X$ is compact then $\operatorname{cl}K_c$ is compact; hence, by Lemma 9.4.5, $\operatorname{ext}\operatorname{cl}K_c\subset K$. The result below subsumes the fact that for compact convex $K$, $K=\operatorname{cl}(\operatorname{ext}K)_c$. It says that any subset of $K$ whose closure contains the extreme points of $K$ can be used to reconstitute $K$.

**Theorem 9.4.6** KREIN–MILMAN III Let $B$ be a subset of the compact convex set $K$ in the LCHS $X$. Then:

(a) $K=\operatorname{cl}B_c$ iff $\operatorname{ext}K\subset\operatorname{cl}B$;

(b) thus, $K=\operatorname{cl}(\operatorname{ext}K)_c$.

**Proof.** We prove only (a). Let $X$ be an LCHS and let $B$ be a subset of the compact convex set $K\subset X$. If $K=\operatorname{cl}B_c$, then, by Lemma 9.4.5, $\operatorname{ext}K=\operatorname{ext}\operatorname{cl}B_c\subset\operatorname{cl}B$.

Conversely, suppose that $\operatorname{ext}K\subset\operatorname{cl}B$. To show that $K=\operatorname{cl}B_c$, we show that the condition of Theorem 9.4.3(b) is satisfied, namely that $\inf f(B)=\min f(K)$ for all real continuous linear functionals $f$ on $X$. By continuity, $\inf f(B)=\inf f(\operatorname{cl}B)$. Since $B\subset K$, $\inf f(B)\ge\min f(K)$. Since $K$ is compact and convex, it follows from Example 9.4.2(c) that $f$ achieves its minimum at an extreme point $x$ of $K$. By hypothesis then $x\in\operatorname{cl}B$. Hence $\min f(K)=f(x)\ge\inf f(\operatorname{cl}B)=\inf f(B)$. ■

Observe that *compact* cannot generally by replaced by *closed and bounded* in Theorem 9.4.6 since there are spaces $X$ (e.g. $c_0$ and $L_1(\mathbf{R})$, Example 9.2.5(c,i)) in which the unit ball $U(X)$ has no extreme points. Nevertheless, there are spaces—$\ell_1$ for example—in which each closed bounded convex subset is the closed convex hull of its extreme points. (See Lindenstrauss [1966] for the statement about $\ell_1$.)

The analytic consequences of the Krein–Milman theorem [9.4.6] are extensive, reaching far into abstract harmonic analysis [Berberian, 1974, p. 293], approximation theory [Singer, 1970, p. 58ff], operator theory [Berberian,

1974, p. 316], ergodic theory [Choquet, 1969c, Chap. 7], and various representation theorems [Choquet, 1969c, Chap. 7; Edwards, 1965, Chap. 10]. It ranks second only to the Hahn–Banach theorem in importance in functional analysis.

How important is local convexity in Krein–Milman theorems? As Kelley [1951] noted, Theorem 9.4.6(b) remains valid in any TVS in which, for any compact convex set $K$ and any point $x$ not in $K$, there is a continuous linear functional whose value at $x$ is smaller than its value at any point of $K$. Kalton [1980] gives an example of a complete metrizable space with trivial dual in which the Krein–Milman theorem holds.

Given any distinct points $x$ and $y$ in an LCHS $X$, there is an $f \in X'$ such that $f(x) \neq f(y)$ [Theorem 7.7.7(a)]. In particular, this is so for any normed space $(X, \|\cdot\|)$. We can sharpen this to the assertion that the unit ball $U(X')$ distinguishes the points of $X$ (just divide $f$ by $\|f\|$). The Krein–Milman theorem permits the following refinement.

**Theorem 9.4.7** Ext $U(X')$ Separates Points For any normed space $X$ and any two distinct points $x$ and $y$ of $X$ there is an extreme point $f$ of the unit ball $U(X')$ of $X'$ such that $f(x) \neq f(y)$.

**Proof.** With notation as above, by the Alaoglu theorem [8.4.1], $U(X')$ is $\sigma(X', X)$-compact, so $U(X') = \mathrm{cl}_{\sigma(X',X)}[\mathrm{ext}\, U(X')]_c$. It suffices to show that there exists $f \in \mathrm{ext}\, U(X')$ that separates $x \neq 0$ from 0, i.e., that $f(x) \neq 0$. Given $x \neq 0$, there exists $g \in U(X')$ such that $g(x) \neq 0$. Since $U(X') = \mathrm{cl}_{\sigma(X',X)}[\mathrm{ext}\, U(X')]_c$, we may approximate $g$ by a convex combination of extreme points $f_1, \ldots, f_n$ of $U(X')$ to any degree of closeness in the $\sigma(X', X)$ topology. Hence for $0 < r < |g(x)|$ there must be a convex combination $\sum_{i=1}^n a_i f_i$ of elements $f_i \in \mathrm{ext}\, U(X')$ such that $|g(x) - \sum_{i=1}^n a_i f_i(x)| < r$; therefore not all the $f_i$ can vanish on $x$. It follows that there must be an $f_i \in \mathrm{ext}\, U(X')$ such that $f_i(x) \neq 0$. ■

Theorem 9.4.7 shows that the dual ball of a normed space has lots of extreme points. The abundance of extreme points of the unit ball is also the theme of Example 9.4.8.

**Example 9.4.8** (a) $C([0,1], \mathbf{R}, \|\cdot\|_\infty)$ Is Not a Dual If the Banach space $X = C([0,1], \mathbf{R}, \|\cdot\|_\infty)$ of continuous real-valued functions on $[0,1]$ were the dual of a normed space $Y$, then the unit ball $U$ of $(C([0,1], \mathbf{R}, \|\cdot\|_\infty))$ would be $\sigma(X, Y)$-compact [Alaoglu, Theorem 8.4.1]. Hence $U = \mathrm{cl}_{\sigma(X,Y)}(\mathrm{ext}\, U)_c$ by a form of the Krein–Milman theorem [9.4.6(b)]. Since $[0,1]$ is connected, the extreme points of $U$ are just $x(t) \equiv 1$ and $y(t) \equiv -1$ by Example 9.2.5(e). The convex hull of $\{x, y\}$, the line segment $[x, y]$, is closed [Theorem 4.4.4(a)], so this would say that any continuous function on $[0,1]$ of norm one could be written as a convex combination of the constants $x$ and $y$ which is ridiculous. It also follows that $C([0,1], \mathbf{R}, \|\cdot\|_\infty)$ is not reflexive. We show in Example

15.5.2 that, for any compact Hausdorff space $T$, $C(T, \mathbf{F}, \|\cdot\|_\infty)$ is reflexive iff $T$ is finite.

(b) INFINITE-DIMENSIONAL BANACH SPACES More generally, no infinite-dimensional normed space $X$ whose unit ball $U$ has only a finite number of extreme points can be the continuous dual of any normed space $Y$. If $X = Y'$ and $\operatorname{ext} U = \{x_1, \ldots, x_n\}$ then $U$ is $\sigma(X, Y)$-compact [Alaoglu] and by Theorem 9.4.6(b), $U = \operatorname{cl}_{\sigma(X,Y)} (\operatorname{ext} U)_c = (\operatorname{ext} U)_c \subset [x_1, \ldots, x_n]$ which implies that $X$ is finite-dimensional. Consequently, *the unit ball of any infinite-dimensional reflexive normed space must have infinitely many extreme points.* ■

## 9.5 THE CHOQUET BOUNDARY

**Notation.** $T$ : a compact Hausdorff space, $\mathbf{F} = \mathbf{R}$ or $\mathbf{C}$.

$C(T, \mathbf{F}, \|\cdot\|_\infty)$ : the Banach space of continuous $\mathbf{F}$-valued functions on $T$.

$e$ : the map $T \to \mathbf{F}$, $t \mapsto 1$.

$X$ : a subspace of $C(T, \mathbf{F})$ that is self-adjoint [$x \in X \Rightarrow \bar{x} \in X$ where $\bar{x}$ denotes the complex conjugate of $x$], contains $e$ and separates points of $T$ in the sense that, given distinct points $s, t \in T$, there exists $x \in C(T, \mathbf{F})$ such that $x(s) \neq x(t)$.

$X'$ : the continuous dual of $X$. When we speak of closure in $X'$ it will always be $\sigma(X', X)$-closure, which we denote by $\operatorname{cl}_\sigma$ in this section.

$U(X), U(X')$ : the norm-unit balls in $X$ and $X'$, respectively.

$SU(X), SU(X')$ : the surfaces (norm one elements) of $U(X)$ and $U(X')$ of $X$ and $X'$, respectively.

$T'$ : the collection of evaluation maps $t'$ determined by the points $t$ of $T$, where for each $x \in X$, $t'(x) = x(t)$.

$\operatorname{cl} B_c$ : the closed convex hull of $B$, i.e., the closure $\operatorname{cl} B_c$ of the convex hull $B_c$ of $B$ ■

We investigate a closed $\sigma(X', X)$-compact subset $S(X)$ of the dual $X'$ called the *state space* of $X$ in this section. In particular, we show that [Theorem 9.5.5]

$$S(X) = \operatorname{cl}_{\sigma(X',X)} (T')_c \text{ and } \operatorname{ext} S(X) \subset T'.$$

Among other things, each $x$ in $X$ achieves its norm on $\operatorname{ch} X$ [Theorem 9.5.9], so that it serves in somewhat the same way as the topological boundary $B = \{a \in \mathbf{C} : |a| = 1\}$ for continuous maps on the closed unit disk which are analytic in its interior. In a finite dimensional space any element $x$ in a convex polygon $K$ decomposes into a convex combination of extreme points. The body of information concerning how elements of a compact convex set $K \subset X$ decompose into extreme points of $K$—how a state can be written as an integral of pure states—is called *Choquet theory*. We briefly introduce it for compact $T$ in this section; the theory for locally compact $T$ is well-developed but harder. Some sources for further exploration are Phelps [1966], Choquet [1969c], Alfsen [1971], and Asimow and Ellis [1980] (cf. Fuchsteiner 1981).

We say that $f \in X'$ is *positive* if $f(x) \geq 0$ whenever $x \geq 0$ in the pointwise ordering on $X$, namely, $x \geq y$ iff for each $t \in T$, $x(t) \geq y(t)$; it is easy to see that *positive* is equivalent to *order preserving.* We abbreviate "$f$ is positive" to $f \geq 0$.

Since the subspace $X$ is self-adjoint, $\operatorname{Re} x$ and $\operatorname{Im} x$ belong to $X$ if $x$ does. If $X$ consists only of real-valued functions then, of course, $X$ is self-adjoint. As we argue below, for any $f \in X'$,

$$f \geq 0 \Rightarrow \operatorname{Re} f(x) = f(\operatorname{Re} x) \qquad (x \in X) \tag{9.1}$$

To see this, suppose $y \in X$ is $\mathbf{R}$-valued and let $w = e + y$. Since $w^2 \geq 0$, $f(w^2) = f(e) + 2f(y) + f(y^2) \geq 0$. Since $f(e)$ and $f(y^2)$ are nonnegative, this implies that $f(y)$ is real. Since $X$ is self-adjoint, any $x \in X$ can be written $x = y + iz$, where $y, z \in X$ are $\mathbf{R}$-valued. Since $f(x) = f(y) + if(z)$ and $f(y)$ and $f(z)$ are real, it follows that $\operatorname{Re} f(x) = f(y) = f(\operatorname{Re} x)$.

We use the following characterization of closed convex hull in Hilbert spaces to prove Theorem 9.5.2.

**Theorem 9.5.1** CONVEX HULL IN HILBERT SPACE If $B$ is a bounded subset of a Hilbert space $X$ then its closed convex hull $H$ is the intersection of all closed balls which contain it.

**Proof.** Let $B$ be a bounded subset of the Hilbert space $X$. Since $B$ is bounded, its convex hull $B_c$ is bounded [Theorem 6.3.3] and so is $\operatorname{cl} B_c = H$ [Theorem 6.3.2(a)]. We show that if $x \notin H$, then there is a closed ball which contains $H$ but excludes $x$. By Theorem 7.8.6(b), for $x \notin H$, there is a continuous $\mathbf{R}$-linear functional $f$ on $X$ such that $f(x) < \inf f(H)$. Choose an orthonormal basis $\{x_s : s \in S\}$ for the null space $N(f)$ of $f$ so that $X = \operatorname{cl}[\{x_s : s \in S\}] \oplus [w]$ for some unit vector $w$ orthogonal to each $x_s$. By replacing $w$ by $-w$ if necessary and since $f$ must be nontrivial, we may suppose that $f(w) > 0$. We may further assume that $f(w) = 1$ by replacing $f$ by $f/f(w)$, if necessary. For each $z \in X$ there exist scalars $a_{s,z}$ $(s \in S)$ and $a_{w,z}$ such that $z = \sum_{s \in S} a_{s,z} x_s + a_{w,z} w$. Since $H$ is bounded, the translate $H - x$ is bounded, so for some positive $K$, $\sum_{s \in S} |a_{s,z} - a_{s,x}|^2 < \sum_{s \in S} |a_{s,z} - a_{s,x}|^2 + |a_{w,z} - a_{w,x}|^2 \leq K^2$ for each $z \in H$. Since $f(x) < \inf f(H)$, $f(x) < f(z)$, and we can choose scalars $c$ and $d$ such that $f(x) = a_{w,x} < c < d \leq a_{w,z} = f(z)$ for each $z \in H$. Let $C(y, r)$ denote the closed solid ball of radius $r > 0$ and center $y$ in $X$. By the argument below, we may choose $p \in \mathbf{N}$ sufficiently large that $H \subset C(\sum_{s \in S} a_{s,x} x_s + pw, p - c)$ and $x \notin C(\sum_{s \in S} a_{s,x} x_s + pw, p - c)$.

For $p > a_{w,z}$,

$$\left\| x - \left( \sum_{s \in S} a_{s,x} x_s + pw \right) \right\|^2 = \|(a_{w,x} - p)w\|^2$$
$$= |p - a_{w,x}|^2 > (p-c)^2$$

Therefore $x \notin C\left(\sum a_{s,x}x_s + pw, p-c\right)$. If $z \in H$, then

$$\begin{aligned}
&\left\|\sum_{s\in S}\left(a_{s,z}x_s + a_{w,z}w\right) - \sum_{s\in S}\left(a_{s,x}x_s + pw\right)\right\| \\
&= \left(\left\|\sum_{s\in S}\left(a_{s,z} - a_{s,x}\right)x_s - \left(p - a_{w,z}\right)w\right\|^2\right)^{1/2} \\
&\le \left(\sum_{s\in S}\left|a_{s,z} - a_{s,x}\right|^2 + (p-d)^2\right)^{1/2} \\
&< [K^2 + (p-d)^2]^{1/2}
\end{aligned}$$

which is less than $p-c$ if $p$ is sufficiently large because $(p-c)^2-(K^2+(p-d)^2) = p^2 - 2pc + c^2 - K^2 - p^2 + 2pd - d^2 = 2p(d-c) - d^2 - K^2$, which is positive for sufficiently large $p$. ■

**Theorem 9.5.2** For $f \in X'$, if $f(e) = 1 = \|f\|$, then $f(x) \in \operatorname{cl} x(T)_c$ for any $x \in X$.

**Proof.** For $a \in \mathbf{F}$ and $r > 0$, let $C(a,r) = \{b \in \mathbf{F} : |b-a| \le r\}$. Since for any $x \in X$, $x(T)$ is compact and $\mathbf{F}$ is complete, $\operatorname{cl} x(T)_c$ is compact [Theorem 4.8.9(c)]. Since $\mathbf{F}$ is a Hilbert space and $\operatorname{cl} x(T)_c$ is bounded, it follows from Theorem 9.5.1 that $\operatorname{cl} x(T)_c = \cap\{C(a,r) : C(a,r) \supset x(T)\}$. For $r > 0$ and $x(T) \subset C(a,r)$ then $|x(t) - a| \le r$ for each $t \in T$ so $\|x - ae\|_\infty = \sup\{|x(t) - a| : t \in T\} \le r$. Hence, with $f$ as in the hypothesis,

$$|f(x) - a| = |f(x) - af(e)| = |f(x - ae)| \le \|f\| \|x - ae\|_\infty \le r$$

If follows that $f(x) \in \operatorname{cl} x(T)_c$. ■

**Theorem 9.5.3** $\|f\| = f(e)$ IFF $f$ POSITIVE For any $f \in X'$, $f(e) = \|f\|$ iff $f$ is positive. Thus, any two of the following conditions imply the third:

(a) $f$ is positive;
(b) $f(e) = 1$;
(c) $\|f\| = 1$.

**Proof.** For $f \in X'$, suppose that $f(e) = \|f\| \neq 0$ and $x \ge 0$. Then $\|(f/\|f\|)\| = 1 = f(e)/\|f\|$. By Theorem 9.5.2, then $f(x)/\|f\| \in \operatorname{cl} x(T)_c$. Since $x \ge 0$, however, $\operatorname{cl} x(T)_c \subset [0,\infty)$ and it follows that $f(x) \ge 0$. Now suppose that $f$ is positive so that by Eq. (9.1), $f(\operatorname{Re} x) = \operatorname{Re} f(x)$ for any $x \in X$. Choose $\|x\|_\infty \le 1$ and $a \in \mathbf{F}$, $|a| = 1$, such that $|f(x)| = af(x)$. Since $\operatorname{Re} ax \le e$,

$$|f(x)| = f(ax) = \operatorname{Re} f(ax) = f(\operatorname{Re} ax) \le f(e)$$

which implies that $\|f\| \leq f(e)$. Since $f(e) \leq |f(e)| \leq \|f\|$, the desired result follows. ■

**Definition 9.5.4** STATE SPACE The *state space* of $X$ is

$$S(X) = \{f \in X' : f \geq 0 \text{ and } f(e) = 1\}$$

The elements of $S(X)$ are called *states* of $X$. The extreme points of $S(X)$ are called the *pure states* of $X$. By Theorem 9.5.3, it follows that $\|f\| = 1$ for every $f \in S(X)$. ■

Evaluation maps $t'$ $(t \in T)$ are clearly states. Some elementary properties of the state space are:

**Theorem 9.5.5** STATE SPACE PROPERTIES

(a) COMPACT $S(X)$ is $\sigma(X', X)$-compact and convex.

(b) EVALUATIONS $S(X) = \text{cl}_{\sigma(X',X)}(T')_c$.

(c) PURE STATES $\text{ext}\, S(X) \subset T'$.

**Proof.** (a) Since $(f+g)(e) = 2$ for any $f, g \in S(X)$, $S(X)$ is not a subspace but it is easy to verify that it is convex. Clearly, any evaluation map $t'$ is a state, so $T' \subset S(X)$. Moreover, since $\langle e, \cdot \rangle$ is a $\sigma(X', X)$-continuous linear functional on $X'$, $H = \{f \in X' : \langle e, f \rangle = 1\}$ is a $\sigma(X', X)$-closed hyperplane in $X'$. Since the conditions $\|f\| \leq 1$ and $f(e) = 1$ imply that $\|f\| = 1$ [Theorem 9.5.3],

$$S(X) = H \cap \{f \in X' : \|f\| = 1\} = H \cap \{f \in X' : \|f\| \leq 1\} = H \cap U(X')$$

As $U(X')$ is convex and $\sigma(X', X)$-compact [Alaoglu, 8.4.1], $S(X)$ is $\sigma(X', X)$-compact.

(b) Since $T' \subset S(X)$, $\text{cl}_{\sigma(X',X)}(T')_c \subset S(X)$. To reverse the inclusion, first note that for any $x \in X$, $x(T) = \langle x, T' \rangle \subset \langle x, \text{cl}_{\sigma(X',X)}(T')_c \rangle$. Since $\langle x, \text{cl}_{\sigma(X',X)}(T')_c \rangle$ is a convex and closed (since it is compact) subset of $\mathbf{F}$, it follows that $\text{cl}\, x(T)_c \subset \langle x, \text{cl}_{\sigma(X',X)}(T')_c \rangle$. For any $f \in S(X)$, $f(e) = 1$ so, by Theorem 9.5.2, $f(x) \in \text{cl}\, x(T)_c$. Therefore

$$\text{Re}\, f(x) \leq \max\{\text{Re}\, h(x) : h \in \text{cl}_{\sigma(X',X)}(T')_c\}$$

If $f$ does not belong to the $\sigma(X', X)$-compact convex set $\text{cl}_{\sigma(X',X)}(T')_c$, then by a consequence of the Hahn–Banach theorem [7.8.6(d)] and the weak representation theorem [8.2.10], there is some $x \in X$ such that $\max\{\text{Re}\, h(x) : h \in \text{cl}_{\sigma(X',X)}(T')_c\} < \text{Re}\, f(x)$, which is contradictory. Therefore $f \in \text{cl}_{\sigma(X',X)}(T')_c$.

(c) $T'$ is a subset of the $\sigma(X', X)$-compact, convex set $S(X)$. Since the map $t \mapsto t'$ of $T$ into $(X', \sigma(X', X))$ is a homeomorphism (as in Example 8.4.10(b)), $T'$ is $\sigma(X', X)$-compact, hence $\sigma(X', X)$-closed. Since $S(X) =$

$\mathrm{cl}_{\sigma(X',X)}(T')_c$ by (b), it follows from the Krein–Milman theorem [9.4.6(a)] that $\mathrm{ext}\, S(X) \subset \mathrm{cl}_{\sigma(X',X)} T' = T'$. ■

Not every point $t'$ of $T'$ need be an extreme point of $S(X)$, i.e., need be a pure state of $X$. We single out those that are for further study.

**Definition 9.5.6** CHOQUET BOUNDARY Those $t \in T$ such that $t' \in \mathrm{ext}\, S(X)$ constitute the *Choquet boundary* $\mathrm{ch}\, X$ of $X$. ■

As we show in Theorem 9.5.12, the points of the Choquet boundary are of sufficient quality and quantity to reconstitute the unit ball of $X'$: $U(X') = \mathrm{cl}_{\sigma(X',X)}[(\mathrm{ch}\, X)']_{bc}$.

With respect to pointwise multiplication $[(xy)(t) = x(t)\, y(t)]$, $C(T, \mathbf{F})$ is an algebra. When speaking of $X$as an algebra, we mean as a subalgebra of $C(T, \mathbf{F})$. A *multiplicative linear functional* or (*algebra*) *homomorphism* of $X$ into $\mathbf{F}$is a linear map $f : X \to \mathbf{F}$ such that $f(xy) = f(x)\, f(y)$ $(x, y \in X)$. Evaluations $t'$are continuous algebra homomorphisms of any subalgebra $X$ of $C(T, \mathbf{F})$ although they may be trivial since $X$, for example, could consist of functions that vanish at certain points of $T$.

**Theorem 9.5.7** HOMOMORPHISMS = PURE STATES If $X$ is an algebra, then $f \in S(X)$ is a nontrivial homomorphism of $X$ iff $f \in \mathrm{ext}\, S(X)$.

**Proof.** Suppose that $X$ is an algebra and $f \in \mathrm{ext}\, S(X)$. We show that $f(xy) = f(x)\, f(y)$ $(x, y \in X)$ in three stages. Let $x \in X$.

(1) Suppose $0 \le y \le e$. Let $g(x) = f(xy) - f(x)\, f(y)$ $(x \in X)$. Since $f(e) = 1$ and $g(e) = 0$, $(f + g)(e) = (f - g)(e) = 1$. We show next that $g = 0$. To see that $f \pm g$ is positive, consider $z \ge 0$. Since $f \ge 0$ and $f(e) = 1$, $(f + g)(z) = f(z) + f(zy) - f(z)\, f(y) = f(z)(1 - f(y)) + f(zy) \ge 0$; similarly, $(f - g)(z) \ge 0$. In other words, $f \pm g \in S(X)$. Since $f \in \mathrm{ext}\, S(X)$ is the midpoint of $[f - g, f + g]$, it follows that $g = 0$.

(2) Real-valued $y$. Since $y$ is bounded, there exist $a, b > 0$ such that $0 \le ay + be \le e$. Hence, by (1),

$$\begin{aligned} &f(x(ay + be)) = f(x)\, f(ay + be) \\ \text{or}\quad &af(xy) + bf(x) = af(x)\, f(y) + bf(x) \quad \text{or} \quad af(xy) = af(x)\, f(y) \end{aligned}$$

(3) Suppose $y$ is arbitrary. We split $y$ into its real and imaginary parts: $y = z + iw$. Since $X$ is self-adjoint, $z, w \in X$. By (2), for any $x \in X$, $f(xy) = f(x(z + iw)) = f(xz) + if(xw) = f(x)\, f(z) + if(x)\, f(w) = f(x)\, f(z + iw) = f(x)\, f(y)$.

To prove the converse, we first demonstrate that for any $f \in S(X)$, $f(x)^2 \le f(x^2)$ for real-valued $x$. Since $X$ is self-adjoint then, by Eq. (9.1), $f \ge 0$ and $x$ real-valued imply that $\mathrm{Re}\, f(x) = f(\mathrm{Re}\, x) = f(x)$, i.e., $f(x)$ is real. Therefore, for any $t \in \mathbf{R}$, $(tx + e)^2 \ge 0$. Hence $0 \le f\left((tx + e)^2\right) =$

$t^2 f(x^2) + 2tf(x) + f(e)$, so the discriminant of this quadratic form must be $\leq 0$, i.e., $4f(x)^2 \leq 4f(x^2) f(e) = 4f(x^2)$.

Now suppose that $f \in S(X)$ is a homomorphism and that $f = (1/2)(g+h)$ for $g, h \in S(X)$. If $x \in X$ is real-valued, since $g, h \geq 0$, it follows from the observation above that

$$\begin{aligned}&(1/2)\, g(x)^2 + (1/2)\, h(x)^2 \leq (1/2)\left[g(x^2) + h(x^2)\right] = f(x^2)\\ &= f(x)^2 = (1/4)\, g(x)^2 + (1/2)\, g(x)\, h(x) + (1/4)\, h(x)^2\end{aligned}$$

Since $(1/2)\, g(x)^2 + (1/2)\, h(x)^2 \leq (1/4)\, g(x)^2 + (1/2)\, g(x)\, h(x) + (1/4)\, h(x)^2$, subtracting the right-hand expression from each side yields

$$(1/4)\, g(x)^2 - (1/2)\, g(x)\, h(x) + (1/4)\, h(x)^2 = [(1/2)(g(x) - h(x))]^2 \leq 0$$

By Eq. (9.1), $g(x)$ and $h(x)$ are real, so $g = h$ on real-valued $x \in X$. Generally, for $z = x + iw \in X$, where $x$ and $w$ are real-valued, $g(z) = g(x) + ig(w) = h(x) + ih(w) = h(z)$, where, as above, we use the fact that $X$ is self-adjoint to infer that $x$ and $w$ belong to $X$. This proves that $g = h$ and consequently that $f \in \operatorname{ext} S(X)$. ■

**Example 9.5.8** CHOQUET BOUNDARY $\operatorname{ch} C(T, \mathbf{F}, \|\cdot\|_\infty) = T$ Each evaluation map $t'$ $(t \in T)$ of $C(T, \mathbf{F}, \|\cdot\|_\infty)$ is a nontrivial homomorphism of $X$, hence is an extreme point of $S(X)$ [Theorem 9.5.7]. Since $\operatorname{ext} S(X) \subset T'$ [Theorem 9.5.5(c)], it follows that $\operatorname{ch} C(T, \mathbf{F}, \|\cdot\|_\infty) = T$. If $X$ is the subalgebra (maximal ideal, actually) of all functions which vanish at $t \in T$, then $\operatorname{ch} X = T \setminus \{t\}$. ■

The structure of the Choquet boundary is quite variable. If $T$ is metrizable, then $\operatorname{ch} X$ is a $G_\delta$-set. Otherwise, it need not even be a Borel set (see, e.g., Stout 1971, pp. 54–55, 138–139). The following characterization of Choquet boundary is often used to define it. (See Sec. 1.6 for definitions of the measure-theoretic terms.) A point $t \in T$ is in the Choquet boundary of $X$ iff $t'(\cdot) = \int_T (\cdot)\, d\mu$ for some probability measure $\mu$ on $T$ implies that $\mu = \mu_t$ where $\mu_t$ denotes the "point mass" concentrated at $t$ [Phelps 1966, p. 38; Larsen 1973b, p. 257]. More generally, $f \in S(X)$ iff there is a probability measure $\mu$ on $T$ such that $f(\cdot) = \int_T (\cdot)\, d\mu$. In this case we say that $\mu$ *represents* $f$ (or is a *representing measure for* $f$) [Asimow and Ellis, 1980, p. 16].

A subset of $T$ on which each $x \in X$ achieves its norm is called a *boundary of* $X$. We prove now that the Choquet boundary is a boundary in this sense.

**Theorem 9.5.9** $\operatorname{ch} X$ IS A BOUNDARY Each $x \in X$ achieves its norm on the Choquet boundary of $X$.

**Proof.** Let $x \in X$. Since $T$ is compact, there is some $t_0 \in T$ such that $|x(t_0)| = \|x\|_\infty$. Now consider the $\sigma(X', X)$-continuous seminorm $|\langle x, \cdot\rangle|$ on $X'$. Since $T' \subset S(X)$, $x(t) = \langle x, t'\rangle \in \langle x, S(X)\rangle$ for all $t \in T$ so

$\|x\|_\infty \leq \max|\langle x, S(X)\rangle|$ ("max" because $S(X)$ is $\sigma(X', X)$-compact by Theorem 9.5.5(a)). On the other hand, for any $h \in S(X) \subset U(X')$, $|h(x)| \leq \|x\|_\infty$, so $\max|\langle x, S(X)\rangle| = \|x\|_\infty$.

Thus $K = \{h \in S(X) : |\langle x, h\rangle| = \max|\langle x, S(X)\rangle| = \|x\|_\infty\}$ is nonempty, convex and $\sigma(X', X)$-compact; hence, by Theorem 9.3.4, it is a convex face of $S(X)$. As a closed convex subset of the $\sigma(X', X)$-compact set $S(X)$, $K$ has extreme points $h$ by the Krein–Milman Theorem [9.4.1]. Since $h \in \operatorname{ext} K$ and $K$ is a face of $S(X)$, it follows from Example 9.3.2(e) that $h \in \operatorname{ext} S(X)$. By Theorem 9.5.5(c), $\operatorname{ext} S(X) \subset T'$, so $h = t'$ for some $t \in \operatorname{ch} X = \{t \in T : t' \in \operatorname{ext} S(X)\}$ and $|h(x)| = \|x\|_\infty = |x(t)|$. ■

**Example 9.5.10** CHOQUET BOUNDARY If $T = [0, 1]$ and

$$X = \{x \in C(T, \mathbf{R}, \|\cdot\|_\infty) : x(1/2) = (1/2)[x(0) + x(1)]\}$$

then $(1/2)' = (1/2)(0' + 1')$. Therefore $1/2 \notin \operatorname{ch} X$. All points $t \in [0, 1] \setminus \{1/2\}$ are in $\operatorname{ch} X$ as can be seen by choosing an $x \in X$ that peaks at $t$ because, by Theorem 9.5.9, each $x \in X$ must achieve its maximum absolute value at a point of $\operatorname{ch} X$. Thus, no point other than $1/2$ can be missing from $\operatorname{ch} X$ and $\operatorname{ch} X = [0, 1] \setminus \{1/2\}$. ■

For the sake of perspective, we consider two other boundaries.

**Definition 9.5.11** (a) SILOV BOUNDARY If there exists a (necessarily unique) minimal closed boundary $\operatorname{Si} X$ for $X$, it is called the *Silov boundary of* $X$ (see Larsen 1973, p. 222ff). ■

(b) BISHOP BOUNDARY A point $t \in T$ is a *peak point* for $X$ if there exists an $x \in X$ such that $|x(t)| = \|x\|_\infty = 1$ and, for all $s \neq t$, $|x(s)| < |x(t)|$. The set $B(X)$ of peak points is called the *Bishop boundary* of $X$. Other names for peak points are *strong boundary points* and *unique maximum points.* ■

If $X$ is a Banach algebra then $\operatorname{Si} X \neq \emptyset$ and for any closed subalgebra $M$ which separates points and contains the identity, we have a sort of Hahn–Banach theorem for homomorphisms in that if a continuous homomorphism $h \in \operatorname{Si} M$ then we can continuously extend $h$ from $M$ to $X$ [Gelfand et al. 1964, pp. 73–80; Naimark 1960, pp. 212–215]. For algebras $X$ such as we have been considering (separating, self-adjoint), $\operatorname{Si} X = T$ [Larsen 1973b, p. 226]. Also, $\operatorname{Si} X = \operatorname{cl}_{\sigma(X', X)} \operatorname{ch} X$ [Larsen 1973b, pp. 236–241]. If $X$ is the Banach algebra of functions continuous on the unit disk $D \subset \mathbf{C}$, analytic in the interior of $D$, $\operatorname{Si} X = \{a \in \mathbf{C} : |a| = 1\}$, the topological boundary of $D$, by the maximum modulus theorem. The Bishop boundary can be empty [Larsen 1973b, pp. 230–232] and generally $B(X) \subset \operatorname{ch} X \subset \operatorname{Si} X$. If $T$ is metrizable and $X$ is a Banach algebra, then $B(X) = \operatorname{ch} X$ [Phelps 1966, p. 53].

**Theorem 9.5.12** UNIT BALL OF $X'$ $U(X') = \operatorname{cl}_{\sigma(X', X)} [(\operatorname{ch} X)']_{bc}$.

**Proof.** Since $\mathrm{ch}\,X$ is a boundary for $X$, for any $x \in X$, there exists $t_0 \in \mathrm{ch}\,X$ such that $|x(t)| = \|x\|_\infty$ by Theorem 9.5.9. Thus, for any $f \in U(X')$, $|f(x)| \leq \|x\|_\infty = |x(t_0)| = |t_0'(x)| = |\langle x, t_0' \rangle|$. Hence, if $x \in [(\mathrm{ch}\,X)']^\circ$, then $|\langle x, f \rangle| \leq 1$ for any $f \in U(X')$; in other words, $[(\mathrm{ch}\,X)']^\circ \subset U(X')^\circ$. By Example 8.3.3, $U(X') = U(X')^{\circ\circ}$. Hence, taking polars, $U(X') \subset \mathrm{cl}_{\sigma(X',X)}[(\mathrm{ch}\,X)']_{bc}$. Since $(\mathrm{ch}\,X)' \subset U(X')$ and $U(X')$ is a $\sigma(X',X)$-closed disk in $X'$, $\mathrm{cl}_{\sigma(X',X)}[(\mathrm{ch}\,X)']_{bc} \subset U(X')$ and the theorem follows. ■

We show next that the extreme points of the unit ball of $C(T, \mathbf{F}, \|\cdot\|_\infty)'$ are unit multiples of evaluation maps, a fact that plays a crucial role in the proof of the Banach–Stone theorem [9.6.2].

**Example 9.5.13** EXT $U(C(T, \mathbf{F}, \|\cdot\|_\infty)') = \{at' : |a| = 1, t \in T\}$ Let $X = C(T, \mathbf{F}, \|\cdot\|_\infty)$. First we show that $\mathrm{ext}\,S(X) \subset \mathrm{ext}\,U(X')$. We accomplish this by showing that $S(X)$ is a convex face of $U(X')$ and using the result of Example 9.3.2(e). To this end let $h \in S(X)$ and suppose that $h = (1/2)(f+g)$ for $f, g \in U(X')$. Since 1 is an extreme point of the unit disk in $\mathbf{F}$, $h(e) = 1 = (1/2)[f(e) + g(e)]$ implies that $f(e) = g(e) = 1$. Hence $\|f\| = \|g\| = 1$ and it follows that $f, g \in S(X)$ by Theorem 9.5.3 and $S(X)$ is a convex face of $U(X')$.

By Example 9.5.8, $\mathrm{ch}\,X = T$, i.e., $T' = \mathrm{ext}\,S(X)$. By Theorem 9.5.12, replacing $\mathrm{ch}\,X$ by $T$, we have $U(X') = \mathrm{cl}_{\sigma(X',X)}(T')_{bc}$. For this to happen, by the Krein–Milman theorem [9.4.6(a)], we must have $\mathrm{ext}\,U(X') \subset \mathrm{cl}_{\sigma(X',X)}(T')_b$. Since $(T')_b$ is the continuous image under scalar multiplication of the compact set $a \in \mathbf{F} : |a| \leq 1 \times T\}$, it is $\sigma(X',X)$-compact; hence $\mathrm{cl}_{\sigma(X',X)}(T')_b = (T')_b$. Therefore, if $f \in \mathrm{ext}\,U(X')$, there exist $a \in \mathbf{F}$, $|a| \leq 1$, and $t \in T$ such that $f = at'$. Since $f \in \mathrm{ext}\,U(X')$, however, $\|f\| = 1$ [Theorem 9.2.4]. Thus $1 = \|f\| = |a|\,\|t'\|$; since $\|t'\| = 1$, $|a| = 1$. With $B = \{at' : |a| = 1, t \in T\}$, we now have $\mathrm{ext}U(X') \subset B$. For $|a| = 1$, the map $A : X' \to X'$, $f \mapsto af$, is a linear isometry of $X'$ onto itself; hence, by Theorem 9.2.3, each $at' \in B$ is an extreme point of $A(U(X')) = aU(X') = U(X')$. In other words, $B \subset \mathrm{ext}\,U(X')$. ■

## 9.6 THE BANACH–STONE THEOREM

**Notation.** $C(T, \mathbf{F}, \|\cdot\|_\infty)$ and $C(S, \mathbf{F}, \|\cdot\|_\infty)$ denote the Banach spaces of continuous maps of the compact Hausdorff spaces $S$ and $T$, respectively, into $\mathbf{F}$ (= $\mathbf{R}$ or $\mathbf{C}$). (By the well-known Gelfand–Naimark theorem, any complex commutative $B^*$-algebra with identity is of this type.) We denote by $e$ the map $e : T \to \mathbf{F}$, $t \mapsto 1$. ■

Clearly there is a correlation between the topological properties of $T$ and Banach space properties of $C(T, \mathbf{F}, \|\cdot\|_\infty)$. If $T$ carries the trivial topology then the only continuous functions are constants, essentially $C(T, \mathbf{F}) = \mathbf{F}$;

if $T$ is discrete then $C(T,\mathbf{F}) = \mathbf{F}^T$. The classical Banach–Stone theorem characterizes the connection between linear isometries of $C(T,\mathbf{F},\|\cdot\|_\infty)$ and $C(S,\mathbf{F},\|\cdot\|_\infty)$ and $S$ and $T$ when $S$ and $T$ are compact. We discuss some variations on that theme in Section 9.9.

If $S$ and $T$ are homeomorphic, then the Banach spaces $C(T,\mathbf{F},\|\cdot\|_\infty)$ and $C(S,\mathbf{F},\|\cdot\|_\infty)$ are linearly isometric: if $h: T \to S$ is a surjective homeomorphism, then the composition map $H: C(S,\mathbf{F},\|\cdot\|_\infty) \to C(T,\mathbf{F},\|\cdot\|_\infty), x \mapsto x \circ h$, is a surjective linear isometry. Thus, homeomorphism of $S$ and $T$ implies linear isometry of $C(T,\mathbf{F},\|\cdot\|_\infty)$ and $C(S,\mathbf{F},\|\cdot\|_\infty)$. The interesting question is the converse: If $C(T,\mathbf{F},\|\cdot\|_\infty)$ is linearly isometric to $C(S,\mathbf{F},\|\cdot\|_\infty)$, must $S$ be homeomorphic to $T$? Using the Mazur–Ulam result, Theorem 9.1.2 and the Gateaux differentiability of $\|\cdot\|_\infty$, Banach [1932, p. 170] proved this to be so for real-valued functions and metrizable $S$ and $T$. Stone [1937b, p. 469] proved it without metrizability by very different methods—essentially using flat faces of the unit ball of $C(S,\mathbf{R},\|\cdot\|_\infty)$; Behrends [1979, p. 139] uses a similar approach. Semadeni [1971, p. 129] also uses Theorem 9.1.2 (for real-valued functions) but—unlike Banach—completes the argument using Kaplansky's theorem [*ibid.*, p.131] that order isomorphism of $C(T,\mathbf{R})$ and $C(S,\mathbf{R})$ implies homeomorphism of $S$ and $T$. Subsequently, Arens and Kelley [1947] proved it using facts about extreme points of the unit ball of $C(T,\mathbf{F},\|\cdot\|_\infty)$ and adjoints. We use their technique to prove what is now known as the *Banach–Stone theorem* in Theorem 9.6.2.

Can *linear isometry* be weakened to *linear homeomorphism*? Absolutely not [Example 9.6.3] but the norm-preservation requirement can be weakened a little. Amir [1965] and Cambern [1966, 1967] showed that $S$ and $T$ are homeomorphic if there is only a surjective linear homeomorphism $A: C(S,\mathbf{R},\|\cdot\|_\infty) \to C(T,\mathbf{R},\|\cdot\|_\infty)$ such that $\|A\| \|A^{-1}\| < 2$ but even if

$$\|A\| \|A^{-1}\| = 2$$

the result can fail [Cambern 1967; see also the discussion after Theorem 9.9.2].

Linear isometries between spaces of continuous functions on compact sets must have the canonical form of Definition 9.6.1(b).

**Definition 9.6.1** WEIGHTED COMPOSITION AND BANACH–STONE MAP Let $S$ and $T$ be topological spaces, let $h: S \to T$ be continuous and let $w \in C(S,\mathbf{F})$. A map $A: C(T,\mathbf{F}) \to C(S,\mathbf{F})$, $Ax(s) = w(s)\,x(h(s))$ $(s \in S)$, is called a (a) *weighted composition map*, $w$ being the *weight* or *weight function*. We write $Ax = w \cdot (x \circ h)$. (b) If $w$ is nonvanishing and $h$ is a surjective homeomorphism, then $A$ is called a *Banach–Stone map*. A scalar-valued function $w$ on $S$ is called *unimodular* if $|w(s)| \equiv 1$. ■

The Banach–Stone theorem shows that linear isometry of $C(S,\mathbf{F},\|\cdot\|_\infty)$ and $C(T,\mathbf{F},\|\cdot\|_\infty)$ not only implies homeomorphism of $S$ and $T$ but also that the linear isometry is a Banach–Stone map with a unimodular weight.

**Theorem 9.6.2** BANACH–STONE THEOREM Let $C(S, \mathbf{F}, \|\cdot\|_\infty) = C(S, \mathbf{F})$ and $C(T, \mathbf{F}, \|\cdot\|_\infty) = C(T, \mathbf{F})$ be the Banach spaces of continuous maps of the compact Hausdorff spaces $S$ and $T$ into $\mathbf{F}$, respectively. If $A : C(T, \mathbf{F}) \to C(S, \mathbf{F})$ is a surjective linear isometry then:

(a) HOMEOMORPHISM There is a surjective homeomorphism $h : S \to T$ and

(b) CANONICAL FORM $A$ is a Banach–Stone map, $Ax = w \cdot (x \circ h)$ for each $x \in C(T, \mathbf{F})$ where the weight $w = Ae \in C(S, \mathbf{F})$ is such that $|w(s)| \equiv 1$.

**Proof.** With notation as above, let $A : C(T, \mathbf{F}) \to C(S, \mathbf{F})$ be a surjective linear isometry and let $A' : C(S, \mathbf{F})' \to C(T, \mathbf{F})'$ be its adjoint. As follows directly from the definitions, the fact that $A$ is a surjective linear isometry implies that $A'$ is a surjective linear isometry when $C(S, \mathbf{F})'$ and $C(T, \mathbf{F})'$ carry their norm topologies. Therefore $A'$ is weak-$*$ continuous by the dual form of Theorem 8.11.3(c), i.e., is continuous when $C(S, \mathbf{F})'$ and $C(T, \mathbf{F})'$ carry their weak-$*$ topologies which we assume that they do for the rest of the proof. $S'$ and $T'$ denote the spaces of evaluation maps determined by the points of $S$ and $T$; as follows from Example 8.4.10(b), $S'$ and $T'$ are $\sigma(X', X)$-compact subsets of $C(T, \mathbf{F})'$ and $C(T, \mathbf{F})'$.

As shown in Example 9.5.13, $\operatorname{ext} U\left(C(S, \mathbf{F})'\right) = \cup_{|a|=1} aS'$. Since $A'$ is a linear isometry, it follows from Theorem 9.2.3 that $A'(\operatorname{ext} U\left(C(S, \mathbf{F})'\right) = A'\left(\cup_{|a|=1} aS'\right) = \operatorname{ext} A'(U(C(S, \mathbf{F})')) = \operatorname{ext} U(C(T, \mathbf{F})' = \cup_{|a|=1} aT'$. Since $A'\left(\cup_{|a|=1} aS'\right) = \cup_{|a|=1} aT'$, for each $s \in S$ there exists $h(s) \in T$ and and a unit scalar $w(s) \in \mathbf{F}$ such that $A's' = w(s)\, h(s)'$. In other words, there exist maps $h : S \to T$ and $w : S \to \mathbf{F}$, $|w(s)| \equiv 1$, such that for each $s \in S$, and $x \in C(T, \mathbf{F})$, $(Ax)(s) = s'(Ax) = A's'(x) = w(s)\, h(s)'(x) = w(s)\, x(h(s))$. It remains to show that $h$ is a homeomorphism and that $w$ is continuous. The continuity of $w$ is easy: Let $e \in C(T, \mathbf{F})$ be the map that is identically 1. Then for each $s$ in $S$, $(Ae)(s) = s'(Ae) = A's'(e) = w(s)h(s)'(e) = w(s)\, e(h(s)) = w(s)$. Therefore $w = Ae$ and $w$ is continuous.

(1) $h$ IS INJECTIVE If $h(s_1) = h(s_2)$, then $h(s_1)' = h(s_2)'$. Since, for any $x \in C(T, \mathbf{F})$, $A's_1'(x) = w(s_1)\, h(s_1)'(x)$ and $A's_2'(x) = w(s_2)\, h(s_2)'(x)$,

$$w(s_1)^{-1} A's_1' = w(s_2)^{-1} A's_2'$$

Applying $(A')^{-1}$ to each side, this implies that $w(s_1)^{-1} s_1' = w(s_2)^{-1} s_2'$. Thus, for each $x \in C(S, \mathbf{F}), |\langle x, s_1' \rangle| = |x(s_1)| = |x(s_2)|$. Since $S$ is a normal Hausdorff space, this implies that $s_1 = s_2$.

(2) $h$ IS SURJECTIVE Since $A'\left(\cup_{|a|=1} aS'\right) = \cup_{|b|=1} bT'$, for each $t \in T$ and $|b| = 1$, there exists $|a| = 1$ and $s \in S$ such that $bt' = A'(as') = aA's' = aw(s)\, h(s)'$. Hence, for each $x \in C(T, \mathbf{F})$, $|x(h(s))| = |x(t)|$. By the normality of $T$, $h(s) = t$.

(3) $h$ IS BICONTINUOUS Since [Example 8.4.10(b)] $S$ and $T$ are homeomorphic to $S'$ and $T'$, respectively, it suffices to show that $h' : S' \to T'$, $s' \mapsto h(s)' = A's'/w(s)$ is a surjective homeomorphism. (1) and (2) show that $h'$

is a bijection. Generally, a continuous 1-1 map between compact Hausdorff spaces is a homeomorphism since it sends closed sets into closed sets. Thus, since $S'$ and $T'$ are compact Hausdorff spaces, it only remains to show that $h'$ is continuous. Since $w$ is never 0, $1/w$ is continuous; since $A'$ is continuous, $A'$ is weak-$*$ continuous by Theorem 8.11.3(c). Identifying $S$ and $S'$ and $T$ and $T'$, the (weak-$*$) continuity of $h' = ((1/w)\,A')|_{S'}$ follows. ■

We can use the Banach–Stone theorem to show that certain normed spaces are not linearly isometric. For example, $\ell_\infty^{\mathbf{R}}(2) = C(\{0,1\}, \mathbf{R}, \|\cdot\|_\infty)$ ($\{0,1\}$ discrete) is not linearly isometric to $\mathbf{R} = C(\{0\}, \mathbf{R}, \|\cdot\|_\infty)$ because $\{0,1\}$ is not homeomorphic to $\{0\}$.

As shown in Example 9.6.3, if the geometric bond between $C(T, \mathbf{F}, \|\cdot\|_\infty)$ and $C(S, \mathbf{F}, \|\cdot\|_\infty)$ is weakened to linear homeomorphism, the homeomorphism between $S$ and $T$ withers:

$C[0,1]$ is linearly homeomorphic to $C([0,1]^2)$ but $[0,1]$ is not homeomorphic to $[0,1]^2$.

**Example 9.6.3** LINEAR ISOMETRY VS. LINEAR HOMEOMORPHISM

(a) $C[0,1]$ IS LINEARLY HOMEOMORPHIC TO $C([0,1]^2)$ Banach [1932, p. 185] remarked that *It is not known if* $C[0,1]$ *is linearly homeomorphic to* $C([0,1]^2)$. Apparently unaware that Banach had raised the question, Miljutin proved a more general result in 1952 [although not published until 1966; see Semadeni 1971, p. 380, for the story], namely that for any uncountable compact metric spaces $S$ and $T$, $C(S, \mathbf{F}, \|\cdot\|_\infty)$ is linearly homeomorphic to $C(T, \mathbf{F}, \|\cdot\|_\infty)$; in particular, for any uncountable compact metric space $T$, $C(T, \mathbf{F}, \|\cdot\|_\infty)$ is linearly homeomorphic to $C[0,1]$, thereby affirmatively settling Banach's question. While we are on the subject, we mention that Cengiz 1978 showed that for locally compact $S$ and $T$, linear homeomorphism of $C(S, \mathbf{F}, \|\cdot\|_\infty)$ and $C(T, \mathbf{F}, \|\cdot\|_\infty)$ implies that $S$ and $T$ have the same cardinality.

(b) $C[0,1]$ IS NOT LINEARLY ISOMETRIC TO $C([0,1]^2)$ If $C([0,1], \mathbf{F}, \|\cdot\|_\infty)$ is linearly isometric to $C([0,1]^2, \mathbf{F}, \|\cdot\|_\infty)$ then, by the Banach–Stone theorem, there is a surjective homeomorphism $h : [0,1] \to [0,1]^2$. This implies the contradictory result that the disconnected space $[0,1] \backslash \{1/2\}$ is homeomorphic to the connected space $[0,1]^2 \setminus \{h(1/2)\}$. ■

What if we weaken the compactness hypothesis on $S$ and $T$ in the Banach–Stone theorem and consider only *bounded* continuous functions $C_b(S, \mathbf{F}, \|\cdot\|_\infty)$ and $C_b(T, \mathbf{F}, \|\cdot\|_\infty)$ on the completely regular Hausdorff spaces $S$ and $T$? For $x \in C_b(T, \mathbf{F})$, let $\beta x$ denote the continuous extension of $x$ to the Stone-Cech compactification $\beta T$ of $T$. The map $x \mapsto \beta x$ establishes a linear isometry between $C_b(T, \mathbf{F}, \|\cdot\|_\infty)$ and $C(\beta T, \mathbf{F}, \|\cdot\|_\infty)$. Hence if $C_b(T, \mathbf{F}, \|\cdot\|_\infty)$ and $C_b(S, \mathbf{F}, \|\cdot\|_\infty)$ are linearly isometric, then so are $C(\beta T, \mathbf{F}, \|\cdot\|_\infty)$ and $C(\beta S, \mathbf{F}, \|\cdot\|_\infty)$. We conclude that:

**Theorem 9.6.4** NONCOMPACT S AND T For completely regular Hausdorff spaces $S$ and $T$, $C_b(T, \mathbf{F}, \|\cdot\|_\infty)$ is linearly isometric to $C_b(S, \mathbf{F}, \|\cdot\|_\infty)$ iff $\beta T$ is homeomorphic to $\beta S$.

Banach [1932, p. 178] (almost) got a result very similar in appearance to the Banach–Stone Theorem for linear isometries $H : L_p[0,1] \to L_p[0,1]$. Lamperti [1958, pp. 461–462] completed Banach's argument and extended the result to the $\sigma$-finite case. Royden [1968, p. 333] has the following result for $L_p[0,1]$:

**Theorem 9.6.5** LINEAR ISOMETRIES OF $L_p[0,1]$ For $p \in [1, \infty) \setminus \{2\}$ and a linear isometry $H : L_p[0,1] \to L_p[0,1]$ there is a Borel measurable map $h$ of $[0,1]$ onto almost all of $[0,1]$ and $w \in L_p[0,1]$ such that for any $x \in L_p[0,1]$, $Hx = w \cdot (x \circ h)$; furthermore $w$ and $h$ are uniquely determined to within a.e. equivalence on the cozero set of $h$.

A topological space is **N**-*compact* if it is homeomorphic to a closed subspace of a power of **N**. The Banach–Stone theorem is about preservation of suprema. Preservation of infima is equivalent to preservation of suprema in a certain situation as shown in Theorem 9.6.6. Although $T$ is assumed to be **N**-compact, note that there is no continuity assumption on $A$.

**Theorem 9.6.6** PRESERVATION OF INFIMA Araujo [2002b, Theorem 4.1] Let $S$ be an **N**-compact Hausdorff space and let $T$ be completely regular Hausdorff space. Let $C(S, \mathbf{F})$ and $C(T, \mathbf{F})$ denote the spaces of continuous maps of $S$ and $T$ into **F**, respectively, and let $A : C(T, \mathbf{F}) \to C(S, \mathbf{F})$ be a surjective linear map. Then the following are equivalent:

(a) PRESERVATION OF INF For every $x \in C(T, \mathbf{F})$, $\inf |x(T)| = \inf |Ax(S)|$.

(b) CANONICAL FORM $A$ is a Banach–Stone map, $Ax = w \cdot (x \circ h)$ for each $x \in C(T, \mathbf{F})$ where the weight $w \in C(S, \mathbf{F})$ is such that $|w(s)| \equiv 1$.

## 9.6.1 The Realcompactification

For any completely regular (Hausdorff ) space $T$, any $x \in C_b(T, \mathbf{R})$ has a unique continuous extension $\beta x$ to $\beta T$. There is a subspace $\upsilon T$ of $\beta T$ such that each $x \in C(T, \mathbf{F})$ has a unique continuous extension $\upsilon x$ to $\upsilon T$; $\upsilon T$ is called by the jawbreaking term *realcompactification* or *repletion* of $T$. There are many ways to describe it. In Sec. 2.4 we discussed the inverse image topology induced on a set $S$ by a function $f : S \to T$ where $T$ is a topological space. It is the weakest topology for $S$ that makes $f$ continuous. The realcompactification can be obtained similarly: There is a weakest uniform structure $\mathcal{C}$ for $C(T, \mathbf{R})$ with respect to which each $x \in C(T, \mathbf{R})$ is *uniformly* continuous. The completion of the uniform space $(T, \mathcal{C})$ is $\upsilon T$ [Beckenstein et al. 1977, p. 22, Gillman and Jerison 1960, p. 225]. Indeed, $\beta T$ can be obtained similarly as the completion of $(T, \mathcal{C}_b)$ where $\mathcal{C}_b$ is the weakest uniform structure for $C(T, \mathbf{R})$ with respect to which each $x \in C_b(T, \mathbf{R})$ is uniformly continuous

[Beckenstein et al. 1977, p. 9, Gillman and Jerison 1960, p. 226]. If $T = \upsilon T$, we say that $T$ is *realcompact.* The points of $\beta T$ are in 1-1 correspondence with the maximal ideals of $C(T, \mathbf{R})$ [Beckenstein et al. 1977, p. 18; Gillman and Jerison 1960, p. 102] whereas the points of $\upsilon T$ are in 1-1 correspondence with the nonzero ring homomorphisms of $C(T, \mathbf{R})$ into $\mathbf{R}$ [Beckenstein et al. 1977, p. 29; Gillman and Jerison 1960, p. 142]. By extending the real and complex parts separately, complex-valued continuous functions can also be extended to $\upsilon T$.

A fact concerning realcompactifications that we need later is listed in Theorem 9.6.7. Its proof can be found in Beckenstein et al. 1977, p. 22, Theorem 1.5-1(b).

**Theorem 9.6.7** REALCOMPACTIFICATION For a completely regular Hausdorff space $T$, $\upsilon T$ consists of those $p \in \beta T$ such that for all sequences $(V_n)$ of neighborhoods of $p$ in $\beta T$, $(\cap_{n \in \mathbf{N}} V_n) \cap T \neq \varnothing$. Thus $p \notin \upsilon T$ iff there is some $G_\delta$-set in $\beta T$ to which $p$ belongs that does not meet $T$.

# 9.7 SEPARATING MAPS

The Banach–Stone theorem [9.6.2] asserts that a surjective linear isometry $H : C(T, \mathbf{F}, \|\cdot\|_\infty) \to C(S, \mathbf{F}, \|\cdot\|_\infty)$ between spaces of continuous functions on compact Hausdorff spaces $S$ and $T$ into $\mathbf{F}$ must be a weighted composition—$Hx = w \cdot (x \circ h)$ for each $x \in C(T, \mathbf{F})$ where the weight $w = He \in C(S, \mathbf{F})$ is such that $|w(s)| \equiv 1$, i.e., a *Banach–Stone* map [Def. 9.6.1(b)]. It is easy to see that such maps satisfy $xy = 0$ implies $HxHy = 0$ $(x, y \in C(T, \mathbf{F}, \|\cdot\|_\infty))$. We call additive maps with this latter property *separating* and investigate them in this and subsequent sections. We show in Theorem 9.7.15 that, for completely regular Hausdorff $S$ and $T$, if $H : C(T, \mathbf{F}) \to C(S, \mathbf{F})$ is biseparating ($H$ and $H^{-1}$ separating), then $\upsilon T$ is homeomorphic to $\upsilon S$; hence, if $S$ and $T$ are realcompact, they are homeomorphic.

## 9.7.1 Definitions and Examples

Let $X$ and $Y$ be linear spaces in which multiplication is possible—spaces of scalar-valued functions, for example—and suppose $H : X \to Y$ is an additive map. If, for $x, y \in X$, $xy = 0$ implies $HxHy = 0$, we say that $H$ is *separating.* For a scalar-valued function $x$ with domain $T$, the *cozero set* of $x$ is $\operatorname{coz} x = \{t \in T : x(t) \neq 0\}$ of $x$. Using this notation, $H$ is separating iff

$$\operatorname{coz} x \cap \operatorname{coz} y = \varnothing \Rightarrow \operatorname{coz} Hx \cap \operatorname{coz} Hy = \varnothing$$

In the vector lattice environment this is: $\inf(|x|, |y|) = 0 \Rightarrow \inf(|Hx|, |Hy|) = 0$. Since elements $x, y$ of a vector lattice such that $\inf(|x|, |y|) = 0$ are called *disjoint* or *orthogonal,* separating maps are also called *disjointness-preserving, orthogonality-preserving* or *d-homomorphisms.*

Linear isometries between spaces of continuous functions on compact sets must be Banach–Stone maps, hence separating [Example 9.7.1(d)]. With the usual identification of almost everywhere equal functions, Banach [1932, p. 175] proved that linear isometries of $L_p[0,1]$ onto itself had to be separating almost everywhere for all $1 \leq p < \infty$, $p \neq 2$. More generally, if $\mu$ is a measure on any measure space, then linear isometries $H : L_p[\mu] \to L_p[\mu]$, $1 \leq p \leq \infty$, $p \neq 2$, onto itself must satisfy

$$xy = 0 \ (\mu\text{-a.e.}) \text{ implies } HxHy = 0 \ (\mu\text{-a.e.}) \qquad (**)$$

As noted before and in Theorem 9.6.5, Lamperti [1958, pp. 461–462; cf. Royden 1968, p. 333] completed Banach's [1932, p. 178] argument characterizing the linear isometries of $L_p[0,1]$ into itself (essentially as weighted compositions) and extended the result to the $\sigma$-finite case. Arendt [1983, p. 200] subsequently called order bounded linear maps between Banach lattices that satisfy $\inf(|x|,|y|) = 0 \Rightarrow \inf(|Hx|,|Hy|) = 0$ *Lamperti operators.* Other cognomens in the $L_p[\mu]$ context are *separation-preserving* operators [Asmar et al. 1991] and *disjoint operators* [Peller 1976].

**Example 9.7.1** SEPARATING MAPS

(a) HOMOMORPHISM IS If $X$ and $Y$ are rings and $H : X \to Y$ is a ring homomorphism then $H$ is separating.

(b) DIFFERENTIATION IS Let $C^1[0,1]$ denote the continuously differentiable functions on $[0,1]$. Then the map $x \mapsto x'$ (the derivative of $x$) of $C^1[0,1]$ into $C[0,1]$ is separating.

(c) INTEGRATION IS NOT The map $H : C([0,1],\mathbf{R}) \to C([0,1],\mathbf{R})$, $x \mapsto \int_0^t x(s)\,ds$ is not separating since it maps "triangles" into functions that are eventually constant.

(d) WEIGHTED COMPOSITION IS Let $S$ and $T$ be topological spaces, suppose $w \in C(S,\mathbf{F})$ and $h : S \to T$ is continuous. Then $H : C(T,\mathbf{F}) \to C(S,\mathbf{F})$, $x \mapsto w \cdot (x \circ h)$, is separating. If $S$ and $T$ are compact and $w$ is unimodular ($|w(s)| \equiv 1$) then $H : C(T,\mathbf{F},\|\cdot\|_\infty) \to C(S,\mathbf{F},\|\cdot\|_\infty)$ is a linear isometry. ■

**Definition 9.7.2** SEPARATING AND SEPARATED Let $C(S,\mathbf{F})$ and $C(T,\mathbf{F})$ be the spaces of continuous functions on the completely regular Hausdorff spaces $S$ and $T$.

(a) SEPARATED SETS Subsets of $S$ or $T$ with disjoint closures are called *separated.*

(b) SEPARATED FUNCTIONS Scalar-valued functions on $S$ or $T$ are *separated* if their cozero sets are separated.

(c) WEAKLY SEPARATING The additive map $H : C(T,\mathbf{F}) \to C(S,\mathbf{F})$ is *weakly separating* if for $x, y \in C(T,\mathbf{F})$, $\operatorname{cl}\operatorname{coz} x \cap \operatorname{cl}\operatorname{coz} y = \varnothing \Rightarrow \operatorname{coz} Hx \cap \operatorname{coz} Hy = \varnothing$.

(d) WEAKLY BISEPARATING If $H : C(T, \mathbf{F}) \to C(S, \mathbf{F})$ is bijective and $H$ and $H^{-1}$ are weakly separating, then $H$ is *weakly biseparating; if* $H$ and $H^{-1}$ are separating then $H$ is *biseparating*. ■

### 9.7.2 Support Map

For the rest of this section through Sec. 9.7.4 we adhere to the following conventions.

**Notation.** $S$,$T$ : completely regular Hausdorff spaces.

$e_W$ : the map $w \mapsto 1$ of a set $W$ into $\mathbf{F}$.

$C(S, \mathbf{F})$, $C(T, \mathbf{F})$ : continuous functions on $S$ and $T$. $C(S, \mathbf{F}, c)$ and $C(T, \mathbf{F}, c)$ denote, respectively, $C(S, \mathbf{F}), C(T, \mathbf{F})$ endowed with their compact-open topologies.

$H : C(T, \mathbf{F}) \to C(S, \mathbf{F})$ : at least a weakly separating map.

$s', t'$ : evaluation maps associated with $s \in S$, $t \in T$, respectively.

For $s \in S$, $s' \circ H$ denotes the composite map sending $x \in C(T, \mathbf{F})$ into $Hx(s)$; if $C(S, \mathbf{F})$ and $C(T, \mathbf{F})$ are TVS and $H$ is linear and weakly continuous, $s' \circ H = H's'$ where $H'$ denotes the adjoint of $H$.

$D = \bigcup \{\operatorname{coz} Hx : x \in C(T, \mathbf{F})\}$. For a surjective map $H : C(T, \mathbf{F}, c) \to C(S, \mathbf{F}, c)$, note that

$$D = \bigcup \{\operatorname{coz} Hx : x \in C(T, \mathbf{F})\} = \cup \{\operatorname{coz} y : y \in C(S, \mathbf{F})\} = S$$

$cT$ : a Hausdorff compactification of $T$.

$h : D \to cT$, $s \mapsto \operatorname{supp} s' \circ H$, the support map of $H$ [Definition 9.7.5, Theorem 9.7.6]. ■

Associated with each weakly separating map $H : C(T, \mathbf{F}) \to C(S, \mathbf{F})$ is a continuous map $h : S \to cT$ called the *support map associated with* $H$. We develop some of its properties in this subsection. To do that we need Dieudonné's result on continuous partitions of the identity. It generalizes the fact that in a normal space if $F$ (closed) $\subset G$ (open), there is an open set $U$ such that $F \subset U \subset \operatorname{cl} U \subset G$.

**Theorem 9.7.3** CONTINUOUS PARTITION OF THE IDENTITY Let $\{G_1, \ldots, G_n\}$ be a finite open cover of the normal space $T$. Then

(a) there exists a closed cover $\{F_1, \ldots, F_n\}$ of $T$ such that $F_i \subset G_i$ for $i = 1, 2, \ldots, n$;

(b) there exist $\{x_1, \ldots, x_n\} \subset C(T, [0, 1])$ such that $\sum_{i=1}^n x_i = e_T$ and $\operatorname{coz} x_i \subset G_i$ for $i = 1, 2, \ldots, n$. The functions $\{x_1, \ldots, x_n\}$ are called a *continuous partition of the identity.*

**Proof.** Let $\{G_1, \ldots, G_n\}$ be a finite open cover of the normal space $T$.

(a) We proceed by induction. For $n = 1$ we have $T = G$ which means $G$ is clopen and the result is clear. For $n = 2$ and $T = G_1 \cup G_2$ it suffices to take

$F_1 = \complement G_2$ and $F_2 = \complement G_1$. Now suppose the theorem holds for open covers consisting of $n-1$ sets for $n \geq 2$. We have $T = \cup_{i=1}^{n} G_i = \left(\cup_{i=1}^{n-1} G_i\right) \cup G_n$. By what we just showed for $n = 2$, there exist closed sets $F \subset \cup_{i=1}^{n-1} G_i$ and $F_n \subset G_n$ such that $F \cup F_n = T$. Consider the open cover $\{G_i \cap F : i = 1, 2, \ldots, n-1\}$ of $F$ in its relative topology. By the induction hypothesis, there exist closed subsets $F_1, \ldots, F_{n-1}$ of $F$ which cover $F$ with $F_i \subset G_i \cap F$ for $i = 1, 2, \ldots, n-1$; hence $\left(\cup_{i=1}^{n-1} F_i\right) \cup F_n = T$.

(b) Let $\{G_1, \ldots, G_n\}$ be a finite open cover of $T$. By (a), there is a closed cover $\{F_1, \ldots, F_n\}$ of $T$ with $F_i \subset G_i$ for $i = 1, 2, \ldots, n$. By Urysohn's lemma, for $i = 1, 2, \ldots, n$, there exist $y_i \in C(T, [0,1])$ such that $y_i = 1$ on $F_i$ and vanishes on $\complement G_i$, i.e., $\operatorname{coz} y_i \subset G_i$. Since $z(t) = \sum_{i=1}^{n} y_i(t) > 0$ for each $t \in T$, we may consider $x_i(t) = y_i(t) / z(t)$ for $i = 1, 2, \ldots, n$. Clearly $\sum_{i=1}^{n} x_i = e_T$ and $\operatorname{coz} x_i \subset G_i$ for $i = 1, 2, \ldots, n$. ■

**Corollary 9.7.4** PARTITION OF IDENTITY ON CLOSED SUBSPACE Given a finite open cover $\{G_1, \ldots, G_n\}$ of a closed subset $F$ of the normal space $T$, there exist $\{x_1, \ldots, x_n\} \subset C(T, [0,1])$ such that $\sum_{i=1}^{n} x_i|_F = e|_F$ and $\operatorname{coz} x_i \subset G_i$ for $i = 1, 2, \ldots, n$.

**Proof.** We use the notation of the statement. Since $\{G_1, \ldots, G_n, \complement F\}$ is an open cover of $T$, by the theorem there are $\{x_1, \ldots, x_{n+1}\} \subset C(T, [0,1])$ such that $\operatorname{coz} x_i \subset G_i$ for $i = 1, 2, \ldots, n$, $\operatorname{coz} x_{n+1} \subset \complement F$ and $\sum_{i=1}^{n+1} x_i = e_T$. It follows that $\sum_{i=1}^{n} x_i|_F = e|_F$. ■

**Definition 9.7.5** VANISHING AND SUPPORT SETS Let $H : C(T, \mathbf{F}) \to C(S, \mathbf{F})$ be weakly separating and let $cT$ be a Hausdorff compactification of $T$. Let $s \in S$.

(a) VANISHING SET An open subset $U$ of $cT$ is a *vanishing set* for $s' \circ H$ if, for any $x \in C(T, \mathbf{F})$, $\operatorname{coz} x \subset U$ implies that $Hx(s) = 0$.

(b) NONVANISHING SET An open subset $U$ of $cT$ is a non*vanishing set* for $s' \circ H$ $(s \in S)$ if there exists $x \in C(T, \mathbf{F})$ such that $\operatorname{coz} x \subset U$ and $Hx(s) \neq 0$.

(c) SUPPORT SET The set $\operatorname{supp} s' \circ H = cT \setminus \cup\{U \subset cT : U$ is a vanishing set for $s' \circ H\}$ is the *support (set)* of $s' \circ H$. ■

Let $D = D(H) = \bigcup \{\operatorname{coz} Hx : x \in C(T, \mathbf{F})\}$. As we show in Theorem 9.7.6, given any weakly separating map $H : C(T, \mathbf{F}) \to C(S, \mathbf{F})$ there is a "support" map $h : D \to cT$ associated with $H$.

**Theorem 9.7.6** SUPPORT MAP Let $H : C(T, \mathbf{F}) \to C(S, \mathbf{F})$ be a weakly separating map and $cT$ be a Hausdorff compactification of $T$. Consider $D = \bigcup \{\operatorname{coz} Hx : x \in C(T, \mathbf{F})\}$, those points $s \in S$ at which $Hx(s) \neq 0$ for some $x \in C(T, \mathbf{F})$ (in effect, we disregard those points $s$ at which $Hx(s) = 0$ for every $x \in C(T, \mathbf{F})$). For $D \neq \varnothing$ and any $s \in D$,

(a) the support set $\operatorname{supp} s' \circ H = cT \setminus \bigcup \{U \subset cT : U$ is a vanishing set for $s' \circ H\}$ is a singleton $\{h(s)\}$ in $cT$; we call $h(s)$ the *H-support point* of $s$. We call the map

$$h : D \to cT, \ s \mapsto \operatorname{supp} s' \circ H$$

the *support map* of $H$.

(b) An open set $U$ is a nonvanishing set for $s' \circ H$ iff $h(s) \in U$. To put it another way, $U$ is a vanishing set for $s' \circ H$ iff $h(s) \notin U$.

**Proof.** We use the notation of the statement. (a) Let $s \in D$. If $\operatorname{supp} s' \circ H = \varnothing$ then $cT = \cup\{U \subset cT : U \text{ is a vanishing set for } s' \circ H\}$; therefore $cT$ is covered by a finite number of vanishing sets $U_1, \ldots, U_n$ for $\operatorname{supp} s' \circ H$. By Theorem 9.7.3, there exists a continuous partition of the identity $\{x_1, \ldots, x_n\} \subset C(cT, \mathbf{F})$ such that $\sum_{i=1}^n x_i = e_{cT}$ and $\operatorname{coz} x_i \subset U_i$ for $i = 1, 2, \ldots, n$. Hence, for any $x \in C(T, \mathbf{F})$, $x = (\sum_{i=1}^n x x_i|_T$. For each $i$, $\operatorname{coz} x x_i = \operatorname{coz} x \cap \operatorname{coz} x_i \subset \operatorname{coz} x_i \subset U_i$ and $U_i$ is a vanishing set for $s' \circ H$. It follows that $H(x x_i)(s) = 0$ for each $x \in C(T, \mathbf{F})$ and therefore that $Hx(s) = \sum_{i=1}^n H(x x_i)(s) = 0$. This contradicts $s \in D = \cup\{\operatorname{coz} Hx : x \in C(T, \mathbf{F})\}$. Having shown that $\operatorname{supp} s' \circ H \neq \varnothing$, suppose that $t$ and $u$ are distinct elements of $\operatorname{supp} s' \circ H$. Since $cT$ is normal, there exist separated open neighborhoods $U$ and $V$ of $t$ and $u$, respectively (i.e., $\operatorname{cl}_{cT} U \cap \operatorname{cl}_{cT} V = \varnothing$). As $t \in \operatorname{supp} s' \circ H$, $U$ cannot be a vanishing set for $s' \circ H$ so there must be some $x \in C(T, \mathbf{F})$ such that $\operatorname{coz} x \subset U$ with $(s' \circ H)(x) = Hx(s) \neq 0$. Similarly, there exists $y \in C(T, \mathbf{F})$ such that $\operatorname{coz} y \subset U$ with $Hy(s) \neq 0$. Now $\operatorname{cl}_T \operatorname{coz} x \cap \operatorname{cl}_T \operatorname{coz} y \subset \operatorname{cl}_{cT} \operatorname{coz} x \cap \operatorname{cl}_{cT} \operatorname{coz} y = \varnothing$ but $HxHy \neq 0$ which contradicts the weakly separating property of $H$ and shows that $\operatorname{supp} s' \circ H$ is a singleton. (b) follows directly from the definitions. ■

We show in Theorem 9.7.8 that the support map is continuous, but first we consider an important example.

**Example 9.7.7** Support Map of a Weighted Composition Let $w \in C(S, \mathbf{F})$ and $g : S \to T$ be continuous. Let $H : C(T, \mathbf{F}) \to C(S, \mathbf{F})$ be the weighted composition defined for each $x \in C(T, \mathbf{F})$ and $s \in S$ as $(Hx)(s) = w(s)\,x(g(s))$. With $D = \cup\{\operatorname{coz} Hx : x \in C(T, \mathbf{F})\}$, we show next that the support map $h$ of $H$ is $g|_D$. First, we show that $D \subset \operatorname{coz} w$. For $s \notin \operatorname{coz} w$, $w(s) = 0$ and $Hx(s) = w(s)\,x(g(s)) = 0$ for all $x \in C(T, \mathbf{F})$; therefore $s \notin D$. For $s \in \operatorname{coz} w$, we show that $\operatorname{supp} s' \circ H = \{g(s)\}$ by showing that an open set $V$ is a vanishing set for $s' \circ H$ iff $g(s) \notin V$ To that end, consider an open set $V$ such that $g(s) \notin V$. If $\operatorname{coz} x \subset V$ then $x(g(s)) = 0$; hence $Hx(s) = w(s)\,x(g(s)) = 0$, i.e., any such $V$ is a vanishing set for $s' \circ H$. Therefore $\operatorname{supp} s' \circ H = \{g(s)\}$ and the support map $h$ of $H$ is $g|_D$. Since $D \subset \operatorname{coz} w$, for any $x \in C(T, \mathbf{F})$, $\operatorname{coz} Hx \cap D = \{s \in D : g(s) \in \operatorname{coz} x\} = g^{-1}(\operatorname{coz} x) \cap D$. Since $h = g|_D$, this implies that $h(\operatorname{coz} Hx) \subset \operatorname{coz} x$ with equality holding if $h$ is surjective. A more general result along these lines is Theorem 9.7.8(c). ■

**Theorem 9.7.8** PROPERTIES OF SUPPORT MAP Let $H : C(T, \mathbf{F}) \to C(S, \mathbf{F})$ be weakly separating and let $cT$ be a Hausdorff compactification of $T$. Consider the support map

$$h : D = \bigcup \{\operatorname{coz} Hx : x \in C(T, \mathbf{F})\} \to cT, \ s \mapsto \operatorname{supp} s' \circ H$$

of $H$. Then:

(a) $h$ is continuous.

(b) For any $x \in C(T, \mathbf{F})$ and any open subset $U$ of $cT$, if $x = 0$ on $U \cap T$ then $Hx = 0$ on $h^{-1}(U)$.

(c) For any $x \in C(T, \mathbf{F})$, $h(\operatorname{coz} Hx) \subset \operatorname{cl}_{cT} \operatorname{coz} x$ (cf. Theorem 9.7.10(b)).

(d) ALTERNATE COMPUTATION OF $h(s)$ For all $s \in D$,

$$h(s) = \bigcap \{\operatorname{cl}_{cT} \operatorname{coz} x : x \in C(T, \mathbf{F}) \text{ such that } Hx(s) \neq 0\}$$

(e) DENSITY $\operatorname{cl}_{cT} h(D) = cT$ iff, for all nonempty open subsets $U$ of $T$, there exists $x \in C(T, \mathbf{F})$ such that $\operatorname{coz} x \subset U$ and $Hx \neq 0$. As proved below, it follows immediately that if $H$ is 1-1 then $\operatorname{cl}_{cT} h(D) = cT$.

(f) SYMMETRIC DESCRIPTION OF WEAKLY SEPARATING If $D = S$ then $H$ is weakly separating iff separated functions $x, y \in C(T, \mathbf{F})$ have separated images $Hx$ and $Hy$.

**Proof.** We use the notation of the statement. (a) Let $s_d$ be a net in $D$ converging to $s \in D$. Since $cT$ is compact, $h(s_d)$ has a convergent subnet—that we continue to denote by $h(s_d)$—convergent to $t \in cT$. If $t \neq h(s)$ then there exist separated open neighborhoods $U$ and $V$ in $cT$ of $h(s)$ and $t$, respectively. As an open neighborhood of $h(s)$, $U$ cannot be a vanishing set for $s' \circ H$ so there must be some $x \in C(T, \mathbf{F})$ with $\operatorname{coz} x \subset U$ such that $Hx(s) \neq 0$. Consequently, there must be an index $d$ such that $Hx(s_d) \neq 0$ and $h(s_d) \in V$. Since $h(s_d) \in V$, there exists $y \in C(T, \mathbf{F})$ such that $\operatorname{coz} y \subset V$ and $Hy(s_d) \neq 0$. Thus $x$ and $y$ are separated but $HxHy \neq 0$ which is contradictory.

(b) Suppose that $U$ is an open subset of $cT$ such that $x = 0$ on $U \cap T$ and that $s \in h^{-1}(U)$. Choose an open neighborhood $V$ of $h(s)$ whose closure is contained in $U$. Since $h(s) \in V$, $V$ is a nonvanishing set for $s' \circ H$ so there exists $y \in C(T, \mathbf{F})$ such that $\operatorname{coz} y \subset V$ and $Hy(s) \neq 0$. Since $x$ and $y$ are separated, and $H$ is weakly separating, it follows that $Hx(s) = 0$.

(c) Suppose that $s \in \operatorname{coz} Hx$ and $h(s) \notin \operatorname{cl}_{cT} \operatorname{coz} x$. Choose an open neighborhood $V$ in $cT$ of $h(s)$ such that $\operatorname{cl}_{cT} V \cap \operatorname{cl}_{cT} \operatorname{coz} x = \varnothing$. Since $h(s) \in V$, $V$ is a nonvanishing set for $s' \circ H$ so there exists $y \in C(T, \mathbf{F})$ such that $\operatorname{coz} y \subset V$ and $Hy(s) \neq 0$. Since $x$ and $y$ are separated and $H$ is weakly separating, it follows that $Hx(s) = 0$, a contradiction. Therefore $h(s) \in \operatorname{cl}_{cT} \operatorname{coz} x$.

(d) For any $x \in C(T, \mathbf{F})$, if $Hx(s) \neq 0$ then $h(s) \in \operatorname{cl}_{cT} \operatorname{coz} x$ by (c). Hence $h(s) \in \cap \{\operatorname{cl}_{cT} \operatorname{coz} x : x \in C(T, \mathbf{F}) \text{ such that } Hx(s) \neq 0\}$. Suppose that $t \neq h(s)$ and let $U$ and $V$ be disjoint neighborhoods in $cT$ of

$h(s)$ and $t$, respectively. Since $h(s) \in U$, there exists $x \in C(T, \mathbf{F})$ such that $\operatorname{coz} x \subset U$ and $Hx(s) \neq 0$. Since $U \cap V = \varnothing$, $t \notin \mathrm{cl}_{cT} \operatorname{coz} x$ and $t \notin \cap\{\mathrm{cl}_{cT} \operatorname{coz} x : x \in C(T, \mathbf{F})$ such that $Hx(s) \neq 0\}$.

(e) If $\mathrm{cl}_{cT}\, h(D) = cT$ then, for each nonempty open set $V$ of $cT$, there is some $s \in D$ such that $h(s) \in V$. Since $V$ is a nonvanishing set for $s' \circ H$, there exists $x \in C(T, \mathbf{F})$ such that $\operatorname{coz} x \subset V \cap T = U$ and $Hx(s) \neq 0$. Conversely, let $V$ be a nonempty open subset of $cT$. Let $W$ be a nonempty open subset of $cT$ such that $\mathrm{cl}_{cT} W \subset V$. By hypothesis, there exists $x \in C(T, \mathbf{F})$ and $s \in S$ such that $\operatorname{coz} x \subset W \cap T = U$ such that $Hx(s) \neq 0$. Hence $s \in D$ and, by (c), $h(s) \in \mathrm{cl}_{cT} \operatorname{coz} x \subset \mathrm{cl}_{cT} W \subset V$. Therefore $V \cap h(D) \neq \varnothing$ and $\mathrm{cl}_{cT}\, h(D) = cT$.

Suppose $H$ is injective and $U$ is a nonempty open subset of $T$. Let $V$ be an open subset of $cT$ such that $U = V \cap T$. Let $t \in U$ and let $W$ be an open neighborhood of $t$ in $cT$ such that $\mathrm{cl}_{cT} W \subset V$. There exists $x \in C(cT, \mathbf{F})$ such that $x(t) = 1$ and $x = 0$ on $\complement W$. For $x = x|_T$, we have $x \neq 0$ and $\operatorname{coz} x \subset W \cap T \subset V \cap T = U$. Since $H$ is 1-1 and $x \neq 0$, $Hx \neq 0$ and the density of $h(D)$ in $cT$ follows.

(f) Sufficiency is clear. Conversely, if $H$ is weakly separating and $x,y \in C(T, \mathbf{F})$ are separated, it follows from (c) that $\mathrm{cl}_{cT}\, h(\operatorname{coz} Hx) \subset \mathrm{cl}_{cT} \operatorname{coz} x$ and $\mathrm{cl}_{cT}\, h(\operatorname{coz} Hy) \subset \mathrm{cl}_{cT} \operatorname{coz} y$. ■

We identify the support map in another important special case below.

**Example 9.7.9** SUPPORT MAP Let $T$ be completely regular, let $S = \upsilon T$, the realcompactification of $T$, and let $cT = \beta T$, the Stone-Cech compactification of $T$. For any $x \in C(T, \mathbf{F})$, let $\upsilon x$ denote the continuous extension of $x$ to $\upsilon T$. Then the linear map $H : C(T, \mathbf{F}) \to C(\upsilon T, \mathbf{F})$, $x \mapsto \upsilon x$, is a surjective ring isomorphism. Since $H$ is multiplicative, it is biseparating. We show below that the support map $h : \upsilon T \to \beta T$ of $H$ is the canonical injection $s \mapsto s$. Suppose $x \in C(T, \mathbf{F})$ and $s \in \upsilon T$ is such that $s \in \operatorname{coz} Hx = \operatorname{coz} \upsilon x$. Since $T$ is dense in $\upsilon T$ there exists a net $(t_d)$ of points in $T$ such that $t_d \to s$. By the continuity of $\upsilon x$ it follows that $\upsilon x(t_d) = x(t_d) \neq 0$ eventually, i.e., that $t_d \in \operatorname{coz} x$ eventually. It follows that $s \in \mathrm{cl}_{\upsilon T} \operatorname{coz} x \subset \mathrm{cl}_{\beta T} \operatorname{coz} x$. Hence $s \in \cap\{\mathrm{cl}_{\beta T} \operatorname{coz} x : x \in C(T, \mathbf{F})$ such that $Hx(s) \neq 0\} = \{h(s)\}$ by Theorem 9.7.8(d). ■

### 9.7.3 Continuity of Weakly Separating Maps

We use the same notation as in Sec. 9.7.2. In order to discuss continuity of weakly separating maps, we endow $C(T, \mathbf{F})$ and $C(S, \mathbf{F})$ with their compact-open topologies which we denote by $C(T, \mathbf{F}, c)$ and $C(S, \mathbf{F}, c)$, respectively [Sec. 5.8].

Let $C^1([0,1], \mathbf{F}, \|\cdot\|_\infty)$ denote the subalgebra of continuously differentiable functions of $C([0,1], \mathbf{F}, \|\cdot\|_\infty)$. As is well known, the differentiation operator $D : C^1([0,1], \mathbf{F}, \|\cdot\|_\infty) \to C([0,1], \mathbf{F}, \|\cdot\|_\infty)$, $x \mapsto x'$, is discontinuous [Example 14.1.2(c)]. It is an example of of a discontinuous separating linear map. This is not an exotic phenomenon. Jarosz [1990] showed that if $T$ is an infinite compact set and $S$ any compact set, there is a discontinuous separating linear map $H : C(T, \mathbf{F}, c) \to C(S, \mathbf{F}, c)$. We show in Theorem 9.7.11 that the only *continuous* separating maps are weighted compositions. In Theorem 9.7.10 we characterize continuity of $s' \circ H$ at one point $s \in S$. It follows from Theorem 9.7.10 that continuity of $s' \circ H$ a *linear* weakly separating map $H$ is equivalent to $h(s) \in T$ and $x(h(s)) = 0 \Rightarrow Hx(s) = 0$ for any $x \in C(T, \mathbf{F}, c)$.

**Theorem 9.7.10** CONTINUITY OF $s' \circ H$ Let

$$h : D = \bigcup \{\operatorname{coz} Hx : x \in C(T, \mathbf{F}, c)\} \to cT, \ s \mapsto \operatorname{supp} s' \circ H$$

be the support map of the weakly separating map $H : C(T, \mathbf{F}, c) \to C(S, \mathbf{F}, c)$. For any $s \in S$, if $s' \circ H$ is continuous on $C(T, \mathbf{F}, c)$ then (a) and (b) hold:

(a) $h(s) \in T$;

(b) for any $x \in C(T, \mathbf{F}, c)$, $x(h(s)) = 0 \Rightarrow Hx(s) = 0$;

(c) Conversely, if $H$ is linear and (a) and (b) hold at $s \in S$ then, for any $x \in C(T, \mathbf{F}, c)$, $Hx(s) = He_T(s)\, x(h(s))$ and $s' \circ H$ is continuous.

**Proof.** With notation as above, suppose that $s' \circ H$ is continuous.

(a) Let $\{K_a : a \in A\}$ denote the compact subsets of $T$ and suppose that $h(s) \notin T$. For a compact set $K_a$ and $t \in K_a$, choose separated [disjoint closures] open neighborhoods in $cT$, $U_t$ of $t$ and $V_t$ of $h(s)$. Since $h(s) \notin U_t$, $U_t$ is a vanishing set for $s' \circ H$ by Theorem 9.7.6(b). Hence $K_a$ is covered by vanishing sets $\{U_t : t \in K_a\}$ for $s' \circ H$ such that $h(s) \notin \operatorname{cl}_{cT} U_t$; therefore there are finitely many vanishing sets $\{U_{t_i} : i = 1, 2, \ldots, n\}$ for $s' \circ H$ that cover $K_a$. With $V_a = \cap_{i=1}^n V_{t_i}$, $\operatorname{cl}_{cT}(\cup_{i=1}^n U_{t_i}) \cap \operatorname{cl}_{cT} V_a = \varnothing$ so each $V_a$ is disjoint from $K_a$. Since each $V_{t_i}$ is a nonvanishing set for $s' \circ H$, so is $V_a$. Therefore, for each $a \in A$, there exists $x_a \in C(T, \mathbf{F}, c)$ such that $\operatorname{coz} x_a \subset V_a$ and $|Hx_a(s)| \geq 1$. Order $A$ by taking, for $a, b \in A$, $a \geq b$ if $K_a \subset K_b$ and consider the net $\{x_a : a \in A\}$. Since $\operatorname{coz} x_a \subset V_a$ for each $a \in A$, $x_a = 0$ on $K_a$. Hence, for any $a \geq b$, the seminorm $p_{K_b}(x_a) = \sup |x_a(K_b)| = 0$. Therefore $x_a \to 0$. Since $|Hx_a(s)| = |(s' \circ H)(x_a)| \geq 1$ for each $a \in A$, it follows that $s' \circ H$ is discontinuous. Thus, the continuity of $s' \circ H$ implies that $h(s) \in T$.

(b) Now suppose that $x(h(s)) = 0$ and let $V(h(s))$ denote the set of open neighborhoods of $h(s)$. Order $V(h(s))$ by taking $U \geq V$ $(U, V \in V(h(s)))$ to mean $U \subset V$. For each $V \in V(h(s))$, choose $y_V \in C(T, \mathbf{F}, c)$ such that $y_V = 1$ on a closed subneighborhood $W$ of $V$ and $y_V = 0$ on $\complement V$. Let $K$ be a compact subset of $T$. If $h(s) \notin K$ then, since $K$ is closed, there exists $V \in V(h(s))$ such that $V \cap K = \varnothing$ which implies that $xy_V = 0$ on $K$. If $h(s) \in K$, given $r > 0$, there exists $V \in V(h(s))$ such that

$$p_K(xy_V) = \sup |xy_V(K)| \leq \sup |x(V)| < r$$

and it follows that the net $xy_V \to 0$. Since $s' \circ H$ is continuous, $(s' \circ H)(xy_V) = H(xy_V)(s) \to 0$. By Theorem 9.7.8(b), since $x = xy_V$ on $V$ $(V \in V(h(s)))$, $H(xy_V)(s) = Hx(s) \to 0$; therefore $Hx(s) = 0$.

(c) Suppose $H$ is linear and (a) and (b) hold at $s \in S$. Then, for any $x \in C(T, \mathbf{F}, c)$, $[(x - x(h(s)))e_T](h(s)) = 0$. Since $H$ linear, $Hx(s) = He_T(s)x(h(s))$ by (b). ■

**Theorem 9.7.11** CONTINUITY OF WEAKLY SEPARATING LINEAR MAPS Let $h : D = \bigcup\{\operatorname{coz} Hx : x \in C(T, \mathbf{F}, c)\} \to cT$, $s \mapsto \operatorname{supp} s' \circ H$, be the support map of the weakly separating linear map $H : C(T, \mathbf{F}, c) \to C(S, \mathbf{F}, c)$. Then the following are equivalent:

(a) $H$ is continuous.

(b) For all $s \in S$, $s' \circ H$ is continuous.

(c) For all $s \in S$, $h(s) \in T$ and $Hx(s) = He_T(s)x(h(s))$ for each $x \in C(T, \mathbf{F}, c)$, i.e., $H$ is a weighted composition. Clearly $He_T(s) \neq 0$ for any $s \in D$.

(d) $h(S) \subset T$ and $\operatorname{coz} x \subset \operatorname{coz} y \Rightarrow \operatorname{coz} Hx \subset \operatorname{coz} Hy$.

**Proof.** We use the notation of the statement. (a) $\Rightarrow$ (b) The continuity of $H$ implies the weak continuity of $H$ by Theorem 8.11.3(c). With $H'$ denoting the adjoint of $H$, the weak continuity of $H$ implies the continuity of $s' \circ H = H's'$ for each $s \in S$ [Theorem 8.10.3].

(b) $\Rightarrow$ (c) Follows from Theorem 9.7.10(c).

(c) $\Rightarrow$ (a) Suppose $(x_d)$ is a net from $C(T, \mathbf{F}, c)$ such that $x_d \to 0$ and let $K$ be a compact subset of $S$. Since $h$ is continuous [Theorem 9.7.8(a)], $h(K)$ is compact and therefore $\sup|x_d(h(K))| \to 0$. Since $He_T(K)$ is bounded, it follows that $\sup|He_T(K)x_d(h(K))| \to 0$, i.e., that $Hx_d \to 0$. Since $H$ is linear, this proves that $H$ is continuous.

(c) $\Rightarrow$ (d) By (b), it follows from Theorem 9.7.10(a) that $h(S) \subset T$. That $\operatorname{coz} x \subset \operatorname{coz} y \Rightarrow \operatorname{coz} Hx \subset \operatorname{coz} Hy$, follows from the fact that, for any $z \in C(T, \mathbf{F})$, $\operatorname{coz} Hz = \operatorname{coz} He_T \cap h^{-1}(\operatorname{coz} z)$.

(d) $\Rightarrow$ (a) We prove the contrapositive. Suppose that $H$ is discontinuous. Then there is some $s \in S$ for which $s' \circ H$ is discontinuous by (b). By Theorem 9.7.10, this means that $h(s) \notin T$ or there is some $x \in C(T, \mathbf{F})$ such that $x(h(s)) = 0$ but $Hx(s) \neq 0$. If $h(s) \notin T$, this negates (d), so suppose $x \in C(T, \mathbf{F})$ is such that $x(h(s)) = 0$ but $Hx(s) \neq 0$. Suppose $He_T(s) = 0$. Then $\operatorname{coz} x \subset \operatorname{coz} e_T = T$ but $\operatorname{coz} Hx \not\subset \operatorname{coz} He_T$ because $s \in \operatorname{coz} Hx$ while $s \notin \operatorname{coz} He_T$, i.e., (d) does not hold. Now suppose $He_T(s) \neq 0$. Since $x(h(s)) = 0$, there exist open neighborhoods $W$ and $V$ of $h(s)$ in $cT$ such that $\operatorname{cl} W \subset V$ such that $\sup|x(V)| < |Hx(s)/He_T(s)|$. Choose $y \in C(cT, \mathbf{R})$, $0 \leq y \leq 1$, such that $y(\operatorname{cl}_{cT} W) = \{1\}$ and $y(\complement V) = \{0\}$. Since $xy = x$ on $W \cap T$, $Hxy = Hx$ on $h^{-1}(W)$ [Theorem 9.7.8(b)], so $Hxy(s) = Hx(s)$. Restrict $y$ to $T$ but continue to denote the restriction by $y$. Consider

$$z = e_T - \frac{He_T(s)}{H(xy)(s)}xy = e_T - \frac{He_T(s)}{Hx(s)}xy$$

We claim that $z$ never vanishes. For $t \notin V \cap T$, $y(t) = 0$ and therefore $z(t) = 1$. For $t \in V \cap T$, since $\sup |x(V)| < |Hx(s)/He_T(s)|$,

$$\left| \frac{He_T(s)}{Hx(s)} (xy)(t) \right| \leq \left| \frac{He_T(s)}{Hx(s)} x(t) \right| < 1$$

Consequently, $|z(t)| \geq \left| 1 - \left| \frac{He_T(s)}{Hx(s)} (xy)(t) \right| \right| > 0$ and $\operatorname{coz} z = T$. Since $s \in h^{-1}(W)$, $Hxy(s) = Hx(s) \neq 0$ and

$$Hz(s) = He_T(s) - \left[ \frac{He_T(s)}{H(xy)(s)} \right] H(xy)(s) = 0$$

Hence $\operatorname{coz} e_T = \operatorname{coz} z = T$ but $\operatorname{coz} He_T \not\subset \operatorname{coz} Hz$. ■

### 9.7.4 Biseparating Maps

We use the same notation as in Sec. 9.7.2. We briefly discussed realcompactifications in Sec. 9.6.1.

We show here [Theorem 9.7.15] that if $H : C(T, \mathbf{F}) \to C(S, \mathbf{F})$ is biseparating then the realcompactifications $\upsilon S$ and $\upsilon T$ of $S$ and $T$ are homeomorphic. Since ring homomorphisms are separating, this generalizes the well-known result that ring isomorphism of $C(T, \mathbf{F})$ and $C(S, \mathbf{F})$ implies homeomorphism of $\upsilon S$ and $\upsilon T$. If $H$ is linear and biseparating and $T$ is realcompact, then $H : C(T, \mathbf{F}, c) \to C(S, \mathbf{F}, c)$ is continuous [Theorem 9.7.16(b)]; consequently, we obtain the automatic continuity result that if $S$ and $T$ are realcompact, then $H$ is bicontinuous.

For a surjective map $H : C(T, \mathbf{F}) \to C(S, \mathbf{F})$, we note that

$$D = \bigcup \{\operatorname{coz} Hx : x \in C(T, \mathbf{F})\} = \cup \{\operatorname{coz} y : y \in C(S, \mathbf{F})\} = S$$

**Theorem 9.7.12** Inverses of Support Maps Let $H : C(T, \mathbf{F}) \to C(S, \mathbf{F})$ be weakly biseparating, let $cT$ and $cS$ denote Hausdorff compactifications of $T$ and $S$, respectively, and let $h : S \to cT$ and $g : T \to cS$ be the support maps of $H$ and $H^{-1}$, respectively. Then:

(a) $h$ is injective on $h^{-1}(T)$ and $g$ is injective on $g^{-1}(S)$;

(b) $g|_{h(S) \cap T} = h^{-1}$ and $h|_{g(T) \cap S} = g^{-1}$.

**Proof.** We use the notation of the statement. (a) We prove only the assertion about $h$. To prove that $h$ is injective on $h^{-1}(T)$, let $s_1$ and $s_2$ be distinct points of $h^{-1}(T)$. Since $cS$ is completely regular, there exist $w, z \in C(cS, \mathbf{F})$ such that $\operatorname{cl}_{cS} \operatorname{coz} w \cap \operatorname{cl}_{cS} \operatorname{coz} z = \varnothing$ and $w(s_1) z(s_2) \neq 0$. Since $H$ is surjective, there exist $x, y \in C(T, \mathbf{F})$ such that $Hx = w|_S$ and $Hy = z|_S$ and we still have $\operatorname{cl}_{cS} \operatorname{coz} Hx \cap \operatorname{cl}_{cS} \operatorname{coz} Hy = \varnothing$. Since $s_1 \in \operatorname{coz} Hx$, $h(s_1) \in h(\operatorname{coz} Hx) \subset \operatorname{cl}_{cT} \operatorname{coz} x$ and, similarly, $h(s_2) \in h(\operatorname{coz} Hy) \subset \operatorname{cl}_{cT} \operatorname{coz} y$ [Theorem 9.7.8(c)]. As $x = H^{-1}Hx$ and $y = H^{-1}Hy$, it also follows from Theorem 9.7.8(c) that $g(\operatorname{coz} x) = g(\operatorname{coz} H^{-1}Hx) \subset \operatorname{cl}_{cS} \operatorname{coz} Hx$ and $g(\operatorname{coz} y) \subset \operatorname{cl}_{cS} \operatorname{coz} Hy$. Since $g$

is continuous, $g(\mathrm{cl}_T \operatorname{coz} x) \subset \mathrm{cl}_{cS} \operatorname{coz} Hx$ and $g(\mathrm{cl}_T \operatorname{coz} y) \subset \mathrm{cl}_{cS} \operatorname{coz} Hy$ which implies that $\mathrm{cl}_T \operatorname{coz} x \cap \mathrm{cl}_T \operatorname{coz} y = \varnothing$ and $h(s_1) \in \mathrm{cl}_{cT} \operatorname{coz} x \cap T = \mathrm{cl}_T \operatorname{coz} x$ and $h(s_2) \in \mathrm{cl}_{cT} \operatorname{coz} y \cap T = \mathrm{cl}_T \operatorname{coz} y$, we conclude that $h(s_1) \neq h(s_2)$.

(b) We show that $g|_{h(S)\cap T} = h^{-1}$. Suppose $s \in S$ is such that $h(s) \in T$. By Theorem 9.7.8(d) applied to $H^{-1}$ and the surjectivity of $H$,

$$\begin{aligned} g(h(s)) &= \bigcap \{\mathrm{cl}_{cS} \operatorname{coz} y : y \in C(S, \mathbf{F}),\ h(s) \in \operatorname{coz} H^{-1}y\} \\ &= \bigcap \{\mathrm{cl}_{cS} \operatorname{coz} Hx : x \in C(T, \mathbf{F}),\ h(s) \in \operatorname{coz} x\} \end{aligned}$$

If $g(h(s)) = s_1 \neq s$ then $s \notin \cap\{\mathrm{cl}_{cS} \operatorname{coz} Hx : x \in C(T, \mathbf{F}),\ h(s) \in \operatorname{coz} x\}$, i.e., there exists $x \in C(T, \mathbf{F})$ such that $h(s) \in \operatorname{coz} x$ but $s \notin \mathrm{cl}_{cS} \operatorname{coz} Hx$. By the complete regularity of $S$, there exists $y \in C(S, \mathbf{F})$ such that $y(s) \neq 0$ and $y = 0$ on $\mathrm{cl}_S \operatorname{coz} Hx$. Choose $z \in C(T, \mathbf{F})$ such that $Hz = y$. Then, since $Hz = y = 0$ on $\mathrm{cl}_S \operatorname{coz} Hx$, $\mathrm{cl}_S \operatorname{coz} Hx \cap \mathrm{cl}_S \operatorname{coz} Hz = \varnothing$. By Theorem 9.7.8(c), $h(\operatorname{coz} Hz) \subset \mathrm{cl}_{cT} \operatorname{coz} z$; therefore, since $s \in \operatorname{coz} Hz$, $h(s) \in \mathrm{cl}_{cT} \operatorname{coz} z$. Since $h(s)$ also belongs to the open set $\operatorname{coz} x$, $\operatorname{coz} x$ must meet $\operatorname{coz} z$. As $H^{-1}$ is weakly separating and $Hx$ and $Hz$ are separated functions, this is a contradiction. ■

Now we need a technical lemma.

**Lemma 9.7.13** Let $H : C(T, \mathbf{F}) \to C(S, \mathbf{F})$ be weakly biseparating, let $cT$ and $cS$ denote Hausdorff compactifications of $T$ and $S$, respectively. Then:

(a) If $\operatorname{coz} x \subset \operatorname{coz} y$ $(x, y \in C(T, \mathbf{F}))$ then $\operatorname{coz} Hx \subset \mathrm{cl}_{cS} \operatorname{coz} Hy$.

(b) Let $W$ be a neighborhood of $s_0 \in S$ and let $U$ be an open neighborhood in $cT$ of $h(s_0)$. Then there exists $x \in C(T, \mathbf{F})$ and a neighborhood $V$ of $s_0$ contained in $W$ such that:

$h(s_0) \in \mathrm{cl}_{cT} \operatorname{coz} x \subset U$,
$\operatorname{coz} Hx \subset W$,
$0 \leq Hx \leq 1$ and $Hx = 1$ on $V$.

**Proof.** We use the notation of the statement. (a) We argue by contradiction. Suppose that $\operatorname{coz} x \subset \operatorname{coz} y$, $s_0 \in \operatorname{coz} Hx$ but $s_0 \notin \mathrm{cl}_{cS} \operatorname{coz} Hy$. Since $\mathrm{cl}_S \operatorname{coz} Hy \subset \mathrm{cl}_{cS} \operatorname{coz} Hy$, it follows that $s_0 \notin \mathrm{cl}_S \operatorname{coz} Hy$. Since $S$ is completely regular and $H$ is surjective, there exists $w \in C(T, \mathbf{F})$ such that $Hw(s_0) \neq 0$ but $Hw = 0$ on $\mathrm{cl}_S \operatorname{coz} Hy$ so that $\mathrm{cl}_S \operatorname{coz} Hy \cap \mathrm{cl}_S \operatorname{coz} Hw = \varnothing$. Since $Hy$ and $Hw$ are separated functions on $S$ and $H^{-1}$ is weakly separating and surjective, it follows from Theorem 9.7.8(f) that $y$ and $w$ are separated. Since $\operatorname{coz} x \subset \operatorname{coz} y$, $x$ and $w$ are separated as well. Therefore $HxHw = 0$. As $Hx(s_0)\,Hw(s_0) \neq 0$, this is a contradiction.

(b) Let $W$ be a neighborhood of $s_0 \in S$ and let $U$ be an open neighborhood in $cT$ of $h(s_0)$. Since $H$ is surjective and $S$ is completely regular, there exists $w \in C(T, \mathbf{F})$ such that $Hw = 1$ on a subneighborhood $V$ (of $s_0$) of $W$ while $Hw = 0$ on $S \setminus W$ (so that $\operatorname{coz} Hw \subset W$). Since $U$ is a neighborhood of $h(s_0)$, it is a nonvanishing set for $s_0' \circ H$ [Theorem 9.7.6(b)] so there exists $z \in C(T, \mathbf{F})$ such that $\mathrm{cl}_{cT} \operatorname{coz} z \subset U$ and $Hz(s_0) \neq 0$. Since $HwHz \in$

$C(S,\mathbf{F})$ and $H$ is surjective, there exists $v \in C(T,\mathbf{F})$, such that $HwHz = Hv$; hence $\operatorname{coz} Hv = \operatorname{coz} Hw \cap \operatorname{coz} Hz \subset W$. Since $\operatorname{coz} Hv \subset \operatorname{coz} Hz$, by (a) applied to $H^{-1}$, $\operatorname{coz} v \subset \operatorname{cl}_{cT} \operatorname{coz} z \subset U$. By the continuity of $Hv$, $\operatorname{coz} Hv$ is a neighborhood of $s_0$. Now let $x \in C(T,\mathbf{F})$ be such that $0 \leq Hx \leq 1$ and $Hx = 1$ on a neighborhood $V$ of $s_0$ contained in $\operatorname{coz} Hv$ while $Hx = 0$ on $S \setminus \operatorname{coz} Hv$. Since $\operatorname{coz} Hx \subset \operatorname{coz} Hv$ it follows from (a) that $\operatorname{coz} H^{-1}Hx = \operatorname{coz} x \subset \operatorname{cl}_{cT} \operatorname{coz} H^{-1}Hv = \operatorname{cl}_{cT} \operatorname{coz} v \subset U$ and therefore that $\operatorname{cl}_{cT} \operatorname{coz} x \subset \operatorname{cl}_{cT} \operatorname{coz} H^{-1}Hv \subset U$. Since [9.7.8(c)] $h(\operatorname{coz} Hx) \subset \operatorname{cl}_{cT}(\operatorname{coz} x)$ and $Hx(s_0) = 1$, it follows that $h(s_0) \in \operatorname{cl}_{cT} \operatorname{coz} x$. ■

We strengthen the hypothesis from *weakly* biseparating to biseparating in Theorems 9.7.14 and 9.7.15 and to *linear* biseparating in Theorem 9.7.16. Theorem 9.7.14 plays a key role in the proof of one of our main results, that the existence of a biseparating map between $C(T,\mathbf{F})$and $C(S,\mathbf{F})$ means that the realcompactifications of $S$ and $T$ are homeomorphic [Theorem 9.7.15].

**Theorem 9.7.14** BISEPARATING MAP $\Rightarrow h(S) \subset \upsilon T$ Let $H : C(T,\mathbf{F}) \to C(S,\mathbf{F})$ be biseparating, let $cT = \beta T$, the Stone-Cech compactification of $T$, and let $h : S \to \beta T$ be the support map of $H$. Then $h(S) \subset \upsilon T$.

**Proof.** We use the notation of the statement. Suppose that $s_0 \in S$ and $h(s_0) \in \beta T \setminus \upsilon T$. Since $h(s_0) \in \beta T \setminus \upsilon T$, by Theorem 9.6.7, there exist open neighborhoods $\{U_n : n \in \mathbf{N}\}$ of $h(s_0)$ in $\beta T$ such that $\operatorname{cl}_{\beta T} U_{n+1} \subset U_n$ for every $n$ and $(\cap_{n\in\mathbf{N}} \operatorname{cl}_{\beta T} U_n) \cap T = \varnothing$. Given the neighborhoods $U_1$ of $h(s_0)$ and $W_1 = h^{-1}(U_1)$ of $s_0$, by Lemma 9.7.13(b), there exists $x_1 \in C(T,\mathbf{F})$ and a neighborhood $V_1 \subset W_1$ of $s_0$ such that

(a$_1$) $h(s_0) \in \operatorname{cl}_{\beta T} \operatorname{coz} x_1 \subset U_1$,
(b$_1$) $\operatorname{coz} Hx_1 \subset W_1$,
(c$_1$) $0 \leq Hx_1 \leq 1$, and $Hx_1 = 1$ on $V_1 \subset W_1$.

Now consider the neighborhoods $U_2 \cap \operatorname{coz} x_1$ of $h(s_0)$ and $W_2 = h^{-1}(U_2) \cap \operatorname{coz} Hx_1 \cap V_1$ of $s_0$. By Lemma 9.7.13(b), there exist $x_2 \in C(T,\mathbf{F})$ and a neighborhood $V_2 \subset W_2$ of $s_0$ such that

(a$_2$) $h(s_0) \in \operatorname{cl}_{\beta T} \operatorname{coz} x_2 \subset U_2$,
(b$_2$) $\operatorname{coz} Hx_2 \subset W_2 \subset V_1$,
(c$_2$) $0 \leq Hx_2 \leq 1$, and $Hx_2 = 1$ on $V_2 \subset W_2$.

For each $n$, let $W_n = h^{-1}(U_n) \cap \operatorname{coz} Hx_{n-1} \cap V_{n-1}$ and continue inductively to generate a sequence $(x_n)$ of elements of $C(T,\mathbf{F})$ and neighborhoods $V_n \subset W_n$ of $s_0$ such that, for $n \geq 2$,

(a$_n$) $h(s_0) \in \operatorname{cl}_{\beta T} \operatorname{coz} x_n \subset U_n$,
(b$_n$) $\operatorname{coz} Hx_n \subset W_n \subset V_{n-1}$,
(c$_n$) $0 \leq Hx_n \leq 1$ and $Hx_n = 1$ on $V_n \subset W_n$.

We now claim that

$$\operatorname{int} \bigcap\nolimits_{n\in\mathbf{N}} V_n = \varnothing \tag{$*$}$$

If $\operatorname{int} \bigcap_{n\in\mathbf{N}} V_n \neq \varnothing$ then, since $H$ is surjective and $S$ is completely regular, there exists $z \in C(T,\mathbf{F})$ such that $\operatorname{coz} Hz \neq \varnothing$ and $\operatorname{coz} Hz \subset \cap_{n\in\mathbf{N}} V_n$. For every $n$ then, by the way the $x_n$ were chosen, $\operatorname{coz} Hz \subset V_n \subset \operatorname{coz} Hx_{n-1}$. By Lemma 9.7.13(a) applied to $H^{-1}$, this means that $\operatorname{coz} z \subset \mathrm{cl}_{\beta T} \operatorname{coz} x_{n-1} \subset \mathrm{cl}_{\beta T} \operatorname{coz} x_n \subset \mathrm{cl}_{\beta T} U_n$ for every $n$; since $(\bigcap_{n\in\mathbf{N}} \mathrm{cl}_{\beta T} U_n) \cap T = \varnothing$ and $\operatorname{coz} z \neq \varnothing$, this is a contradiction.

By ($\mathrm{a}_n$) above, for every $n$, $\operatorname{coz} x_n \subset U_n$, so $\bigcap_{n\in\mathbf{N}} \operatorname{coz} x_n = \varnothing$. Let $t \in T$. Since $(\bigcap_{n\in\mathbf{N}} \mathrm{cl}_{\beta T} U_n) \cap T = \varnothing$ , there must exist $n_0$ such that $t \notin \mathrm{cl}_{\beta T} U_{n_0}$. Let $B$ be a neighborhood in $T$ of $t$ disjoint from $\mathrm{cl}_{\beta T} U_{n_0}$. Since the sets $U_n$ are descending, $B$ is disjoint from $\mathrm{cl}_{\beta T} U_n$ for $n \geq n_0$. Since $\mathrm{cl}_{\beta T} \operatorname{coz} x_n \subset U_n$ for every $n$, it follows that $x_n = 0$ on $B$ for $n \geq n_0$. Consequently, for any scalars $\{a_n : n \in \mathbf{N}\}$, $\sum_{n\in\mathbf{N}} a_n x_n (u) = \sum_{n\leq n_0} a_n x_n (u)$ for any $u \in B$. This implies that $\sum_{n\in\mathbf{N}} a_n x_n \in C(T, \mathbf{F})$; in particular, $x = \sum_{n\in\mathbf{N}} n (x_n - x_{n+1}) \in C(T, \mathbf{F})$. For any $s \in V_{n+1} \subset V_n$, $Hx_n (s) = 1 = Hx_{n+1} (s)$; hence $Hx_{n+1} - Hx_n = 0$ on $V_{n+1}$. This and the facts that $\operatorname{coz} Hx_{n+1} \subset V_n$ and $\operatorname{coz} Hx_n \subset V_{n-1}$ imply that $\operatorname{coz} H (x_n - x_{n+1}) \subset V_{n-1} \setminus V_{n+1}$. Given $r > 0$, choose an open neighborhood $G$ of $s_0$ such that $|Hx (s) - Hx (s_0)| < r$ for each $s \in G$. By $(*)$ , $\operatorname{int} \bigcap_{n\in\mathbf{N}} V_n = \varnothing$ so, for each $k \in \mathbf{N}$, it follows that $V_k \cap G \not\subset \bigcap_{n\in\mathbf{N}} V_n$. Because $(V_n)$ is descending, there exist infinitely many $k$ for which there exists $s_k \in V_k \cap G$ but $s_k \notin V_n$ for $n \geq k + 1$. For $n \in \mathbf{N}$ and $m \geq n + 2$,

$$\operatorname{coz} (Hx_n - Hx_{n+1}) \subset V_{n-1} \setminus V_{n+1} \text{ and } \operatorname{coz} (Hx_m - Hx_{m+1}) \subset V_{m-1} \setminus V_{m+1}$$

Since $V_{n-1} \setminus V_{n+1}$ and $V_{m-1} \setminus V_{m+1}$ are disjoint and $H^{-1}$ is separating, it follows that $\operatorname{coz} (x_n - x_{n+1}) \cap \operatorname{coz} (x_m - x_{m+1}) = \varnothing$. Therefore

$$Hx (s_k) = \sum_{n=1}^{k+1} n (Hx_n - Hx_{n+1}) (s_k) + H \left( \sum_{n\geq k+2} n (x_n - x_{n+1}) \right) (s_k)$$

If $H (k (x_k - x_{k+1}) (s_k)) \neq 0$ then $s_k \in \operatorname{coz} H (x_k - x_{k+1})$ and

$$H \left( \sum_{n\geq k+2} n (x_n - x_{n+1}) \right) (s_k) = 0$$

Since $s_k \in V_k \setminus V_{k+1}$, it follows that $Hx_k (s_k) = 1$ and

$$\begin{aligned} Hx (s_k) &= k (Hx_k - Hx_{k+1}) (s_k) + (k + 1) (Hx_{k+1} - Hx_{k+2}) (s_k) \\ &= k + Hx_{k+1} (s_k) \end{aligned}$$

Since, for each $k$, $0 \leq Hx_k \leq 1$, $Hx$ is unbounded on $G$ which contradicts $|Hx (s) - Hx (s_0)| < r$ for each $s \in G$. This contradiction establishes that $h (s_0) \in \upsilon T$ and completes the proof. ■

For compact Hausdorff spaces $S$ and $T$, Gelfand and Kolmogorov [1939] showed that if the algebraic connection between $C(S, \mathbf{R})$ and $C(T, \mathbf{R})$ is strengthened to ring isomorphism, then the isometry hypothesis can be eliminated: if $C(S, \mathbf{R})$ is *ring* isomorphic to $C(T, \mathbf{R})$, then $S$ and $T$ are homeomorphic and the isomorphism must be a composition map $x \mapsto x \circ h$ where

$h$ is a surjective homeomorphism. The same result holds for complex-valued functions if the ring isomorphism $H$ preserves conjugates ($H\bar{x} = \overline{Hx}$) [Dunford and Schwartz 1958, p. 278]. In the absence of compactness, the result fails. What remains for completely regular Hausdorff spaces $S$ and $T$ is that their realcompactifications $\upsilon S$ and $\upsilon T$ must be homeomorphic as follows from Theorem 9.7.15 since surjective ring isomorphisms are biseparating. The Gelfand–Komolgorov theorem can also be deduced from the Banach–Stone theorem since an algebra isomorphism between $C_b(S, \mathbf{R})$ and $C_b(T, \mathbf{R})$ must be an isometry [Gillman and Jerison 1960, p. 23, Exercises 1I.2 and 1J.6].

**Theorem 9.7.15** BISEPARATING $\implies$ HOMEOMORPHIC REALCOMPACTIFICATIONS If $H : C(T, \mathbf{F}) \to C(S, \mathbf{F})$ is biseparating then $\upsilon T$ is homeomorphic to $\upsilon S$.

**Proof.** We use the notation of the statement. Consider the surjective ring isomorphisms $A : C(\upsilon T, \mathbf{F}) \to C(T, \mathbf{F})$, $x \mapsto x|_T$ and $B : C(S, \mathbf{F}) \to C(\upsilon S, \mathbf{F})$, $y \mapsto \upsilon y$ where $\upsilon y$ represents the extension of $y \in C(S, \mathbf{F})$ to $\upsilon y \in C(\upsilon S, \mathbf{F})$. As is easy to verify, the composition $H^* = BHA : C(\upsilon T, \mathbf{F}) \to C(\upsilon S, \mathbf{F})$ is biseparating. Let $cT = \beta T$, $cS = \beta S$ and let $h^* : \upsilon S \to \beta T$ and $g^* : \upsilon T \to \beta S$ denote the support maps of $H^*$ and $H^{*-1}$, respectively. By Theorem 9.7.14, $h^*(\upsilon S) \subset \upsilon T$ and $g^*(\upsilon T) \subset \upsilon S$. By Theorem 9.7.12, $g^*$ and $h^*$ are injective and are inverses of each other. Since $g^*h^*$ and $h^*g^*$ are identity maps, each is surjective as well. As $h^*$ is a continuous bijection with a continuous inverse, it is a homeomorphism. ■

Theorem 9.7.16 shows that for realcompact $S$ and $T$, linear biseparating maps are homeomorphisms. The "biseparating" hypothesis of Theorem 9.7.16 can often be weakened to "separating" because it is automatic in some cases. If, for example, (1) $S$ is connected or (2) $T$ is 0-dimensional and $S$ is pseudocompact, then a linear separating bijection $H : C(T, \mathbf{F}) \to C(S, \mathbf{F})$ is biseparating [Araujo et al. 1996, Theorems 16 and 17]. For vector lattices $X$ and $Y$, Abramovich and Kitover 2000 extensively discuss when linear separating bijections $H : X \to Y$ are biseparating. For Banach lattices $X$ and $Y$, linear separating bijections are biseparating [Huijsmans and de Pagter 1994, Theorems 2.1, 2.3 and Corollary 2.2; Koldunov 1995, Theorem 3.6]. For an instance in which a linear separating bijection $H : X \to X$, $X$ a normed lattice, is not biseparating, see Abramovich and Kitover 2000, Theorem 13.1.

**Theorem 9.7.16** CASE WHEN LINEAR + BISEPARATING IMPLIES BICONTINUOUS If $H : C(T, \mathbf{F}, c) \to C(S, \mathbf{F}, c)$ is linear and biseparating with support map $h : S \to \beta T$ then:

(a) For $s \in h^{-1}(T)$, $s' \circ H$ is continuous.

(b) If $T$ is realcompact then $H$ is continuous. If $S$ and $T$ are each realcompact, then the support map $h : S \to \beta T$ of $H$ is a homeomorphism and there exists a nonvanishing $w \in C(S, \mathbf{F})$ such that for all $x \in C(T, \mathbf{F})$ and

$s \in S$, $Hx(s) = w(s)\,x(h(s))$; in other words, $H$ is a Banach–Stone map. As a bijective Banach–Stone map is clearly open, $H$ is bicontinuous.

**Proof.** We use the notation of the statement. (a) Suppose that $s \in h^{-1}(T)$. By Theorem 9.7.10(c), if we can show that $x(h(s)) = 0 \Rightarrow Hx(s) = 0$ $(x \in C(T, \mathbf{F}))$ then $s' \circ H$ is continuous so suppose that $x(h(s)) = 0$. By Theorem 9.7.8(b), if $x \in C(T, \mathbf{F})$ vanishes on a neighborhood $U$ of $h(s)$ then $Hx = 0$ on $h^{-1}(U)$, hence on $s$, so the result is proved in this case. Now assume that $x$ does not vanish identically on any neighborhood of $h(s)$. By continuity, there exists a neighborhood $U$ of $h(s)$ on which $|x| \leq 1$. Let $\hat{x} = \inf(|x|, 1)$ and note that $|x| = |\hat{x}|$ on $U$. Consider the nonempty disjoint closed sets $A = \{t \in T : |\hat{x}(t)| \leq 3/4\}$ and $B = \{t \in T : |\hat{x}(t)| \geq 1\}$. By considering the continuous extension $\beta\hat{x}$ of $\hat{x}$ to $\beta T$, it is clear that $A$ and $B$ have disjoint closures in $\beta T$. Consequently, by Urysohn's lemma, there exists $y \in C(\beta T, \mathbf{F})$, $0 \leq y \leq 1$, such that $y(\mathrm{cl}_{\beta T} A) = \{0\}$ and $y(\mathrm{cl}_{\beta T} B) = \{1\}$. With $x_1 = y|_T \in C(T, \mathbf{F})$, we have a function that is 0 on $A$ and 1 on $B$. Let

$$d_n = \left(\frac{1}{4}\right)\left[\frac{1}{n^3} - \frac{1}{(n+1)^3}\right] \qquad (n \in \mathbf{N})$$

For $n \geq 2$, consider the disjoint closed sets

$$L_n = \left\{t \in T : |x(t)| \leq \frac{1}{n^3} - 2d_n\right\} \cup \left\{t \in T : |x(t)| \geq \frac{1}{n^3} + d_{n-1}\right\}$$

$$\text{and } M_n = \left\{t \in T : \frac{1}{n^3} - d_n \leq |x(t)| \leq \frac{1}{n^3}\right\}$$

Similar to what we did above to create the function $x_1$, for each $n \geq 2$, there exists $x_n \in C(T, \mathbf{F})$, $0 \leq x_n \leq 1$, such that

$$x_n(L_n) = \{0\} \text{ and } x_n(M_n) = \{1\}$$

With $D_n = \left(1/n^3 - 2d_n, 1/n^3 + d_{n-1}\right)$, $\complement L_n = |x|^{-1}(D_n)$. For all $n \geq 2$,

$$\frac{1}{(n+1)^3} + d_n < \frac{1}{n^3} - 2d_n$$

so the sets $D_n$ $(n \geq 2)$ are pairwise disjoint. Therefore the sets $\complement L_n = |x|^{-1}(D_n)$ $(n \geq 2)$ are pairwise disjoint as well. Since $\operatorname{coz} xx_n = \operatorname{coz} x \cap \operatorname{coz} x_n \subset \operatorname{coz} x_n \subset \complement L_n$, the cozero sets $\operatorname{coz} xx_n$ are pairwise disjoint open sets and we can define $z_1 = \sum_{n \in \mathbf{N}} xx_n$. For any $n \geq 2$, $xx_n(t) = 0$ if $t \in L_n$ and $|xx_n(t)| \leq 1/n^3 + d_{n-1} \leq 2/n^3$ otherwise. Consequently $|z_1| \leq \sum_{n \in \mathbf{N}} |xx_n(t)| \leq \sum_{n \in \mathbf{N}} (2/n^3)$ and $z_1 \in C(T, \mathbf{F})$ by Weierstrass's $M$-test. Let $z_2 = x - z_1$. As $x(h(s)) = 0$, it follows that

$$z_1(h(s)) = \sum_{n \in \mathbf{N}} x(h(s))\, x_n(h(s)) = 0$$

and therefore that $z_2(h(s)) = x(h(s)) - z_1(h(s)) = 0$.

We show next that $z_2 = x - \sum_{n\in\mathbf{N}} xx_n$ can be written as a sum of continuous functions with disjoint cozero sets. Pick $n \in \mathbf{N}$ and choose $t \in M_n$. Then $(x - xx_n)(t) = x(t) - x(t) = 0$; for $k \neq n$, $xx_k(t) = 0$. Thus, on $M_n$, $z_2 = x - xx_n = 0$ for all $n \geq 2$. Consider the pairwise disjoint open sets

$$G_n = \begin{cases} |x|^{-1}(1/8, \infty), & n = 1 \\ |x|^{-1}\left(1/(n+1)^3, 1/n^3 - d_n\right), & n \geq 2 \end{cases}$$

It is straightforward to verify that $\cup_{n\in\mathbf{N}} G_n = \complement(\cup_{n\in\mathbf{N}} M_n)$. Since $z_2 = 0$ on each $M_n$, $\operatorname{coz} z_2 \subset \cup_{n\in\mathbf{N}} G_n$. With $w_n = z_2 k_{G_n}$ $(n \in \mathbf{N})$, it follows that $z_2 = x - \sum_{n\in\mathbf{N}} xx_n = \sum_{n\in\mathbf{N}} w_n$. Since $\operatorname{coz} x_m \cap G_n = \varnothing$ for $m > n+1$ and all $n$,

$$w_n|_{G_n} = x - xx_n - xx_{n+1}$$

We now show that each $w_n$ is continuous. Let $n \in \mathbf{N}$. Since $w_n$ coincides with the continuous function $z_2$ on $G_n$, $w_n$ is continuous on $G_n$. Since $w_n = 0$ on $\complement G_n$, $w_n$ is continuous on the exterior of $G_n$. It only remains to show that $w_n$ is continuous on the boundary of $G_n$. Fix $n \in \mathbf{N}$ and choose $t \in \operatorname{bdry} G_n$; let $\{t_d\}$ be a net of points from $T$ such that $t_d \to t$. Since $\operatorname{bdry} G_1 \subset M_2$ and $\operatorname{bdry} G_n \subset M_n \cup M_{n+1}$ for $n \geq 2$ and $w_n = z_2 = 0$ on $M_n$, it follows that $w_n(t) = 0 = z_2(t)$. Either $t_d \in \complement G_n$ eventually or $t_d \in G_n$ frequently. In the former case, $w_n(t_d) = 0$ eventually and $w_n(t_d) \to w_n(t) = 0$. In the latter case, $w_n(t_d) = z_2(t_d)$ frequently and it follows that $w_n(t_d) = z_2(t_d) \to z_2(t) = 0$ and establishes the continuity of $w_n$.

We show next that $Hz_1(h(s)) = 0$ and $Hz_2(h(s)) = 0$; since $x = z_1 + z_2$, this yields the desired result that $Hx(s) = 0$.

We claim that $z = \sum_{n\in\mathbf{N}} Hxx_n$ is continuous, consider a point $s \in \cup_{n\in\mathbf{N}} \operatorname{coz} Hxx_n$. Since $H$ is separating, the sets $\{\operatorname{coz} Hxx_n : n \in \mathbf{N}\}$ are pairwise disjoint so there exists a unique $k$ such that $s \in \operatorname{coz} Hxx_k$; hence $z(s) = \sum_{n\in\mathbf{N}} (Hxx_n)(s) = Hxx_k(s)$ and $z$ is continuous at $s$. If $s \notin \operatorname{cl} \cup_{n\in\mathbf{N}} \operatorname{coz} Hxx_n$ then $z(s) = 0$ and any net $s_d \to s$ would eventually have to be in $\complement(\operatorname{cl} \cup_{n\in\mathbf{N}} \operatorname{coz} Hxx_n)$; therefore $z(s_d) = 0$ eventually and $z(s_d) \to z(s)$. Finally, suppose that $s \in \operatorname{bdry} \cup_{n\in\mathbf{N}} \operatorname{coz} Hxx_n$. Since each $\operatorname{coz} Hxx_n$ is open, $s \notin \operatorname{coz} Hxx_n$ for any $n \in \mathbf{N}$. Hence $Hxx_n(s) = 0$ for each $n$ and $z(s) = \sum_{n\in\mathbf{N}} Hxx_n(s) = 0$. If $z$ is discontinuous at $s$ then, for some $r > 0$ and any neighborhood $U$ of $s$, there exists $s_0 \in U$ such that $|z(s_0)| > r$. Since $\{\operatorname{coz} Hxx_n : n \in \mathbf{N}\}$ are pairwise disjoint, there exists a unique $n(s_0) \in \mathbf{N}$ such that $s_0 \in \operatorname{coz} Hxx_{n(s_0)}$ and $|z(s_0)| = |Hxx_n(s_0)| > r$. Since the functions $xx_n$ $(n \in \mathbf{N})$ have disjoint cozero sets, $u = \sum_{n\in\mathbf{N}} nxx_n$ converges. Furthermore, $\sum_{n\in\mathbf{N}} nxx_n$ is continuous because for any $n \in \mathbf{N}$ and any $t \in \operatorname{coz} xx_n \subset \complement L_n$, $|x(t)x_n(t)| \leq 2/n^3$. Hence $\sum_{n\in\mathbf{N}} |nxx_n(t)| \leq 2/n^2$ and continuity follows by the Weierstrass $M$-test. Since the cozero sets of

$xx_{n(s_0)}$ and $\sum_{n\neq n(s_0)} xx_n$ are disjoint and the linear map $H$ is separating,

$$H\left(\sum_{n\in\mathbf{N}} nxx_n\right)(s_0) \tag{$*$}$$

$$= H\left(n(s_0)\,xx_{n(s_0)} + \sum_{n\neq n(s_0)} n(s_0)\left(xx_{n(s_0)}\right)\right)(s_0)$$

$$= n(s_0)\,Hxx_{n(s_0)}(s_0) + H\left(\sum_{n\neq n(s_0)} n(s_0)\left(xx_{n(s_0)}\right)\right)(s_0)$$

$$= n(s_0)\,Hxx_{n(s_0)}(s_0) = n(s_0)\,z(s_0)$$

Since $s \in \operatorname{bdry} \cup_{n\in\mathbf{N}} \operatorname{coz} Hxx_n$, any neighborhood of $s$ must meet

$$\cup_{n\in\mathbf{N}} \operatorname{coz} Hxx_n$$

If some neighborhood of $s$ meets $\operatorname{coz} Hxx_n$ for finitely many values of $n$, then it is straightforward to show that $z$ is continuous at $s$ so we can eliminate that possibility. Finally, suppose that each neighborhood of $s$ meets $\operatorname{coz} Hxx_n$ for infinitely many $n$. Since there must be infinitely many points $s_0 \in W$ such that

$$H\left(\sum_{n\in\mathbf{N}} nxx_n\right)(s_0) = n(s_0)\,|z(s_0)| \geq n(s_0)\,r$$

for arbitrarily large $n(s_0)$, we have a contradiction. Therefore $\sum_{n\in\mathbf{N}} Hxx_n$ is continuous. We show next that $z = \sum_{n\in\mathbf{N}} Hxx_n = Hz_1 = H\left(\sum_{n\in\mathbf{N}} xx_n\right)$. Note that

$$y = Hz_1 - z = H\left(\sum_{n\in\mathbf{N}} xx_n\right) - \left(\sum_{n\in\mathbf{N}} Hxx_n\right)$$

$$= H\left(\sum_{n\geq k} xx_n\right) - \left(\sum_{n\geq k} Hxx_n\right) \quad \text{for any } k \in \mathbf{N}$$

It follows that $\operatorname{coz} y$ is disjoint from $\operatorname{coz} Hxx_n$ for any $n$; this being the case, $\operatorname{coz} y$ is disjoint from $\operatorname{coz} z = \cup_{n\in\mathbf{N}} \operatorname{coz} Hxx_n$ and $yz = 0$. Since $(Hxx_n)\,y = 0$, $xx_nH^{-1}y = 0$ for every $n$ and therefore $\left(H^{-1}y\right)\sum_{n\in\mathbf{N}} xx_n = \left(H^{-1}y\right)z_1 = 0$. Hence

$$0 = yHz_1 = y(z+y) = yz + y^2 = y^2$$

which implies that $y = 0$; therefore $z = Hz_1$.

Now suppose that $Hz_1(s) \neq 0$. Since $H$ is separating and

$$H\left(\sum_{n\in\mathbf{N}} xx_n\right)(s) = z(s) = \sum_{n\in\mathbf{N}} Hxx_n(s)$$

there exists a unique $k$ such that $Hz_1(s) = Hxx_k(s) \neq 0$. By Theorem 9.7.8(c), $h(s) \in \mathrm{cl}_{\beta T} \operatorname{coz} xx_k \subset \mathrm{cl}_{\beta T} \operatorname{coz} x_k$. Since $\operatorname{coz} x_k \subset \complement L_k$ and $|x| \geq 1/k^2 - 2d_n > 0$ on $\complement L_k$, this implies that $x(h(s)) \neq 0$ and contradicts $x(h(s)) = 0$. Hence $Hz_1(s) = 0$. Similarly, $Hz_2(s) = 0$.

(b) Suppose that $T$ is realcompact. By Theorem 9.7.14, for each $s \in S$, $h(s) \in \upsilon T = T$. By (a) and Theorem 9.7.11(a,b), it follows that $H$ is continuous. By Theorem 9.7.15, $h$ is a surjective homeomorphism. The form of $Hx = He \cdot (x \circ h)$ follows from Theorem 9.7.11(c). Since $H$ is onto, $D = \cup\{\operatorname{coz} Hx : x \in C(T, \mathbf{F})\} = \cup\{\operatorname{coz} y : y \in C(S, \mathbf{F})\} = S$, so $w = He$ is nonvanishing. ■

**Example 9.7.17** BISEPARATING $\not\Rightarrow$ CONTINUOUS If $T$ and $S$ are not realcompact, a biseparating linear map need not be continuous.

To illustrate the point, let $T$ be pseudocompact but not realcompact. As such, its realcompactification $\upsilon T$ is compact [Beckenstein et al. 1977, p. 27]; hence $\upsilon T = \beta T$. As noted in Example 9.7.9, the extension map $H$, $x \mapsto \upsilon x$, is a linear biseparating map of $C(T, \mathbf{F}, c)$ onto $C(\upsilon T, \mathbf{F}, c) = C(\beta T, \mathbf{F}, \|\cdot\|_\infty)$. $H$ is discontinuous, however. Suppose $d, r > 0$. Given any proper compact subset $K$ of $T$ and $t \notin K$, there exists $x \in C(T, \mathbf{F})$ such that $x(K) = \{0\}$ while $x(t) = \upsilon x(t) = d + 1$. Hence there is no compact subset $K$ of $T$ such that, for all $x \in C(T, \mathbf{F}, c)$, $\sup|x(K)| \leq r \Rightarrow \sup|\upsilon x(T)| = \|\upsilon x\|_\infty \leq d$. ■

## 9.8 NON-ARCHIMEDEAN THEOREMS

The Banach–Stone theorem [Theorem 9.6.2] tells us that, for compact Hausdorff spaces $S$ and $T$, $C(S, \mathbf{F}, \|\cdot\|_\infty)$ is linearly isometric to $C(T, \mathbf{F}, \|\cdot\|_\infty)$ iff $S$ is homeomorphic to $T$ and that the linear isometry $A : C(T, \mathbf{F}, \|\cdot\|_\infty) \to C(S, \mathbf{F}, \|\cdot\|_\infty)$ is a Banach–Stone map [Definition 9.6.1]: There is a surjective homeomorphism $h : S \to T$ such that for all $x \in C(T, \mathbf{F})$ and $s \in S$, $(Ax)(s) = w(s)\, x(h(s))$ where $w = Ae \in C(S, \mathbf{F})$ and $|w(s)| \equiv 1$. What happens if we replace $\mathbf{F}$ by an ultravalued field $(K, |\cdot|)$? In brief, the result fails: Linear isometries need not be Banach–Stone maps [Example 9.8.6]. We consider necessary and sufficient conditions for a weakly continuous linear bijection to be a Banach–Stone map in Theorem 9.8.3.

**Notation** $(K, |\cdot|)$ denotes a complete nontrivially ultravalued field. Recall that a topological space is 0-dimensional if it has a base of clopen sets. $C(S, K, \|\cdot\|_\infty)$ and $C(T, K, \|\cdot\|_\infty)$ denote the normed spaces of continuous $K$-valued functions on the compact 0-dimensional Hausdorff spaces $S$ and $T$, respectively. For a subset $U$ of $T$ (or $S$), $k_U$ denotes the characteristic function of $U$: $k_U(t) = 1$ for $t \in U$, $k_U(t) = 0$ otherwise. For $t \in T$, $t'$ denotes evaluation at $t$ and, as in Example 8.4.10(a), $\left(T', \sigma\left(C(T, K, \|\cdot\|_\infty)', C(T, K, \|\cdot\|_\infty)\right)\right)$ is homeomorphic to $T$. Terms such as weak topology, adjoint, etc., have formally the same meaning as for real and complex spaces. ■

If $U$ is clopen, $k_U$ is continuous. Not only that, but as we show in Lemma 9.8.1, the linear span of $\{k_U : U \text{ clopen } \subset T\}$—i.e., step functions $\sum_{i=1}^{n} a_i k_{U_i}$, $a_i \in K$—is dense in $C(T, K)$.

**Lemma 9.8.1** DENSITY OF STEP FUNCTIONS IN $C(T, K, \|\cdot\|_\infty)$ Let $T$ be a compact 0-dimensional Hausdorff space and $(K, |\cdot|)$ a nontrivially ultravalued field. Then the linear span of the $K$-valued characteristic functions of clopen subsets of $T$ is dense in $C(T, K, \|\cdot\|_\infty)$.

**Proof.** We use the notation of the statement. Given $x \in C(T, K, \|\cdot\|_\infty)$ and $r > 0$, for each $t \in T$, there exists a clopen neighborhood $V(t, r)$ of $t$ such that $|x(t) - x(s)| < r$ for all $s \in V(t, r)$. As $T$ is compact, a finite number of these, $V(t_i, r)$, $i = 1, 2, \ldots, n$, cover $T$. Let $U_1 = V(t_1, r)$ and $U_2 = V(t_2, r) \setminus U_1$. Note that $U_2$ is clopen and disjoint from $U_1$. By letting $U_3 = V(t_3, r) \setminus (U_1 \cup U_2)$, etc., we can rewrite $\cup_{i=1}^{n} V(t_i, r)$ as a disjoint union $\cup_{i=1}^{n} U_i$ of clopen sets. For $y = \sum_{i=1}^{n} x(t_i) k_{U_i}$, $\|x - y\|_\infty < r$. ■

For linear spaces $X$ and $Y$ over $K$, we define separating map $H : X \to Y$ in formally the same way as in Sec. 9.7.1: namely that $H$ be additive and $\operatorname{coz} x \cap \operatorname{coz} y = \varnothing \Rightarrow \operatorname{coz} Hx \cap \operatorname{coz} Hy = \varnothing$ for all $x$ and $y$ in the domain.

Multiplicative maps ($H(xy) = HxHy$), in particular, evaluation maps and weighted compositions are obviously separating; for other examples, see Example 9.7.1.

**Lemma 9.8.2** EVALUATION MAPS Let $T$ be a compact 0-dimensional Hausdorff space and $(K, |\cdot|)$ a nontrivially ultravalued field. A nontrivial continuous linear functional $f$ on $C(T, K, \|\cdot\|_\infty)$ is a nonzero scalar multiple of an evaluation map iff $f$ is separating on the linear span $M$ of characteristic functions of clopen subsets of $T$, i.e., $k_U k_V = 0$ implies $f(k_U) f(k_V) = 0$ for all clopen subsets $U$ and $V$ of $T$.

**Proof.** We use the notation of the statement. Scalar multiples of evaluation maps are clearly separating on $M$. To prove sufficiency, suppose that $f$ is a nontrivial continuous linear functional on $C(T, K, \|\cdot\|_\infty)$ that is separating on $M$ and let $\mathcal{B}$ denote the clopen subsets $U$ of $T$ such that $f(k_U) \neq 0$. Since $f$ is continuous and nontrivial and $M$ is dense in $C(T, K, \|\cdot\|_\infty)$ [Lemma 9.8.1], it follows that $\mathcal{B} \neq \varnothing$. To see that $\mathcal{B}$ is a filterbase, consider $U, V \in \mathcal{B}$. $U \cap V$ is clopen and $U \setminus (U \cap V)$ and $V \setminus (U \cap V)$ are disjoint clopen sets. By the separating condition, $f\left(k_{U \setminus (U \cap V)}\right) f\left(k_{V \setminus (U \cap V)}\right) = 0$. Since $k_U = k_{U \setminus U \cap V} + k_{U \cap V}$ and $k_V = k_{V \setminus U \cap V} + k_{U \cap V}$, if $f(k_{U \cap V}) = 0$, then $0 \neq f(k_U) = f\left(k_{U \setminus (U \cap V)}\right)$; similarly $f\left(k_{V \setminus (U \cap V)}\right) \neq 0$ which contradicts $f\left(k_{U \setminus (U \cap V)}\right) f\left(k_{V \setminus (U \cap V)}\right) = 0$. Therefore $f(k_{U \cap V}) \neq 0$ and $U \cap V \in \mathcal{B}$.

Since $T$ is compact, $\mathcal{B}$ has a cluster point $t_0 \in \cap \mathcal{B} = \cap \{B : B \in \mathcal{B}\}$ [Theorem 1.1.10(b)]. We show next that $t_0$ is the unique limit of $\mathcal{B}$. Choose $U \in \mathcal{B}$ and consider the map $e : T \to K$, $t \mapsto 1$. Since $e = k_U + k_{\complement U}$ and $f$ is nontrivial, it follows that $f(e) = f(k_U) \neq 0$—indeed, by the same argument it follows that for any clopen set $U$, exactly one of $U, \complement U$ is an element of $\mathcal{B}$. We claim that if $t_0 \in U$ then $f(k_U) \neq 0$—if not, if $f(k_U) = 0$, then $f(k_{\complement U}) \neq 0$ which implies that $\complement U \in \mathcal{B}$ and, since $t_0$ is a cluster point of $\mathcal{B}$, that $t_0 \in \complement U$, a contradiction. Since any clopen neighborhood of $t_0$ belongs to $\mathcal{B}$, $\mathcal{B} \to t_0$; $t_0$ is unique because $T$ is Hausdorff [Theorem 1.1.2(c)]. With $a = f(e)$, we show next that $f = at_0'$ on $M$. For any clopen set $U$, if $t_0 \in U$, then $U \in \mathcal{B}$ and $f(e) = f(k_U) = a = ak_U(t_0) = at_0'(k_U)$. If $t_0 \in \complement U$ then $f(k_U) = 0 = ak_U(t_0) = at_0'(k_U)$. Since $f$ is linear, it follows that $f = at_0'$ on $M$. ■

If $S$ and $T$ are topological spaces and $E$ is a topological field, we say that $A : C(T, E) \to C(S, E)$ is a Banach–Stone map [Definition 9.6.1] if there is a surjective homeomorphism $h : S \to T$ and a nonvanishing $w \in C(S, \mathbf{F})$, such that for all $x \in C(T, E)$ and $s \in S$, $(Ax)(s) = w(s)\, x(h(s))$.

**Theorem 9.8.3** Banach–Stone Maps [Beckenstein and Narici 1987] Let $S$ and $T$ be compact 0-dimensional Hausdorff spaces and $(K, |\cdot|)$ a nontrivially ultravalued field. For a weakly continuous linear bijection $A : C(T, K, \|\cdot\|_\infty) \to C(S, K, \|\cdot\|_\infty)$, the following conditions are equivalent:

(a) $A$ is separating on the linear span $M$ of the characteristic functions of clopen subsets of $T$.

(b) Let $A'$ denote the adjoint of $A$. For each $s \in S$ there exists a nonzero $w(s) \in K$ and point $h(s) \in T$ such that $A's' = w(s)\, h(s)'$.

(c) The map $h : S \to T$ of (b) is a surjective homeomorphism, $w = Ae \in C(S, K)$ and for all $s \in S$ and $x \in C(T, K)$, $Ax(s) = w(s)\, x(h(s))$, i.e., $A$ is a Banach–Stone map.

(d) $A$ is biseparating.

**Proof.** We use the notation of the statement. (a) $\Rightarrow$ (b) Let $s \in S$ and let $U$ and $V$ be disjoint clopen subsets of $T$. Since $A$ is separating on $M$, $\operatorname{coz} Ak_U \cap \operatorname{coz} Ak_V = \varnothing$. Since $A's'(k_U) = Ak_U(s)$ and $A's'(k_V) = Ak_V(s)$, $s$ belongs to at most one of $\operatorname{coz} Ak_U$ and $\operatorname{coz} Ak_V$. Since $A$ is surjective, the continuous linear functional $A's'$ is nontrivial; it is separating on $M$ because $A$ is. It follows from Lemma 9.8.2 that there exist a nonzero $w(s) \in K$ and $h(s) \in T$ such that $A's' = w(s)\, h(s)'$.

(b) $\Rightarrow$ (c) By hypothesis, for each $s \in S$, there exists a nonzero scalar $w(s)$ and point $h(s) \in T$ such that $A's' = w(s)\, h(s)'$. It follows that for each $x \in C(T, K)$,

$$(Ax)(s) = s'(Ax) = (A's')\, x = w(s)\, h(s)'\, x = w(s)\, x(h(s)) \qquad (*)$$

It remains to show that $w$ is continuous and that $h$ is a surjective homeomorphism. Since $S$ is compact and $T$ is Hausdorff, we need only prove that $h$ is a continuous bijection to prove the latter assertion.

Let $e \in C(T, F)$ be the function with constant value 1. By $(*)$, $w(s)\, e(h(s)) = Ae(s) = w(s)$ for each $s \in S$; therefore $w = Ae$ is continuous. To see that $h$ is injective, suppose that $u$ and $v$ are distinct points of $S$ for which $h(u) = h(v)$. Choose a clopen neighborhood $V$ of $v$ such that $u \notin V$. Then $k_V(v) = 1$ and $k_V(u) = 0$. Choose $x \in C(T, F)$ such that $Ax = k_V$. Since $A'u' = w(u)\, h(u)'$ and $A'v' = w(v)\, h(v)'$ and $h(u) = h(v)$,

$$\begin{aligned} \langle x, (1/w(u))\, A'u' \rangle &= \langle x, (1/w(v))\, A'v' \rangle \\ \text{or } (1/w(u))\, \langle Ax, u' \rangle &= (1/w(v))\, \langle Ax, v' \rangle \end{aligned}$$

Thus, $(1/w(u))\, \langle k_V, u' \rangle = (1/w(v))\, \langle k_V, v' \rangle$ so $(1/w(u))\, k_V(u) = 0 = (1/w(v))\, k_V(v) = 1/w(v) \neq 0$. We conclude from this contradiction that $h$ is 1-1.

Next, we show that $h$ is continuous. Note that the weak continuity of $A$ implies the weak-$*$ continuity of $A'$ by Theorem 8.10.5. It is easy to verify that the map $s \mapsto A's'$ from $S$ to $\left(C(T, K, \|\cdot\|_\infty)', \sigma\left(C(T, K, \|\cdot\|_\infty)', C(T, K, \|\cdot\|_\infty)\right)\right)$ is continuous. Since $w$ is nonvanishing, $s \mapsto (1/w(s))\, A's'$ is continuous as well. As $h$ is the composite of the continuous maps (viewing $T'$ as a topological subspace of $\left(C(T, K, \|\cdot\|_\infty)', \sigma\left(C(T, K, \|\cdot\|_\infty)', C(T, K, \|\cdot\|_\infty)\right)\right)$

$$\begin{array}{ccccccc} S & \to & C(T, K, \|\cdot\|_\infty)' & \to & T' & \to & T \\ s & \mapsto & A's' = w(s)^{-1} h(s)' & \mapsto & h(s)' & \mapsto & h(s) \end{array}$$

$h$ is continuous. As such, $h$ is a homeomorphism of $S$ onto $h(T)$. Since $S$ is compact, $h(S)$ is compact and therefore closed in $T$. If $h(S) \neq T$, choose $t \notin h(S)$ and a clopen neighborhood $U$ of $t$ disjoint from $h(S)$. Then for any $s \in S$,

$$0 = k_U(h(s)) = \langle k_U, h(s)' \rangle = \langle k_U, w(s)\, h(s)' \rangle = \langle k_U, A's' \rangle = Ak_U(s)$$

Thus $Ak_U = 0$ even though $k_U \neq 0$ which contradicts the 1-1-ness of $A$. It follows that $h$ is surjective. The implications (c) $\Rightarrow$ (d) $\Rightarrow$ (a) are clear. ■

Although we do not prove it here, in the context of Theorem 9.8.3, a separating linear bijection is bicontinuous [Narici et al. 1994, Corollary 2]. Hence we have another sort of Banach–Stone theorem, namely:

**Theorem 9.8.4** BANACH–STONE WITHOUT ISOMETRY Let $S$ and $T$ be compact 0-dimensional Hausdorff spaces and $(K, |\cdot|)$ a nontrivially ultravalued field. If $A : C(T, K, \|\cdot\|_\infty) \to C(S, K, \|\cdot\|_\infty)$ is a weakly separating linear bijection then $S$ and $T$ are homeomorphic and $A$ is a Banach–Stone map.

As the next example shows, separating linear maps can be discontinuous.

**Example 9.8.5** DISCONTINUOUS LINEAR SEPARATING MAP The subspace $T = \{0\} \cup (1/n : n \in \mathbf{N})$ of $\mathbf{R}$ is a compact ultraregular Hausdorff space. Let $\mathcal{U}$ be an ultrafilter containing the cofinite subsets of $T$. Since $\mathcal{U}$ contains the filter of neighborhoods of 0 in $T$, $\mathcal{U} \to 0$. Choose a nonzero $d \in K$ such that

$|d| < 1$ and define the continuous map $u : T \to K$ by $1/n \mapsto d^n$ $(n > 1)$, $1 \mapsto 1$ and $0 \mapsto 0$. Let $M = \{x \in C(T, K) :$ there exists $U \in \mathcal{U}$ and $c \in K$ such that $xk_U = cuk_U\}$. Scalar multiples of elements in $M$ are also clearly in $M$. If $x, y \in M$ then there are $a, b \in K$ and $U, V \in \mathcal{U}$ such that $xk_U = auk_U$ and $yk_U = buk_U$. Therefore $x + y = (a + b)\, k_{U \cap V}$ and $M$ is a subspace. Let $e \in C(T, K)$ be the map that is identically 1; note that $u \in M, e \notin M$, and let $N$ be an algebraic complement of $M$ in $C(T, K)$ such that $e - u \in N$. Define $f : C(T, K) \to K$ as follows: for $x = m + n \in M + N$, there exists $c \in K$ and $U \in \mathcal{U}$ such that $mk_U = cuk_U$; define $f(x) = c$. Since $e = u + e - u$ and $e - u \in N$, $f(e) = f(u) = 1$. To see that $f$ is well-defined, suppose that $a, b \in K$ and $U, V \in \mathcal{U}$ are such that $mk_U = auk_U$ and $mk_V = buk_V$. We show that $a = b$. Clearly any member of $\mathcal{U}$ must be infinite so there exists $t \in U \cap (V \setminus \{0, 1\})$. Then $au(t)\, k_U(t) = ad^{1/t} = bu(t)\, k_V(t) = bd^{1/t}$ which implies that $a = b$. The linearity of $f$ is clear. To see that it is separating, let $U$ and $V$ be the disjoint cozero sets of $x, y \in C(T, K)$, respectively. If $U \notin \mathcal{U}$ then $\complement U \in \mathcal{U}$ so $xk_{\complement U} = 0 = 0uk_{\complement U}$; hence $f(x) = 0 = f(x)\, f(y)$. If $U \in \mathcal{U}$ then $\complement U \notin \mathcal{U}$ which implies that $V \notin \mathcal{U}$ and, as in the previous case, $f(y) = 0 = f(x)\, f(y)$. ■

In the next example we deal with compact 0-dimensional Hausdorff spaces that have disjoint nonempty homeomorphic clopen subsets. Any discrete doubleton $D$ has this property; so does $D \times T$ where $T$ is a nonempty compact 0-dimensional Hausdorff space with disjoint nonempty homeomorphic clopen subsets.

**Example 9.8.6** A NON-BANACH–STONE LINEAR ISOMETRY Let $T$ be any nonempty compact 0-dimensional Hausdorff space with disjoint homeomorphic nonempty clopen subsets $U$ and $V$ and let $(K, |\cdot|)$ be a nontrivially ultravalued field. Let $h$ map $U$ homeomorphically onto $V$. Choose a nonzero $a \in K$ with $|a| < 1$. Endow $U \cup V$ with its relative topology and define the map $A : C(U \cup V, K, \|\cdot\|_\infty) \to C(U \cup V, K, \|\cdot\|_\infty)$ as follows: For $x \in C(U \cup V, K)$ and $t \in U \cup V$

$$Ax(t) = \begin{cases} ax(t) + x(h(t)), & t \in U \\ x\left(h^{-1}(t)\right) + ax(t), & t \in V \end{cases}$$

We show that $A$ is a surjective linear isometry but not a Banach–Stone map. The linearity of $A$ is clear. As $h$ is a homeomorphism, $Ax \in C(U \cup V, K)$. Since $|\cdot|$ is non-Archimedean, for any $c, d \in K$, $|c| < |d|$ implies that $|c + d| = \max(|c|, |d|)$ [Theorem 1.2.1(a)]. Since $|a| < 1$, it therefore follows that $A$ is an isometry. To see that $A$ is surjective, consider the map $D : C(U \cup V, K) \to C(U \cup V, K)$ defined as follows: For $x \in C(U \cup V, K)$,

$$Dx(t) = \left(a^2 - 1\right)^{-1} \begin{cases} ax(t) - x(h(t)), & t \in U \\ -x(t) + ax\left(h^{-1}(t)\right), & t \in V \end{cases}$$

Since $\left|a^2-1\right|=1$, $D$ is an isometry. Since $AD=DA=I$, the identity map $x\mapsto x$ on $C(U\cup V,K)$, $A$ is surjective. To see that $A$ is not a Banach–Stone map, note that $k_U$ and $k_V$ have disjoint cozero sets but, for any $t\in U$,

$$Ak_U(t)=ak_U(t)+k_U(h(t))=a\neq 0 \text{ and } Ak_V(t)=1$$

Hence $A$ is not separating and is therefore not a Banach–Stone map [Theorem 9.8.3]. We can linearly extend $A$ to a non-Banach–Stone map $B$ of $C(T,K,\|\cdot\|_\infty)$ onto $C(T,K,\|\cdot\|_\infty)$ by defining, for $x\in C(T,K,\|\cdot\|_\infty)$ and $t\in T$,

$$Bx(t)=\begin{cases} x(t), & t\notin U\cup V \\ Ax(t), & t\in U\cup V \end{cases} \quad \blacksquare$$

Araujo and Martínez [1990, Theorem 2] show that if the compact space $T$ has at least two points then there are surjective linear isometries $A: C(T,K,\|\cdot\|_\infty)\to C(T,K,\|\cdot\|_\infty)$ that are not even weighted compositions, let alone Banach–Stone maps; indeed, they show that the collection of linear isometries that are not Banach–Stone maps is dense in the space of linear isometries of $C(T,K,\|\cdot\|_\infty)$ onto itself. Van Rooij [1978, p. 190] shows that if $T$ is an infinite compact ultrametrizable space then $C(T,K,\|\cdot\|_\infty)$ is linearly isometric to the "non-Archimedean Banach" space $(c_0(K),\|\cdot\|_\infty)$ of $K$-valued null sequences. For more on the Banach–Stone theorem in the non-Archimedean setting, see Araujo's impressive investigations 1999a, 2001 and 2002b. Araujo [2002b, Theorem 4.1] proves that preservation of infima is equivalent to preservation of suprema by way of the following result. Note that there is no continuity assumption on $A$.

**Theorem 9.8.7** PRESERVATION OF INFIMA Let $K$ be a complete nontrivially ultravalued field which is not locally compact. Let $S$ be an **N**-compact Hausdorff space and let $T$ be a 0-dimensional Hausdorff space. $C_b(S,K)$ and $C_b(T,K)$ denote the spaces of bounded continuous maps of $S$ and $T$ into $K$, respectively, and let $A: C_b(T,K)\to C_b(S,K)$ be a surjective linear map. Then the following are equivalent:

(a) PRESERVATION OF INF For every $x\in C_b(T,K)$, $\inf|x|(T)=\inf|Ax|(S)$.

(b) CANONICAL FORM $A$ is a Banach–Stone map, $Ax=w\cdot(x\circ h)$ for each $x\in C_b(T,K)$ where the weight $w\in C_b(S,\mathbf{F})$ is such that $|w(s)|\equiv 1$.

In Narici et al. [1977, p. 156] it is shown for *algebra* isometries that:

**Theorem 9.8.8** If $K$ is a complete nontrivially ultravalued field and $S$ and $T$ are compact 0-dimensional Hausdorff spaces then $C(S,K,\|\cdot\|_\infty)$ is algebraically isometric to $C(T,K,\|\cdot\|_\infty)$ (i.e., there exists a surjective norm-preserving algebra isomorphism) iff $S$ is homeomorphic to $T$.

# 9.9 BANACH–STONE VARIATIONS

We discuss some variations on the Banach–Stone theorem [Theorem 9.6.2] in this section.

**Notation.** $S$ and $T$ denote at least completely regular Hausdorff spaces and $C(T, \mathbf{F})$ and $C(S, \mathbf{F})$ denote the spaces of continuous maps on $S$ and $T$, respectively, into $\mathbf{F} = \mathbf{R}$ or $\mathbf{C}$. $C_\infty(S, \mathbf{F})$ and $C_\infty(T, \mathbf{F})$ denote the subspaces of functions that vanish at $\infty$. We denote by $e$ the map $e : T \to \mathbf{F}$, $t \mapsto 1$. A scalar-valued function $w$ on $T$ is called *unimodular* if $|w(t)| \equiv 1$. ■

Some of the questions we consider are:

- For compact $S$ and $T$, if the isometry of $C(T, \mathbf{F}, \|\cdot\|_\infty)$ and $C(S, \mathbf{F}, \|\cdot\|_\infty)$ is strictly into, what connections between $S$ and $T$ can be deduced?
- For compact $S$ and $T$, to guarantee homeomorphism of $S$ and $T$, is linear isometry of a certain subspace of $C(T, \mathbf{F}, \|\cdot\|_\infty)$ and a subspace of $C(S, \mathbf{F}, \|\cdot\|_\infty)$ sufficient?
- If $S$ and $T$ are locally compact and $C_\infty(T, \mathbf{F}, \|\cdot\|_\infty)$ is linearly isometric to $C_\infty(S, \mathbf{F}, \|\cdot\|_\infty)$, is $T$ homeomorphic to $S$? What forms can the linear isometry have?
- For various $S$ and $T$, what if $\mathbf{F}$ is replaced by a Banach space? a Banach lattice?

## 9.9.1 Subspaces

**$S$ and $T$ are locally compact Hausdorff—hence, completely regular—spaces in this subsection.**

As the results quoted here indicate, if sufficiently selective subspaces of $C(T, \mathbf{F}, \|\cdot\|_\infty)$ and $C(S, \mathbf{F}, \|\cdot\|_\infty)$ are closely enough related as normed spaces, we can still deduce homeomorphism of the base spaces $S$ and $T$. Our first selectivity condition is *extreme regularity* [Definition 9.9.1]. As Theorem 9.9.2 shows, if we have something close to isometry between extremely regular subspaces of $C(T, \mathbf{F}, \|\cdot\|_\infty)$ into $C(S, \mathbf{F}, \|\cdot\|_\infty)$, $S$ and $T$ must be homeomorphic.

**Definition 9.9.1** EXTREMELY REGULAR SUBSPACE A closed subspace $X \subset C_\infty(T, \mathbf{F}, \|\cdot\|_\infty)$ is *extremely regular* if for each $t_0 \in T$, each neighborhood $U$ of $t_0$ and each $r \in (0, 1)$ there exists $x \in X$ such that $\|x\|_\infty = |x(t_0)| = 1$ and $|x(t)| < r$ for all $t \in \complement U$. ■

Theorem 9.9.2 differs from the Banach–Stone theorem in three ways: (1) $S$ and $T$ are *locally* compact; (2) We deal with functions that vanish at $\infty$, rather than with all continuous functions and (3) $A$ is not an isometry but is a linear homeomorphism "of small bound."

**Theorem 9.9.2** HOMEOMORPHIC EXTREMELY REGULAR SUBSPACES [Cengiz 1973b] Let $S$ and $T$ be locally compact Hausdorff spaces. If $X \subset C_\infty(T, \mathbf{F}, \|\cdot\|_\infty)$ and $Y \subset C_\infty(S, \mathbf{F}, \|\cdot\|_\infty)$ are extremely regular subspaces and $A : X \to Y$ is a surjective linear homeomorphism such that $\|A\| \|A^{-1}\| < 2$ then $S$ and $T$ are homeomorphic.

This extends earlier results of Amir 1965 (for $\mathbf{R}$-valued functions) and Cambern 1967 concerning surjective isometries

$$A : C_\infty(T, \mathbf{F}, \|\cdot\|_\infty) \to C_\infty(S, \mathbf{F}, \|\cdot\|_\infty)$$

(Cengiz was Cambern's doctoral student.) The number 2 is the largest number for which Theorem 9.9.2 holds: it can fail even if $\|A\| \|A^{-1}\| = 2$ [Cambern 1970; see Exercise 9.109]. The result also shows that Banach's [1932, p. 242] notions of *isometric* and *almost isometric* are equivalent in the context of spaces $C_\infty(T, \mathbf{F}, \|\cdot\|_\infty)$. As to the possibilities for the size of $\|A\| \|A^{-1}\|$, Cambern [1968] showed that if $A$ is a linear homeomorphism of $(c, \|\cdot\|_\infty)$ onto $(c_0, \|\cdot\|_\infty)$ then $\|A\| \|A^{-1}\| \geq 3$ and there exists a linear homeomorphism $A : c \to c_0$ such that $\|A\| \|A^{-1}\| = 3$.

Now consider a weaker notion than extreme regularity.

**Definition 9.9.3** COMPLETELY REGULAR SUBSPACE A closed subspace $X \subset C_\infty(T, \mathbf{F}, \|\cdot\|_\infty)$ is *completely regular* if for each $t_0 \in T$ and any neighborhood $U$ of $t_0$ there exists $x \in X$ such that $\|x\|_\infty = |x(t_0)| = 1$ and $|x(t)| < \|x\|_\infty$ for all $t \in \complement U$. ■

Extreme regularity clearly implies complete regularity but the converse is false [Cengiz 1973a]. It is also clear that $C_\infty(T, \mathbf{F}, \|\cdot\|_\infty)$ is completely regular. For a full description of isometries between completely regular subspaces, see Cengiz 1990.

If we weaken the extreme regularity condition of Theorem 9.9.2 but strengthen $A$ to an isometry, we recover the canonical form of the isometry as a weighted composition with a unimodular weight: In particular, Myers 1948 (for compact $S$ and $T$) and Araujo and Font [1997, Cor. 4.4] showed that:

**Theorem 9.9.4** ISOMETRIC COMPLETELY REGULAR SUBSPACES If $X \subset C_\infty(T, \mathbf{F}, \|\cdot\|_\infty)$ and $Y \subset C_\infty(S, \mathbf{F}, \|\cdot\|_\infty)$ are completely regular subspaces and $A : X \to Y$ is a surjective linear isometry then there is a surjective homeomorphism $h : S \to T$ and a unimodular $w \in C(S, \mathbf{F})$ such that $Ax = w \cdot (x \circ h)$ for each $x \in X$.

Although it can be deduced much more easily using essentially the same proof as the Banach–Stone theorem [Behrends 1979, p. 138], Theorem 9.9.4 yields Corollary 9.9.5.

**Corollary 9.9.5** BANACH–STONE FOR $C_\infty(T, \mathbf{F}, \|\cdot\|_\infty)$ If $A : C_\infty(T, \mathbf{F}, \|\cdot\|_\infty) \to C_\infty(S, \mathbf{F}, \|\cdot\|_\infty)$ is a surjective linear isometry then there is a surjective homeomorphism $h : S \to T$ and a unimodular $w \in C(S, \mathbf{F})$ such that $Ax = w \cdot (x \circ h)$ for each $x \in X$.

Now consider an even weaker separation condition.

**Definition 9.9.6** STRONGLY SEPARATING SUBSPACE A subspace $X \subset C_\infty(T, \mathbf{F}, \|\cdot\|_\infty)$ is *strongly separating* if for any $t \in T$ there exists $x \in X$ such that $x(t) \neq 0$ and for any distinct points $u, v \in T$, there exists $x \in X$ such that $|x(u)| \neq |x(v)|$. ■

Complete regularity clearly implies strongly separating. If strongly separating subspaces $X$ and $Y$ are linearly isometric, then the isometry is a weighted composition but, instead of $S$ and $T$, we can only conclude that the Silov and Choquet boundaries of $X$ and $Y$ are homeomorphic:

**Theorem 9.9.7** ISOMETRIC SUBSPACES [Araujo and Font 1997] If

$$X \subset C_\infty(T, \mathbf{F}, \|\cdot\|_\infty) \text{ and } Y \subset C_\infty(S, \mathbf{F}, \|\cdot\|_\infty)$$

are strongly separating subspaces and $A : X \to Y$ is a surjective linear isometry then there is a surjective homeomorphism $h : \mathrm{Si}\, Y \to \mathrm{Si}\, X$ (Silov boundaries) and a unimodular $w \in C(\mathrm{Si}\, Y, \mathbf{F})$ such that, for every $x \in X$ and $s \in \mathrm{Si}\, Y$, $Ax(s) = w(s) \cdot (x \circ h)(s)$. In addition, $h(\mathrm{Ch}\, Y) = \mathrm{Ch}\, X$, where $\mathrm{Ch}\, X$ denotes the Choquet boundary of $X$.

Araujo and Font show by examples that the result fails if $X$ is not strongly separating and that even though the Silov boundaries of $X$ and $Y$ are homeomorphic, it does not follow that $S$ and $T$ are homeomorphic.

### 9.9.2 Into Isometries

For compact $S$ and $T$, the Banach–Stone theorem asserts that an *onto* linear isometry $A : C_\infty(T, \mathbf{F}, \|\cdot\|_\infty) \to C_\infty(S, \mathbf{F}, \|\cdot\|_\infty)$ is a weighted composition with a surjective *homeomorphism* $h : S \to T$. For an *into* isometry, we have a representation of $A$ as a weighted composition utilizing a *continuous* surjection $h$ defined on a closed subset of $S$.

**Theorem 9.9.8** INTO ISOMETRIES [Holsztynski 1966, p. 134] Let $S$ and $T$ be compact. If $A$ is a linear isometry of $C(T, \mathbf{F}, \|\cdot\|_\infty)$ into $C(S, \mathbf{F}, \|\cdot\|_\infty)$ then there exists a closed subset $S_0$ of $S$, a continuous surjection $h : S_0 \to T$ and $w \in C(S_0, \mathbf{F})$, $|w(s)| \equiv 1$, such that $Ax = w \cdot (x \circ h)$ on $S_0$ for each $x \in C(T, \mathbf{F})$.

Semadeni [1971, p. 391ff.] has a proof of Theorem 9.9.8 in his discussion of linear isometries of $C(S, \mathbf{F})$ into $C_\infty(T, \mathbf{F}, \|\cdot\|_\infty)$ when $T$ is compact and $S$ locally compact. Holsztynski's result figures prominently in the classification of isometric shift operators on $C(T, \mathbf{F})$ [Farid et al. 1994 and Gutek et al.

1991]. Araujo et al. [1996] extend Holsztynski's result to functions that vanish at infinity as follows:

**Theorem 9.9.9** INTO ISOMETRIES Let $S$ and $T$ be locally compact Hausdorff spaces. If $A$ is a linear isometry of a completely regular subspace of $C_\infty(T, \mathbf{F}, \|\cdot\|_\infty)$ into $C_\infty(S, \mathbf{F}, \|\cdot\|_\infty)$ then there exists a closed subset $S_0$ of $S$, a continuous surjection $h : S_0 \to T$ and $w \in C(S_0, \mathbf{F})$, $|w(s)| \equiv 1$, such that $Ax = w \cdot (x \circ h)$ on $S_0$ for each $x \in C(T, \mathbf{F})$.

### 9.9.3 Vector-Valued Functions

**Notation.** $S, T$ completely regular Hausdorff spaces

$X$, $Y$: Banach spaces

$L(X, Y)$: continuous linear maps of $X$ into $Y$ with the strong operator topology [Sec. 11.2].

$C_b(S, Y, \|\cdot\|_\infty)$, $C_b(T, X, \|\cdot\|_\infty)$: bounded continuous maps of $S$ into $Y$ and $T$ into $X$, respectively, normed by

$$\|x\|_\infty = \sup\{\|x(t)\| : t \in T\} \quad (x \in C_b(T, X))$$

with a similar convention for $C_b(S, Y, \|\cdot\|_\infty)$.

$C_\infty(S, Y, \|\cdot\|_\infty)$, $C_\infty(T, X, \|\cdot\|_\infty)$: continuous maps of $S$ into $Y$ and $T$ into $X$, respectively, that vanish at infinity. ■

**Questions.** Suppose $A : C_\infty(T, X, \|\cdot\|_\infty) \to C_\infty(S, Y, \|\cdot\|_\infty)$ is a surjective linear isometry.

(a) Must $S$ be homeomorphic to $T$?

(b) Must $X$ be linearly isometric to $Y$?

(c) What forms may $A$ take?

As the following example illustrates, neither (a) nor (b) has an affirmative answer.

**Example 9.9.10** NON-BANACH–STONE ISOMETRY Consider the compact discrete spaces $\{0\}$ and $\{0,1\}$, let $X = \ell_\infty^{\mathbf{R}}(2)$, $Y = \mathbf{R}$ and let $(a, b) \in \mathbf{R}^2$. For $x \in C(\{0\}, \ell_\infty^{\mathbf{R}}(2))$, $0 \mapsto (a, b)$, consider the surjective linear isometry $H : C(\{0\}, \ell_\infty^{\mathbf{R}}(2), \|\cdot\|_\infty) \to C(\{0,1\}, \mathbf{R}, \|\cdot\|_\infty)$ where $Hx(0) = a$ and $Hx(1) = b$. Since

$$\|Hx\|_\infty = \max(|Hx(0)|, |Hx(1)|) = \max(|a|, |b|) = \|x\|_\infty$$

$H$ is a linear isometry, but $\{0\}$ is not homeomorphic to $\{0,1\}$ and $\ell_\infty^{\mathbf{R}}(2)$ is not linearly isometric to $\mathbf{R}$ (see before Example 9.6.3). ■

Sundaresan [1973, p. 22] shows that for $n \neq 2$, there exist nonhomeomorphic compact Hausdorff spaces $S$ and $T$ such that $C\left(T, \ell_\infty^{\mathbf{R}}(n), \|\cdot\|_\infty\right)$ is linearly isometric to $C\left(S, \ell_\infty^{\mathbf{R}}(n), \|\cdot\|_\infty\right)$.

Recall:

**Definition 9.6.1** BANACH–STONE MAPS Let $S$ and $T$ be topological spaces, let $h : S \to T$ be continuous and let $w \in C(S, \mathbf{F})$. (a) A map $A : C(T, \mathbf{F}) \to C(S, \mathbf{F})$, $Ax(s) = w(s)\, x(h(s))$ $(x \in C(T, \mathbf{F}), s \in S)$, is called a *weighted composition map,* $w$ being the *weight* or *weight function.* We write $Ax = w \cdot (x \circ h)$.

(b) If $w$ is nonvanishing and $h$ a surjective homeomorphism, then $A$ is called a *Banach–Stone map.* ■

Suppose $w \in C(S, \mathbf{F})$. At any $s \in S$ we define $f_{w(s)} \in C(S, \mathbf{F})'$ on $y \in C(S, \mathbf{F})$ as the multiplier map $f_{w(s)}(y) = w(s)\, y(s)$; note that if $w$ is unimodular, then $\left|f_{w(s)}(y)\right| = |y(s)|$. We are considering the functional-valued map $W : S \to C(S, \mathbf{F})'$, $s \mapsto f_{w(s)}$. This motivates the following definition.

**Definition 9.9.11** VECTOR BANACH–STONE MAPS If $W : S \to L(X, Y)$ and $h : S \to T$ are continuous then the map

$$\begin{array}{rccc} H: & C_\infty(T, X, \|\cdot\|_\infty) & \to & C_\infty(S, Y, \|\cdot\|_\infty) \\ & x & \mapsto & (Ws)\left[x(h(s))\right] \quad (s \in S) \end{array}$$

is called:

(a) a *vector Banach–Stone map.*

(b) If $h$ is a surjective homeomorphism and $Ws$ is an isometry for each $s \in S$, then $H$ is a *strong vector Banach–Stone map.* ■

Consider two negative examples.

**Example 9.9.12** NON-VECTOR BANACH–STONE MAPS

(a) NON-VECTOR [Araujo 2004, p. 249] Suppose $T$ is not realcompact, e.g., $T = [0, \Omega) = \{\text{ordinal } \sigma : \sigma < \Omega\}$, the space of ordinals less than the first uncountable ordinal $\Omega$ [Gillman and Jerison 1960, 5.12, Beckenstein et al. 1977, p. 27] and let $X = \ell_2$. Since $\ell_2$ is separable, it is realcompact [Gillman and Jerison 1960, 8.2]. We can therefore continuously extend each bounded continuous map $x : T \to \ell_2$ to a bounded continuous map $vx : vT \to \ell_2$ and thereby define a surjective linear isometry $H : C_b(T, \ell_2, \|\cdot\|_\infty) \to C_b(vT, \ell_2, \|\cdot\|_\infty), x \mapsto vx$, that is clearly not a vector Banach–Stone map.

(b) VECTOR BUT NOT STRONG The map $H$ of Example 9.9.10 is not a strong vector Banach–Stone map since there is only one map $h : \{0,1\} \to \{0\}$ and $h$ is not 1-1. It is a vector Banach–Stone map, however: Let $\mathrm{pr}_1$ and $\mathrm{pr}_2$ denote the projections on the first and second components of $\ell_\infty^{\mathbf{R}}(2)$, respectively, and define $W : \{0,1\} \to L\left(\ell_\infty^{\mathbf{R}}(2), \mathbf{R}\right)$ by taking $W0 = \mathrm{pr}_1$

and $W1 = \mathrm{pr}_2$. Then at each $s \in \{0,1\}$, $(Ws)[x(h(s))] = (Ws)[x(0)] = (Ws)[(a,b)] = Hx(s)$ for each $x \in C\left(\{0\}, \ell_\infty^{\mathbf{R}}(2)\right)$. ■

**Definition 9.9.13** SEPARATING AND BISEPARATING If $H : C(T,X) \to C(S,Y)$ is additive and, for all $x, y \in C(T, \mathbf{F})$,

$$\operatorname{coz} x \cap \operatorname{coz} y = \varnothing \Rightarrow \operatorname{coz} Hx \cap \operatorname{coz} Hy = \varnothing$$

or equivalently, for all $x, y \in C(T, \mathbf{F})$,

$$\text{for all } t \in T,\ \|x(t)\| \, \|y(t)\| = 0 \Rightarrow \text{for all } s \in S,\ \|Hx(s)\| \, \|Hy(s)\| = 0$$

we say that $H$ is *separating; if* $H$ is bijective and $H^{-1}$ is *separating* as well, we say that $H$ is *biseparating.* ■

Biseparating characterizes strong vector Banach–Stone maps in the class of surjective linear isometries for compact $S$ and $T$:

**Theorem 9.9.14** BANACH–STONE AND BISEPARATING

(a) COMPACT SPACES [Hernández et al. 1995, Theorem 4.1] For compact $S$ and $T$, a surjective linear isometry $H : C(T,X) \to C(S,Y)$ is a strong vector Banach–Stone map iff $H$ is biseparating.

(b) REALCOMPACT SPACES [Araujo 2004, Theorem 2.4] Let $X$ and $Y$ be normed spaces (i.e., not necessarily Banach spaces), let $B(X,Y)$ denote the space of all (not necessarily continuous) bijective linear maps of $X$ onto $Y$ and suppose $S$ and $T$ are realcompact. If $H : C(S, Y, \|\cdot\|_\infty) \to C(T, X, \|\cdot\|_\infty)$ is a linear biseparating map or $X$ and $Y$ are infinite-dimensional and $H : C_b(S, Y, \|\cdot\|_\infty) \to C_b(T, X, \|\cdot\|_\infty)$ is a linear biseparating map then there exists a surjective homeomorphism $h : S \to T$ and a map $W : S \to B(X,Y)$ such that for every $s \in S$ and $x$ in the domain of $H$, $Hx(s) = (Ws)(x(h(s)))$.

In regard to dealing with *all* linear biseparating maps in Theorem 9.9.14(b), we note that such maps are usually continuous [Araujo and Jarosz 2003b], but not always as Examples 9.9.15 and 9.7.17 show.

**Example 9.9.15** LINEAR BISEPARATING $\nRightarrow$ CONTINUOUS Let $X = (c_0, \|\cdot\|_\infty)$ and let $S = T = \{t\}$. Consider a Hamel base $\mathcal{B}$ of unit vectors for $c_0$. Let $\mathcal{D} = \{x_n\}$ be a denumerable subset of $\mathcal{B}$ and define $Ax = x$ if $x \in \mathcal{B} \setminus \mathcal{D}$ and $Ax_n = nx_n$ for $x_n \in \mathcal{D}$. Extend the bijective linear map $A$ by linearity to $\bar{A} : C(\{t\}, c_0) \to C(\{t\}, c_0)$. $\bar{A}$ is clearly biseparating but since it is unbounded on the unit ball of $c_0$, it is discontinuous. ■

To state Corollary 9.9.18, we need the notion of strict convexity, something we discuss at length in Sec. 16.1.

**Definition 9.9.16** STRICT CONVEXITY A normed space $X$ is *strictly convex* if the surface $SU(X) = \{x \in X : \|x\| = 1\}$ of the solid unit ball $U$ of $X$ consists entirely of extreme points of $U$.

Two easy examples: the spaces $\ell_p$, $1 < p < \infty$, are strictly convex; $\ell_1$ and $\ell_\infty$ are not.

Jeang and Wong [2003, Theorem 6] show that:

**Theorem 9.9.17** $\ell_\infty(2) \not\subset Y \Rightarrow$ BANACH–STONE MAP Let $S$ and $T$ be locally compact Hausdorff spaces. If $Y$ does not contain a copy of $\ell_\infty(2)$ then every surjective linear isometry $A : C_\infty(T, X, \|\cdot\|_\infty) \to C_\infty(S, Y, \|\cdot\|_\infty)$ is a vector Banach–Stone map.

It follows from this that:

**Corollary 9.9.18** VECTOR-VALUED BANACH–STONE THEOREM Suppose $S$ and $T$ are compact and $A : C(T, X, \|\cdot\|_\infty) \to C(S, Y, \|\cdot\|_\infty)$ is a surjective linear isometry. If $Y$ or its dual $Y'$ is strictly convex then $A$ is a vector Banach–Stone map; if $X$ or $X'$ is also strictly convex then $A$ is a strong vector Banach–Stone map and (therefore) $S$ and $T$ are homeomorphic.

The result of Corollary 9.9.18 for strictly convex $X$ and $Y$ is due to Jerison [1950] (from an article based on his dissertation), that for strictly convex duals to Lau [1975]. It is proved by separating map techniques in Hernández et al. 1995. Cambern [1975] has a stronger version. Cambern [1985] proves a "small bound" version for compact Hausdorff spaces $S$ and $T$ and a uniformly convex [Definition 16.2.1] Banach space $X$ which asserts that $S$ and $T$ are homeomorphic if the linear homeomorphism $A : C(T, X, \|\cdot\|_\infty) \to C(S, X, \|\cdot\|_\infty)$ is such that $\|A\| \|A^{-1}\|$ is smaller than a fixed number which depends on the modulus of rotundity of $X$.

Corollary 9.9.18 also follows from the following extension of Holsztynski's theorem for into isometries [Theorem 9.9.8] to vector-valued functions:

**Theorem 9.9.19** VECTOR-VALUED HOLSZTYNSKI THEOREM [Jeang and Wong 1997, Theorem 1] For locally compact Hausdorff spaces $S$ and $T$, suppose that $Y$ is strictly convex. Then, for any linear isometry

$$A : C_\infty(T, X, \|\cdot\|_\infty) \to C_\infty(S, Y, \|\cdot\|_\infty)$$

there exists a closed subset $S_0$ of $S$, a continuous map $W : S \to L(X, Y)$ and a continuous surjection $h : S_0 \to T$ such that $Ax(s) = (Ws)\, x(h(s))$ for each $s \in S_0$ and $x \in C_\infty(T, X, \|\cdot\|_\infty)$.

We need a little terminology to conclude our discussion of vector-valued Banach–Stone theorems.

**Definition 9.9.20** MULTIPLIERS AND THE CENTRALIZER Let $X$ be a Banach space and $A \in L(X, X)$.

(a) $A$ is called a *multiplier* if each extreme point $f$ of the unit ball $U(X')$ of $X'$ is an eigenvector of the adjoint $A'$. This means that there is a nonvanishing scalar-valued function $w_A$ on $X'$ such that, for each $f \in \operatorname{ext} U(X')$, $A'f = w_A(f) f$.

(b) A multiplier $A$ is an *adjoint* for a multiplier $B$ if $w_A = \bar{w}_B$, the pointwise complex conjugate of $w_B$. (Adjoints don't always exist; they do if $X$ is real.)

(c) The set $Z(X)$ of all multipliers which have an adjoint is called the *centralizer of* $X$. (If $X$ is real, $Z(X)$ is the set of all multipliers.) ■

In many cases the centralizer is trivial ($Z(X) = \{0\}$), e.g., any strictly convex or smooth space [Behrends 1979]. A number of things related to Banach–Stone maps follow if the Banach space(s) have a small centralizer.

**Theorem 9.9.21** ONE-DIMENSIONAL CENTRALIZERS [Araujo 2004, Prop. 3.5 and Theorem 3.6] Suppose $S$ and $T$ are realcompact, the Banach spaces $X$ and $Y$ are infinite-dimensional and $A : C_b(S, Y, \|\cdot\|_\infty) \to C_b(T, X, \|\cdot\|_\infty$ is surjective linear isometry. Then:

(a) $A$ is biseparating.

(b) If $\dim Z(X) = \dim Z(Y) = 1$ then $A$ is a strong vector Banach–Stone map.

We say that a normed space $X$ has the *Banach–Stone property* if for any locally compact Hausdorff spaces $S$ and $T$, linear isometry of $C_\infty(S, X, \|\cdot\|_\infty)$ and $C_\infty(T, X, \|\cdot\|_\infty)$ implies homeomorphism of $S$ and $T$. Neither $\ell_\infty^{\mathbf{R}}(2)$ nor $C([0,1], \mathbf{F}, \|\cdot\|_\infty)$ has the Banach–Stone property [Behrends 1979, p. 143] but every 3-dimensional space does except $\ell_\infty(3)$ [Behrends 1979, pp. 142–147; Cambern 1977]. Moreover, [Behrends 1979]:

**Theorem 9.9.22** TRIVIAL CENTRALIZER ⇒ BANACH–STONE PROPERTY A Banach space with trivial centralizer has the Banach–Stone property.

Araujo [2006] weakened the compactness conditions and showed:

**Theorem 9.9.23** NONCOMPACT BANACH–STONE THEOREM Let $X$ and $Y$ be infinite-dimensional Banach spaces with trivial centralizers and let $S$ and $T$ be realcompact. Then any surjective linear isometry $A : C_b(T, X, \|\cdot\|_\infty) \to C_b(S, Y, \|\cdot\|_\infty)$ is a strong vector Banach–Stone map.

### 9.9.4 Ordered Versions

**Notation.** $S, T$ completely regular Hausdorff spaces.

$X$, $Y$: Banach lattices, hence real vector spaces.

$L(X, Y)$: continuous linear maps of $X$ into $Y$ with the strong operator topology.

$C(S, Y)$, $C(T, X)$: pointwise ordered vector lattices of continuous maps of $S$ into $Y$ and $T$ into $X$, respectively [Example 5.12.1].

$C_b(S,Y)$, $C_b(T,X)$: bounded continuous maps of $S$ into $Y$ and $T$ into $X$, respectively.

$C_\infty(S,Y)$, $C_\infty(T,X)$: continuous maps of $S$ into $Y$ and $T$ into $X$, respectively, that vanish at infinity.

For $x \in C(T,X)$, $z(x) = \{t \in T : x(t) = 0\}$ denotes the zero set of $x$. ■

**Questions.** If $A : C(T,X) \to C(S,Y)$ is a surjective lattice isomorphism,

(a) Must $S$ be homeomorphic to $T$?

(b) Must $X$ be lattice isomorphic to $Y$?

(c) What forms may $A$ take?

As $\ell_\infty^{\mathbf{R}}(2)$ and $\mathbf{R}$ are Banach lattices and the map $H$ of Example 9.9.10 is a lattice isomorphism, neither (a) nor (b) need hold.

If $S$ and $T$ are compact Hausdorff spaces and $C(S,\mathbf{R})$ and $C(T,\mathbf{R})$ are lattice isomorphic then $S$ and $T$ are homeomorphic [Kaplansky 1947; cf. Semadeni 1971, p. 129]. Shirota 1952 and Henriksen 1956 extended Kaplansky's result to noncompact spaces.

Moving to continuous functions $C(T,X)$ taking values in a vector lattice $X$, Cao et al. [2003, Theorem 3.3, Corollary 3.4] proved the following Banach–Stone type theorems in the presence of a nonvanishing condition:

**Theorem 9.9.24** For compact Hausdorff spaces $S$ and $T$ and a real Banach lattice $X$:

(a) If $A : C(T,X) \to C(S,\mathbf{R})$ is a surjective lattice isomorphism such that $z(x) = \varnothing \Rightarrow z(Ax) = \varnothing$ then $S$ and $T$ are homeomorphic and $X$ and $\mathbf{R}$ are lattice isomorphic.

(b) If $C(T,\mathbf{R})$ and $C(S,\mathbf{R})$ are lattice isomorphic then $S$ and $T$ are homeomorphic (since, in this case, the nonvanishing condition of (a) is automatically satisfied).

Chen et al. [2008, Theorems 3 and 4] proved that:

**Theorem 9.9.25** For compact Hausdorff spaces $S$ and $T$, real Banach lattices $X$ and $Y$ and a surjective lattice isomorphism $A : C(T,X) \to C(S,Y)$ such that $z(x) = \varnothing$ iff $z(Ax) = \varnothing$ :

(a) $S$ is homeomorphic to $T$.

(b) REPRESENTATION There exists a continuous map $W : S \to L(X,Y)$ and a surjective homeomorphism $h : S \to T$ such that for each $x \in C(T,X)$ and $s \in S$, $Ax(s) = (Ws)[x(h(s))]$, $Ws$ is a lattice isomorphism of $X$ onto $Y$ for each $s \in S$ and $\|A\| = \sup\{\|Ws\| : s \in S\}$.

## 9.10 EXERCISES

**Notation.** When $X$ is a normed space, $U(X) = \{x \in X : \|x\| \leq 1\}$, the solid unit ball of $X$. ■

9.101 CONVEXITY CRITERION Let $f$ map the vector space $X$ into $\mathbf{R}$. If for all $x,y \in X$, $f((1/2)(x+y)) \leq (1/2)[f(x)+f(y)]$, show that $f$ is convex (defined in Exercise 7.104).

9.102 EXTREME POINTS OF SYMMETRIC SETS If $E$ is a symmetric ($E = -E$) convex subset of a TVS, show that $\operatorname{ext} E$ and $\operatorname{cl} \operatorname{ext} E$ are symmetric as well.

9.103 In $\ell_\infty^{\mathbf{R}}$ let $\{e_n\}$ be the the standard basis vectors and let $K$ be the closed convex hull of $\{0\} \cup \{e_n/n : n \in \mathbf{N}\}$. Show that $K$ is compact but not equal to the convex hull of its extreme points.

9.104 Identify the extreme points of the unit ball of an inner product space.

9.105 Let $X$ be a Banach space. Show that every boundary point of $U(X)$ is an extreme point of $U(X)$ iff, for all $x, y \in X$, $\|x+y\| \neq \|x\| + \|y\|$ unless $x$ and $y$ are linearly dependent.

9.106 For unit vectors $x, y \in X$, a normed space, show that if a point $w \in (x, y)$ has norm one then so does every point in $[x, y]$.

9.107 EXTREME POINTS ARE $G_\delta$ Show that if $E$ is a metrizable compact convex subset of a TVS, then the extreme points of $E$ are a $G_\delta$ set.

9.108 CHOQUET SIMPLEX Let $K$ be a compact convex subset of a real LCHS $X$. Let $(A(K), \|\cdot\|_\infty)$ denote the Banach space of all continuous affine [Def. 9.1.4] maps of $K$ into $\mathbf{R}$.

(a) EVERY COMPACT CONVEX SET IS A STATE SPACE State spaces are generally compact and convex [Theorem 9.5.5(a)]. Show that the map $E$ sending $x \in K$ into the evaluation map $x'$ on $A(K)$ is an affine homeomorphism of $K$ into the state space $S(A(K))$. Let $A(K)$ be ordered pointwise and let its continuous dual $A(K)'$ carry the dual order [Example 5.12.3]. We say that $K$ is a *Choquet simplex* if $A(K)'$ is a vector lattice [Definition 5.12.11].

(b) If $T$ is a compact Hausdorff space and $X$ is a closed subspace of $C(T, \mathbf{R}, \|\cdot\|_\infty)$ which contains constants and separates points, then $X$ is linearly isometric to $A(S(X))$ where $S(X)$ denotes the state space of $X$.

9.109 We consider an example here to show that 2 is the greatest bound for Theorem 9.9.2 [Cambern 1970; Garrido and Jaramillo 2002, pp. 361–362]. Let $S = \{\pm 1/n : n \in \mathbf{N}\} \cup \{0\}$ and $T = \{-1/n : n \in \mathbf{N}\} \cup \{0\} \cup \mathbf{N}$ each in their relative $\mathbf{R}$-topologies. Define $A : C_\infty(T, \mathbf{R}, \|\cdot\|_\infty) \to C_\infty(S, \mathbf{R}, \|\cdot\|_\infty)$ at $x \in C_\infty(T, \mathbf{R}, \|\cdot\|_\infty)$ by

$$Ax(s) = \begin{cases} x(0), & s = 0 \\ x(-1/n) + x(n), & s = 1/n \\ x(-1/n) - x(n), & s = -1/n \end{cases}$$

Show that $A$ is a linear isometry, that $\|A\| \left\|A^{-1}\right\| = 2$ and that $S$ is not homeomorphic to $T$.

9.201 SUPPORTING HYPERPLANES The notion of "supporting hyperplane" [Definition 7.8.9] generalizes the notion of tangent plane to a convex figure in $\mathbf{R}^3$. In the discussion and exercises below, $K$ denotes a convex subset of a real linear space $X$, $f$ is a linear functional on $X$ and $H = \{x \in X : f(x) = t\}$ is a hyperplane in $X$. Recall that $H$ is a supporting hyperplane for $K$ if $K$ meets $H$ and $K$ lies to one side of $H$. A point of $K$ which is in some such hyperplane is called a *support point* of $K$.

(a) Show that $H = \{x \in X : f(x) = a\}$ supports $K$ iff $a$ is the greatest or least element of $f(K)$. A linear functional $g$ such that $\sup g(K) = g(u)$ for some $u \in K$ is called a *support functional* for $K$.

(b) If $K \neq \emptyset$ then $H \cap K$ is a face of $K$.

(c) If $X$ is a TVS and the interior of $K$ is nonempty then any supporting hyperplane $H$ of $K$ is closed and every boundary point $u \in K$ which is in $K$ is a support point of $K$.

(d) The result of this part shows that when $K$ is compact, there are supporting hyperplanes "in any direction." If $X$ is a TVS and $K$ is nonempty and compact, then for any closed hyperplane $H$ in $X$, there is some $x \in X$ such that $x + H$ supports $K$; in other words, given $H$ and $K$, there is a supporting hyperplane for $K$ which is "parallel to $H$."

(e) If $X$ is an LCHS and $K$ a nonempty compact subset, then any closed supporting hyperplane $H$ of $K$ contains an extreme point of $K$.

(f) A *convex set containing only support points*[Rolewicz 1978]. Let $\ell_2(\mathbf{R})$ denote the real Hilbert space of all real-valued functions $x$ on $\mathbf{R}$ which vanish outside a countable subset of $\mathbf{R}$ and such that $\sum_{t \in \mathbf{R}} |x(t)|^2 < \infty$. The inner product of $x, y \in \ell_2(\mathbf{R})$ is $\langle x, y\rangle = \sum_{t \in \mathbf{R}} x(t)y(t)$. Let $K = \{x \in \ell_2(\mathbf{R}) : x(t) \geq 0$ for all $t \in \mathbf{R}$ and $\|x\| \leq 1\}$, the intersection of the positive cone and the unit ball. Show that each point of $K$ is a support point of $K$.

NOTES. The result of (c) shows that closed convex "bodies"—convex sets with nonempty interior—have support points. If we drop the assumption that $K$ has nonempty interior, then the result fails. As shown in Holmes [1975, p. 166], there exist bounded closed convex subsets of Fréchet spaces which have no support points. [This is shown by showing that $f(K)$ is open for each continuous linear functional $f$ on $X$; hence there can be no $u \in K$ such that $\sup f(K) = f(u)$ as would have to occur at a support point by (a).] In a Banach space, however, not only must closed convex sets have support points, the set of support points must be dense in the boundary of the set [Holmes 1975, p. 166]. Moreover, if $K$ is a closed bounded convex subset of a Banach space $X$, then the support functionals of $K$ are dense in $X'$ (in its norm topology) [Holmes 1975, p. 169; Diestel 1975, p. 30]. This last result is the "Bishop-Phelps subreflexivity theorem" [Bishop and Phelps 1963]. One reason for interest in it is its application to questions of reflexivity stemming from James's profound theorem that a real Banach space $X$ is reflexive iff every $f \in X'$ attains its norm on the unit ball $U(X)$ of $X$. For that result and many applications, see James 1957, p. 167, for the separable case and 1964, p. 215 for the general case; cf. also James 1972 and Diestel 1975, p. 12.

9.202 NON-ARCHIMEDEAN KREIN–MILMAN THEOREM [Ajupov 1974]. Let $(K, |\cdot|)$ be a complete discrete ultravalued field and let $X$ be a locally $K$-convex TVS over $K$ [Exercise 4.203]. A subset $E$ of $X$ is called *c-compact* if each filterbase of $K$-convex subsets of $E$ has an adherence point [Exercise 4.204]. A *supporting variety* of a $K$-convex set $A$ is a linear variety $V$ such that

(1) $V$ meets $A$ and

(2) $V$ is the intersection of some family of supporting hyperplanes (usual definition) for $A$.

If $A$ is a $K$-convex subset of $X$ and $\{M_s : s \in S\}$ is the family of all minimal (with respect to set inclusion) supporting varieties of $S$ then $N = \cup_{s \in S}(A \cap M_s)$ is called the *extremal set* of $A$. The following results hold for $K$-convex $c$-compact subsets $A$ of $X$.

(a) Every closed supporting hyperplane of $A$ contains at least one point of the extremal set of $A$.

(b) If $A$ is locally bounded, then $A$ is the closed $K$-convex hull of its extremal set.

(c) If $A$ is compact (not $c$-compact), then $A$ is the closed $K$-convex hull of its extremal set.

9.203 WHEN $U(C_b(T, \mathbf{R}, \|\cdot\|_\infty)) = \text{cl}[\text{ext}\, U(C_b(T, \mathbf{R}, \|\cdot\|_\infty))]_c$ [Goodner 1964]. Let $C_b(T, \mathbf{R}, \|\cdot\|_\infty) = C_b(T, \mathbf{R})$ denote the Banach space of all bounded continuous maps of the Hausdorff space $T$ into $\mathbf{R}$. The point of this exercise is to obtain a necessary and sufficient condition on $T$ for $U(C_b(T, \mathbf{R}))$ to be the closed convex hull of its extreme points. If $T = [0, 1]$, the extreme points of $U(C_b(T, \mathbf{R}))$ are just $e(t) \equiv 1$ and $-e$ [Example 9.2.5(e)], so $U(C_b(T, \mathbf{R}))$ is certainly *not* the closed convex hull of its extreme points. Note too that if $T$ is completely regular, there is no loss of generality in supposing that $T$ is compact since $C_b(T, \mathbf{R}, \|\cdot\|_\infty))$ is linearly isometric to $C(\beta T, \mathbf{R}, \|\cdot\|_\infty))$ where $\beta T$ denotes the Stone-Cech compactification of $T$ [Example 8.4.10]. As usual, "clopen" means closed and open. A *zero set* is a subset of $T$ of the form $x^{-1}(0)$ for some $x$ in $C_b(T, \mathbf{R})$. Prove that $U(C_b(T,\mathbf{R})) = \mathrm{cl}[\mathrm{ext}\, U(C_b(T,\mathbf{R}))]_c$ iff zero sets may be separated by clopen sets, i.e., for each pair of disjoint zero sets $E$ and $F$ of $T$ there exists a clopen set $K \supset E$ such that $K \cap F = \emptyset$.

### Hints to Exercises

9.103 Show that $\mathrm{ext}\, K = \{0\} \cup \{e_n/n : n \in \mathbf{N}\}$ and that $K$ must contain any point of the form $x = \sum_{n\in\mathbf{N}} (t_n/n)\, e_n$ for any sequence $(t_n)$ of nonnegative numbers such that $\sum_{n\in\mathbf{N}} t_n = 1$. After identifying the general form of points in the convex hull of $\mathrm{ext}\, K$, show that if $t_n > 0$ for each $n \in \mathbf{N}$, $\sum_{n\in\mathbf{N}} t_n = 1$, then $\sum_{n\in\mathbf{N}} t_n e_n/n$ cannot be in that convex hull.

9.104 Let $(X, \langle\cdot,\cdot\rangle)$ be an inner product space. Suppose that $\|x\| = 1$ and that $y, z \in U(X)$ are such that $x = (1/2)(y+z)$. Calculate the inner product $\langle x, y\rangle$ and use the Cauchy-Schwarz inequality ($|\langle x, y\rangle| \le \|x\|\,\|y\|$). Conclude that the boundary points of $U(X)$ are the extreme points.

9.106 Let $(X, \langle\cdot,\cdot\rangle)$ be a normed space. Suppose that $0 < a < 1$,

$$\|ax + (1-a)\, y\| = 1$$

and $b \in (a, 1)$. With $z = bx + (1-b)\, y$, $ax + (1-a)\, y = dz + (1-d)\, y$ where $d = a/b < 1$.

9.107 Let $d$ be a metric generating the topology and let $F_n = \{(1/2)(y+z) : y, z \in E,\ d(y, z) \ge 1/n\}$ for each $n \in \mathbf{N}$. Show that each $F_n$ is closed and that a point $x$ of $E$ is not an extreme point of $E$ iff $x \in F_n$ for some $n \in \mathbf{N}$.

9.108 (b) Use Theorem 9.5.12.

9.201 (c) The open convex set $\mathrm{int}\, K$ is disjoint from the linear variety $\{u\} = u + [0]$. Use the geometric form of the Hahn–Banach theorem [7.7.3].

(d) Look at $f(K)$.

(e) Let $\mathcal{F}$ be the family of closed faces $F$ of $K$ which are contained in $H$, ordered by reverse inclusion: $F \le G$ iff $F \supset G$. Conclude that $\mathcal{F}$ has a minimal element.

(f) Let $w \in K$ and choose $s \in \mathbf{R}$ such that $w(s) = 0$. Now consider the evaluation map $s'$ determined by $s$.

9.203 Suppose that zero sets of $T$ may be separated by clopen sets, that $r$ is positive, that $x \in U(C_b(T, \mathbf{R}))$, and let $N$ be the smallest integer $\ge 2/r$. For each integer $n$ such that $0 \le n \le N-1$, let $U_n = \{t \in T : x(t) \ge 1 - nr\}$ and $L_n = \{t \in T : x(t) \le 1 - nr\}$; let $U_N = \{t \in T : x(t) \ge -1\}$

and $L_N = \{t \in T : x(t) \le -1\}$. Show that for $0 \le n \le N$, $U_n$ and $L_n$ are zero sets and that $U_n \cap L_{n+1} = \emptyset$ for $0 \le n \le N+1$. By hypothesis, there exist disjoint clopen sets $P_n$ and $Q_{n+1}$ such that $P_n \supset U_n$, $Q_{n+1} \supset L_{n+1}$ and $P_n \cup Q_{n+1} = T$. Let $Q_0 = P_N = T$. Show that $\cup_{n=0}^N (P_n \cap Q_n) = T$ and if $t \in P_n \cap Q_n$, then $|x(t) - (1 - nr)| < r$. The characteristic function $k_n$ of $P_n \cap Q_n$ is continuous since $P_n \cap Q_n$ is clopen and for $c_n = \pm 1$, $\sum_{n=0}^N c_n k_n$ is an extreme point of $U(C_b(T, \mathbf{R}))$. Let $a_n = nr/2$ for $1 \le n < N$ and let $a_N = 1$. Form the functions $y_0 = \sum_{j=0}^N k_j$,

$$y_1 = (1 - a_1)y_0 + a_1(y_0 - 2k_1) = \sum_{i=0}^{1}(1 - ir)k_i + \sum_{j=2}^{N} k_j$$

$$\vdots$$

$$y_n = (1 - a_n)y_{n-1} + a_n(y_{n-1} - 2k_n) = \sum_{i=0}^{n}(1 - ir)k_i + \sum_{j=n+1}^{N} k_j$$

and

$$y_N = \sum_{i=0}^{N-1}(1 - ir)k_i - k_N$$

The function $y_N$ belongs to $[\operatorname{ext} U(C_b(T, \mathbf{R}))]_c$ and $\|x - y_N\| < r$. Thus $x \in \operatorname{cl}[\operatorname{ext} U(C_b(T, \mathbf{R}))]_c$; therefore $U(C_b(T, \mathbf{R})) = \operatorname{cl}[\operatorname{ext} U(C_b(T, \mathbf{R}))]_c$. Conversely, let $S = \operatorname{cl}[\operatorname{ext} U(C_b(T, \mathbf{R}))]_c$ and let $x^{-1}(0)$ and $y^{-1}(0)$ be disjoint zero sets in $T$. The function $z$ defined as $z(t) = |x(t)| / (|x(t)| + |y(t)|)$ at each $t$ in $T$ has the properties $0 \le z \le 1$, $z(t) = 0$ for $t \in x^{-1}(0)$ and $z(t) = 1$ for $t \in y^{-1}(0)$. Since $z \in S$ there exists $w \in [\operatorname{ext} U(C_b(T, \mathbf{R}))]_c$ such that $\|z - w\|_\infty < 1/2$. Since $w$ can only assume the values $\pm 1$ on $T$ [Example 9.2.5(e)], the sets $H = w^{-1}([-1, 1/2])$ and $K = w^{-1}((1/2, 1])$ are disjoint clopen sets containing $x^{-1}(0)$ and $y^{-1}(0)$, respectively. Since $H \cup K = T$, $T$ is seen to separate zero sets by clopen sets.

# Chapter 10

# Vector-Valued Hahn–Banach Theorems



For TVS $X$ and $Y$, $L(X, Y)$ denotes space of continuous linear maps from $X$ into $Y$.

We consider a continuous extension problem and a norm-preserving extension problem:

- *The Extension Problem for LCS* For what LCS $Y$ can any $A \in L(M, Y)$ defined on any subspace $M$ of any LCS $X$ be extended to $\overline{A} \in L(X, Y)$? We say that such $Y$ have *the extension property.*

- *The Metric Extension Problem for Banach Spaces* For what Banach spaces $Y$ can any $A \in L(M, Y)$ defined on any subspace $M$ of any Banach space $X$ be extended to $\overline{A} \in L(X, Y)$ of the same norm? We say that such $Y$ have the *metric extension property.*

$$\begin{array}{ccc} X & & \\ | & \overset{\overline{A}}{\searrow} & \\ M & \xrightarrow{A} & Y \end{array} \qquad \|\overline{A}\| = \|A\|$$

In the Banach space context, Grothendieck [1953a, p. 169] proved that a Banach space with the extension property is either finite-dimensional or nonseparable. We characterize Banach spaces with the *metric* extension property internally in various ways in terms of intersection properties of balls in Secs. 10.2–10.6. We show [Theorem 10.7.5] that a real Banach space $X$ has the metric extension property iff there is a compact extremally disconnected Hausdorff space $T$ such that $X$ is linearly isometric to the space $C(T, \mathbf{R}, \|\cdot\|_\infty)$ of all continuous maps on $T$. As we note in Theorem 10.8.1(f), the analogous characterization obtains for complex Banach spaces as well.

## 10.1 INJECTIVE SPACES

For a subspace $M$ of a TVS $X$, a continuous linear functional $f : M \to \mathbf{F}$ can be continuously extended to a linear functional $F : X \to \mathbf{F}$ [Corollary 7.3.3]. What if $\mathbf{F}$ is replaced by a TVS $Y$ and $f$ is replaced by $A \in L(M, Y)$? Banach and Mazur [1933], Fichtenholz and Kantorovich [1934], Murray [1937] and others, showed that there are situations in which extensions $\overline{A} \in L(X, Y)$ do not exist. We discuss such a case next.

Recall that a subspace $M$ is complemented in the TVS $X$ iff there is a continuous projection of $X$ onto $M$ [Theorem 4.9.2(a)]. Therefore, since the closed subspace $c_0$ of real null sequences is uncomplemented in the Banach space $\ell_\infty^{\mathbf{R}}$ of bounded real sequences [Example 4.9.7], there is no continuous projection of $\ell_\infty^{\mathbf{R}}$ onto $c_0$. Consider the identity map $I : c_0 \to c_0$, $y \mapsto y$. If $\bar{I} : \ell_\infty^{\mathbf{R}} \to c_0$ is a continuous linear extension of $I$ to $\ell_\infty^{\mathbf{R}}$ then, for any $x \in \ell_\infty^{\mathbf{R}}$, $\bar{I}x \in c_0$; therefore $\bar{I}\left(\bar{I}x\right) = I\left(\bar{I}x\right) = \bar{I}x$, i.e., $\bar{I}\circ\bar{I}=\bar{I}$ and $\bar{I}$ would be a continuous projection of $\ell_\infty$ onto $c_0$. Therefore there can be no continuous extension of $I$ to $\ell_\infty^{\mathbf{R}}$ and $c_0$ does not have the extension property. Although $c_0$ does not have the extension property, if we restrict consideration to **separable** Banach superspaces $X$ of $c_0$, then it does; indeed, for any separable Banach superspace $X$ of $c_0$ there is a continuous projection $P$, $\|P\| \leq 2$, of $X$ onto $c_0$ (see Theorem 10.1.3(b) for why this suffices). As to the existence of projections of norm less than or equal to 2 of $X$ onto $c_0$, see Köthe 1979, p. 21, Diestel 1984, p. 71, or Lindenstrauss and Tzafriri 1977, p. 105.

If there is a continuous projection of $X$ onto $M$ then any $A \in L(M, Y)$ has a continuous linear extension, namely, $\overline{A} = A \circ P_M \in L(X, Y)$. Since [Theorem 4.9.2(a)] there is a continuous projection of $X$ onto $M$ iff $M$ is complemented in $X$ and there are plenty of uncomplemented subspaces [Sec. 7.2], there are many instances in which no continuous projection $P_M$ exists.

**Closed subspace $M$, complete codomain.** Let $Y$ be a TVS. Since it is always possible to continuously extend a linear map $A \in L(M,Y)$ by continuity from $M$ to $\operatorname{cl} M$, we assume that $M$ is closed in our discussion of the extension problem. We also restrict consideration to complete range spaces $Y$ because if $Y$ has the extension property, then $Y$ must be complete—it must be possible to extend the identity map $I : Y \to Y, y \mapsto y$, to $\bar{I} : \widehat{Y} \to Y$ on a completion $\widehat{Y}$ of $Y$. If $(y_s)$ is a (Cauchy) net from $Y$ that converges to $y$ in $\widehat{Y}$, then $\bar{I}y_s = y_s \to y$; since $\bar{I}y_s \to \bar{I}y \in Y$, it follows that $\bar{I}y = y$. Thus, $Y = \widehat{Y}$.

**Definition 10.1.1** INJECTIVE AND $P_1$-SPACES

(a) INJECTIVE An LCS $Y$ is *injective* if, for every LCS $X$ containing $Y$, there is a continuous projection of $X$ onto $Y$.

(b) 1-INJECTIVE A Banach space $Y$ is *1-injective* or a *$P_1$-space* if, for every Banach space $X$ containing $Y$, there is a continuous projection of norm 1 of $X$ onto $Y$.

(c) 1-COMPLEMENTED In view of Theorem 4.9.2(a), an injective Banach space $Y$ has a topological complement in any space in which it is norm embedded; in the special case of a 1-injective space, we say that $Y$ is *1-complemented* in any space in which it is norm embedded. ■

We consider some examples of injective spaces in Theorem 10.1.2; $\ell_\infty(T)$ is as in Example 1.7.5.

**Theorem 10.1.2** INJECTIVE SPACES Let $X$ be a LCS over $\mathbf{F} = \mathbf{R}$ or $\mathbf{C}$ and let $M$ be a subspace of $X$. Then:

(a) $\mathbf{F}^T$ HAS EXTENSION PROPERTY Let $T$ be a set and consider the LCS $\mathbf{F}^T$ of all maps of $T$ into $\mathbf{F}$ with the product topology. Then any $A \in L(M, \mathbf{F}^T)$ has an extension $\overline{A} \in L(X, \mathbf{F}^T)$.

(b) $\ell_\infty(T)$ HAS METRIC EXTENSION PROPERTY For any set $T$ and any $A \in L(M, \ell_\infty(T))$, $A$ has an extension $\overline{A} \in L(X, \ell_\infty(T))$. If $X$ is a normed space, then $\|\overline{A}\| = \|A\|$.

**Proof. Let** $X$ and $M$ be as above. (a) Suppose $A \in L(M, \mathbf{F}^T)$. For each $t \in T$, let $\operatorname{pr}_t$ denote the projection of $\mathbf{F}^T$ onto $\mathbf{F}$. The maps $f_t = \operatorname{pr}_t \circ A$ are therefore continuous linear functionals on $M$; by Corollary 7.3.3, each $f_t$ may be continuously extended to a continuous linear functional $F_t$ on $X$. The desired continuous extension $\overline{A}$ of $A$ is given by the map $\overline{A}x = (F_t(x))$, the continuity of $\overline{A}$ following because $\operatorname{pr}_t \circ \overline{A} = F_t$ is continuous for each $t \in T$.

(b) Suppose $A \in L(M, \ell_\infty(T))$. Since $A$ is continuous, there is a continuous seminorm $p$ on $X$ such that $\|A(\cdot)\|_\infty \leq p(\cdot)$ on $M$ by Theorem 5.7.3. Therefore, for each $t \in T$ and $m \in M$, $|Am(t)| \leq p(m)$ and the linear functionals $f_t(m) = Am(t)$ are continuous on $M$ for each $t \in T$. For each $t \in T$, by Theorem 7.3.2(b), there is a linear extension $F_t$ of $f_t$ on $X$ such that $|F_t| \leq p$. For $x \in X$ and $t \in T$ define $\overline{A}x(t) = F_t(x)$. Since for all $x \in X$

and $t \in T$, $\left|\overline{A}x(t)\right| = |F_t(x)| \le p(x)$, it follows that $\left\|\overline{A}x\right\|_\infty \le p(x)$ and establishes the continuity of $\overline{A}$. If $X$ is a normed space, we may choose $F_t$ to be of the same norm as $f_t$ by the Hahn–Banach theorem [7.4.1] and define $\overline{A}$ as above. Then

$$\begin{aligned}
\left\|\overline{A}\right\| &= \sup_{\|x\|=1} \left\|\overline{A}x\right\|_\infty = \sup_{\|x\|=1} \left(\sup_{t\in T} \left|\overline{A}x(t)\right|\right) = \sup_{t\in T} \left(\sup_{\|x\|=1} |F_t(x)|\right) \\
&= \sup_{t\in T} \|F_t\| = \sup_{t\in T} \|f_t\| = \sup_{t\in T} \left(\sup_{\|x\|=1, x\in M} |f_t(x)|\right) \\
&= \sup_{t\in T} \left(\sup_{\|x\|=1, x\in M} |Ax(t)|\right) = \sup_{\|x\|=1,, x\in M} \left(\sup_{t\in T} |Ax(t)|\right) \\
&= \sup_{\|x\|=1, x\in M} \|Ax\|_\infty = \|A\| \quad \blacksquare
\end{aligned}$$

We show in Theorem 10.1.3 that the extension property is equivalent to injectivity and that the metric extension property implies 1-injectivity; we show in Theorem 10.5.2 that 1-injectivity implies the metric extension property.

**Theorem 10.1.3** INJECTIVE = EXTENSION PROPERTY

(a) EXTENSION PROPERTY IMPLIES INJECTIVE If the LCS $Y$ has the extension property then $Y$ is injective. If the Banach space $Y$ has the metric extension property then it is 1-injective.

(b) INJECTIVE IMPLIES EXTENSION PROPERTY Let $X$ and $Y$ be Banach spaces. If $Y$ is injective then, for any subspace $M \subset X$ and $A \in L(M,Y)$, there exists an extension $\overline{A} \in L(X,Y)$ of $A$.

**Proof.** (a) Suppose the LCS $Y$ has the extension property. Let $M$ be a subspace of the LCS $X$ and let $I : M \to M$, $m \mapsto m$, be the identity map on $M$. Let $\bar{I}$ be its continuous linear extension to $X$. As $\bar{I}\circ\bar{I}=\bar{I}$, $\bar{I}$ is a continuous projection of $X$ onto $M$. If $Y$ is a Banach space with the metric extension property then there exists an extension $\bar{I}$ of $I$ to $X$ such that $\left\|\bar{I}\right\| = \|I\| = 1$.

(b) Suppose the Banach space $Y$ is injective, that $M$ is a subspace of $X$ and $A \in L(M,Y)$. Let $T = \{y' \in Y' : \|y'\| = 1\}$. Now we locate a copy (a linearly isometric image) of $Y$ in $\ell_\infty(T)$. Since $|y'(y)| \le \|y\|$ for all $y' \in T$, we can consider the linear map $B : Y \to \ell_\infty(T)$, $y \mapsto (y'(y))_{y'\in T}$. Since $\|y\| = \sup\{|y'(y)| : y' \in T\} = \|By\|_\infty$ [Theorem 7.4.2(b)], $B$ is an isometry. Let $A' : Y' \to M'$ denote the adjoint of $A$. By the Hahn–Banach theorem [7.4.1], there exists a norm-preserving extension $\overline{A'y'} \in X'$ of $A'y' : M \to \mathbf{F}$ for each $y' \in T$. For $x \in X$ and $y' \in T$, $\left|\overline{A'y'}(x)\right| \le \left\|\overline{A'y'}\right\| \|x\| = \|A'y'\| \|x\| \le \|A'\| \|x\| = \|A\| \|x\|$. Therefore $\left(\overline{A'y'}(x)\right)_{y'\in T} \in \ell_\infty(T)$. The map $U : X \to \ell_\infty(T)$, $x \mapsto \left(\overline{A'y'}(x)\right)_{y'\in T}$ is clearly linear and $\|U\| \le \|A\|$, i.e., $U \in L(X, \ell_\infty(T))$. Identifying $Y$ and $B(Y)$, there exists by hypothesis a continuous projection $P$ of $\ell_\infty(T)$ onto $B(Y)$. It only remains to take $\overline{A} = B^{-1}PU \in L(X,Y)$. $\blacksquare$

We characterize Banach spaces with the metric extension property in Secs. 10.2–10.6.

## 10.2 METRIC EXTENSION PROPERTY

**Notation** $X$ and $Y$ denote Banach spaces over $\mathbf{F}$. $X'$ is the normed dual of $X$ [Example 6.5.3(b)]. $L(X,Y)$ denotes the linear space of continuous linear maps from $X$ into $Y$. ■

**The Problem.** Identify the spaces with the metric extension property, those Banach spaces $Y$ such that any $A \in L(M,Y)$ defined on any closed subspace $M$ of any Banach space $X$ can be extended to $\overline{A} \in L(X,Y)$ with $\|\overline{A}\| = \|A\|$.

$$\begin{array}{ccccl} X & & & & \|\overline{A}\| = \|A\| \\ & \overset{\overline{A}}{} & & & \\ | & \searrow & & & \\ M & \xrightarrow{A} & Y & & Y \text{ fixed, } A, X, M \text{ arbitrary} \end{array}$$

The fields $\mathbf{F} = \mathbf{R}$ or $\mathbf{C}$ are such that linear functionals $f \in L(M, \mathbf{F})$ that satisfy $|f(\cdot)| \leq k\,|f(\cdot)|$ for some $k > 0$ always possess one-dimensional extensions $\bar{f} \in L(M \oplus \mathbf{F}x, \mathbf{F})$, $(x \in X)$ [Theorem 7.3.1]. This is the key inductive step in proving the metric extension property of $\mathbf{F}$ [Theorem 7.4.1]. We show in Theorem 10.2.2 that Banach spaces with the one-dimensional metric extension property have the metric extension property.

**Definition 10.2.1** Immediate 1-Extension Property Let $Y$ be a Banach space and let $M$ be a subspace of a Banach space $X$. Suppose that for any $x \notin M$, any continuous linear map $A : M \to Y$ has a continuous linear extension $\overline{A} : M \oplus \mathbf{F}x \to Y$ such that $\|A\| = \left\|\overline{A}\right\|$. If this is so for any Banach space $X$, any $M$ and $x \notin M$ then we say that $Y$ has the *immediate 1-extension property.* ■

**Theorem 10.2.2** Metric Extension = Immediate 1-Extension The Banach space $Y$ has the metric extension property iff it has the immediate 1-extension property.

**Proof.** We need only show that the immediate 1-extension property of $Y$ implies the metric extension property. To that end, let $M$ be a proper subspace of the normed space $X$ and let $A : M \to Y$ be continuous. We must show that there exists a continuous extension $A\hat{} : X \to Y$ of $A$ of the same norm. To that end let $\mathcal{M}$ be the collection $(N, A_N)$ of subspaces $N$ of $X$ containing $M$ and linear maps $A_N : N \to Y$ such that $\|A_N\| = \|A\|$. Since $Y$ has the immediate 1-extension property, $\mathcal{M}$ is not empty. We order $\mathcal{M}$ by taking $(N, A_N) \leq (N', A_{N'})$ if $N \subset N'$ and $A_{N'}$ extends $A_N$. If $\mathcal{N}$ is a totally ordered subset of $\mathcal{M}$, an upper bound for $\mathcal{N}$ is given by $(N\hat{}, A\hat{})$ where

$N^{\wedge} = \cup_{(N,A)\in\mathcal{N}} N$ and $A^{\wedge}x = A_N x$ if $x \in N$. $A^{\wedge}$ is well-defined because if $x \in N$ and $x \in N'$ then, since $A_{N'}$ extends $A_N$ or vice-versa, $A_{N'}x = A_N x$. It is easy to verify that $A^{\wedge}$ is linear and extends $A$. Next, we show that $(N^{\wedge}, A^{\wedge})$ is an upper bound for $\mathcal{N}$. For any unit vector $x \in N^{\wedge}$, $N$ such that $(N, A) \in \mathcal{N}$ and $x \in N$, $\|A^{\wedge}x\| = \|A_N x\| \leq \|A_N\| \|x\| = \|A\| \|x\|$ so $\|A^{\wedge}\| \leq \|A\|$. Since $M \subset N^{\wedge}$, it follows that $\|A\| \leq \|A^{\wedge}\|$ and therefore that $\|A\| = \|A^{\wedge}\|$. Hence $(A^{\wedge}, N^{\wedge})$ is an upper bound for $\mathcal{M}$. $\mathcal{M}$ is therefore totally ordered and has a maximal element $(\overline{A}, \overline{N})$. If $\overline{N} \neq X$, there exists $z \notin \overline{N}$. Since $Y$ has the immediate extension property, we may extend $\overline{A}$ to a linear map $\overline{A}^{\wedge} : N \oplus \mathbf{F}z \to Y$ of norm $\|\overline{A}^{\wedge}\| = \|A\|$ and thereby contradict the maximality of $\overline{A}$. We conclude that $\overline{N} = X$. ■

We show in Theorem 10.2.3 that the metric extension property of a Banach space $Y$ is characterized by an intersection property (IP) of closed balls. An undesirable feature of this characterization is the presence of a subspace $M$ of a space which has nothing to do with $Y$. We obtain a purely internal characterization of the metric extension property by means of another intersection property of closed balls in Theorem 10.5.2.

**Theorem 10.2.3** METRIC EXTENSION PROPERTY AND INTERSECTIONS OF CLOSED BALLS Let $Y$ be a Banach space and $C(y_0, r) = \{y \in Y : \|y - y_0\| \leq r\}$, $r > 0$. $Y$ has the metric extension property iff for any subspace $M$ of any normed space $X$, any $x \notin M$ and any $A \in L(M, Y)$,

$$\bigcap_{m\in M} C(Am, \|A\| \|x - m\|) \neq \varnothing \tag{IP}$$

**Proof.** With notation as above, suppose the condition holds. By Theorem 10.2.2, it suffices to show that this implies that $Y$ has the immediate 1-extension property. Suppose $x \in X \setminus M$, $A \in L(M, Y)$ and $y \in \bigcap_{m\in M} C(Am, \|A\| \|x - m\|)$. We show that

$$\begin{array}{rccl} \overline{A}: & M \oplus \mathbf{F}x & \to & Y \\ & m + ax & \mapsto & Am + ay \end{array}$$

is a continuous linear extension of $A$ of the same norm. For any nonzero $a \in \mathbf{F}$, and all $m \in M$

$$\begin{aligned} \|\overline{A}(m + ax)\| &= |-a| \|\overline{A}(-m/a - x)\| = |-a| \|A(-m/a) - y\| \\ &\leq |-a| \|A\| \|(-m/a) - x\| = \|A\| \|m + ax\| \end{aligned}$$

Therefore $\overline{A}$ is continuous and $\|\overline{A}\| \leq \|A\|$. Clearly $\|A\| \leq \|\overline{A}\|$ and the condition is sufficient. Conversely, suppose $A \in L(M, Y)$, $x \in X \setminus M$, $\overline{A} : M \oplus \mathbf{F}x \to Y$ is a continuous linear extension of $A$, $\|\overline{A}\| = \|A\|$ and $\overline{A}x = y$. Then, for any $m \in M$, $\|\overline{A}(x - m)\| = \|y - Am\| \leq \|A\| \|x - m\|$ which establishes the necessity of the condition. ■

With the goal of a purely internal characterization of the metric extension property, we consider some other intersection properties next.

## 10.3 INTERSECTION PROPERTIES

**Notation** $X$ and $Y$ denote Banach spaces over $\mathbf{F}$. $C(x,r) = \{y \in X : \|y - x\| \leq r\}$, $r > 0$, and $X'$ denotes the normed dual of $X$. ■

If any finite subcollection of a collection $\mathcal{C}$ of sets has nonempty intersection, we say that $\mathcal{C}$ satisfies the *finite intersection condition.* If $T$ is a compact topological space, for example, and $\mathcal{C}$ is a collection of closed subsets then:

If $\mathcal{C}$ satisfies the finite intersection condition then $\cap\mathcal{C} \neq \varnothing$.

We describe this latter implication by saying that $\mathcal{C}$ has the finite intersection *property.* Compactness of $T$ may be described by saying that every collection of closed subsets of $T$ has the finite intersection property. We use this condition–property scheme in the definition below.

**Definition 10.3.1** $c$-ARY INTERSECTION CONDITION AND PROPERTY Let $\mathcal{C}$ be a collection of closed balls of a normed space $Y$.

(a) $c$-ARY For any cardinal number $c$ we say that $\mathcal{C}$ satisfies the *$c$-ary intersection condition* if any subcollection of cardinality $c$ of them has nonempty intersection; if the $c$-ary intersection condition implies that $\cap\mathcal{C} \neq \varnothing$, we say that $\mathcal{C}$ has the *$c$-ary intersection property.*

(b) If every collection $\mathcal{C}$ of closed balls of $Y$ has the $c$-ary intersection property, we say that $Y$ has the *$c$-ary intersection property.*

(c) BINARY, TERNARY If $c = 2$ or 3, we speak of of the *binary* or *ternary* intersection conditions and properties, respectively. ■

It follows from Theorem 10.3.2 below that $\mathbf{R}$ has the binary intersection property.

**Theorem 10.3.2** BINARY INTERSECTION PROPERTY Any order complete Arch-imedean ordered space $Y$ with an order unit $u$ and the order unit norm [Theorem 5.12.9] has the binary intersection property.

**Proof.** Let $Y$ be as above. As observed in Example 5.12.10, a closed ball $C(x,r)$, $r > 0$, in $Y$ is an order interval

$$[x - ru, x + ru] = \{y \in Y : x - ru \leq y \leq x + ru\}$$

Consider a mutually intersecting family $\mathcal{C} = \{[x_s, y_s] : s \in S\}$ of closed intervals in $Y$. Since $\mathcal{C}$ satisfies the binary intersection condition, each $y_s$ must be $\geq$ each $x_t$ $(s,t \in S)$. Consequently, $\{x_s : s \in S\}$ is majorized by any

$y_t$ and $\{y_s : s \in S\}$ is minorized by any $x_s$. By the completeness of $Y$, there exist $x = \sup_s x_s$ and $y = \inf_s y_s$. Clearly, $x \leq y$, so $\varnothing \neq [x, y] \subset \cap_{s \in S}[x_s, y_s]$. ■

Theorem 10.3.2 enables us to provide many more examples.

**Example 10.3.3** BINARY INTERSECTION PROPERTY

(a) ORDERED SPACES As observed in Example 5.12.10, the pointwise ordered space $\ell_\infty(T, \mathbf{R})$ of bounded $\mathbf{R}$-valued functions on the set $T$ is an order complete Archimedean ordered space with order unit $u(t) \equiv 1$. By Theorem 10.3.2, $\ell_\infty(T, \mathbf{R})$ has the binary intersection property. In particular, $\ell_\infty^{\mathbf{R}}(n)$, $1 \leq n \leq \infty$, has the binary intersection property.

(b) $C(T, \mathbf{R}, \|\cdot\|_\infty)$ The space $C(T, \mathbf{R}, \|\cdot\|_\infty)$ of continuous functions of the compact extremally disconnected Hausdorff space $T$ into $\mathbf{R}$ is order complete by Theorem 5.12.18. Since it has an order unit $u(t) \equiv 1$ and $\|\cdot\|_\infty$ is the order unit norm, $C(T, \mathbf{R}, \|\cdot\|_\infty)$ has the binary intersection property by Theorem 10.3.2. We prove in Sec. 10.7 that this is the most general type of real Banach space with the binary intersection property.

(c) $\ell_p^{\mathbf{R}}(n)$, $1 < p < \infty$, $n \in \mathbf{N}$ DOES NOT Although each $\ell_\infty^{\mathbf{R}}(n)$, $1 \leq n \leq \infty$ has the binary intersection property, $\ell_2^{\mathbf{R}}(2)$ does not—it is easy to construct three mutually intersecting circles whose overall intersection is empty. [Draw one large circle and two pairwise intersecting smaller ones above it in such a way as to leave a gap between their intersection and the larger circle.] For essentially the same reason, none of $\ell_p^{\mathbf{R}}(n)$, $1 < p < \infty$, have the binary intersection property for $n > 1$.

(d) ISOMORPHISM AND HOMEOMORPHISM The $c$-ary intersection property is a metric property, preserved by linear isometry. It is not preserved by linear homeomorphism, however, as $\ell_\infty^{\mathbf{R}}(2)$ is linearly homeomorphic to $\ell_2^{\mathbf{R}}(2)$; the former has the binary intersection property, the latter does not. Indeed, all of the spaces $\ell_p^{\mathbf{R}}(n)$, $n \in \mathbf{N}$, $1 \leq p \leq \infty$, are pairwise linearly homeomorphic but only $\ell_\infty^{\mathbf{R}}(n)$ and $\ell_1^{\mathbf{R}}(n)$ have the binary intersection property. The latter two are the only ones whose unit balls have "sharp corners." ■

When Helly [1912] proved the first version of the fundamental lemma—the one-dimensional extension—to a version of the Hahn–Banach theorem for $C[a, b]$, he observed that a family of closed intervals $\{[a_s, b_s] : s \in S\}$ of $\mathbf{R}$ with the binary intersection property has nonempty intersection. Helly generalized the result to showing that *finite* collections $\{C_1, \ldots, C_k\}$, $k > n$, of convex subsets of $\mathbf{R}^n$ had the $(n+1)$-ary intersection property and presented it to the Wiener Mathematischen Gesellschaft in 1913. He subsequently used it to prove a version of the Hahn–Banach theorem for certain normed spaces [Helly 1921]. While Helly was a POW in Siberia (see Sec. 7.13), Radon [1921] and König [1922] independently discovered Theorem 10.3.5(a) and published it; Helly's proof was not published until 1923. The result of (b) is also Helly's [1930].

First, we prove the following simple lemma.

**Lemma 10.3.4** Given vectors $x_1, \ldots, x_n$, if there exists $1 \le i \le n$ such that $\{x_i - x_j : j \neq i\}$ is linearly dependent then there exist $a_i$, $1 \le i \le n$, not all 0, such that $\sum_{i=1}^n a_i x_i = 0$ and $\sum_{i=1}^n a_i = 0$.

**Proof.** Suppose $\{x_1 - x_j : j \ge 2\}$ is linearly dependent. Then there are $a_j$, $2 \le j \le n$, not all 0 such that $\sum_{j=2}^n a_j (x_1 - x_j) = 0$. It only remains to let $a_1 = -\sum_{j=2}^n a_j$. ■

The finiteness hypothesis of Theorem 10.3.5(a) cannot be dispensed with since, for example, the collection of half-rays $\{[n, \infty) : n \in \mathbf{N}\}$ does not have the binary intersection property in $\mathbf{R}$. Arbitrary collections of *compact* convex subsets of $\mathbf{R}^n$, however, have the $(n+1)$-ary intersection property [Theorem 10.3.5(b)].

**Theorem 10.3.5** HELLY INTERSECTION THEOREM A collection $\mathcal{C}$ of nonempty convex subsets of $\mathbf{R}^n$, $n \in \mathbf{N}$, has the $(n+1)$-ary intersection property if:

(a) FINITELY MANY $\mathcal{C} = \{C_1, C_2, \ldots, C_k\}$, $k > n$, is finite or

(b) ARBITRARILY MANY if the sets in $\mathcal{C}$ are compact.

**Proof.** (a) We prove this by induction on the number $k$ of sets. For $k = n+1$, the conclusion holds by hypothesis so suppose it holds for some positive integer $k \ge n+1$ and consider a collection $\mathcal{C} = \{C_1, C_2, \ldots, C_{k+1}\}$ of $k+1$ convex subsets of $\mathbf{R}^n$ that satisfy the $(n+1)$-ary intersection condition. By the induction hypothesis, for each $1 \le i \le k+1$, there exists $x_i \in \mathbf{R}^n$ such that

$$x_i \in \bigcap_{j=1, j\neq i}^{k+1} C_j$$

Since $k > n$, by Lemma 10.3.4, there exist $a_1, a_2, \ldots, a_{k+1} \in \mathbf{R}$, not all 0, such that $\sum_{i=1}^{k+1} a_i x_i = 0$ and $\sum_{i=1}^{k+1} a_i = 0$. We group $a_1, a_2, \ldots, a_{k+1}$ into positive and negative subsets and reindex them so that $a_1 \ge 0, a_2 \ge 0, \ldots, a_m \ge 0$ and $a_{m+1} < 0, a_{m+2} < 0, \ldots, a_{k+1} < 0$. Since not all of the $a_i$ are 0, it follows that $\sum_{i=1}^m a_i \neq 0$. Let $y = \sum_{i=1}^m a_i x_i / \left(\sum_{i=1}^m a_i\right)$, a convex combination. For $i \le m$, each $x_i \in \bigcap_{j=m+1}^{k+1} C_j$, a convex set; therefore $y \in \bigcap_{j=m+1}^{k+1} C_j$. Similarly, since $\sum_{i=1}^m a_i x_i = \sum_{i=m+1}^{k+1} (-a_i) x_i$ and $\sum_{i=1}^m a_i = \sum_{i=m+1}^{k+1} (-a_i)$,

$$y = \frac{\sum_{i=m+1}^{k+1} (-a_i) x_i}{\sum_{i=m+1}^{k+1} (-a_i)} \in \bigcap_{j=1}^{m} C_j$$

and it follows that $y \in \bigcap_{j=1}^{k+1} C_j$.

(b) Suppose that $\mathcal{C}$ is a collection of nonempty compact subsets of $\mathbf{R}^n$ that satisfies the $(n+1)$-ary intersection condition. By (a), $\mathcal{C}$ satisfies the finite intersection condition. Let $D \in \mathcal{C}$. The collection $\{C \cap D : C \in \mathcal{C}\}$ is then a collection of closed subsets of $D$ that satisfies the finite intersection condition. Since $D$ is compact, $\varnothing \neq \cap \{C \cap D : C \in \mathcal{C}\} \subset \cap \mathcal{C}$. ■

## 10.4 THE CENTER-RADIUS PROPERTY

We investigate another intersection property—the center-radius property—in this section that characterizes the metric extension property. Our first result elaborates on the theme that two circles meet iff their centers are close enough and their radii are big enough.

**Theorem 10.4.1** INTERSECTIONS OF BALLS IN NORMED SPACES Let $x$ and $y$ be elements of the normed space $X$. The closed balls $C(x, r_1)$ and $C(y, r_2)$ $(r_1, r_2 > 0)$ meet iff the distance $\|x - y\|$ between their centers is less than or equal to the sum $r_1 + r_2$ of their radii: $\|x - y\| \leq r_1 + r_2$. If they intersect, they must meet at a point on the line segment $[x, y] = \{ax + (1 - a)y : 0 \leq a \leq 1\}$ joining the centers.

**Proof.** If $C(x, r_1) \cap C(y, r_2) \neq \varnothing$ then, by the triangle inequality, $\|x - y\| \leq r_1 + r_2$. Conversely, if $C(x, r_1) \cap C(y, r_2) = \varnothing$ then for any $z$ such that $\|z - x\| = r_1$, $\|z - y\| > r_2$. This implies that $\|x - y\| > r_1 + r_2$.

Now suppose that $x \neq y$ and $C(x, r_1) \cap C(y, r_2) \neq \varnothing$ so that $\|y - x\| \leq r_1 + r_2$. We may write any $z \in [x, y]$ in the form $z = x + t(y - x), t \in [0, 1]$. For $\|y - x\| < r_1$, then $[x, y] \subset C(x, r_1)$ and $z \in C(y, r_2)$ for $t$ sufficiently close to 1. For $r_1 \leq \|y - x\|$, let $t = r_1 / \|y - x\|$. Then $\|z - x\| = r_1$ and $\|z - y\| = \|(1 - t)(y - x)\| = \|y - x\| - r_1 \leq r_2$. Therefore $z \in C(x, r_1) \cap C(y, r_2)$. ■

Thus, for the closed balls $C(x, r_1)$ and $C(y, r_2)$ $(r_1, r_2 > 0)$ and the scalars $\{a_1, a_2\} = \{1, -1\}$, we have $a_1 + a_2 = 0$ and $\|a_1 x + a_2 y\| = \|x - y\| \leq |a_1| r_1 + |a_2| r_2 = r_1 + r_2$ as a condition for $C(x, r_1)$ and $C(y, r_2)$ to meet. The following notion generalizes this idea.

**Definition 10.4.2** CENTER-RADIUS CONDITION AND PROPERTY Let $X$ be a normed space over $\mathbf{F} = \mathbf{R}$ or $\mathbf{C}$ and let $\mathcal{C} = \{C(x_s, r_s) : s \in S\}$ be a collection of closed balls in $X$.

(a) If for any finite subset $a_1, a_2, \ldots, a_n \in \mathbf{F}$ $(n \in \mathbf{N})$,

$$\sum_{k=1}^{n} a_k = 0 \text{ implies that } \| \sum_{k=1}^{n} a_k x_{s_k} \| \leq \sum_{k=1}^{n} |a_k| r_{s_k}$$

for any $\{s_1, s_2, \ldots, s_n\} \subset S$, then $\mathcal{C}$ satisfies the *center-radius condition* (*CRC*).

(b) If any collection of closed balls that satisfies the center-radius condition has nonempty intersection, we say $X$ has the *center-radius property* (*CRP*). ■

We show in Theorems 10.4.3 and 10.4.5 that **R**, **C** and $\ell_\infty^{\mathbf{R}}(2)$ have the center-radius property, but $\ell_2^{\mathbf{R}}(2)$ and the real space $(c_0, \|\cdot\|_\infty)$ of null sequences do not. The center-radius condition is equivalent to the binary intersection condition in real spaces as we show in Theorem 10.4.3.

**Theorem 10.4.3** CRC = BIC IN REAL SPACES A collection $\mathcal{C} = \{C(x_s, r_s) : s \in S\}$ of closed balls in a real normed space $X$ satisfies the binary intersection condition iff it satisfies the center-radius condition.

**Proof.** CRC $\Rightarrow$ BIC. Suppose that $\mathcal{C}$ satisfies the center-radius condition and consider two balls $C(x, r_1)$ and $C(y, r_2)$ from $\mathcal{C}$. With $a_1 = 1$ and $a_2 = -1$, the center-radius condition implies that $\|x - y\| \le r_1 + r_2$. By Theorem 10.4.1, it follows that the balls meet.

BIC $\Rightarrow$ CRC. Suppose that $\mathcal{C}$ satisfies the binary intersection condition. By Theorem 10.4.1, it follows for $C(x, r_1), C(y, r_2) \in \mathcal{C}$ that $\|x - y\| \le r_1 + r_2$. For scalars $a$ and $b$ such that $a + b = 0$, then $a = -b$ and

$$\|ax + by\| = \|a(x - y)\| \le |a| \|x - y\| \le |a| (r_1 + r_2) = |a| r_1 + |b| r_2$$

which is the center-radius condition for two balls. Now we argue by induction on the number $n$ of elements of the finite subset $J \subset S$. Assume that the center-radius condition is satisfied for sets of $n - 1$ balls and consider balls $C(x_i, r_i)$ $(i = 1, 2, \ldots, n)$ from $\mathcal{C}$ and scalars $a_1, a_2, \ldots, a_n \in \mathbf{F}$, such that $\sum_{i=1}^n a_i = 0$. We may assume that none of the the scalars are 0 and we separate them into positive and negative groups with $a_1, a_2, \ldots, a_m > 0$, and $a_{m+1}, \ldots, a_n < 0$. Since $\sum_{k=1}^n a_k = 0$ we may assume that there exist $i, j$ with $a_i < 0$ and $a_j > 0$ such that $|a_i| \le a_j$. We reindex the scalars so that $|a_{m+1}| \le a_m$; as we will use it below, note that this means that $-a_{m+1} \le a_m$ or $a_m + a_{m+1} \ge 0$. By the induction hypothesis,

$$\left\|\sum_{i=1}^n a_i x_i\right\| = \left\|\sum_{i=1}^m a_i x_i + a_{m+1} x_m - a_{m+1} x_m + \sum_{i=m+1}^n a_i x_i\right\|$$

$$= \left\|\sum_{i=1}^{m-1} a_i x_i + (a_m + a_{m+1}) x_m + \sum_{i=m+2}^n a_i x_i + a_{m+1}(-x_m + x_{m+1})\right\|$$

$$\le \left\|\sum_{i=1}^{m-1} a_i x_i + (a_m + a_{m+1}) x_m + \sum_{i=m+2}^n a_i x_i\right\| + \|a_{m+1}(-x_m + x_{m+1})\|$$

As there are $n - 1$ $x$'s in the expression on the left and their coefficients $\sum_{i=1}^{m-1} a_i + (a_m + a_{m+1}) + \sum_{i=m+2}^n a_i = \sum_{i=1}^n a_i = 0$, it follows from the induction hypothesis that

$$\left\| \sum_{i=1}^{m-1} a_i x_i + (a_m + a_{m+1}) x_m + \sum_{i=m+2}^{n} a_i x_i \right\| \\ \leq \sum_{i=1}^{m-1} a_i r_i + (a_m + a_{m+1}) r_m + \sum_{i=m+2}^{n} (-a_i r_i)$$

and therefore that

$$\left\| \sum_{i=1}^{n} a_i x_i \right\| \\ \leq \sum_{i=1}^{m-1} a_i r_i + (a_m + a_{m+1}) r_m + \sum_{i=m+2}^{n} (-a_i r_i) - a_{m+1} (r_m + r_{m+1}) \\ = \sum_{i=1}^{n} |a_i| r_i \quad \blacksquare$$

In Theorem 10.4.5 we show the equivalence of the ternary intersection condition and the center-radius condition for sets of three balls in **C**. First, we need the following technical result.

**Lemma 10.4.4** Given a complex number $c = |c| e^{i\gamma}$ "between" $a = |a| e^{i\alpha}$ and $b = |b| e^{i\beta}$ in the sense that $\alpha < \gamma < \beta < \pi + \alpha$ then there exist complex numbers $s$ and $t$ that satisfy $s + t = 1$ and $|sa + tb + c| = |s| |a| + |t| |b| + |c|$.

**Proof.** As to the second equation, with $s = \sigma e^{i(\gamma - \alpha)}$ and $t = \tau e^{i(\gamma - \beta)}$, $\sigma, \tau > 0$, we have

$$\begin{aligned} |sa + tb + c| &= \left| \sigma e^{i(\gamma-\alpha)} |a| e^{i\alpha} + \tau e^{i(\gamma-\beta)} |b| e^{i\beta} + |c| e^{i\gamma} \right| \\ &= \left| (\sigma |a| + \tau |b| + |c|) e^{i\gamma} \right| \\ &= \sigma |a| + \tau |b| + |c| = |s| |a| + |t| |b| + |c| \end{aligned}$$

The equation $s + t = 1 = \sigma e^{i(\gamma-\alpha)} + \tau e^{i(\gamma-\beta)}$ is a system of two linear equations in the unknowns $\sigma$ and $\tau$. Since $\alpha < \gamma < \beta < \pi + \alpha$, the ratios below are nonnegative and the solution of the equations is given by

$$\sigma = \frac{\sin(\gamma - \beta)}{\sin(\alpha - \beta)} \text{ and } \tau = \frac{\sin(\alpha - \gamma)}{\sin(\alpha - \beta)} \quad \blacksquare$$

We show next that **C** has the center-radius property and that the center-radius property is equivalent to the ternary intersection property for **C**.

**Theorem 10.4.5** CRC = TERNARY INTERSECTION CONDITION IN **C** Three balls $C(x_i, r_i)$, $i = 1, 2, 3$, in **C** have nonempty intersection iff they satisfy the center-radius condition. Hence, by Helly's intersection theorem [10.3.5(b)], an arbitrary collection of closed balls in **C** has nonempty intersection iff any three of them satisfy the center-radius condition. In other words, the center-radius and ternary intersection properties are equivalent in **C**.

**Proof.** View **C** as a real normed space. If three balls in **C** meet then any two of them meet. Hence, they satisfy the center-radius condition by Theorem 10.4.3. Conversely, still viewing **C** as a real normed space, suppose that $\{C(x_i, r_i) : i = 1, 2, 3\}$ satisfies the center-radius condition. By Theorem 10.4.3, the center-radius condition is equivalent to the binary intersection condition in real normed spaces; hence $B = C(x_1, r_1) \cap C(x_2, r_2) \neq \varnothing$. It follows that if one ball is contained in another, then all three meet, so assume that no ball is a subset of another. Let $S_1$ and $S_2$ denote the circumferences of the balls $C(x_1, r_1) = C_1$ and $C(x_2, r_2) = C_2$ and let $\{x, x'\} = S_1 \cap S_2$; assume that $|x| \leq |x'|$. We may further assume without loss of generality that $x_3 = (0,0)$, and $x_1 = (c, 0)$ with $c > 0$ in Cartesian coordinates. By the symmetry of the problem, we may also assume that $C_2$ is centered in the upper half-plane, at $x_2 = (a, b)$, $b > 0$, say. Since the balls meet pairwise, there exist points $w \in [0, x_2] \cap S_2 \cap C_3$ and $w' \in [0, x_1] \cap S_1 \cap C_3$. Let $d(0, B) = \inf\{|x| : x \in B\}$ and note that $|w| \leq |x|$. We show that $B$ meets $C_3$ by showing that $d(0, B) \leq r_3$. There are two possibilities:

(1) $\arg x < 0$ or $\arg x = \gamma > \arg x_2 = \beta$. In either case we must have (a) $d(0, B) \leq |w| = |x_2| - r_2$ or (b) $d(0, B) \leq |w'| = |x_1| - r_1$.

(a) Since $w \in C_2 \cap C_3 = B \subset C_2$, $|w| = d(0, C_2) \leq d(0, B)$. Hence, $d(0, B) = |w| = |x_3 - x_2| - r_2 \leq r_3 + r_2 - r_2 = r_3$ by the center-radius condition.

(b) $d(0, B) \leq |w'| = |x_3 - x_1| - r_1 \leq r_1 + r_3 - r_1 = r_3$.

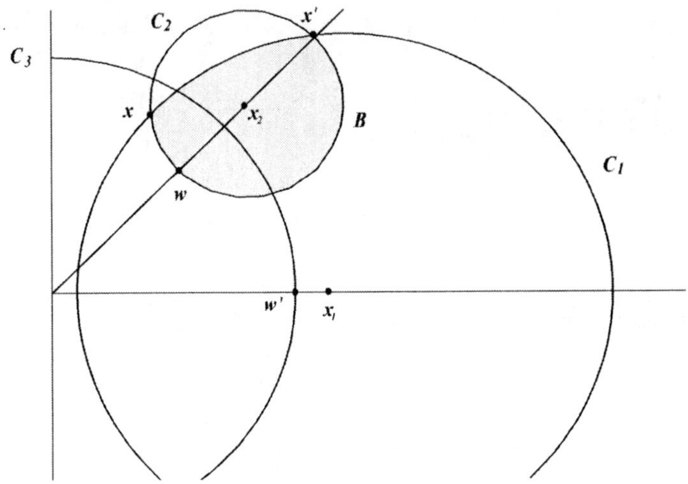


Figure 10.1: Case 1

(2) $0 = \arg x_1 \leq \arg x = \gamma \leq \arg x_2 = \beta$. In this case, $d(x_3, B) = |x_3 - x|$ or $|x_3 - x'|$. Since $x, x' \in B$, $d(x_3, B) \leq d(x_3, x)$ and $d(x_3, B) \leq d(x_3, x')$.

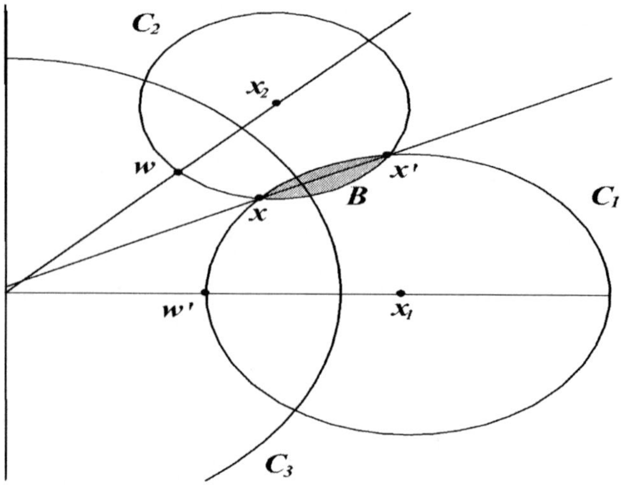


Figure 10.2: Case 2

In this case $x$ is "between" (in the sense of Lemma 10.4.4) $x_1$ and $x_2$. This implies that $-x$ is between $x - x_1$ and $x - x_2$. Hence, by Lemma 10.4.4, there exist complex numbers $s$ and $t$, $s + t = 1$, such that $|s(x - x_1) + t(x - x_2) - x| = |s||x - x_1| + |t||x - x_2| + |x|$. Since $s(x - x_1) + t(x - x_2) - x = -sx_1 - tx_2 = -sx_1 - tx_2 + x_3$, it follows from the center-radius condition that

$$\begin{aligned} |s(x - x_1) + t(x - x_2) - x| &= |s||x - x_1| + |t||x - x_2| + |x| \\ &= |-sx_1 - tx_2 + x_3| \\ &\leq |s| r_1 + |t| r_2 + r_3 \end{aligned}$$

Since $|x - x_1| = r_1$ and $|x - x_2| = r_2$, it follows that $|x| \leq r_3$. ■

## 10.5 METRIC EXTENSION = CRP

We show that the metric extension property is equivalent to the center-radius property in this section.

To show that spaces with the metric extension property have the center-radius property, we need the following lemma which yields a weakened form of the center-radius property. It says that if we start with a collection $\mathcal{C} = \{C(y_s, r_s) : s \in S\}$ of closed balls that satisfy the center-radius condition, we do not get that $\cap \mathcal{C} \neq \varnothing$ but $\cap_{s \in S} C(y_s, \|P\| r_s) \neq \varnothing$ for certain continuous projections $P$, in short, a nonempty intersection of balls in which the radii are enlarged.

**Lemma 10.5.1** WEAKER CRP Let $\mathcal{C} = \{C(y_s, r_s) : s \in S\}$ be a collection of closed balls that satisfy the center-radius condition in the Banach space $(Y, \|\cdot\|)$. Then there exists a normed space $(W, \|\cdot\|)$ in which $Y$ is norm-embedded and, for every continuous projection $P : W \to Y, \cap_{s\in S} C(y_s, \|P\| r_s) \neq \varnothing$.

**Proof.** Let $Y$ and $\mathcal{C}$ be as above, let $W = Y \oplus \mathbf{F}$ and let $r = \inf_{s\in S} r_s$.

CASE 1: $r = 0$. For $y + a \in Y \oplus \mathbf{F}$, define $\|y+a\|' = \|y\| + |a|$. Observe that (1) $\|\cdot\|'$ is clearly a norm and (2) its restriction to $Y$ is $\|\cdot\|$. Since $r = \inf_{s\in S} r_s = 0$ there exists a countable collection $\{r_n\} \subset \{r_s : s \in S\}$ such that $r_n \to 0$. For $\varepsilon > 0$, choose $k$ such that $n \geq k$ implies $r_k < \varepsilon/4$. By the center-radius condition, for $m, n \geq k$,

$$\|y_m - y_n\| \leq \|y_m - y_k\| + \|y_k - y_n\| \leq r_m + r_k + r_k + r_n < \varepsilon$$

This means that $(y_n)$ is Cauchy and therefore has a limit $y$. For any $s \in S$, $\|y_s - y_n\| \leq r_s + r_n$. Taking the limit as $n \to \infty$, this yields $\|y_s - y\| \leq r_s$ for each $s \in S$; in other words $y \in \cap_{s\in S} C(y_s, r_s)$. For any continuous projection $P$ of $Y \oplus \mathbf{F}$ onto $Y$, $\|P\| \geq 1$. Hence $\cap_{s\in S} C(y_s, r_s) \subset \cap_{s\in S} C(y_s, \|P\| r_s)$.

CASE 2: $r > 0$. Let $y + a \in Y \oplus \mathbf{F}$. Let $J$ be a finite subset of $S$ and consider scalars $\{a_j \in \mathbf{F} : j \in J\}$ such that $a = -\sum_{j\in J} a_j$. By taking $y' = y - \sum_{j\in J} a_j y_j$, we have $y + a = y' + \sum_{j\in J} a_j (y_j - 1)$. Define $\|y+a\|' = \inf\left\{\|y'\| + \sum_{j\in J} |a_j| r_j : y + a = y' + \sum_{j\in J} a_j (y_j - 1)\right\}$, the infimum being taken over all such representations (i.e., $a = -\sum_{j\in J} a_j$ for some finite subset $J \subset S$) of $y + a$. To see that $\|\cdot\|'$ is a norm, note first that, for any scalar $c$, $\|c(y+a)\|' = |c| \, \|y+a\|'$. As to the triangle inequality, let $y+a, z+b \in Y \oplus \mathbf{F}$. For any $\varepsilon > 0$, there exist finite subsets $J$ and $K$ of $S$ and sets $\{a_j : j \in J\}$ and $\{b_k : k \in K\}$ of scalars such that $a = -\sum_{j\in J} a_j$ and $b = -\sum_{k\in K} b_k$ with $y + a = y' + \sum_{j\in J} a_j (y_j - 1)$ and $z + b = z' + \sum_{k\in K} b_k (y_k - 1)$ that satisfy

$$\|y'\| + \sum_{j\in J} |a_j| r_j < \|y+a\|' + \frac{\varepsilon}{2} \text{ and } \|z'\| + \sum_{k\in K} |b_k| r_k < \|z+a\|' + \frac{\varepsilon}{2}$$

We now have

$$\begin{aligned} \|y+a+z+b\|' &\leq \left\|(y+z)'\right\| + \textstyle\sum_{j\in J} |a_j| r_j + \sum_{k\in K} |b_k| r_k \\ &\leq \|y'\| + \|z'\| + \textstyle\sum_{j\in J} |a_j| r_j + \sum_{k\in K} |b_k| r_k \\ &< \|y+a\|' + \|z+b\|' + \varepsilon \end{aligned}$$

which establishes the triangle inequality.

To see that $\|\cdot\|'$ extends $\|\cdot\|$, consider $y + a \in Y \oplus \mathbf{F}$ with $a = 0$. Then $y = y' + \sum_{j\in J} a_j (y_j - 1) \in Y$ and $\sum_{j\in J} a_j = -a = 0$; hence $y = y' + \sum_{j\in J} a_j y_j$. Since we can take all the $a_j = 0$ for $j$ from a finite subset $J$ of $S$, $\|y\|$ is one of the values of which $\|y\|'$ is the infimum. Hence $\|y\|' \leq \|y\|$. Since $y = y' + \sum_{j\in J} a_j y_j$ and $Y$ satisfies the center-radius condition,

$$\|y\| \leq \|y'\| + \sum_{j\in J} |a_j| r_j$$

which implies that $\|y\| \leq \|y\|'$ and shows that $\|\cdot\|'$ extends $\|\cdot\|$.

Now suppose that $\|y + a\|' = 0$. If so, given $\varepsilon > 0$, there is a representation $y + a = y' + \sum_{j \in J} a_j (y_j - 1)$ such that

$$\|y'\| + \sum_{j \in J} |a_j| \, r_j < \varepsilon$$

Thus, $\sum_{j \in J} |a_j| \, r_j < \varepsilon$ and therefore $|a| = \left| -\sum_{j \in J} a_j \right| \leq \sum_{j \in J} |a_j| \, r_j < \varepsilon$ which implies that $a = 0$. Since $\|\cdot\|'$ extends $\|\cdot\|$, it follows that $\|y\| = 0$ and that $y = 0$.

Now suppose that $P : Y \oplus \mathbf{F} \to Y$ is a continuous projection of $(Y \oplus \mathbf{F}, \|\cdot\|')$ onto $Y$. To complete the proof, we show that

$$P(0+1) = P(1) \in \bigcap_{s \in S} C(y_s, \|P\| \, r_s)$$

For any $s \in S$, note that

$$\|Py_s - P1\| = \|y_s - P1\| \leq \|P\| \, \|y_s - 1\|' \tag{$*$}$$

For $y_s - 1 = y_s' + (y_s - 1)$ we must have $y_s' = 0$ which implies that $\|y_s - 1\|' \leq r_s$ from the way $\|\cdot\|'$ is defined. It follows from $(*)$ that $\|y_s - P1\| \leq \|P\| \, r_s$. ■

We can now characterize spaces with the metric extension property.

**Theorem 10.5.2** METRIC EXTENSION PROPERTY IFF CRP For a Banach space $Y$, the following are equivalent:

(a) $Y$ is 1-injective;

(b) $Y$ has the center-radius property;

(c) $Y$ has the metric extension property.

**Proof.** (a) $\Rightarrow$ (b) Suppose $Y$ is 1-injective and let $\mathcal{C} = \{C(y_s, r_s) : s \in S\}$ be a family of closed balls that the satisfy the center-radius condition. By Lemma 10.5.1, $Y$ is norm embedded in a certain normed space $W$. Since $Y$ is 1-injective, there is a continuous projection $P$, $\|P\| = 1$, of $W$ onto $Y$. By Lemma 10.5.1, $\cap \{C(y_s, r_s) : s \in S\} \neq \varnothing$ and $Y$ has the center-radius property.

(b) $\Rightarrow$ (c) By Theorem 10.2.3, the metric extension property is equivalent to showing that for any subspace $M$ of any normed space $X$, any $x \notin M$ and any $A \in L(M, Y)$,

$$\bigcap_{m \in M} C(Am, \|A\| \, \|x - m\|) \neq \varnothing \tag{IP}$$

Since $Y$ has the center-radius property, it suffices to show that

$$\{C(Am, \|A\| \, \|x - m\|) : m \in M\}$$

satisfies the center-radius condition. Consider scalars $\{a_i : i = 1, 2, \ldots, n\}$ such that $\sum_{i=1}^n a_i = 0$ and $\{m_1, m_2, \ldots, m_n\} \subset M$. Then

$$\begin{aligned} \left\|\sum_{i=1}^n a_i A m_i\right\| &= \left\|A\left(\sum_{i=1}^n a_i m_i\right)\right\| \\ &\leq \|A\| \left\|\sum_{i=1}^n a_i m_i\right\| \\ &= \|A\| \left\|\sum_{i=1}^n a_i (m_i - x)\right\| \\ &\leq \sum_{i=1}^n |a_i| \|A\| \|m_i - x\| \end{aligned}$$

(c) $\Rightarrow$ (a) Follows from Theorem 10.1.3(a). ■

Since the center-radius and binary intersection conditions are equivalent in real spaces [Theorem 10.4.3] we conclude that:

**Corollary 10.5.3** Real Banach spaces have the metric extension property iff they have the binary intersection property.

**Example 10.5.4** THE METRIC EXTENSION PROPERTY

(a) $\ell_p^{\mathbf{R}}(n)$, $1 < p < \infty$, $n > 1$ DO NOT HAVE THE METRIC EXTENSION PROPERTY Since the real spaces $\ell_p^{\mathbf{R}}(n)$, $1 < p < \infty$, $n > 1$, do not have the binary intersection property [Example 10.3.3(c)], they do not have the metric extension property.

(b) $\ell_\infty^{\mathbf{R}}$ has the binary intersection property by Example 10.3.3(a); therefore it has the metric extension property by Corollary 10.5.3, a fact already established in Theorem 10.1.2(b).

(c) $C(T, \mathbf{R}, \|\cdot\|_\infty)$ HAS METRIC EXTENSION PROPERTY The space $C(T, \mathbf{R}, \|\cdot\|_\infty)$ of $\mathbf{R}$-valued continuous functions of the compact extremally disconnected Hausdorff space $T$ has the binary intersection property [10.3.3(b)], hence has the metric extension property by Corollary 10.5.3. As noted after Theorem 10.8.1, $C([0,1], \mathbf{F}, \|\cdot\|_\infty)$ does not have the metric extension property.

(d) HEREDITY Real $c_0$ is a non-1-injective subspace of the 1-injective space $\ell_\infty^{\mathbf{R}}$ [Theorem 10.1.2(b)] so the metric extension property is not hereditary. Indeed, as we show next, **any real normed space is linearly isometric to a subspace of a Banach space with the metric extension property**. Let $U'$ denote the solid unit ball of the dual $X'$ of a real normed space $X$. Let $\ell_\infty(U', \mathbf{R})$ be the 1-injective space [Theorem 10.1.2(b)] of locally bounded linear maps on $U'$. For each $x \in X$, define $J : X \to \ell_\infty(U', \mathbf{R})$, $x \mapsto \langle x, f \rangle$, $(f \in U')$. Then $J$ is a linear isometry: $\|Jx\|_\infty = \sup_{f \in U'} |f(x)| = \|x\|$ by Theorem 7.4.2(b). Furthermore, every infinite-dimensional injective Banach space has a subspace that is linearly homeomorphic to $\ell_\infty$ [Lindenstrauss and Tzafriri 1977, p.105]. ■

## 10.6 WEAK INTERSECTION PROPERTY

By Theorem 10.5.2 we know that a Banach space has the metric extension property iff it has the center-radius property. We show in Theorem 10.6.2 that the center-radius property is equivalent to an intersection property called the *weak intersection property.*

**Definition 10.6.1** WEAK INTERSECTION PROPERTY (WIP) Let $Y$ be a normed space over $\mathbf{F} = \mathbf{R}$ or $\mathbf{C}$. For each $s \in S$, let $y_s \in Y$ and let $r_s$ be positive. A collection $\mathcal{C} = \{C(y_s, r_s) : s \in S\}$ of closed balls in $Y$ satisfies the weak intersection condition (WIC) if, for any $f$ in the solid unit ball $U'$ of $Y'$,

$$\bigcap \{C(f(y_s), r_s) : s \in S\} \neq \varnothing \text{ (in } \mathbf{F}).$$

If the weak intersection condition implies that $\cap\mathcal{C} \neq \varnothing$, we say that $\mathcal{C}$ has the *weak intersection property* (WIP). If every collection of closed balls in $Y$ has the weak intersection property, then we say that $Y$ has the *weak intersection property* (WIP). ■

The equivalence of the weak intersection property and the metric extension property follows from Theorems 10.5.2 and 10.6.2.

**Theorem 10.6.2** CRP = WIP A normed space $Y$ over $\mathbf{F}$ has the center-radius property iff it has the weak intersection property.

**Proof.** Let $\mathcal{C} = \{C(y_s, r_s) : s \in S\}$ be a collection of closed balls in $Y$.

WIP $\Rightarrow$ CRP Suppose that $Y$ has the weak intersection property and let $\mathcal{C} = \{C(y_s, r_s) : s \in S\}$ satisfy the center-radius condition. First, suppose $Y$ is complex and let $J \subset S$ consist of three elements; consider $\{y_j : j \in J\}$ and consider scalars $\{a_j : j \in J\}$ such that $\sum_{j \in J} a_j = 0$. For any $f \in Y'$ such that $\|f\| \leq 1$, by the center-radius condition,

$$\left|\sum_{j \in J} a_j f(y_j)\right| = \left|f\left(\sum_{j \in J} a_j y_j\right)\right| \leq \left\|\sum_{j \in J} a_j y_j\right\| \leq \sum_{j \in J} |a_j| r_j$$

Since the ternary intersection and center-radius conditions are equivalent in $\mathbf{C}$, [Theorem 10.4.5], it follows that $\cap_{j \in J} C(f(y_j), r_j) \neq \varnothing$. By Helly's intersection theorem [10.3.5], this implies that $\cap_{s \in S} C(f(y_s), r_s) \neq \varnothing$. Since $Y$ has the weak intersection property, it follows that $\cap_{s \in S} C(y_s, r_s) \neq \varnothing$ and shows that $Y$ has the center-radius property. If $Y$ is real, consider a two-element subset $J$ of $S$ and use the equivalence of the binary intersection condition and the center-radius condition of Theorem 10.4.3.

CRP $\Rightarrow$ WIP Suppose that $Y$ has the center-radius property and that $\mathcal{C} = \{C(y_s, r_s) : s \in S\}$ satisfies the weak intersection condition. Since the proofs for the real and complex cases are so similar, we prove it only for the complex case. By the weak intersection condition, for any $f \in Y'$ with $\|f\| \leq 1$, $\cap \{C(f(y_s), r_s) : s \in S\} \neq \varnothing$. Hence, by Theorem 10.4.5, for any 3-element subset $J \subset S$ and scalars $\{a_j : j \in J\}$ such that $\sum_{j \in J} a_j = 0$, it follows that $\left|\sum_{j\in J} a_j f(y_j)\right| \le \sum_{j\in J} |a_j| r_j$. Hence

$$\left|f\left(\sum_{j\in J} a_j y_j\right)\right| = \left|\sum_{j\in J} a_j f(y_j)\right| \le \sum_{j\in J} |a_j| r_j$$

Since [Theorem 7.4.2(b)]

$$\left\|\sum_{j\in J} a_j y_j\right\| = \sup\left\{\left|f\left(\sum_{j\in J} a_j y_j\right)\right| : f \in Y', \ \|f\| \le 1\right\}$$

we conclude that $\left\|\sum_{j\in J} a_j y_j\right\| \le \sum_{j\in J} |a_j| r_j$, i.e., that $\mathcal{C}$ satisfies the center-radius condition. Since $Y$ has the center-radius property, it follows that $\cap\mathcal{C} \neq \varnothing$. ■

## 10.7 REPRESENTATION THEOREM

**Notation.** All the spaces of this section are **real** normed spaces. The dual $X'$ of such a space $X$ carries its weak-$*$ $(= \sigma(X', X))$-topology and the $\sigma(X', X)$-closure of a subset $A$ of $X'$ is denoted $\mathrm{cl}_\sigma A$. The closed unit ball of $X$ or $X'$ is denoted $U(X)$ or $U(X')$, respectively. $C(T, \mathbf{R})$ denotes the Banach space $C(T, \mathbf{R}, \|\cdot\|_\infty)$ of continuous maps of the compact space $T$ into $\mathbf{R}$. ■

In this section we characterize real Banach spaces $Y$ with the metric extension property, those such that for any $A \in L(M, Y)$ of any subspace $M$ of any real normed space $X$ there exists an extension $\overline{A} \in L(X, Y)$ of $A$ with the same norm. We observed in Example 10.5.4(c) that spaces $C(T, \mathbf{R}, \|\cdot\|_\infty)$ of continuous maps of an extremally disconnected compact Hausdorff space $T$ into $\mathbf{R}$ have the metric extension property. We essentially prove the converse in this section. Nachbin [1950] and Goodner [1950] had each proved (in Vols. 68 and 69 of the Transactions of the AMS) that if $Y$ is a real Banach space with the metric extension property *whose unit ball had an extreme point*, then there is a compact extremally disconnected (closures of open sets are open, Sec. 5.12) Hausdorff space $T$ such that $Y$ is linearly isometric to $C(T, \mathbf{R})$. Nachbin [1950, p. 42] conjectured that the unit ball of a space with the metric extension property had to have extreme points, so that to hypothesize their existence was redundant. Kelley [1952] (and also Aronszajn and Panitchpakdi 1956) showed that the assumption was indeed superfluous (cf. Kaufman 1966b and Goodner 1976). Kelley [1952] proved that real Banach spaces with the metric extension property were linearly isometric to some $C(T, \mathbf{R}, \|\cdot\|_\infty)$ of the type just mentioned without the extreme point hypothesis. Since the unit ball of such a space does have extreme points, $e(t) \equiv 1$, for example [Example

9.2.5(e)], this validated Nachbin's conjecture. Goodner [1976] gave a direct proof that the unit ball of a Banach space with the metric extension property has extreme points. We present Kelley's [1952] solution to the problem.

**Remark 10.7.1** Suppose that $A$ is a linear isometry of a Banach space $X$ with the metric extension property into the Banach space $Y$. $A^{-1}$ is then a linear isometry of $A(X)$ onto $X$. Since $X$ has the metric extension property, there is a continuous linear extension $\overline{A^{-1}}$ of $A^{-1}$ to $Y$ such that $\left\|\overline{A^{-1}}\right\| = \left\|A^{-1}\right\| = 1$.

If $X$ is any normed space, the unit ball $U(X')$ of $X'$ is $\sigma(X', X)$-compact by the Alaoglu theorem [8.4.1] and Example 8.3.3 and therefore has extreme points by the Krein-Milman theorem [9.4.1]. Let $S = \mathrm{cl}_\sigma[\mathrm{ext}\, U(X')]$. Since $U(X')$ is symmetric—i.e., $U(X') = -U(X')$—$\mathrm{ext}\, U(X')$ and $\mathrm{cl}_\sigma[\mathrm{ext}\, U(X')]$ are symmetric as well [see Exercise 9.102].

**Definition 10.7.2** TEARING Let $X$ be a real normed space. If $U$ and $V$ are relatively $\sigma(X', X)$-open disjoint subsets of $S = \mathrm{cl}_\sigma[\mathrm{ext}\, U(X')]$ such that $-(U \cup V)$ and $(U \cup V)$ are disjoint and

$$S = \mathrm{cl}_\sigma[-(U \cup V) \cup (U \cup V)] = -(\mathrm{cl}_\sigma U \cup \mathrm{cl}_\sigma V) \cup (\mathrm{cl}_\sigma U \cup \mathrm{cl}_\sigma V)$$

then we say that $U$ and $V$ form a *tearing* of $S$. ■

**Example 10.7.3** TEARING Let $X = \ell_2^{\mathbf{R}}(2)$ so that $X' = \ell_2^{\mathbf{R}}(2)$ as well; since $X$ is finite-dimensional, the norm topology coincides with the weak-$*$ topology on $X'$. $U(X')$ is the solid unit disk in $\ell_2^{\mathbf{R}}(2)$. Its circumference comprises its extreme points: $S = \mathrm{cl}_\sigma\, \mathrm{ext}\, U(X') = SU(X')$. Consider the open arc of the unit circle in the first quadrant, namely

$$U = \left\{(a, b) \in \mathbf{R}^2 : a^2 + b^2 = 1, 0 < a < 1, 0 < b < 1\right\}$$

Let $V$ be the open arc of the circle in the fourth quadrant. It is straightforward to verify that $U$ and $V$ are a tearing of $S$. ■

If $U$ and $V$ form a tearing of $S$, let $T = (\{0\} \times \mathrm{cl}_\sigma U) \cup (\{1\} \times \mathrm{cl}_\sigma V)$, so that $T$ consists of disjoint copies of $\mathrm{cl}_\sigma U$ and $\mathrm{cl}_\sigma V$. We topologize $T$ by taking as an open base sets of the form $\{0\} \times G$, where $G$ is open in $\mathrm{cl}_\sigma U$, and $\{1\} \times W$ with $W$ open in $\mathrm{cl}_\sigma V$. Note that $\{0\} \times \mathrm{cl}_\sigma U$ and $\{1\} \times \mathrm{cl}_\sigma V$ are clopen disjoint subsets of $T$. Since $\mathrm{cl}_\sigma U$ and $\mathrm{cl}_\sigma V$ are closed subsets of the $\sigma(X', X)$-compact set $S \subset U(X')$ and $\{0\} \times \mathrm{cl}_\sigma U$ and $\{1\} \times \mathrm{cl}_\sigma V$ are homeomorphic to $\mathrm{cl}_\sigma U$ and $\mathrm{cl}_\sigma V$, respectively, it follows that

$$T = (\{0\} \times \mathrm{cl}_\sigma U) \cup (\{1\} \times \mathrm{cl}_\sigma V) \text{ is a compact Hausdorff space} \qquad (*)$$

We construct a linear map $H : X \to C(T, \mathbf{R})$ by taking at each $x$ in $X$,

$$(Hx)(0, u) = u(x) \text{ for } u \in \mathrm{cl}_\sigma U \text{ and } (Hx)(1, v) = v(x) \text{ for } v \in \mathrm{cl}_\sigma V \qquad (**)$$

Essentially, $H$ is the canonical embedding $x \mapsto \langle x, \cdot \rangle$ of $X$ into $X''$ and will ultimately yield the linear isometry between $X$ and $C(T, \mathbf{R}, \|\cdot\|_\infty)$ for $T$ as in $(*)$. $H$ is clearly linear. That $Hx \in C(T, \mathbf{R})$ for each $x$ in $X$ follows from the following equivalences: For a net $(u_s)$ in $\text{cl}_\sigma U$,

$$\begin{aligned} (0, u_s) \to (0, u) \quad &\Leftrightarrow \quad u_s \to u \text{ in } \sigma(X', X) \\ &\Leftrightarrow \quad u_s(x) \to u(x) \qquad \text{for each } x \text{ in } X \\ &\Leftrightarrow \quad (Hx)(0, u_s) \to (Hx)(0, u) \quad \text{for each } x \text{ in } X \end{aligned}$$

A similar argument for nets $(v_s)$ in $\text{cl}_\sigma V$ shows that $(1, v_s) \to (1, v)$.

$T'$ denotes the set of evaluation maps of $C(T, \mathbf{R}, \|\cdot\|_\infty) = C(T, \mathbf{R})$ determined by the points $t \in T$, i.e., $t'(x) = x(t)$ for $x \in C(T, \mathbf{R})$. $T$ is homeomorphic to $T'$ endowed with $\sigma(C(T, \mathbf{R})', C(T, \mathbf{R}))$ [Example 8.4.10(a)]. We show next that if $X$ has the metric extension property and $U$ and $V$ form a tearing of $\text{cl}_\sigma[\text{ext}\, U(X')]$, then $X$ is linearly isometric to $C(T, \mathbf{R})$ under the map $H$. In Theorem 10.7.5 we show that tearings exist and that $T$ is extremally disconnected.

**Theorem 10.7.4** $X$ IS LINEARLY ISOMETRIC TO $C(T, \mathbf{R})$ Let the real Banach space $X$ have the metric extension property, let $U$ and $V$ be a tearing of $S = \text{cl}_\sigma \text{ext}\, U(X')$, and let $T$ and $H$ be as in $(*)$ and $(**)$ above. Then:

(a) $H : X \to C(T, \mathbf{R}, \|\cdot\|_\infty) = C(T, \mathbf{R})$ is a surjective linear isometry;

(b) the adjoint $H'$ of $H$ maps $T'$ homeomorphically onto $\text{cl}_\sigma U \cup \text{cl}_\sigma V$;

(c) $\text{cl}_\sigma U \cap \text{cl}_\sigma V = -(\text{cl}_\sigma U \cup \text{cl}_\sigma V) \cap (\text{cl}_\sigma U \cup \text{cl}_\sigma V) = \varnothing$.

**Proof.** We use the notation of the statement and split the proof into several parts.

(1) If $u \in U$ and $v \in V$ are extreme points of $U(X')$, then $H'^{-1}(\{u\}) \cap U(C(T, \mathbf{R})') = \{(0, u)'\}$, where $(0, u)'$ denotes the evaluation map determined by $(0, u)$ and $H'^{-1}(\{v\}) \cap U(C(T, \mathbf{R})') = \{(1, v)'\}$.

PROOF (1) By the Alaoglu theorem [8.4.1], $U(X')$ is $\sigma(X', X)$-compact. For $x$ in $X$ and $z'$ in $X'$ we define $x''(z') = z'(x)$. By Theorem 7.4.2(a), given $x \in X$, there exists $z' \in X'$ of norm 1, such that $z'(x) = \|x\|$; it follows that $x''(U(X')) = [-\|x\|, \|x\|]$. By Theorem 9.3.4, $K = \{z' \in U(X') : x''(z') = \|x\|\}$ is a convex face of $U(X')$. As a $\sigma(X', X)$-closed (convex) subset of $U(X')$, $K$ is $\sigma(X', X)$-compact and therefore has extreme points by the Krein-Milman theorem [9.4.1]. As observed in Example 9.3.2(e), any extreme point $z'$ of $K$ is an extreme point of $U(X')$ as well. Hence $z' \in S = -(\text{cl}_\sigma U \cup \text{cl}_\sigma V) \cup (\text{cl}_\sigma U \cup \text{cl}_\sigma V)$, so $z' \in \text{cl}_\sigma U \cup \text{cl}_\sigma V$ or $z' \in -(\text{cl}_\sigma U \cup \text{cl}_\sigma V)$. Thus

$$\|Hx\|_\infty = \max[\sup_{u \in \text{cl}_\sigma U} |u(x)|, \sup_{v \in \text{cl}_\sigma V} |v(x)|] \geq |z'(x)| = \|x\| \quad (x \in X)$$

Since $\text{cl}_\sigma U \cup \text{cl}_\sigma V \subset U(X')$, however, $\|Hx\|_\infty \leq \|x\|$. Thus $\|Hx\|_\infty = \|x\|$ and $H$ is an isometry which proves part (a).

Let $H'$ denote the adjoint of $H$. By Theorem 8.11.5, $\|H'\| = \|H\| = 1$. For $u \in U$, $v \in V$, and $x \in X$, $H'(0,u)'(x) = (0,u)'Hx = u(x)$ and $H'(1,v)'(x) = v(x)$. Thus

$$H'(0,u)' = u \text{ and } H'(1,v)' = v$$

When $\mathrm{cl}_\sigma U \cap \mathrm{cl}_\sigma V$ is shown to be empty, it will follow that $H'$ is 1-1 on $T'$. We can note now that $H'$ is 1-1 on the sets $(\{0\} \times U)'$ and $(\{1\} \times V)'$.

Next, we show that if $u \in U$, then $H'^{-1}(\{u\}) \cap U(C(T,\mathbf{R})' \cap T' = \{(0,u)'\}$; that is, the only evaluation map in $H'^{-1}(\{u\}) \cap U(C(T,\mathbf{R})')$ is $(0,u)'$. If $(0,w)' \in H'^{-1}(\{u\}) \cap U(C(T,\mathbf{R})' \cap T'$, then $H'(0,w)' = w = u \in \mathrm{cl}_\sigma U$; since $(0,w) = (0,u)$, $(0,w)' = (0,u)'$. Let $v \in \mathrm{cl}_\sigma V$. If $(1,v)' \in H'^{-1}(\{u\}) \cap U(C(T,\mathbf{R})') \cap T'$, then $H'(1,v)' = v \in \mathrm{cl}_\sigma V$. But $H'(0,u)' = u \in U$, so $u = v \in U \cap \mathrm{cl}_\sigma V$. Since $U$ and $V$ are disjoint and open in $S$, however, it follows that $U \cap \mathrm{cl}_\sigma V = \varnothing$. Hence $H'^{-1}(\{u\}) \cap U(C(T,\mathbf{R})') \cap T' = \{(0,u)'\}$. A similar argument shows that $H'^{-1}(\{v\}) \cap U(C(T,\mathbf{R})') \cap T' = \{(1,v)'\}$. Next, we show that we can omit the $T'$.

Suppose that $u \in U$ is an extreme point of $U(X')$. The set $H'^{-1}(\{u\}) \cap U(C(T,\mathbf{R})')$ is a convex face of $U(C(T,\mathbf{R})')$ by Theorem 9.3.5. It now follows from the Krein-Milman theorem [9.4.6] that it is the $\sigma(C(T,\mathbf{R})', C(T,\mathbf{R}))$-closed convex hull of its extreme points and the extreme points of this face of $U(C(T,\mathbf{R})')$ are extreme points of $U(C(T,\mathbf{R})')$. By Example 9.5.13, the extreme points of $U(C(T,\mathbf{R})')$ are evaluation maps $t'$ multiplied by $\pm 1$. By our previous observations, however, the only evaluation map in $H'^{-1}(\{u\}) \cap U(C(T,R)')$ is $(0,u)'$. If $-(1,v)'$ or $-(0,w)'$ belong to this set, application of $H'$ leads to

$$H'[-(0,w)'] = -w = u \in U \cap -\mathrm{cl}_\sigma U \text{ and } H'[-(1,v)'] = -v = u \in U \cap -\mathrm{cl}_\sigma V$$

Since $U$ and $V$ form a tearing of $S$, however, neither of these relationships can hold. Thus, the only extreme point of $U(C(T,\mathbf{R})'$ in $H'^{-1}(\{u\}) \cap U(C(T,\mathbf{R})')$ is $(0,u)'$; by the observation that this set is the closed convex hull of its extreme points, $H'^{-1}(\{u\}) \cap U(C(T,\mathbf{R})') = \{(0,u)'\}$. Similarly, if $v \in V$ is an extreme point of $U(X')$,then $H'^{-1}(\{v\}) \cap U(C(T,\mathbf{R})') = \{(1,v)'\}$ and (1) has been established.

As observed in Remark 10.7.1, since $X$ has the metric extension property and $H$ is an isometry, $H^{-1} : H(X) \to X$ has a continuous extension $G$ to $C(T,\mathbf{R})$ of the same norm.

$$\begin{array}{ccccc} X' & \stackrel{H'}{\longleftarrow} & C(T,\mathbf{R})' & \stackrel{G'}{\longleftarrow} & X' \\ | & & | & & | \\ X & \stackrel{H}{\longrightarrow} & C(T,\mathbf{R}) & \stackrel{G}{\longrightarrow} & X \end{array}$$

(2) $G'$ and $H'$ are linear homeomorphisms and inverses of each other, $G'$ maps $S = \mathrm{cl}_\sigma[\mathrm{ext}\, U(X')]$ onto $T' \cup -T' = \mathrm{ext}\, U(C(T,\mathbf{R})')$ and $H'$ maps $T' \cup -T'$ homeomorphically onto $S$.

PROOF (2) Let $I : X \to X, x \mapsto x$, denote the identity map and note that $I' = (GH)' = H'G'$ is the identity map on $X'$. Since $H^{-1}$ is an isometry, $\|G\| = \|H^{-1}\| = 1$; therefore $\|G'\| = \|G\| = 1$ by Theorem 8.11.5. We consider next the effect of $G'$ on certain extreme points. If $u \in U$ is an extreme point of $U(X')$, then $I'(u) = H'G'(u) = u$ so $G'(u) \in H'^{-1}(\{u\})$; since $\|G'\| = 1$, $G'(u) \in U(C(T,\mathbf{R})') \cap H'^{-1}(\{u\}) = \{(0,u)'\}$; hence $G'(u) = (0,u)'$. Similarly, if $v \in \operatorname{ext} U(X') \cap V$ then $G'(v) = (1,v)'$.

Since $U$ and $V$ form a tearing of $S$, the extreme points of $U(X')$ that lie in $U$ and $V$ together with their images in $-U$ and $-V$ are dense in $S$. Let $\{u_a : a \in A\} \subset U$ and $\{v_b : b \in B\} \subset V$ be the extreme points of $U(X')$ that lie in $U$ and $V$, respectively. As noted in Example 8.4.10(a), $T$ is homeomorphic to $T'$ under the map $t \mapsto t'$ which is the way $G'$ behaves on $\operatorname{ext} U(X') \cap (U \cup V)$. Therefore $G'(\{u_a : a \in A\} \cup \{v_b : b \in B\}) = \{(0,u_a)' : a \in A\} \cup \{(1,v_b)' : b \in B\}$ is dense in $T'$. By the continuity of $G'$ and the compactness of $\operatorname{cl}_\sigma U$, $\operatorname{cl}_\sigma V$ and $T'$, $G'(\operatorname{cl}_\sigma(U \cup V)) = G'(\operatorname{cl}_\sigma U \cup \operatorname{cl}_\sigma V) = T'$ and $G'(-(\operatorname{cl}_\sigma U \cup \operatorname{cl}_\sigma V)) = -T'$. Thus $G'(S) = T' \cup -T'$. The facts that $G'H'(0,u_a)' = G'(u_a) = (0,u_a)'$ and $G'H'(1,v_b)' = G'(v_b) = (1,v_b)'$ for each $a \in A$ and $b \in B$ lead to the conclusion that $G'H' = I$ on a dense subset of $T'$. The continuity of $G'H'$ and compactness of $T'$ mean that $G'H' = I$ on $T'$, therefore on $T' \cup -T'$. But the $\sigma(C(T,\mathbf{R})', C(T,\mathbf{R}))$-closed convex hull of $T' \cup -T'$ is $U(C(T,\mathbf{R})')$, so $G'H' = I'$ on $C(T,\mathbf{R})'$. Since $H'G' = I'$ (see above), $G'$ and $H'$ are homeomorphisms and inverses of each other. Since we have already observed that $G'(S) = T' \cup -T'$, the proof of (2) is complete.

(3) $H'(T') = \operatorname{cl}_\sigma U \cup \operatorname{cl}_\sigma V$.

PROOF (3) This follows from the facts that $H'(\{(0,u_a)' : a \in A\} \cup \{(1,v_b)' : b \in B\}) = \{u_a : a \in A\} \cup \{v_b : b \in B\}$, these last two sets are dense in $\operatorname{cl}_\sigma U \cup \operatorname{cl}_\sigma V$, $T'$ is compact, and $H'$ is continuous. This also proves (b).

(4) $H(X) = C(T,\mathbf{R})$.

PROOF (4) If $H(X) \neq C(T,\mathbf{R})$, since $H$ is an isometry and $X$ is complete, $H(X)$ is a closed proper subspace of $C(T,\mathbf{R})$. Consequently, by Theorem 7.7.5, there exists $f \in C(T,\mathbf{R})'$ such that $f \neq 0$, which vanishes on $H(X)$. Thus, for each $x$ in $X$, $f(Hx) = H'f(x) = 0$. Since, by (2), $H'$ is a homeomorphism, this is contradictory.

(5) $\operatorname{cl}_\sigma U \cap \operatorname{cl}_\sigma V = -(\operatorname{cl}_\sigma U \cup \operatorname{cl}_\sigma V) \cap (\operatorname{cl}_\sigma U \cup \operatorname{cl}_\sigma V) = \varnothing$.

PROOF (5) If $w \in \operatorname{cl}_\sigma U \cap \operatorname{cl}_\sigma V$, then there are nets $(u_a)$ from $U$ and $(v_b)$ from $V$ converging to $w$, with each net consisting of extreme points of $U(X')$. Thus $G'(u_a) = (0,u_a)' \to G'(w)$ and $G'(v_b) = (1,v_b)' \to G'(w)$. Since $T' = (\{0\} \times \operatorname{cl}_\sigma U)' \cup (\{1\} \times \operatorname{cl}_\sigma V)'$, then $G'(w) \in \operatorname{cl}_{\sigma(C(T,\mathbf{R})', C(T,\mathbf{R}))}(\{0\} \times \operatorname{cl}_\sigma U)' \cap \operatorname{cl}_{\sigma(C(T,\mathbf{R})', C(T,\mathbf{R}))}(\{1\} \times \operatorname{cl}_\sigma V)'$ and therefore this intersection is not empty. But since $\{0\} \times \operatorname{cl}_\sigma U$ and $\{1\} \times \operatorname{cl}_\sigma V$ are a clopen disjoint decomposition of $T$, this is contradictory. The fact that $T'$ and $-T'$ are disjoint compact subsets of $U(C(T,\mathbf{R})')$ and an argument similar to the preceding one proves that $(\operatorname{cl}_\sigma U \cup \operatorname{cl}_\sigma V) \cap -(\operatorname{cl}_\sigma U \cup \operatorname{cl}_\sigma V) = \varnothing$ and completes the proof. ■

In view of Theorem 10.7.4(a), to show that any real Banach space $X$

with the metric extension property is linearly isometric to $C(T, \mathbf{R}, \|\cdot\|_\infty)$ for some extremally disconnected compact Hausdorff space $T$, it only remains to show that tearings of $S = \text{cl}_\sigma \operatorname{ext} U(X')$ exist and that the space $T$ of (**) is extremally disconnected.

**Theorem 10.7.5** REAL $X$ WITH METRIC EXTENSION PROPERTY If the real Banach space $X$ has the metric extension property, then there exists a tearing of $S = \text{cl}_\sigma \operatorname{ext} U(X')$ and a compact extremally disconnected Hausdorff space $T$ such that $X$ is linearly isometric to $C(T, \mathbf{R}, \|\cdot\|_\infty)$. By the Banach–Stone theorem [9.6.2], $T$ is unique up to homeomorphism.

**Proof.** Let $\mathcal{W} = \{W_a : a \in A\}$ be the open subsets of $S = \text{cl}_\sigma \operatorname{ext} U(X')$ with the property that $W_a \cap -W_a = \varnothing$. To see that $\mathcal{W}$ is nonempty, let $g \in S$ be such that $\|g\| = 1$ and $w \in X$ such that $|g(w)| = 1$. Then $g \in W = \{f \in X' : |f(w) - g(w)| < 1\} \cap S$ so $W$ is nonempty. If $f \in W \cap -W$, then $|f(w) - g(w)| < 1$ and $|-f(w) - g(w)| < 1$. This implies the contradictory result

$$2 = |-2g(w)| \leq |g(w) - f(w)| + |-f(w) - g(w)| < 2$$

Therefore $\mathcal{W}$ is nonempty. We order $\mathcal{W}$ by inclusion: $W_a \leq W_b$ iff $W_a \subset W_b$. A totally ordered family of these sets will have an upper bound, so, by Zorn's lemma, $\mathcal{W}$ has a maximal element $W$.

We contend that $W \cup -W$ is $\sigma(X', X)$-dense in $S$. If not, let $g \in S \cap \complement(\text{cl}_\sigma (W \cup -W)) = U$. Since $U$ is a relatively $\sigma(X', X)$-open subset of $S = \text{cl}_\sigma \operatorname{ext} U(X')$, $U$ must contain a point of $\operatorname{ext} U(X')$ which we continue to denote by $g$; since $g \in \operatorname{ext} U(X')$, $\|g\| = 1$. The relative $\sigma(X', X)$-openness of $U$ in $S$ implies that there exist $x_i \in X$, $1 \leq i \leq n$, and $r > 0$ such that $V = \{f \in S : |f(x_i) - g(x_i)| < r, 1 \leq i \leq n\} \subset U$. Since $g$ is nontrivial, we may choose $x_0 \in X$ such that $g(x_0) = 1$. Now $G = \{f \in S : |f(x_i) - g(x_i)| < r,\ 0 \leq i \leq n\} \subset V \subset U$ and we may choose $r$ sufficiently small, $r < 1/2$ say, to guarantee that $G \cap -G = \varnothing$. Since $G \subset U$, $G \cap (W \cup -W) = \varnothing$, which implies that

$$(G \cup W) \cap (-(G \cup W)) = \varnothing \tag{$\clubsuit$}$$

Consequently $G \cup W \in \mathcal{W}$. Since $G \cap W = \varnothing$, $G \cup W$ must contain $W$ properly, so ($\clubsuit$) violates the maximality of $W$ in $\mathcal{W}$. Therefore $\text{cl}_\sigma W \cup -\text{cl}_\sigma W = S$.

It is now clear that $W$ and $\varnothing$ form a tearing of $S$ and, with notation as in Theorem 10.7.4, the map $H$ is a linear isometry of $X$ onto $C(T, \mathbf{R})$ by Theorem 10.7.4(a).

It remains to show that $T$ is extremally disconnected, which we effect by the following scheme. By Example 8.4.10(a), $T$ is homeomorphic to

$$(T', \sigma(C(T, \mathbf{R})', C(T, \mathbf{R})))$$

which is homeomorphic to $\text{cl}_\sigma W \cup \text{cl}_\sigma \varnothing = \text{cl}_\sigma W$ by Theorem 10.7.4(b). In the remainder of the argument we show that $\text{cl}_\sigma W$ is extremally disconnected.

Let $D$ be any relatively $\sigma(X',X)$-open subset of $\mathrm{cl}_\sigma W$ and let $E = \mathrm{cl}_\sigma W \cap \complement(\mathrm{cl}_\sigma D)$ so that $D$ and $E$ are disjoint, relatively $\sigma(X',X)$-open subsets of $W$. Since $(D \cup E) \cap -(D \cup E) \subset \mathrm{cl}_\sigma W \cap - \mathrm{cl}_\sigma W = \varnothing$ and $\mathrm{cl}_\sigma[(D \cup E) \cup -(D \cup E)] = \mathrm{cl}_\sigma W \cup - \mathrm{cl}_\sigma W = S$, $D$ and $E$ form a tearing of $S$. Hence, by Theorem 10.7.4(c), $\mathrm{cl}_\sigma D \cap \mathrm{cl}_\sigma E = \varnothing$ and, by the way $D$ and $E$ are defined, the union of the relative $\sigma(X',X)$-closures of $D$ and $E$ in $\mathrm{cl}_\sigma W$ is $\mathrm{cl}_\sigma W$. It follows that the relative $\sigma(X',X)$-closure of $D$ in $\mathrm{cl}_\sigma W$ is open which proves that $\mathrm{cl}_\sigma W$ is extremally disconnected and completes the proof. ■

## 10.8 SUMMARY

We summarize some results about spaces with the metric extension property in this section.

**Theorem 10.8.1** GENERAL CASE A Banach space $Y$ over $\mathbf{F} = \mathbf{R}$ or $\mathbf{C}$ has the metric extension property iff:

(a) $Y$ is 1-injective; [Theorem 10.5.2].

(b) $Y$ is 1-complemented in each Banach space $X$ in which it is norm-embedded [Theorems 4.9.2(a) and 10.5.2].

(c) The identity map $I : Y \to Y$, $y \mapsto y$, can be extended to a linear map of the same norm, 1, to any Banach space $X$ containing $Y$.

(d) $Y$ has the center-radius property [Theorem 10.5.2].

(e) $Y$ has the weak intersection property [Theorem 10.6.2].

(f) [Hasumi 1958] There exists an extremally disconnected [open sets have open closures] compact Hausdorff space $T$ such that $Y$ is linearly isometric to $C(T, \mathbf{F}, \|\cdot\|_\infty)$. By the Banach–Stone theorem [9.6.2], $T$ is unique up to homeomorphism.

Theorem 10.8.1(f) in conjunction with the Banach–Stone theorem implies that $C(T, \mathbf{F}, \|\cdot\|_\infty)$ does not have the metric extension property for any *connected* compact Hausdorff space $T$ since this would imply that $T$ is homeomorphic to a disconnected space. In particular, $C([0,1], \mathbf{R}, \|\cdot\|_\infty)$ does not have the metric extension property.

**Theorem 10.8.2** REAL SPACES A real Banach space $Y$ has the metric extension property iff any of conditions of Theorem 10.8.1 hold as well as iff any of the following conditions hold:

(a) $Y$ has the binary intersection property [Corollary 10.5.3].

(b) $Y$ is linearly isometric to a complete Archimedean ordered vector lattice with order unit with order unit norm. That it suffices for $Y$ to be a complete Archimedean ordered vector lattice with order unit was established in Theorem 10.3.2. The converse follows from Theorems 5.12.18 and 10.7.5 which we state here as (c).

(c) $Y$ is linearly isometric to some $C(T, \mathbf{R}, \|\cdot\|_\infty)$ where $T$ is an extremally disconnected compact Hausdorff space.

Some additional equivalents of the metric extension property for real Banach spaces are listed next. Each is an intersection property of a family of closed balls involving nonempty intersection of the balls with enlarged radii.

(d) [Lindenstrauss 1964] Any family $\mathcal{C} = \{C(y_s, r_s) : y_s \in Y,\ s \in S\}$ of closed balls with the binary intersection property is such that for every $r > 0$, $\cap_{s\in S} C(y_s, (1+r)\, r_s) \neq \varnothing$.

(e) [Davis 1977] Any family $\mathcal{C} = \{C(y_s, 1) : y_s \in Y,\ s \in S\}$ of closed balls with the binary intersection property is such that for every $r > 0$, $\bigcap_{s\in S} C(y_s, 1+r) \neq \varnothing$.

Goodner [1960] showed that imposing the requirement of separability on a space with the metric extension property has drastic consequences—it forces the space to be finite-dimensional; moreover, any real separable space with the metric extension property must be linearly isometric to some $C(T, \mathbf{R}, \|\cdot\|_\infty)$, where $T$ is a finite set with discrete topology.

A point $u$ of the surface $S$ of the unit ball of a normed space $Y$ is a *smooth point* if there is a unique supporting hyperplane at $u$ [Def. 7.8.9]. $Y$ is called *smooth* if it is smooth at each $u \in S$. Akilov [1948] (cf. Goodner 1950, Cor. 4.8) showed that if a real finite-dimensional normed space $Y$ is smooth, then it does not have the metric extension property.

Lindenstrauss [1962; 1963a,b; 1964a,b,c] considered a notion called *ε-extendibility*. A linear map $A$ of one TVS into another is *compact* if there is a neighborhood $U$ of 0 such that $A(U)$ is relatively compact. A real Banach space $Y$ is *ε-extendible* if for any subspace $M$ of any Banach space $X$, any compact map $A$ of $M$ into $Y$ has a compact extension $\overline{A}$ to $X$ such that $\|\overline{A}x\| \leq (1+\varepsilon)\|A\|$. $Y$ has the *metric approximation property* if for each compact subset $K$ of $Y$ and $\varepsilon > 0$, there is a compact map $A : Y \to Y$ with $\|A\| = 1$ such that $\|Ay - y\| < \varepsilon$ for all $y \in K$.

**Theorem 10.8.3** $\varepsilon$–EXTENDIBLE SPACES For a real Banach space $Y$ the following conditions are equivalent:

(a) $Y$ is $\varepsilon$-extendible.

(b) $Y$ has the metric approximation property and for any Banach spaces $Z$ and $\overline{Y}$ where $\overline{Y} \supset Y$, each compact map $A : Y \to Z$ has a compact extension $\overline{A} : \overline{Y} \to Z$ where $\|\overline{A}\| = \|A\|$.

When the unit ball of $Y$ has at least one extreme point, then the following conditions are equivalent:

(c) $Y$ is $\varepsilon$-extendible;

(d) for all Banach spaces $X$ and any subspaces $M \subset N$ with $\dim M = 2$ and $\dim N = 3$ and any $\varepsilon > 0$, each continuous linear map $A : M \to Y$ has a linear extension $\overline{A}$ to $N$ with $\|\overline{A}\| \leq (1+\varepsilon)\|A\|$;

(e) If $C_1, \ldots, C_4$ are pairwise intersecting closed balls in $Y$ whose centers are in a two-dimensional subspace of $Y$, then $C_1 \cap \cdots \cap C_4 \neq \varnothing$.

In addition to the sources cited, see also Herrero 2003, pp. 149f.

**Theorem 10.8.4** REFLEXIVITY [Goodner 1950, Theorem 6.8; Nachbin 1950, Theorem 5] A real Banach space with the metric extension property is reflexive iff it is finite-dimensional.

For real separable spaces, we therefore have:

**Theorem 10.8.5** SEPARABLE SPACES Let $Y$ be a real Banach space with the metric extension property. Then $Y$ is separable if and only if

(a) [Goodner 1960] $Y$ is reflexive.

(b) [Goodner 1950] $Y$ is finite-dimensional.

(c) [Goodner 1960] There exists a finite discrete space $T$ such that $Y$ is linearly isometric to $C(T, \mathbf{R}, \|\cdot\|_\infty)$.

### 10.8.1 Radial Descriptions

We need to define a few terms to state the next characterizations of the metric extension property of real spaces of Theorem 10.8.7.

**Definition 10.8.6** DIAMETER AND RADIUS Let $B$ be a bounded subset of a normed space $X$ and for $x \in X$ and $r > 0$, let $B(x, r) = \{y \in X : \|x - y\| < r\}$.

(a) The *diameter* $d(B)$ of $B$ is $\sup\{\|x - y\| : x, y \in B\}$.

(b) The *radius* $r(B)$ of $B$ is $\inf\{r > 0 : B \subset B(x, r),\ x \in X\}$.

In addition to boundedness, suppose that $B$ is closed and convex for (c) and (d).

(c) $B$ is *diametrically maximal* if for every $x \notin B$, $d(\{x\} \cup B) > d(B)$.

(d) $B$ has *constant width* $d > 0$ if for each $f \in X'$ with $\|f\| = 1$, $\sup f(B - B) = d$. ■

Sets of constant width must be diametrically maximal; the two notions coincide in any two-dimensional space as well as in $\ell_2(n)$ ($n \in \mathbf{N}$) [Eggleston 1965]. They are distinct in certain three-dimensional spaces. It follows from Franchetti 1977, Moreno 2005 and Moreno et. al. 2005 that if $Y = C(T, \mathbf{R}, \|\cdot\|_\infty)$, where $T$ is a compact Hausdorff space, they coincide if and only if $T$ is extremally disconnected; this yields Theorem 10.8.7(b).

**Theorem 10.8.7** RADIAL DESCRIPTIONS A real Banach space $Y$ has the metric extension property iff any of the conditions of Theorem 10.8.2 are satisfied as well as iff:

(a) [Davis 1977] For every bounded subset $B$ of $Y$, the diameter $d(B) = 2r(B)$.

(b) For every closed bounded convex subset $B$ of $Y$, $B$ has constant width iff $B$ is diametrically maximal.

## 10.9 NOTES

Helly's [1921] original argument used his intersection theorem [10.3.5(a)]. Nachbin [1950] returned intersection properties to a position of prominence when he characterized real Banach spaces with the metric extension property by means of the binary intersection property. Some articles from the 1970s in which intersection properties play a major role in subsequent versions of Hahn–Banach type theorems are Davis 1977, Holbrook 1975, Hustad 1973, 1974, and Mira 1982. Ingleton [1952] (cf. Narici et al. [1971, p. 78]) adapted Nachbin's argument to ultranormed spaces over an ultravalued field: If $Y$ is an ultranormed space over an ultravalued field (i.e., the norm and absolute value satisfy $\|x+y\| \leq \max(\|x\|, \|y\|)$, etc.), then $Y$ has the metric extension property iff it has the binary intersection property (see Exercise 10.201).

What normed spaces $X$ have the property that continuous linear maps $A$ of any subspace $M$ of $X$ into any normed space $Y$ always have norm-preserving linear extensions? Kakutani [1941] and Bohnenblust [1942] solved the problem in real and complex spaces, respectively: The real or complex Banach space $X$ has this property iff $\dim X \leq 2$ or $X$ is a Hilbert space!

There are many versions of the Hahn–Banach theorem in which the domain $X$ and codomain $Y$ are preordered linear spaces. The key in Theorem 10.9.1 to the ability to extend dominated linear maps $A : X \to Y$ is order completeness of $Y$.

**Theorem 10.9.1** [Bonnice and Silverman 1966, 1967; To 1971; Ioffe 1981] A real preordered linear space $Y$ (Sec. 5.12) is *p-injective* if for any linear map $A$ of any subspace $M$ of any real linear space $X$, $A : M \to Y$, any sublinear map $p : X \to Y$, $A \leq p|M$, may be extended to a linear map $\overline{A}$ on $X$ which is also dominated by $p$. The ability to extend continuous linear maps in this sense is equivalent to order completeness of $Y$, namely that each majorized subset of $Y$ have a least upper bound [cf. Theorem 9.9.4 and Exercise 10.101].

In the context of linear maps between Banach lattices there is an extensive literature—too much to go into—but we mention Danet and Wong [2002a,b].

## 10.10 EXERCISES

10.101 MONOTONE EXTENSION THEOREMS Let $X$ be a preordered real linear space (Sec. 5.12). We consider positive extensions of a positive linear functional (see below) in various circumstances in this exercise. A map $f : X \to \mathbf{R}$ is called *monotone* if $x \leq y \Rightarrow f(x) \leq f(y)$ for all $x, y \in X$. For linear maps $f$ this is equivalent to $x \geq 0 \Rightarrow f(x) \geq 0$ for all $x \in X$ which is why monotone linear maps are also called *positive*.

(a) COFINAL SUBSPACE A subspace $M$ of $X$ is called *cofinal* if for any $x \geq 0$ in $X$ there exists $m \in M$ such that $m \geq x$. $\mathbf{Z}$, for example, is a cofinal subset of **R**. If $f$ is a positive linear functional defined on a cofinal subspace $M$ of $X$, then $f$ has a positive linear extension to $X$.

(b) ORDER UNIT An element $u \in X$ is an *order unit* if for any $x \in X$ there exists $n \in \mathbf{N}$ such that $-nu \leq x \leq nu$ [Definition 5.12.4]. If $X$ has an order unit $u$ and $M$ is a subspace to which $u$ belongs, show that a positive linear functional defined on $M$ has a positive linear extension to $X$.

(c) Let $P = \{x \in X : x \geq 0\}$. If $f$ is a positive linear functional on a subspace $M$ of $X$, then $f$ has a positive linear extension to $X$ iff there is a convex absorbent set $U$ such that $f \leq 1$ on $M \cap (U - P)$ where $U - P = \{u - p : u \in U,\ p \in P\}$.

10.201 HAHN–BANACH THEOREMS FOR ULTRANORMED SPACES [van Tiel 1965] Let $X$ be a vector space over a nontrivially ultravalued field $(K, |\cdot|)$. "Linear functional on $X$" means a linear map of $X$ into $K$. Suppose that $K$ is spherically complete in the sense that every descending sequence of closed balls has nonempty intersection [Exercise 3.205].

(a) An ultranormed space $(Y, \|\cdot\|)$ over $K$ has the metric extension property if for any continuous linear map $A$ of a subspace $M$ of a ultranormed space $X$ into $Y$, there exists a continuous linear extension $\overline{A}$ of $A$ defined on all of $X$ such that $\sup\{\|Am\| / \|m\| : m \in M,\ m \neq 0\} = \sup\left\{\|\overline{A}x\| / \|x\| : x \in X,\ x \neq 0\right\}$. Show that $Y$ has the metric extension property iff $Y$ is spherically complete. (Note that two balls in an ultranormed space intersect iff one is contained in the other.)

(b) If $M$ is a subspace of a locally $K$-convex space $X$ [Exercise 4.203], $p$ an ultraseminorm [Exercise 5.204] on $X$ and $f$ a linear functional on $M$ satisfying, for some $c > 0$, $|f(m)| \leq cp(m)$ for all $m \in M$—so that $f$ is continuous by Exercise 5.205(b)—then there exists a continuous) linear extension $F$ of $f$ to $X$ such that $|F| \leq cp$ on $X$.

(c) Any continuous linear functional on a linear subspace $M$ of a locally $K$-convex space $X$ has a continuous linear extension to $X$.

(d) If $x$ is a nonzero vector in the locally $K$-convex Hausdorff space $X$, then there exists a continuous linear functional $f$ on $X$ such that $f(x) = 1$.

(e) Let $D$ be a closed $K$-disk [Exercise 4.202] in the locally $K$-convex Hausdorff space $X$ and let $x$ be a vector not in $D$. Then:

i. If the valuation on $K$ is dense, i.e., $\{|a| : a \in K\}$ is dense in the positive reals, then there exists a continuous linear functional $f$ on $X$ such that $\sup |f(D)| \leq 1$ and $f(x) = 1$.

ii. If the valuation on $K$ is discrete, i.e., for some positive $r$, $\{|a| : a \in K\} = \{r^n : n \in \mathbf{Z}\} \cup \{0\}$, then there exists a continuous linear functional $f$ on $X$ such that $\sup |f(D)| \leq 1$ and $|f(x)| > 1$.

(f) If $M$ is a closed linear subspace of the locally $K$-convex Hausdorff space $X$ and $x \notin M$, then there exists a continuous linear functional $f$ on $X$ which vanishes on $M$ and $f(x) = 1$.

# Chapter 11

# Barreled Spaces



*A Principle of Uniform Boundedness for Linear Functionals*: If $(f_n)$ is a sequence of continuous linear functionals on a Banach space $X$ and for each $x \in X$ there exists $M_x$ such that $|f_n(x)| \leq M_x$ for every $n$ then the sequence $(\|f_n\|)$ of norms is bounded. As the latter bound does not depend on any particular point, it is a "uniform" bound. Alternatively, if $\{f_n\}$ is $\sigma(X', X)$-bounded, then it is norm bounded or: the concepts of norm and weak-$*$ boundedness coincide.

Helly [1912, p. 268] established a principle of uniform boundedness for linear functionals on $C[a,b]$; his argument is easily adapted to Banach spaces. Banach space versions were proved by Banach [1922, p. 157], Hahn [1922, p. 6] and Hildebrandt [1923, p. 311].

Banach and Steinhaus [1927] generalized the result to pointwise bounded sequences $(A_n)$ of continuous linear maps between Banach spaces $X$ and $Y$. They used a technique that had been exploited by Lebesgue many times called the "gliding hump" method. Prior to submitting the manuscript for publication, they showed it to S. Saks (1897–1942) who suggested a "category argument" instead of the gliding hump. They followed his advice and changed their argument accordingly. Category arguments have been a powerful tool in Banach space theory ever since.

How can we phrase a principle of uniform boundedness for families $\{A_s \in L(X,Y) : s \in S\}$ of continuous linear maps between locally convex spaces $X$ and $Y$? To do so, we need analogs of pointwise and uniform boundedness. Pointwise boundedness is easy: Require each $\{A_s x : s \in S\}$ to be bounded as in Definition 6.1.1 for each $x \in X$. But what about $\sup_{s \in S} \|A_s\| < \infty$? It happens that the latter condition is equivalent to equicontinuity of $\{A_s : s \in S\}$ so we may rephrase the question as: When is a pointwise bounded family of continuous linear maps equicontinuous? Bourbaki [1953] identified the class of locally convex spaces $X$ for which such an implication holds in $L(X,Y)$ for any LCS $Y$—namely, that every barrel in $X$ (a closed absorbent disk as in Definition 8.8.2) is a neighborhood of 0. We use the catchy term coined by Bourbaki [1950] and call such spaces *barreled*. If $X$ is *barreled* and $Y$ is any LCS, pointwise boundedness implies equicontinuity in $L(X,Y)$ [Theorem 11.9.1]. Conversely, if $X$ is such that pointwise boundedness in $L(X,Y)$ implies equicontinuity for any LCS $Y$, then $X$ is *barreled* [Theorem 11.9.3]. A similar characterization obtains in the absence of local convexity as well (see Exercise 11.201).

We study barreled spaces in this chapter. We also define and investigate Baire spaces, the most important class of barreled spaces—Banach spaces are Baire spaces, for example. As to the stability of barreledness [Sec. 11.12], quotients, locally convex direct sums and products of barreled spaces are barreled. Subspaces generally need not be, an exception being when the subspace is of countable codimension [Theorems 11.12.8 and 11.12.9].

## 11.1 THE SCOTTISH CAFÉ

Mathematics is the most beautiful and most powerful creation of the human spirit. Mathematics is as old as Man.

Stefan Banach

"The book, please." That was all Banach had to say to the waiter at Café Szkocka, The Scottish Café, in Lwów, Poland when he entered. "The book" in which Banach and others wrote problems and solutions and notes is what later became known as *The Scottish Book* (*Księga Szkocka*). The regulars who gathered at The Scottish Café with Banach in the period between the wars, especially on Saturday evenings after the mathematical seminar, included:

> O. Nikodym (1887–1974), H. Steinhaus (1887–1972), W. Orlicz (1903–1990), S. Mazur (1905–1981), M. Kac (1909–1985), S. Ulam (1909–1984),

and, sadly,

> J. Schreier (1909–1942), M. Eidelheit (1911–1943), L. Sternbach (1905–1943), S. Kaczmarz (1895–1939), J. Schauder (1899–1943), H. Auerbach (1901–1942), and A. Łomnicki (1881–1941) [Banach's thesis advisor], and S. Saks (1897–1942).

"Sadly" in regard to the latter because they were all murdered by the Nazis. Ulam in his preface to *The Scottish Book* described the kinetic intellectual energy in the Scottish Café as follows:

> The mathematical life was very intense in Lwów. Some of us met practically every day, informally in small groups, at all times of the day to discuss problems of common interest, communicating to each other the latest work and results.

As Paul Halmos once said, mathematics is not a team science but it is a social science.

Andrzej Turowicz worked at the Kraków branch of the Polish Institute of Mathematical Sciences. Although he continued working there, he became a monk after the war and lived in a nearby cloister. Before the war, he spent three years in Lwów working mainly with Mazur and Eidelheit. Of that period, he described Banach's style of working as:

> [Banach] would spend most of his days in cafés, not only in the company of others but also by himself. He liked the noise and the music. They did not prevent him from concentrating and thinking. There were cases when, after the cafés closed for the night, he would walk over to the railway station where the cafeteria was open around the clock. There, over a glass of beer, he would think about his problems.

Before Café Szkocka, he and some colleagues frequented Café Roma. Annoyed because they wouldn't extend credit to him, Banach suggested that they relocate to Café Szkocka, essentially right in front of the Roma. Steinhaus preferred the more genteel atmosphere and excellent pastry at Ludwik Zalewski's Confectionery at 22 Akademicka Street, but Banach (being Banach) prevailed and Café Szkocka is where they went. Ulam recalled:

> It was difficult to outlast or outdrink Banach during these sessions. We discussed problems proposed right there, often with no solution evident even after several hours of thinking. The next day Banach was likely to appear with several small sheets of paper containing outlines of proofs he had completed.

The group ate and drank and talked and argued and scribbled on the marble tables, to the considerable displeasure of the ownership of the Café who had to scrub them. After a marathon session—one lasted 17 hours—in which they felt they had solved an important problem, they returned the next day to find the precious residue expunged from the table. In part because of this loss, Banach suggested in 1935 that they record the open problems they discussed. His wife Łucja bought a sturdy notebook for them which became *The Scottish Book* [see Mauldin 1981 and other entries listed with it]. When asked for "the book," the waiter retrieved it from its hiding place and brought it to the table along with coffees and a "few" glasses of vodka, one more instance of the dependence of mathematics on the proper mixture of stimulating caffeine and lubricating alcohol. After having decided on a proper formulation, they wrote the problem at the top right of the right hand page, leaving space below and on the following page for subsequent remarks. In a few cases the author(s) offered a prize for the solution. Wavre (Problem 151, 1936) for example, offered a fondue in Geneva to the solver but the best story is about Problem 153 for which Mazur (1936) offered a live goose to the solver. Grothendieck [1955] showed that Problem 153 was equivalent to a problem first formulated by Hildebrandt [1931, pp. 196–197] called *the approximation problem*, namely, given Banach spaces $X$ and $Y$, is every compact linear operator $A : X \to Y$ a limit of operators of finite rank? If $Y$ has a Schauder basis, the answer is affirmative [Theorem 15.9.2]. P. Enflo [1973] showed that there are subspaces of $c_0$ and $\ell_p$, $2 < p < \infty$, that do not have the approximation property. [For a simplified version of Enflo's construction due to Davie, see Lindenstrauss and Tzafriri 1977, p. 87ff.] By Theorem 15.9.2, these (perforce separable) subspaces cannot have Schauder bases, thereby negatively settling *the basis problem*: Does every separable Banach space have a Schauder basis? [The first published reference to the problem is Banach's [1932, p. 111]: *On ne sait pas si tout espace du type (B) séparable admet une base.*] Enflo subsequently went to Warsaw and received the goose from Mazur on live Polish television. His solution to these two problems altered the course of Banach space research forevermore.

By 1939 Banach was a professor at Jan Kazimierz University and President of the Polish Mathematical Society. As a result of the Nazi-Soviet Pact of August 1939, Lwów was occupied by the Soviets from September 1939 until June 1941. On 8 January 1940, they renamed it *Ivan Franko Lviv State University*, the name it bears today. Banach was allowed to retain his post in 1939 and ultimately became Dean of the Faculty of Science. In the same year Banach's father fled from the German armies advancing towards Kraków and came to Lwów while Banach continued his research, textbook writing, lecturing and sessions in the cafés. In 1940 Soviet mathematicians such as Alexandrov, Liusternik and Sobolev sampled the nutrient-rich environment of The Scottish Cafe and made notes in "the book" and Banach attended conferences in the Soviet Union.

Ulam had emigrated to the United States in 1935 but continued to visit Lwów in the summer. His recollection of the summer of 1939 was:

> It seems that in general people were expecting another crisis like that of Munich in the preceding year, but were not prepared for the imminent world war. Mazur, in a discussion concerning such possibilities, suddenly said to me "A world war may break out. What shall we do with *The Scottish Book* and our joint unpublished papers? You are leaving for the United States shortly—and presumably will be safe. In case of a bombardment of the city, I shall put all the manuscripts and The Scottish Book into a case which I shall bury in the ground." We even decided upon a location of this secret hiding place; it was to be near the goal post of a football field outside the city. It is not known to me whether anything of the sort really happened. Apparently the manuscript of *The Scottish Book* survived in good enough shape to have a typewritten copy made, which Professor Steinhaus sent to me ...

The last of the 193 entries—some numerical results dealing with the distribution of the number of matches in a box—was made by Hugo Steinhaus on 31 May 1941, right before the Nazi occupation; he, being not only an intellectual but a Jew (both capital offenses), evaded the Gestapo throughout the occupation in desperate circumstances.

Banach was in Kiev when Germany invaded the Soviet Union on 22 June 1941 and he returned immediately to his family in Lwów. The jackboots arrived in Lwów about a week later, on 30 June 1941. As a continuation of the *Ausserordentliche Befriedungsaktion* started in 1940, the killing in Lwów began immediately. Of the 120,000 Jews of Lwów at the beginning of 1941, only 200–800 were left by 1944. In addition to the Jews and Gypsies, the master plan was to eliminate all Poles who might have the capacity to lead—politicians, artists, sportsmen, intellectuals, priests. Those chosen, some from lists prepared by their Ukrainian students, were transported to killing sites such as the Gestapo prison on Pełczyńska Street, the Brygidki Prison, the former military prison at Zamarstynów and to killing fields surrounding the city. During the night of 3 July–4 July 1941—just three days after the arrival of the Nazis to Lwów—Banach's thesis advisor Antoni Łomnicki and several dozen professors and their families, a total of about 45 people, were arrested by *Einsatzgruppen* for murder by bullet, by bayonet, and by battery (some with a hammer); at least two of them were probably buried alive. The bodies were buried where they fell but several days later the Wehrmacht exhumed and transported them to an unknown location. SS-Hauptsturmführer Hans Krüger commanded the unit supervising the slaughter. Several weeks after the killing in Lwów, his unit was transferred to Stanisławów for more mass murder. He was sentenced to life imprisonment in Germany for the Stanisławów crimes but neither he nor anyone else was ever prosecuted for the Lwów murders. After the war, by Soviet decree, crimes committed east of the Curzon line could not be prosecuted by Polish courts and information on the Lwów atrocities was restricted.

Somehow *The Scottish Book* and Steinhaus survived the war and Banach's son Stefan brought it to Steinhaus in Wroclaw. Steinhaus subsequently sent a copy to Ulam who translated it into English in 1957; he distributed copies to attendees of the International Congress of Mathematicians in Edinburgh in 1958. The original remained with Banach's wife Łucja until her death in 1954, then went to their son Stefan.

Banach was born on 30 March 1892 in Kraków, Galicia, in what was then part of the Austro-Hungarian Empire. Banach's father Stefan Greczek had been drafted into the Austro-Hungarian army. Banach's mother's name is given on his baptismal certificate as Katarzyna Banach but some people believe that his mother was actually some lady of the aristocracy. There is at least one person who claimed he knew who the mother was but was sworn to secrecy. Banach's niece, Monika Waksmundzka-Hajnos [2006] believed that Katarzyna worked for the same officer to whom Greczek was an orderly and that Katarzyna was, in fact, his mother. As Greczek could not marry without permission from the military authorities, the couple separated. The niece also says that Greczek promised Katarzyna that he would take care of the child and never reveal who the mother was. In any case, Greczek took Banach to Ostrowsko, his birth village, and placed him in the care of his mother Antonina. She lived in a small cottage of a type referred to by the locals as a "chicken hut." As Antonina's health began to fail, she went to live with Greczek in Kraków. Because of Antonina's health and with Katarzyna's acquiescence, Banach went to live with Franziska Płowa, a moderately well-to-do laundry owner in Kraków. Greczek paid her for caring for his son. He married two more times and was the father of five more children. One of them, a daughter named Antonina, knew of Banach as a friend of the family. She discovered that he was her half-brother in 1939 at the outbreak of the war when she met him for the first time and he explained what had happened.

Banach had been friendly with Witold Wilkosz since their days at the Henryk Sienkiewicz Gymnasium Number 4 in Kraków. Banach graduated in 1910; Wilkosz had transferred to a different gymnasium and graduated in the same year. Each had an abiding love for mathematics but decided that there was nothing left to do in it. Banach went to Lwów, Galicia (Poland), to study engineering at the Lwów Polytechnic Institute. (*Lwów* is the Polish spelling; the Soviets annexed it after the war and it is now part of Ukraine and spelled *Lviv*; in German, it is *Lemberg*.) Wilkosz studied oriental languages at the Jagiellonian University in Kraków. Banach supported himself by tutoring and other odd jobs. Because of the outbreak of the First World War, he did not graduate and returned to Kraków. He was not conscripted because of his left handedness and poor eyesight. Hugo Steinhaus had no such excuses.

After his military service in the Polish Legion at the beginning of World War I, Steinhaus was living in Kraków in the spring of 1916, waiting to take up a post at the Jan Kazimierz University in Lwów. He liked to walk in the evening. As he related in his memoirs:

> During one such walk [in the Planty Gardens] I overheard the words "Lebesgue measure." I approached the park bench and introduced myself to the two young apprentices of mathematics. They told me they had another companion by the name of Witold Wilkosz, whom they extravagantly praised. The youngsters were Stefan Banach and Otto Nikodym. From then on we would meet on a regular basis, and . . . we decided to establish a mathematical society.

Steinhaus told Banach about a problem he had been working on without success. After a few days Banach had the main idea for a counterexample and Steinhaus and Banach wrote a joint paper which they presented to Stanisław Zaremba for publication. The paper, Banach's first, appeared in the Bulletin of the Kraków Academy in 1918. It was also through Steinhaus that Banach met his wife Łucja Braus. They were married in the mountain resort of Zakopane in 1920. Even though Steinhaus amassed 170 journal publications, he said that Banach was his greatest mathematical discovery.

On Steinhaus's initiative, the Mathematical Society of Kraków was founded in 1919. The first meetings took place in Steinhaus's apartment, much to the annoyance of his landlord. Zaremba chaired the inaugural meeting and was elected its first president. Banach lectured to the Society twice during 1919 and continued to produce high quality research papers. The Mathematical Society of Kraków became the Polish Mathematical Society in 1920.

Through Steinhaus, Banach got an assistantship to Antoni Łomnicki at Polytechnika Lwówska in Lwów in 1920. Banach repeatedly postponed submitting a dissertation on the grounds that he could do better. Ciesielski [2007, p. 2] says that ultimately somebody wrote down Banach's remarks on some problems and this was accepted as his PhD dissertation. Then one day while walking in the hall, Banach was asked to go to a Dean's room to answer some questions and this constituted his examination. As Banach had no university mathematics degree, it was highly unusual to grant him a PhD, but he got it in 1920 from the Jan Kazimierz University in Lwów. The article based on his thesis, *Sur les opérations dans les ensembles abstraits et leur application aux équations intégrales,* was the first display of Banach's incandescent perspicacity. Published in Fundamenta Mathematicae in 1922, it is a landmark in the history of functional analysis—among other things, this is where Banach spaces are first defined. In the same year Banach received his *habilitation*, the qualification that would permit him to mentor doctoral students. He became a full professor in 1924 and remained at the university, which he listed as the University of Lwów, for the rest of his life.

He and Steinhaus founded Studia Mathematica in 1929 and were its first editors. The stated aim of the journal was ". . . to focus on research in functional analysis and related topics."

They started another important publishing venture in 1931, the series called *Monografie Matematczyne.* The original editors were Banach and Steinhaus from Lwów and Knaster, Kuratowski, Mazurkiewicz, and Sierpinski from Warsaw. Volume 1 was the French translation of Banach's *Operacje Liniowe* of 1931, renamed *Théorie des Opérations Linéaires* and became the analyst's vade mecum for the next 30 years. (When it was first published, it was displayed in some Lwów bookshops under "Medical Books.") Volume 7 was Saks's *Theory of the Integral*, Number 15, Banach's *Mechanics.*

In a plenary address at the International Congress of Mathematicians in Oslo in 1936 Banach described the work of the Lwów school and spoke of their plans for further development. Kuratowski spent the years 1927–1934 at the Polytechnika Lwówska and was another important influence on Banach.

During the Nazi occupation of June 1941–July 1944 all universities were closed as Poles were destined to become serfs, doing only agricultural and other menial work in *Die Neuordnung.* Banach and his son survived because they worked at the Rudolf Weigl Bacteriological Institute. There, Banach fed lice with his own blood by placing a box of them over his hand as part of a program about infectious diseases. At one point, Banach was accused of trafficking in German currency and jailed for a few weeks. In the summer of 1944 the Red Army and the Polish Home Army ejected the Wehrmacht from Lwów and the university reopened. Because of the impending annexation of Lwów to Ukraine, the Soviets began to "evacuate" Poles. Banach prepared to leave for Kraków to take up a chair at Jagiellonian University. He was also considered as a possible Minister of Education in Poland. None of this was to be.

The privations of the Nazi occupation had taken a severe toll on Banach's health. He used to tutor the son of the lawyer/businessman Tadeusz Riedl and went to live with the family in their home on Dwernickiego Street in July 1944 for what turned out to be the last 13 months of his life. He met Sobolev outside Moscow during this time but was clearly seriously ill. At a memorial conference for Banach, Sobolev said of this meeting:

> Despite heavy traces of the war years under German occupation, and despite the grave illness that was undercutting his strength, Banach's eyes were still lively. He remained the same sociable, cheerful, and extraordinarily well-meaning and charming Stefan Banach whom I had seen in Lwów before the war. That is how he remains in my memory: with a great sense of humour, an energetic human being, a beautiful soul, and a great talent.

He was diagnosed with lung cancer in January 1945 and died in August of that year. He is buried in Lichakiv cemetery (= Lyczakowski's Necropolis) in Lwów. His tombstone is inscribed (*sic*) "D.R. Stefan Banach." Brailey Sims [2001] describes seeing Banach's yellowed university file at the university in June 2001. On the first page in blue pencil is "File Closed, Died 8 August

1945." It contained Banach's one-page, handwritten CV in Ukrainian from 1945 along with the records of his teaching assignments, promotions, etc., as well as letters of recommendation from Sergei Sobolev and Pavel Aleksandrov. (Imagine submitting a *one-page* CV today?) In a (very) modest tribute, a classroom at the university is dedicated to Banach. It contains a bronze bust and a small oil painting of him atop the blackboard.

A butterfly flapped his wings in the Planty Gardens one evening in 1916; its consequences have been with us ever since.

## 11.2 $\mathcal{S}$-TOPOLOGIES FOR $L(X,Y)$

**Notation.** $L(X,Y)$ denotes the linear space of continuous linear maps of the TVS $X$ into the TVS $Y$. ■

If $X$ and $Y$ are normed spaces, $L(X,Y)$ is a normed space by taking, for $A \in L(X,Y)$ [Example 6.5.3],

$$\|A\| = \sup\{\|Ax\| : \|x\| = 1\}$$

We investigate some other ways to topologize $L(X,Y)$ in this section.

Let $T$ be a set and $Y$ a commutative topological group with neighborhood filter $V(0)$ at 0. Let $F(T,Y)$ denote the commutative group of all functions (with pointwise addition) mapping $T$ into $Y$. In Sec. 2.6 we considered $\mathcal{S}$-topologies on $F(T,Y)$. If $\mathcal{S}$ is any collection of subsets of $T$ then the sets

$$W(S,V) = \{f \in F(T,Y) : f(S) \subset V\} \qquad S \in \mathcal{S},\ V \in V_Y(0)$$

form a subbase at 0 for a group topology $\mathcal{T}_{\mathcal{S}}$, the $\mathcal{S}$-topology, on $F(T,Y)$ [Sec. 2.6, especially Example 2.6.2]. The principal properties of the $W(S,V)$ are listed in Theorem 2.6.1. By Theorem 2.6.3, we know that $\mathcal{T}_{\mathcal{S}}$ is unaffected if $\mathcal{S}$ is enlarged to subsets of finite unions of sets in $\mathcal{S}$ in which case the $W(S,V)$ are a base at 0, rather than just a subbase.

Now replace $T$ and $Y$ by TVS $X$ and $Y$, respectively, and let $V_Y(0)$ denote the filter of neighborhoods of 0 in $Y$. Restrict consideration to the continuous linear maps $L(X,Y)$ of $X$ into $Y$ and let

$$W(S,V) = \{A \in L(X,Y) : A(S) \subset V\} \qquad S \in \mathcal{S},\ V \in V_Y(0)$$

Clearly, for any nonzero scalar $a$, $aW(S,V) = W\left(a^{-1}S,V\right) = W(S,aV)$. Since $U \subset V$ implies that $W(S,U) \subset W(S,V)$ [Theorem 2.6.1(a)], it follows that if $V$ is balanced, so is $W(S,V)$.

When will the $W(S,V)$ be a base at 0 for a vector topology on $L(X,Y)$? If $\mathcal{S}$ is a collection of subsets of $X$ which is closed with respect to the formation of subsets of finite unions and the $V$'s are the balanced neighborhoods of 0 in $Y$, then the $W(S,V)$ form a filterbase of balanced sets for a group topology $\mathcal{T}_{\mathcal{S}}$ on $L(X,Y)$. By Theorem 4.5.1, it is evident that if the $W(S,V)$ are absorbent, then $\mathcal{T}_{\mathcal{S}}$ is a vector topology. We show next that $S \subset X$ is bounded iff $W(S,V)$ is absorbent in $L(X,Y)$. Theorem 11.2.1(a) is the analog of Theorem 8.3.5 in which we showed that, for paired spaces $(X,Y)$, $B \subset X$ is $\sigma(X,Y)$-bounded iff $B^{\circ}$ is an absorbent subset of $Y$.

**Theorem 11.2.1** S BOUNDED $\Longrightarrow$ W(S,V) ABSORBENT Let $X$ and $Y$ be TVS. For $S \subset X$ and $V$ a neighborhood of 0 in $Y$. Then:

(a) $W(S,V) = \{A \in L(X,Y) : A(S) \subset V\}$ is absorbent in $L(X,Y)$ iff $V$ absorbs $A(S)$ for each $A \in L(X,Y)$.

(b) Hence, if $S$ is bounded, then $W(S,V)$ is absorbent in $L(X,Y)$ for every neighborhood $V$ of 0.

**Proof.** We use the notation of the statement. (a) $W(S,V)$ is absorbent in $L(X,Y)$ iff, given any $A \in L(X,Y)$, there exists $r > 0$ such that $|a| \geq r \Rightarrow A \in aW(S,V) = W(S,aV)$ or, equivalently, $A(S) \subset aV$, i.e., $V$ absorbs $A(S)$. (b) If $S$ is bounded then so is $A(S)$ for any $A \in L(X,Y)$ [Theorem 6.4.1]. ■

**Theorem 11.2.2** $\mathcal{S}$-TOPOLOGIES FOR $L(X,Y)$ Let $X$ and $Y$ be TVS, let $B_Y(0)$ denote a base of balanced neighborhoods of 0 in $Y$ and let $\mathcal{S}$ be a collection of bounded subsets of $X$ which is closed with respect to the formation of subsets of finite unions. The sets $W(S,V)$, $S \in \mathcal{S}$, $V \in B_Y(0)$, form a base at 0 for a vector topology $\mathcal{T}_{\mathcal{S}}$ called the *$\mathcal{S}$-topology* on $L(X,Y)$. $\mathcal{T}_{\mathcal{S}}$ is:

(a) Hausdorff if the linear span of $\cup\mathcal{S}$ is dense in $X$ and $Y$ is Hausdorff;

(b) locally convex if $Y$ is. Moreover, if $P$ is a family of continuous seminorms which generates the topology on $Y$ then the seminorms $p_S(A) = \sup p(A(S))$,
$A \in L(X,Y)$, $S \in \mathcal{S}$, $p \in P$, generate $\mathcal{T}_{\mathcal{S}}$.

**Proof.** With notation as above, it follows from the remarks above that $\mathcal{T}_{\mathcal{S}}$ is a vector topology for $L(X,Y)$. (a) Suppose that $Y$ is Hausdorff and that the linear span of $\cup\mathcal{S}$ is dense in $X$. If $A$ is a nonzero element of $L(X,Y)$ then, because of the density of $\cup\mathcal{S}$, there must be some $x \in \cup\mathcal{S}$ such that $Ax \neq 0$. Since $Y$ is Hausdorff, there must be some neighborhood $V$ of 0 in $Y$ such that $Ax \notin V$. Thus, if $x \in S \in \mathcal{S}$, $A \notin W(S,V)$. It follows that $\mathcal{T}_{\mathcal{S}}$ is Hausdorff by Theorem 2.2.2(d).

(b) The local convexity statement follows from two easily verified observations: (1) If $B(0)$ is a base of neighborhoods of 0 in $Y$ then $\{W(S,B) : S \in \mathcal{S}, B \in B(0)\}$ is a neighborhood base at 0 for $\mathcal{T}_{\mathcal{S}}$; and (2) if $B$ is convex, so is $W(S,B)$.

As for the seminorm assertion, if $P$ is a base of continuous seminorms [Definition 5.5.4] for $Y$, then the sets $\overline{V_p} = \{y \in Y : p(y) \leq 1\}$ $(p \in P)$ are a base at 0 for $Y$. Hence the sets $W\left(S, \overline{V_p}\right)$, $S \in \mathcal{S}$, $p \in P$, are a base of neighborhoods of 0 for $\mathcal{T}_{\mathcal{S}}$. Since, for any $p \in P$, $S \in \mathcal{S}$, and $A \in L(X,Y)$,

$$p_S(A) \leq 1 \Leftrightarrow \sup p(A(S)) \leq 1 \Leftrightarrow A(S) \subset \overline{V_p}$$

it follows that $\overline{V}_{ps} = \{A \in L(X,Y) : p_s(A) \le 1\} = W(S, \overline{V}_p)$. ■

As mentioned above, an $\mathcal{S}$-topology is unchanged if $\mathcal{S}$ is replaced by subsets of finite unions of sets in $\mathcal{S}$. Some other modifications which do not alter $\mathcal{T}_{\mathcal{S}}$ are listed in Theorem 11.2.3.

**Theorem 11.2.3** ENLARGEMENTS OF $\mathcal{S}$ Let $X$ and $Y$ be TVS, $\mathcal{S}$ a collection of bounded subsets of $X$ and $L(X,Y)$ the space of continuous linear maps of $X$ into $Y$. Let $\mathcal{T}_{\mathcal{S}}$ be as in Theorem 11.2.2. Then $\mathcal{S}$ may be replaced by any of the following collections $\mathcal{S}'$ without affecting $\mathcal{T}_{\mathcal{S}}$:
(a) subsets of finite unions of sets of $\mathcal{S}$;
(b) scalar multiples $aS$ of sets $S$ in $\mathcal{S}$;
(c) finite sums $S_1 + \cdots + S_n$ of sets $S_1, \ldots, S_n$ in $\mathcal{S}$;
(d) balanced hulls or closures of sets in $\mathcal{S}$; and
(e) if $X$ and $Y$ are LCS, closed balanced convex hulls of sets in $\mathcal{S}$.

**Proof.** We use the notation of the statement. Part (a) follows from Theorem 2.6.3.

(b) Clearly $\mathcal{T}_{\mathcal{S}} \subset \mathcal{T}_{\mathcal{S}'}$. Conversely, by Theorem 6.3.1(c), a scalar multiple of a bounded set is bounded and, for $a \neq 0$ and $V$ a neighborhood of 0 in $Y$, $W(aS, V) = W(S, a^{-1}V)$, a $\mathcal{T}_{\mathcal{S}}$-neighborhood of 0.

(c) Sums of bounded sets are bounded by Theorem 6.3.1(c). Given a neighborhood $V$ of 0 in $Y$, choose a neighborhood $U$ of 0 in $Y$ such that $U+U \subset V$. For $S_1, S_2 \in \mathcal{S}$, note that $W(S_1, U) \cap W(S_2, U) \subset W(S_1 + S_2, V)$.

(d) The balanced hull and the closure of a bounded set $S$ are bounded by Theorem 6.3.2(a,b). Moreover, the balanced hull $S_b$ of $S$ is given by $\cup\{aS : |a| \le 1\}$ by Theorem 4.1.5. If $V$ is balanced and $A(S) \subset V$, then for any $|a| \le 1$, $A(aS) = aA(S) \subset aV \subset V$. Thus $W(S,V) \subset W(S_b, V)$. If $V$ is closed and $A \in W(S,V)$, then $A(\operatorname{cl} S) \subset \operatorname{cl} A(S) \subset V$ by continuity, so $W(S,V) \subset W(\operatorname{cl} S, V)$.

(e) Suppose $X$ and $Y$ are LCS. The closed balanced convex hull of a bounded subset of an LCS is bounded by Theorems 6.3.2 and 6.3.3. Since, for any subset $E$ of any TVS, $\operatorname{cl}(E_{bc}) = (\operatorname{cl} E)_{bc}$, in view of (d), we only have to show that if the sets of $\mathcal{S}$ are replaced by their convex hulls, the $\mathcal{S}$-topology does not change. To this end, let $V$ be a convex neighborhood of 0 in $Y$ and note that if $A \in W(S,V)$ then $A(S_c) \subset V$ or $A \in W(S_c, V)$. ■

If $\mathcal{S}$ is the class of all finite subsets of $X$, the resulting $\mathcal{S}$-topology for $L(X,Y)$ is called the *strong operator topology* or the *topology of pointwise convergence.* Unfortunately, if $Y = \mathbf{F}$—so that $L(X, \mathbf{F}) = X'$—this is the *weak-∗* topology $\sigma(X', X)$. If $\mathcal{S}$ is the class of compact subsets or bounded subsets of $X$, the resulting $\mathcal{S}$-topologies for $L(X,Y)$ are called, respectively, the topologies of *uniform convergence on compact sets* (or the *topology of compact convergence*) and the topology of *uniform convergence on bounded sets* ("topology of bound or bounded convergence" and "bound topology") [cf. Example 6.7.12].

**Example 11.2.4** $\mathcal{S}$-TOPOLOGIES ON NORMED SPACES

(a) UNIFORM TOPOLOGY = BOUND TOPOLOGY ON $L(X,Y)$ Suppose that $X$ and $Y$ are normed spaces and that $L(X,Y)$ is topologized by the uniform norm [Example 6.5.3]: $\|A\| = \sup\{\|Ax\| : \|x\| \leq 1\}$ for $A \in L(X,Y)$. Then the topology $\mathcal{T}_u$ induced by the uniform norm is the topology $\mathcal{T}_b$ of uniform convergence on bounded sets.

Proof. Let $U_X$ and $U_Y$ denote the closed solid unit balls in $X$ and $Y$, respectively and let $r$ be positive. Since $\{A \in L(X,Y) : \|A\| \leq r\} = W(U_X, rU_Y)$, $\mathcal{T}_u \subset \mathcal{T}_b$. Let $B$ be a bounded subset of $X$ and consider $W(B, rU_Y)$, $r > 0$. Since $B$ is bounded, $B \subset dU_X$ for some $d > 0$. Thus $W(B, rU_Y) \supset W(dU_X, rU_Y) = W\left(U_X, d^{-1}rU_Y\right)$ [Theorem 2.6.1(d)] and every $\mathcal{T}_b$-neighborhood of 0 is a uniform neighborhood of 0.

(b) STRONG TOPOLOGY = NORM TOPOLOGY If $X$ is a normed space, there is no distinction between norm bounded and $\sigma(X,X')$-bounded by Mackey's Theorem [8.8.7] on bounded sets. Recall that the strong topology [Example 8.5.5] $\beta(X,X')$ for $X$ is that determined by the polars of all $\sigma(X',X)$-bounded subsets of $X'$. Consequently, $\beta(X',X) = \mathcal{T}_b$ on $X' = L(X,\mathbf{F})$. By (a), $\beta(X',X)$ is the norm topology of Example 6.5.3(b), something already noted in Example 8.8.9. ■

**Example 11.2.5** POLAR VERSUS $\mathcal{S}$-TOPOLOGIES Let $X$ be a TVS.

(a) Polar topologies (Sec. 8.5) are similar to $\mathcal{S}$-topologies, but generally not the same. For polar topologies for $X'$, $\mathcal{S}$ consists of $\sigma(X,X')$-bounded subsets of $X$, not the stronger notion of bounded subsets of $X$. Polar topologies are locally convex; $\mathcal{S}$-topologies need not be.

(b) In TVS $X$ where $\sigma(X,X')$-boundedness and boundedness coincide, as in LCHS, for example [Theorem 8.8.7], every polar topology is an $\mathcal{S}$-topology. ■

**Theorem 11.2.6** EQUICONTINUOUS IMPLIES $\mathcal{T}_{\mathcal{S}}$-BOUNDED Let $X$ and $Y$ be TVS and let $\mathcal{S}$ be a collection of bounded subsets of $X$. A subset $H$ of $L(X,Y)$ is bounded in the $\mathcal{S}$-topology $\mathcal{T}_{\mathcal{S}}$ iff for each neighborhood $V$ of 0 in $Y$, $\cap_{A\in H} A^{-1}(V)$ absorbs each $S \in \mathcal{S}$ or, equivalently, $H(S) = \cup_{A\in H} A(S)$ is a bounded subset of $Y$ for each $S \in \mathcal{S}$. Thus, equicontinuous subsets of $L(X,Y)$ are bounded in any $\mathcal{S}$-topology.

**Proof.** In the notation of the statement, $H \subset L(X,Y)$ is $\mathcal{T}_{\mathcal{S}}$-bounded iff for any $S \in \mathcal{S}$ and any neighborhood $V$ of 0 in $Y$, there exists $r > 0$ such that $|a| \geq r$ implies that $H \subset aW(S,V) = W(S,aV)$ or $H(S) \subset aV$, i.e., that $H(S)$ is bounded. This is equivalent to $S \subset r\cap_{A\in H} A^{-1}(V)$.

If $H$ is equicontinuous then, for any neighborhood $V$ of 0 in $Y$, there is a neighborhood $U$ of 0 in $X$ such that $H(U) \subset V$ or $U \subset \cap_{A\in H} A^{-1}(V)$. Consequently, for each neighborhood $V$ of 0 in $Y$, $\cap_{A\in H} A^{-1}(V)$ is a neighborhood of 0 in $X$ and so absorbs any bounded subset of $X$, hence each $S \in \mathcal{S}$. It follows from the discussion above that $H$ is $\mathcal{T}_{\mathcal{S}}$-bounded. ■

## 11.3 BARRELED SPACES

A barrel, recall from Definition 8.8.2, is a closed absorbent disk.

**Definition 11.3.1** BARRELED A locally convex space $X$ is *barreled* (*tonnelé*) if each barrel in $X$ is a neighborhood of 0. ■

In Theorem 11.2.6 we observed that, for any TVS $X$ and $Y$, an equicontinuous subset $H \subset L(X, Y)$ is bounded in any $\mathcal{S}$-topology for $L(X, Y)$, in particular, $H$ is bounded in the strong operator topology ($\mathcal{S} = \{\{x\} : x \in X\}$), which means that $H(x) = \{Ax : A \in H\}$ is bounded for each $x \in X$. Thus, equicontinuity implies pointwise boundedness. A result in the converse direction is the Banach–Steinhaus theorem or principle of uniform boundedness [Theorem 11.9.1]: If $X$ is barreled and $Y$ an LCS, then pointwise boundedness in $L(X, Y)$ implies equicontinuity. A point of this chapter is to show that the widest class of LCS $X$ for which pointwise boundedness implies equicontinuity in $L(X, Y)$, for any TVS $Y$, is the class of barreled spaces [Theorem 11.9.3].

Any complete pseudometrizable space is barreled [Theorem 11.7.2] but there are plenty of barreled incomplete normed spaces as well as normed spaces that are not barreled (see Examples 11.3.2 and 11.3.3 for two of them). The finest locally convex topology $\mathcal{T}_{lc}$ [Example 5.6.4] for a linear space $X$ has the filterbase $\mathcal{B}$ of all absorbent disks as a base at 0. Thus, if $B$ is a $\mathcal{T}_{lc}$-barrel, $B \in \mathcal{B}$, so any linear space with the finest locally convex topology is barreled. This suggests that stronger topologies are more likely to be barreled than weaker ones [cf. Theorem 11.3.6].

**Example 11.3.2** $(C[0,1], \|\cdot\|_1)$ NOT BARRELED Consider the linear space $X = (C[0,1], \mathbf{F}, \|\cdot\|_1)$ of continuous $\mathbf{F}$-valued functions on the closed interval $[0,1]$ normed by $\|\cdot\|_1 = \int_0^1 |\cdot|\, dt$. We exhibit a barrel $B$ in $X$ which is not a neighborhood of 0. Let $B = \{x \in X : \sup |x[0,1]| \leq 1\}$. That $B$ is absorbent, balanced, and convex is clear. To see that $B$ is closed, let $y \in \operatorname{cl} B$ and $(y_n)$ be a sequence of points from $B$ that is $\|\cdot\|_1$-convergent to $y$. Since $\int_0^1 |y_n - y|\ dt \to 0$, $(y_n)$ must possess a subsequence $(y_{n_k})$ that converges to $y$ pointwise almost everywhere. Since each continuous function $y_{n_k} \in B$, $\sup |y[0,1]| \leq 1$, i.e., $y \in B$ and $B$ is a barrel.

Is $B$ a neighborhood of 0 in $X$? If so, then $B$ would have to contain a ball $C(0,r) = \left\{x \in X : \int_0^1 |x(t)|\, dt \leq r\right\}$, $r > 0$, of functions of small integral. Yet no matter how small a function's integral is, its peak can still be arbitrarily large. Since membership in $B$ involves only a peak restriction, it cannot contain any such $C(0,r)$ and therefore cannot be a neighborhood of 0 in $X$. ■

**Example 11.3.3** $\varphi$ NOT BARRELED Let $\varphi$ denote the space of all sequences whose entries are almost always 0; for $x = (a_n) \in \varphi$, take $\|x\|_\infty = \sup_n |a_n|$ and define $f_n(x) = a_n$ for each positive integer $n$. Let $B = \{x \in \varphi : |f_n(x)| \leq$

$1/n$ for each $n\}$. It is easy to verify that $B$ is a barrel. Can $B$ contain a neighborhood $V = \{x \in P : \|x\|_\infty < r\}$ of 0 ($r > 0$)? For $n$ such that $r/2 > 1/n$ the sequence whose $n$th entry is $r/2$, 0 elsewhere, belongs to $V$ but not $B$. Hence $B$ cannot contain any such neighborhood $V$ and is therefore not a neighborhood of 0. ■

Given a TVS $X$, for $H \subset X'$ to be $\sigma(X', X)$-bounded, each seminorm $p_x(f) = |f(x)|$ ($f \in H$) must be bounded on $H$ for each $x \in X$ [Theorem 6.1.5]; in other words, for each $x$ in $X$, $\langle x, H\rangle = \{\langle x, f\rangle : f \in H\}$ is bounded or $H$ is pointwise bounded. Thus, Theorem 11.3.4 says that pointwise boundedness ($= \sigma(X', X)$-boundedness) of a set of linear functionals on a barreled space implies equicontinuity and characterizes barreled spaces in the class of LCS.

**Theorem 11.3.4** BANACH–STEINHAUS THEOREM FOR FUNCTIONALS An LCS $X$ is barreled iff $\sigma(X', X)$-bounded subsets of $X'$ are equicontinuous. Since equicontinuity generally implies $\sigma(X', X)$-boundedness [Theorem 11.2.6], the equicontinuous and $\sigma(X', X)$-bounded subsets of $X'$ are the same for a barreled space $X$.

**Proof.** Suppose that $(X, \mathcal{T})$ is barreled and that $H$ is a $\sigma(X', X)$-bounded subset of $X'$. Since $H$ is $\sigma(X', X)$-bounded, it is contained in the polar $B^\circ$ of some barrel $B$ in $X$ [Theorem 8.8.3]. Since $X$ is barreled, $B$ is a neighborhood of 0; since $H \subset B^\circ$, equicontinuity of $H$ follows from Theorem 8.6.4(a).

Conversely, suppose that $\sigma(X', X)$-boundedness implies equicontinuity and let $B$ be a barrel in the LCS $X$. Since $B$ is a barrel, $B^\circ$ is $\sigma(X', X)$-bounded by Theorem 8.8.3, hence equicontinuous by hypothesis. Since the polar of an equicontinuous set is a neighborhood of 0 [Theorem 8.6.4(b)], it follows that $B^{\circ\circ}$ is a neighborhood of 0 in $X$. $B^{\circ\circ} = B$ by Theorem 8.8.1(c). ■

Equicontinuity implies more than boundedness.

**Theorem 11.3.5** EQUICONTINUITY IMPLIES STRONG BOUNDEDNESS Let $X$ be an LCHS. If $H$ is an equicontinuous subset of $X'$, then $H$ is $\beta(X', X)$-bounded.

**Proof.** With notation as above, it follows from Theorem 11.2.6 that an equicontinuous subset of $X'$ is bounded in the topology $\mathcal{T}_b$ of uniform convergence on bounded subsets of $X$. By Mackey's theorem [8.8.7] on bounded sets, the bounded subsets of $X$ are the $\sigma(X, X')$-bounded sets so $\mathcal{T}_b = \beta(X', X)$. ■

This yields our first characterization of barreled topologies.

**Theorem 11.3.6** BARRELED TOPOLOGY $= \beta(X, X')$ The LCS $(X, \mathcal{T})$ is barreled iff $\mathcal{T} = \beta(X, X')$.

**Proof.** Suppose that $(X, \mathcal{T})$ is barreled. Since $X$ is locally convex, $\mathcal{T}$ is the topology of uniform convergence on equicontinuous subsets of $X'$ [Theorem 8.6.6]. Since $X$ is barreled, those are the $\sigma(X', X)$-bounded subsets of $X'$ [Theorem 11.3.4]. Thus, $\mathcal{T}$ coincides with the topology $\beta(X, X')$ of uniform convergence on $\sigma(X', X)$-bounded subsets of $X'$.

Conversely, suppose that $\mathcal{T} = \beta(X, X')$. By Theorem 11.3.4, it follows that $X$ is barreled if $\sigma(X', X)$-boundedness implies equicontinuity. If $E \subset X'$ is $\sigma(X', X)$-bounded, $E^\circ$ is a $\beta(X, X')$-neighborhood of 0. Since $E \subset (E^\circ)^\circ$, i.e., $E$ is a subset of the polar of a neighborhood of 0, $E$ is equicontinuous by Theorem 8.6.4. ■

It follows from Theorem 11.3.4 that, for a barreled space $X$, the equicontinuous subsets, the $\sigma(X', X)$-bounded subsets, and the $\beta(X', X)$-bounded subsets of $X'$ are the same.

Continuity implies weak continuity but the converse is generally false [Theorem 8.11.3(c)]. As barreled spaces $X$ carry the strong topology $\beta(X, X')$ and weak continuity implies strong continuity [Theorem 8.11.3(b)] we have:

**Corollary 11.3.7** WEAK CONTINUITY = CONTINUITY IN BARRELED SPACES A linear map $A : X \to Y$ between barreled spaces $X$ and $Y$ is continuous iff it is weakly continuous.

## 11.4 LOWER SEMICONTINUITY

The principal result of this section is the automatic continuity characterization of barreled spaces of Theorem 11.4.3.

**Definition 11.4.1** LOWER SEMICONTINUITY A real-valued function $x$ on a topological space $T$ is *lower semicontinuous* if $x^{-1}(-\infty, a]$ is closed for each $a \in \mathbf{R}$. ■

Continuous functions are lower semicontinuous. So are characteristic functions of open sets; in fact, a set is open iff its characteristic function is lower semicontinuous. If a function is continuous except for a jump discontinuity and it assumes the lower limit at the jump, it is lower semicontinuous. The following lower semicontinuous function is continuous at each irrational: $x : [0, 1] \to \mathbf{R}$ is 0 at each irrational and $-1/q$ at each rational $p/q$ ($p$ and $q$ relatively prime).

**Theorem 11.4.2** LOWER SEMICONTINUOUS PROPERTIES

(a) EXTREMA ON COMPACT SETS A lower semicontinuous function $x : T \to \mathbf{R}$ attains a minimum on a compact space $T$.

(b) SUPREMA If $\{x_s : s \in S\}$ is a family of lower semicontinuous functions with a common domain $T$ such that for each $t \in T$, $x(t) = \sup_{s \in S} x_s(t) < \infty$, then $x = \sup_{s \in S} x_s$ is lower semicontinuous.

**Proof.** We use the notation of the statement. (a) Each of the sets $x^{-1}(-\infty, a]$, $a \in \mathbf{R}$, is closed and the set $\mathcal{B}$ of nonempty ones among them is clearly a filterbase. Since $T$ is compact, $\mathcal{B}$ has a cluster point, $t$ say, which belongs to each $B \in \mathcal{B}$ [Theorem 1.1.10(b)]; $x(t)$ is the desired minimum.

(b) The result follows from the observation that

$$x^{-1}(-\infty, a] = \cap_{s \in S} x_s^{-1}(-\infty, a]$$

■

Note that Theorem 11.4.2 implies that the supremum of a family of continuous functions is lower semicontinuous.

**Theorem 11.4.3** LOWER SEMICONTINUOUS SEMINORMS AND BARRELEDNESS The locally convex space $X$ is barreled iff each lower semicontinuous seminorm on $X$ is continuous.

**Proof.** The essence of the proof lies in showing that the barrels in any TVS $Y$ are the sets of the form $\overline{V}_p = \{y \in Y : p(y) \leq 1\}$ where $p$ is a lower semicontinuous seminorm. To see this, suppose that $B$ is a barrel in $Y$ and let $p$ be its gauge. Since $B$ is closed, $B = \overline{V}_p$ by Theorem 5.3.2. The lower semicontinuity of $p$ follows from the fact that for each $a > 0$, $p^{-1}(-\infty, a] = p^{-1}[0, a] = a\overline{V}_p = aB$. On the other hand, if $p$ is a lower semicontinuous seminorm on $Y$, it is clear that $\overline{V}_p = p^{-1}[0, 1]$ is a barrel.

To return to the original argument, if $X$ is a barreled space and $p$ is a lower semicontinuous seminorm on $X$, then $\overline{V}_p$ is a barrel, therefore a neighborhood of 0, whence [Theorem 5.2.2] the continuity of $p$. Conversely, if each lower semicontinuous seminorm on $X$ is continuous and $B$ is a barrel in $X$, we may write $B$ as $\overline{V}_p$ for some lower semicontinuous—hence continuous—seminorm $p$. The continuity of $p$ implies [Theorem 5.2.2 again] that $B$ is a neighborhood of 0. ■

**Theorem 11.4.4** BARRELED EQUIVALENTS an LCS $(X, \mathcal{T})$ is barreled iff:

(a) [Theorem 11.3.4] $\sigma(X', X)$-bounded (= pointwise bounded) subsets of $X'$ are equicontinuous.

(b) [Theorem 11.3.6] $\mathcal{T} = \beta(X, X')$.

(c) [Theorem 11.4.3] Each lower semicontinuous seminorm on $X$ is continuous.

(d) [Theorem 14.3.1] Each linear map $A$ of $X$ into any locally convex space $Y$ is almost continuous, namely that $\operatorname{cl} A^{-1}(V)$, rather than $A^{-1}(V)$, is a neighborhood of 0 for each neighborhood $V$ of 0 in $Y$.

(e) [Theorem 14.4.3] Each linear map $A$ of any LCS $Y$ onto $X$ is almost open, i.e., for any neighborhood $U$ of 0 in $Y$, $\operatorname{cl} A(U)$ is a neighborhood of 0 in $X$.

# 11.5 RARE SETS

We develop the elementary properties of *rare* sets here. They enter the development because of the role they play in defining classes of spaces (the nonmeager spaces) on which principles of uniform boundedness hold [Theorems 11.6.4, 11.9.1; cf. also Theorem 11.6.8]. Throughout the section $(T, \mathcal{T})$ denotes a topological space.

Early in the development of Lebesgue measure theory, several mathematicians (Baire, Vitali, Lebesgue) attempted to discover the structure of sets of measure 0. It was not merely a case of cardinality for the Cantor perfect set $K$ (all numbers expressible in the form $\sum_{i\in N} n_i/3^i$, where $n_i$ is 0 or 2) was of measure 0, yet had the power of the continuum. The points of this closed set were rather thinly distributed, however: $K$ had no interior. A neighborhood of any of its points had to contain numbers which required 1's in their triadic expansions. Might this low population density provide the denouement to the question of which are the sets of measure 0? The property was abstracted for further investigation.

**Definition 11.5.1** RARE SETS A subset $A$ of a topological space $T$ is called *rare* (*nowhere dense*) if its closure has empty interior: $\operatorname{int} \operatorname{cl} A = \varnothing$. ■

The more descriptive "rare" (due to Bourbaki) is replacing the older term "nowhere dense." Although correct, its translation from the French "rare" might better have been "rarefied" since the desired meaning is "sparsely populated," not "uncommon." As follows from Theorem 11.5.2(b), rare sets $A$ are so thinly distributed that their complements are dense. Before giving some examples, we consider some elementary properties of rare sets.

**Theorem 11.5.2** RARE BASICS In a topological space $T$:

(a) $A \subset T$ has empty interior iff $\complement A$ is dense.

(b) $A \subset T$ is rare iff $\complement(\operatorname{cl} A)$ is dense.

(c) $A \subset T$ is rare iff every nonempty open subset $G$ of $T$ contains a nonempty open subset $U$ that is disjoint from $A$.

**Proof.** We use the notation of the statement. (a) $\operatorname{int} A = \varnothing$ iff every nonempty open subset $G$ of $T$ meets $\complement A$.

(b) Follows immediately from (a).

(c) If $A = \varnothing$, the result is clear so suppose that the rare set $A \neq \varnothing$. Then any nonempty open subset $G$ of $T$ meets the dense subset $\complement(\operatorname{cl} A)$ and the nonempty open set $G \cap \complement(\operatorname{cl} A) \subset \complement A$. Conversely, the condition says that any nonempty open set meets $\complement(\operatorname{cl} A)$ which means that $\complement(\operatorname{cl} A)$ is dense and the result follows from (b). ■

We point out that when we speak of a set $A \subset T$ as being rare, we mean with respect to $T$'s topology, not the topology it induces on $A$—a nonempty set $A$ is never a rare subset of itself, because $\operatorname{int}_A \operatorname{cl}_A A = A$.

**Example 11.5.3** RARE SETS

(a) Any subset of a rare set is rare. The null set is rare; so are singletons in any Hausdorff space as long as the point is not isolated. The only rare subset of a discrete space is $\varnothing$. $\mathbf{R}$ is a rare subset of $\mathbf{R}^2$.

(b) The rationals $\mathbf{Q}$ are not rare in $\mathbf{R}$, but the integers $\mathbf{Z}$ are.

(c) $K = \{(x, y) : xy = 1\}$ is a rare subset of $\mathbf{R}^2$, but its balanced hull is not.

(d) The set $\{1/n : n \in \mathbf{N}\} \cup \{0\}$ is rare in $\mathbf{R}$.

(e) BOUNDARIES OF OPEN OR CLOSED SETS The *boundary*, $\operatorname{bdry} A$, of a set $A$ is the set $(\operatorname{cl} A) \cap (\operatorname{cl} \complement A)$. If $A$ is open, then $\operatorname{bdry} A \subset \complement A$ while a neighborhood of any point in $\operatorname{bdry} A$ must meet $A$ and $\complement A$. Thus $\operatorname{bdry} A$ has no interior. By the same reasoning, if $A$ is closed, $\operatorname{bdry} A \subset A$ and $\operatorname{int} \operatorname{bdry} A = \varnothing$. Two things follow: (1) a closed set is rare iff it coincides with its boundary; (2) boundaries of open or closed sets are rare.

(f) LINEAR SUBSPACES RARE OR DENSE A linear subspace $M$ of a TVS $X$ is rare or dense. Note that the closure of a subspace is again a subspace. If $M$ is not rare, $\operatorname{int} \operatorname{cl} M \neq \varnothing$. By translation, it follows immediately that 0 is an interior point of $\operatorname{cl} M$. Thus $\operatorname{cl} M$ is a neighborhood of 0 and is therefore absorbent. Since $\operatorname{cl} M$ is a subspace, it follows that $\operatorname{cl} M = X$.

(g) FINITE-DIMENSIONAL SUBSPACES RARE In a Hausdorff TVS, finite dimensional subspaces are closed by Theorem 4.10.3(d). Thus, by (f), proper finite-dimensional subspaces of a HTVS are rare. ■

**Theorem 11.5.4** ELEMENTARY PROPERTIES OF RARE SETS Let $T$ be a topological space.

(a) FINITE UNIONS Finite unions of rare sets are rare.

(b) SUBSETS Suppose that $A \subset S \subset T$. If $A$ is rare in $S$, then $A$ is rare in $T$. Conversely, if $S$ is open in $T$ and $A$ is rare in $T$, then $A$ is rare in $S$.

**Proof.** (a) The key fact is that finite intersections of dense open sets are dense. To prove this, suppose that $A$ and $B$ are dense open sets. Thus, if $G$ is a nonempty open set, $G \cap A \neq \varnothing$. As $G \cap A$ is a nonempty open set, $(G \cap A) \cap B = G \cap (A \cap B) \neq \varnothing$ and $A \cap B$ is seen to be dense.

Now suppose that $F$ and $H$ are rare. Since we have to consider closures anyway, we may assume that $F$ and $H$ are closed. To show that $F \cup H$ is rare we note that that $\complement (F \cup H) = \complement F \cap \complement H$. The result follows from Theorem 11.5.2(a).

(b) Suppose that $A \subset S \subset T$ and that $A$ is not rare in $T$. Then there is a nonempty open set $G$ in $T$ that does not meet $\complement \operatorname{cl}_T A$, i.e., such that $G \subset \operatorname{cl}_T A$ and $G \cap S \subset \operatorname{cl}_T A \cap S = \operatorname{cl}_S A$. Since $G \cap A \subset G \cap S$ and $G \cap A \neq \varnothing$ because $G \subset \operatorname{cl}_T A$, this shows that $\operatorname{cl}_S A$ has nonempty interior, i.e., that $A$ is not rare in $S$.

For the converse, suppose that $A$ is rare in $T$, $S$ is open and $G$ is a nonempty open subset of $S$. Since $S$ is open in $T$, so is $G$. Since $A$ is rare in $T$, $G$ must contain a nonempty open subset $U$ of $T$—hence also of $S$—that does not meet $A$. The rareness of $A$ now follows from Theorem 11.5.2(c). ■

# 11.6 MEAGER, NONMEAGER AND BAIRE

As mentioned at the beginning of Sec. 11.5, the motivation to study rare sets was to discover the structure of sets of Lebesgue measure 0. Although it might provide a clue to the structure of such sets, it was obviously only going to be a fragment: The set $\mathbf{Q}$ rationals is of measure 0 but is not a rare subset of $\mathbf{R}$. Still, $\mathbf{Q}$ is a countable union of rare sets. Perhaps this would be the key to sets of measure 0—unbounded sets of measure 0, anyway.

**Definition 11.6.1** MEAGER SETS A *meager (first category)* subset $E$ of a topological space $T$ is one which can be written as a countable union of rare sets or, equivalently, is a subset of such a union. Otherwise, $E$ is called *nonmeager (of the second category)*. ■

The classical terminology for "meager" and "nonmeager" is "of the first category" and "of the second category," respectively. Aside from "meager" being more descriptive, "category" has, as S. Berberian has remarked, "been conscripted for higher service." The empty set is manifestly meager, so to prove a set to be nonmeager automatically shows it to be nonempty. There are some interesting proofs of existence based on this principle—for example, proof of the existence of continuous nowhere differentiable functions (see, e.g., Bachman and Narici 2000, p. 80) by showing that the set of such functions is nonmeager in the Banach space $C([0,1], \mathbf{R}, \|\cdot\|_\infty)$ of continuous functions on $[0,1]$.

**Example 11.6.2** MEAGER AND NONMEAGER SETS (a) As we can write the rationals $\mathbf{Q}$ as a countable union of singletons, $\mathbf{Q}$ is a meager subset of $\mathbf{R}$. Indeed, any countable Hausdorff space without isolated points is meager.

(b) Any topological space which contains an isolated point $t$ is nonmeager, as no set to which $t$ belongs can be rare. Thus discrete spaces are never meager.

(c) Singletons are always nonmeager sub*spaces*. A singleton is a nonmeager sub*set* of a topological space iff the point is isolated.

(d) NONMEAGER SUBSPACES View $\mathbf{Q} \times \mathbf{Q}$ together with the $x$-axis as a subspace of $\mathbf{R}^2$. As the $x$-axis is a rare subset of $\mathbf{R}^2$ and we can enumerate $\mathbf{Q} \times \mathbf{Q}$, $(\mathbf{Q} \times \mathbf{Q}) \cup \mathbf{R}$ is a meager subset of $\mathbf{R}^2$. As Theorem 11.7.2 will show, however, $\mathbf{R}$ is a nonmeager subspace. The point is that the presence of a nonmeager subspace does not imply that the space itself is nonmeager. If, however, a topological space $T$ has an *open* subset $G$ which is a nonmeager subspace, then $T$ is nonmeager—if $T$ is meager and $G$ is open in $T$ then there are rare subsets $\{R_n : n \in \mathbf{N}\}$ of $T$ such that $T = \cup_n R_n$; hence $G =$

$\cup_n (R_n \cap G)$; as $G$ is open in $T$, each $R_n \cap G$ is rare in $G$ and therefore $G$ is meager. ■

To return to the question of meagerness and measure 0 for a moment, Oxtoby and Ulam [1938] proved that for each meager subset $E$ of $[0,1]^n$, $n \in \mathbf{N}$, there is a homeomorphism of $[0,1]^n$ onto itself carrying $E$ into a set of measure 0. On the other hand there are rare—not just meager—sets such as the Cantor-style set of the example below which have positive measure.

**Example 11.6.3** Rare Set of Positive Measure We construct a rare subset of $[0,1]$, similar in style of construction to the classical Cantor set, whose Lebesgue measure is 9/10. The 9/10, as the construction will show, is arbitrary—we can construct a rare subset of $[0,1]$ whose Lebesgue measure is any number less than 1. From the unit interval $I = [0,1]$, delete the open interval $G_{11} = (1/2 - 1/40, 1/2 + 1/40)$ of length $2/40 = (1/2)(1/10)$ centered at 1/2. Let $K_1 = I \setminus G_{11} = [0, .475] \cup [.525, 1] \triangleq F_{11} \cup F_{12}$. Let $G_{21}$ and $G_{22}$ the open intervals of length $(1/8)(1/10)$ centered at the midpoints of $F_{11}$ and $F_{12}$. Let $K_2 = K_1 \backslash (G_{21} \cup G_{22})$. We can now write $K_2$ as the union of four closed intervals: $K_2 \triangleq F_{21} \cup F_{22} \cup F_{23} \cup F_{24}$. In these four closed intervals, delete the open intervals $G_{31}, G_{32}, G_{33}$, and $G_{34}$, each of length $(1/32)(1/10)$, centered at the midpoints of $F_{21}, F_{22}, F_{23}, F_{24}$, respectively. Continue in this way and take $K = \cap_{n \in \mathbf{N}} K_n$. Thus we have deleted an open set from $[0,1]$ of Lebesgue measure equal to $(1/10)(1/2 + 1/4 + 1/8 + \cdots) = 1/10$, so the remaining closed set $K$ is of measure 9/10. $K$ must be rare because if $K$ were to contain an interval $(a,b)$, then $(a,b)$ must be a subset of $K_1$, hence of either $F_{11}$ or $F_{12}$. Assuming it to be a subset $F_{11}$, $(a,b)$ would have to be a subset of $F_{21}$ or $F_{22}$. Continuing in this way, we see that $(a,b)$ must be a subset of each set in a nested sequence of closed intervals whose diameters $2^{-(2n+1)}(1/10)$ go to 0 which cannot happen. Even though $K$ is rare in $[0,1]$ or $\mathbf{R}, K$ is nonmeager in itself by virtue of being compact (closed and bounded) and Hausdorff [Theorem 11.7.3]. ■

**Theorem 11.6.4** Nonmeager LCS Barreled A nonmeager LCS is barreled.

**Proof.** Let $B$ be a barrel in the nonmeager LCS $X$. Since $B$ is absorbent, $X = \cup_{n \in \mathbf{N}} nB$. Since $X$ is nonmeager, there exists $n \in \mathbf{N}$ such that $nB$ has nonempty interior; hence so does $B$. As $B$ is a disk, however, so is $\operatorname{int} B$ [Theorem 4.4.2]; therefore $0 \in \operatorname{int} B$ and $B$ is a neighborhood of 0. ■

The *Baire* spaces, a subcollection of the nonmeager spaces, have very full open sets.

**Definition 11.6.5** Baire Space A topological space is a *Baire space* if each nonempty open set is nonmeager. ■

Complete metric spaces and locally compact Hausdorff spaces are Baire spaces [Theorems 11.7.2 and 11.7.3]; hence Banach spaces are Baire spaces. Since Baire spaces are nonmeager, Baire LCS are barreled.

In the proof of Theorem 11.5.4(b) we showed that the intersection of a finite number of dense open sets is dense. The assertion does not remain valid if we pass to infinitely many sets, however, for the sets $\mathbf{R}\backslash\{x\}$ $(x \in \mathbf{R})$ are each open and dense but their intersection is empty. In Baire spaces a countable intersection of dense open sets is dense as we show next.

**Theorem 11.6.6** EQUIVALENTS OF THE BAIRE CONDITION The following conditions on a topological space $T$ are equivalent:

(a) Countable unions of closed rare sets have no interior.

(b) The intersection of countably many dense open sets is dense.

(c) $T$ is a Baire space.

(d) Complements of meager sets are dense.

**Proof.** The key observation for this proof is that for any set $A$, $\operatorname{int} A = \varnothing$ iff $\complement A$ is dense.

(a) $\Rightarrow$ (b) Suppose that countable unions of closed rare sets have no interior and let $(G_n)$ be a sequence of dense open sets. Then $(\complement G_n)$ is a sequence of closed rare sets. By hypothesis, $\operatorname{int}\left(\cup_n \complement G_n\right) = \varnothing$, so $\complement\left(\cup_n \complement G_n\right) = \cap_n G_n$ is dense.

(b) $\Rightarrow$ (c) Suppose that the intersection of countably many dense open sets is dense. Let $G \subset T$ be a meager open set so that there are rare sets $\{R_n : n \in \mathbf{N}\}$ such that $G = \cup_n R_n \subset \cup_n \operatorname{cl} R_n$. Then $\complement G \supset \cap_n \complement(\operatorname{cl} R_n)$ and since each $\complement(\operatorname{cl} R_n)$ is dense and open, $\cap_n \complement(\operatorname{cl} R_n)$ must be dense by hypothesis. Since $\complement G$ is a closed dense set, it follows that $\complement G = T$ or $G = \varnothing$.

(c) $\Rightarrow$ (d) Suppose that $T$ is a Baire space. Suppose that $A \subset T$ is meager, so $\operatorname{int} A$ must be meager, too. Since $T$ is Baire, $\operatorname{int} A = \varnothing$ and therefore $\complement A$ is dense.

(d) $\Rightarrow$ (a) Suppose that complements of meager sets are dense. Let $(R_n)$ be a sequence of closed rare sets and let $E = \cup_n R_n$. Since $E$ is meager, $\complement E$ is dense by hypothesis. Therefore $\operatorname{int} E = \varnothing$. ■

**Theorem 11.6.7** LOCALLY BAIRE IMPLIES BAIRE If each point in a topological space $T$ has a neighborhood which is a Baire space, then $T$ is a Baire space.

**Proof.** Let $G \subset T$ be a nonempty open subset and let $t \in G$. By hypothesis, $t$ has an open neighborhood $V$ which is a Baire space. If $G$ were meager in $T$, $V \cap G$ would be meager in $V$ and open in $V$, contradicting the Baireness of $V$. ■

By Theorem 11.6.7 it follows that in a TVS $X$, the existence of a Baire neighborhood of 0 implies that $X$ is Baire. In Theorem 11.4.2(a), we observed that a lower semicontinuous function on a compact space is bounded from above. It follows from Theorem 11.6.8 that a lower semicontinuous function on a Baire space $T$ is bounded above on a nonempty open subset $U \subset T$.

**Theorem 11.6.8** PUB FOR LSC FUNCTIONS Let $T$ be a Baire space and $\{f_s : s \in S\}$ a family of lower semicontinuous real-valued functions on $T$ such that, at each $t$ in $T$, $\sup_{s \in S} f_s(t) < \infty$—i.e., $\{f_s : s \in S\}$ is pointwise bounded from above on $T$. Then each nonempty open subset $G$ of $T$ contains a nonempty open subset $U$ on which $\{f_s : s \in S\}$ is bounded from above.

**Proof.** In the notation of the statement, it suffices to prove the theorem for $G = T$. The map $f = \sup_{s \in S} f_s$ is lower semicontinuous by Theorem 11.4.2(b). Decompose $T$ into the union of the closed sets $F_n = \{t \in T : f(t) \leq n\}$, $n \in \mathbf{N}$. Since $T$ is nonmeager, some $F_n$ has a nonempty interior. The theorem now follows. ∎

Not only does Theorem 11.6.8 remain valid for complex-valued functions, but even for functions $\{f_s : s \in S\}$ with values in a normed space: Consider the functions $\|f_s\|$. The theorem can certainly fail if $T$ is not a Baire space, however. Consider the map $f : \mathbf{Q} \to \mathbf{R}$, $p/q \mapsto q$ ($p$ and $q$ relatively prime, $q > 0$). For any $k \in \mathbf{N}$, $f^{-1}(-\infty, k] = \cup_{q=1}^{k} \{p/q : p \in \mathbf{Z},\, p, q \text{ relatively prime}\}$ is a collection of isolated points, hence a closed set. It follows that $f$ is lower semicontinuous. It is unbounded on any nonempty open subset of $\mathbf{Q}$ since any nonempty open subset of $\mathbf{Q}$ must contain rationals with arbitrarily large denominators.

## 11.7 THE BAIRE CATEGORY THEOREM

Osgood showed in 1897 that the intersection of a sequence of dense open sets in $\mathbf{R}$ was dense, that $\mathbf{R}$ was a Baire space. Baire generalized the result to $\mathbf{R}^n$ in 1899. It remains true when $\mathbf{R}^n$ is replaced by a complete metric space, a result known as Baire's "category theorem." The usual proof relies on Cantor's nested sequence characterization of completeness, namely, that a metric space $(X, d)$ is complete iff each decreasing sequence $(F_n)$ of closed sets whose diameters shrink to 0 has nonempty intersection (see, e.g., Bachman and Narici [2000, p. 77ff]). The proof we give of this important result here [Theorem 11.7.2] is due to Choquet [1958, 1969a]. Similar in form to the nested sequence characterization of completeness, it hinges on a space's capacity for admitting a relation among the nonempty open sets which has one open set "strongly" included in another, while "decreasing" sequences still have nonempty intersection. The relation is defined abstractly in Theorem 11.7.1; specific instances are given in Theorems 11.7.2 and 11.7.3. We mention another approach to these results in Exercise 11.105.

**Theorem 11.7.1** CRITERION FOR BAIRENESS A topological space $(T, \mathcal{T})$ is a Baire space if there exists a relation $<$ among the collection $\mathcal{T}^*$ of nonempty open subsets of $T$ such that for $A, B, C, D \in \mathcal{T}^*$,

(a) if $A < B$, then $A \subset B$;

(b) for every $B \in \mathcal{T}^*$, there is an $A \in \mathcal{T}^*$ such that $A < B$;

(c) if $A \subset B < C \subset D$, then $A < D$; and

(d) if $(A_n)$ is a sequence of nonempty open sets such that $A_n > A_{n+1}$ for each $n \in \mathbf{N}$, then $\cap_n A_n \neq \varnothing$.

**Proof.** Suppose that the relation $<$ satisfies (a)–(d) is defined on the nonempty open sets $\mathcal{T}^*$ of the topological space $(T, \mathcal{T})$. If $T$ is not a Baire space, it must contain a nonempty meager open set $G$. Consequently, there must be a sequence $(R_n)$ of closed rare sets whose union contains $G$. We now construct a sequence $(U_n)$ of open subsets of $G$ such that $U_n > U_{n+1}$ and $U_n \cap (R_1 \cup \cdots \cup R_n) = \varnothing$ for each $n \in \mathbf{N}$.

Since $\operatorname{int} R_1 = \varnothing$, certainly $G \not\subset R_1$. Hence $G \cap \complement R_1$ is a nonempty open subset of $G$ which does not meet $R_1$. By (b), there is a nonempty open set $U_1$ such that $U_1 < G \cap \complement R_1$. By (c), $U_1 < G$. Since $U_1$ is a nonempty open set, $U_1 \not\subset (R_1 \cup R_2)$ and there must be a nonempty open set $U_2$ such that $U_2 < U_1$ and $U_2 \cap (R_1 \cup R_2) = \varnothing$. By induction, the sequence $(U_n)$ mentioned above is now seen to exist.

Clearly, $(\cap_n U_n) \cap (\cup_n R_n) = \varnothing$, so $\varnothing = (\cap_n U_n) \cap (\cup_n R_n) \supset (\cap_n U_n) \cap G = \cap_n U_n$ while $\cap_n U_n \neq \varnothing$ by (d) which is a contradiction. ■

We show next that a complete pseudometric space is a Baire space by showing that a relation such as the one of Theorem 11.7.1 may be defined on the nonempty open sets.

**Theorem 11.7.2** BAIRE CATEGORY THEOREM A complete pseudometric space $(T, d)$ is a Baire space, hence is nonmeager.

**Proof.** For any subset $A$ of the complete pseudometric space $(T, d)$, let $d(A) = \sup\{d(s,t) : s, t \in A\}$, the diameter of $A$; define $d^*(A) = \min(1, d(A))$ and note that $A \subset B \Rightarrow d^*(A) \leq d^*(B)$. Let $\mathcal{T}^*$ denote the nonempty open sets of $T$ and suppose $A, B, C, D \in \mathcal{T}^*$. Define $A < B$ if $\operatorname{cl} A \subset B$ and $d(A) \leq (1/2)\, d^*(B)$. To complete the proof, we show that $<$ satisfies conditions (a)–(d) of Theorem 11.7.1. That $A < B \Longrightarrow A \subset B$ is clear, so (a) is satisfied. As for (b), let $s \in G \in \mathcal{T}^*$ and choose $r \in (0,1)$ such that $B(s,r) = \{t \in T : d(s,t) < r\} \subset G$. With $r' = d(B(s,r)) \leq r < 1$ and $U = B(s, r'/4)$, we have $U < G$. As for (c), if $A \subset B < C \subset D$, then $\operatorname{cl} A \subset C \subset D$; since $d(A) \leq (1/2)\, d^*(C) \leq (1/2)\, d^*(D)$, it follows that $A < D$.

We use the completeness of $T$ to show that (d) is satisfied. For this purpose, let $(G_n)$ be a sequence of nonempty open sets, decreasing with respect to $<$. Since, $d^*(G_1) \leq 1$, $d(G_n) \leq 2^{-n+1}$ for each $n \in \mathbf{N}$. Hence, choosing $t_n \in G_n$ for each $n \in \mathbf{N}$ yields a $d$-Cauchy sequence. Since $T$ is complete,

$(t_n)$ has a limit $t$. For $k \geq n$, $t_k \in G_n$, so $t \in \operatorname{cl} G_n$ for each $n \in \mathbf{N}$. Thus $t \in \cap_n \operatorname{cl} G_n \subset \cap_n G_n$ and (d) is satisfied. ■

Since Baire spaces are nonmeager, Theorem 11.7.2 is stronger than the classical Baire category theorem. We note too that a product of complete metric spaces is Baire [Bourbaki 1966b, p. 254, Ex. 17].

In a locally compact Hausdorff space $T$, if $t$ belongs to an open set $U$, there is an open neighborhood $V$ of $t$ such that $\operatorname{cl} V \subset U$. (View $T$ as a subspace of its one-point compactification $T^*$. $T^*$ is normal, hence completely regular. Since complete regularity is hereditary, the complete regularity of $T$ follows.) We mention this for the sake of the result below, another application of Theorem 11.7.1.

**Theorem 11.7.3** LOCALLY COMPACT HAUSDORFF IMPLIES BAIRE Every locally compact Hausdorff space $T$ is a Baire space.

**Proof.** For nonempty open sets $A$ and $B$ of the locally compact Hausdorff space $T$, define $A < B$ if $A$ is relatively compact (i.e., has compact closure) and $\operatorname{cl} A \subset B$. Keeping the comment about complete regularity preceding this theorem in mind, properties (a)–(c) of Theorem 11.7.1 are easy to verify. As for (d), suppose that $(G_n)$ is a sequence of nonempty open sets such that $G_1 > G_2 > \cdots$. The sets $\{G_n : n \geq 2\}$ form a filterbase in the compact space $\operatorname{cl} G_2$. Therefore, by Theorem 1.1.10(b), it has a cluster point $t \in \cap_{n \in \mathbf{N}} \operatorname{cl} G_n \subset \cap_{n \in \mathbf{N}} G_n$. ■

As an easy corollary, we see that a compact Hausdorff space is a Baire space. Thus if $t_n \to t$ in a Hausdorff space, then $\{t_n : n \in \mathbf{N}\} \cup \{t\}$ is a Baire subspace. By Theorem 11.7.3, the locally compact Hausdorff space $\{t_n : n \in \mathbf{N}\}$ is a Baire subspace. Coupled with the remark of Example 11.6.2(a) that countable Hausdorff spaces without isolated points are meager, we have an indirect proof of the fact that the points of a convergent sequence in a Hausdorff space must have an isolated point. We may also infer from Theorem 11.7.3 that $\mathbf{Q}$ is not locally compact, since it is meager.

## 11.8 BAIRE TVS

Since Baire LCS are barreled [Theorem 11.6.4], we see that Fréchet spaces—hence Banach and Hilbert spaces—are barreled by Theorem 11.7.2. Not all barreled spaces are Baire spaces, however, and one of a different kind is given in Example 11.8.1.

**Example 11.8.1** MEAGER BARRELED SPACE Let $X$ be an infinite-dimensional vector space over $\mathbf{F}$ endowed with the finest locally convex topology $\mathcal{T}_{lc}$ [Example 5.6.4], the inverse image topology determined by the class of all seminorms on $X$ or, equivalently, the topology that has the filterbase of all absorbent disks as a base at 0. As noted after Definition 11.3.1, $(X, \mathcal{T}_{lc})$ is barreled. A few more observations about $\mathcal{T}_{lc}$ are now in order.

(a) $\mathcal{T}_{lc}$ IS HAUSDORFF If $x \neq 0$, the map $ax \mapsto a$ defines a nontrivial linear functional $g$ on $\mathbf{F}x$. By extending $\{x\}$ to a Hamel base for $X$ and defining $g$ arbitrarily on the rest of the basis vectors, we get a linear functional $g$ is defined on all of $X$. The seminorm $|g|$ then does not vanish on $x$ and therefore $\mathcal{T}_{lc}$ is a Hausdorff topology by Theorem 5.5.1(a).

(b) ALL LINEAR FORMS ON $X$ ARE CONTINUOUS Given any linear functional $f$ on $X$, $|f|$ is a seminorm.

(c) ALL SUBSPACES OF $X$ ARE CLOSED By an argument like the one in (a), if $x$ is not in the subspace $M$, there exists a linear form $f$ on $X$ which vanishes on $M$ but not on $x$; $f$ is continuous by (b). By Theorem 7.7.7(c), a consequence of the Hahn–Banach theorem, $x \in \operatorname{cl} M$ iff every continuous linear form which vanishes on $M$ vanishes on $x$.

To see that $(X, \mathcal{T}_{lc})$ is meager, let $A = \{x_n\}$ be a denumerable subset of a Hamel base $B$ for $X$. For each $n \in \mathbf{N}$, let $M_n$ be the linear span of $B \cap \complement A$ and $\{x_1, \ldots, x_n\}$. As each $M_n$ is a closed proper subspace, each is a rare subset of $X$ [Example 11.5.3(f)]. Since $X = \cup_{n \in \mathbf{N}} M_n$, $X$ is meager. ■

**Example 11.8.2** COUNTABLE DIMENSION $\Rightarrow$ MEAGER Let $X$ be a Hausdorff TVS and suppose that $\{x_n : n \in \mathbf{N}\}$ is a Hamel base for $X$. For each $n$ the linear span $M_n$ of $\{x_1, \ldots, x_n\}$ is closed by Theorem 4.10.3(d). Since each $M_n$ is rare [Example 11.5.3(g)] and $X = \cup_{n \in \mathbf{N}} M_n$, this means that $X$ is meager. Hence any infinite-dimensional Baire space has uncountable dimension. In particular, by the Baire category theorem [Theorem 11.7.2] any infinite-dimensional complete pseudometrizable TVS is of uncountable dimension. ■

Suppose $X$ is an infinite-dimensional normed space. As such, $X'$ is infinite-dimensional as well [Theorem 8.4.6] and therefore the $\sigma(X, X')$-neighborhoods of 0 are unbounded [Example 6.1.7(c)]; hence $\sigma(X, X')$ is not normable [see the discussion following Theorem 6.2.1]. But when is $\sigma(X, X')$ metrizable? The non-metrizability of the weak topology for infinite-dimensional normed spaces of Example 11.8.3 was discovered by Wehausen [1938].

**Example 11.8.3** $\sigma(X, X')$ METRIZABLE IFF $X$ FINITE-DIMENSIONAL If $X$ is a finite-dimensional HTVS then $(X, \sigma(X, X'))$ is linearly homeomorphic to $\ell_\infty(n)$ [Theorem 4.10.3(c)] so $\sigma(X, X')$ is metrizable. If $X$ is an infinite-dimensional normed space, then $X'$ is a Banach space, hence a Baire space, and therefore the dimension of $X'$ is uncountable [Example 11.8.2]. We show next that $\sigma(X, X')$ is not first countable, hence not metrizable [Theorem 4.8.3]. To see that $\sigma(X, X')$ is not first countable, suppose that $(V_n)$ is a denumerable base of $\sigma(X, X')$-neighborhoods of 0. Each $V_n$ is the polar of a finite subset $F_n$ of $X'$. Since $F = \cup_{n \in \mathbf{N}} F_n$ is denumerable and $X'$ is of nondenumerable dimension, the linear span of $F$ is a proper subset of $X'$. Hence, there is some $g \in X'$ which is linearly independent of $F$. Thus, no matter what $n \in \mathbf{N}$ you choose, $\cap_{f \in F_n} f^{-1}(0) \not\subset g^{-1}(0)$ [Theorem 1.4.1(f)]. This means that there must be some $x \in \cap_{f \in F_n} f^{-1}(0)$ and $g \in X'$ such that $g(x) > 1$. Thus $x \in F_n{}^\circ = V_n$ but $x \notin \{g\}^\circ$, so $V_n \not\subset \{g\}^\circ \in \sigma(X, X')$ for any $n$ and $(V_n)$ cannot be a base at 0 for $\sigma(X, X')$. ■

**Example 11.8.4** BARRELED $\not\Rightarrow$ BAIRE Let $\varphi$ be the space of all "finite" sequences $(t_n)$ of complex numbers, i.e., $(t_n)$ such that $t_n = 0$ for almost all $n$, with the finest locally convex topology. Since $\varphi$ is of denumerable dimension, it cannot be a Baire space by Example 11.8.2. It is barreled by the result of Example 11.8.1. ■

Nonmeager topological spaces need not be Baire [Example 11.8.5] but a nonmeager TVS must be [Theorem 11.8.6].

**Example 11.8.5** NONMEAGER $\not\Rightarrow$ BAIRE Obviously, Baire spaces are nonmeager. To see that nonmeager spaces need not be Baire, consider the metric space $\mathbf{Q} \cup (0,1)$. Since it contains the open nonmeager subspace $(0,1)$, it is nonmeager [Example 11.6.2(d)]. Since it has meager open sets such as $\{t \in \mathbf{Q} : t > 1\}$, it is not a Baire space. Note also that this is an incomplete metric space which is nonmeager. ■

**Theorem 11.8.6** NONMEAGER TVS IS BAIRE A TVS $X$ is a Baire space iff it is nonmeager. Thus, by Theorem 11.6.4, a Baire LCS is barreled.

**Proof.** Since a Baire space is nonmeager, we need only show that a nonmeager TVS is Baire. To this end, let $X$ be a TVS which is not a Baire space. As such, it must have a meager neighborhood $V$ of 0. Hence $X = \cup_n nV$ and $X$ is meager. ■

The result of Theorem 11.8.8 is even stronger, but first we need the following lemma.

**Lemma 11.8.7** Let $A$ and $B$ be subsets of a topological space. If $B$ is closed and $\text{int}\, A = \varnothing$, then $\text{int}\,(A \cup B) = \text{int}\, B$.

**Proof.** Suppose that $B$ is closed, $\text{int}\, A = \varnothing$ and $t \in \text{int}\, A \cup B$. Let $G$ be an open neighborhood of $t$ such that $G \subset A \cup B$. Since $G \subset A \cup B$, $G \cap \complement B \subset A$. Since $\text{int}\, A = \varnothing$, $G \cap \complement B = \varnothing$, i.e., $G \subset B$. ■

**Theorem 11.8.8** BAIRE IFF NO RARE $B$ SUCH THAT $X = \cup_{n \in \mathbf{N}} nB$ A TVS $X$ is Baire iff there is no rare set $B$ for which $X = \cup_{n \in \mathbf{N}} nB$.

**Proof.** If the TVS $X$ is Baire, it is nonmeager so there cannot be a rare set $B$ for which $X = \cup_n nB$. Conversely, suppose that $X$ is not a Baire space. As such, it must be meager by Theorem 11.8.6, so let $(B_n)$ be a sequence of rare sets which covers $X$. Since not all the $B_n$ can be empty, i.e., since there are nonempty rare sets, the topology on $X$ cannot be the indiscrete topology $\{X, \varnothing\}$. Let $U$ be a proper balanced neighborhood of 0 and choose a closed balanced neighborhood $V$ of 0 such that $V + V \subset U$. Note that $V = V \cap (\cup_{n\in\mathbf{N}} B_n) = \cup_{n\in\mathbf{N}} (V \cap B_n)$. We show that

$$B = \bigcup_{n\in\mathbf{N}} \left(\frac{1}{n}(V \cap B_n)\right)$$

is a rare set such that $X = \cup_{n\in\mathbf{N}} nB$.

To see that $X = \cup_{n\in\mathbf{N}} nB$, note that for any $n \in \mathbf{N}$, $V \cap B_n \subset nB$. Thus, for any $k \in \mathbf{N}$, $kV = k[\cup_{n\in\mathbf{N}} (V \cap B_n)] \subset k[\cup_{n\in\mathbf{N}} nB] \subset \cup_{n\in\mathbf{N}} nB$. Since $V$ is absorbent, $\cup_{k\in\mathbf{N}} kV = X$ and the statement follows. To see that $B$ is rare, suppose that $y \in \operatorname{int}(\operatorname{cl} B)$ and let $W$ be an open neighborhood of $y$ such that $W \subset \operatorname{cl} B$. Since $B = \cup_{n\in\mathbf{N}} \left(n^{-1}(V \cap B_n)\right)$, for any $k \in \mathbf{N}$,

$$\operatorname{cl} B = \operatorname{cl}\left(\cup_{n<k} n^{-1}(V \cap B_n)\right) \cup \operatorname{cl}\left(\cup_{n\geq k} n^{-1}(V \cap B_n)\right)$$

Since each $V \cap B_n$ is rare, $\cup_{n<k} n^{-1}(V \cap B_n)$ is rare; therefore, by Lemma 11.8.7, $W \subset \operatorname{cl}\left[\cup_{n\geq k} n^{-1}(V \cap B_n)\right]$. Since $V$ is balanced, $\cup_{n\geq k} n^{-1}(V \cap B_n) \subset k^{-1}V$ and $W \subset k^{-1}V$. Choose $x \notin U$ and $r > 0$ so that $\pm rx \in W - y$, so that $2rx = (y + rx) - (y - rx) \in W - W \subset k^{-1}V + k^{-1}V \subset k^{-1}U$. Since $2krx \in U$ and $U$ is balanced, for $k$ such that $2kr \geq 1$, $x = (1/2kr)(2krx) \in U$ which contradicts $x \notin U$. ■

Theorem 11.8.8 is strengthened a little further in Theorem 11.8.9. In the course of that argument we construct a balanced set by taking the *balanced core* of a certain set, the balanced core of a set $A$ being the largest balanced subset $B$ of $A$ (see Sec. 4.1).

**Theorem 11.8.9** BAIRE IFF NO RARE BALANCED ABSORBENT SET A TVS $X$ that does not have the indiscrete topology is a Baire space iff there are no rare balanced absorbent sets in $X$.

**Proof.** By Theorem 11.8.6, if the TVS $X$ is a Baire space there cannot be a rare subset $B$ such that $X = \cup_{n\in\mathbf{N}} nB$, hence no rare balanced absorbent set. We argue the converse for complex vector spaces only and just mention what happens in the real case which is much easier to prove. We suppose that $X$ is not a Baire space and construct a rare balanced absorbent subset.

If $X$ is not a Baire space, it is meager by Theorem 11.8.6, so let $(R_n)$ be a sequence of closed rare sets which covers $X$. Let $U$ be a closed balanced proper neighborhood of 0. Choose a closed balanced neighborhood $V$ of 0 such that $V + V \subset U$. For each positive integer $n$, let $A_n = V \cap R_n$ so that $(A_n)$ is a sequence of closed rare sets which covers $V$. We combine and rotate the $A_n$ as follows: For each $n \in \mathbf{N}$, let

$$B_n = \cup_{k=0}^{n-1} e^{2k\pi i/n} (A_1 \cup \cdots \cup A_n)$$

[In the real case, take $B_n = (A_1 \cup \cdots \cup A_n) \cup (-1)(A_1 \cup \cdots \cup A_n)$.] Using $1/n$ as an astringent, as in the proof of Theorem 11.8.8, we use contractions of $B_n$ to form the rare absorbent set $B = \cup \{1/n)B_n : n \in \mathbf{N}\}$. The balanced core of $B$ is the set we ultimately seek. The argument of Theorem 11.8.8 *mutatis mutandis* yields the rareness of $B$. The remainder of this argument is devoted to proving that $B$ is absorbent.

For any $y \in X$, it follows from the Baire category theorem [11.7.2] that $\mathbf{C}y$ is a Baire subspace of $X$. Since $V \cap \mathbf{C}y$ is a relative neighborhood of 0 in $\mathbf{C}y$, then it must be nonmeager in $\mathbf{C}y$. Since the sets $\{A_n \cap \mathbf{C}y : n \in \mathbf{N}\}$ cover $V \cap \mathbf{C}y$, one of them, $A_p$ say, must have nonempty interior in the relative topology on $V \cap \mathbf{C}y$, i.e., there must exist $b \in \mathbf{C}$ and $r > 0$ such that

$$C(b,r)\, y \subset A_p \qquad \text{where} \qquad C(b,r) = \{a \in \mathbf{C} : |a-b| \le r\} \qquad (*)$$

Since the map $t \mapsto e^{it}$ is uniformly continuous on $[0, 2\pi]$, there must be an integer $q \ge p$ such that $|s-t| \le 2\pi/q$ implies $\left|e^{is} - e^{it}\right| |b| \le r/2$. We show next that if $a = |a|\, e^{it}$ is such that $|a|$ is sufficiently close to $|b|$, $ay$ belongs to $\cup_{k=0}^{n-1} e^{i2k\pi/n} A_p \subset B_n$ for sufficiently large $n$. Let $b = |b|\, e^{is}$ and suppose that $||a| - |b|| \le r/2$. Then $|a| = |b| \pm u$ for some $u \in [0, r/2]$. For $n \ge q$ and $s, t \in [0, 2\pi]$, since $|s-t| \le 2\pi$, there exists $k$, $0 \le k \le n-1$, such that $|t - (2k\pi + ns)/n| \le 2\pi/n$. Since $n \ge q$, $2\pi/n \le 2\pi/q$. Hence $\left|e^{it} - e^{i(2k\pi+ns)/n}\right| |b| \le r/2$. Therefore

$$\begin{aligned} \left|a - be^{i2\pi k/n}\right| &= \left|(|b| \pm u)\, e^{it} - |b|\, e^{i(2k\pi+ns)/n}\right| \\ &\le \left|e^{it} - e^{i(2k\pi+ns)/n}\right| |b| + r/2 \le r \end{aligned}$$

so $ay \in C\left(e^{i2\pi k/n} b, r\right) y \subset \cup_{k=0}^{n-1} e^{2ki\pi/n} A_p \subset B_n$ whenever $||a| - |b|| \le r/2$. Now choose $N \ge q$ so that $|b| / (N+1) \le r/2$. For any $d \in \mathbf{C}$ such that $|d| \le |b|/N$, there exists $n \ge N$ such that $|b|/(n+1) \le |d| \le |b|/n$. Since

$$\begin{aligned} (1/n)(|b| - r/2) &= |b|/n - (r/2)(1/n) \le |b|/n - (|b|/(n+1))(1/n) \\ &= |b|/(n+1) \le |d| \le |b|/n \le (1/n)(|b| + r/2) \end{aligned}$$

it follows that $|n|d| - |b|| \le r/2$ and therefore that $dy \in (1/n) B_n$. So for any $d$ such that $0 \le |d| \le |b|/N$ (since 0 is clearly in $B$), $dy \in B$ and $B$ is seen to be absorbent. The balanced core of $B$ is the desired rare balanced absorbent set. ■

### 11.8.1 Baire Variations

Barreled spaces have been defined by prescribing that certain absorbent disks be neighborhoods of 0. Bornological spaces [Definition 13.2.3] are similarly defined as those LCS in which every bornivorous disk is a neighborhood of 0. Theorem 11.8.9, first proved by Saxon [1974], affords an analogous description of linear Baire spaces as those TVS (not necessarily LCS) in which each *absorbent balanced* set is a neighborhood of *some point*.

Consider the following variations on the theme of "Baire space":

An LCHS $X$

1. has property (S) if its dual $X'$ is $\sigma(X', X)$-sequentially complete.

2. has property (C) if every $\sigma(X', X)$-bounded subset of $X'$ is $\sigma(X', X)$-relatively countably compact.

3. is *$\omega$-barreled* if every countable $\sigma(X', X)$-bounded subset of $X'$ is equicontinuous (sometimes called "$\sigma$-barreled").

4. is *barreled* if every $\sigma(X', X)$-bounded subset of $X'$ is equicontinuous [Theorem 11.3.4].

5. is *quasi-Baire* if $X$ is barreled and not the union of an increasing sequence of rare subspaces.

6. is *Baire-like* if $X$ is not the union of an increasing sequence of rare balanced convex sets.

7. is *suprabarreled* if for any increasing sequence $(M_n)$ of subspaces whose union is $X$, one of them must be barreled and dense in $X$.

8. is *unordered Baire-like* if $X$ is not the union of an arbitrary sequence of rare disks.

9. is *Baire* if $X$ is not the union of an increasing sequence of rare sets [Theorem 11.8.8].

The implications read from (9) up to (1) and none of them is reversible as shown by arguments and examples in Saxon 1972, Saxon and Levin 1971, Levin and Saxon 1971, Todd and Saxon 1973, Dierolf et al. 1978, and Valdivia 1981. It is also shown in these articles that (1)–(6) are equivalent in the class of metrizable spaces and (4)–(6) are equivalent in the wider class of spaces not containing a copy of $\varphi$, the space of sequences of complex numbers which are almost always 0 endowed with the finest locally convex topology.

It is not known whether the product of Baire TVS is a Baire space. Oxtoby 1961, however, assuming the continuum hypothesis, has shown that it is *not* generally true for completely regular topological spaces. Todd and Saxon 1973 and Todd 1980 show that products of unordered Baire-like spaces are again unordered Baire-like.

## 11.9 BANACH–STEINHAUS THEOREM

$L(X, Y)$ denotes the linear space of continuous linear maps of the TVS $X$ into the TVS $Y$. $\mathcal{T}_p$ denotes the strong operator topology (= topology of pointwise convergence) on $L(X, Y)$ as discussed in Sec. 11.2. To say that a subset $H$ of $L(X, Y)$ is "pointwise" bounded means that $H(x)$ is a bounded subset of $Y$ for each $x$ in $X$ or, equivalently, that $H$ is $\mathcal{T}_p$-bounded.

The main result of this section is Theorem 11.9.1, the Banach–Steinhaus theorem or principle of uniform boundedness for barreled spaces $X$ which asserts that pointwise bounded subsets $H \subset L(X,Y)$ are equicontinuous for any LCS $Y$. Theorem 11.9.3 shows that the barreled spaces $X$ are the widest class of LCS for which such a principle holds in $L(X,Y)$ for all LCS $Y$.

**Theorem 11.9.1** BANACH–STEINHAUS THEOREM FOR LCS. For any LCS $Y$ and barreled space $X$, if $H \subset L(X,Y)$ is pointwise bounded, then $H$ is equicontinuous.

**Proof.** In the notation of the statement, let $H \subset L(X,Y)$ be pointwise bounded. Given a neighborhood $V$ of 0 in $Y$, we must find a neighborhood $U$ of 0 in $X$ such that $H(U) = \cup_{A \in H} A(U) \subset V$. Since $Y$ is locally convex, it has a base of barrels at 0 [Theorem 4.4.6], so we may assume that $V$ is a barrel. By continuity and linearity, each $A^{-1}(V)$ is a closed disk [Theorems 4.1.8(c) and 4.2.4(b)]. Since these properties are stable under intersection, $U = \cap_{A \in H} A^{-1}(V)$ is a closed disk as well. If we can show that $U$ is absorbent, $U$ will be a barrel, hence a neighborhood of 0 in $X$, and the equicontinuity of $H$ will follow. But absorbency of $U$ means, given $x \in X$, $ax \in U = \cap_{A \in H} A^{-1}(V)$ or, equivalently, $aH(x) \subset V$ for all sufficiently small scalars $a$. Since $H$ is pointwise bounded, the condition is satisfied and the theorem follows. ■

A Banach space $X$ is nonmeager by the Baire category theorem [11.7.2], hence barreled by Theorem 11.6.4. Therefore if $Y$ is any normed space and $\{A_s : s \in S\} \subset L(X,Y)$ is such that $\sup_{s \in S} \|A_s x\| < \infty$ for each $x \in X$, $\{A_s\}$ is equicontinuous by Theorem 11.9.1. Therefore it is bounded in the topology of uniform convergence on bounded sets [Theorem 11.2.6]. For normed spaces, this means that $\{A_s\}$ is bounded in the norm topology on $L(X,Y)$ [Example 11.2.4], i.e., $\sup_{s \in S} \|A\| < \infty$. In summary:

**Theorem 11.9.2** BANACH–STEINHAUS FOR NORMED SPACES If $\{A_s : s \in S\}$ is a collection of continuous linear maps of a Banach space $X$ into a normed linear space $Y$ such that at every $x \in X$, $\sup_{s \in S} \|A_s x\| < \infty$, then $\sup_{s \in S} \|A_s\| < \infty$.

The following converse shows that in the class of LCS the principle of uniform boundedness will not stretch beyond the barreled spaces.

**Theorem 11.9.3** PUB IMPLIES BARRELED Let $X$ be an LCS and let $Y$ be a TVS not consisting of just the zero vector nor carrying the indiscrete topology. If pointwise boundedness implies equicontinuity in $L(X,Y)$ then $X$ is barreled.

**Proof.** We use the notation of the statement. To prove that $X$ is barreled, we show that each $\sigma(X', X)$-bounded subset $B$ of $X'$ is equicontinuous [Theorem 11.3.4]. To that end, let $y$ be a vector in $Y$ not in $\mathrm{cl}\{0\}$. For each $f \in B$, consider the map $yf : X \to Y$, $x \mapsto f(x)y$. Clearly $yf \in L(X,Y)$.

Since $B$ is $\sigma(X', X)$-bounded, $yB = \{yf : f \in B\}$ is a pointwise bounded subset of $L(X, Y)$, hence equicontinuous by hypothesis. To see that $B$ is equicontinuous, let $W$ be a balanced neighborhood of 0 in $Y$ to which $y$ does not belong. Since $yB$ is equicontinuous, there is a neighborhood $V$ of 0 in $X$ such that $B(V)y = (\cup_{f \in B} f(V))y \subset W$. We claim that $B(V) \subset \{a \in \mathbf{F} : |a| \leq 1\}$, i.e., that $B \subset V^{\circ}$; this implies that $B$ is equicontinuous by Theorem 8.6.4(a). Since $B(V)y \subset W$, if there exists $a \in B(V)$ such that $|a| \geq 1$, then $ay \in W$. Since $W$ is balanced, $y = (1/a)ay \in W$, contradicting the way in which $W$ was chosen. Consequently $B(V) \subset \{a \in \mathbf{F} : |a| \leq 1\}$. ■

**Theorem 11.9.4** POINTWISE LIMITS Let $X$ be barreled, $Y$ an LCS and $F(X, Y)$ denote the space of all maps of $X$ into $Y$. If $(A_s)$ is a pointwise bounded net from $L(X, Y)$ that converges to $A \in F(X, Y)$ pointwise [i.e., $A_s x \to Ax$ for each $x$ in $X$ or, equivalently, $A_s \to A$ in the topology of pointwise convergence of $F(X, Y)$], then $A \in L(X, Y)$ and $A_s \to A$ in the compact-open topology.

**Proof.** We use the notation of the statement. That such a limit $A$ is linear is straightforward to show. Since $(A_s)$ is pointwise bounded, it is equicontinuous by the Banach–Steinhaus theorem [11.9.1]. The pointwise closure $\mathrm{cl}_p\{A_s : s \in S\}$ of the equicontinuous subset $\{A_s : s \in S\} \subset L(X, Y)$ is equicontinuous by an argument similar to that of Lemma 5.10.3. Since $A \in \mathrm{cl}_p\{A_s : s \in S\}$, $A$ is continuous. An argument similar to that of Lemma 5.10.2 shows that the point-open and compact-open topologies coincide on equicontinuous subsets of $L(X, Y)$; hence $A_s \to A$ in the compact-open topology. ■

If we restrict consideration to pointwise convergent sequences in Theorem 11.9.4 then we can omit "pointwise bounded" from the hypothesis since pointwise convergent sequences are automatically pointwise bounded.

Some variants of the Banach–Steinhaus theorem that do not involve convexity are the subject of Theorem 11.9.5; the idea is to substitute nonmeager for barreled. Some other versions which do not involve local convexity are given in Exercise 11.201.

**Notation.** Let $E$ be a family of continuous linear maps of the TVS $X$ into the TVS $Y$. For subsets $C$ of $X$ and $D$ of $Y$,

$$E(C) = \cup\{A(C) : A \in E\} \qquad E^{-1}(D) = \cap\{A^{-1}(D) : A \in E\} \qquad ■$$

**Theorem 11.9.5** BANACH–STEINHAUS WITHOUT CONVEXITY Let $E$ be a collection of continuous linear maps of the TVS $X$ into the TVS $Y$. Then:

(a) PUB If $E$ is pointwise bounded on a nonmeager set $B \subset X$ then $E$ is equicontinuous and $E$ is pointwise bounded on $X$. In particular, if $X$ is a Baire space and $E$ is pointwise bounded on $X$, then $E$ is equicontinuous.

(b) POINTWISE LIMITS If $X$ is nonmeager and the pointwise bounded net $(A_s)_{s\in S}$ of continuous linear maps of $X$ into $Y$ converges pointwise to the map $A$, then $A$ is a continuous linear map.

**Proof.** (a) Let $E$, $X$ and $Y$ be as above and suppose that $E$ is pointwise bounded on a nonmeager set $B \subset X$. Given a neighborhood $V$ of 0 in $Y$, we produce a neighborhood $U$ of 0 in $X$ such that $E(U) \subset V$. Choose a closed balanced neighborhood $W$ of 0 such that $W + W \subset V$. For any $x \in B$, $E(x)$ is bounded so there exists $n \in \mathbf{N}$ such that $E(x) \subset nW$. Hence $x/n \in E^{-1}(W) = C$. Therefore $B \subset \cup_{n\in\mathbf{N}} nC$. Since $C$ is closed, the nonmeagerness of $B$ implies that some $nC$—hence $C$ itself—has a nonempty interior. Suppose that $x \in \operatorname{int} C$ and that $U$ is a neighborhood of 0 such that $x+U \subset C$. For any $A \in E$, $Ax+A(U) \subset A(C)$ or $A(U) \subset A(C)-Ax$. Since $x \in C$, $Ax \in W$ and therefore $A(C) - Ax \subset W - W \subset V$, i.e., $E(U) \subset V$.

Since $E$ is equicontinuous, $E$ is bounded in any $\mathcal{S}$-topology by Theorem 11.2.6. Hence $E$ is bounded in the topology of pointwise convergence and therefore $E(x)$ is bounded at each $x \in X$.

(b) Let $A$ and $X$ be as in the statement. Obviously, such an $A$ is linear. Since $\{A_s : s \in S\}$ is bounded at each point of $X$, it is equicontinuous by (a). Thus, given a closed neighborhood $V$ of 0 in $Y$, there is a neighborhood $U$ of 0 in $X$ such that, for every index $s$, $A_s(U) \subset V$, which implies that $A(U) \subset \operatorname{cl} V = V$ and establishes the continuity of $A$. ■

**Theorem 11.9.6** BANACH–STEINHAUS VARIANT Let $E$ be a family of continuous linear maps of the HTVS $X$ into the TVS $Y$. If $E$ is pointwise bounded on a convex compact subset $K$ of $X$, then $E(K) = \cup\{A(K) : A \in E\}$ is bounded.

**Proof.** We use the notation of the statement. Let $W$ be a neighborhood of 0 in $Y$. Choose a closed balanced neighborhood $U$ of 0 in $Y$ such that $U + U \subset W$. Let $B = E^{-1}(U)$. For any $x \in K$, $E(x)$ is bounded, so for some $p \in \mathbf{N}$, $E(x) \subset pU$; hence $x/p \in E^{-1}(U) = B$. It follows that $K = \cup_n (K \cap nB)$. Since $K$ is compact, $K$ is a Baire space by Theorem 11.7.3. Thus, since $K$ and $B$ are closed, there must be some $q \in \mathbf{N}$ such that $K \cap qB$ has a nonempty interior in $K$. Let $w \in \operatorname{int}_K (K \cap qB)$ and choose an open neighborhood $V$ of 0 such that $(w + V) \cap K \subset qB$. Since $\{w + nV : n \in \mathbf{N}\}$ is an open cover of $K$ and $K$ is compact, there exists an integer $j > 1$ such that $K \subset w + jV$. Now, for any $x$ in $K$, $y = (1 - 1/j)\,w + (1/j)\,x \in K$ because $K$ is convex. Also $y - w = (1/j)\,(x - w) \in V$ by the way $j$ was chosen. Thus $y \in (w + V) \cap K \subset qB$. Since $B = E^{-1}(U)$, $E(qB) \subset qU$. Since $x = jy - (j-1)\,w \in jqB - (j-1)\,qB$, we have, for any $A \in E$, $Ax \in jqU - (j-1)\,qU \subset jqU + jqU \subset jqW$. The boundedness of $E(K)$ is now established. ■

## 11.10 A DIVERGENT FOURIER SERIES

Where does the Fourier series of a continuous function converge pointwise? The question has stimulated science (not just mathematics) for two centuries. Attempts to solve it have led to diverse important discoveries.

In 1829 Dirichlet proved that the Fourier series of a *piecewise smooth* continuous function on $[-\pi, \pi]$ converged to the function at every point. At the end of the article he commented that he believed that the Fourier series of any continuous function converged to the function at every point. Such greats as Riemann, Weierstrass and Dedekind expressed similar beliefs over the next 40 years. They were all wrong. In 1876 P. du Bois-Reymond gave an example of a continuous function whose Fourier series did not converge at the points of a dense subset of $[-\pi, \pi]$. Fejér simplified du Bois-Reymond's construction in 1909 and gave other examples of continuous functions whose Fourier series diverged at certain points. Hardy and Rogosinski [1950] give others. As a consequence of these results and Kolmogorov's 1926 proof that there is a function $x \in L_1[-\pi, \pi]$ whose Fourier series diverged everywhere [Bary 1964, p. 455, Katznelson 1968, p. 59], it was believed by many that it was only a matter of time until it was shown that there were *continuous* functions whose Fourier series diverged everywhere. The consensus was wrong again. The Swedish mathematician Lennart Carleson [1966] proved the very deep result that the Fourier series of a continuous function $x$ converges pointwise to $x$ almost everywhere.

For the spaces $X = L_p[-\pi, \pi]$, $1 \leq p < \infty$, and $X = C\left([a, b], \|\cdot\|_\infty\right)$ (and others), there is the following dichotomy:

> Either there exists $x \in X$ whose Fourier series diverges everywhere or the Fourier series of every $x \in X$ converges almost everywhere [Katznelson 1968, p. 59].

As noted above, Kolmogorov showed that $L_1[-\pi, \pi]$ was of the first type. Lusin had conjectured in 1915 that $L_2[-\pi, \pi]$ was of the latter type, that Fourier series of $L_2$ functions converged almost everywhere. Carleson confirmed Lusin's conjecture in 1966; then Hunt [1968] extended Carleson's result to all the $L_p[-\pi, \pi]$ for $p > 1$. Kahane and Katznelson [Katznelson 1968, p. 58f] showed that on any set $E \subset [-\pi, \pi]$ of measure 0 there was a continuous function on $[-\pi, \pi]$ whose Fourier series diverged at every point of $E$. For the results mentioned above as well as many other profound discoveries, Carleson received the Abel Prize from The Norwegian Academy of Science and Letters in May 2006.

In an intriguing application of the Banach–Steinhaus theorem [11.9.2], we present Kolmogorov's argument here on the existence of a continuous function whose Fourier series diverges at a certain point.

**Theorem 11.10.1** There exists a real-valued continuous function on $[-\pi, \pi]$ whose Fourier series diverges at a point.

**Proof.** Let $X = C([-\pi,\pi], \mathbf{R}, \|\cdot\|_\infty)$ and let $Y = \mathbf{R}$. Let $D_0(t) = 1$ and for $n \in \mathbf{N}$ the $n$th *Dirichlet kernel* is defined at $t \in [-\pi,\pi]$ as the continuous function

$$D_n(t) = 1 + \sum_{k=1}^{n} 2\cos kt = \begin{cases} \frac{\sin(n+1/2)t}{\sin(t/2)}, & t \neq 0 \\ 2n+1, & t = 0 \end{cases},$$

Consider the linear functionals

$$f_n(x) = \frac{1}{2\pi}\int_{-\pi}^{\pi} x(t) D_n(t)\, dt,\ x \in C([-\pi,\pi], \mathbf{R})$$

As in Theorem 7.4.4, each integral operator $f_n$ is continuous and $\|f_n\| = v(D_n)$, the variation of $D_n$, which happens to be $\frac{1}{2\pi}\int_{-\pi}^{\pi} |D_n(t)|\, dt$ [Bachman et al. 2000, p. 131]. The $n$th partial sum of the Fourier series for $x$ is given by

$$s_n(t) = \frac{1}{2\pi}\int_{-\pi}^{\pi} x(u) D_n(t-u)\, du$$

Since $D_n$ is an even function, $f_n(x) = s_n(0)$. If we can show that $\|f_n\| \to \infty$, it will follow from the Banach–Steinhaus theorem [11.9.2] that there is some continuous function $x$ such that $|f_n(x)| = |s_n(0)| \to \infty$, i.e., that the Fourier series for $x$ diverges at $t = 0$. Since $\sin(t/2) \leq t/2$ for $0 \leq t \leq \pi$, for any $n \in \mathbf{N}$,

$$\begin{aligned} \|f_n\| &= \frac{1}{2\pi}\int_{-\pi}^{\pi} |D_n(t)|\, dt \\ &= \frac{1}{\pi}\int_0^{\pi} \left|\frac{\sin(n+1/2)t}{2\sin(t/2)}\right| dt \\ &\geq \frac{1}{\pi}\int_0^{\pi} \frac{|\sin(n+1/2)t|}{t}\, dt \end{aligned}$$

Making the substitution $u = (n+1/2)t$, we see that for every $n$,

$$\frac{1}{2\pi}\int_{-\pi}^{\pi} |D_n(t)|\, dt \geq \frac{1}{\pi}\int_0^{(n+1/2)\pi} \frac{|\sin u|}{u}\, du \tag{11.1}$$

But consider the function $|\sin u|/u$ for $u > 0$ with triangles inscribed in each lobe. As the peaks of the triangles occur at odd multiples of $\pi/2$, the altitudes of the triangles are $2/(2n-1)\pi$. Thus, summing the areas of the triangles,

$$\int_0^{\infty} \frac{|\sin u|}{u}\, du \geq \sum_{n\in\mathbf{N}} \frac{1}{2}\pi \frac{2}{(2n-1)\pi} = \infty$$

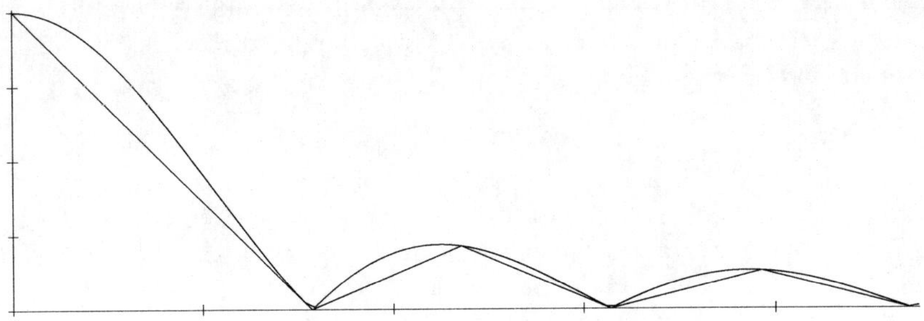

Hence, the terms on the right of Ineq. (11.1) go to infinity as $n \to \infty$ and the argument is complete. ■

## 11.11 INFRABARRELED SPACES

We introduce and briefly discuss infrabarreled spaces here. Among the reasons for considering such spaces are results such as "an LCHS is reflexive iff it is semireflexive and infrabarreled" [Theorem 15.2.6]. Some other variations of barreledness are discussed at the end of the section.

**Definition 11.11.1** (a) BORNIVOROUS A subset $D$ of a TVS is *bornivorous* or a *bornivore* if it absorbs bounded sets, i.e., for any bounded set $B$, there is a positive number $r$ such that $B \subset aD$ for $|a| \geq r$.

(b) INFRABARRELED An LCS $X$ is *infrabarreled* (*quasi-barreled, evaluable*) if every bornivorous barrel is a neighborhood of 0. ■

Obviously, barreledness implies infrabarreledness. Normed spaces need not be barreled—$(C([0,1], \|\cdot\|_1)$ is not barreled, for example [Example 11.3.2]—but they are infrabarreled as we show in Theorem 11.11.2. More generally, every bornological (every bornivorous disk is a neighborhood of 0) space is infrabarreled.

**Theorem 11.11.2** NORMED IMPLIES INFRABARRELED In a normed space $X$, every bornivore is a neighborhood of 0; consequently, any normed space is infrabarreled.

**Proof.** Let $C(0,r) = \{x \in X : \|x\| \leq r\}$, $r > 0$. If $B$ is bornivorous then there is an $a > 0$ such that $C(0,r) \subset aB$; equivalently, $C(0, r/a) \subset B$. ■

We have already seen [Theorem 8.8.3] that $\sigma(X', X)$-boundedness in an LCS $X$ is equivalent to being a subset of the polar of a barrel. The analogous result for $\beta(X', X)$-boundedness follows. Recall that $\beta(X', X)$ is the polar topology generated by the set of all $\sigma(X, X')$-bounded sets [Example 8.5.5].

**Theorem 11.11.3** STRONG BOUNDEDNESS AND POLARS OF BARRELS Let $X$ be an LCHS. A subset $H$ of $X'$ is $\beta(X', X)$-bounded iff $H$ is contained in the polar of a bornivorous barrel $B$ in $X$.

**Proof.** We use the notation of the statement. To prove sufficiency, we show that the polar of each bornivorous subset $B$ of $X$ is $\beta(X', X)$-bounded, i.e., that $B^{\circ}$ is absorbed by the polar of any $\sigma(X, X')$-bounded subset $E$ of $X$. Since the bounded sets are the same in any topology of the dual pair [Theorem 8.8.7], such an $E$ is bounded in the original topology on $X$. Since $B$ is bornivorous, there is an $r > 0$ such that $E \subset rB$. Hence $B^{\circ} \subset rE^{\circ}$ and therefore $B^{\circ}$—and any subset thereof—is $\beta(X', X)$-bounded.

Necessity. Suppose that $H$ is $\beta(X', X)$-bounded. Consequently, $H$ is $\sigma(X', X)$-bounded. Therefore $H^{\circ}$ is a $\sigma(X, X')$-barrel by Theorems 8.3.5 and 8.3.6(a), hence a barrel in the original topology on $X$ by Theorem 8.8.1(a). Since $H$ is $\beta(X', X)$-bounded, $H$ must be absorbed by the polar of any $\sigma(X, X')$-bounded subset $B$ of $X$: There must be an $r > 0$ such that $H \subset rB^{\circ}$, which implies that $rH^{\circ} \supset B^{\circ\circ} \supset B$. It follows that $H^{\circ}$ is bornivorous and it only remains to observe that $H \subset H^{\circ\circ}$. ■

An LCS $X$ is barreled iff $\sigma(X', X)$-boundededness implies equicontinuity [Theorem 11.3.4]. For infrabarreled spaces, we have:

**Theorem 11.11.4** INFRABARRELED IFF $\beta(X', X)$-BOUNDEDNESS IMPLIES EQUICONTINUITY The LCHS $X$ is infrabarreled iff each $\beta(X', X)$-bounded subset of $X'$ is equicontinuous. [We observed in Theorem 11.3.5 that equicontinuity implies $\beta(X', X)$-boundedness.]

**Proof.** Suppose that $X$ is infrabarreled and that $H$ is a $\beta(X', X)$-bounded subset of $X'$. By Theorem 11.11.3, there is a bornivorous barrel $B$ in $X$ such that $H \subset B^{\circ}$. Since $X$ is infrabarreled, $B$ is a neighborhood of 0 and $H$ is equicontinuous by Theorem 8.6.4(a).

Conversely, suppose that $\beta(X', X)$-boundedness implies equicontinuity and let $B$ be a bornivorous barrel in $X$. Since $B$ is bornivorous, $B^{\circ}$ is $\beta(X', X)$-bounded by Theorem 11.11.3, hence equicontinuous. The polar of an equicontinuous set is a neighborhood of 0 by Theorem 8.6.6 so $B^{\circ\circ}$ is a neighborhood of 0. It only remains to observe that $B^{\circ\circ} = B$ by Theorem 8.8.1(c). ■

A relatively compact subset of a TVS is totally bounded [discussion before Theorem 3.4.1], hence bounded [Theorem 6.1.2]; hence a relatively $\sigma(X', X)$-compact set is $\sigma(X', X)$-bounded. For LCHS $X$, the bounded sets are the same in any topology of a dual pair [Theorem 8.8.7]; hence relatively $\sigma(X', X)$-compact sets are $\tau(X', X)$-bounded. We show next that relatively $\sigma(X', X)$-compact *convex* sets are even $\beta(X', X)$-bounded.

**Theorem 11.11.5** Let $X$ be an LCHS. Every $\sigma(X', X)$-relatively compact convex subset $H$ of $X'$ is $\beta(X', X)$-bounded [cf. Exercise 11.109].

**Proof.** We use the notation of the statement. Let $H$ be a $\sigma(X', X)$-relatively compact convex subset of $X'$; $\text{cl}_{\sigma(X',X)} H$ is convex by Theorem 4.4.2(g) and $\left(\text{cl}_{\sigma(X',X)} H\right)_{bc} = E$ is $\sigma(X', X)$-compact by a slight modification of the argument used in Theorem 4.4.4 (use $\{a_1, \ldots, a_n\}$ such that $\sum_{i=1}^{n} |a_i| \leq 1$ instead of $\sum_{i=1}^{n} a_i = 1$). We show next that $E$ is absorbed by the polar $B^\circ$ of any $\sigma(X, X')$-bounded set $B$. Such a $B$ must be a $\tau(X, X')$-bounded by Theorem 8.8.7. Hence, since $E$ is a $\sigma(X', X)$-compact disk, there must be an $r > 0$ such that $B \subset rE^\circ$ which implies that $H \subset E \subset rB^\circ$ and the result follows. ■

We showed in Theorem 11.3.6 that barreled spaces $X$ carry $\beta(X, X')$.

**Theorem 11.11.6** INFRABARRELED SPACES CARRY $\tau(X, X')$ Let $(X, \mathcal{T})$ be an LCHS. If $X$ is infrabarreled, then $\mathcal{T} = \tau(X, X')$.

**Proof.** Let the LCHS $(X, \mathcal{T})$ be infrabarreled. By Theorem 8.7.4, $\mathcal{T} \subset \tau(X, X')$. Conversely, if $B$ is a $\sigma(X', X)$-compact disk of $X'$, it is $\beta(X', X)$-bounded by Theorem 11.11.5. Since $X$ is infrabarreled, $B$ is equicontinuous by Theorem 11.11.4. Since [Theorem 8.6.6] $\mathcal{T}$ is the polar topology generated by the equicontinuous subsets of $X'$, $B^\circ \in \mathcal{T}$. ■

**Example 11.11.7** CLASS OF NONBARRELED SPACES If $(X, \mathcal{T})$ is an infinite-dimensional normed space then $\sigma(X, X')$ is strictly weaker than $\tau(X, X')$; consequently, $(X, \sigma(X, X'))$ is not infrabarreled, hence not barreled.

**Proof.** Suppose $(X, \mathcal{T})$ is an infinite-dimensional normed space. Then $\sigma(X, X')$ is not metrizable [Example 11.8.3]. Hence $\sigma(X, X')$ is strictly weaker than the norm topology. Since $\sigma(X, X') \subset \mathcal{T} \subset \tau(X, X')$ [Theorem 8.7.4], $\sigma(X, X')$ is strictly weaker than $\tau(X, X')$ as well. Therefore, $(X, \sigma(X, X'))$ is not infrabarreled by Theorem 11.11.6. ■

The following variations on "barrel" and "barreled" are in current use. Another is mentioned in Exercise 11.201.

**Definition 11.11.8** OTHER BARRELS An absorbent subset $B$ of a TVS $X$ is a

(a) *barrel* if it is a closed disk;

(b) *d-barrel* if it is the intersection of countably many closed disked neighborhoods of 0;

(c) *σ-barrel* if it is an intersection $\bigcap_{n \in \mathbf{N}} \{x : p_n(x) \leq r_n\}$ of denumerably many closed semiballs where each $p_n$ is a $\sigma(X, X')$-continuous seminorm and each $r_n > 0$. ■

**Definition 11.11.9** OTHER KINDS OF BARRELEDNESS A locally convex space is:

(a) *barreled* if each barrel is a neighborhood of 0;

(b) *d-barreled* if each $d$-barrel is a neighborhood of 0;

(c) *$\sigma$-barreled* if each $\sigma$-barrel is a neighborhood of 0;

(d) *infrabarreled* (*evaluable*) if each bornivorous barrel is a neighborhood of 0;

(e) *d-infrabarreled* (*d-evaluable*) if each bornivorous $d$-barrel is a neighborhood of 0;

(f) *$\sigma$-infrabarreled* (*d-evaluable*) if each bornivorous $\sigma$-barrel is a neighborhood of 0. ■

Call an LCS $(X, \mathcal{T})$ *Mackey* if $\mathcal{T} = \tau(X, X')$. We have the following irreversible relations between the various kinds of barreledness [Schmets 1976, p. 7ff]:

$$\begin{array}{ccccc}
\text{barreled} & \Longrightarrow & d\text{-barreled} & \Longrightarrow & \sigma\text{-barreled} \\
\Downarrow & & \Downarrow & & \Downarrow \\
\text{infrabareled} & \Longrightarrow & d\text{-infrabarreled} & \Longrightarrow & \sigma\text{-infrabarreled} \\
\Downarrow & & & & \\
\text{Mackey} & & & &
\end{array}$$

# 11.12 PERMANENCE PROPERTIES

The major result of this section is that subspaces of countable codimension of barreled spaces are barreled [Theorem 11.12.9]. Dieudonné [1952, 1981b, p. 346] first proved that *finite*-codimensional subspaces $M$ of barreled spaces $X$ are barreled. The proof is simplified by showing [Theorem 11.12.7(a)] that if $B$ is a barrel in a finite-codimensional subspace $M$, there is a barrel $B'$ in $X$ such that $B = B' \cap M$. The barreledness of subspaces of *countable* codimension was obtained independently by Valdivia [1971b] and Saxon and Levin [1971]. The argument given in Theorem 11.12.9 is a simplification due to Webb [1981]. Levin and Saxon [1971] consider some other inheritance properties of subspaces of countable codimension.

**Theorem 11.12.1** DENSE BARRELED SUBSPACE $\Longrightarrow$ BARRELED If a TVS $X$ has a dense barreled subspace then $X$ is barreled. Thus the completion of a barreled space is barreled. (This result is strengthened in Exercise 11.115.)

**Proof.** Let $M$ be a dense barreled subspace of the TVS $X$. If $S$ is a dense subset of a topological space $T$ then the closure $\text{cl}_T V$ in $T$ of a neighborhood $V$ in $S$ is a neighborhood in $T$ [Exercise 11.101]. Since $M$ is barreled it is locally convex; since closures of convex sets are convex [Theorem 4.4.2(g)], it follows that $X$ is locally convex. If $B$ is a barrel in $X$, then $B \cap M$ is a barrel in $M$, therefore a neighborhood of 0 in $M$. By the remark above, $\text{cl}\,(B \cap M)$

is a neighborhood of 0 in $X$. Since $\mathrm{cl}\,(B \cap M) \subset B$, $B$ is a neighborhood of 0 in $X$. ■

Not every dense subspace of a barreled space is barreled, however, as shown in Example 11.12.5.

We already know that the finest locally convex topology is barreled [see after Def. 11.3.1]. Theorem 11.12.2 is a stronger result that immediately yields the barreledness and infrabarreledness of quotients and locally convex direct sums of barreled and infrabarreled spaces, respectively [Theorem 11.12.3].

**Theorem 11.12.2** DIRECT IMAGE LC TOPOLOGIES BARRELED For each $i \in I$, let $A_i$ be a surjective linear map from a TVS $X_i$ into a vector space $X$. Let $\mathcal{T}$ be the direct image locally convex topology for $X$ generated by $\{A_i : i \in I\}$ of Example 4.11.3(b). Then:

(a) BARRELED If each $X_i$ is barreled, then $(X, \mathcal{T})$ is barreled.

(b) INFRABARRELED If each $X_i$ is infrabarreled, then $(X, \mathcal{T})$ is infrabarreled.

**Proof.** We use the notation of the statement. (a) As discussed in Example 4.11.3(b), a base at 0 for the direct image locally convex topology $\mathcal{T}$ is given by the filterbase of all absorbent disks $B \subset X$ such that each $A_i^{-1}(B)$ is a neighborhood of 0 in $X_i$. If $B$ is a barrel in $(X, \mathcal{T})$ then each $A_i^{-1}(B)$ is absorbent, balanced, and convex by Theorems 4.1.8(c) and 4.2.4(b); $A_i^{-1}(B)$ is closed because $A_i$ is continuous. Thus each $A_i^{-1}(B)$ is a barrel in $X_i$ and therefore a neighborhood of 0 in $X_i$. It follows that $B$ is a $\mathcal{T}$-neighborhood of 0 and proves part (a).

(b) If $B$ is a bornivorous barrel in $X$, then so is each $A_i^{-1}(B)$. Since each $X_i$ is infrabarreled, each $A_i^{-1}(B)$ is a neighborhood of 0 in $X_i$. ■

Since quotient [Examples 4.7.2 and 4.11.3(b)] and locally convex direct sum topologies [Example 4.11.4] are each direct image locally convex topologies, Theorem 11.12.3 follows immediately from Theorem 11.12.2.

**Theorem 11.12.3** QUOTIENTS AND LC DIRECT SUMS Quotients and locally convex direct sums of barreled or infrabarreled spaces are barreled or infrabarreled, respectively.

To have it near its closest relatives, we record the result about products here. For a proof, see Robertson and Robertson [1973, p. 94] or Jarchow [1981, p. 223].

**Theorem 11.12.4** PRODUCTS Any product of barreled or infrabarreled spaces is barreled or infrabarreled, respectively.

**Example 11.12.5** NONBARRELED DENSE SUBSPACE OF A BARRELED SPACE Endow $\mathbf{R}^{\mathbf{N}}$ with the product topology and let $M$ be the subspace of all bounded sequences. As a power of complete spaces, $\mathbf{R}^{\mathbf{N}}$ is complete [Example 3.2.3]. Let $\mathrm{pr}_n$ denote the projection $(a_j) \mapsto a_n$ of $\mathbf{R}^{\mathbf{N}}$ onto $\mathbf{R}$. Since the product topology is determined by the countable family of seminorms $|\mathrm{pr}_n|$, $n \in \mathbf{N}$, it is metrizable by Theorem 5.6.1 (and the discussion before). Therefore $\mathbf{R}^{\mathbf{N}}$ is nonmeager by the Baire category theorem [11.7.2]. As a product of LCS, it is locally convex [Theorem 5.5.3], hence barreled [Theorem 11.6.4]. Moreover, since any basic open set in $\mathbf{R}^{\mathbf{N}}$ meets $M$, $M$ is dense in $\mathbf{R}^{\mathbf{N}}$; $M$ is not barreled because $B = \{(t_n) \in X : \sup_n |t_n| \leq 1\}$ is a barrel in $M$ but not a neighborhood of 0. (By Theorems 11.12.8 and 11.12.9, it will follow that $M$ cannot be of countable codimension.) ■

Example 11.12.5 shows that barreled spaces may have subspaces that are not barreled. The fact that Fréchet spaces are generally barreled allows us to generate a class of such spaces—namely, a metric completion of a nonbarreled metrizable LCS. Arbitrary subspaces of barreled spaces need not be barreled, but what about *closed* subspaces? Here, too, the answer is generally negative. Each LCHS $X$ is linearly homeomorphic to a subspace of a product $Y$ of Banach spaces [Theorem 5.11.4(b)]. Since Banach spaces are barreled, $Y$ is barreled [Theorem 11.12.4]. Thus if there is a *complete* LCHS which is not barreled, we have an example of a closed, nonbarreled subspace of a barreled space. Such an example follows.

**Example 11.12.6** Complete Nonbarreled Space Let $T$ be a nondenumerable set and let $X = \{x \in \mathbf{R}^T : x(t) = 0 \text{ for almost all } t\}$ endowed with the relative topology induced by the box topology on $\mathbf{R}^T$ [Exercise 4.112]. As noted in Exercise 4.112(c), this topology makes $X$ into a complete LCS. It is straightforward to verify that $B = \{x \in X : \sum_{t \in T} |x(t)| \leq 1\}$ is a barrel. It is not a neighborhood of 0 by the following argument. Given the basic box neighborhood of 0, $U = \Pi_{t \in T} [-r_t, r_t]$ where each $r_t$ is a positive number, there is a finite subset $H \subset T$ such that $\sum_{t \in H} r_t > 1$—if not, then for all $n \geq 2$, the set $A_n = \{t \in T : r_t > 1/n\}$ must have fewer than $n$ elements which implies that $\{r_t : t \in T\} = \cup_{n \in \mathbf{N}} A_n$ is denumerable. Hence, let $H$ be a finite subset such that $\sum_{t \in H} r_t > 1$. The function $x$ such that $x(t) = r_t$ for $t \in H$, $x(t) = 0$ otherwise, is an element of $U \cap X$ but not of $B$. Since $B$ cannot contain any $U \cap X$, $B$ is not a neighborhood of 0. ■

Having mentioned some negative results about subspaces of barreled spaces, we now turn toward the positive side. The results below show that sufficiently large subspaces of barreled spaces—large in the sense that they are of small codimension—are barreled. It is easy to verify that the projection $B \cap M$ of a barrel $B$ in a TVS $X$ onto a subspace $M$ is a barrel. The key to showing that a finite-codimensional subspace of a barreled or infrabarreled space is barreled or infrabarreled, respectively, is the following result that asserts that all barrels in subspaces are of this type. Lemma 12.1.2 is similar: if $U$ is a disked neighborhood of 0 in $M$—with no hypothesis of finite codimensionality—then there is a disked neighborhood $V$ of 0 in $X$ such that $U = V \cap M$.

**Theorem 11.12.7** BARRELS, ETC., IN FINITE-CODIMENSIONAL SUBSPACES Let $M$ be a finite-codimensional subspace of the LCS $X$. If $B$ is a

(a) barrel in $M$, there is a barrel $B'$ in $X$ such that $B = B' \cap M$;

(b) bornivorous disk in $M$, there is a bornivorous disk $B'$ in $X$ such that $B = B' \cap M$;

(c) bornivorous barrel in $M$, there is a bornivorous barrel $B'$ in $X$ such that $B = B' \cap M$.

**Proof.** We use the notation of the statement. (a) If we can prove the assertion for subspaces of codimension 1, it will suffice: By an induction argument, it will then hold for subspaces of finite codimension. So assume that $M$ is of codimension 1 and let $B$ be a barrel in $M$. Let $W$ be the linear span $[\mathrm{cl}_X B]$ of the closure of $B$ in $X$. Since $B$ is absorbent in $M$, $M \subset [\mathrm{cl}_X B] = W$. Consider two possibilities for $W$:

(1) If $W = X$, any $y \in X$ can be written as $y = \sum_{i=1}^{n} a_i x_i$ ($a_i \in \mathbf{F}$, $x_i \in \mathrm{cl}_X B$). Since $B$ is a disk, so is $\mathrm{cl}_X B$ and therefore, for $|c| \geq \sum_{i=1}^{n} |a_i|$, $y \in c(\mathrm{cl}_X B)$ [Theorem 4.2.9], i.e., $\mathrm{cl}_X B$ is absorbent, and $\mathrm{cl}_X B$ is a barrel in $X$. Since $B$ is closed in $M$, $(\mathrm{cl}_X B) \cap M = B$ and we may take $B'$ to be $\mathrm{cl}_X B$.

(2) If $W \neq X$, then $B$ must be closed in $X$ (not just $M$). Choose $x \in X$ such that $X = W \oplus \mathbf{F}x = M \oplus \mathbf{F}x$. Let $D = \{a \in \mathbf{F} : |a| \leq 1\}$ and note that $Dx$ is a closed disk in $X$. We now show that $B' = B + Dx$ is the desired barrel. Since $B' \cap M = (B + Dx) \cap M = B$, all that remains to be verified is that $B'$ is a barrel in $X$.

First note that $B'$ is closed because it is the sum of a closed and a compact set [Theorem 4.4.1(a)]. Generally, finite sums of disks are disks, so $B'$ is a disk. To see that $B'$ is absorbent, suppose that $w \in X = M \oplus \mathbf{F}x = [B] \oplus \mathbf{F}x$. Choose $u \in [B]$ and a scalar $t$ such that $w = u + tx$. Since $B$ is absorbent in $M$, then $au \in B$ for scalars $a$ of sufficiently small magnitude; hence $aw = au + atx \in B + Dx = B'$ and the proof of (a) is complete.

(b) As in the proof of (a), it suffices to consider subspaces of codimension 1. Choose $x \in X$ such that $X = M \oplus \mathbf{F}x$ and let $B$ be a bornivorous disk in $M$. Let $P$ be the projection $m + tx \mapsto m$ of $X = M \oplus \mathbf{F}x$ onto $M$. There are two possibilities for $P$: It is locally bounded (maps bounded sets into bounded sets) or it is not. If it is locally bounded, then $P^{-1}(B)$ is a bornivorous disk in $X$, as is straightforward to verify. Moreover, $P^{-1}(B) \cap M = \{m \in M : m \in B\} = B$, so $B' = P^{-1}(B) = B \oplus \mathbf{F}x$ is the desired bornivorous disk in $X$.

Now suppose that $P$ is not locally bounded. By Theorem 6.1.4(b) then there is a bounded sequence $(x_n)$ in $X$ such that $(m_n) = (Px_n)$ is not bounded. For $x_n = m_n + t_n x$ ($m_n \in M$, $t_n \in \mathbf{F}$), then $(t_n)$ must be unbounded, too—if not, then $(Px_n) = (m_n) = (x_n - t_n x)$ would be bounded. By extracting a subsequence, if necessary, we may suppose that $|t_n| \geq 1$ for every $n$ and that $|t_n| \to \infty$. Let $H$ be the balanced convex hull of $\{x, x_1, \ldots\}$ and let

$y_n = -Px_n/t_n = -m_n/t_n$. For every $n \in \mathbf{N}, y_n \in M$ and

$$y_n - x = -m_n/t_n - x = -x_n/t_n \in (1/t_n)\, H \subset H \qquad (\clubsuit)$$

Let $\mathcal{B}$ denote the class of all bounded disks in $X$ containing the bounded set $H$. For a set to be bornivorous in $X$, it suffices for it to absorb the sets in $\mathcal{B}$, for if $J$ absorbs each set in $\mathcal{B}$ and $L$ is any bounded set in $X$, $(L \cup H)_{bc}$ is bounded and is in $\mathcal{B}$. As such, $J$ must absorb it and therefore $J$ absorbs $L$.

Let $D \in \mathcal{B}$. Since $D \cap M$ must be bounded in $M$ and $B$ is bornivorous in $M$, there must be $r_D > 0$ such that $2r_D\,(D \cap M) \subset B$. Let $G = (\cup_{D \in \mathcal{B}} r_D D)_{bc}$. Since $G$ absorbs each set in $\mathcal{B}$, $G$ is bornivorous. We now show that

$$G \cap M \subset B \qquad (*)$$

Each $w \in G \cap M$ is a finite sum of the form $\sum_i a_i\,(m_i + b_i x)$ where $a_i$ and $b_i$ are scalars such that $\sum_i |a_i| \leq 1$, $\sum_i a_i b_i = 0$ (since $x \notin M$), $m_i \in M$, and $m_i + b_i x \in r_{D_i} D_i$. For each $i$, $H \subset D_i$; hence, since $|b_i/t_n| < r_{D_i}$ for sufficiently large $n$,

$$(b_i/t_n)\, H \subset (b_i/t_n)\, D_i \subset r_{D_i} D_i$$

By this inclusion, and since $m_i + b_i x \in r_{D_i} D_i$ for each $i$ and $(\clubsuit)$, for sufficiently large $n$,

$$m_i + b_i y_n = m_i + b_i x + b_i\,(y_n - x) \in r_{D_i} D_i + (b_i/t_n)\, H \subset (2r_{D_i} D_i) \cap M \subset B$$

Since $\sum_i a_i b_i = 0$, $\sum_i a_i\,(m_i + b_i x) = \sum_i a_i\,(m_i + b_i y_n) \in B$, which proves $(*)$.

Let $B' = (G \cup B)_{bc}$. Since $G$ is bornivorous, so is $B'$. Obviously, $B'$ is a disk. That $B' \cap M \subset B$ follows from $(*)$ and the fact that $(G \cup B) \cap M = (G \cap M) \cup (B \cap M) \subset B$. Since $B \subset (G \cup B)_{bc} \cap M = B'$, the proof of (b) is complete.

(c) Let $B$ be a bornivorous barrel in $M$. As such, $B$ is a closed bornivorous disk in $M$. By (b), there is a bornivorous disk $D$ in $X$ such that $B = D \cap M$. Therefore $B' = \mathrm{cl}_X\, D$ is the desired bornivorous barrel in $X$. ∎

By Theorem 11.12.7(a,c), it is clear that:

**Theorem 11.12.8** SUBSPACE OF FINITE CODIMENSION A finite-codimensional subspace of a barreled or infrabarreled space is barreled or infrabarreled, respectively.

**Theorem 11.12.9** SUBSPACES OF DENUMERABLE CODIMENSION A subspace $M$ of denumerable codimension of a barreled space $X$ is barreled.

**Proof.** Let $X$ and $M$ be as in the statement. Since $M$ is of denumerable codimension, there is an increasing sequence $(M_n)$ of subspaces of $X$ such that $M_1 = M$, the codimension of each $M_n$ in $M_{n+1}$ is 1 and $X = \cup_n M_n$ (as were constructed in Example 11.8.1, for example). Given a barrel $B$ in $M$, by Theorem 11.12.7(a), there is a sequence $(B_n)$ of barrels, starting with $B_1 = B$ in $M$, and $B_n$ in $M_n$ such that $B_{n+1} \cap M_n = B_n$ for each $n$. Clearly $V = \cup_n B_n$ is an absorbent disk in $X$ so $\operatorname{cl} V$ is a barrel in $X$. Since $X$ is barreled, $\operatorname{cl} V$ is a neighborhood of 0. In the remainder of the argument we show that $\operatorname{cl} V \subset 2V$, from which it follows that $V$ is a neighborhood of 0. Since $B = V \cap M$, as a routine verification will show, it follows that $B$ is a neighborhood of 0 in $M$.

For $z \notin 2V = \cup_n 2B_n$, clearly $z \notin 2B_n$ for any $n$. If we view each $M_n$ as a real vector space, there exists a real continuous linear functional $f_n$ on $M_n$ such that $\sup f_n(2B_n) < f_n(z)$ by a consequence [Theorem 7.8.6(b)] of the Hahn–Banach theorem. Clearly, we may suppose that $f_n(z) = 2$, so that $\sup f_n(B_n) \leq 1$ for every $n$; since $B_n$ is balanced, $\sup |f_n(B_n)| \leq 1$. By Corollary 7.3.3 we may assume that each $f_n$ has been extended to a member of $X'$.

We now contend that $\{f_n : n \in \mathbf{N}\}$ is $\sigma(X', X)$-bounded on $X = \cup_n M_n$. To see this, choose $x \in \cup_n M_n$ and $n \in \mathbf{N}$ such that $x \in M_n$. Since $B_n$ is absorbent in $M_n$, it follows that $tx \in B_n$ for some $t > 0$. Since the barrels $B_n$ are increasing, it follows that $|f_k(x)| \leq 1/t$ for all $k \geq n$. Thus, for all $j \in \mathbf{N}$, $|f_j(x)| \leq \max(1/t, |f_1(x)|, \ldots, |f_{n-1}(x)|)$. Having shown that $\{f_n\}$ is $\sigma(X', X)$-bounded, there exists a barrel $U$ in $X$ such that $\{f_n\} \subset U^\circ$ [Theorem 8.8.3]. Since $X$ is barreled, $U$ must be a neighborhood of 0 in $X$; hence $U^\circ$ is $\sigma(X', X)$-compact by the Alaoglu theorem [8.4.1]. Consequently, $(f_n)$ has a $\sigma(X', X)$-cluster point $f \in X'$ [Theorem 1.1.10(b)]. We now verify that (1) $f(z) = 2$ and (2) $f \in V^\circ = (\operatorname{cl} V)^\circ$.

(1) PROOF THAT $f(z) = 2$: Given $r > 0$, the $\sigma(X', X)$-neighborhood $V(f, z, r) = \{g \in X' : |f(z) - g(z)| \leq r\}$ must contain some $f_n$. It follows that $|f(z) - f_n(z)| = |f(z) - 2| \leq r$. Since $r$ is arbitrary, $f(z) = 2$.

(2) PROOF THAT $f \in V^\circ$: Given $r > 0$, suppose that $x \in V = \cup_n B_n$. The $\sigma(X', X)$-neighborhood $V(f, x, r)$ must contain $f_n$ for infinitely many $n$. Choose $n$ such that $f_n \in V(f, x, r)$ and $x \in B_n$. Since $\sup f_n(B_n) \leq 1$, $|f(x)| \leq |f_n(x)| + r \leq 1 + r$. It follows that $|f(x)| \leq 1$.

By the continuity of $f$, it is now clear that $z \notin \operatorname{cl} V$. Therefore, $\operatorname{cl} V \subset 2V$ and the theorem is proved. ■

## 11.13 INCREASING SEQUENCES OF DISKS

The algebraic closure $\operatorname{acl} A$ of a subset $A$ of a vector space $X$ is the union of $A$ and those points which are linearly accessible from $A$—those $x \in X$ for which there exists $y \in A$ such that the line segment $[y, x) \subset A$ [Exercise 4.111]. Theorem 11.13.2 shows that if an increasing sequence $(D_n)$ of disks satisfies a certain condition, the topological closure of $\cup_{n \in \mathbf{N}} D_n$ coincides with its algebraic closure. We use it to prove the completeness of strict inductive limits of complete spaces in Theorem 12.1.10. We place it here because of the features the proof shares with Theorem 11.12.9.

**Theorem 11.13.1** Algebraic Closure of a Disk  If $D$ is a disk in a vector space $X$, then $\operatorname{acl} D = \cap_{r>0}(1+r)D$.

**Proof.** Let $D$ be a disk. Note that for all $0 < a < b$, $(1+a)D \subset (1+b)D$ and suppose that $x \in \cap_{r>0}(1+r)D$. For any $r > 0$ and $t = 1/(1+r)$, then $tx \in D$, i.e., $[0,x) \subset D$ so $x \in \operatorname{acl} D$. Conversely, suppose that $r > 0$ and that $x \in \operatorname{acl} D$. Let $y \in D$ be such that $[y,x) \subset D$ and choose $t \in [0, r/(1+r)]$. Since $ty + (1-t)x \in D$ and $-y \in D$, it follows that, for any $c \in [0,1]$,

$$c(-y) + (1-c)[ty + (1-t)x] = [-c + (1-c)t]y + (1-c)(1-t)x \in D$$

For $c = t/(1+t)$, the coefficient $-c + (1-c)t$ of $y$ above is 0 and therefore $(1-c)(1-t)x = (1-t)/(1+t)x \in D$. Since $t \le r/(1+r)$, $1/(1+r) \le (1-t)/(1+t)$. Since $D$ is a disk, it follows that $[1/(1+r)]x \in D$ and therefore that $x \in (1+r)D$. ■

**Remark.** $\mathcal{B}+V(0)$. If $\mathcal{B}$ is a Cauchy filterbase in a TVS and $V(0)$ is the filter of neighborhoods of 0 then the filterbase $\mathcal{B}+V(0) = \{B+V : B \in \mathcal{B}, V \in V(0)\}$ is a Cauchy filterbase by the following argument: Let $V \in V(0)$ and choose a balanced $W \in V(0)$ such that $W+W+W \subset V$. Pick $B \in \mathcal{B}$ such that $B - B \subset W$. Then $(B+W)-(B+W) \subset W+W+W \subset V$. ■

**Theorem 11.13.2** Closures of Increasing Sequences of Disks  Let $X$ be a locally convex space with neighborhood filter $V(0)$ at 0, and let $(D_n)$ be an increasing sequence of disks of some subspace $M$ of $X$ such that

$$\text{for any } f_n \in D_n^{\circ},\ \{f_n : n \in \mathbf{N}\} \text{ is equicontinuous on } M \qquad (*)$$

Then:

(a) $\operatorname{cl}(\cup_n D_n) = \cap_{r>0}(1+r)(\cup_n \operatorname{cl} D_n)$, the algebraic closure of $\cup_n \operatorname{cl} D_n$ or, equivalently,

(b) if $\mathcal{B}$ is a Cauchy filterbase in $\cup_n D_n$, then, given any $r > 0$, $\mathcal{B}+V(0)$ induces a Cauchy filterbase on $(1+r)D_m$ for some positive integer $m$.

**Remark.** If $(D_n)$ is an increasing sequence of subspaces of $M$, then each $D_n^{\circ} = \{0\}$, so the sequence $\{f_n\}$ of $(*)$ is just $\{0\}$. As $\{0\}$ is equicontinuous, $(*)$ is trivially satisfied.

**Proof.** We use the notation of the statement. View the LCS $X$ as a real vector space.

$(*) \Rightarrow$ (a) To say that $y \in \operatorname{acl} C$, the algebraic closure of the set $C$, means that there exists $x \in C$ such that $[x,y) \subset C$. Hence $y \in \operatorname{cl} C$ and $\operatorname{acl} C \subset \operatorname{cl} C$. In particular, $\operatorname{acl}(\cup_n \operatorname{cl} D_n) \subset \operatorname{cl}(\cup_n \operatorname{cl} D_n)$. By general topological considerations, $\cup_n \operatorname{cl} D_n \subset \operatorname{cl}(\cup_n D_n)$, so $\operatorname{cl}(\cup_n \operatorname{cl} D_n) \subset \operatorname{cl}(\cup_n D_n)$. It follows that $\operatorname{acl}(\cup_n \operatorname{cl} D_n) \subset \operatorname{cl}(\cup_n D_n)$.

To show that $\operatorname{cl}(\cup_n D_n) \subset \operatorname{acl}(\cup_n \operatorname{cl} D_n)$, we show that $\operatorname{cl}(\cup_n D_n) \subset (1+r)(\cup_n \operatorname{cl} D_n)$ for each $r > 0$ [Theorem 11.13.1]. We argue by contrapositive. Suppose that for some $r > 0$, $x \notin (1+r)(\cup_n \operatorname{cl} D_n)$. By the Hahn–Banach theorem [Theorem 7.8.6(b)], for each $n$ there exists a real linear functional $f_n \in D_n^\circ$ such that $f_n(x) = 1 + r$. By hypothesis, $\{f_n : n \in \mathbf{N}\}$ is therefore equicontinuous on $M$. If $\operatorname{cl} M = X$, then, as we show next,

$$S = \{f_n : n \in \mathbf{N}\} \text{ is equicontinuous on } X \qquad (**)$$

Since $S \subset X'$ is equicontinuous on $M$, then given $d > 0$, there is a neighborhood $V$ of 0 in $X$ such that $|f_n(v)| < d$ for each $f_n \in S$ and $v \in V$. By the continuity of $f_n$, $|f_n(v)| \leq d$ for each $f_n \in S$ and $v \in \operatorname{cl}_X V$. By the density of $M$, $\operatorname{cl}_X V$ is a neighborhood of 0 in $X$ [Exercise 11.101] and therefore $S$ is equicontinuous on $X$.

Now, returning to the original argument, since $\{f_n : n \in \mathbf{N}\}$ is equicontinuous on $X$, it is $\sigma(X', X)$-relatively compact by Theorem 8.6.5 and therefore has a $\sigma(X', X)$-cluster point $f$. Since $f_n(x) = 1 + r$ for every $n$, as in the proof of (1) (proof that $f(z) = 2$) in Theorem 11.12.9, $f(x) = 1 + r$. As in (2) of that same argument, $f \in (\cup_n D_n)^\circ = [\operatorname{cl}(\cup_n D_n)]^\circ$. We conclude that $x \notin \operatorname{cl}(\cup_n D_n)$ which proves (a) when $M$ is dense in $X$.

If $\operatorname{cl} M \neq X$, the previous argument shows that $\{f_n : n \in \mathbf{N}\}$ is equicontinuous on $\operatorname{cl} M$. By Theorem 8.6.3 and the fact that the continuous seminorms on $M$ are restrictions to $M$ of continuous seminorms on $X$, there is a continuous seminorm $p$ on $X$ such that for each positive integer $n$, $|f_n| \leq p$ on $M$. By a consequence of the Hahn–Banach theorem [7.3.2(b)], we may assume that $|f_n| \leq p$ on all of $X$ for each $n \in \mathbf{N}$. In other words, using Theorem 8.6.3 again, we may assume that $\{f_n : n \in \mathbf{N}\}$ is equicontinuous on $X$. The same argument as follows $(**)$ above shows that $x \notin \operatorname{cl}(\cup_n D_n)$ and completes the proof of (a).

(a) $\Rightarrow$ (b) Let $r > 0$ be given and let the LCS $Y$ be a completion of $M$ [Theorem 5.11.5] so that a Cauchy filterbase $\mathcal{B}$ in $\cup_n D_n$ converges to some $x \in Y$. It follows that $x \in \operatorname{cl}_Y(\cup_n D_n) = \cap_{r>0} (1+r)(\cup_n \operatorname{cl}_Y D_n)$. Hence, for some positive integer $m$, $x \in (1+r) \operatorname{cl}_Y D_m$.

Let $U \in \mathcal{V}(0)$ and choose a balanced neighborhood $V$ of 0 such that $V + V \subset U$. Since $\mathcal{B} \to x$, there exists $B^* \in \mathcal{B}$ such that $B^* \subset x + V$. For any $B \in \mathcal{B}$, there exists $B' \in \mathcal{B}$ such that $B' \subset B^* \cap B$. Hence $B' \subset B^* \subset x + V$. Since $V$ is balanced, this implies that $x \in B' + V$, so $x + V \subset B' + U \subset B + U$. Since $x \in (1+r) \operatorname{cl}_Y D_m$, $x + V$ must meet $(1+r)D_m$; hence, so must $B + U$; in other words, each set $B + U$, $(B \in \mathcal{B},\ U \in \mathcal{V}(0))$, meets $(1+r)D_m$, so we can consider the trace [Example 1.1.3(d)] of $\mathcal{B} + \mathcal{V}(0)$ on $(1+r)D_m$. As noted in the Remark before this theorem, $\mathcal{B} + \mathcal{V}(0)$ is a Cauchy filterbase; so, therefore, is its trace on $(1+r)D_m$.

(b) $\Rightarrow$ (a) Let $r > 0$ be given. Let $x \in \operatorname{cl}(\cup_n D_n)$ and consider the filterbase $\mathcal{B} = \{(x + V) \cap (\cup_n D_n) : V \in \mathcal{V}(0)\}$ of neighborhoods of $x$ in $\cup_n D_n$. Since

$\mathcal{B} \to x$, $\mathcal{B}$ is Cauchy. Moreover, the filterbase $\mathcal{B} + V(0)$ induces $\mathcal{B}$ again on $\cup_n D_n$. So, by hypothesis, $\mathcal{B}$ is a Cauchy filterbase in $(1+r)D_m$ for some positive integer $m$. Hence $x \in (1+r)\operatorname{cl} D_m$ and, as argued in the proof of $(*) \Rightarrow (a)$ above, it follows that $\operatorname{cl}(\cup_n D_n)$ is the algebraic closure of $\cup_n \operatorname{cl} D_n$. ■

## 11.14 EXERCISES

11.101 TOPOLOGICAL RESULT If $S$ is a dense subset of a topological space $T$ and $V$ is a relative neighborhood of $t \in S$ in $S$, then $\operatorname{cl}_T V$ is a neighborhood of $t$ in $T$.

11.102 RARE SETS

(a) In any topological group, a subgroup which is closed and not open is rare.

(b) A disk $D$ in a TVS is not rare iff $\operatorname{cl} D$ is a neighborhood of 0.

11.103 NORMED SPACES NEED NOT CARRY $\beta(X, X')$ Show that a normed space $X$ need not carry $\beta(X, X')$. Note that the norm topology on $X'$ must be $\beta(X', X)$ [Example 8.8.9].

11.104 NONMEAGER SPACES A topological space $T$ is nonmeager iff every countable family of dense open sets has nonempty intersection [cf. Theorem 11.6.6].

11.105 BAIRE SPACES (a) Let $\mathcal{T}^*$ denote the nonempty open sets of the topological space $(T, \mathcal{T})$. Show that $T$ is a Baire space if there exists a function $f : \mathcal{T}^* \to \mathcal{T}^*$ which is:

(1) *reducing* in the sense that $f(G) \subset G$ for each $G$ in $\mathcal{T}^*$, and

(2) if $(G_n)$ is a sequence of nonempty open sets such that $f(G_n) \supset G_{n+1}$ for every $n$, then $\cap_n G_n \neq \varnothing$.
Use this to show that:

(a) $(0, 1)$ is a Baire space in its usual topology.

(b) A locally countably compact (each point has a countably compact neighborhood) regular Hausdorff space is Baire.

(c) A *pseudocompact* topological space $T$ is one on which each continuous $\mathbf{R}$-valued function is bounded. Show that a completely regular pseudocompact space is Baire.

11.106 INCOMPLETE SPACE Show that $C([0,1], \mathbf{F}, \|\cdot\|_1)$ is not complete.

11.107 EQUICONTINUITY AND COMPACTNESS Let $X$ be an LCHS. We already know that if $X$ is barreled then a subset $E$ of $X'$ is relatively $\sigma(X', X)$-compact iff $E$ is equicontinuous [Theorem 11.3.4].

(a) Show that an LCHS $X$ carries $\tau(X, X')$ iff each relatively $\sigma(X', X)$-compact *convex* subset of $X'$ is equicontinuous.

(b) For any LCHS $X$, a relatively $\sigma(X', X)$-compact *convex* subset of $X'$ is $\beta(X', X)$-bounded.

11.108 WEAK CAUCHY IMPLIES BOUNDED Let $X$ be a normed space and let $(x_n)$ be a $\sigma(X, X')$-Cauchy sequence. Show that $(x_n)$ is bounded in the norm topology on $X$.

11.109 CASE WHEN $\sigma(X', X)$-BOUNDED IMPLIES $\beta(X', X)$-BOUNDED If $X$ is a quasi-complete [closed bounded subsets are complete, Exercise 6.109] LCHS, then a $\sigma(X', X)$-bounded subset $H \subset X'$ is $\beta(X', X)$-bounded.

11.110 BOUNDED AND EQUICONTINUOUS SUBSETS OF $L(X, Y)$ Let $X$ and $Y$ be HTVS, let $\mathcal{S}$ be a collection of bounded subsets of $X$, and let $H$ be a subset of the TVS $L(X, Y)$ of all continuous linear maps of $X$ into $Y$ endowed with the $\mathcal{S}$-topology [Sec. 11.2]. Then (a)–(c) are equivalent.

(a) $H$ is bounded.

(b) For each neighborhood $V$ of 0 in $Y$, $\cap\{A^{-1}(V) : A \in H\}$ absorbs each $S$ in $\mathcal{S}$.

(c) For each $S$ in $\mathcal{S}$, $\cup\{A(S) : A \in H\}$ is bounded in $Y$.

Statements (d)–(f) are also mutually equivalent (to each other, not to the statements above).

(d) $H$ is equicontinuous.

(e) For each neighborhood $V$ of 0 in $Y$, $\cap\{A^{-1}(V) : A \in H\}$ is a neighborhood of 0 in $X$.

(f) For each neighborhood $V$ of 0 in $Y$ there is a neighborhood $U$ of 0 in $X$ such that $\cup\{A(U) : A \in H\} \subset V$.

(g) If $X$ and $Y$ are LCHS and $H$ is an equicontinuous subset of $L(X, Y)$ then the restrictions to $H$ of the topology of pointwise convergence [$\mathcal{S}$ consists of all singletons in $X$] and the topology of precompact convergence [$\mathcal{S}$ consists of the totally bounded subsets of $X$] are the same.

11.111 $\varphi$ NOT BARRELED Show that the space $(\varphi, \|\cdot\|_\infty)$ of finite sequences is not barreled by showing that the continuous linear functionals $f_n$

$$f_n((a_j)) = na_n, \quad n \in \mathbf{N}, (a_j) \in \varphi$$

are pointwise bounded but not equicontinuous.

11.112 If $H$ is a subset of the dual $X'$ of a barreled LCHS $X$, show that the following statements are equivalent:

(a) $H$ is equicontinuous.

(b) $H$ is relatively $\sigma(X', X)$-compact.

(c) $H$ is $\beta(X', X)$-bounded.

(d) $H$ is $\sigma(X', X)$-bounded.

11.113 BARRELED CRITERION The LCS $(X, \mathcal{T})$ is barreled iff any locally convex topology $\mathcal{T}^*$ which has a base of $\mathcal{T}$-closed neighborhoods of 0 is weaker than $\mathcal{T}$.

11.114 INFRABARRELED SPACES If $X$ is an LCHS, then the following statements are equivalent:

(a) $X$ is infrabarreled.

(b) Each $\beta(X', X)$-bounded subset of $X'$ is equicontinuous.

(c) Each locally bounded [maps bounded sets into bounded sets] lower semicontinuous seminorm $p$ on $X$ is continuous.

11.115 COMPLETION OF AN INFRABARRELED SPACE If the LCHS $X$ has a dense infrabarreled subspace $M$ then $X$ is barreled. Thus the completion of an infrabarreled LCHS is barreled [cf. Theorem 11.12.1].

11.116 BANACH–STEINHAUS THEOREM FOR INFRABARRELED SPACES [cf. Theorem 11.9.1] If $X$ is an infrabarreled LCHS and $Y$ is any LCHS, then any $\beta(X, X')$-bounded subset $H$ of $L(X, Y)$ is equicontinuous.

11.117 QUASI-COMPLETE INFRABARRELED SPACE A quasi-complete [closed bounded subsets are complete; Exercise 6.109] infrabarreled LCHS is barreled.

11.118 $X$ BARRELED IMPLIES $(X', \sigma(X', X))$ QUASI-COMPLETE The dual $X'$ of a barreled space $(X, \mathcal{T})$ is $\sigma(X', X)$-quasi-complete.

11.119 DISCONTINUOUS LINEAR FORMS ON BAIRE SPACES Show that a discontinuous linear form must exist on an infinite-dimensional Hausdorff Baire TVS $X$.

11.120 INCOMPLETE METRIZABLE BAIRE SPACE In the Baire space $X = \ell_p$, $1 \leq p < \infty$, for each $n \in \mathbf{N}$, let $e_n = (\delta_{jn})$. Extend $\{e_n : n \in \mathbf{N}\}$ to a Hamel base $B$ for $X$ and let $(x_n)$ be a sequence of distinct elements of $B \setminus \{e_n : n \in \mathbf{N}\}$. Let $M_k$ be the linear span of $B \setminus \{x_n : n > k\}$. Show that some $M_k$ is nonmeager and incomplete.

11.201 BARRELEDNESS WITHOUT CONVEXITY Let $(X, \mathcal{T})$ be an HTVS. A closed balanced subset $B$ of $X$ is called an *ultrabarrel* if there exists a sequence $(B_n)$ of closed balanced absorbent sets such that for every $n$, $B_{n+1} + B_{n+1} \subset B_n$ with $B_1 + B_1 \subset B$. Generally, a sequence $(B_n)$

of absorbent balanced sets such that $B_{n+1} + B_{n+1} \subset B_n$ for every $n$ is called a *string*; each $B_n$ is called a *knot* of the string. If each $B_n$ is closed, it is called a *closed string.* Thus, an ultrabarrel is a closed balanced set $B$ for which there exists a closed string $(B_n)$ such that $B_1 + B_1 \subset B$, in which case we say that $B$ *contains the string.* If we do not require that $B$ be closed nor that the string be closed, then $B$ is called a *suprabarrel.*

(a) A barrel is an ultrabarrel, as is any closed balanced neighborhood of 0.

(b) An ultrabarrel need not be a barrel.

$(X, \mathcal{T})$ is called *ultrabarreled* if any vector topology $\mathcal{T}^*$ for $X$ which has a base of $\mathcal{T}$-closed neighborhoods of 0 at 0 is weaker than $\mathcal{T}$ (see Exercise 11.113).

(c) If $X$ is a Baire space, $X$ is ultrabarreled.

(d) $X$ is ultrabarreled iff each ultrabarrel is a neighborhood of 0.

(e) BANACH–STEINHAUS THEOREM If $H$ is a pointwise bounded collection of continuous linear maps of the ultrabarreled space $X$ into the HTVS $Y$, then $H$ is equicontinuous.

(f) CONVERSE TO BANACH–STEINHAUS If $X$ is such that for any HTVS $Y$, any pointwise bounded subset $H$ of $L(X, Y)$ is equicontinuous, then $X$ is ultrabarreled.

(g) A locally convex ultrabarreled space $X$ is barreled, but not conversely.

(h) ASSOCIATED ULTRABARRELED TOPOLOGY Let $(X, \mathcal{T})$ be a TVS and consider the collection of all vector topologies $\{\mathcal{T}_s : s \in S\}$ on $X$ which are stronger than $\mathcal{T}$ and which are ultrabarreled (the finest vector topology for $X$ of Exercise 4.117 is one such topology). Show that $\mathcal{T}_b = \inf\{\mathcal{T}_s : s \in S\} = \cap\{\mathcal{T}_s : s \in S\} = \sup\{\mathcal{T}' : \mathcal{T}'$ is a vector topology for $X$ such that $\mathcal{T}' \subset \mathcal{T}_s$ for every $s \in S\}$ is an ultrabarreled topology for $X$ which is stronger than $\mathcal{T}$.

11.202 NON-ARCHIMEDEAN BARRELED SPACES Let $(X, \mathcal{T})$ be a locally $K$-convex space [Exercises 4.203 and 5.205] over a nontrivially ultravalued field $K$. A closed absorbent $K$-disk in $X$ is called a *$K$-barrel* and $X$ is called *$K$-barreled* if each $K$-barrel is a neighborhood of 0.

(a) $X$ is $K$-barreled if any locally $K$-convex topology $\mathcal{T}^*$ on $X$ which has a base of $\mathcal{T}$-closed neighborhoods of 0 is weaker than $\mathcal{T}$.

(b) If $X$ is a Baire space then $X$ is $K$-barreled.

(c) If $X$ is $K$-barreled, then each non-Archimedean lower semicontinuous seminorm [Exercise 5.204] on $X$ is continuous; the converse holds if the valuation on $K$ is discrete.

Say that a subset $E$ of $X$ is *bornivorous* if it absorbs each bounded set $B$ in the sense that $B \subset aE$ for sufficiently large $|a|, (a \in K)$. $X$ is called $K$*-infrabarreled* if each bornivorous $K$-barrel is a neighborhood of 0.

(d) If $X$ is $K$-infrabarreled then each locally bounded (bounded sets into bounded sets) lower semicontinuous ultraseminorm on $X$ is continuous. If $K$ is discretely valued, then the converse holds as well.

In (e)–(g), $Y$ denotes a locally $K$-convex space.

(e) BANACH–STEINHAUS THEOREM If $X$ is $K$-barreled, then each pointwise bounded subset $H$ of $L(X,Y)$ is equicontinuous.

(f) If $X$ is $K$-infrabarreled, then strongly bounded subsets $H$ of $L(X,Y)$ are equicontinuous.

(g) If $X$ is $K$-barreled, $Y$ is Hausdorff and $(A_s)$ is a net in $L(X,Y)$ which converges pointwise to a map $A$ of $X$ into $Y$, then $A \in L(X,Y)$ and $A_s \to A$ uniformly on each totally bounded subset of $X$.

As shown in Prolla [1982, pp. 94, 106], every ultrabarreled space $X$ over $K$ [see Exercise 11.201] is $K$-barreled, but not conversely.

### Hints to Exercises

11.101 Let $U$ be an open neighborhood of $t$ in $T$ such that $V = U \cap S$. Since $U$ is open, for any subset $B$ of $T$, $U \cap \mathrm{cl}_T B \subset U \cap \mathrm{cl}_T(U \cap B)$. Hence $\mathrm{cl}_T V = \mathrm{cl}_T(U \cap S) \supset U \cap \mathrm{cl}_T S = U$.

11.103 Use Theorem 11.3.6.

11.105 Use Theorem 11.7.1 for (a).

11.106 Use the result of Example 11.3.2.

11.108 Let $X'$ carry its norm topology, which is the same as $\beta(X', X)$ by Example 11.2.4(b). As $\mathbf{F}$ is complete, so is $X'$ [see Theorem 13.2.13], hence barreled. View $\{x_n\}$ now as a $\sigma(X'', X')$-bounded subset of $X''$.

11.109 If $H$ is $\sigma(X', X)$-bounded then there is some barrel $B$ in $X$ such that $H \subset B^\circ$ by Theorem 8.8.3. If $E \subset X$ is any bounded set, so is its closed disked hull $D$ [Theorems 6.3.2 and 6.3.3]; $D$ is therefore complete by hypothesis. Hence $B$ absorbs $D$ by Theorem 8.8.6, i.e., $B$ is a bornivorous barrel. It follows that $H$ is $\beta(X', X)$-bounded by Theorem 11.11.3.

11.111 See Example 11.3.3.

11.112 See Theorems 11.3.4, 8.6.5, and 11.2.6.

11.113 NECESSITY Given a $\mathcal{T}^*$-neighborhood $V$ of 0, choose a $\mathcal{T}$-closed $\mathcal{T}^*$-neighborhood $W$ of 0 such that $W \subset V$. Let $U$ be a disked $\mathcal{T}^*$-neighborhood of 0 which is contained in $W$. The $\mathcal{T}$-closure of $U$ is a $\mathcal{T}$-barrel, hence a $\mathcal{T}$-neighborhood of 0.

SUFFICIENCY The set of all $\mathcal{T}$-barrels forms a base at 0 for a locally convex topology $\mathcal{T}^*$; $\mathcal{T}^*$ is weaker than $\mathcal{T}$ by hypothesis. Thus, since any $\mathcal{T}$-barrel $B$ is a $\mathcal{T}^*$-neighborhood of 0, $B$ is also a $\mathcal{T}$-neighborhood of 0.

11.114 For the equivalence of (a) and (b), use Theorem 11.11.4; for that of (a) and (c), note that $\{x \in X : p(x) \leq 1\}$ is a bornivorous barrel iff $p$ is a locally bounded lower semicontinuous seminorm [cf. the proof of Theorem 11.4.3].

11.115 Let $B$ be a barrel in $X$. Clearly, $B \cap M$ is a barrel in $M$. Let $E$ be a bounded subset of $M$. Its closed disked hull $H$ in $X$ is complete and bounded [Theorems 6.3.2 and 6.3.3]. By Theorem 8.8.6, $B$ absorbs $H$; therefore, $B \cap M$ absorbs $E$ in $M$. Hence $B \cap M$ is a bornivorous barrel in $M$ and therefore a neighborhood of 0 in $M$. Thus $\mathrm{cl}_X(B \cap M) = B$ is a neighborhood of 0 in $X$ by Exercise 11.101.

11.116 Let $V$ be a closed disked neighborhood of 0 in $Y$ and let $B = \cap\{A^{-1}(V) : A \in H\}$. Show that $B$ is a closed disk. Show also that $B$ is bornivorous because $H$ is strongly bounded. Conclude that $B$ is a neighborhood of 0 in $X$ from which the equicontinuity of $H$ follows.

11.117 Let $H$ be a $\sigma(X', X)$-bounded subset of $X'$. Since $X$ is a quasi-complete LCHS, $H$ is $\beta(X', X)$-bounded by Exercise 11.109. By Theorem 11.11.4, $H$ is therefore equicontinuous. Hence $X$ is barreled by Theorem 11.3.4.

11.118 Let $B$ be a closed bounded subset of $(X', \sigma(X', X))$ so that $B^\circ$ is a $\beta(X, X')$-neighborhood of 0 in $X$, hence a $\mathcal{T}$-neighborhood of 0 by Theorem 11.3.6. Thus $B^{\circ\circ}$ is $\sigma(X', X)$-compact by the Alaoglu theorem [8.4.1]. The completeness of $B$ follows from its being a $\sigma(X', X)$-closed subset of $B^{\circ\circ}$ and Theorem 3.1.11.

11.119 If $X$ is pseudometrizable, the result follows from Theorem 5.6.2. But otherwise, suppose that $H = B \cup \{x_n : n \in \mathbf{N}\}$ is a Hamel base for $X$ and let $M_k$ be the linear span of $H \setminus \{x_k\}$, $k \in \mathbf{N}$. As the $M_k$ cover $X$, there must be some $j$ such that $\mathrm{cl}\, M_j = X$, i.e., $M_j$ is not closed. The linear form $f$ obtained by defining $f$ to be 1 on $x_j$ and 0 on the remaining basis vectors is discontinuous by Theorem 5.7.7(a). Must the subspaces $M_j$ be linearly homeomorphic to each other?

11.120 To show that $M_k$ is incomplete, show that it is dense in $X$. Since $X = \cup_k M_k$ and $X$ is a complete metric space, at least one of the $M_k$ must be nonmeager.

11.201 (a) Consider $\{2^{-n}B : n \in \mathbf{N}\}$ when $B$ is a barrel.

(c) Let $\mathcal{T}^*$ be a vector topology for $X$ which has a base of $\mathcal{T}$-closed neighborhoods of 0 and let $V$ be a $\mathcal{T}^*$-neighborhood of 0. Choose a $\mathcal{T}^*$-neighborhood $W$ of 0 which is $\mathcal{T}$-closed, balanced, and such that $W + W \subset V$. Since $X = \cup_{n\in\mathbf{N}} nW$ and $X$ is a Baire space, there exists $n \in \mathbf{N}$ such that the $\mathcal{T}$-interior of $nW$—hence of $W$—is not empty. For any $\mathcal{T}$-interior point $x$ of $W$, $-x + W$ is a $\mathcal{T}$-neighborhood of 0 and $-x + W \subset W + W \subset V$, so $V$ is a $\mathcal{T}$-neighborhood of 0, i.e., $\mathcal{T}^* \subset \mathcal{T}$.

(d) Suppose that $X$ is ultrabarreled and $B$ is an ultrabarrel. The collection of knots of all strings that are contained in $B$ together with $B$ then determine a vector topology $\mathcal{T}^*$ for $X$.

Conversely, suppose that each $\mathcal{T}$-ultrabarrel is a neighborhood of 0 and let $V$ be a $\mathcal{T}^*$-neighborhood of 0. There exists a balanced $\mathcal{T}$-closed $\mathcal{T}^*$-neigborhood $U$ of 0 such that $U + U \subset V$. Continue by induction to show that $V$ is a $\mathcal{T}$-ultrabarrel.

(e) If $V$ is a closed balanced neighborhood of 0 in $Y$, then $\cap\{A^{-1}(V) : A \in H\}$ is an ultrabarrel in $X$.

(f) Waelbroeck's argument can be found in Waelbroeck [1971] as well as Prolla [1982, p. 63], Husain and Khaleelulla [1978, p. 239], and Adasch et al. [1978, p. 38].

(g) A barrel $B$ in $X$ is an ultrabarrel. As for the false converse, see Husain and Khaleelulla [1978, p. 98] or Robertson [1958].

11.202 (a) See Exercise 11.113.

(c) If $p$ is a lower semicontinuous non-Archimedean seminorm, then $\overline{V}_p = \{x \in X : p(x) \leq 1\}$ is a $K$-barrel (cf. Exercise 5.204(b)). By (a), $\overline{V}_p$ is a neighborhood of 0 which implies that $p$ is continuous. Conversely, if the valuation on $K$ is discrete and $V$ is a $K$-barrel in $X$, let $p$ be its gauge [Exercise 5.204(c)]. If $r > 0$ is a generator of $|K^*| = \{|d| : d \neq 0\} = \{r^n : n \in \mathbf{Z}\}$, for any $a > 0$ there exists $k \in \mathbf{Z}$ such that $\{x \in X : p(x) \leq a\} = \{x \in X : p(x) \leq r^k\} = b^k\overline{V}_p$ where $|b| = r$. Since $\{x \in X : p(x) \leq a\}$ is closed, $p$ is lower semicontinuous, hence continuous, and it follows that $V$ is a neighborhood of 0.

(e) Let $V$ be a $K$-convex neighborhood of 0 in $Y$. We show that $W = \cap\{A^{-1}(V) : A \in H\}$ is a neighborhood of 0 in $X$. Since $V$ is $K$-convex, $V + V \subset V$ which implies that $V$ is clopen, hence closed and therefore $W$ is closed. $W$ is clearly $K$-convex; it is absorbent because $H$ is pointwise bounded. Thus $W$ is a $K$-barrel.

(f) Proceed as in (e), but note that $W$ is bornivorous because $H$ is strongly bounded.

(g) The set $\{A_s\}$ is equicontinuous. As in Lemmas 5.10.2 and 5.10.3, the topologies of pointwise convergence and uniform convergence on totally bounded sets induce the same topology on equicontinuous subsets of $L(X, Y)$.

# Chapter 12

# Inductive Limits



The Dirac delta function $\delta$ which was 0 everywhere but 0 on $\mathbf{R}$ but still satisfied $\int_{-\infty}^{\infty} \delta(t)\, dt = 1$ served physicists and engineers well, although it affronted and puzzled contemporary mathematicians. The comment of L. Schwartz [1966a, p. 3] is typical: "Les ingénieurs 1'utilisent systématiquement, chacun avec sa conception personelle, avec la conscience plus ou moins tranquille." It and its symbolic calculus relatives' efficacy provoked mathematicians into probing the cause of the rumbling. The ultimate success of Sobolev [1936] and Schwartz [1966a] undoubtedly provided smug comfort to some complacent engineers and physicists.

Although $\delta$ was spoken of as a function and used in some outrageous pure manipulations as such, its "integral property," namely,

$$\text{for } any\ x, \quad \int_{-\infty}^{\infty} x(t)\, \delta(t-a)\, dt = x(a)$$

was ultimately perceived as all that was needed. That, however, is merely the evaluation map, $x \mapsto x(a)$, at $a$, a perfectly decent linear functional, provided that the $x$'s at least form a vector space. The derivatives (!) of the delta function were amenable to a similar, sensible, interpretation.

Let $D(\mathbf{R})$ denote the linear space of infinitely differentiable maps of $\mathbf{R}$ into $\mathbf{R}$ which vanish outside some compact set. There is a vector topology for $D(\mathbf{R})$ called the *strict inductive limit* topology. That topology for $D(\mathbf{R})$ and other spaces on which it conveniently fits is the main subject of this chapter. The continuous dual $D(\mathbf{R})'$ of $D(\mathbf{R})$ is called a space of *distributions* and comfortably houses Dirac's delta function and its relatives. $D(\mathbf{R})$ is a special kind of *LF-space* ("LF" for "limit of Fréchet spaces") and all the results about them here were first proved in a classic 1949 paper of Dieudonné and Schwartz,

"La dualité dans les espaces $\mathcal{F}$ et $\mathcal{LF}$." The continuous duals of another kind of LF-space, continuous functions with compact support [Example 12.1.6], yield the spaces of *Radon measures* which are also of great importance in contemporary mathematics.

We consider a more general inductive limit in Sec. 12.2, one that we touched on in Example 4.11.3. We say more about it now especially for the sake of its utility in understanding bornological and ultrabornological spaces in the next chapter: Every such space can be expressed as the inductive limit of normed or Banach spaces [Theorems 13.2.10, 13.2.11]. In that connection and others, inductive limits and LF-spaces play an important role in Chap. 14 on closed graph theorems [Theorems 14.7.2 and 14.7.3, for example] and webbed spaces.

## 12.1 STRICT INDUCTIVE LIMITS

**Notation.** Throughout the section $(X_n)$ is an increasing sequence of LCS such that each $X_n$ is a subspace of $X_{n+1}$ and $X = \cup_n X_n$ is a vector space. $\mathcal{D}$ denotes the collection of all disks $D$ in $X$ such that $D \cap X_n$ is a neighborhood of 0 in $X_n$ for each $n \in \mathbf{N}$. ■

Is there a locally convex topology for $X$ which makes each $X_n$ a topological subspace of $X$? There is [Theorem 12.1.3(a)]. We set forth the main properties of this topology, the strict inductive limit topology, and also mention an important class of spaces, the LF-spaces, which carry it.

For each $n \in \mathbf{N}$, let $I_n : X_n \to X$, $x \mapsto x$, be the canonical injection of $X_n$ into $X$. The direct image locally convex topology $\mathcal{T}_i$ for $X$ induced by $\{I_n : n \in \mathbf{N}\}$ is the finest locally convex topology for $X$ with respect to which each $I_n$ is continuous [Example 4.11.3(b)]; it has as a base at 0 the class of all *absorbent* disks $D \subset X$ such that $D \cap X_n$ is a neighborhood of 0 in $X_n$ for each $n \in \mathbf{N}$. In the situation under consideration, we can omit "absorbent"—each disk $D$ such that $D \cap X_n$ is a neighborhood of 0 in $X_n$ for each $n \in \mathbf{N}$ is absorbent because any $x \in X$ must belong to some $X_n$ and since $D \cap X_n$ is a neighborhood of 0 in $X_n$, there is some $t > 0$ such that $x \in t\,(D \cap X_n) \subset tD$. The collection $\mathcal{D}$ of all disks $D$ in $X$ such that $D \cap X_n$ is a neighborhood of 0 in $X_n$ for each $n \in \mathbf{N}$ forms a base at 0 for the direct image locally convex topology $\mathcal{T}_i$ on $X$ which is called the *strict inductive limit topology* for $X$ ["determined by the maps $(I_n)$"] in this context. We say that $(X, \mathcal{T}_i)$ is the *strict inductive limit* of $(X_n)$ and write $X = \text{str ind}_n X_n$.

The word "strict" is reserved for denumerable increasing collections $(X_n)$ of LCS. In the next section we consider the inductive limit of arbitrarily many spaces.

**Theorem 12.1.1** BASE AT 0 For $X = \text{str ind}_n X_n$, the disked hulls

$$(\cup_{n \in \mathbf{N}} U_n)_{bc}$$

in $X$ of sets of the form $\cup_{n\in\mathbf{N}}U_n$ where each $U_n$ is a neighborhood of 0 in $X_n$ form a base at 0 for the strict inductive limit topology $\mathcal{T}_i$.

**Proof.** We use the notation of the statement. For each $n \in \mathbf{N}$, let $U_n$ be a neighborhood of 0 in $X_n$. Clearly $D = (\cup_n U_n)_{bc}$ is an absorbent disk in $X$ whose intersection with any particular $X_k$ contains $U_k$ so $D \in \mathcal{D}$. Conversely, for any $D \in \mathcal{D}$ and any $n \in \mathbf{N}$, $D \cap X_n \subset D$; hence $\cup_{n\in\mathbf{N}} D \cap X_n \subset D$. Since $D$ is a disk, $[\cup_n (D \cap X_n)]_{bc} \subset D$. ■

We put Theorem 12.1.2 to immediate use in Theorem 12.1.3 to show that the strict inductive limit topology induces the original topology on each of the component spaces $X_n$.

**Theorem 12.1.2** Disked Neighborhood of 0 in Subspace Let $M$ be a subspace of the LCS $X$. If $U$ is a disked neighborhood of 0 in $M$ then there is a disked neighborhood $V$ of 0 in $X$ such that $U = V \cap M$. If $M$ is closed and $x \notin U$ then $V$ can be chosen such that $x \notin V$.

**Proof.** Let $M$ be a subspace of the LCS $X$. If $U$ is a disked neighborhood of 0 in $M$, there is some neighborhood $W$ of 0 in $X$ such that $U = W \cap M$ but $W$ need not be convex. As $X$ is locally convex, $W$ must contain some disked neighborhood $W_1$ of 0; thus $W_1 \cap M \subset U$. Let $V = (W_1 \cup U)_c$, the convex hull of $W_1 \cup U$. With this convex binding, we show that $U = V \cap M$. Clearly, $U \subset V \cap M$. If, conversely, $z \in V \cap M$, since $W_1$ and $U$ are disks, there exist $w \in W_1$ and $u \in U$ such that $z = tw + (1-t)\,u$ for some $t \in [0,1]$ [Theorem 4.2.3(b)]. If $t = 0$, then $z \in U$ and we are done. If $t > 0$, then $w = (1/t)\,[z-(1-t)\,u] \in M$. Consequently, $w \in W_1 \cap M \subset W \cap M = U$. Since $U$ is convex, $z \in U$.

With notation as above, suppose that $M$ is closed and $x \notin U = V \cap M$. If $x \in M$ then $x \notin V$. If $x \notin M$, then there exists a disked neighborhood $N$ of 0 in the LCS $X$ such that $(x+N) \cap M = \varnothing$. Let $W_2 = W_1 \cap N$ so $(x+W_2) \cap M = \varnothing$. Now let $V = (W_2 \cup U)_c$. Could $x \in V$? If so, then there exists $w \in W_2$, $u \in U$, and $t \in [0,1]$ such that $x = tw+(1-t)\,u$ which implies that $x - tw = (1-t)\,u \in (x+W_2) \cap M$ which contradicts $(x+W_2) \cap M = \varnothing$. Finally, since $W_2 \subset W_1$, $(W_2 \cup U)_c \cap M \subset U$. The reverse inclusion follows from the fact that $U \subset M$. ■

**Theorem 12.1.3** Properties If $X = \text{str ind}_n X_n$, then:

(a) each $X_n$ is a topological subspace of $X$;

(b) if each $X_n$ is Hausdorff, so is $X$;

(c) if each $X_n$ is closed in $X_{n+1}$ then each $X_n$ is closed in $X$.

**Proof.** Let $X = \text{str ind}_n X_n$. (a) We show that any disked neighborhood $U_n$ of 0 in $X_n$ may be written in the form $U \cap X_n$ for some disked neighborhood $U$ of 0 in $X$. We argue by induction. For $n \in \mathbf{N}$, let $U_n$ be a disked neighborhood of 0 in $X_n$. By Theorem 12.1.2 there is a disked neighborhood $U_{n+1}$ of 0 in

$X_{n+1}$ such that $U_{n+1} \cap X_n = U_n$. Likewise, for each $k \in \mathbf{N}$, there is a disked neighborhood $U_{n+k}$ of 0 in $X_{n+k}$ such that

$$U_{n+k} \cap X_{n+k-1} = U_{n+k-1}$$

Clearly $U = \cup_{k \in \mathbf{N}} U_{n+k}$ is a disk in $X$ and $U \cap X_{n+m} = U_{n+m}$ for $m = 0, 1, \ldots$. For $p < n$, $U \cap X_p = U \cap (X_n \cap X_p) = U_n \cap X_p$ which is a neighborhood of 0 in $X_p$. Thus $U \cap X_m$ is a neighborhood of 0 in $X_m$ for each $m \in \mathbf{N}$. Therefore, $U$ is a neighborhood of 0 in the strict inductive limit topology and (a) is established.

(b) For $x \in X$, if $x \neq 0$, there must be some $n \in \mathbf{N}$ and neighborhood $V_n$ of 0 in $X_n$ such that $x \in X_n$ and $x \notin V_n$. By (a) there is a neighborhood $V$ of 0 in $X$ such that $V_n = V \cap X_n$. Clearly, $x$ cannot belong to $V$.

(c) Given $n \in \mathbf{N}$, suppose that $x \notin X_n$. There is some $k > n$ such that $x \in X_k$. As $X_n$ must be closed in $X_k$ (by induction, from the hypothesis) there exists a neighborhood $U_k$ of $x$ in $X_k$ such that $U_k \cap X_n = \varnothing$. By (a) there is a neighborhood $U$ of $x$ in $X$ such that $U \cap X_k = U_k$. Since $U \cap X_n = U \cap (X_k \cap X_n) = U_k \cap X_n = \varnothing$, it follows that $X_n$ is closed in $X$. ■

**Definition 12.1.4** SUPPORT Let $T$ be a topological space. For any map $x : T \to \mathbf{F}$, $\operatorname{cl}\{t \in T : x(t) \neq 0\}$ is called the *support* of $x$. ■

The most important kind of strict inductive limit is that which arises when each $X_n$ is a Fréchet space. Then $X = \operatorname{str\,ind}_n X_n$ is called an *LF-space*. LF-spaces are generally meager since the closed proper linear subspaces $X_n$ must have empty interior [Example 11.5.3(f)]. Note also that since each $X_n$ is complete, it is a closed subspace of $X_{n+1}$ for every $n$. We consider two important LF-spaces next. In the LF-space $D(\mathbf{R})$ of Example 12.1.5, differentiation is a continuous linear map [see after Theorem 13.2.9].

**Example 12.1.5** TEST FUNCTIONS AND DISTRIBUTIONS Let $D(\mathbf{R})$ denote the linear space (with pointwise operations) of infinitely differentiable functions $x : \mathbf{R} \to \mathbf{R}$ with compact support. $D(\mathbf{R})$ is a very narrow class of functions, to be sure (can you think of a member?), but it does consist of more than just 0. The function $x$ below, for instance, is a nontrivial member.

$$x(t) = \begin{cases} \exp\left(\dfrac{1}{t^2 - a^2}\right), & |t| < a \ (a > 0) \\ 0, & |t| \geq a \end{cases}$$

We topologize $D(\mathbf{R})$ as a strict inductive limit by considering an increasing sequence $(K_n)$ of compact subsets of $\mathbf{R}$ which cover $\mathbf{R}$ such that $K_n \subset \operatorname{int} K_{n+1}$ for each $n \in \mathbf{N}$. ($K_n$ could be $[-n, n]$, for example.) Let $D_n$ be the subspace of $D(\mathbf{R})$ consisting of those functions whose support is in $K_n$. $D_n$ is then topologized by the family of seminorms $\{p_m : m \in \mathbf{N}\}$ where, for $x \in D_n$,

$$p_m(x) = \sup\left\{\left|x^{(j)}(K_n)\right| : 0 \leq j \leq m\right\}$$

Each $D_n$ is metrizable since its topology is determined by a denumerable family of seminorms—indeed, each is a Fréchet space [see (a) below] and each $D_n$ is a closed subspace of $D_{n+1}$ [see (b)]. Clearly, $D(\mathbf{R}) = \cup_n D_n$. When endowed with the strict inductive limit topology determined by the $D_n$, $D(\mathbf{R})$ is called a space of *test functions.* Its continuous dual is called a space of *distributions.* Each continuous map $y : \mathbf{R} \to \mathbf{R}$ determines a distribution by way of the map $x \mapsto \int_{-\infty}^{\infty} x(t)\, y(t)\, dt$.

(a) EACH $D_n$ IS FRÉCHET Recall the following analytic fact: If $(x_n)$ is a sequence of functions which converges uniformly to $x$ on some closed interval $[a, b]$ and whose continuous first derivatives $x_n'$ converge uniformly to $y$, then $y = x'$.

To prove that $D_k$ is a Fréchet space, note that if $(x_n)$ is a Cauchy sequence in $D_k$, then $(p_0(x_n)), (p_1(x_n)), \ldots$ are Cauchy sequences as well; in other words, the sequences $(x_n), (x_n'), (x_n''), \ldots$ are uniform Cauchy sequences. By the result quoted above, it follows that $(x_n)$ converges to an infinitely differentiable function whose support, obviously, is also contained in $K_k$.

(b) $D_n$ IS CLOSED IN $D_{n+1}$ Clearly, $D_n$ is a topological subspace of $D_{n+1}$. As each $D_n$ is complete, each is closed. ■

With some minor modifications, the statements above remain valid if $\mathbf{R}$ is replaced by any open subset $G$ of $\mathbf{R}^n$. The space $D(G)$ then denotes the space of functions which have partial derivatives of all orders and compact support. Any such $D(G)$ with strict inductive limit topology [formed, analogously, by taking an increasing cover $(K_n)$ of compact subsets of $G$] is also called a space of test functions and its continuous dual a space of distributions.

**Example 12.1.6** RADON MEASURES Let $G$ be an open subset of $\mathbf{R}^k$ ($k \in \mathbf{N}$) in its usual topology. (a) There exists an increasing sequence $(K_n)$ of compact subsets of $G$ whose union is $G$ and such that $K_n \subset \operatorname{int} K_{n+1}$ for each $n$ in $\mathbf{N}$. If $G = \mathbf{R}^k$, the closed balls $C(0, n)$ of radius $n \in \mathbf{N}$ about 0 are such a sequence. If $G \neq \mathbf{R}^k$, let $A_n$ denote the points of $G$ of distance $\geq 1/n$ ($n \in \mathbf{N}$) away from the boundary of $G$. The $A_n$ are closed but need not be bounded. The sets $K_n = A_n \cap C(0, n)$ then have the desired properties.

Let $C_c(G, \mathbf{F})$ denote the linear space of continuous $\mathbf{F}$-valued functions on $G$ with compact support. For $n \in \mathbf{N}$, let

$$X_n = \left\{x \in C(G, \mathbf{F}) : x\left(\complement K_n\right) = \{0\}\right\}$$

The Banach space $(X_n, \|\cdot\|_\infty)$ is a topological subspace of $X_{n+1}$. Note that $C_c(G, \mathbf{F}) = X = \cup_n X_n$ and consider $X = \operatorname{str\,ind}_n X_n$. As is easy to verify, the strict inductive limit topology does not depend on the sets $K_n$—any increasing sequence $(H_n)$ of compact subsets of $G$ that cover $G$ will produce the same topology. Thus we may speak of the strict inductive limit topology for $X$ without reference to a particular increasing sequence of compact sets. A member of the continuous dual of $X$ is called a *Radon measure on $G$.* If a

Radon measure $f$ is positive [i.e., $x \geq 0 \Rightarrow f(x) \geq 0$] then $f$ is called a *positive* Radon measure.

(b) For each $n$, $X_n$ is closed in $X_{n+1}$ by the following argument: If $x \in X_{n+1} \backslash X_n$, there is some $t \in K_{n+1} \backslash K_n$ such that $x(t) \neq 0$. As each function of $X_n$ must vanish at $t$, no sequence of points from $X_n$ can converge to $x$ and $x \notin \operatorname{cl} X_n$. ■

**Theorem 12.1.7** TOPOLOGICAL PROPERTIES For $X = \text{str ind}_{n \in \mathbf{N}} X_n$, suppose that each $X_n$ is a closed subspace of $X_{n+1}$. Then:

(a) BOUNDEDNESS A subset $B$ of $X$ is bounded or totally bounded iff there exists $k \in \mathbf{N}$ such that $B \subset X_k$ and $B$ is bounded or totally bounded, respectively, in $X_k$.

(b) CONVERGENCE A sequence $x_n \to x$ in $X$ iff for some $k$ in $\mathbf{N}$, $x \in X_k$, each $x_n \in X_k$ and $x_n \to x$ in $X_k$.

(c) COMPACTNESS If each $X_n$ is Hausdorff, a subset $B$ of $X$ is compact iff there is some $k \in \mathbf{N}$ such that $B$ is a compact subset of $X_k$.

**Proof.** Let $X = \text{str ind}_n X_n$ and suppose that each $X_n$ is a closed subspace of $X_{n+1}$. (a) If $B$ is bounded in some $X_k$ and $U$ is a neighborhood of 0 in $X$, then $B$ is absorbed by $U \cap X_k$, hence by $U$. To prove the converse, we show that if $B$ is not contained in any $X_k$, then $B$ is unbounded in $X$. To accomplish this, we use the criterion of Theorem 6.1.4(a) and show that such a $B$ contains a sequence $(x_n)$ such that $x_n/n \nrightarrow 0$.

Since $B$ is not contained in any $X_k$, there exists $x_1 \in B$ such that $x_1 \notin X_1$. Let $i_2$ be the least positive integer such that $x_1 \in X_{i_2}$ choose $x_2 \in B \backslash X_{i_2}$ and let $i_3$ be the least positive integer such that $x_2 \in X_{i_3}$. Thus, by induction, there exists an increasing sequence $(i_k)$ of positive integers and elements $x_k \in B$ such that $x_k \notin X_{i_k}$ but $x_k \in X_{i_{k+1}}$. For any disked neighborhood $W_1$ of 0 in $X_1$, since $x_1 \notin X_1$, $x_1 \notin W_1$. Consequently, by Theorem 12.1.2, there exists a disked neighborhood $W_2$ of 0 in $X_{i_2}$ such that $W_1 = W_2 \cap X_1$ and $x_1 \notin W_2$. Since $x_2 \notin X_{i_2}$, $x_2/2 \notin W_2$. Since neither $x_1$ nor $x_2/2$ belong to $W_2$, there exist disked neighborhoods $V_3$ and $V_3'$ of 0 in $X_{i_3}$ such that $x_1 \notin V_3$ and $x_2/2 \notin V_3'$ and $W_2 = V_3 \cap X_{i_2} = V_3' \cap X_{i_2}$ [Theorem 12.1.2]. Letting $W_3 = V_3 \cap V_3'$, then $W_2 = W_3 \cap X_{i_2}$ and neither $x_1$ nor $x_2/2$ belong to $W_3$. Continuing in this fashion, we see that there exists an increasing sequence of disked neighborhoods $W_k$ of 0 in $X_{i_k}$ such that, for any $k \in \mathbf{N}$, $x_k/k \notin W_n$ for any $n$. Consider the disk $W = \cup_{n \in \mathbf{N}} W_n$ in $X$. If, for some $k \in \mathbf{N}$, $\{x_k/k, \dots\} \subset W$, however, we have the contradictory result that $x_k/k \in W_n$ for some $n$. Therefore $\{x_k/k, \dots\} \not\subset W$ for any $k$. Since each $W_n$ is a disk, $W$ is a disk, hence a neighborhood of 0 in $X$ [Theorem 12.1.1]; it follows that $x_n/n \nrightarrow 0$ and therefore that $B$ is unbounded.

If $B$ is totally bounded in $X_k$ and $V$ is a neighborhood of 0 in $X$ then there exist $x_1, \dots, x_n$ in $X_k$ such that $B \subset \cup_{i=1}^n (x_i + (V \cap X_k)) \subset \cup_{i=1}^n (x_i + V)$.

Conversely, if $B$ is totally bounded in $X$, it is bounded [Theorem 6.1.2]. Consequently, there exists $k \in \mathbf{N}$ such that $B \subset X_k$. To show that it is totally bounded in $X_k$, we show that it is Cauchy bounded in $X_k$ [Definition 3.3.4] which is equivalent to total boundedness by Theorem 3.3.5. To this end let $I$ be a denumerable subset of $B$ and let $V_k$ be a neighborhood of 0 in $X_k$. Let $V$ be a neighborhood of 0 in $X$ such that $V_k = V \cap X_k$. As $B$ is Cauchy bounded in $X$, there exist distinct points $x, y \in I \subset X_k$ such that $x - y \in V$. Since $X_k$ is a subspace, $x - y \in V \cap X_k = V_k$.

(b) The condition clearly suffices. The converse follows from the fact that if $x_n \to x$ then $\{x_n : n \in \mathbf{N}\} \cup \{x\}$ is compact, hence bounded, and therefore contained in some $X_k$ by (a).

(c) The sufficiency of the condition is clear. If $B \subset X$ is compact, it is bounded and therefore contained in some $X_k$. If $(G_s)$ is an open cover of $B$ from $X_k$, there exist open subsets $W_s$ of $X$ such that $G_s = W_s \cap X_k$ for each index $s$. The desired result now follows from the fact that a finite number of the $W$'s must cover $B$. ■

**Theorem 12.1.8** LF NONMETRIZABLE If $X = \text{str ind}_n X_n$ and each $X_n$ is a closed proper subspace of $X_{n+1}$ then $X$ is not pseudometrizable. Hence LF-spaces are not pseudometrizable.

**Proof.** We use the notation of the statement. To show that $X$ is not pseudometrizable, we show that no countable collection $(U_n)$ of balanced neighborhoods of 0 in $X$ can be a base at 0 [Theorem 4.8.3]. If $(U_n)$ is a base at 0, we may suppose that it is decreasing since $U_1, U_1 \cap U_2, \ldots, U_1 \cap \cdots \cap U_n, \ldots$ would also be a base of balanced neighborhoods of 0. Note that for each $n \in \mathbf{N}, U_n$ cannot be a subset of $X_n$ because each $U_n$ is absorbent in $X$ and $X_n \neq X$. Thus, for every $n$, we may choose an element $x_n$ in $U_n$, not in $X_n$. Since the $x_n$'s cannot be in any one $X_k$, they comprise an unbounded set [Theorem 12.1.7(a)]; yet since the $U_n$'s are decreasing, for any $j \in \mathbf{N}$, $x_n \in U_n \subset U_j$ for $n \geq j$ while $U_j$ must absorb $x_1, \ldots x_{j-1}$—in other words, any $U_j$ absorbs the $x_n$'s. Thus, if $(U_n)$ were a base at 0, $\{x_n : n \in \mathbf{N}\}$ would be a bounded set which it is not. ■

Utilizing Theorem 12.1.8, we show next that strict inductive limit topologies are rather strong topologies.

**Example 12.1.9** FINENESS OF INDUCTIVE LIMIT TOPOLOGY (a) Consider the normed space $(\varphi, \|\cdot\|_\infty)$ of all finite sequences of real scalars, i.e., real sequences $(t_n)$ such that $t_n = 0$ for almost all $n$. Under the obvious identification, $\ell_\infty^{\mathbf{R}}(n)$ $(n \in \mathbf{N})$ is an increasing sequence of subspaces whose union is $\varphi$ and each $\ell_\infty^{\mathbf{R}}(n)$ is a closed subspace of $\ell_\infty^{\mathbf{R}}(n+1)$.

How does the strict inductive limit topology $\mathcal{T}_i$ on $\varphi$ compare to the $\|\cdot\|_\infty$-topology $\mathcal{T}_\infty$ on $\varphi$?

Recall that the finest locally convex topology $\mathcal{T}_{lc}$ for $\varphi$ is that which has the filterbase of all absorbent disks as a base of neighborhoods of 0 [Example 5.6.4]. We show that $\mathcal{T}_i = \mathcal{T}_{lc}$ by the following argument. Since $\mathcal{T}_i$ is locally convex, $\mathcal{T}_i \subset \mathcal{T}_{lc}$. Conversely, if $D$ is an absorbent disk in $\varphi$, then $D \cap \ell_\infty^{\mathbf{R}}(n)$ is a disk in $\ell_\infty^{\mathbf{R}}(n)$; it is absorbent because $D$ is. As an absorbent disk in $\ell_\infty^{\mathbf{R}}(n)$, $D \cap \ell_\infty^{\mathbf{R}}(n)$ is a basic neighborhood of 0 in the finest locally convex topology for $\ell_\infty^{\mathbf{R}}(n)$. But as all Hausdorff linear topologies coincide on a finite-dimensional space [Theorem 4.10.3(b)], $D \cap \ell_\infty^{\mathbf{R}}(n)$ is a $\mathcal{T}_\infty$-neighborhood of 0 in $\ell_\infty^{\mathbf{R}}(n)$. Thus *any* absorbent disk in $\varphi$ induces a $\mathcal{T}_\infty$-neighborhood of 0 in $\ell_\infty^{\mathbf{R}}(n)$, so $\mathcal{T}_{lc} \subset \mathcal{T}_i$.

As $\mathcal{T}_\infty$ is a locally convex topology for $\varphi$, $\mathcal{T}_\infty \subset \mathcal{T}_i$. Since $\mathcal{T}_\infty$ is metrizable and $\mathcal{T}_i$ is not [Theorem 12.1.8 or Example 5.6.4], it follows that $\mathcal{T}_i$ is strictly finer than $\mathcal{T}_\infty$. Yet each induces the same topology on each of the subspaces $\mathbf{R}^n$.

(b) Let $C_c(\mathbf{R}, \mathbf{R})$ denote the linear space of all continuous maps of $\mathbf{R}$ into $\mathbf{R}$ with compact support. View $C_c(\mathbf{R}, \mathbf{R})$ as a strict inductive limit $C_c(\mathbf{R}, \mathbf{R}) = \text{str ind}_n X_n$ as in Example 12.1.6 with the strict inductive limit topology $\mathcal{T}_i$. Two other topologies for $C_c(\mathbf{R}, \mathbf{R})$ are the sup norm topology $\mathcal{T}_\infty$ induced by $\|\cdot\|_\infty$ and the compact-open topology $\mathcal{T}_c$ [Section 5.8]. Clearly, $\mathcal{T}_c \subset \mathcal{T}_\infty \subset \mathcal{T}_i$. Furthermore, $\mathcal{T}_\infty$ is strictly finer than $\mathcal{T}_c$ since to be small on some compact set does not imply smallness everywhere. $\mathcal{T}_\infty$ is metrizable and $\mathcal{T}_i$ is not [Theorem 12.1.8]. Hence $\mathcal{T}_i$ is strictly finer than $\mathcal{T}_\infty$. Despite this, $\mathcal{T}_i$ and $\mathcal{T}_c$ still induce the same topology on each of the subspaces $X_n$. ■

It follows from Theorem 12.1.10 that the LF-spaces $D(\mathbf{R})$ and $C_c(G, \mathbf{R})$ of Examples 12.1.5 and 12.1.6 are complete.

**Theorem 12.1.10** COMPLETENESS If each $X_n$ is complete then $X = \text{str ind}_n X_n$ is complete; hence LF-spaces are complete.

**Proof.** Suppose that each $X_n$ is complete and $X = \text{str ind}_n X_n$. As observed in the remark after the statement of Theorem 11.13.2, an increasing sequence of subspaces such as $(X_n)$ satisfies condition $(*)$ of Theorem 11.13.2. Hence, by Theorem 11.13.2, if $Y$ is a completion of $X$, the closure of $X = \cup_n X_n$ in $Y$ is the algebraic closure of $\cup_n \text{cl}_Y X_n = \cup_n X_n$ which is just $X$ again since it is a subspace. ■

If $U$ is a subset of a TVS, $\text{cl}\, U$ can be so much bigger than $U$ that statements such as $(1/2)\,\text{cl}\, U \subset U$ are generally false. Consider $U = (0,1) \cup \mathbf{Q} \subset \mathbf{R}$, for example. But $U$ is not convex. What happens if $U$ is a disk in an LCS? The subspace $P$ of polynomials is a disk in the normed space $X = C([0,1], \mathbf{F}, \|\cdot\|_\infty)$ and $\text{cl}\, P = X$ (by the Stone-Weierstrass theorem, Theorem 16.5.7) so even the disk condition does not suffice. We show in Theorem 12.1.11, however, that $(1/2)\,\text{cl}\, D \subset D$ for disked neighborhoods $D$ of 0.

**Theorem 12.1.11** If $D$ is a disked neighborhood of 0 in a TVS then $(1/2)\,\text{cl}\, D \subset D$.

**Proof.** If $D$ is any convex set then $D + D = 2D$ by Theorem 4.2.5. Suppose that $D$ is a disked neighborhood of 0 and that $y \in \operatorname{cl} D$. Since $D$ is a neighborhood of 0, then there exists $z \in (y + D) \cap D$. Since $z \in y + D$, $z - y \in D$. Since $z \in D$ and $D$ is a disk, $y \in D + D = 2D$. Therefore $(1/2) \operatorname{cl} D \subset D$. ■

As we will use it in proving Theorem 12.1.12, we recall:

**Theorem 11.13.2** Let $X$ be a locally convex space with neighborhood filter $V(0)$ at 0 and let $(D_n)$ be an increasing sequence of disks of some subspace $M$ of $X$ such that

$$\text{for } f_n \in D_n^{\circ},\ \{f_n : n \in \mathbf{N}\} \text{ is equicontinuous on } M. \tag{$*$}$$

Then

(a) $\operatorname{cl}(\cup_n D_n)$ is the algebraic closure of $\cup_n \operatorname{cl} D_n = \cap_{r>0} (1 + r) (\cup_n \operatorname{cl} D_n)$ or, equivalently,

(b) if $\mathcal{B}$ is a Cauchy filterbase in $\cup_n D_n$, then, given any $r > 0$, $\mathcal{B} + V(0)$ induces a Cauchy filterbase on $(1 + r) D_m$ for some positive integer $m$.

**Theorem 12.1.12** BARRELED SPACES If $(X, \mathcal{T})$ is a barreled LCHS and $(X_n)$ is an increasing sequence of subspaces such that $X = \cup_n X_n$ then $X$ is the strict inductive limit of the $X_n$, i.e., $\mathcal{T}$ is the strict inductive limit topology $\mathcal{T}_i$.

**Proof.** In the notation of the statement, since $X$ is an LCS, there is a base of disked $\mathcal{T}$-neighborhoods $D$ of 0 in $X$. Since the $X_n$ are subspaces of $X$, each such $D$ meets each $X_n$ in a neighborhood of 0 so $\mathcal{T} \subset \mathcal{T}_i$. To show that $\mathcal{T}_i \subset \mathcal{T}$, we show that if $D$ is a disk in $X$ which meets each $X_n$ in a neighborhood of 0, then $D$ is a $\mathcal{T}$-neighborhood of 0.

A disk $D$ which meets each $X_n$ in a neighborhood of 0 must be absorbent in $X = \cup_n X_n$; hence $\operatorname{cl}_X D$ is a barrel in $X$. In the remainder of the proof we show that $(1/8) \operatorname{cl}_X D \subset D$ from which it follows that $D$ is a $\mathcal{T}$-neighborhood of 0. Let $x \in (1/2) \operatorname{cl}_X D$ so that there exists a Cauchy filterbase $\mathcal{B}$ in $(1/2)D$ such that $\mathcal{B} \to x$. Since $X$ is barreled and $D = \cup_n (D \cap X_n)$ is absorbent, it follows that given any $y \in X$ there is some $t_y > 0$ such that $y \in t_y D$ and therefore that for some $n \in \mathbf{N}$, $y \in t_y (D \cap X_n)$. Since the disks $D \cap X_n$ are increasing, $k \geq n \Rightarrow y \in t_y (D \cap X_k)$. Hence, for $k \geq n$ and $f_k \in (D \cap X_k)^{\circ}$, $|f_k(y)| \leq t_y$; therefore, for any sequence $f_n \in (D \cap X_n)^{\circ}$, there is some $K_y > 0$ such that $|f_n(y)| \leq K_y$ for all $n \in \mathbf{N}$ i.e., $\{f_n : n \in \mathbf{N}\}$ is pointwise bounded on $X$. Since $X$ is barreled, it follows from the Banach-Steinhaus theorem [11.9.1] that $\{f_n : n \in \mathbf{N}\}$ is equicontinuous on $X$, i.e., satisfies $(*)$ of Theorem 11.13.2. Therefore, with $r = 1$ and $V(0)$ denoting the filter of neighborhoods of 0 in $X$, there is some $m \in \mathbf{N}$ such that $\mathcal{B} + V(0)$ induces a Cauchy filterbase on $2(D \cap X_m)$. Since the disks are increasing, we may suppose that $m$ is big enough so that $x \in X_m$. Since $\mathcal{B} \to x$, it is straightforward to verify that $\mathcal{B} + V(0) \to x$ as well; therefore $x \in \operatorname{cl}_{X_m} 2(D \cap X_m)$. By Theorem

12.1.11, since $D \cap X_m$ is a disked neighborhood of 0 in $X_m$, $\mathrm{cl}_{X_m}(D \cap X_m) \subset 2(D \cap X_m)$. Therefore $x \in 4(D \cap X_m)$ and $(1/2)\,\mathrm{cl}_X D \subset \cup_n 4(D \cap X_n) = 4D$. As $(1/8)\,\mathrm{cl}_X D$ is a barrel, the desired result follows. ■

# 12.2 INDUCTIVE LIMITS OF LCS

We have already looked at something more general than the strict inductive limit topology, namely, the direct image locally convex topology mentioned in Examples 4.11.3(b) and 5.11.7, particular instances of which are quotients of LCS [Example 4.7.2] and locally convex direct sums [Example 4.11.4]. Another term for "direct image locally convex topology" is "inductive limit topology." To have it nearby, we restate the definition here.

**Definition 12.2.1** INDUCTIVE LIMITS Let $\{X_s : s \in S\}$ be a family of LCS and $X$ a vector space. Suppose that for each $s \in S$ that $A_s : X_s \to X$ is a surjective linear map and that the linear span of $\cup_{s\in S} A_s(X_s)$ is $X$. In this case, the finest locally convex topology for $X$ with respect to which each $A_s$ is continuous, the direct image locally convex topology, is called the *inductive limit topology* for $X$ ["determined by $\{A_s : s \in S\}$"]. We write $X = \mathrm{ind}_{s\in S} X_s$ and say that $X$ is the *inductive limit* of $\{X_s : s \in S\}$. ■

A base at 0 for the inductive limit topology is given by the collection of disks $D$ in $X$ such that $A_s^{-1}(D)$ is a neighborhood of 0 in $X_s$ for each $s \in S$. The requirement that $\cup_{s\in S} A_s(X_s)$ span $X$ is not essential. Its loss entails taking *absorbent* disks $D$ in $X$ such that each $A_s^{-1}(D)$ is a neighborhood of 0 in each $X_s$ as a base at 0 for the inductive limit topology.

The inductive limit topology is the finest locally convex topology for $X$ such that the *inverse image* topology determined by $A_s$ on $X_s$ [Example 4.7.1] is coarser than the original topology on $X_s$ for each $s \in S$. Two significant differences between inductive limit topologies and *strict* inductive limit topologies are: (1) In the case when each $X_s$ is a subspace of $X$ and each $A_s$ is the canonical injection of $X_s$ into $X$, the inductive limit topology may induce a strictly weaker topology on $X_s$ than its original topology, and (2) the inductive limit topology may not be Hausdorff even if each $X_s$ is a Hausdorff space. (Consider the quotient topology on $X/M$ when $M$ is a nonclosed subspace of a Hausdorff LCS.)

**Theorem 12.2.2** CONTINUITY With notation as in Definition 12.2.1, let $X$ carry the inductive limit topology determined by the maps $A_s : X_s \to X$ $(s \in S)$ and let $A$ be a linear map of $X$ into an LCS $Y$. Then $A$ is continuous iff for each $s \in S$, $A \circ A_s$ is continuous.

**Proof.** $A$ is continuous iff for each disked neighborhood $D$ of 0 in $Y$,
$A^{-1}(D)$ is a neighborhood of 0 in $X$;
$\Leftrightarrow A_s^{-1}(A^{-1}(D))$ is a neighborhood of 0 in $X_s$ for each $s \in S$;

$\Leftrightarrow A \circ A_s$ is continuous for each $s \in S$. ■

As follows from Theorem 11.12.2, inductive limits of barreled and infrabarreled spaces are barreled and infrabarreled, respectively.

**Theorem 12.2.3** INDUCTIVE LIMITS AND DIRECT SUMS If $X = \operatorname{ind}_{s \in S} X_s$ determined by the maps $A_s : X_s \to X$ $(s \in S)$ then $X$ is linearly homeomorphic to a quotient by a closed subspace of the locally convex direct sum $\oplus_{s \in S} X_s$.

**Proof.** Let $X = \operatorname{ind}_{s \in S} X_s$ determined by the maps $A_s : X_s \to X$ $(s \in S)$ and consider the map $A : \oplus_{s \in S} X_s \to X$, $(x_s) \mapsto \sum_{s \in S} A_s x_s$. For each $s \in S$, let $I_s$ denote the canonical injection of $X_s$ into $\oplus_{s \in S} X_s$ and note that $\oplus_{s \in S} X_s$ carries the inductive limit topology determined by the canonical injections $\{I_s : s \in S\}$ [Example 4.11.4]. Note also that $A \circ I_s = A_s$ for each $s \in S$. By Theorem 12.2.2, the continuity of $A$ is equivalent to that of $A \circ I_s$ for each $s \in S$. Thus, since each $A_s$ is continuous, $A$ is continuous. Since $\cup_{s \in S} A_s (X_s)$ spans $X$, $A$ is onto. Let $N$ denote the null space of $A$ and $\dot{A}(x + N) = Ax$ $(x \in N)$.

$$\begin{array}{ccccc} X_s & \xrightarrow{A_s} & X & \xrightarrow{\dot{A}^{-1}} & (\oplus_{s \in S} X_s)/N \\ & {\scriptstyle I_s} \searrow & & \nearrow {\scriptstyle k} & \\ & & \oplus_{s \in S} X_s & & \end{array}$$

To show that $X$ is linearly homeomorphic to $\oplus_{s \in S} X_s / N$, by Theorem 4.7.4 it only remains to show that $\dot{A}^{-1}$ is continuous. Let $k$ denote the canonical map $x \mapsto x + N$ of $\oplus_{s \in S} X_s$ onto $(\oplus_{s \in S} X_s)/N$. Then $\dot{A}^{-1} \circ A_s$ and $k \circ I_s$ each map $x \in X_s$ into $x + N$. In order words, $\dot{A}^{-1} \circ A_s = k \circ I_s$. Since $k \circ I_s$ is continuous for each $s$, the continuity of $\dot{A}^{-1}$ follows by Theorem 12.2.2. ■

## 12.3 EXERCISES

12.101 MEAGERNESS An LF-space is meager.

12.102 MACKEY TOPOLOGY A Fréchet or LF-space carries its Mackey topology $\tau(X, X')$.

12.103 $D(K)$ NOT NORMABLE Let $K$ be a compact subset of $\mathbf{R}$ and let $D(K)$ denote the linear space of infinitely differentiable functions on $K$ topologized by the family of seminorms

$$p_n(x) = \sup\left\{\left|x^{(k)}(K)\right| : 0 \le k \le n\right\}, \quad n \in \mathbf{N}, x \in D(K)$$

This is a Fréchet space [Example 12.1.5]. Show that it is not normable.

12.104 Let $X$ be a vector space, let $\{X_s : s \in S\}$ be a family of LCS and $A_s : X_s \to X$ $(s \in S)$ a family of surjective linear maps but do not assume that the linear span of $\cup_{s \in S} A_s(X_s)$ is $X$. Let $X$ carry the direct image locally convex topology $\mathcal{T}$ determined by $\{A_s : s \in S\}$. Show that a base at 0 for $\mathcal{T}$ is given by the set of *absorbent* disks $D$ such that each $A_s^{-1}(D)$ is a neighborhood of 0 in $X_s$.

12.105 Give examples of inductive limits $X = \text{ind}_{s \in S} X_s$ for which

(a) $X$ is not Hausdorff, even though each $X_s$ is, and

(b) the topology on some $X_s$ as a subspace of the inductive limit is strictly weaker than its original topology.

12.106 STRONG DUAL OF AN LF-SPACE If $X$ is an LF-space, show that $(X', \beta(X', X))$ is metrizable only if there exist Banach spaces $X_n$ $(n \in \mathbf{N})$ such that $X = \text{str ind}_n X_n$.

12.107 EQUICONTINUITY Let $X$ be a vector space, let $\{X_s : s \in S\}$ be a family of LCS and $A_s : X_s \to X$ $(s \in S)$ a family of surjective linear maps. Suppose that the $X$ carries the direct image locally convex topology determined by the maps $(A_s)$. If $E$ is a set of linear maps of $X$ into the LCS $Y$, show that $E$ is equicontinuous iff $E \circ A_s$ is equicontinuous for each $s \in S$.

12.108 ITERATED INDUCTIVE LIMITS Show that an inductive limit of inductive limits is an inductive limit, i.e., if $Y = \text{ind}_{s \in S} Y_s$ and each $Y_s = \text{ind}_t Y_{st}$, then $Y = \text{ind}\, Y_{st}$.

12.109 SEPARABILITY The inductive limit of countably many separable spaces is separable.

12.110 NON-ARCHIMEDEAN STRICT INDUCTIVE LIMITS Let $K$ be a nontrivially ultravalued field and let $(X_n)$ be an increasing sequence of non-Archimedean Fréchet spaces [Exercise 5.205(c)] such that each $X_n$ is a topological linear subspace of $X_{n+1}$. The collection $\mathcal{B}$ of $K$-disks [Exercise 4.202] $V$ of $X = \cup_n X_n$ such that each $V \cap X_n$ is a neighborhood of 0 in $X_n$ defines a locally $K$-convex topology $\mathcal{T}$ on $X$ for which $\mathcal{B}$ is a base at 0. It is the finest locally $K$-convex topology for $X$ for which each of the canonical injections of $X_n$ into $X$, $x \mapsto x$, is continuous. $\mathcal{T}$ is called the inductive limit topology for $X$ and $X$ the strict inductive limit of the $X_n$; $X$ is then called a *non-Archimedean LF-space*. Show that:

(a) $X$ is a complete nonmetrizable space.

(b) A subset $B$ of $X$ is bounded iff, for some $n \in \mathbf{N}$, $B$ is a bounded subset of $X_n$.

(c) A subset $E$ of $X$ is closed in $X$ iff $E \cap X_n$ is closed in each $X_n$.

(d) If $Y$ is locally $K$-convex [Exercise 4.203] then the linear map $A : X \to Y$ is continuous iff each restriction $A|_{X_n}$ is continuous.

(e) $X$ is $K$-barreled in the sense that each closed absorbent $K$-disk is a neighborhood of 0.

(f) If $Y$ is a metrizable locally $K$-convex space or a non-Archimedean LF-space and $Z$ a locally $K$-convex space, then a linear map $A : Y \to Z$ is continuous iff $A$ maps bounded sets into bounded sets.

12.201 DIRECT SUMS Let $\{X_s : s \in S\}$ be a family of locally convex spaces and let $\oplus_{s\in S}X_s$ be their locally convex direct sum [Example 4.11.4]. Let $I_s : X_s \to \Pi_{s\in S}X_s$ be the canonical injection sending $x$ into that element of the product which has $x$ as its $s$th entry and 0's everywhere else.

(a) Each $I_s : X_s \to I_s(X_s)$ is a linear homeomorphism; if each $X_s$ is Hausdorff, then $I_s(X_s)$ is closed in $\Pi_{s\in S}X_s$.

(b) FINENESS The locally convex direct sum topology on $\oplus_{s\in S}X_s$ is finer than the product topology; they coincide if $S$ is finite. The locally convex direct sum topology induces the original topology on each $X_s$.

(c) SEPARATION $\oplus_{s\in S}X_s$ is Hausdorff iff each $X_s$ is; in that case each $X_s$ is closed in $\oplus_{s\in S}X_s$.

(d) BOUNDED SETS A subset $B$ of a Hausdorff locally convex direct sum $\oplus_{s\in S}X_s$ is bounded or totally bounded iff $B$ is a bounded or totally bounded, respectively, subset of a finite sum $Y = \oplus_{s\in H}X_s$, i.e., where $H$ is a finite subset of $S$.

12.202 PROJECTIVE LIMITS For each $s$ in an index set $S$, let $A_s$ be a linear map of a vector space $X$ into an LCS $X_s$ such that for each nonzero vector $x \in X$, there is some $s \in S$ such that $A_s x \neq 0$. When $X$ carries the inverse image topology determined by $(A_s)$ [Example 4.7.1] it is called the *projective limit* of the spaces $(X_s)$ determined by the maps $(A_s)$.

(a) CONTINUITY Let $Y$ be an LCS and let $X$ be as above. A linear map $A : Y \to X$ is continuous iff for each $s \in S$, $A_s \circ A$ is continuous.

(b) BOUNDED SETS A subset $B$ of $X$ is bounded or totally bounded iff each $A_s(B)$ is bounded or totally bounded, respectively.

(c) DUAL OF AN INDUCTIVE LIMIT For each $s \in S$, let $Y_s$ be an LCHS and let $\mathcal{S}_s$ be a collection of $\sigma(Y_s, Y_s')$-bounded sets. Suppose that each $Y_s'$ carries the $\mathcal{S}_s$-topology [Section 8.5]. Let $Y$ be the inductive limit of $(Y_s)$ determined by the maps $(A_s)$ and suppose that $Y$ is Hausdorff. Let $\mathcal{S}$ be the set of finite unions of sets of $\{A_s(\mathcal{S}_s) : s \in S\}$. The continuous dual $Y'$ of $Y$ with the $\mathcal{S}$-topology is the projective limit of the spaces $Y'_s$ determined by the adjoints $A'_s$ of the maps $A_s$ $(s \in S)$.

(d) DUAL OF A DIRECT SUM The dual of the locally convex direct sum $\oplus_{s\in S}X_s$ is the product $\Pi_{s\in S}X'_s$ of the duals. If each $X_s$ is Hausdorff and each $X'_s$ carries an $\mathcal{S}_s$-topology [as in (c)], then the product topology on $\Pi_{s\in S}X'_s$ is the $\mathcal{S}$-topology determined by the set $\mathcal{S}$ of finite unions of $\cup_{s\in S}\mathcal{S}_s$.

(e) DUAL OF A PRODUCT Let $(X_s)$ be a family of LCS. The dual of the product $\Pi_{s\in S}X_s$ is the locally convex direct sum $\oplus_{s\in S}X'_s$.

(f) PRODUCTS OF BARRELED SPACES A product of Hausdorff barreled spaces is barreled.

12.203 MONTEL SPACES An LCHS $X$ is *semi-Montel* if bounded subsets of $X$ are relatively compact. A barreled semi-Montel space is called *Montel.*

(a) EXAMPLES Every finite-dimensional Hausdorff TVS is Montel. If a normed space is Montel, it must be finite-dimensional. Although it is not easy to show, the space $D(\mathbf{R})$ of test functions of Example 12.1.5 is Montel, as is its strong dual, the space of distributions [Horvath 1966, pp. 236, 241].

(b) BOUNDED SUBSETS If $B$ is a bounded subset of a semi-Montel space $X$ then the topology $B$ receives as a subspace of $X$ is the same as that induced by $\sigma(X, X')$.

(c) CLOSED SUBSPACES AND PRODUCTS Any closed subspace of a semi-Montel space and any product of semi-Montel spaces is semi-Montel.

(d) STRICT INDUCTIVE LIMITS If $X = \operatorname{str\,ind}_n X_n$ where each $X_n$ is semi-Montel and closed in $X_{n+1}$ then $X$ is semi-Montel.

## Hints to Exercises

12.106 If $(V_n)$ is a denumerable base at 0 in $(X', \beta(X', X))$ then $(V_n^\circ)$ is a base of bounded sets in $X$ in the sense that any bounded subset of $X$ is contained in one of them.

12.201 (b) Let the locally convex direct sum and product topologies be denoted by $\mathcal{T}_d$ and $\mathcal{T}_p$, respectively. Each canonical injection $I_s$ is continuous when $\oplus_{s\in S}X_s$ carries $\mathcal{T}_p$, so $\mathcal{T}_p$ is weaker than $\mathcal{T}_d$. Suppose that $S$ is an infinite index set and that $U_s$ is a proper disked neighborhood of 0 in $X_s$ for each $s$. Let $\mathrm{pr}_s$ denote the projection of $\oplus_{s\in S}X_s$ onto $X_s$. Show that $\cap_{s\in S}\mathrm{pr}_s^{-1}(U_s)$ is a neighborhood of 0 in the locally convex direct sum topology but not the product topology. Use the result of (a) to show that the locally convex direct sum topology induces the original topology on each $X_s$.

(c) The fact that the locally convex direct sum induces the original topology on each $X_s$ shows that if $\oplus_{s\in S}X_s$ is Hausdorff, then so is each $X_s$. For the converse, use the facts that the product of Hausdorff spaces is Hausdorff and that the locally convex direct sum topology is finer than the product.

(d) If $B$ is bounded in $\oplus_{s\in S}X_s$ then so is each $\mathrm{pr}_s(B)$; clearly, $B \subset \oplus_{s\in S}\,\mathrm{pr}_s(B)$. Suppose that there are infinitely many indices $s$ for which $\mathrm{pr}_s(B) \neq \{0\}$ so that there exists a sequence $(s_n)$ of indices and nonzero $x_n \in \mathrm{pr}_{s_n}(B)$. Choose disked neighborhoods $U_{s_n}$ of 0 in $X_{s_n}$ such that $x_n \notin nU_{s_n}$. For $s \neq s_n$, let $U_s = X_s$ and consider the disked hull $U$ of $(\cup_{n\in\mathbf{N}}U_{s_n}) \cup (\cup_{s\neq s_n}U_s)$. $U$ is a neighborhood of 0 in the locally convex direct sum topology and $\mathrm{pr}_{s_n}(U) \subset U_{s_n}$ for every $n$ so $B \not\subset nU$ for any $n$.

12.202 (b) If $B$ is bounded in $X$, so is each $A_s(B)$. If $U_s$ is a disked neighborhood of 0 in $X_s$, then $U_s$ absorbs $A_s(B)$; hence $A_s^{-1}(U_s)$ absorbs $B$. What is a base at 0 in $X$?

(c) Each adjoint $A_s'$ maps $Y'$ into $Y_s'$. Suppose $f \in Y_s'$. If, for each $s$, $\langle A_s x_s, f\rangle = \langle x_s, A_s' f\rangle = 0$ for each $x_s \in X_s$, then $f$ vanishes on $\cup_s A_s(X_s)$ which spans $X$; hence $\cap_s A_s'^{-1}(0) = \{0\}$. The projective limit topology on $X'$ is the weakest topology for which $\{A_s'^{-1}(K_s^\circ) : K_s \in \mathcal{S}_s\}$, $s \in S$, are neighborhoods of 0. Since $A_s'^{-1}(K_s^\circ) = A_s(K_s)^\circ$, the projective limit topology is the desired $\mathcal{S}$-topology.

(d) For a linear functional $f$ on $\oplus_{s\in S}X_s$, let $f_s$ be the restriction of $f$ to $X_s$. Let $\mathrm{pr}_s$ denote the projection of $\oplus_{s\in S}X_s$ onto $X_s$. Show that the map $f \mapsto \sum_s f_s(\mathrm{pr}_s(\cdot))$ establishes an algebraic isomorphism between $(\oplus_{s\in S}X_s)'$ and $\Pi X_s'$. Show that the adjoint $I_s'$ of the canonical injection $I_s$ of $X_s$ into $\oplus_{s\in S}X_s$ is the projection $\mathrm{pr}_s'$ of $\Pi X_s'$ onto $X_s'$ and use the result of (c).

(e) Let $f$ be a continuous linear functional on the product which is bounded on the neighborhood $\cap_{s\in H}\,\mathrm{pr}_s^{-1}(U_s)$ where $H$ is finite and each $U_s$ is a disked neighborhood of 0 in $X_s$. Then $f$ vanishes on $X_s$ for $s \notin H$. Let $f_s$ be the restriction of $f$ to $X_s$ and show that $f = \sum_{s\in H} f_s$ and therefore that $f \in \oplus_{s\in S}X_s'$.

(f) By (e), the dual of $\Pi_{s\in S}X_s = X$ is $\oplus_{s\in S}X_s'$. Suppose that each $X_s'$ carries $\sigma(X_s', X_s)$. Let $\mathcal{T}'$ denote the locally convex direct sum topology for $\oplus_{s\in S}X_s'$. By (d), $(\oplus_{s\in S}X_s', \mathcal{T}')' = \Pi_{s\in S}X_s$. If $B$ is a $\sigma(X', X)$-bounded set, it is $\mathcal{T}'$-bounded by Theorem 8.8.7. Therefore, by Exercise 12.201(d), there is a finite set $H$ such that $B \subset \oplus_{s\in H}B_s'$ where each $B_s'$ is a $\sigma(X_s', X_s)$-bounded subset of $X_s'$. Since each $X_s$ is barreled, each $B_s'$ equicontinuous. Therefore $B$ is equicontinuous and $X$ is barreled.

12.203 (b) It suffices to consider *closed* bounded subsets $B$. Now consider the identity map of (the compact set) $B$ in its subspace topology into (the Hausdorff space) $B$ with its induced $\sigma(X, X')$ topology.

(c) Let $B$ be a bounded subset of the product $\Pi_s X_s$ of the semi-Montel spaces $\{X_s : s \in S\}$. Let $\text{pr}_s$ denote the projection of $\Pi_s X_s$ onto $X_s$. Note that $B \subset \Pi_s \, \text{pr}_s(B)$.

(d) Suppose that $B$ is a closed bounded subset of $X$. As such, by Theorem 12.1.7(a), $B$ is contained in some $X_n$.

# Chapter 13

# Bornological Spaces



Continuous linear maps are always locally bounded in the sense that they map bounded sets into bounded sets [Theorem 6.4.1]. If the domain is pseudometrizable, the converse holds: Local boundedness implies continuity [Theorem 6.5.2]. The purpose of this chapter is to identify the class of LCS $X$ for which all locally bounded linear maps into any LCS $Y$ are continuous. Mackey [1946] first singled out this class of spaces, which we call bornological, the term coined by Bourbaki.

A special type of bornological space, the ultrabornological space (Grothendieck's [1955, p. 17] "espace du type $(\beta)$") is also studied here. It is closely analogous to the bornological space and we have developed the properties of the two simultaneously so as to highlight the analogy. Ultrabornological spaces are the domain for a rather general closed graph theorem [Theorem 14.7.3].

## 13.1 BANACH DISKS

In Sec. 13.2 we introduce and discuss bornological and ultrabornological spaces. Each has an important characterization as an inductive limit of normed spaces [Theorems 13.2.10 and 13.2.11]. We develop the basic properties of those component normed spaces and some other collateral material here.

The linear span $[D]$ of a disk $D$ is easy to describe: It is just $\cup_{n\in\mathbf{N}} nD$, since if $x = \sum_{i=1}^{k} a_i d_i \in [D]$ and $n > \sum_{i=1}^{k} |a_i|$, then $x \in nD$ [Theorem 4.2.9]. It is thus clear that $D$ is absorbent in $[D]$. Consequently, we may consider the gauge $p_D$ of $D$ in $[D]$.

**Notation.** If $D$ is a disk in a vector space $X$, the seminormed space $([D], p_D)$ is denoted $X_D$. ■

Note that the set of all positive multiples of $D$ is a base at 0 in $X_D$.

If $B$ and $D$ are disks and $B \subset D$, then $p_D \leq p_B$ on $[B]$ as follows directly from properties of infima. Consequently, the topology $X_B$ gets as a subspace of $X_D$ is weaker than the topology on $X_B$.

Having discussed $X_D$ when $D$ is any disk, we consider next the case of most interest, the case when $D$ is a *bounded* disk. If $D$ is a bounded disk in a TVS $X$, then each neighborhood $U$ of 0 contains some positive multiple $rD$ of $D$. Hence $U \cap X_D \supset rD$ and $X_D$'s topology is seen to be finer than the topology $X_D$ inherits as a subspace of $X$.

**Definition 13.1.1** (a) BANACH DISKS If $D$ is a bounded disk in the TVS $X$ and $X_D$ is a Banach space then $D$ is called a *Banach disk.*

(b) INFRABOUNDED MAP A linear map $A : X \to Y$ between LCS $X$ and $Y$ is *infrabounded* if it maps Banach disks into bounded disks. ■

As follows from Theorem 13.1.2, any closed bounded disk in a Banach space is a Banach disk.

**Theorem 13.1.2** BANACH DISKS Let $D$ be a bounded disk in a HTVS $X$. Then:

(a) the gauge $p_D$ of $D$ is a norm;

(b) if $D$ is sequentially complete, then $X_D$ is a Banach space, i.e., $D$ is a Banach disk.

**Proof.** Let $D$ and $X$ be as above. (a) Let $x$ be a nonzero element of $X_D$ and choose a neighborhood $U$ of 0 in $X$ such that $x \notin U$. Since $D$ is bounded, $U \supset rD$ for some $r > 0$ so $x \notin rD$. Hence $p_D(x) \geq r > 0$.

(b) Suppose that $D$ is sequentially complete and let $(x_n)$ be a Cauchy sequence in $X_D$. By taking a subsequence if necessary, we may assume that $x_{n+1} - x_n \in 2^{-n}D$ for each $n \in \mathbf{N}$. Since $X_D$'s topology is finer than the subspace topology on $X_D$, $(x_n)$ is a Cauchy sequence in $X$ as well. Since, for any $n$,

$$\begin{aligned} x_n &= x_1 + (x_2 - x_1) + \cdots + (x_n - x_{n-1}) \\ &\in x_1 + 2^{-1}D + \cdots + 2^{-(n-1)}D \subset x_1 + D \end{aligned}$$

and $D$ (hence $x_1 + D$) is sequentially complete, there exists $x \in x_1 + D \subset X_D$ such that $x_n \to x$ in the subspace topology. To see that $p_D(x_n - x) \to 0$ as well, fix $k \geq 2$. Then for each $j \geq k$,

$$\begin{aligned} x_j - x_k &= (x_j - x_{j-1}) + (x_{j-1} - x_{j-2}) + \cdots + (x_{k+1} - x_k) \\ &\in \left(2^{-(j-1)} + 2^{-(j-2)} + \cdots + 2^{-k}\right) D \subset 2^{-(k-1)}D \end{aligned}$$

As $D$ is sequentially closed, it follows that $x - x_k \in 2^{-(k-1)}D$ and therefore $p_D(x - x_n) \to 0$. ■

## 13.2 BORNOLOGICAL SPACES

Bornological spaces are those LCS $X$ for which local boundedness of a linear map $A : X \to Y$, where $Y$ is any LCS, is equivalent to continuity [Theorem 13.2.7(a)]. A smaller class, the ultrabornological spaces, is treated simultaneously because of the close analogy with bornological spaces.

A set $D$ in a TVS is *bornivorous* if $D$ absorbs all bounded sets $B$ in the sense that $B \subset aD$, $a \in \mathbf{F}$, for all sufficiently large $|a|$. In an LCS the disked hull of a bounded set is bounded by Theorems 6.3.2 and 6.3.3. Hence in an LCS, a set $D$ is bornivorous if it absorbs all bounded *disks* $B$. Neighborhoods of 0 are bornivorous by the definition of boundedness but generally there may be other bornivorous sets; there may even be other bornivorous disks (see Example 13.2.8(c)).

**Definition 13.2.1** INFRABORNIVOROUS A disk $D$ is *infrabornivorous* if it absorbs all Banach disks. ■

**Theorem 13.2.2** INFRABORNIVOROUS DISKS In an LCHS $X$ a disk is infrabornivorous iff it absorbs all compact disks, i.e., is "compactivorous."

**Proof.** By Theorem 13.1.2(b), it follows that a compact disk $D$ in an LCHS $X$ is a Banach disk; therefore an infrabornivorous disk is compactivorous.

Conversely, suppose that $D \subset X$ is a compactivorous disk and let $B$ be a Banach disk. If $D$ does not absorb $B$ then for each $n \in \mathbf{N}$ there exists $x_n \in B$ such that $x_n \notin n^2 D$. Since $x_n/n \in (1/n) B$, it follows that $p_B (x_n/n) \leq 1/n$ and $x_n/n \to 0$ in $X_B$. Hence $C = \{x_n/n : n \in \mathbf{N}\} \cup \{0\}$ is compact in $X_B$. Since the Banach space $X_B$ is a complete LCHS, the closed disked hull $\operatorname{cl} C_{bc}$ of $C$ is also compact [Theorem 4.8.9(c)]. Since (as noted in Sec. 13.1) the norm topology of $X_B$ is finer than the subspace topology, $\operatorname{cl} C_{bc}$ is compact in $X$; hence $D$ absorbs $\operatorname{cl} C_{bc}$ and for some $k \in \mathbf{N}$, $x_n/n \in kD$ for every $n$. Hence, for $n > k$, $x_n/n \in nD$ which contradicts $x_n \notin n^2 D$. ■

**Definition 13.2.3** BORNOLOGICAL AND ULTRABORNOLOGICAL An LCS $X$ is:

(a) *bornological* if each bornivorous disk is a neighborhood of 0;

(b) *ultrabornological* if each infrabornivorous disk is a neighborhood of 0.

■

Before giving examples, we prove an external characterization of bornological and ultrabornological spaces in Theorem 13.2.7. First, we establish some useful preliminary results.

**Theorem 13.2.4** Let $A$ be a linear map of a vector space $X$ into a vector space $Y$. If $B$ is balanced in $X$ and $D$ balanced in $Y$, then $D$ absorbs $A(B)$ iff $A^{-1}(D)$ absorbs $B$.

**Proof.** In the notation of the statement, if $A^{-1}(D)$ absorbs $B$, then $B \subset rA^{-1}(D)$ for sufficiently large $r > 0$. Hence $A(B) \subset rA\left(A^{-1}(D)\right) \subset rD$.

Conversely, if $D$ absorbs $A(B)$, then $A(B) \subset rD$ for sufficiently large $r > 0$. Thus $B \subset A^{-1}(A(B)) \subset rA^{-1}(D)$. ■

Local boundedness of a linear map $A$ may now be characterized in terms of the behavior of $A^{-1}$.

**Theorem 13.2.5** $A^{-1}$ FOR LOCALLY BOUNDED $A$ Let $A$ be a linear map of the LCS $X$ into the LCS $Y$. Then the following are equivalent:

(a) $A$ is locally bounded;

(b) $A$ maps bounded disks into bounded disks;

(c) $A^{-1}$ takes bornivorous disks into bornivorous disks.

**Proof.** Let $A$, $X$, and $Y$ be as in the statement. (a) $\Leftrightarrow$ (b) We need only prove that (b) $\Rightarrow$ (a). If $B$ is any bounded subset of $X$, its disked hull $B_{bc}$ is bounded by Theorems 6.3.2 and 6.3.3. Thus $A(B)$ is a subset of the bounded set $A(B_{bc})$.

(b) $\Rightarrow$ (c) Let $D$ be a bornivorous disk in $Y$ and $B$ a bounded disk in $X$. $A(B)$ is bounded, so $D$ absorbs $A(B)$. By Theorem 13.2.4, $A^{-1}(D)$ absorbs $B$ and $A^{-1}(D)$ is seen to be bornivorous.

(c) $\Rightarrow$ (b) Let $B$ be a bounded disk in $X$ and $U$ any disked neighborhood of 0 in $Y$. Since $U$ is bornivorous so is $A^{-1}(U)$, by hypothesis. Hence $A^{-1}(U)$ absorbs $B$ or, equivalently, $U$ absorbs $A(B)$ [Theorem 13.2.4 again]. ■

**Theorem 13.2.6** $A^{-1}$ FOR INFRABOUNDED $A$ Let $A$ be a linear map of the LCS $X$ into the LCS $Y$. Then $A$ maps Banach disks into bounded disks (i.e., $A$ is *infrabounded*) iff $A^{-1}$ takes bornivorous disks into infrabornivorous disks.

**Proof.** Let $A$, $X$, and $Y$ be as in the statement. Suppose that $A$ maps Banach disks into bounded disks and let $D$ be a bornivorous disk in $Y$. If $B$ is a Banach disk in $X$, then $A(B)$ is bounded. Therefore $D$ absorbs $A(B)$ which is equivalent to $A^{-1}(D)$ absorbing $B$ [Theorem 13.2.4]. Thus $A^{-1}(D)$ is infrabornivorous.

Conversely, suppose that $A^{-1}$ takes bornivorous disks into infrabornivorous disks and let $B$ be a Banach disk in $X$. To see that $A(B)$ is bounded, let $U$ be a disked neighborhood of 0 in $Y$. $U$ is bornivorous, so $A^{-1}(U)$ absorbs $B$ [Theorem 13.2.4]. It follows that $U$ absorbs $A(B)$ and $A(B)$ is seen to be bounded. ■

**Theorem 13.2.7** CONTINUITY AND BORNOLOGICITY Let $X$ be an LCS.

(a) $X$ is bornological iff for any LCS $Y$, each locally bounded linear map $A : X \to Y$ is continuous;

(b) If $X$ is ultrabornological then each infrabounded linear map $A$ of $X$ into any LCS $Y$ is continuous. If $X$ is Hausdorff and infrabounded linear maps $A$ of $X$ into any LCS $Y$ are continuous then $X$ is ultrabornological.

**Proof.** Let $X$ be an LCS. (a) Suppose that $X$ is bornological, $A$ is locally bounded and $U$ is a disked neighborhood of 0 in $Y$. Since $U$ is bornivorous, $A^{-1}(U)$ is a bornivorous disk by Theorem 13.2.5(c). Hence $A^{-1}(U)$ is neighborhood of 0 and $A$ is continuous.

Conversely, suppose that locally bounded linear maps are continuous and let $D$ be a bornivorous disk in $X$. Let $p_D$ be the gauge of $D$ and consider the identity map $I : X \to (X, p_D)$, $x \mapsto x$. To see that $I$ is locally bounded, let $B$ be a bounded disk in $X$. Since $D$ is bornivorous, there exists $r > 0$ such that $B \subset rD \subset r\{x \in X : p_D(x) \leq 1\}$ [Theorem 5.3.2] and $B$ is seen to be bounded in $(X, p_D)$. Thus, by hypothesis, $I$ is continuous. Since $D \supset V_{p_D}$, it follows that $D = I^{-1}(D)$ is a neighborhood of 0 in $X$.

(b) If $X$ is ultrabornological then the continuity of infrabounded linear maps $A$ of $X$ into any LCS $Y$ follows from Theorem 13.2.6. As for the converse, let $B = \{a \in \mathbf{F} : |a| \leq 1\}$ and note that, for any $x \in X$, $Bx = \{x\}_{bc}$ is a Banach disk by Theorems 13.2.2 and 13.1.2(b). Thus, if $D$ is an infrabornivorous disk in $X$, it must absorb $Bx$, hence $x$. In other words, infrabornivorous disks must be absorbent. The remainder of the argument is now virtually identical to that used in (a). ■

**Example 13.2.8** BORNOLOGICITY

(a) NORMED IMPLIES BORNOLOGICAL The closed balls $C(0, r)$, $r > 0$, are a base of bounded disked neighborhoods of 0 in any normed space $X$. Consequently, if $D$ is a bornivorous disk, $D$ contains a neighborhood of 0, $C(0, r)$ for some $r > 0$. Hence $D$ is a neighborhood of 0 and therefore $X$ is bornological. Since we have already encountered non-barreled normed spaces (in Examples 11.3.2 and 11.3.3, for example), it follows that BORNOLOGICAL DOES NOT IMPLY BARRELED. Clearly, bornological implies infrabarreled, however.

(b) PSEUDOMETRIZABLE IMPLIES BORNOLOGICAL Local boundedness implies continuity on pseudometrizable spaces [Theorem 6.5.2]. Hence, by Theorem 13.2.7(a), a pseudometrizable LCS is bornological.

(c) NONBORNOLOGICAL SPACES If $X$ is an infinite-dimensional normed space, then $\sigma(X, X')$ is strictly coarser than $\tau(X, X')$ [Example 11.11.7]. Consequently, there must be a disked $\tau(X, X')$-neighborhoods $U$ of 0 which is not a $\sigma(X, X')$-neighborhoods of 0. Since the family of bounded sets is the same for any topology of the dual pair [Theorem 8.8.7], $U$ is bornivorous in $(X, \sigma(X, X'))$ and therefore $(X, \sigma(X, X'))$ is not bornological.

Another way to argue the point is to observe that the identity map $x \mapsto x$ from $(X, \sigma(X, X'))$ to $(X, \tau(X, X'))$ is locally bounded [Theorem 8.8.7] but not continuous and invoke Theorem 13.2.7(a).

(d) FRÉCHET SPACES ARE ULTRABORNOLOGICAL The closed unit ball $U$ of a Fréchet space $X$ is sequentially complete, hence a Banach disk by

Theorem 13.1.2(b). Thus, if $B$ absorbs all Banach disks, $B$ absorbs $U$ and $B$ is therefore a neighborhood of 0. Thus the spaces $D_n \subset D(\mathbf{R})$ (test functions) and the $X_n$ (continuous functions with compact support) of Examples 12.1.5 and 12.1.6, respectively, are ultrabornological.

(e) FINEST LOCALLY CONVEX TOPOLOGY Let $X$ be any vector space over $\mathbf{R}$ or $\mathbf{C}$ with its finest locally convex topology [Example 5.6.4], that having the family of all absorbent disks as a base at 0. As a bornivorous disk is absorbent, $X$ is bornological.

(f) ASSOCIATED BORNOLOGICAL SPACE Let $(X, \mathcal{T})$ be an LCS. The collection of all bornivorous disks of $X$ satisfies the conditions of the basis theorem [4.5.2] for a locally convex topology $\mathcal{T}_b$ on $X$. Clearly $(X, \mathcal{T}_b)$ is bornological. It is called the *associated bornological space* and $\mathcal{T}_b$ the *associated bornological topology*. As any disked $\mathcal{T}$-neighborhood $U$ of 0 is bornivorous, each such $U$ is a $\mathcal{T}_b$-neighborhood of 0 so $\mathcal{T} \subset \mathcal{T}_b$. Hence $\mathcal{T}_b$-bounded sets are $\mathcal{T}$-bounded. A $\mathcal{T}$-bounded set $B$, however, is absorbed by exactly the same class of disks in $(X, \mathcal{T}_b)$ as in $(X, \mathcal{T})$ so the class of bounded sets is the same for each topology. $\mathcal{T}_b$ is the finest locally convex topology for $X$ with the same bounded sets as the original topology.

(g) ASSOCIATED ULTRABORNOLOGICAL SPACE If in (f), instead of taking the bornivorous disks as a base at 0, we take the infrabornivorous disks, we get an ultrabornological topology $\mathcal{T}_{ub}$ called the *associated ultrabornological topology*. $(X, \mathcal{T}_{ub})$ is called the *associated ultrabornological space.* ■

The following result generates a wide class of bornological and ultrabornological spaces.

**Theorem 13.2.9** DIRECT IMAGE LOCALLY CONVEX TOPOLOGY Let $X$ be a vector space and let $\{X_s : s \in S\}$ be a family of bornological or ultrabornological spaces and for each $s \in S$, let $A_s : X_s \to X$ be a surjective linear map. Then $X$ equipped with the direct image locally convex topology for $X$ determined by $\{A_s : s \in S\}$ is bornological or ultrabornological, respectively. Hence, inductive limits of bornological and ultrabornological spaces are bornological and ultrabornological, respectively.

**Proof.** Let $A_s$, $X_s$ and $X$ be as in the statement. Recall from Example 4.11.3(b) that the direct image locally convex topology $\mathcal{T}$ for $X$ is the finest locally convex topology for $X$ with respect to which each of the maps $A_s$ is continuous; those disks $D$ in $X$ for which every $A_s^{-1}(D)$ is a neighborhood of 0 in $X_s$ form a base at 0 for $\mathcal{T}$. Since each $A_s$ is continuous, each is locally bounded by Theorem 6.4.1. Thus, if $D$ is bornivorous disk in $X$, then each $A_s^{-1}(D)$ is a bornivorous disk by Theorem 13.2.5(c). Since each $X_s$ is bornological, each $A_s^{-1}(D)$ is a neighborhood of 0 in $X_s$. Therefore, $D$ is a $\mathcal{T}$-neighborhood of 0 and $X$ is seen to be bornological.

Now suppose that each $X_s$ is ultrabornological and let $D$ be an infrabornivorous disk in $X$. To show that $D$ is a $\mathcal{T}$-neighborhood of 0, we show that each $A_s^{-1}(D)$ is an infrabornivorous disk—hence a neighborhood of 0—in each of the ultrabornological spaces $X_s$. Equivalently, by Theorem 13.2.2, we must show that $A_s^{-1}(D)$ is compactivorous. To this end, let $B$ be a compact disk in $X_s$. As its continuous image $A_s(B)$ is a compact disk in $X$, $D$ absorbs $A_s(B)$. Hence $A_s^{-1}(D)$ absorbs $B$ by Theorem 13.2.4. ■

Thus, quotients and locally convex direct sums of bornological spaces are bornological, hence inductive limits of normed and pseudometrizable LCS are bornological; the LF-spaces spaces $D(\mathbf{R})$ and $C_c(G, \mathbf{F})$ of Examples 12.1.5 and 12.1.6 are ultrabornological. Using this, we can now easily prove that differentiation is a continuous linear map of $D(\mathbf{R})$ onto $D(\mathbf{R})$ by showing that the differentiation operator $D : D(\mathbf{R}) \to D(\mathbf{R})$ is locally bounded. To this end let $B$ be a bounded subset of $D(\mathbf{R})$ and let $D_n$ be as in Example 12.1.5: $D(\mathbf{R}) = \text{str ind}_n D_n$. As such, there is some $n \in \mathbf{N}$ such that $B \subset D_n$ by Theorem 12.1.7(a). Therefore, each of the seminorms

$$p_m(x) = \sup\left\{\left|x^{(j)}(K_n)\right| : 0 \le j \le m\right\}, \quad m \in \mathbf{N}$$

which define the topology on $D_n$ is bounded on $B$. Clearly, $D(B) \subset D_n$ and each $p_n$ is bounded on $D(B)$ as well. Compare this to the result of Example 14.1.2(c), a case in which differentiation is discontinuous.

As LF-spaces are never metrizable [Theorem 12.1.8], they constitute a class of nonmetrizable bornological spaces.

**Theorem 13.2.10** BORNOLOGICAL EQUIVALENTS The following statements about an LCHS $X$ are equivalent:

(a) $X$ is bornological.

(b) $X$ is an inductive limit of normed spaces. More specifically, let $\mathcal{B}$ denote the bounded or closed bounded disks of $X$ and for each $B \in \mathcal{B}$, let $X_B$ denote the normed space of Sec. 13.1. If $X$ is bornological, then $X = \text{ind}_{B \in \mathcal{B}} X_B$ determined by the canonical injections $I_B : X_B \to X$, $x \mapsto x$ $(B \in \mathcal{B})$.

(c) Each locally bounded seminorm on $X$ is continuous [cf. Theorem 11.4.3].

(d) $X$ carries the Mackey topology $\tau(X, X')$ and each locally bounded linear functional on $X$ is continuous.

**Proof.** $X$ denotes an LCHS. We show that (a) $\Rightarrow$ (b) $\Rightarrow$ (c) $\Rightarrow$ (a), then (a) $\Leftrightarrow$ (d).

(a) $\Rightarrow$ (b) Let $\mathcal{B}$ denote the class of bounded disks in $X$. For $B \in \mathcal{B}$, let $X_B = ([B], p_B)$ be the normed space of Sec. 13.1. Let $\mathcal{T}$ be the norm topology on $X$ and let $\mathcal{T}_i$ be the inductive limit topology determined by the normed spaces $X_B$ and canonical injections $I_B : X_B \to X$, $x \mapsto x$, $B \in \mathcal{B}$. As $\mathcal{T}$ on $X_B$ is finer than the topology $X_B$ inherits as a subspace of $(X, \mathcal{T})$, each of the canonical injections $I_B : X_B \to (X, \mathcal{T})$ is continuous. Hence $\mathcal{T} \subset \mathcal{T}_i$.

If $U$ is a basic $\mathcal{T}_i$-neighborhood of 0 in $X$, i.e., if $U$ is a disk in $X$ whose intersection with each $X_B$ is a neighborhood of 0 in $X_B$, then for any $B \in \mathcal{B}$, there exists $r_B > 0$ such that $r_B B \subset U \cap X_B \subset U$. Thus $U$ is a bornivorous disk in $X$. Since $X$ is bornological, $U$ is a $\mathcal{T}$-neighborhood of 0. Hence $\mathcal{T}_i \subset \mathcal{T}$.

The same exact argument shows that the family of *closed* bounded disks of $X$ could also have been used. If $X$ were not Hausdorff, the result would still hold but the $X_B$'s would be seminormed rather than normed spaces.

(b) $\Rightarrow$ (c) If $X$ is an inductive limit of normed (or seminormed) spaces, then $X$ is bornological by Theorem 13.2.9 and Example 13.2.8(a). If $p$ is a locally bounded seminorm on $X$ (maps bounded sets into bounded sets), then $\overline{V}_p = \{x \in X : p(x) \leq 1\}$ is a bornivorous disk because, for any bounded subset $B \subset X$, there exists $r > 0$ such that $r^{-1}B \subset \overline{V}_p$. Therefore $\overline{V}_p$ is a neighborhood of 0 in $X_B$, hence a neighborhood of 0 in $X$. The continuity of $p$ now follows from Theorem 5.2.2.

(c) $\Rightarrow$ (a) Suppose $B$ is a bornivorous disk in $X$ with gauge $p_B$ and that $D \subset X$ is bounded. Since there exists $r > 0$ such that $D \subset rB$, $\sup p_B(D) \leq r$. Hence $p_B$ is locally bounded, therefore continuous by hypothesis. The continuity of $p_B$ implies that $B$ is a neighborhood of 0 [Theorem 5.3.3(b)].

(a) $\Rightarrow$ (d) If $X$ is bornological then each bornivorous disk—hence every bornivorous barrel—is a neighborhood of 0; therefore $X$ is infrabarreled [Definition 11.11.1]. Infrabarreled spaces carry the Mackey topology by Theorem 11.11.6. The continuity of locally bounded linear functionals on $X$ follows from Theorem 13.2.7(a).

(d) $\Rightarrow$ (a) Let $X$ carry the Mackey topology $\tau(X, X')$ and suppose that each locally bounded linear functional on $X$ is continuous. Let $\mathcal{T}_b$ be the associated bornological topology on $X$ [Example 13.2.8(f)]; as noted there, $\tau(X, X') \subset \mathcal{T}_b$. To show that $\tau(X, X') \supset \mathcal{T}_b$, we show that $(X, \mathcal{T}_b)$ and $(X, \tau(X, X'))$ have the same continuous dual. It will then follow from the Mackey-Arens theorem [8.7.4] that $T_b \subset \tau(X, X')$.

Since $\tau(X, X') \subset \mathcal{T}_b$, each $\tau(X, X')$-continuous linear functional on $X$ is $\mathcal{T}_b$-continuous. Thus it only remains to show that each $\mathcal{T}_b$-continuous linear functional on $X$ is $\tau(X, X')$-continuous. Let $p_B$ denote the gauge of the bornivorous disk $B \subset X$. Since $\mathcal{T}_b$ is generated by the saturated family of seminorms $p_B$ where $B$ is a bornivorous disk in $X$, the $\mathcal{T}_b$-continuity of a linear functional $f$ on $X$ means that $|f| \leq p_B$ for some bornivorous disk $B$ [Theorem 5.7.3]. As noted in the proof of (c) $\Rightarrow$ (a), $p_B$ is locally bounded which implies that $f$ is a locally bounded linear functional and therefore $\tau(X, X')$-continuous by hypothesis. ■

By merely replacing "bounded disk" by "Banach disk" in the proof of (a) $\Rightarrow$ (b)—that a space is bornological iff it is the inductive limit of normed spaces—of Theorem 13.2.10, we obtain:

**Theorem 13.2.11** ULTRABORNOLOGICAL SPACES If the LCHS $X$ is ultrabornological, then $X$ is the inductive limit of Banach spaces. In particular, $X$ is the inductive limit of $\{X_B : B \text{ is a Banach disk in } X\}$. Thus, in view of Theorem 13.2.9 (and Example 13.2.8(d)), an LCHS is ultrabornological iff it can be written as an inductive limit of Banach spaces.

As an immediate consequence we have:

**Theorem 13.2.12** SEQUENTIALLY COMPLETE + BORNOLOGICAL $\Rightarrow$ ULTRABORNOLOGICAL If the LCHS $X$ is sequentially complete and bornological, then it is ultrabornological.

**Proof.** Let $\mathcal{B}$ denote the set of closed bounded disks of the sequentially complete bornological LCHS $X$. Since $X$ is bornological, $X = \text{ind}_{B \in \mathcal{B}} X_B$ [Theorem 13.2.10(b)]. Since $X$ is sequentially complete, so is each $B$. Thus each $B$ is a Banach disk by Theorem 13.1.2(b). As Banach spaces are ultrabornological [Example 13.2.8(d)], the desired result follows from Theorem 13.2.9. ■

Recall that an infrabarreled space is an LCS in which each bornivorous barrel is a neighborhood of 0; consequently bornological spaces are infrabarreled. It is also clear that ultrabornological spaces are barreled since (1) Banach spaces are barreled [Theorems 11.7.2 and 11.6.4], and (2) inductive limits of barreled spaces are barreled [Theorem 11.12.2]. Hence ultrabornological spaces $X$ carry $\beta(X, X')$ by Theorem 11.3.6.

Bornological does not imply barreled [Example 13.2.8(a)]. Barreled does not imply bornological either, but this is a very deep result. Nachbin [1954] and Shirota [1954] showed independently that a space $C(T, \mathbf{R}, c)$ of continuous real-valued functions on a completely regular Hausdorff space $T$ with compact-open topology is bornological iff $T$ is realcompact. Schmets and De Wilde [1971] sharpened this to: $C(T, \mathbf{R}, c)$ is *ultra*bornological iff $T$ is realcompact. Barreledness of $C(T, \mathbf{R}, c)$ may also be characterized by a condition on $T$ and there are nonrealcompact spaces that satisfy it. Thus there are barreled spaces that are not bornological. For the details, aside from the original papers cited above, see Beckenstein et al. [1977, pp. 100, 103], Schmets [1976, p. 66], or Weir [1975, p. 155].

The result below generalizes the fact the dual $X'$ of a normed space $X$ is complete in its norm topology.

**Theorem 13.2.13** $(X', \beta(X', X))$ COMPLETE If $X$ is bornological then $(X', \beta(X', X))$ is complete.

**Proof.** Let $X$ be bornological. We show that any Cauchy net $(f_s)$ in $(X', \beta(X', X))$ converges to its "pointwise" limit $f$.

To define $f$, choose $x \in X$ and a $\sigma(X, X')$-bounded set $D$ to which $x$ belongs. Since $(f_s)$ is a $\beta(X', X)$-Cauchy net, given $r > 0$, there exists $t'$ such that $f_s - f_t \in rD^\circ$ for $s, t \geq t'$. Hence $|f_s(x) - f_t(x)| \leq r$ for $s, t \geq t'$. Thus $(f_s(x))$ is a Cauchy net in $\mathbf{F}$. We define $f(x)$ to be the limit of $(f_s(x))$; $f$ is obviously linear.

It remains to show that $f$ is continuous and that $f_s \to f$ with respect to $\beta(X', X)$. To show that $f$ is continuous, we show that $f$ is locally bounded and use the bornologicity of $X$. To this end, let $D \subset X$ be $\sigma(X, X')$-bounded. For sufficiently large indices $s$ and $t$, $f_s - f_t \in D^\circ$; thus, for any $x \in D$, $|f_s(x) - f_t(x)| \leq 1$. Since each $f_t \in X'$, each $f_t$ is locally bounded [Theorem 6.4.1]. Fixing $t$, for sufficiently large $s$ and $x \in D$,

$$|f_s(x)| \leq 1 + |f_t(x)| \leq 1 + \sup_{x \in D} |f_t(x)| \doteqdot r \qquad (*)$$

Taking the limit, it follows that $f(D)$ is bounded and the continuity of $f$ follows from Theorem 13.2.7(a).

Last, to show that $f_s \to f$ with respect to $\beta(X', X)$, let $D$ be a $\sigma(X, X')$-bounded set in $X$. Since $(f_s)$ is a $\beta(X', X)$-Cauchy net, given $r > 0$, there exists $t'$ such that $f_s - f_t \in rD^\circ$ for $s \geq t$. Given any $x$ in $D$, we may choose an index $s \geq t'$ such that $|f(x) - f_s(x)| \leq r$. Hence, for $t \geq t'$, $|f(x) - f_t(x)| \leq 2r$ for each $x$ in $D$, i.e., $f - f_t \in 2rD^\circ$, which yields the desired convergence statement. ■

Theorem 11.12.7(b) says that if $B$ is a bornivorous disk in a finite-codimensional subspace $M$ of the LCS $X$ then there is a bornivorous disk $B'$ in $X$ such that $B = B' \cap M$. It follows immediately that a finite-codimensional subspace $M$ of a bornological space is bornological. Generally, however, even a closed subspaces of a bornological space need not be bornological. By Theorem 13.2.12 we know that sequentially complete bornological spaces are ultrabornological and, by the discussion after Theorem 13.2.12, ultrabornological spaces are barreled. Köthe [1983, p. 384] gives an example of a closed subspace of a bornological space which is complete, therefore sequentially complete, but not barreled—hence not bornological. Countable products of bornological spaces are bornological but for uncountable products, the situation is more complex. To describe what happens, we mention that an *Ulam measure* on a set $S$ is $\{0, 1\}$-valued countably additive set function $m$ defined on the power set $2^S$ of $S$ which vanishes on each singleton of $S$ but is such that $m(S) = 1$. Such measures may not even exist; if they do, the cardinality of the set $S$ would have to be immense. The result for arbitrary products of bornological spaces is the *Mackey-Ulam* theorem: The product of a family $\{X_s : s \in S\}$ of bornological spaces is bornological iff $S$ does not admit an Ulam measure [Köthe 1983, p. 392; de Wilde 1978, p. 40; Jarchow 1981, p. 281]. Thus, for all practical purposes, the product of bornological spaces is bornological. The case of denumerable products is discussed in Exercise 13.203.

| LCS | Disk which must be neighborhood of 0 |
|---|---|
| barreled | barrel |
| infrabarreled | bornivorous barrel |
| bornological | bornivorous |
| ultrabornological | infrabornivorous (absorbs compact disks) |

**Table 13.2.14**

## 13.3 EXERCISES

13.101 Give an example of a bounded disk which is not absorbent in $\mathbf{R}^2$.

13.102 Show that an absorbent disk $D$ in an LCS $X$ is:

(a) bornivorous iff its gauge is locally bounded,

(b) infrabornivorous iff its gauge is infrabounded.

13.103 Let $X$ and $Y$ be LCS. Let $X_b$ be the associated bornological space of $X$ [Example 13.2.8(f)], i.e., $X$ retopologized by taking all bornivorous disks as a base at 0. Show that a linear map $A : X \to Y$ is locally bounded iff $A : X_b \to Y$ is continuous.

13.104 INDUCTIVE LIMITS Let $\mathcal{B}$ be a family of bounded disks in the LCHS $X$. For each $B \in \mathcal{B}$, let $X_B$ be as in Sec. 13.1 and suppose that $\cup\{X_B : B \in \mathcal{B}\}$ spans $X$. Show that a disk $D \subset X$ is a neighborhood of 0 in $\text{ind}_{B\in\mathcal{B}} X_B$ iff $D$ absorbs each $B \in \mathcal{B}$. Using this, show that a compactivorous disk $D$ in an ultrabornological space is a neighborhood of 0.

13.105 ULTRABORNOLOGICAL SPACES Let $\mathcal{K}$ denote the set of compact disks in an LCS $X$. For $K \in \mathcal{K}$, $X_K$ is as in Sec. 13.1. Show that $X$ is ultrabornological iff $X = \text{ind}_{K\in\mathcal{K}} X_K$ determined by the canonical injections $I_K : K \to X$, $x \mapsto x$ $(K \in \mathcal{K})$.

13.106 FINITE PRODUCTS Show that a finite product of bornological or ultrabornological spaces is bornological or ultrabornological, respectively.

13.107 ASSOCIATED SPACES (a) Each continuous linear map $A$ of the bornological space $X$ into the LCHS $Y$ remains continuous when $Y$ is replaced by the associated bornological space $Y_b$ of Example 13.2.8(f).

(b) Each continuous linear map $A$ of the ultrabornological space $X$ into the LCHS $Y$ remains continuous when $Y$ is replaced by the associated ultrabornological space $Y_{ub}$ of Example 13.2.8(g).

13.108 This extends the result of Theorem 13.2.13 that, for any bornological space $X$, $(X', \beta(X', X))$ is complete. Let $X, Y$ be LCS and let the linear space $L(X, Y)$ of all continuous linear maps of $X$ into $Y$ carry the topology of uniform convergence on bounded sets defined after Theorem 11.2.3. Show that if $Y$ is complete, then so is $L(X, Y)$.

13.109 If $X$ is a metrizable LCS, for each bounded set $B$ of $X$ there is a bounded disk $B'$ such that the topology induced by $X$ on $B$ and that induced by $X_{B'}$ [as in Sec. 13.1] on $B$ coincide.

13.110 In a metric LCS, every bornivore is a neighborhood of 0.

13.111 If $X$ is bornological, then each convex bornivore is a neighborhood of 0.

13.112 If each locally bounded linear map of the LCS $X$ into any seminormed space is continuous, then $X$ is bornological. Show too that "seminormed space" may be replaced by "Banach space."

13.113 SEQUENTIAL DESCRIPTION OF LOCALLY BOUNDED LINEAR MAPS For any TVS $X$ and $Y$, a linear map $A : X \to Y$ is locally bounded iff it maps null sequences into bounded sequences [cf. Exercise 13.201(d)].

13.114 WHEN SEQUENTIAL CONTINUITY IMPLIES CONTINUITY Let $A$ be a linear map of the bornological space $X$ into the LCS $Y$. If $A$ is sequentially continuous, then $A$ is continuous.

13.115 Let $A$ be a continuous linear map of the TVS $X$ into the pseudometrizable TVS $Y$. Show that if $A^{-1}$ takes bounded sets into bounded sets, then $A$ is relatively open—i.e., if $G \subset X$ is open then $A(G)$ is an open subset of $A(X)$.

13.116 Let $D$ be a disk in a vector space $X$. Its gauge $p_D$ in $X_D$ is a norm in $X_D$ [as in Sec. 13.1] iff $D$ does not contain a nonzero subspace of $X$.

13.201 MACKEY CONVERGENCE AND BORNOLOGICAL SPACES Let $X$ be an LCHS. A sequence $(x_n)$ in $X$ is *Mackey convergent* to $x$, $x_n \to_M x$, if there exists a bounded disk $B$ such that $x_n \to x$ in $X_B$ [as in Sec. 13.1]. This sort of convergence means that there exists a sequence $(r_n)$ of real numbers such that $r_n \to \infty$ while $r_n(x_n - x)$ remains bounded in $X$. (Proof?) It shifts dependence on $X$'s topology to just the dual pair as it depends only on the weakly bounded subsets of $X$ for its definition. As the spaces $X_B$ carry finer topologies than that induced by $X$, Mackey convergence implies convergence. A sufficient condition for the converse is given in (a).

(a) If $X$ is metrizable then its convergent sequences are Mackey convergent.

(b) Give an example of a convergent sequence that is not Mackey convergent.

(c) A disk $D$ in $X$ is bornivorous iff it absorbs all Mackey convergent sequences. On the basis of this result, give another description of bornological spaces.

(d) SEQUENTIAL DESCRIPTION OF LOCALLY BOUNDED MAPS A linear map $A$ of $X$ into the LCS $Y$ is locally bounded iff:

(1) $A$ maps Mackey convergent sequences into Mackey convergent sequences, or

(2) $A$ maps Mackey convergent sequences of $X$ into bounded subsets of $Y$.

Use these results to give another description of bornological spaces, analogous to that of Theorem 13.2.7(a)

13.202 FAST CONVERGENCE AND ULTRABORNOLOGICAL SPACES Let $X$ be an LCHS. A sequence $(x_n)$ in $X$ is *fast convergent* to $x$ if there is a Banach disk $B$ such that $x_n \to x$ in $X_B$ [as in Sec. 13.1]. We write $x_n \to_F x$. Clearly, fast convergence implies Mackey convergence.

(a) Show that $x_n \to_F x$ iff there exists a compact disk $K$ such that $x_n \to x$ in $X_K$.

(b) If $x_n \to_F x$, there exists $r_n \to \infty$ such that $r_n(x_n \to x) \to_F 0$.

(c) If $X$ is a Fŕechet space, its convergent sequences are fast convergent (cf. Exercise 13.201(a)).

(d) A disk $D$ in $X$ is infrabornivorous iff it absorbs all fast convergent sequences.

(e) SEQUENTIAL DESCRIPTION OF INFRABOUNDED MAPS A linear map $A$ of the LCHS $X$ into the LCHS $Y$ is infrabounded iff it is bounded on compact disks of $X$ or on the fast convergent null sequences of $X$, or if $A$ maps fast convergent null sequences into fast convergent null sequences. Use these results to give another description of ultrabornological spaces analogous to that of Theorem 13.2.7(b).

13.203 DENUMERABLE PRODUCTS (a) Let $(X_n)$ be a sequence of LCS and let $A$ be a linear map of $\Pi_{n \in \mathbf{N}} X_n$ into the Banach space $Y$. If $A$ is locally bounded, show that for all but finitely many $n$, the restriction $A_n$ of $A$ to the (appropriately identified) subspace $X_n$ of $X$ is 0. (b) Show that a denumerable product of bornological spaces is bornological.

13.204 NONCONVEX BORNOLOGICAL SPACES Let $X$ be a TVS. If every locally bounded linear map of $X$ into any TVS $Y$ is continuous, we call $X$ *quasi-bornological.* By Theorem 6.5.2, any pseudometrizable TVS is quasi-bornological. For the sake of developing an internal characterization of these spaces, we introduce the notion of "bornivorous string." As mentioned in Exercise 11.201, a string is a sequence $(U_n)$ of absorbent balanced sets $U_n$ called knots such that $U_{n+1} + U_{n+1} \subset U_n$ for every $n$. If each knot is bornivorous, we call $(U_n)$ a *bornivorous string.*

(a) The set of all knots of all bornivorous strings of a TVS $(X, \mathcal{T})$ is a base at 0 for a vector topology $\mathcal{T}_{qb}$ called the *associated quasi-bornological topology.* Show that $\mathcal{T}_{qb}$ is the finest vector topology which is stronger than $\mathcal{T}$ and has the same bounded sets as $\mathcal{T}$.

(b) $X$ is quasi-bornological iff any knot in any bornivorous string is a neighborhood of 0.

(c) $(X, \mathcal{T})$ is quasi-bornological iff $\mathcal{T} = \mathcal{T}_{qb}$.

(d) Let $\mathcal{T}_c$ be the finest locally convex topology on $(X, \mathcal{T})$ which is coarser than $\mathcal{T}$. Show that if $(X, \mathcal{T})$ is quasi-bornological, then $(X, \mathcal{T}_c)$ is bornological.

(e) $X$ is quasi-bornological iff for any complete metrizable TVS $Y$, any locally bounded linear map of $X$ into $Y$ is continuous.

(f) Any locally convex quasi-bornological space is bornological.

(g) A bornological space need not be quasi-bornological. Iyahen [1968] introduced quasi-bornological spaces, calling them "ultrabornological." We use "quasi-bornological" to prevent confusion with the ultrabornological spaces of Definition 13.2.3. For stability properties of quasi-bornologicity, see Adasch et al. [1978, p. 62ff.]

13.205 NON-ARCHIMEDEAN BORNOLOGICAL SPACES Let $X$ be a locally $K$-convex space [Exercises 4.203 and 5.205] over a nontrivially ultravalued field $K$. A subset $E$ of $X$ is called bornivorous if $E$ absorbs each bounded subset $B$ of $X$ (i.e., $B \subset aE$ for sufficiently large $|a|$ where $a \in K$). $X$ is called *$K$-bornological* if each bornivorous $K$-disk is a neighborhood of 0.

(a) $X$ is $K$-bornological iff every locally bounded linear map $A$ of $X$ into any locally $K$-convex space $Y$ is continuous.

(b) If $X$ is metrizable, then $X$ is $K$-bornological.

(c) In the terminology of Exercise 13.204, suppose that any knot in any bornivorous string in $X$ is a neighborhood of 0; call such an $X$ *$K$-quasi-bornological.* Show that a $K$-quasi-bornological space must be $K$-bornological but that the converse is false.

### Hints to Exercises

13.105 A compact disk is a Banach disk so ind $\{X_K : K \in \mathcal{K}\}$ is ultrabornological. If $X$ is ultrabornological and $X = \text{ind}\{X_B : B \text{ Banach disk}\}$, let $\mathcal{T}$ denote this inductive limit topology and let $\mathcal{T}'$ denote the topology of ind$\{X_K : K \in \mathcal{K}\}$. Each compact disk is a Banach disk [Theorem 13.1.2], so $\mathcal{T} \subset \mathcal{T}'$. Conversely, let $D$ be a compactivorous disk and use Theorem 13.2.2.

13.108 Let $(A_s)$ be a Cauchy net in $L(X, Y)$. Since $Y$ is complete, the "pointwise" limit $A$, $Ax = \lim A_s x$ exists for every $x \in X$; $A$ is clearly linear. To show that $A$ is locally bounded, let $B$ be any bounded subset of $X$, let $q$ be any continuous seminorm on $Y$, and show that $q(A(B))$ is bounded. Since $(A_s)$ is Cauchy, given any $r > 0$, $q(A_s x - A_t y) < r$ for sufficiently large $s$ and $t$ and any $x, y \in B$. Hence $q(Ax - A_t y) \leq r$ for sufficiently large $t$. Since $A_t$ is continuous, $A_t(B)$ is bounded, so there exists $k$ such that $q(A_t x) \leq k$ for every $x$ in $B$. Conclude that $q(Ax) \leq k + r$ for each $x \in B$.

13.109 Let $(U_n)$ be a base of disked neighborhoods of 0 in $X$ and let $(t_n)$ be a sequence of positive numbers such that $B \subset \cap_{n\in\mathbf{N}} t_n U_n$. $B' = \cap_{n\in\mathbf{N}} 2^n t_n U_n$ is bounded and for each $x \in B$, $r > 0$ and $k \in \mathbf{N}$,

$$\left(x + r\bigcap_{n=1}^{k} 2^n t_n U_n\right) \cap B \subset x + (B - B) \cap r\bigcap_{n=1}^{k} 2^n t_n U_n$$
$$\subset x + r\bigcap_{n\in\mathbf{N}} 2^n t_n U_n = x + rB' \quad \text{if } r2^{k+1} \geq 1$$

13.111 If $D$ is a convex bornivore then for any bounded disk $B$ there exists $r_B > 0$ such that $r_B B \subset D$. Let $\mathcal{B}$ denote the set of bounded disks and show that the convex hull $(\cup_{B\in\mathcal{B}} r_B B)_c \subset D$ and $(\cup_{B\in\mathcal{B}} r_B B)_c$ is a neighborhood of 0 in $X$.

13.112 Let $D$ be a bornivorous disk in $X$, let $p_D$ be its gauge, and consider the identity map $x \mapsto x$ of $X$ into the seminormed space $(X, p_D)$. For the second part, consider a completion of $(X, p_D)/p_D^{-1}(0)$.

13.113 If $A$ is not locally bounded, there is a bounded set $B$ such that $A(B)$ is unbounded. Hence there is a neighborhood $V$ of 0 in $Y$ and points $x_n$ in $B$ such that $Ax_n \notin n^2 V$ for each $n$. Now $x_n/n \to 0$ but $(A(x_n/n))$ is not bounded.

13.114 Use the result of Exercise 13.113.

13.115 Use Theorem 6.5.2.

13.201 (a) It suffices to consider $x_n \to 0$. Let $B$ be the disked hull of $\{x_n\}$ and let $(U_n)$ be a decreasing base of disked neighborhoods of 0 in $X$. Choose an increasing sequence $(n_k)$ of positive integers such that $x_{n_k} \in k^{-1}U_k$ $(k \in \mathbf{N})$. For $n_k \leq n < n_{k+1}$, $kx_n \in B$.

(c) If $D$ is bornivorous and $x_n \to_M x$ then $D$ absorbs $\{x_n\} \cup \{x\}$. Conversely, if $D$ is not bornivorous, there is some bounded disk $B$ which $D$ does not absorb. Hence, for every $n$, there exists $x_n \in B$ such that $x_n \notin n^2 D$. Since $x_n/n \in (1/n)B$, $x_n/n \to_M 0$.

(d) If $A$ is locally bounded and $x_n \to_M 0$, there is some bounded disk $B$ such that for each $r > 0$, $x_n \in rB$ eventually; hence $Ax_n \in rA(B)$ eventually. Therefore a locally bounded map maps Mackey convergent sequences into Mackey convergent sequences, hence into bounded sets. If $A$ is not locally bounded, there is a bounded set $B$ such that $A(B)$ is not bounded. As such, there must be a disked neighborhood $V$ of 0 in $Y$ and points $x_n \in B$ such that $Ax_n \notin n^2 V$; now note that $x_n/n \to_M 0$, while $\{A(x_n/n)\}$ is unbounded.

13.202 (a) It suffices to consider null sequences, so suppose that $x_n \to 0$ in $X_B$ for some Banach disk $B$. The disked hull $K$ of $\{x_n\} \cup \{0\}$ is compact in $X_B$, hence in $X$ and $x_n \to 0$ in $X_K$.

(c) See hint to Exercise 13.201(a).

(d) See hint to Exercise 13.201(c).

(e) See hint to Exercise 13.201(d).

13.203 (a) Argue by contradiction. Create a bounded sequence $(x_n)$ in $X$ such that $(Ax_n)$ is unbounded.

(b) By Exercise 13.112, it suffices to show that any locally bounded linear map $A$ of $\Pi_{n \in \mathbf{N}} X_n$ into a Banach space $Y$ is continuous. By (a), there exists a finite set $H \subset \mathbf{N}$ such that if $U$ is the unit ball of $Y$, then $A^{-1}(U) = \Pi_{n \in H} A_n^{-1}(U) \times \Pi_{n \notin H} X_n$ where $A_n$ denotes the restriction of $A$ to $X_n$.

13.204 (b) SUFFICIENCY Suppose that the condition holds, let $Y$ be a TVS, and let $A : X \to Y$ be a locally bounded linear map. Given a balanced neighborhood $W$ of 0 in $Y$, choose a string $(U_n)$ of neighborhoods of 0 in $Y$ with $U_1 = W$. Then, since $A$ is locally bounded, $\{A^{-1}(U_n) : n \in \mathbf{N}\}$ is a bornivorous string in $X$. By hypothesis, each knot in such a string is a neighborhood of 0; hence $A^{-1}(W)$ is a neighborhood of 0.

NECESSITY If $(X, \mathcal{T})$ is quasi-bornological, show that any $\mathcal{T}$-bounded set $B$ is $\mathcal{T}_{qb}$-bounded. Now consider the identity map from $(X, \mathcal{T})$ onto $(X, \mathcal{T}_{qb})$.

(e) See Iyahen [1968] (the result is Iyahen's) or Adasch et al. [1978, p. 61].

(g) See Robertson [1958].

13.205 (a) NECESSITY Show that $A^{-1}$ takes $K$-convex neighborhoods of 0 in $Y$ into bornivorous $K$-disks.

SUFFICIENCY Let $\mathcal{B}$ be the set of bornivorous $K$-disks in $X$. Show that for every $B \in \mathcal{B}$, $B + B \subset B$ and for $0 < |a| < 1$, $aB \in \mathcal{B}$, so that $\mathcal{B}$ is a base of neighborhoods of 0 for a locally $K$-convex topology $\mathcal{T}_b$ for $X$ [cf. Example 13.2.8(f)]. Now consider the identity map from $X$ onto $(X, \mathcal{T}_b)$.

(c) Suppose that $X$ is $K$-quasi-bornological and let $B$ be a bornivorous $K$-disk. For $0 < |a| < 1$, $\{a^{n-1}B : n \in \mathbf{N}\}$ is a bornivorous string, from which it follows by hypothesis that $B$ is a neighborhood of 0. For the false converse, see Prolla [1982, p. 110].

# Chapter 14

# Closed Graph Theorems



The *graph* $G(A)$ of a map $A : X \to Y$ is the set $\{(x, Ax) : x \in X\}$. This chapter is concerned with pairs $(X, Y)$ of TVS for which linear maps with closed graphs in $X \times Y$ must be continuous. When this happens, we say we have a *closed graph theorem.*

The earliest closed graph theorem [Banach 1932, p. 41, Th. 7] was proved for complete metrizable TVS $X$ and $Y$. Although the assumptions on $X$ and $Y$ are symmetric, they were utilized differently. The completeness and metrizability were used directly on the range side, but only by way of the Baire category theorem [11.7.2] in the domain—essentially the same proof shows that if $X$ is a Baire (= nonmeager) TVS and $Y$ a complete pseudometrizable TVS, a closed graph theorem holds.

After some preliminary things about maps with closed graphs in general, we obtain some closed graph theorems in Sec. 14.3. After some discussion of an equivalent dual result—the open mapping theorem—we work toward a significant generalization of the closed graph theorem, de Wilde's closed graph theorem for webbed spaces, in Sec. 14.7, namely that a *sequentially* closed linear map of an *ultrabornological* (= inductive limit of Banach spaces) space into a *webbed* space [Sec. 14.6] must be continuous.

In Sec. 14.9 we discuss without proofs some other closed graph theorems such as Pták's and some for topological groups.

## 14.1 MAPS WITH CLOSED GRAPHS

In this section $S$ and $T$ denote topological spaces; $f$ is a map from $S$ into $T$. The set $G(f) = \{(s, f(s)) : s \in S\}$ is called the *graph of f*. We say that $f$ is a *closed map* if $G(f)$ is a closed subset of $S \times T$ in the product topology. Continuity usually ($T$ Hausdorff) implies closedness but even closed *linear* maps do not have to be continuous [Example 14.1.2(b)]. We discuss some basic descriptions of closed maps and connections between continuity and closedness in this section. We begin with a useful description of closedness.

**Theorem 14.1.1** Criterion for Closed Graph Let $f$ map the topological space $S$ into the topological space $T$; $f$ has a closed graph iff for any net $(x_r)$ in $S$

$$x_r \to x \text{ and } f(x_r) \to y \Rightarrow y = f(x)$$

If the condition only holds for sequences, we say that $f$ and $G(f)$ are *sequentially closed.*

**Remarks.** An important difference between closedness and continuity is that for continuous maps $f$, if $x_r \to x$ then $f(x_r)$ *must* converge. If $S$ and $T$ are first countable spaces, sequential closedness is the same as closedness.

**Proof.** We use the notation of the statement. If $G(f)$ is closed, $x_r \to x$, and $f(x_r) \to y$, then obviously $(x, y) \in \operatorname{cl} G(f) = G(f) \Leftrightarrow y = f(x)$. Conversely, suppose that the condition holds and $(x, y) \in \operatorname{cl} G(f)$. If so, there is a net $(x_r, f(x_r))$ from $G(f)$ such that $(x_r, f(x_r)) \to (x, y)$. By the continuity of projections, $x_r \to x$ and $f(x_r) \to y$. By the condition, $y = f(x)$ and therefore $(x, y) \in G(f)$. ■

A continuous map $f$ which does not have a closed graph is the map of $\mathbf{R}$ with its usual topology into $\mathbf{R}_t$, $\mathbf{R}$ with the trivial topology sending 0 into 1 and everything else into 0. Clearly $f$ is continuous. Consider a basic open set $U = (a, b) \times \mathbf{R} \subset \mathbf{R} \times \mathbf{R}_t$, $a < 0, b > 0$, to which $(0, 0)$ belongs. Since $(0, 1) \in U$, $(0, 0) \in \operatorname{cl} G(f)$. Since $(0, 0) \notin G(f)$, $f$ is not closed. If however the range space $T$ is Hausdorff, continuous maps must be closed: To use the criterion of Theorem 14.1.1, suppose that $x_r \to x$ and $f(x_r) \to y$. If $f$ is continuous, $f(x_r)$ must converge to $f(x)$. Since limits are unique in a Hausdorff space, $f(x) = y$.

Two instances in which closedness does not imply continuity are given in Example 14.1.2.

**Example 14.1.2** (a) CLOSED BUT DISCONTINUOUS. The map of $\mathbf{R}$ into $\mathbf{R}$ (with its usual topology in each case) sending 0 into 0 and $x$ into $1/x$ otherwise has a closed graph but is discontinuous.

(b) DISCONTINUOUS LINEAR MAP WITH CLOSED GRAPH Let $\mathbf{R}_d$ denote $\mathbf{R}$ with the discrete topology. Then the identity map $I : \mathbf{R} \to \mathbf{R}_d$, $x \mapsto x$, is clearly linear and discontinuous. It is closed, however, for if $x_n \to x$ in $\mathbf{R}$ and $Ix_n = x_n \to y$ in $\mathbf{R}_t$ then $x_n = y$ eventually. Therefore $y = x = Ix$.

(c) DISCONTINUOUS LINEAR MAP WITH CLOSED GRAPH Let

$$X = C\left([0,1], \mathbf{R}, \|\cdot\|_\infty\right)$$

be the space of continuous functions of $[0,1]$ into $\mathbf{R}$. Let $C_1 \subset C\left([0,1], \mathbf{R}, \|\cdot\|_\infty\right)$ be the subspace of elements with continuous derivative. Consider the differentiation operator $D : C_1 \to C\left([0,1], \mathbf{R}, \|\cdot\|_\infty\right)$, $x \mapsto x'$, the derivative of $x$. Clearly, $D$ is linear. The collection $x_n$, $x_n(t) = t^n$ $(n \in \mathbf{N})$ is contained in the unit ball of $C_1$, but its image $D(C_1) = \{nt^{n-1} : n \in \mathbf{N}\}$ is unbounded. Since $D$ does not map bounded sets into bounded sets, $D$ is discontinuous [Theorem 6.4.1]. To see that $D$ is closed, we use the criterion of Theorem 14.1.1 for sequences: Suppose that $x_n \to x$ $(x_n, x \in C_1)$ and $Dx_n = x_n' \to y$. As convergence in $C_1$ is uniform, $(x_n)$ and $(x_n')$ are uniformly convergent sequences. By standard theorems of analysis, this implies that $Dx = x' = \lim_n x_n' = y \in C\left([0,1], \mathbf{R}, \|\cdot\|_\infty\right)$. ■

As Example 14.1.2 shows, closedness does not generally imply continuity. When additional conditions are imposed, it may, as Theorem 14.1.3 demonstrates.

**Theorem 14.1.3** CLOSED IMPLIES CONTINUOUS FOR COMPACT CODOMAIN Let $f$ map the topological space $S$ into the compact space $T$. If $f$ is closed, then $f$ is continuous.

**Proof.** We use the notation of the statement. We show that $f$ is continuous by showing that, for any closed subset $K \subset T$, $f^{-1}(K)$ is closed. To this end, let $x \in \operatorname{cl} f^{-1}(K)$ and let $(x_r)$ be a net in $f^{-1}(K)$ such that $x_r \to x$. Since $K$ is compact, $(f(x_r))$ has a convergent subnet $f(x_u) \to y \in K$ [Theorem 1.1.10(b)]. Since $x_u \to x$ and $f$ is closed, $y = f(x) \in K$ or $x \in f^{-1}(K)$. ■

## 14.2 CLOSED LINEAR MAPS

**Notation.** $X$ and $Y$ are LCHS and $A : X \to Y$ is a linear map in this section unless otherwise noted. ■

As in Sec. 8.10, let $A'$ denote the linear map of the dual $Y'$ of $Y$ into the algebraic dual $X^*$ of $X$ defined by taking $\langle x, A'y' \rangle = \langle Ax, y' \rangle$ for all $x \in X$ and $y' \in Y'$. By Theorem 8.10.3, $A(Y') \subset X'$ iff $A$ is weakly continuous [i.e., continuous when $X$ and $Y$ carry $\sigma(X, X')$ and $\sigma(Y, Y')$, respectively]. We may characterize the subspace

$$D(A') = \{y' \in Y' : A'y' \in X'\}$$

as follows:

**Theorem 14.2.1** Let $X$ and $Y$ be LCHS and let $A : X \to Y$ be linear. Then $D(A') = \cup\{A(V)^\circ : V$ is a disked neighborhood of 0 in $X\}$.

**Proof.** We use the notation of the statement. Let $V(0)$ denote the filter of neighborhoods of 0 in $X$. For $y' \in D(A')$, we must show that there is a disk $V \in V(0)$ such that $y' \in A(V)^\circ$. From the continuity of $A'y'$ on $X$, there must be a disk $V$ in $V(0)$ such that $\sup|A'y'(V)| \leq 1$. In other words, for all $v$ in $V$, $|\langle v, A'y'\rangle| = |\langle Av, y'\rangle| \leq 1$, i.e., that $y' \in A(V)^\circ$. To complete the proof, we only need to reverse the steps. ■

Theorem 14.2.2 shows that certain conditions on $D(A')$ may imply that $A$ is continuous.

**Theorem 14.2.2** *$X$* MACKEY AND *$D(A') = Y' \Rightarrow A$* CONTINUOUS Let $X$ and $Y$ be LCHS and let $A : X \to Y$ be linear. If $X$ carries its Mackey topology $\tau(X, X')$, the polar topology generated by $\sigma(X', X)$-compact disks of $X'$, and $D(A') = Y'$, then $A$ is continuous. (Any bornological space carries its Mackey topology by Theorem 13.2.10(d).)

**Proof.** We use the notation of the statement. $D(A') = Y'$ implies that $A$ is weakly continuous by Theorem 8.10.3, hence Mackey continuous by Theorem 8.11.3(a). Since $Y$'s topology is coarser than $\tau(Y, Y')$ by the Mackey-Arens theorem [8.7.4], $A$ is continuous. ■

A "size" condition on $D(A')$ may also be used to describe closedness of a linear map.

**Theorem 14.2.3** *$A$* CLOSED IFF *$D(A')$* IS *$\sigma(Y', Y)$*-DENSE For LCHS $X$ and $Y$, the linear map $A : X \to Y$ is closed iff $D(A')$ is $\sigma(Y', Y)$-dense in $Y'$.

**Proof.** Let $A, X$ and $Y$ be as above. Suppose that $D(A')$ is $\sigma(Y', Y)$-dense in $Y'$ and that $(x_0, y_0) \in X \times Y \setminus G(A)$. We show that $(x_0, y_0) \notin \operatorname{cl} G(A)$. Since $y_0 - Ax_0 \neq 0$, there exists $y' \in Y'$ such that $\langle y_0 - Ax_0, y'\rangle \neq 0$. By the $\sigma(Y', Y)$-density of $D(A')$ in $Y'$, we may assume that $y' \in D(A')$. Recall that generally $(X \oplus Y)' = X' \oplus Y'$ [Exercise 8.113]. Viewing $(x, Ax)$ as an element of $X \oplus Y$ and $(-A'y', y')$ as an element of $(X \oplus Y)'$, for every $x \in X$,

$$\begin{aligned} 0 &= \langle x, -A'y' + A'y'\rangle = \langle x, -A'y'\rangle + \langle Ax, y'\rangle \\ &= \langle (x, Ax), (-A'y', y')\rangle \end{aligned}$$

Hence $(-A'y', y')$ vanishes on $G(A)$ and therefore also on $\operatorname{cl} G(A)$. To prove that $(x_0, y_0) \notin \operatorname{cl} G(A)$, we show that $\langle (x_0, y_0), (-A'y', y') \rangle \neq 0$. We have

$$\begin{aligned}\langle (x_0, y_0), (-A'y', y') \rangle &= \langle x_0, -A'y' \rangle + \langle y_0, y' \rangle \\ &= \langle x_0, -A'y' \rangle + \langle y_0 - Ax_0 + Ax_0, y' \rangle \\ &= \langle x_0, -A'y' \rangle + \langle x_0, A'y' \rangle + \langle y_0 - Ax_0, y' \rangle \\ &= \langle y_0 - Ax_0, y' \rangle \neq 0\end{aligned}$$

Conversely, suppose that $A$ is closed. To demonstrate the $\sigma(Y', Y)$-density of $D(A')$, we show that $D(A')^\circ = \{0\}$. To that end, suppose that $z \in D(A')^\circ$. If $z \neq 0$, $(0, z) \notin G(A)$ because $A$ is linear. Since $G(A)$ is a closed subspace of $X \oplus Y$, there exists [Theorem 7.7.5] a continuous linear functional $(x', y') \in (X \oplus Y)'$ which vanishes on $G(A)$ but $\langle (0, z), (x', y') \rangle = \langle z, y' \rangle = 2$. Since $(x', y')$ vanishes on $G(A)$, for every $x \in X$, $0 = \langle (x, Ax), (x', y') \rangle = \langle x, x' \rangle + \langle Ax, y' \rangle$. Hence $A'y' = -x'$, which implies that $y' \in D(A')$ and leaves us with the contradictory statements $z \in D(A')^\circ$ and $\langle z, y' \rangle = 2$ for $y' \in D(A')$. Therefore $D(A')^\circ = \{0\}$, which implies that $D(A')^{\circ\circ} = \operatorname{cl}_{\sigma(Y', Y)} D(A') = Y'$ by the bipolar theorem [8.3.8]. ■

**Theorem 14.2.4** $A$ CLOSED $\Rightarrow A^{-1}(0)$ CLOSED If $A$ is a closed linear map from the TVS $X$ into the HTVS $Y$, then $A^{-1}(0)$ is closed.

**Proof.** We use the notation of the statement. Since $Y$ is a Hausdorff space, $X \times \{0\}$ is a closed subset of $X \times Y$. Hence, since $G(A)$ is closed, $A^{-1}(0) \times \{0\} = G(A) \cap (X \times \{0\})$ is closed in $X \times Y$. Since the map $f : X \to X \times Y$, $x \mapsto (x, 0)$, is continuous (a homeomorphism of $X$ into $X \times Y$, actually), then $f^{-1}(A^{-1}(0) \times \{0\}) = A^{-1}(0)$ is closed in $X$. ■

In inner product spaces, "adjoint" has a different meaning, as explained in Example 14.2.5.

**Example 14.2.5** "ADJOINTS" IN INNER PRODUCT SPACES ARE CLOSED Let $(X, \langle \cdot, \cdot \rangle)$ and $(Y, \langle \cdot, \cdot \rangle)$ be inner product spaces and let $A : X \to Y$ be *any* map (not even linear, necessarily). Given $y \in Y$ there may be an element $z \in X$ such that $\langle Ax, y \rangle = \langle x, z \rangle$ for every $x \in X$. The set of $y \in Y$ for which this is true is denoted $D^*$ and we define the *adjoint* of $A$ to be the map $A^* : D^* \to X$, $y \mapsto z$. To see that $A^*$ is well defined, suppose that $\langle Ax, y \rangle = \langle x, w \rangle = \langle x, z \rangle$ for every $x \in X$. Then $z - w$ is orthogonal to each vector in $X$ and so must be 0. Even though $A$ may not be linear, $A^*$ is a closed linear map. As to the linearity, let $u, v \in D^*$, $A^*u = y$, and $A^*v = z$. For any scalars $a$ and $b$ and $x \in X$, then $\langle Ax, au + bv \rangle = \bar{a} \langle Ax, u \rangle + \bar{b} \langle Ax, v \rangle = \bar{a} \langle x, y \rangle + \bar{b} \langle x, z \rangle = \langle x, ay + bz \rangle$. Therefore, $D^*$ is a subspace and $A^*$ is linear. To verify that $A^*$ is closed, suppose that $y_n \to y$ ($y_n \in D^*$) and $A^*y_n = z_n \to z$. For every $n$, $\langle Ax, y_n \rangle = \langle x, z_n \rangle$. Since the inner product is continuous, this implies that $\langle Ax, y \rangle = \langle x, z \rangle$. Thus, $y \in D^*$ and $A^*y = z$. ■

## 14.3 CLOSED GRAPH THEOREMS

If a pair $(X, Y)$ of TVS is such that a closed linear map from $X$ to $Y$ must be continuous, we say that a "closed graph theorem holds for the pair."

The main results of this section are roughly parallel to the generalizations [Theorem 11.9.1 and 11.9.5(b)] of the Banach-Steinhaus theorem, another early victory for category arguments. Viewing those theorems in the context of "for what conditions on the spaces $(X, Y)$ does a Banach-Steinhaus theorem hold?" we obtained two results—one for when $X$ and $Y$ were each locally convex and one for when they were not necessarily locally convex. In the locally convex case, $X$ had to be barreled; without it, $X$ had to be nonmeager. For closed graph theorems there is a similar bifurcation: assuming constantly that $Y$ is a complete pseudometrizable TVS, then "$X$ barreled" or "$X$ nonmeager" activates a closed graph theorem, respectively, in the locally convex and nonlocally convex cases [Theorem 14.3.4].

A notion which plays an important role in proving the closed graph theorems of this section is that of *almost continuity* of a linear map $A : X \to Y$, namely that $\operatorname{cl} A^{-1}(V)$, rather than $A^{-1}(V)$, is a neighborhood of 0 for each neighborhood $V$ of 0 in $Y$. In certain cases ($X$ barreled in the locally convex case, $X$ nonmeager otherwise) linear maps are automatically almost continuous. Our first closed graph theorems show that an almost continuous closed linear map is continuous. If $X$ is ultrabornological and $Y$ is "webbed" [see Sec. 14.7], there is an inverse notion of almost continuity—namely, that $\operatorname{cl} A^{-1}(W)$ must be a neighborhood of 0 for certain sets $W$ of the web—that yields a closed graph theorem [14.7.3].

**Theorem 14.3.1** BARRELED $\Longrightarrow$ ALMOST CONTINUOUS If $X$ is barreled then every linear map $A$ of $X$ into any locally convex space $Y$ is almost continuous.

**Proof.** In the notation of the statement, $\operatorname{cl} A^{-1}(B)$ is a barrel in $X$ for any barrel $B$ in $Y$. Since $Y$ is locally convex, there is a base of barrels $B$ at 0. Consequently, if $X$ is barreled, any linear map is almost continuous. ■

**Theorem 14.3.2** ABSORBENT SETS IN NONMEAGER SPACES ARE NONMEAGER If $E$ is an absorbent subset of the nonmeager (= Baire) TVS $X$ then $E$ is nonmeager [cf. Theorem 11.8.8].

**Proof.** Suppose that $E$ is absorbent subset of the TVS $X$ and that $(R_n)$ is a sequence of closed sets such that $E \subset \cup_n R_n$. For each positive integer $k$, $kE \subset \cup_{n\in\mathbf{N}} kR_n$ so $X = \cup_{k\in\mathbf{N}} kE = \cup_{k\in\mathbf{N}} \cup_{n\in\mathbf{N}} kR_n$. Since $X$ is nonmeager, there must be integers $n$ and $k$ such that $\operatorname{int} kR_n \neq \emptyset$, which implies that $\operatorname{int} R_n \neq \emptyset$ and $E$ is seen to be nonmeager. ■

Theorem 14.3.3 is the analog of Theorem 14.3.1 for nonlocally convex spaces.

**Theorem 14.3.3** BAIRE $\Longrightarrow$ ALMOST CONTINUOUS If $X$ is a Baire TVS and $Y$ any TVS, then any linear map $A : X \to Y$ is almost continuous.

**Proof.** We use the notation of the statement. Given a neighborhood $U$ of 0 in $Y$, choose a neighborhood $V$ of 0 in $Y$ such that $V - V \subset U$. Since $V$ is a neighborhood of 0, $A^{-1}(V)$ is absorbent in $X$; hence it must be nonmeager by Theorem 14.3.2. Since $A^{-1}(V)$ cannot be rare, $\text{cl}\, A^{-1}(V)$ is a neighborhood of some point. Hence the algebraic difference $\text{cl}\, A^{-1}(V) - \text{cl}\, A^{-1}(V)$ is a neighborhood of 0 in $X$. Since

$$\begin{aligned}\text{cl}A^{-1}(U) &\supset \text{cl}A^{-1}(V - V) = \text{cl}\left[A^{-1}(V) - A^{-1}(V)\right] \\ &\supset \text{cl}A^{-1}(V) - \text{cl}A^{-1}(V),\end{aligned}$$

it follows that $\text{cl}\, A^{-1}(U)$ is a neighborhood of 0 in $X$. ■

We can now prove our first closed graph theorems.

**Theorem 14.3.4** CLOSED GRAPH THEOREMS Let $X$ be a TVS and $Y$ a complete pseudometrizable linear space. Then any closed almost continuous linear map $A : X \to Y$ is continuous. By Theorems 14.3.3 and 14.3.1, respectively, it follows that:

(a) BAIRE SPACES If $A$ is a closed linear map of the Baire TVS $X$ into the complete pseudometrizable TVS $Y$, then $A$ is continuous.

(b) LOCALLY CONVEX SPACES If $A$ is a closed linear map of the barreled space $X$ into the complete pseudometrizable LCS $Y$, then $A$ is continuous.

**Proof.** We use the notation of the statement. Let $\{V_i : i \in \mathbf{N}\}$ be a base of closed balanced neighborhoods of 0 in $Y$ such that for each $i \in \mathbf{N}$, $V_{i+1} + V_{i+1} \subset V_i$. Since $A$ is almost continuous, there is a countable family $(U_i)$ of balanced neighborhoods of 0 in $X$ such that $U_i \subset \text{cl}\left(A^{-1}(V_i)\right)$ and $U_{i+1} + U_{i+1} \subset U_i$ for each $i \in \mathbf{N}$. We demonstrate continuity by showing that $U_i \subset A^{-1}(V_{i-1})$ for each $i \geq 2$. To do this, fix $i \geq 2$ and choose $x_i \in U_i$. We show $x_i \in A^{-1}(V_{i-1})$ by showing that there is some $z \in V_{i-1}$ such that $(x_i, z) \in \text{cl}\, G(A) = G(A)$. For any subset $B$ of a TVS and any neighborhood $V$ of 0,

$$\text{cl}\, B \subset B + V \tag{$*$}$$

Hence $U_i \subset \text{cl}\, A^{-1}(V_i) \subset A^{-1}(V_i) + U_{i+1}$. It follows by induction that for all $k \geq i$ there exist $x_k \in U_k$ and $y_k \in A^{-1}(V_k)$ such that $x_k = y_k + x_{k+1}$. Applying $A$ and summing both sides for $i \leq k \leq n$,

$$Ax_i - Ax_{n+1} = \sum_{k=i}^{n} Ay_k \in V_i + \cdots + V_n \subset V_{i-1}$$

The sequence of partial sums of $\sum_{k \geq i} Ay_k$ is therefore Cauchy, so it converges to some $z$ in the closed neighborhood $V_{i-1}$. Since $x_{n+1} \in U_{n+1} \subset$

$\operatorname{cl} A^{-1}(V_{n+1})$, $\sum_{k=i}^{n} y_k = x_i - x_{n+1} \in x_i + \operatorname{cl} A^{-1}(V_{n+1})$. For any neighborhood $U$ of 0, $\operatorname{cl} A^{-1}(V_{n+1}) \subset A^{-1}(V_{n+1}) + U$. Hence, for any $n \geq i$, $\sum_{k=i}^{n} y_k \in x_i + A^{-1}(V_{n+1}) + U$ and:

(1) there exists $w \in A^{-1}(V_{n+1})$ such that $\sum_{k=i}^{n} y_k - w \in x_i + U$.

Since $\sum_{k=i}^{n} Ay_k - z = -\sum_{k \geq n+1} Ay_k \in V_{n+1} + V_{n+2} + \cdots \subset V_n$, then:

(2) $A(\sum_{k=i}^{n} y_k - w) - z = \sum_{k=i}^{n} Ay_k - z - Aw \in V_n + V_{n+1} \subset V_{n-1}$.

Hence the basic neighborhood $(x_i + U) \times (z + V_{n-1})$ of $(x_i, z)$ contains the point $(\sum_{k=i}^{n} y_k - w, A(\sum_{k=i}^{n} y_k - w)) \in G(A)$, so $(x_i, z) \in \operatorname{cl} G(A) = G(A)$. Thus $Ax_i = z \in V_{i-1}$, so $U_i \subset A^{-1}(V_{i-1})$ and $A$ is seen to be continuous. ■

## 14.4 OPEN MAPPING THEOREMS

An *open* map sends open sets into open sets. For $m > n$, for example, projections $\operatorname{pr}_n : \ell_2^{\mathbf{R}}(m) \to \ell_2^{\mathbf{R}}(n)$ are open linear maps. An open *linear* map $A : X \to Y$, $X$ and $Y$ TVS, must be onto because $A(X)$ is an open linear subspace of $Y$. The subject of the closed graph theorem [14.3.4] is when a closed linear map $A$ takes open sets into open sets inversely. In this section we investigate when $A$ does this directly. For open mapping theorems, we look for completeness in the domain, Baire or barreledness in the codomain. The first open mapping theorem was proved by Banach [1929, p. 238], who proved that continuous linear isomorphisms between Banach spaces must be open maps. Schauder [1930] also proved it for Banach spaces but, significantly, used a category argument. This made it possible to generalize the result to complete metric linear spaces, which Banach did using Schauder's argument [Banach 1932, pp. 41–42]. It was a (relatively) early triumph for "the category argument."

We have already defined almost continuity of a linear map $A : X \to Y$, namely that $\operatorname{cl} A^{-1}(V)$, rather than $A^{-1}(V)$, is a neighborhood of 0 for each neighborhood $V$ of 0 in $Y$. We define *almost open* analogously:

**Definition 14.4.1** Almost Open Linear Map Let $X$ and $Y$ be TVS. A linear map $A : X \to Y$ is *almost* (*nearly*) *open* if for any neighborhood $U$ of 0 in $X$, $\operatorname{cl} A(U)$ is a neighborhood of 0 in $Y$. [Many authors require that $\operatorname{cl} A(U)$ be a neighborhood of 0 in $A(X)$ rather than $Y$.]

Clearly, if a map $f$ is bijective then $f$ is almost open iff $f^{-1}$ is almost continuous, a fact we use in the proof of the open mapping theorem.

Having defined a notion dual to almost continuity, we now obtain a dual—the open mapping theorem—to the closed graph theorem [14.3.4]. The *Satz von der Gebietsinvarianz* below was first proved by Schauder [1930, p. 6].

**Theorem 14.4.2** Open Mapping Theorem If $(X, d)$ is a complete pseudometrizable TVS and $Y$ is any HTVS then any closed almost open linear surjection $A : X \to Y$ is open.

**Proof.** We use the notation of the statement. Let $N = A^{-1}(0)$ and consider the bijective map $\dot{A} : X/N \to Y$, $\dot{x} = x + N \mapsto Ax$. The idea of the proof is to apply the closed graph theorem to $\dot{A}^{-1}$.

Since $A$ is closed, so is $N$ by Theorem 14.2.4. Consequently, $X/N$ is a complete pseudometrizable space by Theorem 3.7.4. The quotient pseudometric [Example 2.7.5] $\dot{d}$ on $X/N$ is defined for $x, y \in X$ by $\dot{d}(\dot{x}, \dot{y}) = \inf\{d(x - y + z, 0) : z \in N\}$. A base at 0 for the quotient topology on $X/N$ is given by sets of the form $V + N$ where $V$ is a neighborhood of 0 in $X$. For any such set, $\operatorname{cl} \dot{A}(V + N) = \operatorname{cl} A(V)$ is a neighborhood of 0 in $Y$ by hypothesis so $\dot{A}$ is almost open. Since $\dot{A}$ is bijective, $\dot{A}^{-1}$ is almost continuous. We now show that $\dot{A}$ is closed.

To this end, so that the criterion of Theorem 14.1.1 may be applied, suppose that $\dot{x}_n \to \dot{x}$ and $\dot{A}\dot{x}_n \to y$. Since $\inf\{d(x_n - x + z, 0) : z \in N\} \to 0$, there exists a subsequence $(x_{n_k})$ of $(x_n)$ and a sequence $(z_k)$ from $\mathbf{N}$ such that for each positive integer $k$, $d(x_{n_k} - x + z_k, 0) < 1/k$. Hence $x_{n_k} + z_k \to x$ while $A(x_{n_k} + z_k) = Ax_{n_k} = \dot{A}\dot{x}_{n_k} \to y$. Since $A$ is closed, $y = Ax = \dot{A}\dot{x}$ and therefore $\dot{A}$ is closed.

Since the graph of $\dot{A}$ is closed, so is the graph of $\dot{A}^{-1}$. Hence, since $\dot{A}$ is almost open, $\dot{A}^{-1}$ is a closed almost continuous linear map of the TVS $Y$ into the complete pseudometrizable space $X/N$. The continuity of $\dot{A}^{-1}$ now follows from the closed graph theorem [14.3.4]; thus, $\dot{A}$ is open. Since $\dot{A}(\dot{U}) = A(U)$ for any subset $U$ of $X$, the openness of $A$ follows. ■

Two situations that produce almost openness, the inverse versions of Theorems 14.3.1 and 14.3.3, are given in Theorems 14.4.3 and 14.4.4. In conjunction with the open mapping theorem above, these yield two particular open mapping theorems.

**Theorem 14.4.3** Almost Open Maps and Barreled Spaces If $Y$ is barreled then each surjective linear map $A$ of any LCS $X$ onto $Y$ is almost open.

**Proof.** Suppose that $A$ is a linear map of the LCS $X$ onto the barreled space $Y$. Since $X$ is locally convex, it must have a base of neighborhoods of 0 which are barrels [Theorem 4.4.6]. Since $A$ is onto, if $B$ is any barrel in $X$, $\operatorname{cl} A(B)$ is a barrel in $Y$, hence a neighborhood of 0. ■

**Theorem 14.4.4** Almost Open Maps and Baire Spaces A linear map $A$ of a TVS $X$ onto a Baire (= nonmeager) TVS $Y$ is almost open.

**Proof.** We use the notation of the statement. For any balanced neighborhood $V$ of 0 in $X$, $A(V)$ is an absorbent subset of $Y$ because $A$ is onto. $A(V)$ is therefore nonmeager by Theorem 14.3.2, hence not rare. Since $\operatorname{cl} A(V)$ is a balanced nonempty set, $\operatorname{int} \operatorname{cl} A(V)$ is balanced; therefore $0 \in \operatorname{int} \operatorname{cl} A(V)$ which is what we wanted to prove. ■

Combining the previous three results, we have:

**Theorem 14.4.5** OPEN MAPPING THEOREMS Let $A$ be a closed linear map of a complete pseudometrizable space $X$ onto a TVS $Y$. if

(a) $Y$ is a Baire space, or

(b) $X$ is locally convex and $Y$ is barreled,

then $A$ is open.

If $A$ is continuous and $Y$ Hausdorff in either of the situations Theorem 14.4.5 [or Theorem 14.4.2], then $A$ must be open since continuous maps are closed when the range is Hausdorff [see after Theorem 14.1.1].

The following result is a variant of Theorem 14.4.5(a). To compensate for slightly relaxed conditions on the codomain $Y$, we strengthen the assumption on $A$.

**Theorem 14.4.6** OPEN MAPPING THEOREM FOR CONTINUOUS MAP Let $A$ be a continuous linear map of a complete pseudometrizable TVS $X$ into a Hausdorff TVS $Y$. If $A(X)$ is a nonmeager subset of $Y$, then:

(a) $A$ is surjective,

(b) $A$ is an open map, and

(c) $Y$ is a complete pseudometrizable space.

**Proof.** Suppose that $A$ is a continuous linear map of a complete pseudometrizable TVS $X$ into a Hausdorff TVS $Y$ and that $A(X)$ is a nonmeager subset of $Y$. Since $A$ is continuous and $Y$ is Hausdorff, $N = A^{-1}(0)$ is closed; hence $X/N$ is a complete pseudometrizable space by Theorem 4.8.7. Let $\dot{A} : X/N \to Y$ be the map $\dot{x} = x + N \mapsto Ax$. Since $\dot{A}$ is continuous and $Y$ is Hausdorff, $\dot{A}$ is closed. Since $A(X) = \dot{A}(X/N)$ is a Baire space, $\dot{A}$ is an open map by Theorem 14.4.5(a). Thus $\dot{A}$ is a linear homeomorphism. Since a linear homeomorphism is a unimorphism [Theorem 3.5.1] and unimorphisms preserve completeness [Theorem 3.5.6], $A(X)$ is complete. In particular, $A(X)$ is closed since $Y$ is Hausdorff by Theorem 3.1.9(b). Linear subspaces must be rare or dense by Example 11.5.3(f). Since $A(X)$ is a nonmeager subset of $Y$, it cannot be rare; hence $\text{cl}_Y A(X) = Y$. Thus $\text{cl}_Y A(X) = A(X) = Y$. Since $\dot{A}$ is open, it is evident that $A$ is open as well. To see that $Y$ is pseudometrizable, note that if $(U_n)$ is a countable base of neighborhoods of 0 in $X$, then $(A(U_n))$ is a base at 0 in $Y = A(X)$ by the openness and continuity of $A$. ∎

We deduced the open mapping theorem [14.4.2] here from the closed graph theorem [14.3.4] by applying the closed graph theorem to $\dot{A}^{-1}$. The exact same technique may be used to deduce the closed graph theorem from the open mapping theorem.

## 14.5 APPLICATIONS

Let $(X, \langle\cdot,\cdot\rangle)$ be a Hilbert space and let $A : X \to X$ be linear. Define the adjoint $A^*$ of $A$ as in Example 14.2.5 and suppose that $D^* = \{y \in Y : \exists z \in X \text{ such that } \langle Ax, y\rangle = \langle x, z\rangle\ \forall x \in X\} = X$. If $A$ is *self-adjoint,* i.e., $A = A^*$ or $\langle Ax, y\rangle = \langle x, Ay\rangle$ for all $x, y \in X$, then $A$ must be closed [Example 14.2.5]. For a Hilbert space, version (a) or (b) of the closed graph theorem [14.3.4] applies and asserts that $A$ is continuous. Thus, self-adjoint maps on a Hilbert space must be continuous. (To appreciate the strength of the closed graph theorem, try to prove this directly.)

The bounded inverse theorem below has interesting applications to questions of uniqueness of norm topologies, some of which are discussed after the theorem. It shows, in particular, that a continuous linear bijection between Fréchet spaces must be a homeomorphism. The first version is due to Banach [1932, p. 41, Th. 5].

**Theorem 14.5.1** BOUNDED INVERSE THEOREM Let $X$ be a complete pseudo-metrizable TVS and $Y$ a Baire HTVS. If $A : X \to Y$ is a continuous linear bijection, then $A^{-1}$ is continuous, i.e., $A$ is a linear homeomorphism.

**Proof.** In the notation of the statement, since $A$ is continuous and $Y$ Hausdorff, $A$ must have a closed graph, as observed in Sec. 14.1. It follows from the open mapping theorem [14.4.5(a)] that $A$ is open, i.e., $A^{-1}$ is continuous. ■

If $\mathcal{T}$ and $\mathcal{T}'$ are comparable ($\mathcal{T} \subset \mathcal{T}'$ or $\mathcal{T}' \subset \mathcal{T}$) metric vector topologies for a vector space $X$ and $X$ is complete with respect to each of them, then they must be equal. To see this, apply Theorem 14.5.1 to the identity map. Other results along these lines are: (1) A homomorphism of a Banach algebra into a semisimple (the intersection of its maximal ideals is $\{0\}$) commutative Banach algebra is automatically continuous; and (2) any semisimple commutative Banach algebra has a unique norm topology [Rickart 1960, pp. 70–76]. Thus, for example, the sup norm $\|\cdot\|_\infty$ is the "only" Banach algebra norm for the algebra $C([0,1], \mathbf{F}, \|\cdot\|_\infty)$ of continuous scalar-valued functions on $[0,1]$.

**Example 14.5.2** CLOSED SUBSPACES OF DIFFERENTIABLE FUNCTIONS Let $Y$ be the Banach space $C([0,1], \mathbf{R}, \|\cdot\|_\infty)$ of continuous real-valued functions on $[0,1]$.

(a) If the subspace $C^1$ of continuously differentiable functions were a closed subspace of $Y$, it too would be a Banach space. Consequently, the differential operator $D : C_1 \to C([0,1], \mathbf{R}, \|\cdot\|_\infty)$, $x \mapsto x'$, since it is closed by Example 14.1.2(c), would be continuous by the closed graph theorem [14.3.4]. We have already observed [Example 14.1.2(c)], however, that it is not. (Of course, it follows easily from the Stone-Weierstrass theorem that $C^1$ is not closed in $C([0,1], \mathbf{R}, \|\cdot\|_\infty)$.)

(b) Suppose that $M \subset Y$ is *some* closed subspace of continuously differentiable functions such that the differential operator $D : M \to Y$, $x \mapsto x'$, is continuous. If $U$ is the closed unit ball of $M$, then for each $u \in U$, $\|Du\|_\infty = \|u'\|_\infty \le \|D\| \|u\|_\infty \le \|D\|$. In other words, the functions in $U$ have uniformly bounded derivatives which, by the mean value theorem, implies that $U$ is equicontinuous. Since $\|u\|_\infty \le 1$ for each $u \in U$, it follows from Ascoli's theorem [5.10.4] that $U$ is relatively compact in $Y$. Therefore $U = (\text{cl}_Y U) \cap M$ is compact in $M$ and $M$ is locally compact. Therefore, by Theorem 4.10.4, $M$ is finite-dimensional. ■

## 14.6 WEBBED SPACES

If $X$ and $Y$ are TVS and $A : X \to Y$ a linear map, let us say we have "a closed graph theorem" when we posit enough conditions on $A, X$, and $Y$ to ensure that closedness of $A$ implies continuity. Our basic closed graph theorem [14.3.4] requires that the range space be a complete pseudometrizable space. The pseudometrizability is a countability condition on $Y$ [Theorem 4.8.3]. We consider another, weaker, countability condition on $Y$—that it possess a "web." In compensation, we impose stronger conditions on the domain space (Fréchet space, ultrabornological space) and get closed graph theorems. Among the spaces that we show to possess webs are Fréchet spaces and strong duals $(X', \beta(X', X))$ of metrizable locally convex spaces $X$.

The notion of a web (*réseau*) is due to de Wilde [1969a, 1978]; webs are also investigated and discussed in Robertson and Robertson [1973], Robertson [1972], Köthe [1979], and Jarchow [1981]. It was inspired by a closed graph theorem of Schwartz [1966b], the "Borel graph theorem" [14.9.2]. Schwartz's proof was measure-theoretic but Martineau [1966a,b; 1968] (cf. Treves [1967, App.] and Hogbe-Nlend [1973]) reproved it without measure theory, using instead category and the idea of a "sifting" or "sieve" (Sec. 14.9) of a Souslin space [de Wilde 1978; Bourbaki 1966b]. The notion of sieve motivated de Wilde to invent the web. We discuss a special kind of web here for our introductory sample.

A *web* in an LCHS $X$ (Hausdorff so that we may freely use duality theory) consists of sequences of disks arranged in strata. In the first stratum there is a sequence $(W_n)$ of disks whose union absorbs $X$. In the second stratum, let $(W_{1j})$ be a sequence of disks of $(1/2)\,W_1$ whose union absorbs $W_1$. Suppose that there is such a sequence $(W_{ij})$ for each $W_i$ in the first stratum. This sequence of sequences comprises the second stratum:

$$\begin{array}{lllll} W_1 & W_2 & W_3 & \cdots & \bigcup_i W_i \text{ absorbs } X \\ (W_{1j}) & (W_{2j}) & (W_{3j}) & \cdots & \forall i,j\ W_{ij} \subset (1/2)\,W_i \text{ and } \bigcup_j W_{ij} \text{ absorbs } W_i \end{array}$$

To each set $W_{ij}$ in the second stratum, we suppose that there is associated a sequence $(W_{ijk})$ of disks, each of which is a subset of $(1/2)\,W_{ij}$ whose union over $k$ absorbs $W_{ij}$. The sequences $(W_{ijk})$ of this type constitute the third stratum and so on. The various absorbency conditions enunciated above are collectively referred to as the *absorbency requirement.* A *strand* of the web is a sequence of the form $W_i, W_{ij}, W_{ijk}, \ldots$, one from each stratum, each new set from a sequence directly "below" the set above it. Any particular strand can be reindexed so we can speak of a "strand $(S_n)$." If one vector $x_n$ is chosen from each disk $S_n$ in a strand, we demand that $\sum_n x_n$ converge; we refer to this as the *convergence requirement.* The convergence requirement immediately implies that strands $(S_n)$ contain arbitrarily small sets in the sense that if $V$ is a neighborhood of 0, there must be $n \in \mathbf{N}$ such that $S_n \subset V$. (*Proof:* Suppose that $V$ is a neighborhood of 0 and that no $S_n$ is a subset of $V$. Then for each $n \in \mathbf{N}$ there must be $x_n \in S_n$ such that $x_n \notin V$. Hence $x_n \nrightarrow 0$ and $\sum_n x_n$ is not convergent.) In summary:

A *web* is a stratified collection of disks in an LCHS satisfying the absorbency and convergence requirements. An LCHS $X$ on which a web can be defined is called a *webbed space.*

To verify that a collection of sets is a web, usually the most difficult thing to verify is the convergence requirement. In the three examples of webbed spaces to follow, Theorem 14.6.1 is helpful in this regard.

**Theorem 14.6.1** CAUCHY SEQUENCE IN A BOUNDED DISK If $(x_k)$ is a sequence from a bounded disk $D$ in a TVS $X$ then the partial sums of $\sum_k 2^{-k}x_k$ are Cauchy; thus, if $X$ is *sequentially* complete then $\sum_k 2^{-k}x_k$ converges.

**Proof.** We use the notation of the statement. Let $U$ be a balanced neighborhood of 0. Choose $a > 0$ such that $D \subset aU$ and $k \in \mathbf{N}$ such that $a < 2^{k-1}$. With $(x_k)$ as above, let $s_n = \sum_{k=1}^{n} 2^{-k}x_k$. For all positive integers $p$,

$$\begin{aligned} s_{k+p} - s_{k-1} &= \left(1/2^k\right)x_k + \left(1/2^{k+1}\right)x_{k+1} + \cdots + \left(1/2^{k+p}\right)x_{k+p} \\ &= \left(1/2^k\right)x_k + \left(1/2^k\right)\left((1/2)\,x_{k+1} + \cdots + (1/2^p)\,x_{k+p}\right) \\ &\in \left(1/2^k\right)D + \left(1/2^k\right)D = \left(1/2^{k-1}\right)D \subset U \end{aligned}$$

by Theorems 4.2.9 and 4.2.5. ■

Banach spaces $X$ possess a very simple web. If $U$ denotes the closed unit ball of $X$, the disks

$$\begin{array}{cccc} U & U & U & \cdots \\ (1/2)\,U & (1/2)\,U & (1/2)\,U & \cdots \\ (1/4)\,U & (1/4)\,U & (1/4)\,U & \cdots \\ & \cdots\cdots & & \end{array}$$

constitute a web. The absorbency requirement is obviously satisfied. Since $U$ is a bounded disk and $X$ is complete, the convergence requirement is seen to be satisfied by Theorem 14.6.1.

As shown next, it is possible to define a web in a Fréchet space which has essentially just one strand. By way of a converse to Theorem 14.6.2, de Wilde [1978, p. 72] shows webbed *Baire* spaces to be Fréchet spaces.

**Theorem 14.6.2** Every Fréchet Space Is Webbed.

**Proof.** Let $(U_n)$ be a base of closed disked neighborhoods of 0 in a Fréchet space such that $U_{n+1} + U_{n+1} \subset U_n$ [or, equivalently, $U_{n+1} \subset (1/2)\, U_n$] for every $n \in \mathbf{N}$. Now consider

$$\begin{array}{cccc} U_1 & U_1 & U_1 & \cdots \\ U_2 & U_2 & U_2 & \cdots \\ U_3 & U_3 & U_3 & \cdots \\ & \cdots\cdots & & \end{array}$$

The absorbency requirement is clearly satisfied. If $x_j \in U_j$ for each $j$, the partial sums of $\sum_{j\in\mathbf{N}} x_j$ are easily seen to form a Cauchy sequence since, for any $p \in \mathbf{N}$, $x_n + \cdots + x_{n+p} \in U_n + \cdots + U_{n+p} \subset (1/2)\, U_{n-1} + (1/4)\, U_{n-1} + \cdots + \left(1/2^{p-1}\right) U_{n-1} \subset U_{n-1}$, from which the convergence requirement is seen to be satisfied. ■

In the case of the strict inductive limit $X$ of metrizable LCS $\{X_n : n \in \mathbf{N}\}$, $(X', \beta(X', X))$ is webbed [Theorem 14.6.4] and the strands are rather simple, not neighborhoods of 0 but certain $\sigma(X', X)$-compact sets. The following observation is helpful in Theorems 14.6.4 and 14.6.5.

**Lemma 14.6.3** if $\mathcal{B}$ is a base at 0 for a TVS $X$, then $\cup_{B\in\mathcal{B}} B^{\circ} = X'$.

**Proof.** If the linear functional $f \in X'$ and $f \notin \cup_{B\in\mathcal{B}} B^{\circ}$ then for each $B \in \mathcal{B}$, there is $x_B \in B$ such that $|f(x_B)| > 1$. We now have the filterbase $\mathcal{B} \to 0$ in $X$ but $f(\mathcal{B}) \nrightarrow 0$, and $f$ is discontinuous. ■

**Theorem 14.6.4** Strong Duals of Metrizable Spaces Are Webbed If $X$ is a metrizable LCS, then its strong dual $(X', \beta(X', X))$ is webbed.

**Proof.** Let $\{B_n : n \in \mathbf{N}\}$ be a base at 0 in the metrizable LCS $X$. As the web in $X'$, take

$$\begin{array}{cccc} B_1^{\circ} & B_2^{\circ} & B_3^{\circ} & \cdots \\ (1/2)\, B_1^{\circ} & (1/2)\, B_2^{\circ} & (1/2)\, B_3^{\circ} & \cdots \\ (1/4)\, B_1^{\circ} & (1/4)\, B_2^{\circ} & (1/4)\, B_3^{\circ} & \cdots \\ & \cdots\cdots & & \end{array}$$

Each $B_n^{\circ}$ is a disk and the absorbency requirement follows from Lemma 14.6.3. To verify the convergence requirement, we wish to invoke Theorem 14.6.1. $X$ is metrizable, therefore bornological [Example 13.2.8(b)] and the strong dual of a bornological space is complete [Theorem 13.2.13]. To see that each $B_n^{\circ}$

is $\beta(X', X)$-bounded, let $B$ be a $\sigma(X, X')$-bounded subset of $X$ so that $B^\circ$ is a basic $\beta(X', X)$-neighborhood of 0. For such a $B$, since boundedness $= \sigma(X, X')$-boundedness in dual pairs by Theorem 8.8.7, given any basic neighborhood $B_n$ of 0, $B \subset aB_n$ for sufficiently large $a > 0$. Hence $B_n^\circ \subset aB^\circ$ and the $B_n^\circ$ $(n \in \mathbf{N})$ are seen to be $\beta(X', X)$-bounded disks. The convergence requirement now follows from Theorem 14.6.1. ■

Theorem 14.6.4 is generalized in Theorem 14.6.5. The web there is more complex than that of a Fréchet Space or the strong dual of a metrizable space.

**Theorem 14.6.5** Strong Duals of Inductive Limits of Metrizable Spaces If $X = \operatorname{str\,ind}_n X_n$ is the strict inductive limit of metrizable LCS $\{X_n : n \in \mathbf{N}\}$ then $(X', \beta(X', X))$ is webbed.

**Proof.** Let $X$ be as above and let $(A_n)$ be a base at 0 for $X_1$, $(B_n)$ a base at 0 for $X_2$, $(C_n)$ for $X_3$, and so on. As the web, take

$$\begin{array}{cc}
A_1^\circ & A_2^\circ \\
(1/2)\left(A_1^\circ \cap B_j^\circ\right) & (1/2)\left(A_2^\circ \cap B_j^\circ\right) \\
(1/4)\left(A_1^\circ \cap B_1^\circ \cap C_k^\circ\right), (1/4)\left(A_1^\circ \cap B_2^\circ \cap C_k^\circ\right), \ldots & \cdots\cdots \\
\cdots\cdots & \cdots\cdots
\end{array}$$

Every set is a disk. That the absorbency requirement is satisfied follows from Lemma 14.6.3.

To verify the convergence requirement, we use Theorem 14.6.1 as we did in Theorems 14.6.2 and 14.6.4. Note that $X$ is bornological because it is the inductive limit of bornological spaces [Theorem 13.2.9] and bornological spaces have strongly complete duals by Theorem 13.2.13—hence $(X', \beta(X', X))$ is complete. To invoke Theorem 14.6.1, it suffices to show that sets of the form $\cap_{j\in\mathbf{N}} U_j^\circ$, where each $U_j$ is a neighborhood of 0 in $X_j$, are $\beta(X', X)$-bounded disks. In order to do this, note that the balanced convex hull of $\cup_{j\in\mathbf{N}} U_j$ is a neighborhood of 0 in the strict inductive limit topology on $X$ [Theorem 12.1.1]. Hence its polar $\cap_{j\in\mathbf{N}} U_j^\circ$, aside from being a disk in $X'$, is equicontinuous by Theorem 8.6.4, hence $\beta(X', X)$-bounded by Theorem 11.3.5. ■

## 14.7 CLOSED GRAPH THEOREMS

The closed graph theorem [14.3.4(b)] enables us to deduce continuity of a linear map $A$ in the following situation for LCS:

$$\text{barreled} \xrightarrow{A \text{ closed}} \text{complete pseudometrizable LCS}$$

The results of this section (due to de Wilde 1969a, 1978) yield a closed graph theorem for

$$\text{ultrabornological} \xrightarrow{A \text{ sequentially closed}} \text{webbed}$$

The conditions on $A$ and the codomain have been weakened, while a stronger one is used on the domain side. For a fixed class of codomain spaces (Banach spaces, say, or finite-dimensional spaces), how much relaxation is possible on the domain side for a closed graph theorem to hold? We discuss some possibilities in Sec. 14.8. We show, for example, that if a closed graph theorem holds for an LCHS $X$ into any Banach space, then $X$ must be barreled.

**Theorem 14.7.1** CLOSED GRAPH THEOREM: FRÉCHET $\overset{\text{seq closed}}{\rightarrow}$ WEBBED A sequentially closed linear map $A$ of a Fréchet space $X$ into a webbed space $Y$ is continuous.

**Proof.** We use the notation of the statement. Let $\{U_k : k \in \mathbf{N}\}$ be a base of disks at 0 for $X$ for which $2U_{k+1} \subset U_k$ for all $k \in \mathbf{N}$ and let $\mathcal{W} = \{W_1, W_2, \ldots, W_{11}, W_{12}, \ldots\}$ be a web for $Y$. The union of the sets $W_n$ of the first stratum is absorbent in $Y$; so therefore is $A^{-1}(\cup_n W_n)$ in $X$. Since $X$ is nonmeager, $\cup_n A^{-1}(W_n)$ is nonmeager by Theorem 14.3.2. Since countable unions of meager sets are meager, some $A^{-1}(W_n)$ must be nonmeager. Since $\cup_j W_{nj}$ absorbs $W_n$, $A^{-1}(\cup_j W_{nj})$ absorbs $A^{-1}(W_n)$. Hence $\cup_j A^{-1}(W_{nj})$ is nonmeager and there must be some $j$ such that $A^{-1}(W_{nj})$ is nonmeager. In other words, by induction, there is a strand $(S_n)$ of $\mathcal{W}$ such that $A^{-1}(S_n)$ is nonmeager for every $n$. Since each $A^{-1}(S_n)$ is nonmeager, none of them can be rare: Each $\mathrm{cl}(A^{-1}(S_n))$ has nonempty interior. Since the $S_n$ are disks, $0 \in \mathrm{int}\,\mathrm{cl}\, A^{-1}(S_n)$ for every $n$ [Exercise 4.106(d)]. Thus, there exist neighborhoods $(U_{k_n})$ of 0 such that for every $n$, $U_{k_n} \subset \mathrm{cl}\, A^{-1}(S_n)$. As observed in Sec. 14.6, strands must contain arbitrarily small sets: Given any neighborhood $V$ of 0 in $Y$, there is $n \in \mathbf{N}$ such that $S_n \subset V$. Hence $A$ is almost continuous in the sense that each $\mathrm{cl}\, A^{-1}(V)$ is a neighborhood of 0. The rest of the argument shows that the "cl" can be removed.

Let $V$ be a closed neighborhood of 0 in $Y$ and choose a member $S_{m-1}$ of the strand such that $S_{m-1} \subset V$. We show that $U_{k_m} \subset A^{-1}(V)$. As in $(*)$ in the proof of Theorem 14.3.4, it follows that, for every $n$, $U_{k_n} \subset \mathrm{cl}\, A^{-1}(S_n) \subset A^{-1}(S_n) + U_{k_{n+1}}$. Hence for any $x_m \in U_{k_m}$, there exist sequences $x_n \in U_{k_n}$, for $n \geq m$, and $y_n \in A^{-1}(S_n)$, for $n \geq m$, such that $x_n = y_n + x_{n+1}$ for $n \geq m$. For $k \geq m$, then

$$x_m + x_{m+1} + \cdots + x_k = y_m + x_{m+1} + y_{m+1} + x_{m+2} + \cdots + y_k + x_{k+1}$$

or

$$x_m = \sum_{n=m}^{k} y_n + x_{k+1}$$

Since the filterbase $(U_{k_n}) \to 0$, $x_{k+1} \to 0$ and it follows that $\sum_{n \geq m} y_n = x_m$. Since $Ay_n \in S_n$ for all $n \geq m$ and $(S_n)$ is a strand, it follows from the convergence requirement that $\sum_{n \geq m} Ay_n$ converges to some $z \in Y$. Moreover, for all $p \in \mathbf{N}$,

$$\begin{aligned} Ay_m + Ay_{m+1} + \cdots + Ay_{m+p} &\in S_m + S_{m+1} + \cdots + S_{m+p} \\ &\subset (1/2)S_{m-1} + (1/4)S_{m-1} + \cdots + (1/2^{m+p})S_{m-1} \subset S_{m-1} \end{aligned}$$

so $\sum_{n \geq m} Ay_n = z \in \operatorname{cl} S_{m-1} \subset \operatorname{cl} V = V$. Since $G(A)$ is sequentially closed, $z = Ax_m$. Hence $x_m \in A^{-1}(V)$, so $U_{k_m} \subset A^{-1}(V)$ and the proof is complete. ■

We want to extend the result just obtained to the case where the domain is ultrabornological, hence an inductive limit of Banach spaces [Theorem 13.2.11]. To this end, for a given LCHS $Y$, let $D_s(Y)$ (the "domains" for $Y$) be the class of all LCHS $X$ for which a sequentially closed linear map of $X$ into $Y$ must be continuous. We prove now that $D_s(Y)$ is closed with respect to the formation of inductive limits, from which it follows that a closed graph theorem holds for ultrabornological spaces into webbed spaces by Theorem 14.7.1

**Theorem 14.7.2** INDUCTIVE LIMITS OF "CLOSED GRAPH" DOMAINS Let $Y$ be an LCHS and let $D_s(Y)$ be as defined above. Then inductive limits of elements in $D_s(Y)$ also belong to $D_s(Y)$.

**Proof.** We use the notation of the statement. Let $\{X_s : s \in S\}$ be a family of members of $D_s(Y)$, and let $X = \operatorname{ind}_{s \in S} X_s$ be the inductive limit determined by a family $\{A_s : s \in S\}$ of linear maps [Definition 12.2.1]. Let $A : X \to Y$ be a sequentially closed linear map. By Theorem 12.2.2, to show that $A$ is continuous we must show that $A \circ A_s$ is continuous for each $s \in S$. Since each $X_s \in D_s(Y)$, this reduces to showing that each $A \circ A_s$ is sequentially closed. To do this, suppose that $x_n \to x$ in $X_s$ and that $(A \circ A_s)x_n \to y$. By Theorem 14.1.1, we must show that $y = (A \circ A_s)x$.

Since $X = \operatorname{ind}_{s \in S} X_s$, each $A_s$ is continuous; therefore $x_n \to x$ implies that each $A_s x_n \to A_s x$. This, together with the fact that $A(A_s x_n) \to y$ and $A$ being sequentially closed, implies that $y = A(A_s x)$. It follows that $A \circ A_s$ is sequentially closed and the proof is complete. ■

As a consequence of Theorems 14.7.1 and 14.7.2, we have:

**Theorem 14.7.3** CLOSED GRAPH THEOREM ULTRABORNOLOGICAL SPACES A sequentially closed linear map of an ultrabornological space into a webbed space is continuous.

It is *not* more general to take an inductive limit of Fréchet spaces (as opposed to Banach spaces) in Theorem 14.7.3 for the following reasons: (1) Fréchet spaces are ultrabornological by Example 13.2.8(d); (2) inductive limits of ultrabornological spaces are ultrabornological by Theorem 13.2.9; (3) therefore, an inductive limit of Fréchet spaces is ultrabornological, i.e., an inductive limit of Banach spaces.

## 14.8 LIMITS ON THE DOMAIN SPACE

By analogy with the way $D_s(Y)$ was defined in Sec. 14.7 for an LCHS $Y$, let $D(Y)$ stand for the class of LCHS $X$ for which a closed linear map $A : X \to Y$ must be continuous. For a class $\mathcal{Y}$ of LCHS $Y$, let $\mathcal{D}(\mathcal{Y}) = \cap_{Y\in\mathcal{Y}} D(Y)$. We obtain $\mathcal{D}(\mathcal{Y})$ for certain $\mathcal{Y}$'s in this section. Further results along these lines may be found in de Wilde [1978, pp. 20–24].

**Theorem 14.8.1** $\mathcal{D}$(FINITE-DIMENSIONAL) = ALL LCHS The class of LCHS $X$ such that a closed linear map of $X$ into any finite-dimensional LCHS $Y$ is continuous is the class of all LCHS.

**Proof.** Let $Y$ be a finite-dimensional LCHS, $X$ any LCHS, and $A$ a closed linear map of $X$ into $Y$. Since $Y$ is finite-dimensional, we may assume that it is a normed space by Theorem 4.10.3(c). If $A$ is not continuous, there exists a closed ball $C(0, r)$, $r > 0$, about 0 in $Y$ such that for any neighborhood $U$ of 0 in $X$, there exists $x \in U$ such that $\|Ax\| > r$. By renorming $Y$, replacing $\|\cdot\|$ by $(1/r)\|\cdot\|$ if necessary, we may assume that $r = 1$. For any neighborhood $U$ of 0 in $X$, $B_U = \{x/\|Ax\| : x \in U, \|Ax\| > 1\} \neq \varnothing$. As $U$ runs through the neighborhoods of 0 in $X$, $(B_U)$ is seen to be a filterbase in $X$; since $B_U \subset U$ for each $U$, $(B_U) \to 0$. Since $Y$ is finite-dimensional, its unit ball is compact [Theorem 4.10.4]; so, therefore, is its surface $S = \{y \in Y : \|y\| = 1\}$. Since $A((B_U))$ is a filterbase in $S$, $A(B_U))$ must have a cluster point $y \in S$ [Theorem 1.1.10(b)]. Thus $(0, y) \in \operatorname{cl} G(A) = G(A)$ which implies the contradictory result that $y = A0 = 0$. ■

**Theorem 14.8.2** $\mathcal{D}$(ALL LCHS) = ($X$, FINEST LC TOPOLOGY) If every closed linear map of the LCHS $(X, \mathcal{T})$ into any LCHS $Y$ is continuous, then $X$ carries the finest locally convex topology $\mathcal{T}_{lc}$ [Example 5.6.4]. Conversely, if $X$ carries $\mathcal{T}_{lc}$, then any linear map—closed or not—of $X$ into any LCHS is continuous.

**Proof.** Let $(X, \mathcal{T})$ be an LCHS and suppose that every closed linear map of $(X, \mathcal{T})$ into any LCHS $Y$ is continuous. Since $(X, \mathcal{T})$ is Hausdorff, the diagonal—the graph of the identity map—of $(X, \mathcal{T}) \times (X, \mathcal{T})$ must be closed. Consequently, the identity map $x \mapsto x$ of $(X, \mathcal{T})$ into $(X, \mathcal{T}_{lc})$ must also be closed, therefore continuous by hypothesis. This implies that $\mathcal{T}_{lc} \subset \mathcal{T}$ and proves the first part of the theorem.

Conversely, let $Y$ be an LCHS, let $\mathcal{T}_{lc}$ be the finest locally convex topology for the linear space $X$ and consider a linear map $A : (X, \mathcal{T}_{lc}) \to Y$. For any disked neighborhood $U$ of 0 in $Y$, $A^{-1}(U)$ is an absorbent disk in $X$, hence a $\mathcal{T}_{lc}$-neighborhood of 0. ■

Finally, we present Wilansky's [1976] proof of a result of Mahowald [1961].

**Theorem 14.8.3** $\mathcal{D}$(BANACH SPACES) = BARRELED SPACES The class of LCHS $X$ for which a closed linear map of $X$ into any Banach space $Y$ must be continuous is the class of Hausdorff barreled spaces.

**Proof.** If the LCHS $X$ is barreled, then a closed linear map of $X$ into any Banach space must be continuous by Theorem 14.3.4(b). Conversely, let $X$ be an LCHS for which closed linear maps of $X$ into any Banach space must be continuous and let $B$ be a barrel in $X$. Let $B^\circ$ carry its relative weak-$*$ topology $\sigma(X', X)$ and consider the Banach space $C_b(B^\circ, \mathbf{R}, \|\cdot\|_\infty)$ of continuous bounded maps of $B^\circ$ into $\mathbf{R}$. Consider the linear map

$$A : X \to C_b(B^\circ, \mathbf{R}, \|\cdot\|_\infty),\ x \mapsto \langle x, \cdot\rangle$$

where, for $x \in X$ and $x' \in B^\circ$, $(Ax)(x') = \langle x, x'\rangle$. We show next that $A$ is closed.

Suppose that $x_s \to x$ in $X$ and $Ax_s \to y$ in $C_b(B^\circ, \mathbf{R}, \|\cdot\|_\infty)$. For any $x' \in B^\circ$, $\langle x_s, x'\rangle \to \langle x, x'\rangle$ by the continuity of $x'$. Thus $Ax_s \to Ax$ in the topology $\mathcal{T}_p$ of pointwise convergence [Example 5.9.1] on $C_b(B^\circ, \mathbf{R})$. Since $Ax_s \to y$ in the $\|\cdot\|_\infty$-topology on $Y$, $Ax_s \to y$ in the weaker topology $\mathcal{T}_p$. As $\mathcal{T}_p$ is Hausdorff, limits are unique so $y = Ax$ and $A$ is closed. Since $C_b(B^\circ, \mathbf{R}, \|\cdot\|_\infty)$ is a Banach space, $A$ is continuous by hypothesis. The inverse image under $A$ of the unit ball of $C_b(B^\circ, \mathbf{R}, \|\cdot\|_\infty)$ consists of those $x \in X$ for which $\|Ax\|_\infty \leq 1$, i.e., those $x$ for which $|\langle x, x'\rangle| \leq 1$ for every $x' \in B^\circ$ which is precisely $B^{\circ\circ} = B$ by the bipolar theorem [8.8.1(c)]. Since $A$ is continuous, $B$ must be a neighborhood of 0 and $X$ is barreled. ■

## 14.9 OTHER CLOSED GRAPH THEOREMS

Banach's closed graph theorem [Banach 1932, p. 41, Th. 7] was for closed linear maps from one complete metric TVS into another. We considered a variation of it in Theorem 14.7.3 in which the domain was ultrabornological and the range webbed. There are many other versions and we discuss some of them here without proofs. Some good sources for results like these are Köthe 1979, de Wilde 1978, Robertson and Robertson 1973, and Husain 1965.

Let us take Theorem 14.3.4(b) as our basic closed graph theorem for locally convex spaces, namely: If $X$ is barreled and $Y$ is a complete pseudometrizable LCS, then closed linear maps of $X$ into $Y$ are continuous. It is reasonable to ask what happens if $Y$ is just complete, dropping the pseudometrizability. It turns out that the theorem fails in this case, showing that something stronger than completeness is needed. One such stronger condition—*$B_r$-completeness*—was developed by Pták [1953, 1958] (cf. Husain 1965, Robertson and Robertson 1973 and Köthe 1979):

**Theorem 14.9.1** PTÁK'S CLOSED GRAPH THEOREM Every closed linear map from a barreled Hausdorff space into a *$B_r$-complete* space (defined below) is continuous.

For any neighborhood $U$ of 0 in the LCHS $X$, $U^\circ$ is $\sigma(X',X)$-compact [Alaoglu, 8.4.1]. Hence, for any $\sigma(X',X)$-closed $F \subset X'$, $F \cap U^\circ$ is $\sigma(X',X)$-compact. A subset $S \subset X'$ is called *nearly* (*almost*) *closed* if $S \cap U^\circ$ is $\sigma(X',X)$-compact for each neighborhood $U$ of 0 in $X$. The LCHS $X$ is called *Pták* (*fully complete, B-complete*) if each nearly closed *subspace* of $X'$ is closed. If each nearly closed $\sigma(X',X)$-dense subspace $M$ of $X'$ is $\sigma(X',X)$-closed (equivalently, $\text{cl}_{\sigma(X',X)} M = X'$), then $X$ is called *$B_r$-complete* or *infra-Pták.*

Dieudonné and Schwartz 1949 proved that closed linear maps $A : X \to Y$ between strict inductive limits of Fréchet spaces are continuous. The "strict" was removed by Köthe 1950 and an even more general result was obtained by Grothendieck [1955, p. 13, Th. B], who proved it in the case when $X$ was ultrabornological and $Y$ a strict inductive limit of Fréchet spaces when the linear map was only *sequentially* closed. Grothendieck conjectured at the time that weaker conditions on the codomain should suffice, a conjecture that de Wilde's results (Sec. 14.7) have borne out.

As mentioned in the discussion at the beginning of Sec. 14.6, de Wilde was motivated to invent the web by Schwartz's "Borel graph theorem" which we state below. The pertinent definitions follow the statement.

**Theorem 14.9.2** BOREL GRAPH THEOREM [Schwartz 1966b; cf. de Wilde 1978, p. 136; and Treves 1967, p. 549] Let $X$ be ultrabornological and $Y$ be an LCHS which is also a Souslin space. Let $A : X \to Y$ be a linear map whose graph is a sq-Borel subset of $X \times Y$. Then $A$ is continuous.

A *Souslin space* is a regular Hausdorff space which is the continuous image of a separable complete metric space. The *sq-Borel* sets of a topological space are those of the $\sigma$-algebra (stable under complementation and countable union) generated by the sequentially closed sets. Webs begin to enter the picture because:

> A regular Hausdorff space is a Souslin space iff it admits a sieve [de Wilde 1978, p. 126].

and a *sieve* (*crible* in French) on a topological space $T$ is a collection of subsets of $T$ arranged in strata (cf. Sec. 14.6)

$$\begin{array}{cccc} S_1 & S_2 & S_3 & \cdots \\ (S_{1j}) & (S_{2j}) & (S_{3j}) & \cdots \\ (S_{11k}),(S_{12k}),\ldots & (S_{21k}),(S_{22k}),\ldots & (S_{31k}),(S_{32k}),\ldots & \cdots \\ & \cdots\cdots & & \end{array}$$

so that (1) each set in the sieve contains any set directly beneath it, (2) each set in the sieve is the union of all sets directly beneath it, and (3) defining a *strand* of the sieve to be a sequence of sets of the sieve, one selected from each stratum with each new set selected from a sequence directly below the preceding one, any strand must be a convergent filterbase.

Since the proof of our basic closed graph theorem [14.3.4] did not involve scalar multiplication, we have:

**Theorem 14.9.3** CLOSED GRAPH THEOREM TOPOLOGICAL GROUPS I If $G$ is a topological group and $H$ a complete pseudometrizable topological group, then any closed almost continuous homomorphism of $G$ into $H$ is continuous.

The inverse result, i.e., the corresponding open mapping theorem, is that (with $G$ and $H$ as above) if $f : H \to G$ is an almost open [meaning that $\operatorname{cl} f(V)$ is a neighborhood of 0 in $f(H)$ for each neighborhood $V$ of 0 in $G$] homomorphism with closed graph, then $f$ is continuous.

De Wilde [1978, p. 143] obtained the following generalization of Theorem 14.9.3.

**Theorem 14.9.4** CLOSED GRAPH THEOREM TOPOLOGICAL GROUPS II Let $G$ and $H$ be topological groups with $G$ a Baire space and $H$ a Souslin space. Then any homomorphism of $G$ into $H$ whose graph is a sq-Borel set is continuous.

### 14.9.1 Webs without Convexity Conditions

If $X$ is a nonlocally convex TVS, there is also a notion of web. Instead of disks, the $W$'s are balanced sets such that for all $k$ and $j$, $W_{kj} + W_{kj} \subset W_k$, etc., with the same absorbency and convergence requirements. As has happened previously [in Theorem 14.3.4, for example], Baire replaces barreled in the absence of local convexity: Specifically, as an analog of Theorem 14.7.3 we have:

**Theorem 14.9.5** CLOSED GRAPH WITHOUT CONVEXITY If the TVS $X$ is an inductive limit of Baire TVS and $Y$ is a webbed TVS (as defined above) then every closed linear map of $X$ into $Y$ is continuous.

For a proof, see Jarchow [1981, p. 92]; compare Exercise 14.204 and the hint thereto.

## 14.10 EXERCISES

14.101 WHEN A MAP IS CLOSED (a) Show that the linear map $A$ of the TVS $(X, \mathcal{T})$ into the HTVS $(Y, \mathcal{T}')$ has a closed graph iff there exists a Hausdorff vector topology $\mathcal{T}^*$ for $Y$ which is weaker than $\mathcal{T}'$ and $A$ is a continuous map of $(X, \mathcal{T})$ into $(Y, \mathcal{T}^*)$.

(b) Let $X$ and $Y$ be TVS with neighborhood bases $\mathcal{V}_X(0)$ and $\mathcal{V}_Y(0)$, respectively, at 0. The linear map $A : X \to Y$ is closed iff $\cap\{A(U) + V : U \in \mathcal{V}_X(0), V \in \mathcal{V}_Y(0)\} = \{0\}$.

14.102 Give an example of an almost continuous (open) linear map that is not continuous (open).

14.103 Show that a strand of a web on an LCHS $X$ forms a base at 0 for a topology for $X$ which makes it a topological group. Is it a base at 0 for a TVS topology?

14.104 CLOSED CANONICAL MAP Let $X$ and $Y$ be LCHS, $A : X \to Y$ a closed linear map, $N = A^{-1}(0)$, and $\dot{A}$ the associated canonical map $\dot{x} \mapsto Ax$ of $X/N$ into $Y$. Show that $G(A)$ is closed iff $G(\dot{A})$ is closed.

14.105 CLOSED GRAPH THEOREM FOR INFRABARRELED SPACES Show that a closed locally bounded linear map $A$ of an infrabarreled space $X$ into a complete pseudometrizable LCS $Y$ is continuous.

14.106 DISCONTINUOUS CLOSED MAP Let $\mathcal{T}$ be a Hausdorff vector topology on a vector space $X$ and let $\mathcal{T}^*$ be a strictly finer one. Show that the identity map from $(X, \mathcal{T})$ onto $(X, \mathcal{T}^*)$ is a closed discontinuous map.

14.107 COMPLEMENTS (a) Let $M$ and $N$ be algebraically complementary subspaces of the TVS $X$ (i.e., $X = M + N$ and $M \cap N = \{0\}$). Show that the projection $P$ on $M$ along $N$ [Definition 4.9.1] is closed iff $M$ and $N$ are closed.

(b) Using (a) and the closed graph theorem [14.3.4(a)] show that algebraically complementary subspaces $M$ and $N$ of a complete pseudometrizable TVS are topologically complementary (Sec. 4.9).

14.108 OPEN MAPPING THEOREM FOR ULTRABARRELED SPACES A closed linear map $A$ of a complete pseudometrizable space $X$ onto an ultrabarreled space $Y$ (defined in Exercise 11.201) is open.

14.109 CLOSED GRAPH THEOREM FOR ULTRABARRELED SPACES A closed linear map $A$ of an ultrabarreled space $X$ (see Exercise 11.201) into a complete pseudometrizable TVS $Y$ is continuous. (A converse holds as well, an analog of Theorem 14.8.3: Specifically, if any closed linear map of the TVS $X$ into any complete pseudometrizable TVS $Y$ is continuous, then $X$ must be ultrabarreled [Husain and Khaleelulla 1978, pp. 88–89].)

14.110 NON-ARCHIMEDEAN CLOSED GRAPH THEOREMS (a) Show that the basic closed graph and open mapping theorems [14.3.4 and 14.4.2] hold when the underlying field is a nontrivially ultravalued field $K$.

Now suppose that $X$ and $Y$ are locally $K$-convex spaces (Exercises 4.203 and 5.205) and that $A : X \to Y$ is a linear map.

(b) Show that if $Y$ is $K$-barreled [Exercise 11.202] and $A$ is onto, then $A$ is almost open. If, on the other hand, $X$ is $K$-barreled, then $A$ is almost continuous.

Using (a) and (b), show that:

(c) A closed linear map of a complete pseudometrizable locally $K$-convex space onto a $K$-barreled space is open.

(d) A closed linear map of a $K$-barreled space into a complete pseudometrizable locally $K$-convex space is continuous.

14.201 STABILITY OF WEBS (a) SUBSPACES Every sequentially closed subspace of a webbed space is webbed.

(b) SEQUENTIALLY CONTINUOUS LINEAR IMAGES If the HTVS $Y$ is the image of a webbed space under a sequentially continuous linear map then $Y$ is webbed.

(c) QUOTIENTS Hausdorff quotients of webbed spaces are webbed.

(d) COARSER TOPOLOGIES Webbed spaces remain webbed under any coarser locally convex Hausdorff topology.

(e) PRODUCTS Countable products of webbed spaces are webbed.

14.202 OPEN MAPPING THEOREM FOR WEBBED SPACES The technique of Theorem 14.4.6 may be used to deduce the open mapping theorems (a) and (b) below once the result of Exercise 14.201(c) has been established.

(a) A continuous linear map of a webbed space $X$ onto an ultrabornological space $Y$ is open.

(b) A closed linear map $A$ of the webbed space $X$ onto a nonmeager subspace $A(X)$ of an LCHS $Y$ is open and $A(X) = Y$.

14.203 STABILITY OF THE RANGE Given an LCHS $X$, let $R(X) = \{Y : Y$ is an LCHS and the closed graph theorem holds for $(X, Y)\}$.

(a) CONTINUOUS LINEAR IMAGES Let $A : X \to Y$ be a linear map. If $A(Y)$ is an LCHS and $Y \in R(X)$, then $A(Y) \in R(X)$.

(b) COARSER TOPOLOGIES, CLOSED SUBSPACES, QUOTIENTS If $(Y, \mathcal{T}) \in R(X)$ and $\mathcal{T}' \subset \mathcal{T}$, where $\mathcal{T}'$ is a locally convex Hausdorff topology, then $(Y, \mathcal{T}') \in R(X)$. If $Y \in R(X)$ and $M$ is a closed subspace of $Y$, then $M \in R(X)$ and $Y/M \in R(X)$.

14.204 CLOSED GRAPH THEOREM FOR INDUCTIVE LIMITS OF BAIRE SPACES In Theorem 14.7.1 it was shown that a sequentially closed linear map of a Fréchet space into a webbed space is continuous. From Theorem 14.7.2 it then followed that a sequentially closed linear map of an inductive limit of Fréchet spaces, i.e., an ultrabornological space, into a webbed space is continuous. Modify the proof of Theorem 14.7.1 to show that if "Fréchet" is weakened to "Baire LCHS" while strengthening the requirement on $A$ from sequentially closed to closed, then: A closed linear map $A$ of a Baire LCHS $X$ into a webbed space $Y$ is continuous.

### Hints to Exercises

14.105 Show that $A$ is almost continuous by Theorem 13.2.5; then use Theorem 14.3.4.

14.108 Show that $\operatorname{cl} A(V)$, where $V$ is a closed balanced neighborhood of 0, is an ultrabarrel. Hence, in the situation of the exercise, $A$ is almost open. Now use the basic open mapping theorem, Theorem 14.4.2.

14.109 Show, for any linear map $T$ and any ultrabarrel $B$, that $\operatorname{cl} T^{-1}(B)$ is an ultrabarrel. Thus the map $A$ of the exercise is almost continuous. Now use the basic closed graph theorem, Theorem 14.3.4.

14.201 (a) Letting $M$ be the subspace, consider the sets $W \cap M$ as $W$ ranges through the web.

(b) If $A$ is the map, consider the sets $A(W)$ as $W$ ranges through the web.

(c) Use (b).

14.204 As in the proof of Theorem 14.7.1, get a strand $(S_n)$ of the web on $Y$ such that each $A^{-1}(S_n)$ is nonmeager. Then choose disked neighborhoods $U_n$ of 0 in $X$ ($n \in \mathbf{N}$) such that $U_{n+1} + U_{n+1} \subset U_n \subset \operatorname{cl} A^{-1}(S_n)$. Hence, for every $n$, $U_n \subset \operatorname{cl} A^{-1}(S_n) \subset A^{-1}(S_n) + U_{n+1}$. Given a closed neighborhood $V$ of 0 in $Y$, choose $S_{m-1}$ such that $S_{m-1} \subset V$. For the sake, ultimately, of showing that $U_m \subset A^{-1}(V)$, pick $x_m \in U_m$. For $n \geq m$, choose $x_n \in U_n$ and $y_n \in A^{-1}(S_n)$ such that $x_n = y_n + x_{n+1}$. Thus, for $k \geq m$,

$$x_m = \sum_{n=m}^{k} y_n + x_{k+1}$$

In this case, however [unlike the proof of Theorem 14.7.1], it does not follow that $x_{k+1} \to 0$ since the neighborhoods $(U_n)$ do not form a neighborhood base at 0. For all $n \geq m$, $y_{n+1} = x_{n+1} - x_{n+2} \in U_{n+1} + (1/2)U_{n+1} \subset U_n$, so for any neighborhood $U$ of 0 in $X$,

$$x_{n+1} + y_{n+1} \in U_{n+1} + U_n \subset U_{n-1} \subset \operatorname{cl} A^{-1}(S_{n-1}) \subset A^{-1}(S_{n-1}) + U$$

Hence, for all $k > m + 1$, there exists $z_k \in A^{-1}(S_k)$ such that $x_{k+1} + y_{k+1} - z_{k-1} = u \in U$. Thus

$$x_m - \sum_{n=m}^{k} y_n + x_{k+1} = u + z_{k-1} - y_{k+1}$$

Therefore, $x_m - \sum_{n=m}^{k} y_n - z_{k-1} + y_{k+1} \in U$. Since $(S_n)$ is a strand, $(Az_{k-1})$ and $(Ay_{k+1})$ are null sequences while $(\sum_{n=m}^{k} Ay_n)$ converges to some $z$ in $S_{m-1} \subset V$. Thus $A(\sum_{n=m}^{k} y_n - z_{k-1} + y_{k+1}) \to z$. Hence $(x_m, z) \in \mathrm{cl}\, G(A) = G(A)$, so $z = Ax_m$ or $x_m \in A^{-1}(V)$; hence $U_m \subset A^{-1}(V)$. It follows that $A$ is continuous.

# Chapter 15

# Reflexivity



Let $X$ be a normed space. If $X$ is finite-dimensional then its normed dual $X'$ is just $X$. Consequently, $X'' = X$ as well. One of the first infinite-dimensional normed spaces to be investigated was $\ell_2$ and here too $\ell_2'' = \ell_2$; likewise $L_2[a,b]'' = L_2[a,b]$ and $\ell_p'' = \ell_p$ $(1 < p < \infty)$. Did this always happen? Hahn [1927, p. 219] observed that you could locate a norm-isomorphic copy of $X$ in $X''$ by means of the *canonical embedding*

$$J : X \to X'',\ x \mapsto \langle x, \cdot \rangle$$

first discussed in Sec. 8.4 and we frequently identify $J(X)$ and $X$. As noted in Theorem 8.4.5, not only is $J$ a linear isomorphism, it is an isometry. By Goldstine's theorem [8.4.7], $X$ is $\sigma(X'', X')$-dense in $X''$ but when is $J$ *surjective*? or, identifying $X$ and $J(X)$ as normed spaces, when is $X = X''$? Helly [1921, p. 80] had already noted that the Banach space $(c, \|\cdot\|_\infty)$ of convergent sequences was a proper subset of $c''$, i.e., that $J$ is not always surjective. Hahn [1927, p. 220] singled out those normed spaces $X$ for which $X = X''$ and called them "regular spaces" (*reguläre Räume*). Mercifully, mathematics has been spared this $10^{11}$th use of "regular"; we use Lorch's [1939] *reflexive* instead.

Consider some examples.

**Example 15.0.1** REFLEXIVE NORMED SPACES

(a) The unit ball $U$ of an infinite-dimensional Banach space $X$ has infinitely many extreme points [Example 9.4.8(b)]; hence if $U$ has only a finite number, $X$ is not reflexive. Since the unit balls of $c_0$ and $L_1(\mathbf{R})$ have no extreme points [Example 9.2.5(c,i)], and $C([0,1], \mathbf{R}, \|\cdot\|_\infty)$ has only two [Example 9.2.5(e)], they are not reflexive. As we note after Example 15.8.5, $U$ having an infinite number of extreme points does not suffice for reflexivity.

(b) Any Hilbert space is reflexive. [Example 15.7.1].

(c) $\ell_p$, $1 < p < \infty$, is reflexive [Example 15.7.2].

(d) By (a), $c_0$ is not reflexive. We show in Example 15.7.3(a,b) that $c_0$ is not reflexive directly, by showing that $c_0'' = \ell_\infty$.

(e) If a normed space $X$ contains a linearly homeomorphic copy of $c_0$ then $X$ is not reflexive. Roughly, $c_0'' \subset X''$. ■

When we leave normed spaces for TVS $(X, \mathcal{T})$ the situation is much more delicate because, unlike the situation for normed spaces, $X'$ does not come equipped with a "natural" topology. Things are further complicated by the fact that, for LCHS $X$, if $X'$ carries any topology $\mathcal{T}$ such that $\sigma(X', X) \subset \mathcal{T} \subset \tau(X', X)$, then $X'' = X$.

**Problem 15.0.2** To investigate reflexivity of a TVS $X$, how should we topologize $X'$?

We deal with the issue for LCHS in our consideration of reflexivity in this chapter. The most significant results are about Banach spaces, some of the most exciting discoveries of latter 20th century functional analysis.

**Notation for the Chapter** $X$ is at least an LCHS, so $(X, X')$ is a dual pair. $J$ denotes the canonical embedding. Note that if $Jx = 0$ then $f(x) = 0$ for all $f \in X'$; since $(X, X')$ is a dual pair, this means that $x = 0$ so $J$ is generally a linear isomorphism and we often identify $X$ and $J(X) \subset X''$.

$X'$ stands for the strong dual $(X', \beta(X', X))$ of $X$.

For a subset $B' \subset X'$, we sometimes compute its polar in $X''$ rather than $X$; in this case we denote it by $(B')^d$ ("$d$" as in "dual"):

$$(B')^d = \{x'' \in X'' : \sup |x''(B')| \leq 1\}$$

When topologized, $X''$ is endowed with $\beta(X'', X')$, the topology determined by sets of the form $(B')^d$ where $B'$ is a $\sigma(X', X'')$-bounded subset of $X'$. As we show in Theorem 15.2.1(a), $\sigma(X', X'')$-boundedness is the same as $\beta(X', X)$-boundedness so $\beta(X'', X')$ may be described as the topology determined by sets of the form $(B')^d$ where $B'$ is $\beta(X', X)$-bounded. ■

## 15.1 REFLEXIVITY BASICS

For any normed space $X$, the norm topology on $X'$ is the strong topology $\beta(X', X)$, the polar topology generated by all $\sigma(X, X')$-bounded subsets of $X$ [Example 8.8.9]. Motivated by that, we define:

**Definition 15.1.1** BIDUAL Let $X$ be an LCHS. With $X' = (X', \beta(X', X))$, we take $X'' = (X', \beta(X', X))'$ and call it the *bidual* of $X$. In other words, the bidual is the dual of the strong dual $(X', \beta(X', X))$. We call $X''$ endowed with $\beta(X'', X')$ the *strong* bidual of $X$. ■

We suppose $X'$ and $X''$ to be equipped with these strong topologies throughout the chapter.

Suppose $(X, \mathcal{T})$ is an LCHS and $x \in X$. Is $Jx : (X', \beta(X', X)) \to \mathbf{F}$, $x' \mapsto \langle x, x' \rangle$, continuous? Since $Jx : (X', \sigma(X', X)) \to \mathbf{F}$ is continuous and $\sigma(X', X) \subset \beta(X', X)$, it follows that $Jx : (X', \beta(X', X)) \to \mathbf{F}$ is continuous, i.e., that $J(X) \subset X''$. If $J$ is surjective we say that $X$ is *semireflexive.* If $J$ is a surjective homeomorphism or, equivalently,

$$\beta(X'', X') \cap X = \mathcal{T}$$

we say that $X$ is *reflexive.*

It is immediate that reflexivity is preserved by linear homeomorphism.

The strong dual $(X', \beta(X', X))$ of any bornological space $X$ is complete [Theorem 13.2.13]. Thus, if a normed—hence bornological—space $X$ is reflexive, it is a Banach space. If $(X, \mathcal{T})$ is a normed space, $J$ is an isometry [Theorem 8.4.5]; hence $\beta(X'', X') \cap X = \mathcal{T}$, so:

> To show that a normed space is reflexive, it is only necessary to show that $J$ is onto—that $X$ is semireflexive.

The polar of the unit ball $U$ of any TVS $X$ is $\sigma(X', X)$-compact [Alaoglu, 8.4.1]. A principal result of this chapter is:

Theorem 15.5.1. A Banach space $X$ is reflexive iff $U$ is $\sigma(X, X')$-compact.

## 15.2 REFLEXIVE SPACES

Theorem 15.2.1(a) enables us to describe $\beta(X'', X')$ in a simpler way (see after the theorem). Theorem 15.2.1(b) is a cousin of Theorems 8.3.5 and 11.2.1, relating boundedness to absorbency of a polar.

**Theorem 15.2.1** $\beta(X', X)$-BOUNDEDNESS Let $(X, X')$ be a dual pair. Then:

(a) the bounded sets of $X'$ are the same with respect to $\beta(X', X)$, $\sigma(X', X'')$, and $\tau(X', X'')$;

(b) a subset $D \subset X'$ is $\beta(X',X)$-bounded iff $D^\circ \subset X$ absorbs each $\sigma(X,X')$-bounded disk of $X$.

**Proof.** Let $(X,X')$ be a dual pair. (a) The topologies $\sigma(X',X'')$, and $\tau(X',X'')$ are topologies of the pair $(X',X'')$ as is $\beta(X',X)$ by the way $X''$ is defined; hence the bounded sets of all three topologies are the same by Mackey's theorem [8.8.7].

(b) Suppose $D \subset X'$. Let $B \subset X$ be a $\sigma(X,X')$-bounded disk. $D^\circ$ absorbs $B$ iff for some $r > 0$, $rB \subset D^\circ$ and this is equivalent to $(1/r)B^\circ \supset D^{\circ\circ} \supset D$. As the sets $B^\circ$ are a base at 0 for $\beta(X',X)$, it follows that $D$ is $\beta(X',X)$-bounded iff $D^\circ$ absorbs all $\sigma(X,X')$-bounded disks $B$. ■

In Theorems 15.2.2 and 15.2.3, we use the following notation: For $B' \subset X'$, we sometimes compute the polar of $B'$ in $X''$ rather than $X$; we denote this polar by $(B')^d$ ("$d$" as in "dual"):

$$(B')^d = \{x'' \in X'' : \sup|x''(B')| \le 1\}$$

By Theorem 15.2.1(a), $\beta(X'',X')$ may be described as the polar topology determined by sets of the form $(B')^d$ where $B' \subset X'$ is $\beta(X',X)$-bounded.

**Theorem 15.2.2** STRENGTH OF $\beta(X'',X')$ If $(X,\mathcal{T})$ is an LCHS then $\mathcal{T} \subset \beta(X'',X') \cap X$.

**Proof.** Let $(X,\mathcal{T})$ be an LCHS. Since any $\mathcal{T}$-closed, disked neighborhood $U$ of 0 absorbs $\mathcal{T}$-bounded subsets of $X$, it absorbs $\sigma(X,X')$-bounded disks. Since $U = U^{\circ\circ}$ [Theorem 8.8.1(c)], it follows that $U^\circ \subset X'$ is $\beta(X',X)$-bounded by Theorem 15.2.1(b). Since $\beta(X',X)$-bounded $= \sigma(X',X'')$-bounded [Theorem 15.2.1(a)], it follows that $U^{\circ d} \in \beta(X'',X')$. By the bipolar theorem [8.3.8], $U^{\circ d} \cap X = U^{\circ\circ} = \mathrm{cl}_{\sigma(X,X')} U$. Since the closure of a convex set is the same in any topology of the dual pair [Theorem 8.8.1(b)], $\mathrm{cl}_{\sigma(X,X')} U = \mathrm{cl}_{\mathcal{T}} U = U$. Thus $U$ is a $\beta(X'',X') \cap X$-neighborhood of 0 and $\mathcal{T} \subset \beta(X'',X') \cap X$. ■

Recall that an LCS is infrabarreled if every bornivorous barrel is a neighborhood of 0.

**Theorem 15.2.3** $J$ HOMEOMORPHISM IFF $X$ INFRABARRELED For an LCHS $(X,\mathcal{T})$, the canonical embedding $J : X \to X'' = (X'', \beta(X'',X'))$ is a homeomorphism iff $X$ is infrabarreled.

**Proof.** Let $(X,\mathcal{T})$ be an LCHS. Since $\mathcal{T} \subset \beta(X'',X') \cap X$ [Theorem 15.2.2], $J$ is generally a relatively open map so it only remains to show that $J$ is continuous iff $X$ is infrabarreled. By Theorem 11.11.4, $X$ is infrabarreled iff every $\beta(X',X)$-bounded set is equicontinuous. If $B$ is $\beta(X',X)$-bounded then, given any basic $\beta(X',X)$-neighborhood $V$ of 0, $B \subset rV$ for sufficiently large

$r > 0$. Since subsets of equicontinuous sets are equicontinuous, it suffices to show that continuity of $J$ is equivalent to the equicontinuity of basic $\beta(X', X)$-neighborhoods $V$ of 0. A basic $\beta(X', X)$-neighborhood of 0 is the polar $U^\circ$ of a $\sigma(X, X')$-closed, bounded disk $U \subset X$. A basic $\beta(X'', X')$-neighborhood of 0 is therefore of the form $U^{\circ d} = \{f \in X'' : \sup |f(U^\circ)| \leq 1\}$. Hence, $J$ is continuous iff $J^{-1}(U^{\circ d}) = \{x \in X : \sup |x(U^\circ)| \leq 1\} = U^{\circ\circ} = U$ (by the bipolar theorem, 8.8.1(c)) is a neighborhood of 0. By Theorem 8.6.4(b), this is equivalent to the equicontinuity of $U^\circ$. ■

By Theorem 15.2.3, a reflexive space is infrabarreled. We strengthen this to "a reflexive space is barreled" in Theorem 15.2.6. James [1951] constructed a non-reflexive Banach space $V$, the *James space,* with the odd property that it is linearly isometric to $V''$ but $J$ is not the linear isometry—to be reflexive, it is $J$ that must be the isometry. $V$ is of codimension 1 in $V''$ however [Lindenstrauss and Tzafriri 1977, p. 25], and this motivated Civin and Yood [1957] to define *quasi-reflexive* spaces as those Banach spaces $X$ such that $X''/X$ is finite-dimensional. Herman and Whitley [1967] call a Banach space *somewhat reflexive* if every infinite-dimensional subspace contains an infinite-dimensional *reflexive* subspace. (Quasi-reflexivity implies somewhat reflexivity.) James [1974] constructed a space called the *James tree,* which has the property that every infinite-dimensional subspace contains a copy of $\ell_2$!

A normed space $X$ is finite-dimensional iff each closed bounded set is compact [Theorem 4.10.4]. We characterize semireflexivity by means of a weak Heine-Borel property in Theorem 15.2.4(b), generalizing Banach's [1932, p. 189, Th. 13] result for separable Banach spaces.

**Theorem 15.2.4** SEMIREFLEXIVITY An LCHS $(X, \mathcal{T})$ is semireflexive iff either of the following conditions is satisfied.

(a) $\beta(X', X) = \tau(X', X)$ (or, equivalently, $\beta(X', X)$ is a topology of the dual pair $(X', X)$).

(b) Every bounded subset of $X$ is contained in a $\sigma(X, X')$-compact set or, equivalently, every closed bounded subset of $X$ is $\sigma(X, X')$-compact.

**Proof.** Let $(X, \mathcal{T})$ be an LCHS. (a) If $\beta(X', X) = \tau(X', X)$—a topology of the dual pair $(X', X)$—then $X'' = (X', \beta(X', X))' = X$ by the Mackey-Arens theorem [8.7.4] and $X$ is semireflexive. The converse is clear.

(b) Suppose $X$ is semireflexive so that $\beta(X', X) = \tau(X', X)$ by (a). If $B$ is a $\sigma(X, X')$-bounded set, then $B^\circ \in \beta(X', X) = \tau(X', X)$. Thus there must be a $\sigma(X, X')$-compact disk $K$ such that $K^\circ \subset B^\circ$. Hence $B \subset B^{\circ\circ} \subset K^{\circ\circ} = K$ and every $\sigma(X, X')$-bounded set [hence every bounded set by Theorem 8.8.7] is contained in a $\sigma(X, X')$-compact set.

Conversely, suppose that every bounded subset of $X$ is contained in a $\sigma(X, X')$-compact set. To show that $X$ is semireflexive, it suffices to show that $\beta(X', X) = \tau(X', X)$ by (a). To this end, let $B \subset X$ be $\mathcal{T}$-bounded. By the condition, there exists a $\sigma(X, X')$-compact set $K$ containing $B$. Hence $D = \mathrm{cl}_{\sigma(X,X')} B$ is a $\sigma(X, X')$-compact disk containing $B$. Therefore $B^\circ \supset D^\circ \supset K^\circ$ and $B^\circ$ is a $\tau(X', X)$-neighborhood of 0, i.e., $\beta(X', X) \subset \tau(X', X)$. ■

By Theorem 11.3.6, $(X', \beta(X', X))$ is barreled iff $\beta(X', X) = \beta(X', X'')$. If $X$ is semireflexive, however, then $X'' = X$; hence $\beta(X', X) = \beta(X', X'')$. Consequently:

**Corollary 15.2.5** If $X$ is semireflexive then $X'$ is barreled.

**Theorem 15.2.6** REFLEXIVITY EQUIVALENTS For any LCHS $(X, \mathcal{T})$, the following statements are equivalent:

(a) $X$ is reflexive;

(b) $X$ is barreled and semireflexive;

(c) $X$ is infrabarreled and semireflexive.

**Proof.** Let $(X, \mathcal{T})$ be an LCHS. (a) $\Rightarrow$ (b). Since $(X, \mathcal{T})$ is reflexive, $X$ is semireflexive and therefore infrabarreled by Theorem 15.2.3. To show that $X$ is barreled, we show that $\mathcal{T} = \beta(X, X')$ [Theorem 11.3.6]. To that end, let $D \subset X'$ be $\sigma(X', X)$-bounded—hence $\tau(X', X)$-bounded. Since $X$ is semireflexive, $\tau(X', X) = \beta(X', X)$ [Theorem 15.2.4(a)]; hence $D$ is $\beta(X', X)$-bounded. Since $X$ is infrabarreled, $D$ is equicontinuous by Theorem 11.11.4. As any topology for an LCS $X$ is determined by polars of the equicontinuous subsets of $X'$ [Theorem 8.6.6], the basic $\beta(X, X')$-neighborhood $D^\circ$ of 0 is a $\mathcal{T}$-neighborhood of 0; therefore $\mathcal{T} = \beta(X, X')$.

Clearly (b) $\Rightarrow$ (c). To see that (c) $\Rightarrow$ (a), since $X$ is infrabarreled, the canonical embedding $J$ is a homeomorphism by Theorem 15.2.3; since $X$ is semireflexive, $J$ is onto. ■

Since neither $(C[0,1], \|\cdot\|_1)$ nor $\varphi$ is barreled [Examples 11.3.2 and 11.3.3], neither is reflexive.

**Theorem 15.2.7** CLOSED SUBSPACE OF REFLEXIVE BANACH SPACE A closed subspace $M$ of a reflexive Banach space $X$ is reflexive.

**Proof.** We use the notation of the statement. Since $M$ is a normed space, we need only show that $M$ is semireflexive. We use the criterion of Theorem 15.2.4(b) and show that any bounded subset $B$ of $M$ is contained in a $\sigma(M, M')$-compact set. $B$ being bounded in $M$ implies that $B$ is bounded in $X$. Since $X$ is reflexive, $B \subset K$ for some $\sigma(X, X')$-compact set $K \subset X$ by Theorem 15.2.4(b). Clearly $K \cap M$ is $\sigma(X, X') \cap M$-compact. By the continuous extendibility of continuous linear functionals [Corollary 7.3.3], $\sigma(X, X') \cap M = \sigma(M, M')$. Thus $B$ is a subset of the $\sigma(M, M')$-compact set $K \cap M$. ■

As the Banach space $c_0$ of null sequences is not reflexive [Example 15.0.1(a)] and $c_0$ is a closed subspace of $\ell_\infty$, it follows that $\ell_\infty$ is not reflexive.

Reflexivity of a Banach space was the first property to be recognized as a *three-space property* in the following sense: Let $M$ be a closed subspace of a Banach space $X$. If two of $X$, $M$ and $X/M$ are reflexive, then so is the third, a result due to Krein and Smulian [1940, p. 575]. For more on three-space properties generally, see Castillo and González 1997. Our next result demonstrates the close connection between reflexivity of a space and reflexivity of its dual.

**Theorem 15.2.8** $X$ Reflexive Almost $\Leftrightarrow X'$ Reflexive An LCHS $X$ is reflexive iff $X'$ is reflexive and $\beta(X', X)$ is a topology of the dual pair $(X', X)$.

**Proof.** Suppose the LCHS $X$ is reflexive. Then $(X')'' = (X'')' = X'$ so $X'$ is semireflexive. Furthermore, by Theorem 15.2.4(a), $\beta(X', X) = \tau(X', X)$ so $\beta(X', X)$ is a topology of the pair $(X', X)$. Since $\beta(X', X'') = \beta(X', X)$, the topology of $X'$, $X'$ is barreled [Theorem 11.3.6]. Since $X'$ is barreled and semireflexive, it is reflexive by Theorem 15.2.6(b).

Now suppose that $X'$ is reflexive and $\beta(X', X)$ is a topology of the dual pair $(X', X)$. As $\beta(X', X)$ is a topology of the dual pair $(X', X)$, $X'' = (X', \beta(X', X))' = X$ and $X$ is semireflexive. We observed in Corollary 15.2.5 that semireflexive spaces have a barreled strong dual. Therefore, since $X'$ is reflexive, $(X'', \beta(X'', X'))$ is barreled. Since $X'' = X$, $\beta(X'', X') = \beta(X, X')$ so which implies that $(X, \beta(X, X'))$ is barreled [Theorem 11.3.6]. Since $\mathcal{T} \subset \beta(X, X')$, $(X, \mathcal{T})$ is barreled. Since $X$ is semireflexive and barreled, it is reflexive [Theorem 15.2.6(b)] ■

We cannot omit the condition that $\beta(X', X)$ be a topology of the pair $(X', X)$ to show that reflexivity of $X'$ implies reflexivity of $X$. Consider the subspace $\varphi$ of $\ell_2$ consisting of sequences which are eventually 0. Clearly $\varphi$ is dense in $\ell_2$. Thus $\varphi' = \ell_2' = \ell_2$ [Example 15.7.1] and $\varphi'' = \ell_2' = \ell_2$. Two things follow: $\varphi$ is not semireflexive and $\beta(\varphi', \varphi) = \beta(\ell_2, \varphi)$ is not a topology of the pair $(\ell_2, \varphi)$: if it were, then $(\ell_2, \beta(\ell_2, \varphi))' = \varphi$ whereas a little computation and the Cauchy-Schwarz inequality show that $(\ell_2, \beta(\ell_2, \varphi))' = \ell_2$. Since $(\varphi', \beta(\varphi', \varphi'')) = (\ell_2, \beta(\ell_2, \ell_2))$ is barreled, $\varphi$ is an example of a space $X$ where $(X', \beta(X', X))$ is barreled but $X$ is not semireflexive.

## 15.3 WEAK-STAR CLOSED SETS

In the Krein–Smulian theorem [15.3.9] we obtain a necessary and sufficient condition for a convex subset $F$ of the dual of a Banach space $X$ to be $\sigma(X', X)$-closed—namely that $F \cap E$ is $\sigma(X', X)$-closed for every $\sigma(X', X)$-closed equicontinuous disk $E \subset X'$. Its immediate consequences include criteria for (1) a subspace of $X'$ to be $\sigma(X', X)$-closed and (2) for a $\sigma(X, X')$-compact set $K$ to be metrizable in its relative weak topology $\sigma(X, X') \cap K$.

Recall that equicontinuity implies relative $\sigma(X', X)$-compactness [Theorem 8.6.5] and that equicontinuous sets are bounded in any topology of the dual pair $(X', X)$ [Theorem 11.2.6].

**Definition 15.3.1** EQUICONTINUOUS WEAK-$*$ TOPOLOGY Let $X$ be an LCHS over $\mathbf{F}$. The *equicontinuous weak-$*$ topology* $e\sigma(X', X)$ is the strongest topology on $X'$ that induces $\sigma(X', X) \cap E$ on each equicontinuous subset $E \subset X'$. Thus $e\sigma(X', X)$ consists of all $B \subset X'$ such that $B \cap E \in \sigma(X', X) \cap E$ for each equicontinuous set $E \subset X'$. ∎

Since a translate $f + E$ of an equicontinuous set $E$ is equicontinuous, the $e\sigma(X', X)$-neighborhoods of $f \in X'$ are of the form $f + V$ where $V$ is an $e\sigma(X', X)$-neighborhood of 0. Generally, however, $e\sigma(X', X)$ is not a linear topology. Komura [1962] exhibited an LCHS $X$ and an $e\sigma(X', X)$-neighborhood $U$ of 0 such that $V + V \not\subset U$ for any $e\sigma(X', X)$-neighborhood $V$ of 0. Hence addition is not $e\sigma(X', X)$-continuous in this case. However [Theorem 15.3.6], if $X$ is a metrizable LCHS, then $e\sigma(X', X)$ is a polar topology—indeed, if $X$ is a Fréchet space, $e\sigma(X', X) = \tau(X', X)$, a topology of the dual pair $(X', X)$ [Cor. 15.3.7].

Recall that a subset $B$ of a TVS $X$ is totally bounded (= precompact) if for any neighborhood $V$ of 0 there exist $x_1, \ldots, x_n \in X$ such that $B \subset \cup_{i=1}^{n}(x_i + V)$. Totally bounded sets are bounded [Theorem 6.1.2], hence weakly bounded. Thus [Theorem 8.5.1] polars of totally bounded sets generate a vector topology on $X'$ which we discuss now.

**Definition 15.3.2** UNIFORM CONVERGENCE ON PRECOMPACT SETS Let $X$ be an LCHS over $\mathbf{F}$ and $X'$ its dual. Let $\{P_s : s \in S\}$ be the precompact subsets of $X$. The collection $\{P_s^{\circ} : s \in S\}$ forms a base of neighborhoods of 0 in $X'$ for a vector topology $p(X', X)$ called the *topology of precompact convergence* on $X'$. ∎

Since finite sets are precompact, $\sigma(X', X) \subset p(X', X)$ and $p(X', X)$ is Hausdorff.

**Theorem 15.3.3** $p(X', X) \subset e\sigma(X', X)$ Let $X$ be an LCHS over $\mathbf{F}$. If $E \subset X'$ is equicontinuous then $p(X', X) \cap E = \sigma(X', X) \cap E$. Hence $p(X', X) \subset e\sigma(X', X)$.

**Proof.** Let $(X, \mathcal{T})$ be an LCHS. Since, as observed above, $\sigma(X', X) \subset p(X', X)$, if $E \subset X'$ is equicontinuous then $\sigma(X', X) \cap E \subset p(X', X) \cap E$. We show that $p(X', X) \cap E \subset \sigma(X', X) \cap E$ by showing that a $p(X', X)$-neighborhood $(g + P^{\circ}) \cap E$ ($P$ precompact) of $g \in E$ contains a relative $\sigma(X', X)$-neighbor-hood $(g + \{x_1, \ldots, x_n\}^{\circ}) \cap E$ of $g$ for some $x_1, \ldots, x_n \in X$. Since $E$ is equicontinuous, $E - g$ is equicontinuous. Therefore there exists a neighborhood $W$ of 0 in $X$ such that $\sup_{f \in E} |(f - g)(W)| \leq 1/2$. Since $P$ is precompact, there exist $x_1, \ldots, x_n \in X$ such that $P \subset \cup_{i=1}^{n}(x_i + W)$.

Suppose that $f \in E \cap (g + (1/2)\{x_1, \ldots, x_n\}^\circ) \in \sigma(X', X) \cap E$. Then $(f - g) \in (1/2)\{x_1, \ldots, x_n\}^\circ$. For any $x \in P \subset \cup_{i=1}^n (x_i + W)$, there exists $i \in \{1, 2, \ldots, n\}$ and $w \in W$ such that $x = x_i + w$. Hence $|(f-g)(x)| = |(f-g)(x_i) + (f-g)(w)| \le 1/2 + 1/2 = 1$. Hence $f - g \in P^\circ$ and $E \cap (g + (1/2)\{x_1, \ldots, x_n\}^\circ) \subset E \cap (g + P^\circ)$. ■

We now show that $e\sigma(X', X)$ has a base of balanced absorbent neighborhoods of $0 \in X'$.

**Theorem 15.3.4** BASE OF $e\sigma(X', X)$-BALANCED ABSORBENT NEIGHBORHOODS Let $X$ be an LCHS over $\mathbf{F}$. Then $e\sigma(X', X)$ has a base of balanced absorbent neighborhoods of 0.

**Proof.** Let $(X, \mathcal{T})$ be an LCHS. We show first that if $W$ is an $e\sigma(X', X)$-neighborhood of 0, the balanced core $B(W) = \cap_{|a| \ge 1} aW$ [Sec. 4.1] of $W$ is also an $e\sigma(X', X)$-neighborhood of 0. For any equicontinuous subset $E \subset X'$, the balanced core $B(E)$ of $E$ is equicontinuous. Since $W$ is an $e\sigma(X', X)$-neighborhood of 0, $W \cap B(E)$ contains a $\sigma(X', X) \cap B(E)$ neighborhood of 0, say $\{x_1, \ldots, x_n\}^\circ \cap B(E)$, where $x_1, \ldots, x_n \in X$. Since $\{x_1, \ldots, x_n\}^\circ \cap B(E)$ is a balanced subset of $W$, $\{x_1, \ldots, x_n\}^\circ \cap B(E) \subset B(W)$. Hence $B(W)$ is an $e\sigma(X', X)$-neighborhood of 0. Since $B(W) \subset W$, the result follows.

To show that there is a base of absorbent $e\sigma(X', X)$-neighborhoods of 0, let $W$ be a balanced $e\sigma(X', X)$-neighborhood of 0 and let $f \in X'$. The balanced hull $E = \{f\}_b$ is an equicontinuous subset of $X'$ to which $f$ belongs. Since $W$ is an $e\sigma(X', X)$-neighborhood of 0, there exist $x_1, \ldots, x_n \in X$ such that $\{x_1, \ldots, x_n\}^\circ \cap E \subset W$. Since $\{x_1, \ldots, x_n\}^\circ$ is absorbent, there exists $r \in (0, 1]$ such that $af \in \{x_1, \ldots, x_n\}^\circ$ for $|a| \le r$. Since $W$ is balanced, $af \in \{x_1, \ldots, x_n\}^\circ \cap E \subset W$ for $|a| \le r$ and $W$ is absorbent. ■

We show next that when $X$ is metrizable, $e\sigma(X', X)$ is a vector topology; in fact, $e\sigma(X', X) = p(X', X)$. First, we establish the following technical lemma.

**Lemma 15.3.5** Let $(X, \mathcal{T})$ be a metrizable LCHS and let $W \subset X'$ be an $e\sigma(X', X)$-open neighborhood of 0. Let $\{V_n : n \in \mathbf{N}\}$ be a descending base of $\mathcal{T}$-closed, disked neighborhoods of 0 with $V_1 = X$. For each $k \in \mathbf{N}$ there exists a finite set $F_k \subset V_k$ such that, with $A_n = \cup_{k=0}^{n-1} F_k$ $(F_0 = \{0\})$, $A_n^\circ \cap V_n^\circ \subset W$.

**Proof.** Let $(X, \mathcal{T})$ be a metrizable LCS and let $W \subset X'$ be an $e\sigma(X', X)$-open neighborhood of 0. Since $V_1 = X$, $V_1^\circ = \{0\}$. Thus $A_1^\circ \cap V_1^\circ = F_0^\circ \cap \{0\} = X \cap \{0\} = \{0\} \subset W$. Now suppose $F_k$ has been found for $k = 1, \ldots, n-1$, so that $A_n^\circ \cap V_n^\circ = \left(\cup_{k=0}^{n-1} F_k\right)^\circ \cap V_n^\circ \subset W$ or $A_n^\circ \cap V_n^\circ \cap \complement W = \varnothing$. Suppose that there is no finite set $F_n$ satisfying the desired condition. Then, for any finite set, $F_n \subset V_n$, $(A_{n+1})^\circ \cap V_{n+1}^\circ \cap \complement W = (A_n \cup F_n)^\circ \cap V_{n+1}^\circ \cap \complement W = A_n^\circ \cap F_n^\circ \cap V_{n+1}^\circ \cap \complement W \ne \varnothing$.

Since $V_{n+1}^{\circ}$ is equicontinuous [8.6.4], $\complement W \cap V_{n+1}^{\circ}$ is $\sigma(X', X)$-closed in $V_{n+1}^{\circ}$. Since $V_{n+1}^{\circ}$ is $\sigma(X', X)$-compact [Alaoglu, 8.4.1], $\complement W \cap V_{n+1}^{\circ}$ is $\sigma(X', X)$-compact. Since there is no finite set $F_n \subset V_n$ such that $A_n^{\circ} \cap F_n^{\circ} \cap V_{n+1}^{\circ} \cap \complement W \neq \varnothing$ then, for any finite subsets $F$ and $G$ of $V_n$,

$$\begin{aligned}&(A_n \cup F \cup G)^{\circ} \cap V_{n+1}^{\circ} \cap \complement W \\ &= \left[A_n^{\circ} \cap F^{\circ} \cap V_{n+1}^{\circ} \cap \complement W\right] \cap \left[A_n^{\circ} \cap G^{\circ} \cap V_{n+1}^{\circ} \cap \complement W\right] \neq \varnothing\end{aligned}$$

Thus the family of $\sigma(X', X)$-closed subsets $A_n^{\circ} \cap F^{\circ} \cap V_{n+1}^{\circ} \cap \complement W$, where $F$ is a finite subset of $V_n$, of the $\sigma(X', X)$-compact set $V_{n+1}^{\circ} \cap \complement W$ has a nonempty intersection. Hence, for the family $\{\{x\} : x \in V_n\}$ of singletons,

$$\bigcap_{x \in V_n} A_n^{\circ} \cap \{x\}^{\circ} \cap V_{n+1}^{\circ} \cap \complement W \neq \varnothing$$

But $A_n^{\circ} \cap (\cap_{x \in V_n} \{x\}^{\circ}) \cap V_{n+1}^{\circ} \cap \complement W = A_n^{\circ} \cap V_n^{\circ} \cap V_{n+1}^{\circ} \cap \complement W \subset A_n^{\circ} \cap V_n^{\circ} \cap \complement W = \varnothing$ and we have arrived at a contradiction. ■

**Theorem 15.3.6** BANACH–DIEUDONNÉ THEOREM If $X$ is a metrizable LCS then $e\sigma(X', X) = p(X', X)$, i.e., the strongest topology on $X'$ which induces the $\sigma(X', X)$-topology on each equicontinuous subset of $X'$ is the topology of uniform convergence on precompact subsets of $X$.

**Proof.** Let $X$ be a metrizable LCHS. Since $p(X', X) \subset e\sigma(X', X)$ [Theorem 15.3.3] and $e\sigma(X', X)$-neighborhoods of $f \in X'$ are translates of $e\sigma(X', X)$-neighborhoods of 0 (discussion following Definition 15.3.1), it is only necessary to show that if $W$ is an $e\sigma(X', X)$-open neighborhood of 0, there exists a precompact set $P$ such that $P^{\circ} \subset W$. Let $F_n$, $A_n$, and $V_n$ be as in Lemma 15.3.5 and let $P = \cup_{k \in \mathbf{N}} F_k$. Since $F_k \subset V_k$ and $\{V_n : n \in \mathbf{N}\}$ is a descending base of neighborhoods of 0 in $X$, it follows that $P$ is precompact by the following argument: If $V \subset X$ is a neighborhood of 0, then there exists $n$ such that $V_n \subset V$. Since $F_k \subset V_k \subset V_n \subset V$ for all $k \geq n$, $V$ contains all but a finite number of vectors $x_1, \ldots, x_m$ from $P = \cup_{k \in \mathbf{N}} F_k$. Thus $P \subset \cup_{i=1}^{m} (x_i + V)$ and $P$ is precompact. Since $A_n \subset P$ and $\overset{\circ}{A_n} \cap \overset{\circ}{V_n} = \left(\cup_{k=0}^{n-1} F_k\right)^{\circ} \cap \overset{\circ}{V_n} \subset W$ for all $n$ [Lemma 15.3.5], $P^{\circ} \cap \overset{\circ}{V_n} \subset \overset{\circ}{A_n} \cap \overset{\circ}{V_n} \subset W$ for all $n \in \mathbf{N}$. Thus $P^{\circ} \cap \left(\cup_{n \in \mathbf{N}} \overset{\circ}{V_n}\right) = P^{\circ} \cap X' = P^{\circ} \subset W$. ■

For any LCHS $(X, \mathcal{T})$, $p(X', X)$ is generated by polars of sets $\text{cl}_{\sigma(X, X')} P_{bc}$ where $P \subset X$ is precompact. By Theorem 8.8.1(b), $\text{cl}_{\sigma(X, X')} P_{bc} = \text{cl}_{\mathcal{T}} P_{bc}$. If $(X, \mathcal{T})$ is Fréchet then each $\text{cl}_{\mathcal{T}} P_{bc}$ is compact [Theorems 3.4.1 and 4.8.9(b)] which implies that $p(X', X) \subset \tau(X', X)$; since every compact set is precompact, it follows that $p(X', X) = \tau(X', X)$. By Theorem 15.3.6, $p(X', X) = e\sigma(X', X) = \tau(X', X)$ and $e\sigma(X', X)$ is a topology of the dual pair $(X', X)$. We summarize this below for ease of future reference.

**Corollary 15.3.7** If $X$ is a Fréchet space then $p(X', X) = e\sigma(X', X) = \tau(X', X)$.

**Theorem 15.3.8** $e\sigma(X', X)$-CLOSED Let $X$ be a metrizable LCS over $\mathbf{F}$. The set $F \subset X'$ is $e\sigma(X', X)$-closed $= p(X', X)$-closed iff $F \cap D$ is $\sigma(X', X)$-closed for every $\sigma(X', X)$-closed equicontinuous disk $D \subset X'$.

**Proof.** Let $X$ be a metrizable LCHS. If $F$ is $e\sigma(X', X)$-closed, then by the definition of $e\sigma(X', X)$, $F \cap E$ is $\sigma(X', X)$-closed for every $\sigma(X', X)$-closed equicontinuous subset $E$ of $X'$, hence for every $\sigma(X', X)$-closed equicontinuous disk $D$.

Conversely, suppose that $F \cap D$ is $\sigma(X', X)$-closed for every equicontinuous $\sigma(X', X)$-closed disk $D \subset X'$. Then if $M \subset X'$ is equicontinuous, $M^{\circ\circ} = \text{cl}_{\sigma(X', X)} M_{bc}$ is equicontinuous by Theorems 8.6.1 and 8.6.2. Thus $M^{\circ\circ}$ is an equicontinuous $\sigma(X', X)$-closed disk. By hypothesis, $F \cap M^{\circ\circ}$ is $\sigma(X', X)$-closed. Since $M \subset M^{\circ\circ}$, $F \cap M$ is closed in $M$ for the topology $\sigma(X', X) \cap M$. As $M$ is an arbitrary equicontinuous subset of $X'$, $F$ is $e\sigma(X', X)$-closed. ■

Now for an alternative description of what it means for a convex subset of a Fréchet space to be weak-$*$ closed.

**Theorem 15.3.9** KREIN–SMULIAN THEOREM Let $X$ be a Fréchet space over $\mathbf{F}$. A convex set $F \subset X'$ is $\sigma(X', X)$-closed iff $F$ is $e\sigma(X', X)$-closed.

**Proof.** Let $X$ be a Fréchet space. Since $\sigma(X', X) \subset e\sigma(X', X)$, if $F$ is $\sigma(X', X)$-closed, then $F$ is $e\sigma(X', X)$-closed. Now suppose that $F$ is $e\sigma(X', X)$-closed. Then $F$ is $\tau(X', X)$-closed [Corollary 15.3.7]. Since the closed convex subsets of $X'$ are the same for any topology of the dual pair $(X', X)$ [Theorem 8.8.1(a)], $F$ is $\sigma(X', X)$-closed. ■

As we show next, $\sigma(X', X)$-closedness of a subspace of a Banach space is equivalent to $\sigma(X', X)$-closedness of its unit ball.

**Theorem 15.3.10** $\sigma(X', X)$-CLOSED SUBSPACES OF $X'$ Let $X$ be a normed space over $\mathbf{F}$, let $U(X')$ denote the unit ball of $X'$ and let $M$ be a subspace of $X'$. Then $M$ is $\sigma(X', X)$-closed iff the unit ball $U(M)$ is $\sigma(X', X)$-closed.

**Proof.** Let $X$ be a normed space. Since equicontinuous sets are $\sigma(X', X)$-bounded [Theorem 8.6.5], they are norm-bounded by Mackey's theorem [8.8.7]. Hence every $\sigma(X', X)$-closed equicontinuous disk $D$ of $X'$ is contained in $nU(X')$ for some $n \in \mathbf{N}$. Thus $D \cap M = [D \cap nU(X')] \cap M = D \cap nU(M)$. Thus if $U(M)$ is $\sigma(X', X)$-closed, $D \cap M$ is $\sigma(X', X)$-closed. Thus, $M$ is $e\sigma(X', X)$-closed, therefore $\sigma(X', X)$-closed by Theorem 15.3.9. Conversely, suppose that $M$ is $\sigma(X', X)$-closed, that $f \in \text{cl}_{\sigma(X', X)} U(M)$ and that $(f_s)$ is a net from $U(M)$ that is $\sigma(X', X)$-convergent to $f$. Since $M$ is $\sigma(X', X)$-closed, $f \in M$ and for every $x \in U(X)$, the unit ball of $X$, $f_s(x) \to f(x)$.

For $x \in U(X)$ and every $s$, $|f_s(x)| \leq 1$ which implies that $|f(x)| \leq 1$ and therefore that $\|f\| \leq 1$; thus $f \in U(M)$. ■

**Theorem 15.3.11** METRIZABILITY OF WEAKLY COMPACT SETS Let $K \subset X$ be a $\sigma(X, X')$-compact subset of a normed space $(X, \|\cdot\|)$ over $\mathbf{F}$. If there exists a countable total subset $B = \{f_n \in X' : n \in \mathbf{N}, \|f_n\| = 1\}$ of $X'$, then $\sigma(X, X') \cap K$ is metrizable.

**Proof.** We use the notation above. Let $K \subset X$ be $\sigma(X, X')$-compact. The metrizability of $K$ is clear if $X$ is finite-dimensional so suppose that $X$ is infinite-dimensional and that $B = \{f_n \in X' : n \in \mathbf{N}, \|f_n\| = 1\}$ is a denumerable total subset of $X'$. For $x \in X$, let $p(x) = \sum_{n\in\mathbf{N}} 2^{-n} |f_n(x)|$. Because $\|f_n\| = 1$ for all $n \in \mathbf{N}$ and $B$ is total, it follows that $p$ is a norm and $p(\cdot) \leq \|\cdot\|$. Since $K$ is $\sigma(X, X')$-compact, $K$ is $\sigma(X, X')$-bounded [Theorem 6.1.2]. Since the bounded sets are the same for any topology of the pair $(X, X')$ [Theorem 8.8.7], $K$ is norm-bounded. Assuming, with no loss of generality, that $\|x\| \leq 1$ for all $x \in K$, it follows that $|f_n(x)| \leq 1$ for all $f_n \in B$ and $x \in K$. For $r > 0$, choose $k$ big enough so that $\sum_{n\geq k} 2^{-n} < r$. Hence $K \cap r\{f_1, \ldots, f_{k-1}\}^\circ \subset \{x \in K : p(x) < 2r\}$. Therefore, as we may assume that $0 \in K$, every $p$-neighborhood of 0 contains a $\sigma(X, X') \cap K$-neighborhood of 0. This implies that the identity map $I : (K, \sigma(X, X') \cap K) \to (K, p)$, $x \mapsto x$, is continuous. But since $K$ is $\sigma(X, X')$-compact, $I$ maps closed sets into closed sets, hence is open. Therefore $I$ is a homeomorphism and $\sigma(X, X') \cap K$ is the topology on $K$ induced by $p$. ■

## 15.4 EBERLEIN–SMULIAN THEOREM

In this section we study weakly compact subsets of an LCHS $X$ and $X'$. We prove in Theorem 15.4.8 that in a normed space $X$, $\sigma(X, X')$-compact $=$ $\sigma(X, X')$-countably compact $= \sigma(X, X')$-sequentially compact. In Theorem 15.5.1 we show that a Banach space is reflexive iff its closed unit ball is weakly compact.

The Bolzano-Weierstrass property of $\mathbf{R}$ is commonly formulated as (1) every bounded infinite subset of $\mathbf{R}$ has a limit point or (2) every bounded sequence of $\mathbf{R}$ has a convergent subsequence. These motivated the following definitions [Sec. 1.1.3].

**Definition 15.4.1** TYPES OF COMPACTNESS Let $T$ be a Hausdorff topological space. Then:

(a) $T$ is *sequentially compact* if every sequence in $T$ has a subsequence which converges to an element of $T$;

(b) $T$ is *countably compact* if every infinite subset of $T$ has a limit point $x$ (neighborhoods of $x$ contain infinitely many points of $T$) in $T$. ■

Sequential compactness implies countable compactness and they are equivalent in a first countable space. In any Hausdorff space countable compactness is equivalent to the statement that every denumerable open cover has a finite subcover [Engelking 1977, p. 258]. The following implications hold in any Hausdorff space $T$.

| | | compact | | | |
|---|---|---|---|---|---|
| | 2nd ctble $\nearrow$ | | $\searrow$ | | |
| ctble cover finite subcover | | $\leftrightarrows$ | ctbly compact | $\leftrightarrows$ 1st ctble | seq compact |

**Compactness in Hausdorff Spaces**

If $T$ is second countable, then every infinite cover of $T$ has a denumerable subcover; hence if the Hausdorff Space $T$ is countably compact, then $T$ is compact. Consequently, all notions of compactness considered here are equivalent in a second-countable Hausdorff space. If $T$ is a countably compact metric space, then $T$ is separable [Engelking 1977, p. 320, Th. 4.1.17], hence second-countable. In summary, *these notions of compactness are equivalent in a metric space.* Even though weak topologies on infinite-dimensional normed spaces $X$ are far from metrizable—$\sigma(X, X')$ is first countable iff $X$ is finite-dimensional [Example 11.8.3]—the various notions of $\sigma(X, X')$-compactness are equivalent which is the content of the Eberlein–Smulian theorem [15.4.8].

**Lemma 15.4.2** TOTAL SUBSETS Given a countable subset $B$ of a normed space $X$, there exists a countable subset $\{f_n : n \in \mathbf{N}\} \subset X'$ which is total on the closed linear span $M$ of $B$ and $M$ is separable.

**Proof.** The result is clear if $B$ is finite, so consider a denumerable subset $B = \{x_n \in X : n \in \mathbf{N}\}$ of a normed space $X$ and let $M$ be the closed linear span of $B$. A dense subset of $M$ is given by the denumerable set $D = \{\sum_{i=1}^{n} a_i x_i \in X : n \in \mathbf{N},\, a_i \in \mathbf{Q}+i\mathbf{Q}\} = \{y_k : k \in \mathbf{N}\}$ so $M$ is separable. By the norm-preserving Hahn–Banach theorem [7.4.1], there exists a family $\{f_n : n \in \mathbf{N}\} \subset X'$ such that $f_n(y_n) = \|y_n\|$ and $\|f_n\| = 1$ for each $n \in \mathbf{N}$. This family $\{f_n : n \in \mathbf{N}\}$ is a total family of linear functionals on $M$ for if $0 \neq x \in M$, there exists $y_n \in D$ such that $\|x - y_n\| < \|x\|/4$. Then $|f_n(x) - f_n(y_n)| = |f_n(x) - \|y_n\|| \leq \|f_n\| \|x - y_n\| = \|x - y_n\| < \|x\|/4$. Therefore $\|y_n\| > 3\|x\|/4$ and $|f_n(x)| > \|y_n\| - \|x\|/4 > 3\|x\|/4 - \|x\|/4 = \|x\|/2 \neq 0$. Thus $\{f_n : n \in \mathbf{N}\}$ is total on $M$. ■

**Theorem 15.4.3** WEAK COMPACTNESS $\Rightarrow$ WEAK SEQUENTIAL COMPACTNESS If $K$ is a $\sigma(X, X')$-compact subset of a normed. space $X$ then every sequence $(x_n)$ in $K$ has a subsequence which is $\sigma(X, X')$-convergent to an element of $K$.

**Proof.** We use the notation of the statement. Consider a sequence $(x_n)$ from the $\sigma(X, X')$-compact subset $K$ of $X$. If the set $\{x_n : n \in \mathbf{N}\}$ is finite, $(x_n)$ clearly has a $\sigma(X, X')$-convergent subsequence so suppose that $\{x_n : n \in \mathbf{N}\}$ is infinite and let $M$ be its norm-closed linear span. Let $D \subset X'$ be a denumerable total subset on the separable space $M$ [Lemma 15.4.2]. By the Hahn–Banach continuous extension theorem [Corollary 7.3.3], $\sigma(X, X') \cap M = \sigma(M, M')$. Since $M$ is a norm-closed [therefore, $\sigma(X, X')$-closed by Theorem 8.8.1(a)] subspace of $X$, $K \cap M$ is $\sigma(X, X') \cap M = \sigma(M, M')$-compact. Since $\{f_n : n \in \mathbf{N}\}$ is total on $M$, $(K \cap M) \cap \sigma(M, M') = \sigma(X, X') \cap M$ is metrizable [Theorem 15.3.11]. Since $K \cap M$ is $\sigma(M, M')$-compact, it is countably $\sigma(M, M')$-compact and therefore $\{x_n : n \in \mathbf{N}\}$ has a limit point $x \in K$. The metrizability of $\sigma(M, M') \cap K$ implies that there is a subsequence $(x_{n_k})$ of $(x_n)$ convergent to $x$ with respect to $\sigma(X, X') \cap M = \sigma(M, M')$. Thus $x_{n_k} \to x$ in the $\sigma(X, X')$-topology and $K$ is $\sigma(X, X')$-sequentially compact. ■

We show next that weakly countably compact sets are weakly bounded.

**Theorem 15.4.4** $\sigma(X, X')$-COUNTABLY COMPACT $\Rightarrow$ $\sigma(X, X')$-BOUNDED A countably $\sigma(X, X')$-compact subset $K$ of a normed space $X$ is $\sigma(X, X')$-bounded.

**Proof.** Let $X$ be a normed space and suppose that $K \subset X$ is $\sigma(X, X')$-countably compact but not $\sigma(X, X')$-bounded. If for any $f \in X'$ there exists $r > 0$ such that $K \subset r\{f\}^\circ$ then for any finite subset $\{f_1, \ldots, f_n\} \subset X'$ there exists $r > 0$ such that $K \subset r\{f_1, \ldots, f_n\}^\circ$. Hence there exists $f \in X'$ such that for each $n \in \mathbf{N}$ there exists $x_n \in K$ such that $|f(x_n)| > n$. Let $x$ be a weak limit point of $\{x_n\}$. Then, for any $f \in X'$, the weak neighborhood $V(x, f, 1) = \{v \in X : |f(x) - f(v)| < 1\}$ must contain infinitely many $x_n$ which implies the contradictory result that $n < |f(x_n)| \leq |f(x)| + 1$ for infinitely many $n$. Thus $K$ is $\sigma(X, X')$-bounded. ■

Now we can establish the first equivalence.

**Theorem 15.4.5** $\sigma(X, X')$-SEQUENTIALLY COMPACT = $\sigma(X, X')$-COUNTABLY COMPACT A subset $K$ of the normed space $X$ is $\sigma(X, X')$-sequentially compact iff it is $\sigma(X, X')$-countably compact.

**Proof.** Let $X$ be a normed space. As discussed at the beginning of this section, sequential compactness generally implies countable compactness. Conversely, suppose that $K \subset X$ is $\sigma(X, X')$-countably compact. If $K$ is finite, any sequence from $K$ must contain a constant subsequence so $K$ is weakly sequentially compact. Therefore suppose that $K$ is infinite and consider a denumerable subset $\{x_n\}$ of $K$. We show that $\{x_n\}$ has a weakly convergent subsequence. By Lemma 15.4.2, there exists a total set $H = \{f_n : n \in \mathbf{N}\} \subset X'$ on the norm-closure $M$ of the linear span of $\{x_n\}$. Since $K$ is countably $\sigma(X, X')$-compact, $\{x_n\}$ has a $\sigma(X, X')$-limit point $x$. Since $M$ is norm-closed, it is weakly closed [Theorem 8.8.1(a)] so $x \in M$. For $r > 0$, the $\sigma(X, X')$-neighborhood $V(x, f_1, r) = \{w \in X : |f_1(w) - f_1(x)| < r\}$ contains $x_n$ for infinitely many $n$. Consequently, $(x_n)$ has a subsequence $(y_n)$ such that $|f_1(y_n) - f_1(x)| < 1/n$ for every $n$ and $f_1(y_n) \to f_1(x)$. Similarly $(y_n)$ has a subsequence $(z_n)$ such that $f_2(z_n) \to f_2(x)$. By repeating the process and choosing the sequence $(v_n)$ of "diagonal" elements $y_1, z_2, w_3, \ldots$, a subsequence of $(x_n)$,

$$\lim_n f_k(v_n) = f_k(x) \text{ for each } k \in \mathbf{N} \tag{1}$$

It remains to show that (1) holds for all $f \in X'$. If not, then for some $f \in X'$ and $r > 0$, there exists a subsequence $(u_n)$ of $(v_n)$ such that

$$|f(u_n) - f(x)| \geq r \text{ for each } n \tag{2}$$

By the $\sigma(X, X')$-countable compactness of $K$, $(u_n)$ has a $\sigma(X, X')$-limit point $u \in M$ and, as in (1), there exists a subsequence $(u_n^*)$ of $(u_n)$ such that $\lim_n f_k(u_n^*) = f_k(u)$ and

$$\lim_n f_k(u_n^*) = f_k(u) \text{ for each } k \in \mathbf{N} \tag{3}$$

Eqs. (1) and (3) imply that $f_k(u - x) = 0$ for every $f_k \in H$. Since $H$ is total on $M$, this implies that $u = x$. This means that $\lim_n f(u_n^*) = f(x)$ which is incompatible with (2) and proves the theorem. ■

With $J$ denoting the canonical embedding of $X$ into $X''$, we identify $J(X)$ and $X$ and show in Theorem 15.4.6 that the $\sigma(X'', X')$-closure of a relatively weakly countably compact subset of a normed space $X$ lies in $X$. This is instrumental in proving Theorem 15.4.8.

**Theorem 15.4.6** $K$ $\sigma(X, X')$-COUNTABLY COMPACT $\Rightarrow \mathrm{cl}_{\sigma(X'', X')} K \subset X$
For any relatively countably $\sigma(X, X')$-compact subset $K$ of a normed space $X$, $\mathrm{cl}_{\sigma(X'', X')} K \subset X$.

**Proof.** Let $K$ be a relatively countably $\sigma(X, X')$-compact subset of a normed space $X$. We show that for $f \in \mathrm{cl}_{\sigma(X'', X')} K$, there exists $x \in X$ such that $f = Jx$. For a finite-dimensional subspace $M$ of $X'$, $B = \{g \in M : \|g\| = 1\}$ is compact [Theorem 4.10.4]. Hence $B$ is totally bounded and there exists a finite $(1/4)$-net $\{g_1, \ldots, g_n\} \subset B$ for $B$. Choose $h \in B$ and $i$ such that $\|h - g_i\| \leq 1/4$. Since each $\|g_i\| = 1$, there are unit vectors $\{x_1, \ldots, x_n\} \subset X$ such that $|g_i(x_i)| \geq 3/4$ for each $i$. Hence, for any $i$, $|h(x_i)| \geq |g_i(x_i)| - |(h - g_i)(x_i)| \geq 3/4 - 1/4 = 1/2$, it follows that $\max_i |h(x_i)| \geq 1/2$. Thus for any $h$ from a finite-dimensional subspace $M$, there exist unit vectors $x_1, \ldots, x_n \in X$ such that

$$\max_i |h(x_i)| \geq (1/2) \|h\| \tag{1}$$

For $g_1 \in SU(X') = \{g \in X' : \|g\| = 1\}$ and $f \in \mathrm{cl}_{\sigma(X'',X')} K$, the $\mathrm{cl}_{\sigma(X'',X')}$-neighborhood $V(f, g_1, 1)$ must contain a point $x_1 \in K$, i.e., $x_1 \in K$ such that $|Jx_1(g_1) - f(g_1)| = |g_1(x_1) - f(g_1)| \leq 1$. Since the linear span $[f, Jx_1 - f]$ of $\{f, Jx_1 - f\}$ is a finite-dimensional subspace of $X''$, by (1) there exists $n(2) \in \mathbf{N}$ and $\{g_2, \ldots, g_{n(2)}\} \subset SU(X')$ such that

$$\text{for any } u \in [f, Jx_1 - f], \ \max\{|u(g_i)| : i = 2, \ldots, n(2)\} \geq (1/2)\|u\|$$

Since the weak neighborhood $V(f, g_1, g_2, \ldots, g_{n(2)}, 1/2)$ must contain a point $x_2 \in K$, we have

$$\begin{aligned} &\max\{|Jx_2(g_i) - f(g_i)| : i = 1, \ldots, n(2)\} \\ &= \max\{|g_i(x_2) - f(g_i)| : i = 1, \ldots, n(2)\} \leq 1/2 \end{aligned} \tag{2}$$

Repeating the process, there exists $n(3) \in \mathbf{N}$ and $\{g_{n(2)+1}, \ldots, g_{n(3)}\} \subset SU(X')$ such that $\max\{|u(g_i)| : i = n(2)+1, \ldots, n(3)\} \geq (1/2)\|u\|$ for each $u \in [f, Jx_1 - f, Jx_2 - f]$. As above, there exists $x_3 \in K$ such that

$$\max\{|g_i(x_3) - f(g_i)| : i = 1, \ldots, n(3)\} \leq 1/3 \tag{3}$$

Choose countable subsets $\{x_n\} \subset K$ and $\{g_n\} \subset SU(X')$ in this way and suppose that $\{x_n\}$ is infinite. Since $K$ is relatively countably $\sigma(X, X')$-compact, there exists a $\sigma(X, X')$-limit point $x \in X$ of $\{x_n\}$. Since the norm closure $\mathrm{cl}[\{x_n\}]$ of $[\{x_n\}]$ is $\sigma(X, X')$-closed, $x \in \mathrm{cl}[\{x_n\}]$. Since $J$ is an isometry, $Jx \in \mathrm{cl}[Jx_1, \ldots, Jx_n, \cdots] \subset \mathrm{cl}[f, Jx_1 - f, Jx_2 - f, \ldots, Jx_n - f, \ldots] = Z$. By the way we chose things, for any $u \in [f, Jx_1 - f, \ldots, Jx_n - f, \ldots]$, $\sup_i |u(g_i)| \geq (1/2)\|u\|$. Consequently, this same inequality is satisfied if $u \in Z$. Hence $\sup_i |(Jx - f)(g_i)| = \sup_i |g_i(x) - f(g_i)| \geq (1/2)\|Jx - f\|$. If we can show that $g_i(x) = f(g_i)$ for all $i \in \mathbf{N}$, then it will follow from the preceding inequality that $\|Jx - f\| = 0$ and prove the theorem. If $n(m) < k$, then $m \leq n(m) < n(k)$ and therefore, as in (3), $|g_i(x_k) - f(g_i)| \leq 1/k$ for $i = 1, \ldots, n(k)$. Consequently,

$$|g_i(x_k) - f(g_i)| \leq 1/k < 1/m \text{ for } i = 1, \ldots, n(m) \text{ and } n(m) < k \tag{4}$$

Thus, for $i = 1, \ldots, n(m)$,

$$|g_i(x) - f(g_i)| \leq |g_i(x) - g_i(x_k)| + |g_i(x_k) - f(g_i)| \leq |g_i(x - x_k)| + 1/m \tag{5}$$

Since $x$ is a weak limit point of $\{x_n\} \subset X$, there exists $k \in \mathbf{N}$ such that $|g_i(x_k - x)| < 1/m$ for $i \leq n(m)$ and $k > n(m)$. Thus, by (5), $|g_i(x) - f(g_i)| \leq 2/m$ for $i \leq n(m)$. Hence, taking the limit as $m \to \infty$, we conclude that $g_i(x) = f(g_i)$ for all $i \in \mathbf{N}$ and therefore that $Jx = f$. If $\{x_n\}$ is finite, any $x \in \{x_n\}$ which repeats infinitely many times may be used in the previous argument and the proof is done. ■

**Theorem 15.4.7** COUNTABLY $\sigma(X, X')$-COMPACT $\Rightarrow$ $\sigma(X, X')$-COMPACT A relatively countably $\sigma(X, X')$-compact subset $K$ of a Banach space $X$ is relatively $\sigma(X, X')$-compact.

**Proof.** Let $X$ be a Banach space and suppose that $K$ is relatively countably $\sigma(X, X')$-compact. Then $K$ is $\sigma(X, X')$-bounded [Theorem 15.4.4]. As the bounded sets are same in any topology of the pair $(X, X')$ [Theorem 8.8.7], $K$ is norm-bounded and therefore, since $J$ is an isometry, $\mathrm{cl}_{\sigma(X'',X')} J(K)$ is norm-bounded; hence it is relatively $\sigma(X'', X')$-compact by Example 8.3.3 $[U(X')^\circ = U(X'')]$ and the Alaoglu theorem [8.4.1]. Since $\mathrm{cl}_{\sigma(X'',X')} J(K) \subset J(X)$ [Theorem 15.4.6] and $J(X) \cap \sigma(X'', X') = \sigma(X, X')$ it follows that $\mathrm{cl}_{\sigma(X'',X')} K = \mathrm{cl}_{\sigma(X,X')} J(K)$ is $\sigma(X, X')$-compact. ■

We call the result below the Eberlein–Smulian theorem—there are some variations on this theme in use. Smulian [1940, p. 435] showed that weak countable compactness implies weak sequential compactness. Eberlein [1947, p. 53] showed that weak countable compactness implies weak compactness.

**Theorem 15.4.8** COMPACT = SEQUENTIALLY COMPACT = COUNTABLY COMPACT For any $\sigma(X, X')$-closed subset $K$ of a normed space $X$, the following are equivalent:

(a) $K$ is $\sigma(X, X')$-compact;
(b) $K$ is $\sigma(X, X')$-countably compact;
(c) $K$ is $\sigma(X, X')$-sequentially compact.

**Proof.** Let $X$ be a normed space. The implications (a) $\Rightarrow$ (b) $\Rightarrow$ (c) are generally true in a first countable space. That (c) $\Rightarrow$ (b) follows from Theorem 15.4.5, (b) $\Rightarrow$ (a) from Theorem 15.4.7. ■

Theorem 15.4.9 on weak-$*$ compactness is useful in proving James's [1957, p. 167] theorem on reflexivity of separable Banach spaces, namely that a separable Banach space $X$ is reflexive iff for every $f \in X'$ there exists a unit vector $x \in X$ such that $f(x) = \|f\|$.

**Theorem 15.4.9** WHEN $\sigma(X', X)$-COMPACT $\Rightarrow$ $\sigma(X', X)$-SEQUENTIALLY COMPACT Let $X$ be a separable Banach space and $X'$ its dual. If $K \subset X'$ is $\sigma(X', X)$-compact, then $K$ is $\sigma(X', X)$-sequentially compact.

**Proof.** Let $X$ be a separable Banach space and $X'$ its dual. Let $K \subset X'$ be $\sigma(X', X)$-compact. If $X$ is finite-dimensional, then $X$ is linearly homeomorphic to $\ell_\infty(n)$; since $\ell_\infty(n)$ is first countable, the result follows. If $X$ is infinite-dimensional and separable, then there exists a denumerable dense subset $\{x_n\}$ of unit vectors. Since $\{Jx_n\}$ is clearly a total subset of $X'$ and $K$ is $\sigma(X', X)$-compact by hypothesis, $K \cap \sigma(X', X)$ is metrizable [Theorem 15.3.11]—hence first countable—and therefore the notions of $\sigma(X', X)$-compactness and sequential $\sigma(X', X)$-compactness coincide. ■

## 15.5 REFLEXIVITY OF BANACH SPACES

Banach [1932, p. 189, Th. 13] obtained the following criterion for reflexivity:

> If every bounded sequence of a *separable* Banach space $X$ has a weakly convergent subsequence then $X$ is reflexive.

In other words, if the unit ball $U$ of $X$ is sequentially $\sigma(X, X')$-compact, then $X$ is reflexive. By Theorem 15.4.8, we know that it is unnecessary to distinguish between the various kinds of $\sigma(X, X')$-compactness on $U$. The Alaoglu theorem [8.4.1] provides a result in the converse direction: It says that the unit ball $U(X')$ of the dual $X'$ of a normed. space $X$ is $\sigma(X', X)$-compact. Hence if a Banach space is reflexive—in which case $X$ is a dual—its unit ball must be $\sigma(X, X')$-compact. We prove next that reflexivity of a Banach space is equivalent to weak compactness of its unit ball. We put it to immediate use in Example 15.5.2 to show that $C(T, \mathbf{F}, \|\cdot\|_\infty)$ is reflexive iff $T$ is finite.

**Theorem 15.5.1** REFLEXIVE $\Longleftrightarrow$ $U$ WEAKLY COMPACT The Banach space $X$ is reflexive iff its closed unit ball $U$ is weakly compact.

**Proof.** Let $X$ be a Banach space. If $U$ is $\sigma(X, X')$-compact then for any $r > 0$, $rU = \{x \in X : \|x\| \leq r\}$ is $\sigma(X, X')$-compact. Since every bounded set is contained in $rU$ for some $r > 0$, $X$ is semireflexive [Theorem 15.2.4(b)]. As $X$ is barreled, $X$ is reflexive [Theorem 15.2.6(a,b)]. Conversely, if $X$ is reflexive, then $U \subset K$ for some $\sigma(X, X')$-compact set $K$ by Theorem 15.2.4(b). Since $U$ is a norm-closed convex set, $U$ is $\sigma(X, X')$-closed [Theorem 8.8.1], hence $\sigma(X, X')$-compact. ■

Since finite-dimensional TVS are locally compact, it follows immediately that finite-dimensional spaces are reflexive [Theorem 4.10.4]. As discussed in Example 9.4.8(a), $C([0,1], \mathbf{R}, \|\cdot\|_\infty)$ is not reflexive. The following application of Theorem 15.5.1 characterizes reflexivity of $C(T, \mathbf{F}, \|\cdot\|_\infty)$.

**Example 15.5.2** $C(T, \mathbf{F}, \|\cdot\|_\infty)$ REFLEXIVE $\Leftrightarrow$ T FINITE Consider the Banach space $X = C(T, \mathbf{F}, \|\cdot\|_\infty)$ of continuous maps of the compact Hausdorff space $T$ into $\mathbf{F}$. If $T$ is finite, with $n$ elements say, then $C(T, \mathbf{F}, \|\cdot\|_\infty)$ is linearly isometric to $\ell_\infty(n)$. As $\ell_\infty(n)$ is locally compact with respect to $\sigma(\ell_\infty(n), \ell_\infty(n)')$, it is reflexive by Theorem 15.5.1.

Now suppose that $T$ is infinite and that $\{t_n\} \subset T$ is denumerable. As $T$ is compact, $\{t_n\}$ has a limit point $t^*$. If $t^* \in \{t_n\}$, remove it from $\{t_n\}$ and reindex the set. By Urysohn's lemma, for each $n \in \mathbf{N}$ there exists $x_n \in C(T, [0,1])$ such that $x_n(\{t_1, \ldots, t_n\}) = \{1\}$ and $x_n(t^*) = 0$. Thus $\|x_n\|_\infty = 1$ for all $n \in \mathbf{N}$ and $\{x_n\}$ is a bounded subset of $C(T, \mathbf{F}, \|\cdot\|_\infty)$. If $X$ is reflexive, its unit ball is weakly compact by Theorem 15.5.1. Hence, by the Eberlein–Smulian theorem [15.4.8], there exists a weakly convergent subsequence $y_n \to x \in C(T, \mathbf{F}, \|\cdot\|_\infty)$ of the bounded sequence $(x_n)$. For each $t \in T$, the evaluation map $t'$ $[t'(z) = z(t)$ for all $z \in C(T, \mathbf{F}, \|\cdot\|_\infty)]$ is a continuous linear functional on $X = C(T, \mathbf{F})$ with $X$ carrying either the norm or $\sigma(X, X')$-topology. Therefore, since $y_n \to x$ in the $\sigma(X, X')$-topology,

$y_n(t) \to x(t)$ for all $t \in T$. But $y_n(t_i) = 1$ for $n \geq i$ implies that $x(t_i) = 1$ for all $i \in \mathbf{N}$. Also, $y_n(t^*) = 0$ for all $n \in \mathbf{N}$ implies that $x(t^*) = 0$. Since $t^*$ is a limit point of $\{t_n\}$, a subsequence converges to $t^*$ so this contradicts the continuity of $x$. Thus, when $T$ is infinite, $C(T, \mathbf{F})$ is not reflexive. [Cf. Exercise 15.102.] ■

The extreme points of the unit ball $U$ of $C(T, \mathbf{F}, \|\cdot\|_\infty)$ are in 1-1 correspondence with the clopen subsets of $T$ [Example 9.2.5(e)]. If $T$ is an infinite zero-dimensional compact Hausdorff space, there are infinitely many clopen sets, hence infinitely many extreme points of $U$. Nevertheless $C(T, \mathbf{F}, \|\cdot\|_\infty)$ is not reflexive by Example 15.5.2; cf. Example 15.0.1(a).

## 15.6 NORM-ATTAINING FUNCTIONALS

The preceding results are useful in proving James's profound theorem that a Banach space $X$ is reflexive iff for every $f \in X'$ there exists a unit vector $x \in X$ such that $f(x) = \|f\|$ [James 1957, p. 167, for the separable case and 1964, p. 215, for the general case; cf. also James 1972]. There is no "easy" proof of James's theorem. In addition to James's papers, proofs in special cases can be found in Megginson 1998 and Holmes 1975.

**Definition 15.6.1** NORM-ATTAINING FUNCTIONALS Let $X$ be a Banach space and $f \in X'$. If there exists a unit vector $x \in X$ such that $f(x) = \|f\|$ then $f$ is called *norm-attaining* and $x$ a *maximal element* for $f$. ■

The maximal element $x$ can also be defined by requiring that $|f(x)| = \|f\|$. In this case, if $f(x) = re^{it}$, then $v = e^{-it}x$ maximizes $f$. By Theorem 7.4.2(a), any unit vector $x$ is a maximal element for some $f \in X'$. For $f = 0$, any unit vector is a maximal element for $f$.

**Example 15.6.2** A NON-NORM-ATTAINING FUNCTIONAL For $(a_n) \in \ell_1$, consider the continuous linear functional [Example 15.7.3(a)] $f : c_0 \to \mathbf{F}$, $(a_n) \mapsto \sum_{n \in \mathbf{N}} 2^{-n} a_n$. Note that $\|f\|_1 = \sum_{n \in \mathbf{N}} 2^{-n} = 1$. For any $x = (a_n)$ in the unit ball of $c_0$ there exists $N \in \mathbf{N}$ such that $n \geq N$ implies $|a_n| \leq 1/2$. Therefore

$$\begin{aligned} |f(x)| &= \left| \sum_{n \in \mathbf{N}} a_n 2^{-n} \right| \leq \left| \sum_{n \leq N} a_n 2^{-n} \right| + \left| \sum_{n > N} a_n 2^{-n} \right| \\ &\leq \left| \sum_{n \leq N} 2^{-n} \right| + \left| \sum_{n > N} \left( \frac{1}{2} \right) 2^{-n} \right| = 1 - 2^{-N} + 2^{-(N+1)} < 1 \end{aligned}$$

Hence $f$ does not attain its norm. ■

Suppose that $x$ and $y$ are unit vectors in $X$ which are maximal elements for $f \in X'$. If, for some scalar $b$, $y = bx$, then $f(x) = f(y) = \|f\|$ immediately implies that $b = 1$. Thus distinct maximal elements must be linearly independent. But can there be distinct maximal elements?

**Example 15.6.3** DISTINCT MAXIMAL ELEMENTS Let $X$ be the Banach space $C(T, \mathbf{F}, \|\cdot\|_\infty)$ of continuous maps of the compact Hausdorff space $T$ into $\mathbf{F}$. Let $t \in T$ be disjoint from the closed set $K \subset T$. By Urysohn's lemma, there exist $x \in C(T, [0,1])$ such that $x(t) = 1$ and $x(K) = \{0\}$ and $y \in C(T, [1/2, 1])$ such that $y(t) = 1$, $y(K) = \{1/2\}$. Thus $x$ and $y$ are distinct maximal elements for the evaluation map $t' \in C(T, \mathbf{F})'$ where $t'(z) = z(t)$ for all $z \in C(T, \mathbf{F})$.

The question of uniqueness of maximal element is significant in approximation theory. For the remainder of this discussion, we investigate only existence. We begin by showing that every continuous linear functional on a reflexive Banach space has maximal elements.

**Theorem 15.6.4** REFLEXIVE IMPLIES MAXIMAL ELEMENTS If $X$ is a reflexive Banach space then any $f \in X'$ is norm-attaining.

**Proof.** If the Banach space $X$ is reflexive, the unit ball $U$ of $X$ is $\sigma(X, X')$-compact by Theorem 15.5.1. Let $f \in X'$, $f \neq 0$. Since $f$ is $\sigma(X, X')$-continuous, there exists a nonzero $u \in U$ such that $|f(u)| = \max\{|f(x)| : x \in U = \|f\|\}$. Since $U$ is convex and compact, the upper semicontinuous convex function $|f|$ assumes its maximum at an extreme point of $U$ [Example 9.4.2(d)]; hence $\|u\| = 1$. ■

It follows from Example 15.6.2 that $c_0$ is not reflexive. The converse of Theorem 15.6.4 is the theorem of James mentioned above. We now show that the question of whether $\mathbf{F}$ is $\mathbf{R}$ or $\mathbf{C}$ is irrelevant with regard to the existence of maximal elements and therefore, ultimately, to reflexivity.

**Theorem 15.6.5** MAXIMAL ELEMENTS FOR REAL AND COMPLEX FUNCTIONALS Let $X$ be a complex Banach space, let $f \in X'$, and let $g = \operatorname{Re} f$. Then:

(a) $g$ is a continuous real linear functional;
(b) $f(x) = \operatorname{Re} f(x) - i \operatorname{Re} f(ix) = g(x) - ig(ix)$;
(c) $\|f\| = \|g\|$;
(d) $x$ is a maximal element for $f$ iff $x$ is a maximal element for $g$.

**Proof.** We use the notation of the statement. Parts (a) and (b) are Theorems 5.7.7 and 1.4.2, respectively, restated. As to (c), for any unit vector $x$, $|g(x)| \leq |f(x)|$ so $\|g\| \leq \|f\|$. However, if $x_n \in X$ are unit vectors and $|f(x_n)| \to \|f\|$ then, for $r_n = |f(x_n)|$, there exist $t_n \in [0, 2\pi)$ such that

$f(x_n) = r_n e^{it_n}$ $(n \in \mathbf{N})$ and $f(e^{-it_n} x_n) = r_n = g(e^{it_n} x_n) \to \|f\|$ and $\|f\| = \|g\|$.

(d) If $x$ is a maximal element for $f$, then $f(x) = \|f\| = \operatorname{Re} f(x) = g(x) = \|g\|$ by (c) and $x$ is a maximal element for $g$. Conversely, if $x$ is a maximal element for $g$, then $g(x) = \|g\| = \|f\|$; since $\|f\| = g(x) = \operatorname{Re} f(x) \leq |f(x)| \leq \|f\|$, it follows that $|f(x)| = \|f\|$ and $x$ is a maximal element for $f$. ■

As a consequence of the previous result, proofs of James's theorem are usually restricted to real Banach spaces.

Phelps [1957] considered a weaker notion than reflexivity—not that every continuous linear functional on a Banach space $X$ attained its norm but only that $X'$ contain a dense subset of them. He called such spaces *subreflexive.* Bishop and Phelps [1961] then proved the *Bishop–Phelps subreflexivity theorem,* that all Banach spaces are subreflexive. See also the Notes to Exercise 9.201.

## 15.7 PARTICULAR DUALS

We identify the duals of a few particular spaces in this section. We observe that any Hilbert space $X$ is isometrically conjugate-isomorphic to its dual, that is, there is a bijective norm-preserving additive map $A : X \to X'$ that satisfies $A(ax) = \overline{a}Ax$ $(a \in \mathbf{F},\ x \in X)$. It immediately follows that Hilbert spaces are reflexive.

In Example 15.7.2 we show that if $p, q \in \mathbf{R}$, $p, q > 1$, and $1/p + 1/q = 1$, then $\ell_p' = \ell_q$. Hence $\ell_p'' = \ell_q' = \ell_p$ and the $\ell_p$ spaces are reflexive. In Example 15.7.3 we show that $c_0' = \ell_1$ and $\ell_1' = \ell_\infty$. Hence $c_0$ is not reflexive.

**Example 15.7.1** HILBERT SPACES SELF-DUAL AND REFLEXIVE The Riesz representation theorem for Hilbert spaces [Bachman and Narici 2000, p. 209] states that any continuous linear functional $f$ on a Hilbert space $(X, \langle \cdot, \cdot \rangle)$ is of the form $f = \langle \cdot, x \rangle$ for some unique $x \in X$. Moreover, as is straightforward to verify, the map $A : X \to X'$, $x \to \langle \cdot, x \rangle$, is a norm-preserving additive bijection with the property $A(ax) = \overline{a}Ax$ for any $a \in \mathbf{F}$ and $x \in X$. It follows that $X' = X$ as sets and that $X$ and $X''$ are linearly isometric. For more details, see Bachman and Narici 2000, p. 212, Theorem 12.6. ■

In the Banach sequence spaces $X$ of Examples 15.7.2 and 15.7.3, $\{e_n\}$ denotes the standard basis vectors, sequences $e_n$ which are 1 at the $n$th location, 0 elsewhere. The technique for characterizing the dual $X'$ is, given $f \in X'$, compute $(f(e_n))$ and show that, for any $x = (a_n) \in X$, $f(x) = \sum_{n \in \mathbf{N}} a_n f(e_n)$.

**Example 15.7.2** $\ell_p' = \ell_q$ FOR P $> 1$ AND $1/$P $+ 1/$Q $= 1$ Consider the Banach space $\ell_p$, $p > 1$, $1/p + 1/q = 1$. We show that for any $f \in \ell_p'$, $(f(e_n)) \in$

$\ell_q$, that for $x = (a_n) \in \ell_p$, $f(x) = \sum_{n\in\mathbf{N}} a_n f(e_n)$, that $\|f\| = \|(f(e_n))\|_q$ and that each $(b_n) \in \ell_q$ determines a continuous linear functional on $\ell_p$, namely $(a_n) \mapsto \sum_{n\in\mathbf{N}} a_n b_n$.

Recall the Hölder inequality: If $(a_n) \in \ell_p$ and $(b_n) \in \ell_q$, then

$$\sum_{n\in\mathbf{N}} |a_n b_n| \leq \left(\sum_{n\in\mathbf{N}} |a_n|^p\right)^{1/p} \left(\sum_{n\in\mathbf{N}} |b_n|^q\right)^{1/q} \tag{$*$}$$

and the signum function: for $a \in \mathbf{F}$ $\operatorname{sgn} a = 0$ and $\operatorname{sgn} a = a/|a|$ otherwise. For $a \neq 0$, $|\operatorname{sgn} a| = 1$; for all $a$, $a \operatorname{sgn} \bar{a} = |a|$. For $x = (a_n) \in \ell_p$, $x = \sum_{n\in\mathbf{N}} a_n e_n$ where the convergence is with respect to $\|\cdot\|_p$. Let $f \in \ell_p'$ and $f(e_n) = c_n$ $(n \in \mathbf{N})$. By continuity, $f(x) = \sum_{n\in\mathbf{N}} a_n c_n$. We now show that $(c_n) \in \ell_q$ and that the map $A : \ell_p' \to \ell_q$, $f \mapsto (c_n)$, is a linear isometry of $\ell_p'$ onto $\ell_q$. To begin, we show that $(c_n) \in \ell_q$. This is clear if $(c_n) = 0$ so assume that $(c_n) \neq 0$.

Choose $n$ large enough that $w = (d_1, \ldots, d_n, 0, \ldots, 0, \ldots) \neq 0$ where $d_k = |c_k|^{q-1} \operatorname{sgn} \bar{c}_k$ $(1 \leq k \leq n)$. Since $1/p + 1/q = 1$, $p(q-1) = q$ and it follows that $|d_k|^p = |c_k|^{p(q-1)} = |c_k|^q$ for $1 \leq k \leq n$. Thus $f(w) = f(\sum_{i=1}^n d_i e_i) = \sum_{i=1}^n |c_i|^{q-1} (\operatorname{sgn} \bar{c}_i) f(e_i) = \sum_{i=1}^n |c_i|^{q-1} (\operatorname{sgn} \bar{c}_i) c_i = \sum_{i=1}^n |c_i|^q = |f(w)|$. Since $|f(w)| \leq \|f\| \|w\|_p$, it follows that

$$|f(w)| = \sum_{i=1}^n |c_i|^q \leq \|f\| (\sum_{i=1}^n |c_i|^{(q-1)p})^{1/p} = \|f\| (\sum_{i=1}^n |c_i|^q)^{1/p}$$

Thus, dividing by $(\sum_{i=1}^n |c_i|^q)^{1/p}$, $(\sum_{i=1}^n |c_i|^q)^{1-1/p} = (\sum_{i=1}^n |c_i|^q)^{1/q} \leq \|f\|$. Therefore the bounded increasing sequence $(\sum_{i=1}^n |c_i|^q)^{1/q}$ converges; hence $(c_n) \in \ell_q$ and $\|(c_n)\|_q \leq \|f\|$. Therefore $A$ maps $\ell_p'$ into $\ell_q$. That $A$ is a linear isomorphism is clear.

To show that $A$ is an isometry, let $x = \sum_{n\in\mathbf{N}} a_n e_n \in \ell_p$. Then $f(x) = \sum_{n\in\mathbf{N}} a_n c_n$. By the Hölder inequality $(*)$,

$$|f(x)| \leq \sum_{n\in\mathbf{N}} |a_n c_n| \leq (\sum_{n\in\mathbf{N}} |a_n|^p)^{1/p} (\sum_{n\in\mathbf{N}} |c_n|^q)^{1/q}$$

so $\|f\| \leq \|(c_n)\|_q$. Therefore $\|f\| = \|(c_n)\|_q = \|Af\|_q$. It is straightforward to show that if $(c_n) \in \ell_q$ and $f$ is the linear functional defined at $(a_n) \in \ell_p$ by $f((a_n)) = \sum_{n\in\mathbf{N}} a_n c_n$, then $f \in \ell_p'$ and $Af = (c_n)$; hence $A$ is onto. ■

Note that if $p = 2$, $\ell_p$ is a Hilbert space and this case was discussed in Example 15.7.1. We show that the duals of $c_0$ and $\ell_1$ are $\ell_1$ and $\ell_\infty$, respectively, in Example 15.7.3.

**Example 15.7.3** THE DUALS OF $c_0$ AND $\ell_1$

(a) $c_0' = \ell_1$ Consider the Banach space $(c_0, \|\cdot\|_\infty)$ of null complex sequences. For $x = (a_n) \in c_0$, $x = \sum_{n\in\mathbf{N}} a_n e_n$ with respect to $\|\cdot\|_\infty$. For $f \in$

$c_0'$, $f(\sum_{n\in\mathbf{N}} a_n e_n) = \sum_{n\in\mathbf{N}} a_n f(e_n)$. The vectors $x_n = \sum_{i=1}^n \left(\operatorname{sgn} \overline{f(e_i)}\right) e_i \in c_0$ and since $|f(x_n)| = \sum_{i=1}^n |f(e_i)| \leqq \|f\| \|x_n\|_\infty = \|f\|$, it follows that $(f(e_n)) \in \ell_1$. Since

$$\left| f(\sum_{n\in\mathbf{N}} a_n e_n) \right| \leq \max_{n\in\mathbf{N}} |a_n| (\sum_{n\in\mathbf{N}} |f(e_n)|) = \left\| \sum_{n\in\mathbf{N}} a_n e_n \right\|_\infty (\sum_{n\in\mathbf{N}} |f(e_n)|)$$

it follows that $\|f\| \leq \sum_{n\in\mathbf{N}} |f(e_n)| = \left\|\sum_{n\in\mathbf{N}} f(e_n) e_n\right\|_1$. For $f \neq 0$, $\|x_n\|_\infty = 1$ eventually and $|f(x_n)| \to \sum_{n\in\mathbf{N}} |f(e_n)|$ as $n \to \infty$; hence

$$\|f\| \geq \sum_{n\in\mathbf{N}} |f(e_n)|$$

and therefore $\|f\| = \left\|\sum_{n\in\mathbf{N}} f(e_n) e_n\right\|_1$. As $A : c_0' \to \ell_1$, $f \mapsto (f(e_n))$ is clearly linear, $A$ is a linear isometry. To see that $A$ is surjective, suppose $(b_n) \in \ell_1$ and define $f$ at $(a_n) \in c_0$ by $f((a_n)) = \sum_{n\in\mathbf{N}} b_n a_n$. Since $\left|\sum_{n\in\mathbf{N}} b_n a_n\right| \leq \max_{n\in\mathbf{N}} |a_n| (\sum_{n\in\mathbf{N}} |b_n|) = \|(a_n)\|_\infty (\sum_{n\in\mathbf{N}} |b_n|)$, it follows that $f \in c_0'$ and $A$ is surjective.

(b) $\ell_1' = \ell_\infty$ As noted in Example 1.7.5, $(\ell_\infty, \|\cdot\|_\infty)$ is a Banach space. If $f \in \ell_1'$, then for $x = (a_n) \in \ell_1$, $x = \sum_{n\in\mathbf{N}} a_n e_n$ and $f(\sum_{n\in\mathbf{N}} a_n e_n) = \sum_{n\in\mathbf{N}} a_n f(e_n)$. Let $f(e_n) = b_n$ $(n \in \mathbf{N})$. For every $n$, $|f(e_n)| = |b_n| \leq \|f\| (\|e_n\|_\infty) = \|f\|$; hence $(f(e_n)) = (b_n) \in \ell_\infty$ and $\|(b_n)\|_\infty \leq \|f\|$. Moreover

$$\begin{aligned} \left| f\left(\sum_{n\in\mathbf{N}} a_n e_n\right)\right| &= \left|\sum_{n\in\mathbf{N}} a_n f(e_n)\right| = \left|\sum_{n\in\mathbf{N}} a_n b_n\right| \\ &\leq \sup_{n\in\mathbf{N}} |b_n| (\sum_{n\in\mathbf{N}} |a_n|) = \|(b_n)\|_\infty \|(a_n)\|_1 \end{aligned}$$

Thus $\|f\| \leq \|(b_n)\|_\infty$ and it follows that

$$A : \ell_1' \to \ell_\infty, \ f \mapsto (f(e_n))$$

is an into linear isometry. Consider $(b_n) \in \ell_\infty$ and $\sum_{n\in\mathbf{N}} a_n e_n \in \ell_1$. Define $f : \ell_1 \to \mathbf{F}$ by taking $\left|f(\sum_{n\in\mathbf{N}} a_n e_n)\right| = \left|\sum_{n\in\mathbf{N}} a_n b_n\right|$. We then have $\left|f(\sum_{n\in\mathbf{N}} a_n e_n)\right| \leq \|(b_n)\|_\infty \|(a_n)\|_1$; thus $f \in \ell_1'$ and $A$ is onto.

Since $c_0' = \ell_1$ and $\ell_1' = \ell_\infty$, $c_0$ is not reflexive. In Sec. 15.8 (see after Example 15.8.4) we show that $\ell_1$ is not reflexive either.

(c) $\beta(\ell_1, c_0)$ is not a topology of the pair $(\ell_1, c_0)$. Since $\ell_1$ is a Banach space, it is barreled and therefore its topology is $\beta(\ell_1, \ell_1')$ [Theorem 11.3.6] which is $\beta(\ell_1, \ell_\infty)$ by (b). Since $\ell_1$ is the dual of $c_0$, its topology is $\beta(\ell_1, c_0)$ by Example 8.8.9. As $(\ell_1, \beta(\ell_1, c_0))' = \ell_\infty \neq c_0$, $\beta(\ell_1, c_0)$ is not a topology of the pair $(\ell_1, c_0)$. ∎

As $c_0$ is an uncomplemented subspace of $\ell_\infty^{\mathbf{R}}$ [Example 4.9.7], we see that a Banach space can be uncomplemented in its bidual. Duals higher than $X''$ have generally been neglected but Dixmier [1948, p. 1066] showed that for any Banach space $X$, $X'$ is complemented in $X'''$.

## 15.8 SCHAUDER BASES

The Riesz-Fischer theorem in modern parlance states that for any orthonormal sequence $(x_n)$ in a Hilbert space $X$ and sequence $(a_n)$ of scalars, $\sum_{n\in\mathbf{N}} a_n x_n < \infty$ iff $(a_n) \in \ell_2$ [Bachman et al. 2000, p. 104; Bachman and Narici 2000, p. 148]. At the time of its discovery [Riesz 1907, Fischer 1907 in the same volume of C. R. Acad. Sci.], it was a statement about the linear isometry of $L_2[a,b]$ and $\ell_2$. It inspired further investigation, first, into orthonormal bases (essentially the method that Riesz used to prove his version) in Hilbert spaces and subsequently into "Schauder bases" [Def. 15.8.1] in Banach spaces. Schauder's [1927, p. 52] first proof of his fixed point theorem was for a Banach space with a Schauder basis and it may have motivated him to invent the notion. Unlike a Hamel basis, a Schauder basis is not just a set; it is an *ordered* set.

**Definition 15.8.1** SCHAUDER BASES If $X$ is a Banach space, $\{x_n : n \in \mathbf{N}\} \subset X$ is a *Schauder basis* (or *base*) for $X$ if for each $x \in X$ there exists a unique sequence $(a_n)$ of scalars such that $x = \sum_{n\in\mathbf{N}} a_n x_n$; the Schauder basis $\{x_n : n \in \mathbf{N}\}$ is:

(a) *unconditional* if, for any permutation $\pi$ of $\mathbf{N}$, $x = \sum_{n\in\mathbf{N}} a_{\pi(n)} x_{\pi(n)}$;

(b) *shrinking* if for each $f \in X'$, with $X_n = \text{cl}[\{x_k : k \geq n\}]$, it follows that $\|f|X_n\| = \sup\{|f(x)| : \|x\| = 1,\ x \in X_n\} \to 0$;

(c) *boundedly complete* if $\sup_n \|\sum_{i=1}^n a_i x_i\| < \infty$ implies that $\sum_{n\in\mathbf{N}} a_n x_n < \infty$. ■

As linear combinations with rational coefficients of elements from a Schauder basis are dense in $X$, a Banach space with a Schauder basis is separable.

The existence of a Schauder basis for a Banach space $X$ allows us to convert elements $x = \sum_{n\in\mathbf{N}} a_n x_n$ into sequences $(a_n)$. In practice, this is not usually much of an advantage. Some tangible benefits of the presence of a Schauder basis are the use made of them by Kadets [1967] in proving that all separable infinite-dimensional Banach spaces are homeomorphic to each other by reducing the general case to the case where the space has a Schauder basis.

**Definition 15.8.2** Let $X$ be a Banach space with Schauder basis $\{x_n : n \in \mathbf{N}\}$ and let $x = \sum_{n\in\mathbf{N}} a_n x_n$.

(a) NATURAL PROJECTION For $n \in \mathbf{N}$ the linear map

$$P_n : X \to X,\ \sum_{i\in\mathbf{N}} a_i x_i \mapsto \sum_{i=1}^n a_i x_i$$

is called the $n$th *natural projection* associated with the basis $\{x_n : n \in \mathbf{N}\}$.

(b) MONOTONE BASIS If for every $x = \sum_{n\in\mathbf{N}} a_n x_n$ the sequence

$$\left(\left\|\sum_{i=1}^n a_i x_i\right\|\right) = (\|P_n x\|)$$

is increasing then $\{x_n\}$ is called *monotone.* ■

If $X$ has a Schauder basis then, for any $x \in X$, $P_n x \to x$. The standard basis $\{e_n\}$ is a monotone basis for $c_0$ and $\ell_p$ $(1 \le p < \infty)$. Since $\ell_\infty$ is not separable, there is no Schauder basis for $\ell_\infty$. As was observed by Banach [1932, p. 111], any Banach space $X$ with a Schauder basis $\{x_n\}$ can be renormed with an equivalent norm $\|\cdot\|^*$ with respect to which $\{x_n\}$ is monotone as we demonstrate now. If $x = \sum_{n\in\mathbf{N}} a_n x_n \in X$, then it is straightforward to verify that: (1) $\|x\|^* = \sup_n \|\sum_{i=1}^n a_i x_i\|$ is a norm on $X$ with respect to which $\{x_n\}$ is monotone; (2) $\|\cdot\| \le \|\cdot\|^*$; and (3) $(X, \|\cdot\|^*)$ is a Banach space. Consequently, the identity map $I : (X, \|\cdot\|^*) \to (X, \|\cdot\|)$, $x \mapsto x$, is continuous. By the open mapping theorem [14.4.6], it follows that $\|\cdot\|$ and $\|\cdot\|^*$ are equivalent norms which implies that there exists $K > 0$ such that $\|\cdot\|^* \le K\|\cdot\|$. Therefore for each $m \in \mathbf{N}$, $\|\sum_{i=1}^m a_i x_i\|^* = \sup_{1\le n\le m} \|\sum_{i=1}^n a_i x_i\| \le K\|\sum_{i=1}^m a_i x_i\|$. Hence, for any $n < m$,

$$\left\|\sum_{i=1}^n a_i x_i\right\| \le K \left\|\sum_{i=1}^m a_i x_i\right\| \tag{$*$}$$

For $x = \sum_{n\in\mathbf{N}} a_n x_n \in X$, since $\|P_n x\| = \|\sum_{i=1}^n a_i x_i\| \le \|x\|^* \le K\|x\|$, it follows that each $P_n : X \to X$ is continuous and that the *basis constant* $\sup_n \|P_n\| \le K$.

**Theorem 15.8.3** SCHAUDER BASIS CONDITION Let $X$ be a Banach space. The nonzero vectors $\{x_n : n \in \mathbf{N}\}$ are a Schauder basis iff $\mathrm{cl}\,[\{x_n\}] = X$ and there exists $K > 0$ such that

$$\left\|\sum_{i=1}^n a_i x_i\right\| \le K \left\|\sum_{i=1}^m a_i x_i\right\| \quad \text{for all } n, m \in \mathbf{N},\ m > n, \text{ and } (a_i) \in \mathbf{F}^m \tag{$*$}$$

**Remark.** By $(*)$, $\sum_{n\in\mathbf{N}} a_n x_n = 0$ implies that $a_n = 0$ for every $n$; hence the coefficients $a_n$ are unique and $\{x_n : n \in \mathbf{N}\}$ is linearly independent.

**Proof.** We have already established the necessity of $(*)$ before the theorem. Conversely, let $\{x_n : n \in \mathbf{N}\}$ be a set of nonzero vectors such that $\mathrm{cl}\,[\{x_n : n \in \mathbf{N}\}] = X$ and satisfy $(*)$ for some $K > 0$. We may assume that $\|x_n\| = 1$ for every $n$. For $1 \le i \le n$, let $f_i$ be the linear functional (the *coefficient functional*) on $[\{x_n : 1 \le i \le n\}]$ defined by $f_i(x_j) = \delta_{ij}$, the Kronecker delta. For $1 \le i \le n$, by $(*)$,

$$\left\|f_i\left(\sum_{k=1}^n a_k x_k\right)\right\| = |a_i| = \left\|\sum_{k=1}^i a_k x_k - \sum_{k=1}^{i-1} a_k x_k\right\| \le 2K \left\|\sum_{k=1}^n a_k x_k\right\|$$

Thus $\|f_i\| \le 2K$ on $[\{x_n\}]$ so $f_i$ is continuous on $[\{x_n\}]$ and we may assume that $f_i \in X'$ by the Hahn–Banach theorem [Corollary 7.3.3]. Given $x \in X$, for each $m \in \mathbf{N}$, let $y_m = x - \sum_{i=1}^{m} f_i(x)x_i$. We show next that $y_m \in \mathrm{cl}[\{x_n : n > m\}]$ for every $m$ and that $\|y_m\| \to 0$.

Given $m \in \mathbf{N}$, $y_m \in X = \mathrm{cl}[\{x_n : n \in \mathbf{N}\}]$, so there exist $z_k \in [\{x_n : n \in \mathbf{N}\}]$ ( $k \in \mathbf{N}$) such that $z_k \to y_m$. Each $z_k$ is of the form $\sum_{i \in J_k} a_i^k x_i$ for a finite subset $J_k \subset \mathbf{N}$ and scalars $\{a_i^k : a_i^k = 0 \text{ for } i \notin J_k\}$. Thus $y_m = \lim_k \sum_{i \in J_k} a_i^k x_i$. Since, for $j = 1, \ldots, m$, $f_j(y_m) = f_j(x - \sum_{i=1}^{m} f_i(x)x_i) = f_j(x) - f_j(x) = 0$, it follows that $0 = f_j(y_m) = \lim_k f_j\left(\sum_{i \in J_k} a_i^k x_i\right) = \lim_k a_j^k$. Consequently $y_m \in \mathrm{cl}[x_n : n > m]$.

By the density of $[\{x_n : n \in \mathbf{N}\}]$, for $r > 0$ there exists $n(r) \in \mathbf{N}$ and $a_1, \ldots, a_{n(r)} \in \mathbf{F}$ such that

$$\left\| x - \sum_{i=1}^{n(r)} a_i x_i \right\| < r \tag{1}$$

For $m > n(r)$, let $a_m = 0$ and choose scalars $b_{m+1}, \ldots, b_{k(m)}$ [since $y_m \in \mathrm{cl}[x_n : n > m]$] such that

$$\left\| y_m - \sum_{m+1}^{k(m)} b_i x_i \right\| = \left\| x - \sum_{i=1}^{m} f_i(x)x_i - \sum_{m+1}^{k(m)} b_i x_i \right\| < r \tag{2}$$

By (1) and (2), adding and subtracting $x$, for $m > n(r)$, it follows that

$$\left\| \sum_{i=1}^{m} (f_i(x)x_i + \sum_{m+1}^{k(m)} b_i x_i - \sum_{i=1}^{n(r)} a_i x_i \right\| = \left\| \sum_{i=1}^{m} (f_i(x) - a_i)x_i + \sum_{m+1}^{k(m)} b_i x_i \right\| < 2r \tag{3}$$

By $(*)$ and $(3)$, since $m > n(r)$,

$$\left\| \sum_{i=1}^{m} (f_i(x) - a_i)x_i \right\| \le K \left\| \sum_{i=1}^{m} (f_i(x) - a_i)x_i + \sum_{i=m+1}^{k(m)} b_i x_i \right\| \le 2Kr \tag{4}$$

Hence

$$\begin{aligned} \left\| \sum_{i=1}^{m} f_i(x)x_i - \sum_{i=1}^{n(r)} a_i x_i + y_m \right\| &= \left\| \sum_{i=1}^{m} f_i(x)x_i - x + x - \sum_{i=1}^{n(r)} a_i x_i + y_m \right\| \\ &= \left\| -y_m + x - \sum_{i=1}^{n(r)} a_i x_i + y_m \right\| \\ &= \left\| x - \sum_{i=1}^{n(r)} a_i x_i \right\| < r \quad (\text{by } (1)) \end{aligned}$$

i.e.,

$$\left\| \sum_{i=1}^{m} (f_i(x) - a_i)x_i + y_m \right\| < r \tag{5}$$

By (4) and (5), for $m > n(r)$,

$$\|y_m\| = \left\| y_m + \sum_{i=1}^{m} (f_i(x) - a_i)x_i - \sum_{i=1}^{m} (f_i(x) - a_i)x_i \right\| \le r + 2Kr$$

Thus $y_m \to 0$, $x = \sum_{i \in \mathbf{N}} f_i(x)x_i$, and $\{x_n : n \in \mathbf{N}\}$ is a Schauder basis. ■

We consider some examples next.

**Example 15.8.4** SCHAUDER BASES (a) The Banach space $(\ell_\infty, \|\cdot\|_\infty)$ of bounded sequences of scalars has no Schauder basis because a space with a Schauder basis must be separable and $\ell_\infty$ is not [Bachman and Narici 2000, p. 83].

In parts (b)–(e) $\{e_n\}$ represents the standard basis.

(b) Consider the Banach space $(\ell_1, \|\cdot\|_1)$ of absolutely summable sequences. Clearly $[\{e_n\}]$ is dense in $\ell_1$. For $m > n$ and $(a_i) \in \mathbf{F}^m$,

$$\left\| \sum_{i=1}^{n} a_i e_i \right\|_1 = \sum_{i=1}^{n} |a_i| \le \left\| \sum_{i=1}^{m} a_i e_i \right\|_1$$

so we may take the constant $K$ of Theorem 15.8.3 to be 1. Not only is $\{e_n\}$ a Schauder basis for $\ell_1$, it is an unconditional basis since absolutely convergent series can be summed in any order. It is boundedly complete because $\sup\{\|\sum_{i=1}^{n} a_i e_i\|_1 : n \in \mathbf{N}\} = \sup\{\sum_{i=1}^{n} |a_i| : n \in \mathbf{N}\} < \infty$ implies that $\sum_{n \in \mathbf{N}} a_n e_n \in \ell_1$. Recall that $\ell_1' = \ell_\infty$ [Example 15.7.3(b)] and consider $f = (1) \in \ell_\infty$. For all $n \in \mathbf{N}$, $f(e_n) = 1$ and $\|e_n\|_1 = 1$. Hence $\|f|_{X_n}\| = 1$ for all $n \in \mathbf{N}$ and $\{e_n\}$ is not a shrinking basis for $\ell_1$.

The space $\ell_1$ is critical in determining whether or not a Schauder basis for a Banach space $X$ is shrinking. James [1950] showed that if $\{x_n : n \in \mathbf{N}\}$ is a Schauder basis for $X$, it is a shrinking basis iff $X$ does not contain a subspace which is linearly homeomorphic to $\ell_1$.

(c) For the space $(c_0, \|\cdot\|_\infty)$ of null sequences, $\{e_n\}$ clearly satisfies the conditions of Theorem 15.8.3 and so is a Schauder basis. It is an unconditional basis but not boundedly complete since $(x_n) = (\sum_{i=1}^{n} e_i)$ is a sequence of unit vectors in $c_0$ while $\sum_{n \in \mathbf{N}} e_n$ diverges. Since $c_0' = \ell_1$ [Example 15.7.3(a)], it is clear that $\{e_n\}$ is a shrinking basis for $c_0$: For $f = (a_n) \in \ell_1$, $\|f|X_n\| = \sup \sum_{k \ge n} |a_k| \to 0$.

(d) In Example 15.7.2 we showed that for $p > 1$, $\ell_p' = \ell_q$—for $f \in \ell_p'$ and any $(a_n) \in \ell_p$, $f((a_n)) = \sum_{n \in \mathbf{N}} a_n f(e_n)$. From this and the Hölder inequality (stated in Example 15.7.2), it follows using elementary arguments pertaining to infinite sums that $\{e_n\}$ is an unconditional, shrinking, boundedly complete basis for $\ell_p$.

(e) With $e = (1)$, $\{e\} \cup \{e_n\}$ is a Schauder basis for the Banach space $(c, \|\cdot\|_\infty)$ of convergent sequences—if $x = (a_n) \in c$ and $a_n \to a$ then $x = ae + \sum_{n \in \mathbf{N}} (a_n - a) e_n$. ■

James [1950] showed that if $X$ has a Schauder basis, $X$ is reflexive iff the basis is shrinking and boundedly complete. Hence, as already observed in Example 15.8.4(d), the standard basis for the reflexive spaces $\ell_p$ $(p > 1)$ must be shrinking and boundedly complete. As bases for $c_0$ and $\ell_1$ do not have both these properties, neither is reflexive [cf. Example 15.7.3.]

We construct a Schauder basis for $C[0,1]$ in Example 15.8.5.

**Example 15.8.5** Basis for $C[0,1]$ We show that elementary sawtooth functions form a Schauder basis for $C([0,1], \mathbf{R}, \|\cdot\|_\infty) = C[0,1]$. Let $\{t_n : n \in \mathbf{N}\} = \{n/2^m : n, m \geq 0,\ n \leq 2^m\} \subset [0,1]$ be the binary numbers. We index them as follows: $\{t_0, t_1, t_2, t_3, \ldots\} = \{0, 1, 1/2, 1/4, 3/4, 1/8, \ldots\}$. Let $\{x_n : n \geq 0\}$ be the following functions in $C[0,1]$: $x_0 = 1, x_1 = t$, and for $n \geq 2$, $x_n$ is as below:

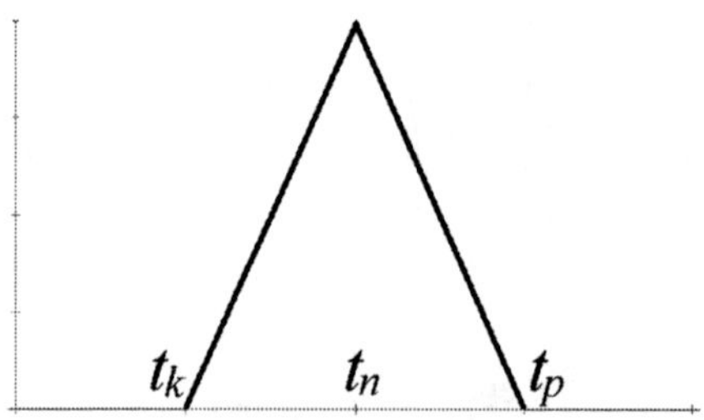


Figure 15.1: $x_n$

where $t_k$ and $t_p$, $k, p < n$, are the closest binary numbers on either side of $t_n$, that is, the closest binary numbers to $t_n$ that precede $t_n$ in the indexed list. Note that we can have $t_p < t_k$ or $t_k < t_p$. For example, $x_4$ is:

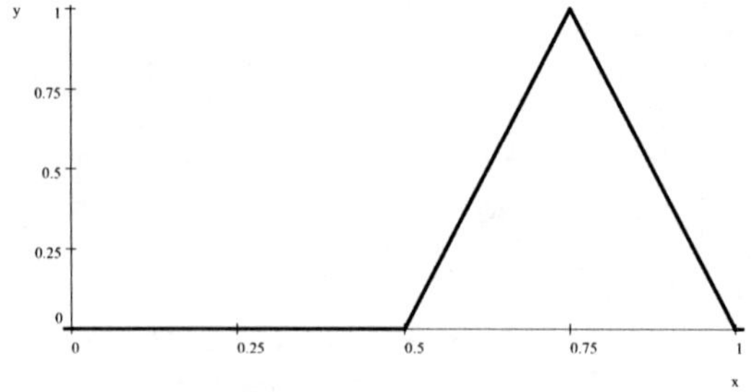


Figure 15.2: $x_4$

For $x \in C[0,1]$, we show that there exists a unique sequence $(a_n)_{n\geq 0}$ such that $x = \sum_{n\geq 0} a_n x_n$. We choose the numbers $a_i$ $(0 \leq i \leq n)$ so that $y_n = \sum_{i=0}^{n} a_i x_i$ meets $x$ at the points $t_0, \ldots, t_n$, thus producing a "polygonal approximation" to $x$. The facts that $x$ is uniformly continuous on the closed bounded interval $[0,1]$ and that the binary numbers are dense in $[0,1]$ guarantee the uniform convergence of $y_n$ to $x$. Since $x_n(0) = 0$ for $n \geq 1$, it follows that $a_0 = x(t_0) = x(0)$ and $y_0 = x(0)$. In order that $y_1 = a_0 x_0 + a_1 x_1$ meet the graph of $x$ at both $t_0$ and $t_1$, $a_1$ must satisfy

$$x(1) = a_0 x_0(1) + a_1 x_1(1) = x(0) \cdot 1 + a_1 \cdot 1$$

Thus $a_1 = x(1) - x(0)$ and $y_1 = x(0) + (x(1) - x(0))\, t$. Likewise, $a_2 = x(t_2) - a_0 - a_1 x_1(t_2)$. Suppose $a_0, \ldots, a_{n-1}$ satisfy $y_{n-1}(t_j) = x(t_j)$, $0 \leq j \leq n-1$ and $a_n$ satisfies

$$a_n x_n(t_n) = a_n = x(t_n) - \sum_{i=0}^{n-1} a_i x_i(t_n) \qquad (*)$$

Since $x_n(t_j) = 0$ for $j < n$, $(*)$ also suffices to make $y_n(t_j) = x(t_j)$ for all $0 \leq j \leq n-1$. The uniqueness of the $a_n$ follows from the fact that in order for $\sum_{i=0}^{n} b_i x_i$ to converge to $x$ on $[0,1]$, the partial sums $w_m = \sum_{i=0}^{m} b_i x_i$ must be equal to $x$ at the points $t_0, \ldots, t_m$. If not, if $w_m(t_i) - x(t_i) \neq 0$ for some $m$ and some $i \leq m$, then, since $x_n(t_i) = 0$ for all $n > m \geq i$, then $w_n(t_i) - x(t_i) = w_m(t_i) - x(t_i) \neq 0$ for $n > m$; hence $w_n(t_i) \nrightarrow x(t_i)$. ■

Simply because the linear span of the linearly independent set $\{x_n : n \in \mathbf{N}\}$ is dense in $X$, it does not follow that $\{x_n : n \in \mathbf{N}\}$ is a Schauder basis. To be a Schauder basis we require that every $x \in X$ can be written as a convergent series $\sum_{n\in\mathbf{N}} a_n x_n$ where the $a_n$ are unique. The linear span of $\{t^n : n \geq 0\}$—i.e., polynomials—is dense in $C[0,1]$ by the Stone-Weierstrass theorem [16.5.7]. But not every $x \in C[0,1]$ can be written in the form $x = \sum_{n\in\mathbf{N}} a_n t^n$—i.e., as a power series with positive radius of convergence—for this would imply that every $x \in C[0,1]$ is differentiable at 0.

We show in Theorem 15.8.7 that every infinite-dimensional Banach space has an infinite-dimensional subspace with a Schauder basis. First [Lemma 15.8.6], we need to establish that if $M$ is a finite-dimensional subspace of an infinite-dimensional normed space $X$, there is a unit vector $x \in X$ such that $\|bx + m\| \geq (0.9)\,\|m\|$ for all $m \in M$ and $b \in \mathbf{F}$. In other words, adding scalar multiples of $x$ to $m$ cannot significantly diminish the size of $bx + m$ compared to $m$.

**Lemma 15.8.6** FINITE-DIMENSIONAL SUBSPACES Let $M$ be a finite-dimensional subspace of an infinite-dimensional normed space $X$. Then for any $a > 0$ there exists a unit vector $x \in X$ such that $\|m\| \leq (1+a)\,\|m + bx\|$ for all $b \in \mathbf{F}$ and $m \in M$.

**Proof.** Suppose that $a > 0$ and $M$ is a finite-dimensional subspace of an infinite-dimensional normed space $X$. Since $M$ is locally compact [Theorem 4.10.4], $SU(M) = \{m \in M : \|m\| = 1\}$ is compact and therefore totally bounded [Theorem 3.4.2]. Thus, there exist $m_1, \ldots, m_k \in SU(M)$ such that for each $m \in SU(M)$ there exists $m_i$ with $\|m - m_i\| \leq a/(1+a)$. By Theorem 7.4.2(a), there exist $f_1, \ldots, f_k \in X'$ such that $f_i(m_i) = \|m_i\| = 1$ while $\|f_i\| = 1$ for all $i$. Since $X$ is infinite-dimensional, it follows that $X'$ is infinite-dimensional [Theorem 8.4.6]. If $\cap_{i=1}^k N(f_i) = \{0\}$ then, since $\{0\} \subset N(f)$ for any $f \in X'$, this would imply that $f$ is a linear combination of $f_1, \ldots, f_k$ [Theorem 1.4.1(f)]. Thus $\cap_{i=1}^k N(f_i) \neq \{0\}$ and there exists a unit vector $x \in X$ such that $f_i(x) = 0$ for all $i$. For $m \in SU(M)$, choose $i$ such that $\|m - m_i\| \leq a/(1+a)$. Then, for any $b \in \mathbf{F}$, $\|m + bx\| \geq \|m_i + bx\| - \|m - m_i\| \geq \|m_i + bx\| - a/(1+a) \geq |f_i(m_i + bx)| - a/(1+a) = 1 - a/(1+a) = 1/(1+a)$. Hence $(1+a)\|m + bx\| \geq 1$ for all $m \in SU(M)$. Therefore, for all nonzero $m \in M$, $(1+a)\|(1/\|m\|)m + bx\| \geq 1$ or $(1+a)\|m + b\|m\|x\| \geq \|m\|$. Since $b$ is arbitrary, the result follows. ■

**Theorem 15.8.7** Every infinite-dimensional Banach space has an infinite-dimensional subspace with a Schauder basis.

**Proof.** Let $X$ be an infinite-dimensional Banach space. We show that there exists a linearly independent set $\{x_n : n \in \mathbf{N}\}$ that satisfies $(*)$ of Theorem 15.8.3, i.e., that there exists $K > 0$ such that

$$\left\|\sum_{i=1}^{n} a_i x_i\right\| \leq K \left\|\sum_{i=1}^{m} a_i x_i\right\| \text{ for } n, m \in \mathbf{N},\ m > n, \text{ and } (a_i) \in \mathbf{F}^m \qquad (*)$$

Suppose $0 < a < 1$ and let $K = 1 + a$. We construct inductively a sequence $(c_n)$ of positive numbers such that $\Pi_{n\in\mathbf{N}}(1 + c_n) \leq 1 + a$. Let $c_1 = a/2$ and suppose that $c_1, \ldots, c_n$ are positive numbers such that $b_n = \Pi_{i=1}^n (1 + c_i) < 1 + a$. Since $b_n < 1 + a$, we can choose $c_{n+1} > 0$ so that $b_{n+1} = b_n(1 + c_{n+1}) < 1 + a$. The sequence $(b_n)$ is a bounded increasing sequence and $\Pi_{n\in\mathbf{N}}(1 + c_n) = \lim_n b_{n+1}\Pi_{i=1}^n(1 + c_i) \leq 1 + a$. For a nonzero vector $x_1$, let $M = \mathbf{F}x_1$. For any sequence $(a_n)$ of scalars, by Lemma 15.8.6, there exists a sequence $(x_n)$ of linearly independent vectors satisfying $\|\sum_{i=1}^n a_i x_i\| \leq (1 + c_{n+1})\|\sum_{i=1}^n a_i x_i + a_{n+1}x_{n+1}\|$ for every $n$. Thus, for $m > n$, $\|\sum_{i=1}^n a_i x_i\| \leq (1 + c_{n+1})(1 + c_{n+2}) \cdots (1 + c_m)\|\sum_{i=1}^m a_i x_i\| \leq (1 + a)\|\sum_{i=1}^m a_i x_i\|$. Therefore $\{x_n : n \in \mathbf{N}\}$ is a Schauder basis for $\mathrm{cl}[\{x_n : n \in \mathbf{N}\}]$. ■

Having a Schauder basis is not a hereditary property: Enflo [1973] showed that each of the reflexive Banach spaces $\ell_p$, $p > 2$—which have a Schauder basis [Example 15.8.4(d)]—has a subspace that does not have a Schauder basis.

## 15.9 APPROXIMATION PROPERTIES

As discussed in Sec. 11.1, it remained open for many years whether every separable Banach space has a Schauder basis. Existence of a Schauder basis relates to the *approximation property:*

**Definition 15.9.1** APPROXIMATION PROPERTY Let $X$ be a Banach space and let $I : X \to X$ denote the identity map $x \mapsto x$. We say that $X$ has the *approximation property* if given any compact set $K \subset X$ and $r > 0$ there exists $A \in L(X, X)$ of finite rank (= finite-dimensional range) such that $\|x - Ax\| = \|(I - A)x\| < r$ for all $x \in K$. ■

Thus, in a Banach space with the approximation property, the identity map may be approximated by operators of finite rank on compact sets. Grothendieck [1955, p. 165], incidentally, lists nine equivalent formulations of the approximation property.

If $X$ and $Y$ are Banach spaces, $A \in L(X, Y)$ is *compact* if $\operatorname{cl} A(U)$ is compact in $Y$ where $U$ is the closed unit ball of $X$; in other words, $A$ maps bounded sets into relatively compact sets. The simplest compact operators are those of finite rank (such as the natural projections of Def. 15.8.2, for example) and even norm limits $A$ of operators $\{A_n\}$ of finite rank are compact, i.e., $\|A - A_n\| \to 0$ with respect to the uniform norm for $L(X, Y)$. The converse question is called the approximation *problem,* namely, if $A \in L(X, Y)$ is compact, is there a sequence $A_n \in L(X, Y)$ of operators of finite rank such that $A_n \to A$?

**Theorem 15.9.2** SCHAUDER BASIS $\Rightarrow$ APPROXIMATION PROPERTY A Banach space $X$ with a Schauder basis $\{x_n : n \in \mathbf{N}\}$ has the approximation property. In particular, $P_n \to I$, the identity map, where $P_n$ ( $n \in \mathbf{N}$ ) denotes the natural projection associated with $\{x_n : n \in \mathbf{N}\}$.

**Proof.** Let $\{x_n : n \in \mathbf{N}\}$ be a Schauder basis for a Banach space $X$ and let $\{P_n\}$ be the associated natural projections. By Theorem 15.8.3, there exists $K > 0$ such that $\|\sum_{i=1}^{n} a_i x_i\| \leq K \|\sum_{i=1}^{m} a_i x_i\|$ for $n, m \in \mathbf{N}$, $m > n$, and $(a_i) \in \mathbf{F}^m$; hence $\|P_n\| \leq K$ for any $n$. A compact subset $B$ of $X$ is totally bounded [Theorem 3.4.2]. Hence, for any $r > 0$ and $x \in B$, there exist $y_1, \ldots, y_m \in B$ ($m \in \mathbf{N}$) such that $\|x - y_i\| < r$ for some $i \in \{1, 2, \ldots, m\}$. Since $X$ has a Schauder basis, $\lim_n P_n y_i = y_i$ for every $i$. Hence, for sufficiently large $n$, $\|P_n y_i - y_i\| < r$; therefore, for any $x \in B$, there exists $i$ such that $\|x - P_n x\| \leq \|x - y_i\| + \|y_i - P_n y_i\| + \|P_n y_i - P_n x\| \leq r + r + \|P_n\| \, \|y_i - x\| \leq r + r + Kr = (K + 2)\, r$. Thus $I$ is approximated by $\{P_n\}$ on $B$. ■

As Enflo [1973] showed, there are subspaces of $c_0$ and $\ell_p$, $2 < p < \infty$, that do not have the approximation property. [For a simplified version of the construction due to Davie, see Lindenstrauss and Tzafriri 1977, Chap. 2.]

By Theorem 15.9.2, these subspaces cannot have a Schauder basis. As these are separable spaces [$\{e_n : n \in \mathbf{N}\}$ is a Schauder basis for each by Example 15.8.4(c,d)], this negatively settled *the basis problem*: Does every separable Banach space have a Schauder basis?

We show in Theorem 15.9.3 that if $Y$ has a Schauder basis then, for any compact operator $A \in L(X,Y)$, $P_n A \to A$ where $P_n$ denotes the natural projection associated with the basis [Definition 15.8.2].

**Theorem 15.9.3** FINITE RANK OPERATORS APPROXIMATE COMPACT OPERATORS Let $X$ and $Y$ be Banach spaces, suppose that $Y$ has a Schauder basis and let $P_n$ denote the natural projection associated with the basis. Then for any compact $A \in L(X,Y)$, the finite rank operators $P_n A \to A$.

**Proof.** In the notation of the statement, let $U$ denote the unit ball of $X$. Since $A$ is compact, $\operatorname{cl} A(U)$ is compact in $Y$. Since $Y$ has a Schauder basis, by Theorem 15.9.2, for all $r > 0$ there exists $n \in \mathbf{N}$ such that $\|P_n y - y\| < r$ for all $y \in \operatorname{cl} A(U)$. Thus $\|P_n(Ax) - Ax\| = \|(P_n A - A)x\| < r$ for all $x \in U$. Consequently, $\|P_n A - A\| \leq r$ and the rank of $P_n A$ is clearly $\leq n$. ■

For the approximation property we demand that there be an operator $A \in L(X,X)$ of finite rank that approximates the identity. If we impose a constraint on the bound of $A$, then we have:

**Definition 15.9.4** BOUNDED APPROXIMATION PROPERTY A Banach space $X$ has the *bounded approximation property* if there exists a constant $b \geq 1$ such that, for every $r > 0$ and compact subset $K \subset X$, there is a finite rank operator $A \in L(X,X)$, $\|A\| \leq b$, such that $\|Ax - x\| \leq r$. ■

Since the natural projections $P_n$ of Theorem 15.9.2 are such that $\|P_n\| \leq K$ for any $n$, Banach spaces with Schauder bases have the bounded approximation property. For a separable Banach space $X$, the bounded approximation property is equivalent to $X$ being linearly homeomorphic to a complemented subspace of a space with a Schauder basis [Pelczynski 1971, p. 239; Johnson et al. 1971, p. 503].

## 15.10 EXERCISES

15.101 BANACH–BOURBAKI Prove that if $X$ is a normed space then $X$ is reflexive iff the closed unit ball $U(X)$ of $X$ is $\sigma(X,X')$-compact. That is, show that in Theorem 15.5.1 it is not necessary to assume that $X$ is complete

15.102 EXTREME POINTS OF UNIT BALL OF $C(T,\mathbf{F},\|\cdot\|_\infty)$ Let $(T,d)$ be an infinite compact ultrametric space [i.e., $d(t_1,t_2) \leq \max(d(t_1,t_3), d(t_2,t_3))$ for all $t_1,t_2,t_3,z \in T$].

(a) Show that if $d(t_1, t_3) > d(t_2, t_3)$, then $d(t_1, t_2) = d(t_1, t_3)$.

(b) Show that the sets $\{t \in T : d(t_0, t) = r\}$, $r > 0$, are clopen.

(c) Show that for all $n \in \mathbf{N}$ there exist clopen sets $U_n \subset T$ such that $U_{n+1} \subsetneq U_n$.

(d) Use (c) to construct a denumerable linearly independent set of extreme points of the closed unit ball of $C(T, \mathbf{F}, \|\cdot\|_\infty)$.

(e) Show that (c) and (d) hold if $T$ is any infinite 0-dimensional compact Hausdorff space. [A space is 0-dimensional if there is a basis of clopen sets for the topology.]

15.103 $\sigma(X, X')$-COMPACT SETS AND POSITIVE DISTANCE Let $X$ be a Banach space and let $B$ be a bounded $\sigma(X, X')$-closed subset of $X$.

(a) Show that if $B.$ is $\sigma(X, X')$-compact and $F$ is $\sigma(X, X')$-closed and $B.\cap F = \varnothing$, then $d(B., F) = \inf\{\|x - y\| : x \in B., \ y \in F\} > 0$.

(b) Show that if $X$ is reflexive, then when $F \subset \complement U$ (where $U$ denotes the closed unit ball of $X$) is $\sigma(X, X')$-closed, there exists $r > 0$ such that $\|x\| > 1 + r$ for all $x \in F$.

15.104 NATURAL PROJECTIONS EQUICONTINUOUS Let $\{x_n\}$ be a Schauder basis for a Banach space $X$. Show that the natural projections

$$\{P_n : n \in \mathbf{N}\}$$

[Def. 15.8.2] are equicontinuous.

15.105 DIRECT SUMS AND PRODUCTS Show that the locally convex direct sum and product of semireflexive LCS is semireflexive.

15.106 Show that a Banach space $X$ is reflexive iff $X'$ is reflexive.

15.107 SEPARABILITY Show that if the dual $X'$ of the normed space $X$ is separable then so is $X$.

### Hints to Exercises

15.101 Let $\hat{X}$ be a completion of $X$. $(\hat{X}, X')$ is a dual pair and $U(\hat{X}) = \text{cl}_n U(X)$ where $\text{cl}_n$ denotes closure in $\hat{X}$ in the norm topology. Use the fact that $\sigma\left(\hat{X}, X'\right) \cap U(X) = \sigma(X, X') \cap U(X)$ and the compactness of $U(X)$ to conclude that $U(X) = U(\hat{X})$ or $X = \hat{X}$.

15.102 (c) Since $T$ is infinite and compact, there must be a limit point $t_0 \in T$. The sets can be selected from the neighborhoods of $t_0$.

(d) Let

$$x_n(t) = \begin{cases} 1, & t \in U_n \\ -1, & t \in \complement U_n \end{cases}$$

Suppose that $\sum_{i=1}^{n} a_i x_i = 0$. Choosing points $t_j \in U_j \setminus U_{j+1}$ and evaluating the sum at $t_j$, we obtain the equations

$$a_1 - a_2 - a_3 - \cdots - a_n = 0 \tag{1}$$

$$a_1 + a_2 - a_3 + \cdots - a_n = 0 \tag{2}$$

$$\vdots$$

$$a_1 + a_2 + a_3 + \cdots + a_{n-1} - a_n = 0 \tag{n-1}$$

$$a_1 + a_2 + a_3 + \cdots + a_{n-1} + a_n = 0 \tag{n}$$

whose only solution is the trivial solution.

15.103 (a) If $B$ is $\sigma(X, X')$-compact and there are sequences $(x_n)$ of points in $B$, and $(y_n)$, of points in $F$, such that $\|x_n - y_n\| \to 0 = d(B, F)$, we can assume [Theorem 15.4.8] that $(x_n)$ converges to $x \in B$ with respect to $\sigma(X, X')$. Consequently $y_n \to x$ with respect to $\sigma(X, X')$ and $x \in B \cap F$.

# Chapter 16

# Norm Convexities and Approximation



**Notation for Secs. 16.1–16.3.**

$X$, normed space over $\mathbf{F}$.

$U(X) = \{x \in X : \|x\| \leq 1\}$, closed unit ball of $X$.

$SU(X) = \{x \in X : \|x\| = 1\}$, the surface of $U(X)$.

$[x, y] = \{tx + (1-t)y : 0 \leq t \leq 1\}$, the line segment joining $x$ and $y$.

$\operatorname{ext} U(X)$, the extreme points of $U(X)$.

$d(x, M) = \inf_{m \in M} \|x - m\|$, the distance from $x \in X$ to $M \subset X$.

In Sections 16.1 and 16.2 we consider two geometric properties of normed spaces—strict and uniform convexity, attributes that ensure a certain roundness of the unit ball of $X$. Inner product spaces, for example, are uniformly convex and (hence) strictly convex; the spaces $\ell_1(2)$ and $\ell_\infty(2)$ with their pointed unit balls are neither. Strict and uniform convexity depend intrinsically on the norm, not the topology induced by it. Properties of this type are called *metric properties*. We consider the connection of these notions to "best approximation" in Sec. 16.3, the ability to find a vector in a subspace closest to a vector not in it.

Hahn–Banach extensions of the same norm are not generally unique [Example 16.4.4] but they are in Hilbert spaces. We consider general criteria for uniqueness of norm-preserving extensions in Secs.16.4.2 and 16.4.3. In Sec. 16.5 we consider approximation concerning density of certain subspaces; in particular we prove the Stone–Weierstrass theorem about the density of polynomials in $C(T, \mathbf{F}, \|\cdot\|_\infty)$.

## 16.1 STRICT CONVEXITY

Strictly convex spaces $X$ are those whose unit ball $U(X)$ is "round" in the sense that its surface $SU(X)$ consists entirely of extreme points. We investigate strict convexity in this section and get some equivalent descriptions of it, e.g., it is equivalent to each continuous nontrivial linear functional $f$ on $X$ having at most one maximal element.

**Definition 16.1.1** STRICTLY CONVEX A normed space $X$ is *strictly convex* or *strictly normed* or *rotund* if the surface $SU(X) = \{x \in X : \|x\| = 1\}$ of the unit ball consists entirely of extreme points—$SU(X) = \operatorname{ext} U(X)$; we also say that the norm is a *strict norm* or that the norm is *strictly subadditive*. ■

Clearly, in a strictly convex space $X$, every closed ball $C(x, r)$ ($x \in X$, $r > 0$) is such that its surface $\{y \in X : \|y - x\| = r\}$ consists only of extreme points. Moreover, $X$ is strictly convex iff $SU(X)$ contains no nontrivial (i.e., $x \neq y$) line segments $[x, y]$ [Theorem 16.1.3]. It is easy to see that $\ell_2(2)$ is strictly convex and $\ell_\infty(2)$ [Example 16.1.6] and $\ell_1(2)$ are not. Fréchet [1925, pp. 39–40] considered strictly convex spaces—although he called them by the distinctly wacky name *metric spaces*—and observed that Hilbert spaces are strictly convex [Theorem 16.1.5]. The notion of strict convexity was rediscovered by Clarkson [1936, p. 104] and Krein [1938, pp. 183–184]; see also Akhieser 1956, p. 11] and Smulian [1939, p. 404]. We already encountered strictly convex spaces in connection with a vector-valued Banach-Stone theorem [Corollary 9.9.18].

**Theorem 16.1.2** WHEN $SU(X)$ CONTAINS INTERVALS Let $x$ and $y$ be unit vectors in a normed space $X$. If a point in the segment $(x, y)$ lies in $SU(X)$ then $[x, y] \subset SU(X)$.

**Proof.** Let $X$ be a normed space, suppose that $x, y \in SU(X)$ and that, for some $a \in (0, 1)$, $\|ax + (1 - a)y\| = 1$. Further, suppose that there is some $b \in (a, 1)$ such that $\|bx + (1 - b)y\| < 1$. With $z = bx + (1 - b)y$, $ax +$

$(1-a)\,y = dz + (1-d)\,y$ where $d = a/b < 1$. Then $\|ax + (1-a)\,y\| \leq d\,\|z\| + (1-d)\,\|y\| < d + (1-d) = 1$ which is a contradiction. The proof for $b \in (0,a)$ is essentially the same. ■

**Theorem 16.1.3** STRICT CONVEXITY IFF $[x,y] \not\subset SU(X)$ A normed space $X$ is strictly convex iff the surface $SU(X)$ of the unit ball contains no nontrivial line segments.

**Proof.** Certainly, if $X$ is strictly convex, $SU(X)$ contains no nontrivial line segments. If $X$ is *not* strictly convex, there exist unit vectors $x, y \in X$ and $0 < a < 1$ such that $\|ax + (1-a)\,y\| = 1$. By Theorem 16.1.2, $[x,y] \subset SU(X)$. ■

**Theorem 16.1.4** STRICT CONVEXITY EQUIVALENTS For a normed space $X$ the following are equivalent:

(a) $X$ is strictly convex, i.e. $SU(X) = \operatorname{ext} U(X)$.

(b) For distinct $x, y \in SU(X)$, $[x,y] \not\subset SU(X)$.

(c) The midpoint of the line segment joining two distinct points $x, y \in SU(X)$ lies in the interior of $U(X)$: $\left\|\left(\frac{1}{2}\right)(x+y)\right\| = \left(\frac{1}{2}\right)\|x+y\| < 1$.

(d) For any nonzero $x$ and $y$, $\|x+y\| = \|x\| + \|y\|$ implies that there exists $a > 0$ such that $y = ax$.

**Proof.** Let $X$ be a normed space. The equivalence of (a) and (b) is Theorem 16.1.3.

(b) $\Rightarrow$ (c) If, for distinct $x, y \in SU(X)$, $\left\|\left(\frac{1}{2}\right)(x+y)\right\| = 1$ then $\left(\frac{1}{2}\right)(x+y) \in SU(X)$. This implies that $[x,y] \subset SU(X)$ by Theorem 16.1.2 and contradicts (b). That (c) $\Rightarrow$ (b) is clear.

(d) $\Rightarrow$(a) Suppose that (d) holds. We use the extreme point characterization of Theorem 9.2.2(b) and show that if $z$ is the midpoint of $[x,y]$, $x, y \in SU(X)$, then $x = y$. If $x, y \in SU(X)$ and $z = \left(\frac{1}{2}\right)(x+y) \in \operatorname{ext} U(X)$ then $\|x+y\| = 2 = \|x\| + \|y\|$. By hypothesis, this means that there exists $a > 0$ such that $y = ax$. Hence $2 = \|x + ax\| = 1 + a$ which implies that $a = 1$ and $y = x$.

(a) $\Rightarrow$ (d) Suppose that $SU(X) = \operatorname{ext} U(X)$ and that $x$ and $y$ are nonzero vectors such that $\|x+y\| = \|x\| + \|y\|$ with $\|x\| \leq \|y\|$. By (c), if $x/\|x\| \neq y/\|y\|$ then $\|x/\|x\| + y/\|y\|\| < 2$. But

$$\begin{aligned}
\left\|\frac{x}{\|x\|} + \frac{y}{\|y\|}\right\| &\geq \left\|\frac{x}{\|x\|} + \frac{y}{\|x\|}\right\| - \left\|\frac{y}{\|x\|} - \frac{y}{\|y\|}\right\| \\
&= \frac{\|x+y\|}{\|x\|} - \|y\|\left(\frac{1}{\|x\|} - \frac{1}{\|y\|}\right) \\
&= \frac{\|x\| + \|y\|}{\|x\|} - \|y\|\left(\frac{1}{\|x\|} - \frac{1}{\|y\|}\right) \\
&= 1 + \frac{\|y\|}{\|x\|} - \frac{\|y\|}{\|x\|} + 1 = 2
\end{aligned}$$

which is contradictory; therefore $x/\|x\| = y/\|y\|$ or $y = (\|y\|/\|x\|)\,x$. ■

By Theorem 16.1.4(d), since equality holds in the Minkowski inequality [Example 1.7.8] for $p > 1$ iff one vector is a scalar multiple of the other, the $L_p$ spaces are strictly convex for $p > 1$.

**Theorem 16.1.5** INNER PRODUCT SPACES ARE STRICTLY CONVEX An inner product space $(X, \langle\cdot,\cdot\rangle)$ is strictly convex.

**Proof.** Let $(X, \langle\cdot,\cdot\rangle)$ be an inner product space. If $x, y \neq 0$ and $\|x+y\| = \|x\| + \|y\|$ then $\|x+y\|^2 = \langle x+y, x+y\rangle = \|x\|^2 + \langle x,y\rangle + \langle y,x\rangle + \|y\|^2 = (\|x\| + \|y\|)^2$. Thus $2\|x\|\,\|y\| = 2\operatorname{Re}\langle x,y\rangle$ and $\operatorname{Re}\langle x,y\rangle = \|x\|\,\|y\|$. Since $|\langle x,y\rangle| \leq \|x\|\,\|y\| = \operatorname{Re}\langle x,y\rangle \leq |\langle x,y\rangle|$, it follows that $|\langle x,y\rangle| = \operatorname{Re}\langle x,y\rangle$ and $\operatorname{Im}\langle x,y\rangle = 0$. Hence $|\langle x,y\rangle| = \langle x,y\rangle = \|x\|\,\|y\|$. Since $\langle x,y\rangle = \|x\|\,\|y\|$, it is easy to verify that

$$\left\langle x - \frac{\|x\|}{\|y\|}y, x - \frac{\|x\|}{\|y\|}y \right\rangle = \|x - \frac{\|x\|}{\|y\|}y\|^2 = 0$$

which implies that $x - (\|x\|/\|y\|)\,y = 0$ or $y = (\|y\|/\|x\|)\,x$ The result follows from Theorem 16.1.4. ■

We show next that there can be equivalent norms, one of which is strict but the other is not. This means that strict convexity is a metric property in that it depends on the norm, not the topology induced by it.

**Example 16.1.6** EQUIVALENT NORMS, ONE STRICT, ONE NOT For any $n \in \mathbf{N}$, $\ell_2(n)$ is strictly convex because it is an inner product space [Theorem 16.1.5] but $\ell_\infty(n)$ is not: for $x = (1, 1, 0, \ldots, 0)$ and $y = (1, 0, \ldots, 0)$, $\|x+y\|_\infty = 2 = \|x\|_\infty + \|y\|_\infty$ but $y$ is not a scalar multiple of $x$. Since the spaces are finite-dimensional, the norms are equivalent [Theorem 4.10.3(b)]. ■

Recall that a unit vector $x \in X$ such that $f(x) = \|f\|$ is a *maximal element* for $f \in X'$ [Def. 15.6.1].

**Theorem 16.1.7** STRICT CONVEXITY AND MAXIMAL ELEMENTS A normed space $X$ is strictly convex iff each continuous nontrivial linear functional $f$ on $X$ has at most one maximal element.

**Proof.** Suppose that $X$ is strictly convex and that $f \in X'$, $f \neq 0$, has maximal elements $x$ and $y$. We may assume without loss of generality that $\|f\| = 1$. Hence $f(x) = f(y) = 1 = \|x\| = \|y\|$. Thus $f(x+y) = f(x) + f(y) = \|x\| + \|y\|$. Since $f(x+y) = \|x\| + \|y\| \leq \|x+y\|$, this implies that $\|x+y\| = \|x\| + \|y\|$. Since $X$ is strictly convex, there exists $a > 0$ such that $y = ax$. Hence $a = \|y\|/\|x\| = 1$ or $y = x$.

Now suppose that $X$ is not strictly convex. As a consequence of Theorem 15.6.5(d), we need only prove the theorem for $\mathbf{F} = \mathbf{R}$ and we show that there exists a continuous real linear functional on $X$ with distinct maximal elements. Since $X$ is not strictly convex, there exist distinct $x, y \in SU(X)$ such that the line segment $[x, y] \subset SU(X)$ [Theorem 16.1.3]. Hence for all $c \in [0,1]$, $\|cx + (1-c)y\| = 1$. It follows that for any $a, b \geq 0$, not both 0,

$$\|ax + by\| = (a+b)\left\|\frac{ax}{a+b} + \frac{by}{a+b}\right\| = a + b$$

Similarly, for $a, b \geq 0$, not both 0, and $a > b$, $\|ax - by\| = a - b$. We define a real linear functional $f$ on the linear span $M$ of $x$ and $y$ by $f : M \to \mathbf{R}$, $ax + by \mapsto a + b$ $(a, b \in \mathbf{F})$. For $a$ and $b$ not both 0, we compute $\|f\|$ by considering the following cases:

$$\begin{array}{ll}
\dfrac{|f(ax+by)|}{\|ax+by\|} = \dfrac{a+b}{a+b} = 1 & a, b \geq 0 \\
\dfrac{|f(ax-by)|}{\|ax-by\|} = \dfrac{|a-b|}{\|ax-by\|} \leq \dfrac{a-b}{a-b} = 1 & a, b \geq 0,\ b < a \\
\dfrac{|f(ax-by)|}{\|ax-by\|} = \dfrac{|a-b|}{\|ax-by\|} \leq \dfrac{b-a}{b-a} = 1 & a, b \geq 0,\ a < b \\
\dfrac{|f(-ax-by)|}{\|-ax-by\|} = \dfrac{|-a-b|}{a+b} = \dfrac{a+b}{a+b} = 1 & a, b \geq 0
\end{array}$$

Thus $\|f\| = 1$ and, by the Hahn–Banach theorem [7.4.1], $f$ may be extended to $F \in X'$ where $\|F\| = 1$. Since $F(x) = F(y) = \|x\| = \|y\| = 1$, $x$ and $y$ are distinct maximal elements for $F$. ■

## 16.2 UNIFORM CONVEXITY

Consider two points $x$ and $y$ on the circumference of the unit circle in $\mathbf{R}^2$. The midpoint $\left(\frac{1}{2}\right)(x+y)$ of the chord joining $x$ and $y$ is close to the rim iff $x$ and $y$ are close to each other; furthermore, the points $x$ and $y$ are irrelevant: the distance $1 - \left\|\left(\frac{1}{2}\right)(x+y)\right\|$ from $\left(\frac{1}{2}\right)(x+y)$ to the circumference depends only on the distance between $x$ and $y$. In connection with his work on vector measures, Clarkson [1936] abstracted this idea to normed spaces. It is a property with surprising consequences—reflexivity, for example [Theorem 16.2.6].

**Definition 16.2.1** UNIFORM CONVEXITY A normed space $X$ is uniformly convex if, for any sequences $(x_n)$ and $(y_n)$ of unit vectors,

$$\left\|\frac{x_n + y_n}{2}\right\| \to 1 \Rightarrow \|x_n - y_n\| \to 0 \quad ■$$

As we show in Examples 16.2.2 and 16.2.3, $\ell_\infty(2)$ is not uniformly convex but $\ell_2(2)$ is even though their norms are equivalent. As is the case for strict convexity, uniform convexity depends intrinsically on the norm, not the topology it induces.

**Example 16.2.2** $\ell_\infty(2)$ NOT UNIFORMLY CONVEX Let $x_n = (0,1)$, $y_n = (1, 1-1/n)$, $n \in \mathbf{N}$. Then $\|x_n\|_\infty = \|y_n\|_\infty = 1$ for every $n$. The vectors $\left(\frac{1}{2}\right)(x_n + y_n) = (1/2, 1-1/2n)$ approach the surface of the unit ball but $y_n - x_n = (1, -1/n) \nrightarrow 0$. ■

**Example 16.2.3** INNER PRODUCT SPACES ARE UNIFORMLY CONVEX In an inner product space $(X, \langle\cdot,\cdot\rangle)$, if $\|x_n\| = \|y_n\| = 1$ for every $n$ and $\|x_n + y_n\| \to 2$ then $\|x_n + y_n\|^2 = \langle x_n + y_n, x_n + y_n\rangle \to 4$. Hence, by the parallelogram law, $\|x_n - y_n\|^2 = 4 - \|x_n + y_n\|^2 \to 0$. ■

$L_p[0,1]$ and $\ell_p$ $(p > 1)$ are also uniformly convex, a matter we discuss in Exercise 16.201.

The following characterization of uniform convexity is immediate.

**Theorem 16.2.4** UNIFORM CONVEXITY EQUIVALENTS A normed space $X$ is uniformly convex if, for any $r > 0$, there exists $d = d(r) > 0$ such that, for all $x, y \in SU(X)$,

(a) $1 - \left\|\left(\frac{1}{2}\right)(x+y)\right\| < d$ implies $\|x - y\| < r$, or, equivalently,

(b) $\|x - y\| \geq r$ implies $1 - \left\|\left(\frac{1}{2}\right)(x+y)\right\| \geq d$.

Uniformly convex spaces are a subclass of the strictly convex spaces.

**Theorem 16.2.5** UNIFORM IMPLIES STRICT A uniformly convex normed space $X$ is strictly convex. Conversely, if $X$ is a strictly convex finite-dimensional space then $X$ is uniformly convex.

**Proof.** Suppose that $X$ is uniformly convex. To show that $X$ is strictly convex, we use the extreme point criterion of Theorem 16.1.4(c): We show that, for unit vectors $x$ and $y$, $\|(1/2)(x+y)\| = 1 \Rightarrow x = y$. By the uniform convexity of $X$, given $r > 0$, there exists $d > 0$ such that $\|x - y\| < r$. If $\|(1/2)(x+y)\| = 1$, however, then $(1/2)\|x+y\| > 1 - d$ for all $d > 0$. Hence $x = y$.

Now suppose that $X$ is finite-dimensional, strictly convex and $r > 0$. Since $X$ is finite-dimensional, the surface $SU(X) = \{x \in X : \|x\| = 1\}$ of the unit ball is compact; therefore $SU(X) \times SU(X)$ is a compact subset of $X \times X$ in the product topology. Consequently, the closed subset $K = \{(x,y) \in SU(X) \times SU(X) : \|x - y\| \geq r\}$ is compact as well. The continuous map defined on $X \times X$ by $f(x,y) = 1 - (1/2)\|x+y\|$ therefore assumes a minimum $d = f(x_0, y_0) = \min f(K)$ on $K$. As $X$ is strictly convex, $(1/2)\|x+y\| < 1$ for distinct $x, y \in K$ [Theorem 16.1.4(c)]. Since the elements $x, y \in K$ are distinct, $(\|x-y\| \geq r > 0)$, this implies that $f(x,y) > 0$ for all $(x,y) \in K$, in particular $d = f(x_0, y_0) > 0$. Thus $\|x - y\| \geq r$ implies that $f(x,y) = 1 - (1/2)\|x+y\| \geq d$ and $X$ is uniformly convex [Theorem 16.2.4(b)]. ■

In Theorem 16.2.6 we present Milman's [1938] result that uniformly convex Banach spaces are reflexive, the proof of which gestated through Kakutani [1939, p. 172] and Pettis [1939]. Since inner product spaces are uniformly convex [Example 16.2.3], this yields another proof that Hilbert spaces are reflexive.

**Theorem 16.2.6** Uniformly Convex Spaces Are Reflexive A uniformly convex Banach space $X$ is reflexive.

**Proof.** Let $X$ be a uniformly convex Banach space. As noted at the end of Sec. 15.1, to show that a Banach space is reflexive, it suffices to demonstrate that $X$ is semireflexive. To do this it suffices to show that if $z \in SU(X'')$ then $z \in X$. By Goldstine's theorem [8.4.7], $X$ is $\sigma(X'', X')$-dense in $X''$ so there exists a net $\{x_s : s \in S\}$ of points from $X$ such that $x_s \to z$ with respect to $\sigma(X'', X')$. Hence, for each $f \in SU(X')$,

$$|\langle f, 2z\rangle| = \lim_{s,t\in S} |\langle f, x_s + x_t\rangle| \leq \lim_{s,t\in S} \|x_s + x_t\| = \|2z\| = 2$$

Therefore $\lim_{s,t\in S} \|(1/2)(x_s + x_t)\| = 1$. Since $X$ is uniformly convex, $\lim_{s,t\in S} \|x_s - x_t\| = 0$, i.e., $(x_s)$ is a Cauchy net in the Banach space $X$. Its limit $z$ therefore belongs to $X$. ■

For uniformly convex normed spaces $X$, the argument above shows that $\sigma(X, X')$-convergence of a sequence $x_n \to x$ from $SU(X)$ implies norm convergence. Therefore:

**Theorem 16.2.7** If $X$ is a uniformly convex normed space, the norm and weak topologies coincide on the surface $SU(X)$ of the closed unit ball of $X$.

We summarize some results about convexities in the table below.

| | | |
|---|---|---|
| $SU(X) = \operatorname{ext} U(X)$ | $\Longleftrightarrow$ | $\exists \leq 1$ best approximation to any $x \notin M$ for each $M \subset X$ [Theorem 16.3.15] |
| $\Updownarrow$ | | $\Updownarrow$ |
| strictly convex | | $\exists \leq 1$ maximal element for each $f \in X'$ [Theorem 16.1.7] |
| $\Updownarrow$ | | |
| $[x, y] \not\subset SU(X)$ for distinct $x, y \in SU(X)$ | | uniformly convex |
| $\Updownarrow$ | | |
| $\|x + y\| = \|x\| + \|y\|$ iff $y = ax$ for $a > 0$ | $\Longleftarrow$<br>$\Longrightarrow$<br>if $\dim X < \infty$ | $\|(x_n + y_n)/2\| \to 1 \Rightarrow \|x_n - y_n\| \to 0$ |

## 16.3 BEST APPROXIMATION

Let $M$ be a subset of the normed space $X$. We say that a *best approximation to* $x \in X$ *from* $M$ is an element $m_0 \in M$ closest to $x$, i.e., $m_0 \in M$ such that $\|x - m_0\| = \inf\{\|x - m\| : m \in M\}$. For example, $(0,0,0)$ is the best approximation to $(0,0,1) \in \ell_2^{\mathbf{R}}(3)$ from $M = \ell_2^{\mathbf{R}}(2)$. If $x \in M$ then $x$ is the best approximation to itself. If $x \in \operatorname{cl} M$ then there is a sequence $(m_n)$ from $M$ which converges to $x$ so there is no best approximation to $x$ from $M$, just as there is no best approximation to $\sqrt{2}$ from $\mathbf{Q}$. For these reasons, we only consider closed subspaces $M$ and the ability to approximate $x \notin M$.

We investigate:

16.1 Which normed spaces $X$ and subsets $M \subset X$ are such that for each $x \notin M$ there exists a best approximation $m_0 \in M$ to $x$?

16.2 When is $m_0$ unique?

16.3 For which normed spaces $X$ will Questions 1 and 2 be true for all closed subspaces $M$ of $X$?

**Definition 16.3.1** If $M$ is a subset of a metric space $(T, d)$ and $t \in T$, we call $d(t, M) = \inf\{d(t, m) : m \in M\}$ the *distance* from $t$ to $M$. ■

In this notation, we restate the definition of best approximation.

**Definition 16.3.2** BEST APPROXIMATION Let $M$ be a closed subset of the normed space $X$ and suppose $x \notin M$. We say that $m_0 \in M$ is a *best approximation to* $x$ *from* $M$ or that $m_0$ is an *$M$-minimal distance vector for* $x$ if

$$\|x - m_0\| = \inf\{\|x - m\| : m \in M\} = d(x, M) \quad ■$$

If best approximations exist, they need not be unique: What is the best approximation to the origin from the surface $SU\left(\ell_2^{\mathbf{R}}(3)\right)$ of the unit ball $U$ of $\ell_2^{\mathbf{R}}(3)$? For a finite-dimensional subspace $M$ of an inner product space $X$, we show in Theorem 16.3.3 that unique $M$-minimal distance vectors exist for any $x \notin M$; indeed, we can even say what the best approximation is. We generalize the existence statement to finite-dimensional subspaces of any normed space in Theorem 16.3.6.

**Theorem 16.3.3** FINITE-DIMENSIONAL SUBSPACES OF INNER PRODUCT SPACES Let $\{x_1, \ldots, x_n\}$ be an orthonormal subset of the inner product space $(X, \langle\cdot,\cdot\rangle)$. Then the best approximation to any $x \in X$ from $M = [x_1, x_2, \ldots, x_n]$ (linear span) is given by $\sum_{i=1}^{n} \langle x, x_i\rangle x_i$, the "sum of its projections" $\langle x, x_i\rangle x_i$ on the subspaces $[x_i]$.

**Proof.** Let $(X, \langle \cdot, \cdot \rangle)$ be an inner product space space and let $M \subset X$ denote the linear span of the orthonormal set $\{x_1, \ldots, x_n\}$. For any scalar $a$,

$$\begin{aligned} \|x - ax_1\|^2 &= \|x\|^2 - \bar{a} \langle x, x_1 \rangle - a \langle x_1, x \rangle + a\bar{a} \\ &= \|x\|^2 - \langle x, x_1 \rangle \overline{\langle x, x_1 \rangle} + \langle x, x_1 \rangle \overline{\langle x, x_1 \rangle} \\ &\quad - \bar{a} \langle x, x_1 \rangle - a \langle x_1, x \rangle + a\bar{a} \\ &= \|x\|^2 - |\langle x, x_1 \rangle|^2 + |\langle x, x_1 \rangle - a|^2 \end{aligned}$$

The last expression is clearly minimized when $a = \langle x, x_1 \rangle$. When $ax_1$ is replaced by $\sum_{i=1}^n a_i x_i$, we get more terms like the ones in the expression above. We get

$$\left\| x - \sum_{i=1}^n a_i x_i \right\|^2 = \|x\|^2 - \sum_{i=1}^n |\langle x, x_i \rangle|^2 + \sum_{i=1}^n |\langle x, x_i \rangle - a_i|^2$$

Clearly the best approximation to $x$ is obtained by setting $a_i = \langle x, x_i \rangle$ for each $i$. The best approximation to $x$ by vectors in $M$ is therefore $m_0 = \sum_{i=1}^n \langle x, x_i \rangle x_i$. ■

Finite-dimensional subspaces of any HTVS are closed [Theorem 4.10.3(d)]. It follows from Theorem 16.3.4 that a unique best approximation to any $x \in X$ from a *complete* subspace $M$ of an inner product space $X$ exists no matter what the dimension of $M$ is.

**Theorem 16.3.4** BEST APPROXIMATION FROM COMPLETE SUBSETS Let $M$ be a complete convex subset of the inner product space $X$. Then, for any $x \notin M$:

(a) there is a unique best approximation $m_0 \in M$ to $x$;

(b) if $M$ is a subspace then $x - m_0 \perp M$ and $m_0$ is the only point in $M$ such that $x - m_0 \perp M$.

**Proof.** (a) With $M$ as above, we seek $m_0 \in M$ such that $\|x - m_0\| = d(x, M) = d$. We create a sequence of vectors $m_n \in M$ that almost have this property, then take a limit. Since $d(x, M)$ is an infimum, for each $n \in \mathbf{N}$ there exists $m_n \in M$ such that

$$d \leq \|x - m_n\| < d + 1/n \tag{1}$$

Since the $m_n$ are close to $x$, they are close to each other; indeed $(m_n)$ is Cauchy. To verify this, consider a "parallelogram" with sides $x - m_n$ and $x - m_k$ and diagonals $2x - m_n - m_k$ and $m_k - m_n$, the sum and the difference. It follows from the parallelogram law that

$$2\|x - m_n\|^2 + 2\|x - m_k\|^2 = \|2x - m_n - m_k\|^2 + \|m_k - m_n\|^2 \tag{2}$$

As $M$ is convex, $\left(\frac{1}{2}\right)(m_n + m_k) \in M$ so, for any $n$ and $k$,

$$\left\| x - \frac{1}{2}(m_n + m_k) \right\| \geq d \tag{3}$$

Since $\|x - m_n\| < d + 1/n$, $\|x - m_n\|^2 < d^2 + 2d/n + 1/n^2$. For $r > 0$ we can choose $N$ to simultaneously satisfy $2d/n < r/2$ and $1/n^2 < r/2$ for $n \geq N$ so that $\|x - m_n\|^2 \leq d^2 + r$ for $n \geq N$. It therefore follows from (1)–(3) that, for $n, k \geq N$,

$$\|m_k - m_n\|^2 \leq \left(2d^2 + 2r\right) + \left(2d^2 + 2r\right) - 4d^2 = 4r$$

Hence $(m_n)$ is Cauchy. Since $M$ is complete, there exists $m_0 \in M$ such that $m_n \to m_0$. By the continuity of the norm it follows from Inequality (1) that $\|x - m_0\| = d$. Suppose $m \in M$ is such that $\|x - m\| = d$. Since $(m_0 + m)/2 \in M$, the parallelogram law implies that

$$\begin{aligned} \|m - m_0\|^2 &= 2\|x - m_0\|^2 + 2\|x - m\|^2 - \|2x - m_0 - m\|^2 \\ &= 2d^2 + 2d^2 - 4\left\|x - \tfrac{1}{2}(m_0 + m)\right\|^2 \\ &\leq 4d^2 - 4d^2 = 0 \end{aligned}$$

and it follows that $m = m_0$.

(b) Now suppose that $M$ is a subspace and $x$ and $m_0 are$ as above. By the uniqueness of $m_0$, for any nonzero $m \in M$ and any nonzero scalar $a$,

$$\|x - (m_0 + am)\|^2 > d^2$$

By expanding the left-hand term and using the fact that $\|x - m_0\| = d$, it follows that

$$|a|^2 \|m\|^2 - 2\,\mathrm{Re}\,\bar{a}\,\langle x - m_0, m\rangle > 0$$

Suppose that $\langle x - m_0, m\rangle \neq 0$ and let $a = \langle m, x - m_0\rangle b$ where $b \neq 0$ is real. The above inequality becomes

$$b^2 |\langle x - m_0, m\rangle|^2 \|m\|^2 - 2b\,|\langle x - m_0, m\rangle|^2 > 0$$

which implies that $b^2\|m\|^2 - 2b > 0$ for all real nonzero $b$. For $b = 1$ then $\|m\|^2 > 2$ for any $m \in M$ which is impossible. We conclude that $\langle x - m_0, m\rangle = 0$, i.e., that $x - m_0 \perp M$.

Finally, suppose that $m' \in M$ also has the property that $x - m' \perp M$. Then, for any $m \in M$,

$$\langle x - m_0, m\rangle = 0 = \langle x - m', m\rangle$$

This implies that $\langle m' - m_0, m\rangle = 0$ for every $m \in M$. Therefore $m' - m_0 \perp m' - m_0$ and this yields the uniqueness of $m_0$. ■

We categorize the possibilities about existence and uniqueness of best approximations as follows.

**Definition 16.3.5** PROXIMINAL AND CHEBYSHEV SUBSPACES Let $M$ be a closed subspace of a normed space $X$. Then:

(a) AT LEAST ONE $M$ is *proximinal* if for all $x \notin M$ there exists $m_0 \in M$ such that $\|x - m_0\| = \inf_{m \in M} \|x - m\| = d(x, M)$;

(b) AT MOST ONE $M$ is *semi-Chebyshev* if, for each $x \notin M$, there is at most one $m_0 \in M$ satisfying $\|x - m_0\| = d(x, M)$ (there may be none for some $x \notin M$);

(c) EXACTLY ONE $M$ is *Chebyshev* if for each $x \notin M$, there is a unique $m_0 \in M$ such that $\|x - m_0\| = d(x, M)$. ■

By Theorem 16.3.3, a finite-dimensional subspace of an inner product space is Chebyshev. We generalize this in Theorem 16.3.6.

**Theorem 16.3.6** FINITE-DIMENSIONAL $\Rightarrow$ PROXIMINAL If $M$ is a finite-dimensional subspace of a normed space $X$ then:

(a) $M$-minimal distance vectors $m_0 \in M$ exist for every $x \notin M$ and

(b) if $X$ is strictly convex then the $m_0$ of (a) is unique, i.e., $M$ is Chebyshev.

**Proof.** Let $X$ be a normed space. (a) Let $M \subset X$ be finite-dimensional and suppose $x \notin M$. As $M$ is finite-dimensional, $M$ is locally compact [Theorem 4.10.4]. Clearly $d(x, M) \leq d(x, 0) = \|x\|$. Hence, for any $m \in M$, $\|x - m\| \leq \|x\|$ and it follows from the triangle inequality that $\|m\| \leq 2\|x\|$. Hence to find $m_0 \in M$ such that $\|x - m_0\| = d(x, M)$, we need only consider candidates from the compact set $K = \{m \in M : \|m\| \leq 2\|x\|\}$. Since $K$ is compact, there exists $m_0 \in K$ such that $\|x - m_0\| = \min_{m \in K} \|x - m\| = \min_{m \in M} \|x - m\| = d(x, M)$.

(b) Suppose that $m_0$ and $m_{00}$ are $M$-minimal distance vectors for $x \notin M$ : $\|x - m_0\| = \|x - m_{00}\| = d(x, M)$. For a basis $\{m_1, \ldots, m_n\}$ for $M$, suppose that $m_0 = \sum_{i=1}^n a_i m_i$ and $m_{00} = \sum_{i=1}^n b_i m_i$ $(a_i, b_i \in \mathbf{F})$. Since

$$\begin{aligned} d(x, M) &\leq \left\| x - \sum_{i=1}^n \frac{a_i + b_i}{2} m_i \right\| = \left\| \frac{x}{2} + \frac{x}{2} - \sum_{i=1}^n \frac{a_i + b_i}{2} m_i \right\| \\ &\leq \frac{1}{2} \left\| x - \sum_{i=1}^n a_i m_i \right\| + \frac{1}{2} \left\| x - \sum_{i=1}^n b_i m_i \right\| \\ &= (1/2)\, d(x, M) + (1/2)\, d(x, M) = d(x, M) \end{aligned}$$

it follows that

$$\left\| x - \sum_{i=1}^n \frac{(a_i + b_i)}{2} m_i \right\| = \left\| \frac{x}{2} - \sum_{i=1}^n \frac{a_i}{2} m_i \right\| + \left\| \frac{x}{2} - \sum_{i=1}^n \frac{b_i}{2} m_i \right\|$$

Since $X$ is strictly convex, there exists $a > 0$ such that $x - \sum_{i=1}^n a_i m_i = a(x - \sum_{i=1}^n b_i m_i)$ which implies that $(1 - a)x \in M$. Since $x \notin M$, $a = 1$. Since $\{m_1, \ldots, m_n\}$ is linearly independent, $a_i = b_i$ for $i = 1, \ldots, n$; therefore $m_0 = m_{00}$. ■

It follows from Theorem 16.3.15 that a closed subspace of a uniformly convex Banach space is Chebyshev.

**Theorem 16.3.7** UNIFORMLY CONVEX $\Rightarrow$ UNIQUE BEST APPROXIMATION Let $M$ be a closed convex subset of a uniformly convex Banach space $X$. Then for each $x \notin M$ there exists a unique $m_0 \in M$ such that $\|x - m_0\| = \inf\{\|x - m\| : m \in M\} = d(x, M)$. Hence a closed subspace of a uniformly convex Banach space is Chebyshev.

**Proof.** Let $d(x, M) = r > 0$. Consider the closed convex set $(1/r)(x - M) = \{(1/r)(x - m) : m \in M\} = L$. Clearly $y \in L \Rightarrow \|y\| \geq 1$. Thus, we are asking: Is there a unique $y \in L$ such that $\|y\| = 1$. Since $d(x, M) = r$, there must be a sequence $y_n \in L$ such that $\|y_n\| \to 1$. With $d_n = 1/\|y_n\|$, it follows that, for any $n, m \in \mathbf{N}$, adding and subtracting $y_n + y_m$,

$$\left\|\frac{d_n y_n + d_m y_m}{2}\right\| \geq \frac{1}{2}\|y_n + y_m\| - \frac{1}{2}(1 - d_n)\|y_n\| - \frac{1}{2}(1 - d_m)\|y_m\| \quad (*)$$

Since $L$ is convex, $(1/2)(y_n + y_m) \in L$ so $(1/2)\|y_n + y_m\| \geq 1$. Since $X$ is uniformly convex, there exists $d > 0$ such that for all $u, v \in SU(X)$, $1 - \|(u + v)/2\| < d \Rightarrow \|u - v\| < r$ [Theorem 16.2.4(a)]. Choose $k \in \mathbf{N}$ such that $\|y_n\| < 1 + d$ for $n \geq k$ and suppose that $m, n \geq k$. Since $\|d_n y_n\| = \|d_m y_m\| = 1$, and $\|y_m\|, \|y_n\| < 1 + d$, $(*)$ implies that

$$\begin{aligned} 1 &\geq (1/2)\|d_n y_n + d_m y_m\| \\ &\geq \tfrac{1}{2}\|y_n + y_m\| - \tfrac{1}{2}\|y_n\| + \tfrac{d_n}{2}\|y_n\| - \tfrac{1}{2}\|y_m\| + \tfrac{d_m}{2}\|y_m\| \\ &= \tfrac{1}{2}\|y_n + y_m\| - \tfrac{1}{2}\|y_n\| + \tfrac{1}{2} - \tfrac{1}{2}\|y_m\| + \tfrac{1}{2} \\ &\geq 2 - \tfrac{1}{2}\|y_n\| - \tfrac{1}{2}\|y_m\| \\ &> 2 - (1/2)(1 + d) - (1/2)(1 + d) = 1 - d \end{aligned}$$

Therefore, for $n, m \geq k$, $1 \geq (1/2)\|d_n y_n + d_m y_m\| > 1 - d$. Hence

$$\|d_n y_n - d_m y_m\| < r \; for \; n, m \geq k$$

and $(d_n y_n)$ is a Cauchy sequence. Since $X$ is a Banach space, there exists $y \in \operatorname{cl} L = L$ such that $d_n y_n \to y$. Since $\|d_n y_n\| = 1$ for every $n$, $\|y\| = 1$. Thus there exists $m_0 \in M$ such that $y = (1/r)(x - m_0)$ and $1 = \|y\| = \|(1/r)(x - m_0)\|$, i.e., $\|x - m_0\| = r = d(x, M)$.

We demonstrate the uniqueness of $m_0$ by proving the uniqueness of $y$. Suppose that $z \in L$ is such that $\|z\| = 1$. Since $(z + y)/2 \in L$, $1 \leq \|(z + y)/2\| \leq \|z\|/2 + \|y\|/2 = 1$. Thus $\|(z + y)/2\| = 1 > 1 - d$ for all $d > 0$; therefore, by Theorem 16.2.4(a), $\|z - y\| < r$ for any $r > 0$. It follows that $z = y$. ■

**Definition 16.3.8** ORTHOGONALS For a subset $M$ of a normed space $X$, recall [Def. 8.9.1] that the *orthogonal* of $M$ is given by $M^\perp = \{f \in X' : \langle M, f\rangle = \{0\}\}$. For $N \subset X'$, $N^\perp = \{x \in X : \langle x, N\rangle = \{0\}\}$. For a subspace $N \subset X'$ we define for $x \in X$

$$\|x\|_N = \sup\{|f(x)| : f \in N \cap U(X')\}$$

Since $\|x\| = \sup\{|f(x)| : f \in U(X')\}$, $\|\cdot\|_N \leq \|\cdot\|$. ■

A trivial observation that we use several times is that, for any linear functional $f$, $f \in N(f)^{\perp}$ where $N(f) = f^{-1}(0)$. The quality of being proximinal is not rare as our next result and its corollary illustrate.

**Theorem 16.3.9** $M^{\perp}$ IS PROXIMINAL For any subspace $M$ of a normed space $X$ over $\mathbf{F}$,

(a) $M^{\perp}$ is proximinal in $X'$ and

(b) for each $f \in X' \setminus M$, $d(f, M^{\perp}) = \|f|_M\|$.

**Proof.** Let $M$ be a subspace of the normed space $X$ and let $f \in X' \setminus M$. For any $g \in M^{\perp}$,

$$\begin{aligned}\|f|_M\| &= \sup\{|f(m)| : m \in U(M) = U(X) \cap M\} \\ &= \sup\{|(f-g)(m)| : m \in U(M)\} \\ &\leq \sup\{|(f-g)(x)| : x \in U(X)\} = \|f-g\|\end{aligned}$$

which implies that $\|f|_M\| \leq d(f, M^{\perp})$. By the norm-preserving Hahn–Banach extension theorem [7.4.1], there exists an extension $h \in X'$ of $f$ such that $\|h\| = \|f|_M\|$. Hence $f - h \in M^{\perp}$ and $\|f|_M\| = \|h\| = \|f - (f-h)\| \geq d(f, M^{\perp})$. We conclude that $\|f|_M\| = d(f, M^{\perp}) = \|f - (f-h)\|$ and that $f - h$ is a best approximation to $f$ from $M^{\perp}$. ■

Since $M = M^{\perp\perp}$ for any $\sigma(X', X)$-closed—hence norm closed—subspace $M \subset X'$ [Theorem 8.9.2(e)], we have the following corollary:

**Corollary 16.3.10** Any $\sigma(X', X)$-closed subspace $M$ of the dual $X'$ of a normed space is proximinal.

We can now relate a minimal distance vector for a subspace $M$ to a maximal element $f \in M^{\perp}$ as follows:

**Theorem 16.3.11** MINIMAL DISTANCE VECTORS Let $M$ be a closed subspace of a normed space $X$ and suppose $x \notin M$ and $m_0 \in M$. Then the following are equivalent:

(a) $m_0$ is an $M$-minimal distance vector for $x$;

(b) there exists $f \in M^{\perp} \cap SU(X')$ such that $f(x - m_0) = \|x - m_0\|$ or, equivalently, $\frac{x - m_0}{\|x - m_0\|}$ is a maximal element for $f$;

(c) $\|x - m_0\|_{M^{\perp}} = \|x - m_0\|$.

**Proof.** We use the notation of the statement. (a) $\Rightarrow$ (b) Suppose that $m_0$ is an $M$-minimal distance vector for $x \notin M$. For $a \in \mathbf{F}$ and $m \in M$, define $g : M \oplus \mathbf{F}x$, $m + ax \mapsto a\|x - m_0\|$. Clearly, $g$ is a continuous linear functional on $M \oplus \mathbf{F}x$, $g(x - m_0) = \|x - m_0\|$ and $g$ vanishes on $M$. Since $m_0$ is $M$-minimal for $x$, for every $a \neq 0$ and $m \in M$, $|g(ax - m)| = |a|\,\|x - m_0\| \leq |a|\,\|x - a^{-1}m\| = \|ax - m\|$; hence $\|g\| = 1$. We extend $g$ to a continuous linear functional $f$ of norm 1 on $X$ by the norm-preserving Hahn–Banach extension theorem [7.4.1].

(b) $\Rightarrow$ (c) Suppose $f \in SU(X') \cap M^{\perp}$ is such that $f(x - m_0) = \|x - m_0\|$. Since $f \in M^{\perp}$,

$$\begin{aligned} \|x - m_0\| &= |f(x - m_0)| \leq \|x - m_0\|_{M^{\perp}} \\ &= \sup\left\{|g(x - m_0)| : g \in M^{\perp} \cap U(X)\right\} \end{aligned}$$

Since, generally, $\|x - m_0\|_{M^{\perp}} \leq \|x - m_0\|$, $\|x - m_0\|_{M^{\perp}} = \|x - m_0\|$.

(c) $\Rightarrow$ (a) Suppose $\|x - m_0\|_{M^{\perp}} = \|x - m_0\|$. Then for every $f \in U(X') \cap M^{\perp}$, $|f(x - m)| \leq \|x - m\|$ for all $m \in M$ and since $f$ vanishes on $M$, for every $m \in M$,

$$\begin{aligned} \|x - m_0\| = \|x - m_0\|_{M^{\perp}} &= \|x - m\|_{M^{\perp}} \\ &= \sup\left\{|f(x)| = |f(x - m)| : f \in M^{\perp} \cap U(X')\right\} \leq \|x - m\| \end{aligned}$$

Therefore $m_0$ is an $M$-minimal distance vector for $x$. ■

For $f \in SU(X')$, let $N(f)$ denote the null space of $f$. Clearly $f \in N(f)^{\perp} \cap SU(X')$. If $x$ is a maximal element for $f$, then 0 is an $N(f)$-minimal distance vector for $x \notin M$ by Theorem 16.3.11(a,b). We summarize this below for future reference.

**Corollary 16.3.12** Let $X'$ be the dual of a normed space $X$. Suppose $f \in SU(X')$ and $N(f)$ denotes the null space of $f$. Then $x$ is a maximal element for $f$ iff 0 is an $N(f)$-minimal distance vector for $x$.

Next, we make some connections between best approximation for the null space $N(f)$ of a continuous linear functional $f$ and maximal elements for $f$ on the way to showing that every subspace of $X$ is semi-Chebyshev iff $X$ is strictly convex [Theorem 16.3.15].

**Theorem 16.3.13** MAXIMAL ELEMENTS Let $N(f)$ be the null space of a continuous linear functional $f$ on a normed space $X$. For any $g \in X'$, note that $g \in N(f)^{\perp} \Leftrightarrow N(f) \subset N(g) \Leftrightarrow g = af$ for some $a \in \mathbf{F}$; if $g = af$ and $a \neq 0$, then $N(f) = N(g)$. If $f$ has a maximal element then any scalar multiple $g = af$, $a \in \mathbf{F}$, has a maximal element.

**Proof.** Let $X$ be a normed space, let $f \in X'$ and let $x \in X$ be a unit vector such that $f(x) = \|f\|$. Suppose $g = af$ for some nonzero $a \in \mathbf{F}$, the result being trivial for $a = 0$. Then $\|g\| = |a| \cdot \|f\|$ and $g((\overline{a}/|a|)x) = |a| f(x) = |a| \|f\|$; therefore $(\overline{a}/|a|)x$ is a maximal element for $g$. ■

**Theorem 16.3.14** PROXIMINAL NULL SPACE Let $X$ be a normed space and let $f$ be a nontrivial continuous linear functional with null space $N(f)$. Then:

(a) $N(f)$ is proximinal iff $f$ has a maximal element and
(b) $N(f)$ is Chebyshev iff $f$ has a unique maximal element.

**Proof.** We use the notation of the statement. (a) Suppose that $0 \neq f \in X'$ has a maximal element $w \in X$. By Theorem 16.3.13, we may assume that $\|f\| = 1 = f(w) = \|w\|$. For any $x \notin N(f)$, there exists a nonzero $a \in \mathbf{F}$ and $y \in N(f)$ such that $x = aw + y$. Now $(\overline{a}/|a|)f(x-y) = (\overline{a}/|a|)f(aw) = |a| = \|aw + y - y\| = \|x - y\|$. Let $g = (\overline{a}/|a|)f$. Since $g(x-y) = \|x-y\|$, $y$ is an $N(g)$-minimal distance vector for $x$ [Theorem 16.3.11(a,b)]. Since $N(f) = N(g)$, $y$ is an $N(f)$-minimal distance vector for $x$ as well. Since $x$ is arbitrary, $N(f)$ is proximinal.

Conversely, suppose that $N(f)$ is proximinal, that $x \notin N(f)$ and that $m_0 \in M$ is an $N(f)$-minimal distance vector for $x$. By Theorem 16.3.11(a,b), there exists $g \in N(f)^{\perp} \cap SU(X')$ such that $g(x - m_0) = \|x - m_0\|$; hence $(x - m_0)/\|x - m_0\|$ is a maximal element for $g$. Since $N(f) \subset N(g)$, it follows from Theorem 16.3.13 that $f$ has a maximal element.

(b) Suppose that $N(f)$ is Chebyshev. Since $f \neq 0$, it suffices to consider $f \in SU(X')$. Since $N(f)$ is Chebyshev, $f$ has a maximal element by Theorem 16.3.11. If $x$ and $y$ are maximal elements for $f$ then $f(x) = f(y) = 1$. Hence $y - x = m \in N(f)$. Since $f(y) = \|y\|$, $0$ is an $N(f)$-minimal distance vector for $y$ [Corollary 16.3.12]. Since $f(y-m) = f(x) = 1 = \|x\| = \|y - m\|$, $m$ is an $N(f)$-minimal distance vector for $y$ as well. Since $N(f)$ is Chebyshev, $m = 0$ and $x = y$.

Conversely, suppose that $N(f)$ is not Chebyshev. Then, for some $x \notin N(f)$, there exist distinct $m_0, m_1 \in N(f)$ such that $\|x - m_0\| = \|x - m_1\| = d(x, N(f))$. By Theorem 16.3.11(a,b), there exists $g \in N(f)^{\perp} \cap SU(X')$ such that $g(x) = \|x - m_0\| = \|x - m_1\| = d(x, N(f))$. Since $g \in N(f)^{\perp}$, $N(f) \subset N(g)$ and there exists $a \in \mathbf{F}$ such that $g = af$. Since $g \neq 0$, $a \neq 0$ and therefore $N(f) = N(g)$. Since $m_0, m_1 \in N(f) = N(g)$, $g(x - m_0) = g(x - m_1) = g(x) = \|x - m_0\| = \|x - m_1\|$. Since $m_0 \neq m_1$, it follows that $(x - m_0)/\|x - m_0\|$ and $(x - m_1)/\|x - m_1\|$ are distinct maximal elements for $g$, hence also distinct maximal elements for $f = a^{-1}g$. ■

We show next that all subspaces of $X$ are semi-Chebyshev iff $X$ is strictly convex.

**Theorem 16.3.15** Strictly Convex iff All Subspaces Semi-Chebyshev Let $X$ be a normed space. Then:

(a) If $X$ is strictly convex then every closed convex subset of $X$ is semi-Chebyshev.

(b) Conversely, if every one-dimensional subspace $\mathbf{F}x$ $(x \neq 0)$ is semi-Chebyshev then $X$ is strictly convex.

Consequently, $X$ is strictly convex iff every closed subspace is semi-Chebyshev.

**Proof.** (a) Let $M$ be a closed convex subset of a strictly convex space $X$, $x \in X$, and $m_1, m_2 \in M$ be best approximations to $x$. Then $d \doteqdot d(x, M) = \|x - m_1\| = \|x - m_2\|$. Since $d \leq \|x - (m_1 + m_2)/2\|$, it follows that $2d \leq \|2x - (m_1 + m_2)\| \leq 2d$. This implies that $d = \|x - (m_1 + m_2)/2\|$, i.e., that $(m_1 + m_2)/2$ is a point on the surface of the closed ball $C(x, d)$. It follows from the strict convexity of $X$ that $m_1 = m_2$ [Theorem 16.1.4(c)].

(b) If $X$ is not strictly convex, there exists $f \in SU(X')$ with distinct maximal elements $x$ and $y$ [Theorem 16.1.7]. If $M = \mathbf{R}(x - y)$ then $f \in M^{\perp}$ since $f(x - y) = 1 - 1 = 0$. By Theorem 16.3.11, since $f(y) = 1 = \|y\|$, 0 is an $M$-minimal distance vector for $y$. Since $f(y - (y - x)) = f(x) = 1 = \|x\| = \|y - (y - x)\|$, $y - x$ is a nonzero $M$-minimal distance vector for $y$. Therefore $M$ is not semi-Chebyshev. ■

### 16.3.1 Best Approximation in $C(T, \mathbf{F}, \|\cdot\|_{\infty})$

In Theorems 16.3.16 and 16.3.17 we characterize minimal distance vectors using extreme points of the closed unit ball $U(X')$ of $X'$. Extreme points of $U(X')$ exist as a consequence of the Alaoglu and Krein-Milman theorems [8.4.1 and 9.4.1]. We use those results to obtain a characterization of minimal distance vectors in $C(T, \mathbf{F}, \|\cdot\|_{\infty})$ [Theorem 16.3.18].

**Theorem 16.3.16** MINIMAL DISTANCE VECTORS AND EXTREME POINTS Let $M$ be a closed subspace of the normed space $X$ and $m_0$ an $M$-minimal distance vector for $x \notin M$. Then there exists $f_0 \in SU(X')$ with the following properties:

(a) $f_0$ is an extreme point of the unit ball $U(X')$ of $X'$;

(b) $\operatorname{Re} f_0(m_0) \geq 0$;

(c) $f_0(x - m_0) = \|x - m_0\|$.

**Proof.** We use the notation of the statement. Let $m_0$ be an $M$-minimal distance vector for $x \in X$. By Theorem 16.3.11(a,b),

$$K = \{f \in SU(X') : f(x - m_0) = \|x - m_0\|\} \neq \varnothing$$

$K$ is a $\sigma(X', X)$-closed convex subset of the $\sigma(X', X)$-compact set $U(X')$ [Alaoglu, 8.4.1]; therefore it has extreme points by the Krein-Milman theorem [9.4.1]. Moreover, $K$ is a face of $U(X')$ by the following argument. Suppose $f \in K$ is a proper convex combination, $f = af_1 + (1 - a)f_2$, of $f_1, f_2 \in U(X')$ and some $a \in (0, 1)$. Then

$$\|x - m_0\| = f(x - m_0) = af_1(x - m_0) + (1 - a)f_2(x - m_0)$$

Since $\|x - m_0\|$ is an extreme point of the circle $C(0, \|x - m_0\|) \subset \mathbf{F}$ and $f_1(x - m_0), f_2(x - m_0) \in C(0, \|x - m_0\|)$, it follows from Theorem 9.2.2(c) that $\|x - m_0\| = f_1(x - m_0) = f_2(x - m_0)$, i.e., that $f_1, f_2 \in K$. Consider the map $h : K \to \mathbf{F}$, $f \mapsto f(m_0)$. When $K$ carries the $\sigma(X', X) \cap K$ topology,

$h$ is clearly continuous. Thus $h(K) \subset \mathbf{F}$ is compact; it is easy to verify that it is also convex. Clearly $0 \in h(K)$ because, by Theorem 16.3.11(a,b), there exists $f \in K \cap M^{\perp}$ and therefore $f(m_0) = 0$. Hence $r = \sup \operatorname{Re} h(K) \geq 0$. As $h(K)$ is closed, there exists $a_1 \in h(K)$ such that $r = \operatorname{Re} a_1$. When $\mathbf{F} = \mathbf{C}$, let $L$ be the vertical line in $\mathbf{F}$ that intersects the real axis at $(r, 0)$. Let $b = \sup \{\operatorname{Im} a : a \in h(K), \operatorname{Re} a = r\}$. Since $h(K)$ is closed, $b = \operatorname{Im} a_2$ for some $a_2 \in h(K)$ for which $\operatorname{Re} a_2 = r$. The point $a_2 = r + ib$ is clearly an extreme point of $h(K)$. For some $g \in K$, $a_2 = h(g) = g(m_0) = r + ib$. Now $h^{-1}(\{r + ib\})$ is a closed convex face of $K$ and therefore has an extreme point $f_0$ [Krein-Milman, 9.4.1]; $f_0$ is an extreme point of $K$ by Example 9.3.2(e). Likewise, since $K$ is a face of $U(X')$, $f_0$ is an extreme point of $U(X')$. If $F = \mathbf{R}$, $r = \max h(K) \geq 0$. In this case $r$ itself is an extreme point of $h(K)$ and the rest of the argument goes through unchanged. ■

In Theorem 16.3.17 we consider the existence of a family of continuous linear functionals satisfying conditions similar to those of Theorem 16.3.16.

**Theorem 16.3.17** MINIMAL DISTANCE VECTORS AND EXTREME POINTS Let $M$ be a closed subspace of the normed space $X$. Then $m_0$ is an $M$-minimal distance vector for $x \notin M$ iff for each $m \in M$ there exists $f_m \in SU(X')$ such that

(a) $f_m \in \operatorname{ext} U(X')$;
(b) $\operatorname{Re} f_m(m_0 - m) \geq 0$; and
(c) $f_m(x - m_0) = \|x - m_0\|$.

**Proof.** Let $M$ be a closed subspace of the normed space $X$. Suppose $m_0$ is an $M$-minimal distance vector for $x \notin M$. Then, for any $m \in M$, $m_0 - m$ is an $M$-minimal distance vector for $x - m$ because for any $m' \in M$,

$$\|(x - m) - (m_0 - m)\| = \|x - m_0\| \leq \|(x - m) - m'\|$$

Hence, by Theorem 16.3.16, for each $m \in M$ there exists $f_m \in X'$ such that:

(a) $f_m$ is an extreme point of $U(X')$;
(b) $\operatorname{Re} f_m(m_0 - m) \geq 0$; and
(c) $f_m((x - m) - (m_0 - m)) = f_m(x - m_0) = \|(x - m) - (m_0 - m)\| = \|x - m_0\|$.

Conversely, suppose that $\{f_m \in X' : m \in M\}$ satisfies (a)–(c). Then, for each $m \in M$,

$$\begin{aligned} \|x - m_0\| &= f_m(x - m_0) \leq f_m(x - m_0) + \operatorname{Re} f_m(m_0 - m) \\ &= \operatorname{Re} f_m(x - m) \leq |f_m(x - m)| \leq \|x - m\| \end{aligned}$$

Hence $\|x - m_0\| = d(x, M)$. ■

What are the $M$-minimal distance vectors in $C(T, \mathbf{F}, \|\cdot\|_{\infty})$? First, recall [Example 9.5.13] that the extreme points of $U\left(C(T, \mathbf{F}, \|\cdot\|_{\infty})'\right)$ are evaluation maps $t'$ at points $t \in T$ multiplied by scalars $a \in \mathbf{F}$ of magnitude 1.

**Theorem 16.3.18** MINIMAL DISTANCE VECTORS IN $C(T, \mathbf{F}, \|\cdot\|_\infty)$ Let $T$ be a compact Hausdorff space and let $M$ be a closed subspace of $X = C(T, \mathbf{F}, \|\cdot\|_\infty)$. Then $m_0 \in M$ is an $M$-minimal distance vector for $x \notin M$ iff:

(a) for each $m \in M$ there exists $t_m \in T$ such that

$$\operatorname{Re}\left[\overline{(x(t_m) - m_0(t_m))}m(t_m)\right] \geq 0 \text{ and}$$

(b) $|x(t_m) - m_0(t_m)| = \|x - m_0\|_\infty$.

**Proof.** We use the notation of the statement. By Theorem 16.3.17, $m_0$ is an $M$-minimal distance vector for $x$ iff, for each $m \in M$, there exists $t_m \in T$ and $a_m \in \mathbf{F}$ $(|a_m| = 1)$ such that:

(a) $\operatorname{Re} a_m [t'_m(m_0 - m)] = \operatorname{Re} a_m [m_0(t_m) - m(t_m)] \geq 0$ and

(b) $a_m [x(t_m) - m_0(t_m)] = \|x - m_0\|_\infty$.

Since $|a_m| = 1$ and $a_m [x(t_m) - m_0(t_m)] = \|x - m_0\|_\infty > 0$,

$$a_m = \frac{\overline{x(t_m) - m_0(t_m)}}{|x(t_m) - m_0(t_m)|}$$

Substituting this value for $a_m$ into (a) and (b), we get that for each $m \in M$ there exists $t_m \in T$ such that:

(a′) $\operatorname{Re} \overline{(x(t_m) - m_0(t_m))}(t'_m(m_0 - m)) \geq 0$ and

(b′) $|x(t_m) - m_0(t_m)| = \|x - m_0\|_\infty$.

But as $m_0 \in M$, $\{m_0 - m : m \in M\} = M$; therefore, by Theorem 16.3.17, $m_0$ is an $M$-minimal distance vector for $x$ iff for all $m \in M$ there exists $t_m \in T$ such that

(a″) $\operatorname{Re}[\overline{(x(t_m) - m_0(t_m))}m(t_m)] \geq 0$ and

(b″) $|x(t_m) - m_0(t_m)| = \|x - m_0\|_\infty$. ■

## 16.4 UNIQUENESS OF HB EXTENSIONS

Let $f$ be a linear functional defined on a subspace $M$ of a vector space $X$. We proved the existence of dominated and norm-preserving extensions of $f$ in Theorems 7.3.2 and 7.4.1. We discuss uniqueness of such extensions in this section. As we shall see, the existence of unique norm-preserving extensions is closely linked to the smoothness of $X$.

### 16.4.1 Dominated Extensions

Let $M$ be a subspace of a vector space $X$. In proving the dominated Hahn–Banach extension theorem [7.3.2(b)], we showed that the ability to extend a real linear functional $f : M \to \mathbf{R}$ dominated by a sublinear functional $p$, $f \leq p$, to a real linear functional $F : M \to \mathbf{R}$, $F \leq p$, implied that you could extend a complex linear functional $f : M \to \mathbf{C}$, $|f| \leq p$, $p$ a seminorm to a complex linear functional $F : M \to \mathbf{R}$, $|F| \leq p$. The technique: Extend $\operatorname{Re} f$

and use the fact that $f(x) = \operatorname{Re} f(x) - i \operatorname{Re} f(ix)$ [Theorem 1.4.2]. Clearly, if $f$ has a an extension $F$ with $|F| \leq p$ then $\operatorname{Re} f$ is extended by $\operatorname{Re} F$ and $\operatorname{Re} f \leq p$; $F$ is unique iff $\operatorname{Re} F$ is. We summarize these observations below:

**Theorem 16.4.1** DOMINATED COMPLEX EXTENSIONS Let $M$ be a subspace of a vector space $X$ over $\mathbf{C}$, let $p : X \to \mathbf{R}$ be a seminorm and $f : M \to \mathbf{C}$ a linear functional such that $|f| \leq p$. Then $f$ has a unique linear extension $F : X \to \mathbf{C}$ such that $|F| \leq p$ iff $\operatorname{Re} f$ has a unique linear extension $G : X \to \mathbf{R}$ such that $G \leq p$.

In the proof of the key lemma to the Hahn–Banach theorem, the one-dimensional dominated Hahn–Banach extension theorem [7.3.1, specifically Ineq. (7.1)], to extend a real linear functional $f$ from a subspace $M$ of a real vector space $X$ to $M + \mathbf{R}x$ $(x \notin M)$, we chose a number $c$ such that

$$\sup_{m \in M} [-p(-x - m) - f(m)] \leq c \leq \inf_{m \in M} [p(x + m) - f(m)]$$

There is only one choice for $c$ iff

$$\sup_{m \in M} [-p(-x - m) - f(m)] = \inf_{m \in M} [p(x + m) - f(m)] \tag{16.1}$$

In other words, the extension to $M \oplus \mathbf{R}x$ is unique iff Eq. (16.1) is satisfied. This leads to the following criteria for there to be a unique dominated extension of a linear functional defined on a subspace.

**Theorem 16.4.2** UNIQUE DOMINATED EXTENSION Let $f$ be a linear functional defined on a subspace $M$ of a vector space $X$. Then:

(a) REAL If $X$ is real, $p$ is a sublinear functional and $f \leq p$ on $M$ then $f$ has a unique extension $F$ to $X$, $F \leq p$, iff for every $x \notin M$

$$\sup_{m \in M} [-p(-x - m) - f(m)] = \inf_{m \in M} [p(x + m) - f(m)] \tag{16.2}$$

(b) COMPLEX If $X$ is complex, $p$ is a seminorm and $|f| \leq p$ on $M$ then $f$ has a unique extension $F$ to $X$, $|F| \leq p$, iff, for every $x \notin M$,

$$\sup_{m \in M} [-p(-x - m) - \operatorname{Re} f(m)] = \inf_{m \in M} [p(x + m) - \operatorname{Re} f(m)] \tag{16.3}$$

**Proof.** (a) With $f$ and $M$ as above, suppose that $F$ is an extension of $f$ to $X$ with $F \leq p$. Then, for $x \notin M$ and any $m \in M$, $F(x) = F(x + m) - F(m) \leq p(x + m) - F(m)$ which implies that $F(x) \leq \inf_{m \in M} [p(x + m) - f(m)]$. Similarly, $\sup_{m \in M} [-p(-x - m) - f(m)] \leq F(x)$ and therefore

$$\sup_{m \in M} [-p(-x - m) - f(m)] \leq F(x) \leq \inf_{m \in M} [p(x + m) - f(m)]$$

Hence if Eq. (16.2) holds for every $x \notin M$, there is only one choice for $F(x)$ for any $x \notin M$, namely

$$F(x) = \sup_{m \in M} [-p(-x-m) - f(m)] = \inf_{m \in M} p(x+m) - f(m)$$

hence only one dominated extension of $f$ to $X$. If Eq. (16.2) does not hold for some $x \notin M$, there is more than one choice for $F(x)$, hence more than one extension of $f$ to $M \oplus \mathbf{R}x$ as in Theorem 7.3.1 and, ultimately, more than one extension of $f$ to to $X$.

(b) Follows from Theorem 16.4.1. ■

Bandyopadhyay and Roy 2003 characterize when a single linear functional dominated by a sublinear functional $p$ on a subspace $M$ of a real vector space $X$ has a unique extension to $X$ dominated by $p$ in terms of nested sequences of "$p$-balls" in a quotient space; by considering the canonical embedding of $M$ in its bidual $M''$, they characterize unique extendibility of elements of $M'$ in terms of sequences from $M$.

### 16.4.2 Norm-Preserving Extensions

We consider uniqueness of extensions of continuous linear functionals $f : M \to \mathbf{F}$ defined on a subspace $M$ of a normed space $X$ in this section. We only consider norm-preserving extensions because without the preservation-of-norm requirement, the answer is simple: there are no unique extensions—there are always continuous extensions of greater norm [Theorem 7.4.3]. The passage from the real to the complex case is easy because a continuous complex linear functional $f$ defined on a subspace of a normed space has a unique extension of the same norm iff its real part does—for every $x \in X$, $f(x) = \operatorname{Re} f(x) - i \operatorname{Re} f(ix)$ and $\|f\| = \|\operatorname{Re} f\|$ by Theorems 1.4.2 and 5.7.7(d), respectively.

Consider the subspace $M = \mathbf{R} \subset \ell_2^{\mathbf{R}}(2)$ and the linear functional $f(a, 0) = a$ defined on $M$. Let $y \in \mathbf{R}^2$ be a unit vector of angle $\beta \neq 0, \pi$ with the $x$-axis. The subspace $N = \mathbf{R}y$ is a topological complement of $M$ [Theorem 4.9.5]. For any such $N$, the projection $P_M$ on $M$ along $N$ is a continuous extension of $f$ and $f \circ P_M$ is a continuous extension of $f$ of norm $|\csc \beta|$, so there are infinitely many continuous extensions of $f$ but only one of the same norm ($\|f\| = 1$), namely when $\beta = \pm\pi/2$ (when $N = M^{\perp}$); this, incidentally, yields the extension of $f$ of minimal norm, the smallest value of $|\csc \beta|$.

**Definition 16.4.3** UNIQUE EXTENSION PROPERTY We say that a closed subspace $M$ of a normed space $X$ has the *unique extension property* if every $f \in M'$ has a unique extension $F \in X'$ of the same norm. If every closed subspace has the unique extension property, we say that $X$ has the *unique extension property*. ■

There are more Banach spaces with the unique extension property than you might think—any $L_p$, $1 < p < \infty$, for example has the unique extension property. In the following examples of spaces that lack the unique extension property, note that the unit ball has sharp corners.

**Example 16.4.4** NON-UNIQUE EXTENSIONS (a) TWO EXTENSIONS For the subspace $M = \mathbf{R}$ of $\ell_1^{\mathbf{R}}(2)$ or $\ell_\infty^{\mathbf{R}}(2)$ and the continuous linear functional $f(a,0) = a$ defined on $M$, distinct extensions of the same norm are given by $F(a,b) = a + b$ and $G(a,b) = a - b$.

(b) INFINITELY MANY Consider the subspace $M$ of constant functions of the Banach space $C([0,1], \mathbf{F}, \|\cdot\|_\infty)$ of $\mathbf{F}$-valued continuous functions on $[0,1]$ and the continuous linear functional $f : M \to \mathbf{F}$, $x \mapsto x(0)$. Clearly $\|f\| = 1$. For any $t \in [0,1]$, the evaluation map $t' : C[0,1] \to \mathbf{F}$, $x \mapsto x(t)$, extends $f$ and is of norm 1. ■

Sharp corners of the unit ball do not eliminate the possibility of unique extensions. For example, the subspace $c_0$ of $\ell_\infty$ has the unique extension property [Example 16.4.11(e)] and, for $X = C(T, \mathbf{F}, \|\cdot\|_\infty)$ where $T$ is a compact Hausdorff space then, for each $t \in T$, the maximal ideal $M_t = \{x \in C(T, \mathbf{F}, \|\cdot\|_\infty) : x(t) = 0\}$ has the unique extension property [Example 16.4.11(b)].

**Example 16.4.5** HILBERT SPACE HAS UNIQUE EXTENSION Suppose $f$ is a continuous linear functional defined on a closed subspace $M$ of a Hilbert space $(X, \langle\cdot,\cdot\rangle)$. Any closed subspace $M$ of a Hilbert space $(X, \langle\cdot,\cdot\rangle)$ is complemented by its orthogonal complement $M^\perp = \{m^\perp \in X : \langle m, m^\perp\rangle = 0$ for all $m \in M\}$ [Bachman and Narici 2000, p. 172]. By Theorem 4.9.2(a) the orthogonal projection $P_M : X = M \oplus M^\perp \to M$, $m + m^\perp \mapsto m$, is continuous; hence $F = f \circ P_M$ is a continuous extension of $f$. Therefore, by the Riesz representation theorem (quoted in Example 15.7.1), there exist unique $m_0 \in M$ and $m_0^\perp \in M^\perp$ such that $F(\cdot) = \langle\cdot, m_0 + m_0^\perp\rangle$ and $\|F\| = \|m_0 + m_0^\perp\|$. Since $0 = F(m_0^\perp) = \|m_0^\perp\|^2$, it follows that $m_0^\perp = 0$. Hence $\|F\| = \|m_0\| = \|f\|$. If $G$ is any extension of $f$ of the same norm, then $G(\cdot) = \langle\cdot, m_0 + m^*\rangle$ where (since $G$ extends $f$) $m_0$ is as above and $m^* \in M^\perp$. Since $f = \langle\cdot, m_0\rangle$. $\|f\| = \|m_0\| = \|G\|$. Since $\|G\|^2 = \|m_0\|^2 + \|m^*\|^2$, $m^* = 0$ and $G(\cdot) = \langle\cdot, m_0\rangle = F(\cdot)$. ■

The situation for certain subspaces of $\ell_p$, $1 < p < \infty$, is similar. Let $\{e_n\}$ be the standard basis for $\ell_p$, let $S$ be a subset of $\mathbf{N}$ and let $f$ be a continuous linear functional defined on the closed linear span $M$ of $\{e_n : n \in S\}$. By Example 15.7.2, for $q = p/(p-1)$, $\|f\|^q = \sum_{n\in S} |f(e_n)|^q$. Given $(a_n) \in \ell_p$ and an extension $F((a_n)) = \sum_{n\in S} a_n f(e_n) + \sum_{n\notin S} a_n F(e_n)$ of $f$ of the same norm then $\|f\|^q = \|F\|^q$ implies that $F(e_n) = 0$ for all $n \notin S$.

Unique extendibility is generally related to "smoothness" of the unit ball $U(X)$ meaning that at each point $x \in SU(X)$, there is a unique supporting hyperplane at $x$ [Def. 7.8.9]. The first hint at the connection is Theorem 16.4.6.

**Theorem 16.4.6** ATTAINS NORM AT SMOOTH POINT $\Rightarrow$ UNIQUE EXTENSION If a nonzero continuous linear functional $f$ defined on a closed subspace $M$ of the normed space $X$ over $\mathbf{F}$ attains its norm at a smooth point $u$, $\|u.\| = 1$, of $U(X)$ then $f$ has a unique continuous extension of the same norm to $X$.

**Proof.** Let $M$ be a closed subspace of the normed space $X$ and let $f$ be a nonzero element of $M'$ and $u$ be a unit vector such that $f(u) = \|f\|$. We may assume that $\|f\| = 1$. Extend $f$ to $F \in X'$ where $\|F\| = 1$ by the Hahn–Banach theorem [7.4.1]. Since $u$ is a smooth point and $H = F^{-1}(1)$ clearly supports $U(X)$ at $u$, $F$ is unique by Theorem 7.8.11. ■

The seminal result characterizing the unique extension property is that of Phelps 1960a [Theorem 16.4.7 below], a theorem that is remarkable for its simplicity. Phelps referred to what we call the *unique extension property* of a subspace as *property* $U$; Lima [1983], Sullivan [1977] and others call such subspaces *Hahn–Banach smooth.*

**Theorem 16.4.7** $M$ HAS UNIQUE EXTENSION IFF $M^{\perp}$ CHEBYSHEV A closed subspace $M$ of a normed space $X$ over $\mathbf{F}$ has the unique extension property iff $M^{\perp}$ is Chebyshev. Consequently, $X$ has the unique extension property iff for each closed subspace $M$ of $X$, $M^{\perp}$ is Chebyshev.

**Proof.** Let $X$ be a normed space. *Necessity.* Let $M$ be a closed subspace of $X$ with the unique extension property. $M^{\perp}$ is proximinal by Theorem 16.3.9(a) so we need only show that best approximations are unique. To that end, suppose $f \in X' \setminus M^{\perp}$ has distinct best approximations $g_1, g_2 \in M^{\perp}$, let $h_1 = f - g_1$ and $h_2 = f - g_2$. Since $h_1 = f = h_2$ on $M$, they are distinct extensions of $h_1|_M = h_2|_M$ of the same norm—$\|f - g_1\| = \|f - g_2\| = d(f, M^{\perp})$. Since $M$ has the unique extension property, this is a contradiction.

*Sufficiency.* Let $M$ be a closed subspace, suppose that $M^{\perp}$ is Chebyshev and consider $f \in M'$. We may assume that $f \neq 0$ and that $\|f\| = 1$. Suppose $g, h \in SU(X') = \{\varphi \in X' : \|\varphi\| = 1\}$ are continuous extensions of $f$ of the same norm. Clearly $g - h \in M^{\perp}$. Since $d(g, M^{\perp}) = \|g|_M\| = \|h|_M\|$ [Theorem 16.3.9(b)],

$$1 = \|g\| = \|g - 0\| = \|g - (g - h)\| \geq d(g, M^{\perp}) = \|g|_M\| = 1$$

It follows that 0 and $g - h$ are best approximations to $g$ from $M^{\perp}$. Since $M^{\perp}$ is Chebyshev, $g = h$. ■

Phelps's theorem was generalized by Park [1993].

It follows from Theorem 16.3.4 that every closed subspace of a Hilbert space $X$ is Chebyshev. Since $X$ is conjugate-linearly isometric to $X'$ for

Hilbert spaces, the same is true for $X'$. Therefore this is another way of seeing that Hilbert spaces have the unique extension property. Since every closed subspace of a uniformly convex space is Chebyshev [Theorem 16.3.7], any Banach space with a uniformly convex dual has the unique extension property.

Taylor [1939] proved that if the dual $X'$ of a Banach space $X$ is strictly convex then $X$ has the unique extension property. He proved the converse for reflexive $X$. Foguel [1958] removed the reflexivity, thereby showing that the unique extension property was equivalent to strict convexity of the dual. We use Phelps's theorem to prove this in Theorem 16.4.8.

**Theorem 16.4.8** UNIQUE EXTENSION IFF STRICTLY CONVEX DUAL A normed space $X$ over $\mathbf{F}$ has the unique extension property iff $X'$ is strictly convex.

**Proof.** Suppose that $X'$ is strictly convex and that $M$ is a closed subspace of $X$. By Theorem 16.3.9(a), $M^{\perp}$ is proximinal. Since $X'$ is strictly convex, each closed subspace is semi-Chebyshev [Theorem 16.3.15(a)]; therefore $M^{\perp}$ is Chebyshev. Hence $M$ has the unique extension property by Phelps's theorem [16.4.7]. Conversely, suppose that $X$ has the unique extension property. We use the criterion of Theorem 16.3.15(b) to show that $X'$ is strictly convex by showing that every 1-dimensional subspace $M = \mathbf{F}f$, $f \neq 0$, of $X'$ is semi-Chebyshev. Clearly $M^{\perp} = N(f)$. Since $X$ has the unique extension property and $M$ is closed, $M^{\perp\perp} = M$ is Chebyshev by Phelps's theorem [16.4.7]. ■

### 16.4.3 HB-Subspaces

$X$ denotes a Banach space over $\mathbf{F}$ in this subsection. $L(X, X)$ and $K(X, X)$ denote, respectively, the spaces of all continuous linear operators and compact operators $A \in L(X, X)$ [i.e., $\operatorname{cl} A(U(X))$ is compact] of $X$ into $X$.

A sufficient condition for a subspace to have the unique extension property is that it be an HB-subspace; an important subclass of the HB-subspaces is collection of the $M$-ideals.

**Definition 16.4.9** HB-SUBSPACES AND M-IDEALS Let $M$ be a closed subspace of a Banach space $X$ such that there is a continuous (norm-one, actually) projection $P$ on $X'$ such that $P^{-1}(0) = M^{\perp}$.

(a) HB-SUBSPACES $M$ is an *HB-subspace* if for each $f \in X'$,

$$f \neq Pf \Rightarrow \|Pf\| < \|f\| \text{ and } \|f - Pf\| \leq \|f\| \tag{HB}$$

(b) M-IDEALS $M$ is an *M-ideal*.if, instead of (HB),

$$f \neq Pf \Rightarrow \|f\| = \|f - Pf\| + \|Pf\| \tag{M}$$

Clearly (M) $\Rightarrow$ (HB) so every $M$-ideal is an HB-subspace. $M$-ideals were invented by Alfsen and Effros [1972; cf. Ando 1973, p. 395]. They are explored at length in Behrends 1979 and Harmand et al. 1993.

**Theorem 16.4.10** HB-SUBSPACE $\Rightarrow$ UNIQUE EXTENSION If $M$ is an HB-subspace of the Banach space $X$ then $M$ has the unique extension property.

**Proof.** Let $M$ be an HB-subspace of the Banach space $X$ and let $P$ be a projection on $X'$ that satisfies (HB). We first show that $\|P\| = 1$. Since $P$ is a projection, $P^2 = P$ and therefore $\|P\| \geq 1$. If there exists $f \in X'$, $\|f\| = 1$, such that $\|Pf\| > 1$ and $f \neq Pf$ then $1 < \|Pf\| < \|f\| = 1$, a contradiction. Thus, for any $f$, $\|f\| = 1$, such that $\|Pf\| > 1$, $f = Pf$ which yields the contradiction $1 = \|f\| = \|Pf\| > 1$. Therefore $\|P\| = 1$. Now suppose that $F \in X'$ is a norm-preserving extension of $f \in M'$. Since $P^2 = P$, $F - PF \in P^{-1}(0) = M^{\perp}$ so, for any $m \in M$, $F(m) = PF(m)$. Hence

$$\|F\| = \|f\| = \|F|_M\| = \sup_{m \in SU(M)} |PF(m)| \leq \|PF\|$$

If $F \neq PF$, then $\|PF\| < \|F\|$ which is contradictory; hence $F = PF$ for any norm-preserving extension $F$ of $f$. If $G \in X'$ is a norm-preserving extension of $f$ then $F - G \in M^{\perp} = P^{-1}(0)$ so $PF = PG = G = F$. ■

Hennefeld [1979] introduced the notion of HB-subspaces and asked if they were the only subspaces with the unique extension property. Oja [1984] showed that they were not—for $X = \mathbf{R}^2$ normed by

$$\|(a, b)\| = \max(|a|, |a + b|/2)$$

the subspace $\mathbf{R}$ has the unique extension property but is not an HB-subspace. Oja [1997] gets some equivalent conditions for $M$ to be an HB-subspace and also shows that $X$ is an HB-subspace of its bidual whenever the compact operators $K(X, X)$ form an HB-subspace of $L(X, X)$.

Clearly $\{0\}$ and $X$ are $M$-ideals of a Banach space $X$, but here are some nontrivial examples. Since each $M$-ideal is an HB-subspace, each has the unique extension property.

**Example 16.4.11** M-IDEALS (a) If $X$ is a B*-algebra then any closed 2-sided ideal in $X$ is an $M$-ideal [Smith and Ward 1978, p. 347].

(b) By (a), spaces $C(T, \mathbf{F}, \|\cdot\|_\infty)$ of continuous functions on a compact set $T$ have an abundance of $M$-ideals: For any $t \in T$, the maximal ideal $M_t = \{x \in C(T, \mathbf{F}) : x(t) = 0\}$ is an $M$-ideal.

(c) If $T$ is locally compact and Hausdorff, the $M$-ideals of $C_\infty(T, \mathbf{F}, \|\cdot\|_\infty)$, the continuous scalar-valued functions on $T$ that vanish at infinity, are precisely $M_F = \{x \in C_\infty(T, \mathbf{F}) : x(F) = \{0\}\}$ where $F$ is closed in $T$ [Alfsen and Effros 1972, p. 139; Behrends 1979, p. 40].

(d) For $X = L_2(\mu)$, for some measure $\mu$, the subspace $K(X, X)$ of compact operators is an $M$-ideal in $L(X, X)$.

(e) The subspace $c_0$ of null sequences is an $M$-ideal in $\ell_\infty$ [Harmand et al. 1993, Proposition 1.12].

(f) A strictly convex or smooth Banach space does not contain any nontrivial $M$-ideals [Holmes et al. 1975; Behrends 1978, p. 265]. Hence even though every subspace of the smooth space $\ell_p$ has the unique extension property, none of them is an $M$-ideal. ■

Because M-ideals are proximinal [Alfsen and Effros 1972, p. 120], they play an important role in approximation theory.

**Internal characterization of M-ideals** $M$-ideals may be characterized internally in various ways by intersection properties of balls. For real $X$, the closed subspace $M$ is an $M$-ideal iff $M$ satisfies the *3-ball property*, namely that if three open balls $B_1, B_2, B_3 \subset X$ have nonempty intersection and each meets $M$ then $M \cap (\cap_{i=1}^3 B_i) \neq \varnothing$ [Alfsen and Effros 1972; Behrends 1979, p. 46f.]. Behrends [1991] proved it for subspaces of complex spaces. The 3-ball property characterizes $M$-ideals, but the 2-ball property does not [Alfsen and Effros 1972, pp. 98, 120–122].

**Internal characterizations of unique extension property** We have characterized subspaces with the unique extension property so far in terms involving the dual space. Lima [1983] obtained the following internal characterization.

**Theorem 16.4.12** UNIQUE EXTENSION PROPERTY INTERNALLY A closed subspace $M$ of a Banach space $X$ has the unique extension property iff for each $r > 0$, $x \in X$ and sequence $(m_n)$ from $M$ such that for all $n \in \mathbf{N}$,

$$\|m_1\| \leq 1 + r \text{ and } \|m_{n+1} - m_n\| \leq 1 + r/2^{n+1}$$

there exists $m \in M$ and $n_0 \in \mathbf{N}$ such that $\|x - m \pm m_{n_0}\| \leq n_0 + 2r - r/2^{n_0}$.

Oja and Põldvere [1996, p. 291] show that $M$ has the unique extension property iff for every $x \in X$ and every increasing sequence of open balls $B_n$ with centers in $M$ and infinitely increasing radii such that $0 \in B_1$, there exists $m \in M$ such that $\pm(x + m) \in \cup_{n \in \mathbf{N}} B_n$.

For the sake of another internal characterization, consider the following condition.

**Definition 16.4.13** TWO-BALL SEQUENCE PROPERTY Let $M$ be a closed subspace of a Banach space. Consider increasing sequences of open balls $B_n = B(m_n, r_n)$ and $D_n = B(y_n, s_n)$ such that for each $n \in \mathbf{N}$, $B_n \subset B_{n+1}$, $D_n \subset D_{n+1}$, $m_{n+1} - m_n \in M$, $y_{n+1} - y_n \in M$, and $\lim_n r_n = \lim_n s_n = \infty$. If

$$(\cap_{n \in \mathbf{N}} B_n) \cap (\cap_{n \in \mathbf{N}} D_n) \neq \varnothing \Rightarrow M \cap (\cap_{n \in \mathbf{N}} B_n) \cap (\cap_{n \in \mathbf{N}} D_n) \neq \varnothing$$

then $M$ satisfies the *2-ball sequence property*. ■

Oja and Põldvere show [1999, p. 1258] show that a closed subspace $M$ has the unique extension property iff $M$ satisfies the 2-ball sequence property. Hence a Banach space $X$ has the unique extension property iff every closed subspace satisfies the 2-ball sequence property; equivalently this characterizes those $X$ with strictly convex dual [Theorem 16.4.8]. Costara and Popa [2001] give further examples of subspaces with the unique extension property.

## 16.5 STONE–WEIERSTRASS THEOREM

**Notation.** $T$ is a completely regular Hausdorff space. $C(T, \mathbf{R}, c)$ denotes the pointwise ordered lattice of real-valued continuous functions on $T$ endowed with the compact-open topology. $C(T, \mathbf{F}, c)$ stands for $\mathbf{F}$-valued functions with compact-open topology. For a compact subset $K$ of $T$ and $x \in C(T, \mathbf{F})$, $p_K(x) = \sup |x(K)|$. ■

In one of the great landmarks of analysis, Weierstrass [1885] proved that continuous functions on $[0, 1]$ could be uniformly approximated by polynomials. Stone [1937b, pp. 467–468] significantly generalized Weierstrass's result. We prove our version of the Stone–Weierstrass theorem on the density of self-adjoint subalgebras of $C(T, \mathbf{F}, c)$ in Theorem 16.5.7. In his excellent book on Banach spaces of bounded continuous functions, Semadeni [1971, pp. 118–119] discusses some of the history of the theorem and many of its generalizations.

The ability to approximate $x(t) = |t|$ on a closed interval by polynomials [Theorem 16.5.1] is a key to the proof of the Stone–Weierstrass theorem [16.5.7].

**Theorem 16.5.1** APPROXIMATION OF $x(t) = |t|$ For every $0 < r < 1$ and $k > 0$, there exists a polynomial $p \in C([-k, k], \mathbf{R}, c)$ such that $\sup_{t \in [-k,k]} |p(t) - |t|| < r$.

**Proof.** Let $r$ and $k$ be as above. If we can find a polynomial $p \in C([-1, 1], \mathbf{R})$ such that $|p(t) - |t|| < r$ for $|t| \leq 1$ then $||p(t/k)| - |t/k|| \leq r$ for $|t| \leq k$. Hence $|kp(t/k) - |t|| \leq rk$ for $|t| \leq k$. Thus, it suffices to assume that $k = 1$.

We start with the polynomial $p_0 = 0$ and inductively generate polynomials $p_n(t)$ that are uniformly closer to $x(t) = |t|$. For each $n \geq 0$, let $p_{n+1}(t) = p_n(t) + (1/2)\left[t^2 - p_n(t)^2\right]$. The first three polynomials are sketched with $|t|$ in the following diagram.

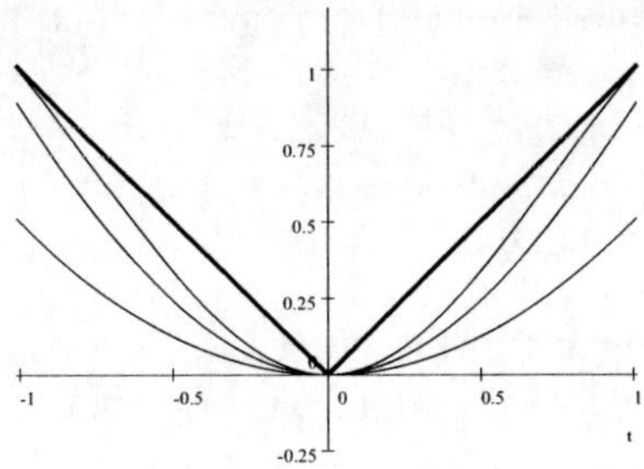


$|t|$, $p_1(t)$, $p_2(t)$ and $p_3(t)$

We prove by induction that $0 \leq p_n(t) \leq |t|$ for all $n \in \mathbf{N}$ and $|t| \leq 1$. This is clear for $n = 0$, so suppose that $0 \leq p_n(t) \leq |t|$ for $n \in \mathbf{N}$. This implies that $p_n(t)^2 \leq t^2$, so $p_{n+1}(t) = p_n(t) + (1/2)\left[t^2 - p_n(t)^2\right] \geq 0$. Since $p_n(t) \leq |t|$ then $0 \leq (1/2)(|t| + p_n(t)) \leq |t|$. Hence, for $|t| \leq 1$,

$$\begin{aligned} |t| - p_{n+1}(t) &= |t| - p_n(t) - (1/2)\left[|t|^2 - p_n(t)^2\right] \\ &= (|t| - p_n(t))\left[1 - (1/2)(|t| + p_n(t))\right] \geq 0 \end{aligned}$$

Therefore $0 \leq p_{n+1}(t) \leq |t|$ for $|t| \leq 1$ and shows that $0 \leq p_n(t) \leq |t|$ for all $n \in \mathbf{N}$.

Next, we show that

$$|t| - p_n(t) \leq |t|(1 - |t|/2)^n \text{ for all } n \in \mathbf{N} \text{ and } |t| \leq 1 \tag{$*$}$$

This is clear for $n = 0$ so for $n \in \mathbf{N}$ suppose that $|t| - p_n(t) \leq |t|(1 - |t|/2)^n$ for $|t| \leq 1$. Then

$$\begin{aligned} |t| - p_{n+1}(t) &= |t| - p_n(t) - (1/2)\left[t^2 - p_n(t)^2\right] \\ &= (|t| - p_n(t))\left[1 - (1/2)(|t| + p_n(t))\right] \end{aligned}$$

Since $p_n(t) \geq 0$, $1 - (1/2)(|t| + p_n(t)) \leq 1 - |t|/2$. This, plus the induction hypothesis, $0 \leq |t| - p_n(t) \leq |t|(1 - |t|/2)^n$, yields

$$\begin{aligned} |t| - p_{n+1}(t) &= (|t| - p_n(t))\left[1 - (1/2)(|t| + p_n(t))\right] \\ &\leq |t|(1 - |t|/2)^n (1 - |t|/2) = |t|(1 - |t|/2)^{n+1} \end{aligned}$$

which establishes $(*)$.

To complete the proof, suppose that $|t| \leq 1$. Since $p_n(t) \geq 0$ for every $n$, $|t| - p_n(t) \leq |t|$. Hence, by $(*)$, for $0 < r \leq |t| \leq 1$,

$$0 \leq |t| - p_n(t) \leq |t|(1 - |t|/2)^n \leq (1 - |t|/2)^n \leq (1 - r/2)^n$$

For sufficiently large $n$, $(1 - r/2)^n < r$. Thus $0 \leq |t| - p_n(t) < r$ for sufficiently large $n$ and $|t| \leq 1$. ■

Recall that a preordered linear space $X$ which contains the supremum (and therefore the infimum) of each pair of its elements is called a (vector) lattice.

Generally, a subalgebra of $C(T, \mathbf{R})$ is not a lattice—the sup and inf of two polynomials, for example, is not usually a polynomial—but, as we show next, a *closed* subalgebra of $C(T, \mathbf{R}, c)$ is a lattice.

**Theorem 16.5.2** CLOSED SUBALGEBRA IS A LATTICE For any completely regular Hausdorff space $T$, a closed subalgebra $X$ of $C(T, \mathbf{R}, c)$ is a lattice.

**Proof.** We use the notation of the statement. Suppose that $x \in X$ and that $K \subset T$ is compact. Let $k = p_K(x) = \sup |x(K)|$. Hence $x(t) \in [-k, k] \subset \mathbf{R}$ for all $t \in K$. Given $r \in (0, 1)$, there exists a polynomial $p$ such that $|p(x(t)) - |x(t)|| < r$ for all $t \in K$ [Theorem 16.5.1]. Therefore $p_K(p(x) - |x|) \leq r$. Since $X$ is a subalgebra, $p(x) \in X$; since $K$ is arbitrary, it follows that $|x| \in \operatorname{cl} X = X$, i.e., that $x \in X \Rightarrow |x| \in X$. Now, using Stone's marvelous observation about sups and absolute values, for any $x, y \in X$, $\sup(x, y) = [(x + y) + |x - y|]/2 \in X$ and $\inf(x, y) = [(x + y) - |x - y|]/2 \in X$. Thus, $X$ is a lattice. ■

**Definition 16.5.3** SELF-ADJOINT ALGEBRAS Let $T$ be a completely regular Hausdorff space and let $X$ be a subalgebra of $C(T, \mathbf{C})$. For $x \in C(T, \mathbf{C})$ $\operatorname{Re} x$ denotes the function whose value at $t \in T$ is $\operatorname{Re} x(t)$ with a similar convention for $\operatorname{Im} x$; clearly, $\operatorname{Re} x$ and $\operatorname{Im} x$ are continuous. We define $\operatorname{Re} X$ to be $\{\operatorname{Re} x : x \in X\}$. By the *conjugate* $\bar{x}$ of $x \in C(T, \mathbf{C})$, we mean $\bar{x} = \operatorname{Re} x - i \operatorname{Im} x$. We say that $X$ is *self-adjoint* if $x \in X \Rightarrow \bar{x} \in X$. ■

We have the following decomposition of self-adjoint algebras.

**Theorem 16.5.4** SELF-ADJOINT ALGEBRAS Let $T$ be a completely regular Hausdorff space and let $X$ be a self-adjoint subalgebra of $C(T, \mathbf{C})$. Then:

(a) REAL PARTS $\operatorname{Re} X$ is a subalgebra of $C(T, \mathbf{R})$ and $X = \operatorname{Re} X + i \operatorname{Re} X$.

(b) CLOSURE $\operatorname{cl} \operatorname{Re} X = \operatorname{Re} \operatorname{cl} X$ and $\operatorname{cl} X = \operatorname{cl} \operatorname{Re} X + i \operatorname{cl} \operatorname{Re} X$.

**Proof.** We use the notation of the statement. (a) For $a, b \in \mathbf{R}$ and $x, y \in X$, $ax + by \in X$; hence $\operatorname{Re}(ax + by) = a \operatorname{Re} x + b \operatorname{Re} y \in \operatorname{Re} X$ and therefore $\operatorname{Re} X$ is an $\mathbf{R}$-subspace of $C(T, \mathbf{R})$. Since $X$ is self-adjoint, $x\bar{y} \in X$ and therefore $(\operatorname{Re} x)(\operatorname{Re} y) = (\operatorname{Re} xy + \operatorname{Re} x\bar{y})/2 = \operatorname{Re}(xy + x\bar{y})/2 \in X$. Thus $\operatorname{Re} X$ is a subalgebra of $C(T, \mathbf{R})$. Since, for any $x \in X$, $\operatorname{Re} x = (x + \bar{x})/2 \in X$, it follows that $\operatorname{Re} X \subset X$ and therefore $\operatorname{Re} X + i \operatorname{Re} X \subset X$. Since $x = \operatorname{Re} x + i \operatorname{Re}(-ix) \in \operatorname{Re} X + i \operatorname{Re} X$, it follows that $X = \operatorname{Re} X + i \operatorname{Re} X$.

(b) It is straightforward to verify that $\operatorname{cl} \operatorname{Re} X = \operatorname{Re} \operatorname{cl} X$. Likewise, if $X$ is self-adjoint then so is $\operatorname{cl} X$. The result now follows from (a). ■

As a precursor to the Stone–Weierstrass theorem, we prove the Kakutani-Stone Theorem [16.5.5]. It reduces the problem of confirming a property for all compact subsets of $T$ to verification for two-point sets. We use this surprising fact to characterize the dense subalgebras of $C(T, \mathbf{F}, c)$ in the Stone–Weierstrass theorem.

**Theorem 16.5.5** CLOSURES OF SUBLATTICES OF $C(T, \mathbf{R}, c)$ Let $T$ be a completely regular Hausdorff space and let $X$ be a sublattice of $C(T, \mathbf{R}, c)$. Then for any $y \in C(T, \mathbf{R}, c)$, $y \in \operatorname{cl} X$ iff for all $s, t \in T$ and $r > 0$ there exists $x \in X$ such that $p_{\{s,t\}}(y - x) < r$.

**Proof.** We use the notation of the statement. Of course, if $y \in \operatorname{cl} X$, since $K = \{s, t\}$ is a compact subset of $T$, the condition is satisfied.

We show next that if $y \in C(T, \mathbf{R})$ satisfies the condition then $y \in \operatorname{cl} X$. Fix $s \in T$. By the condition, for each $r > 0$ and $t \in T$, there exists $x_t \in X$ such that $|y(t) - x_t(t)| < r$ and $|y(s) - x_t(s)| < r$. For $t \in T$, let $V(t) = \{u \in T : x_t(u) > y(u) - r\}$ and let $K$ be a compact subset of $T$. By the condition, the open sets $\{V(t) : t \in K\}$ cover $K$. As $K$ is compact, there exist $t_1, \ldots, t_n \in T$ such that $K \subset \cup_{i=1}^{n} V(t_i)$. Since $X$ is a lattice, $y_s = \sup\{x_{t_i} : i = 1, \ldots, n\} \in X$. For any $u \in K$, there exists $i$ such that $u \in V(t_i)$. Hence $y_s(u) \geq x_{t_i}(u) > y(u) - r$. Since $x_t(s) < y(s) + r$ for all $t \in T$, $y_s(s) < y(s) + r$. Thus, for each $s \in T$, there exists $y_s \in X$ such that

$$y_s(s) < y(s) + r \text{ and } y_s(u) > y(u) - r \text{ for all } u \in K \qquad (*)$$

By $(*)$, $s \in W(s) = \{u \in T : y_s(u) < y(u) + r\}$. Hence $\{W(s) : s \in K\}$ is an open cover of $K$; therefore there exist $s_i$, $i = 1, \ldots, k$, such that $K \subset \cup_{i=1}^{k} W(s_i)$. Since $X$ is a lattice, $z = \inf\{y_{s_i} : i = 1, \ldots, k\} \in X$. For $t \in K$, $t \in W(s_i)$ for some $i$ so $z(t) \leq y_{s_i}(t) < y(t) + r$. By the right side of $(*)$, $z(t) > y(t) - r$; therefore $|y(t) - z(t)| < r$ for any $t \in K$ and $p_K(y - z) = \sup |(z - y)(K)| \leq r$. We conclude that $y \in \operatorname{cl} X$. ■

**Definition 16.5.6** SEPARATING ALGEBRAS Let $T$ be a completely regular Hausdorff space and let $X$ be a subalgebra of $C(T, \mathbf{F})$. We say that $X$ *separates points of* $T$ (is a *separating subalgebra*) if for any distinct points $s, t \in T$, there exists $x \in X$ such that $x(s) \neq x(t)$. $X$ is *nonvanishing* if for each $t \in T$ there exists $x \in X$ such that $x(t) \neq 0$. ■

If we multiply elements $(a, b), (c, d) \in \mathbf{R}^2$ pointwise then $\mathbf{R}^2$ is an algebra (indeed, $\mathbf{R}^2 = C(\{1, 2\}, \mathbf{R})$, $\{1, 2\}$ discrete), a fact that we use in the proof of Theorem 16.5.7. Note that if $X$ is a subalgebra of $\mathbf{R}^2$ and $(1, 2) \in X$ then $(1, 2)^2 = (1, 4) \in X$. Since $(1, 2)$ and $(1, 4)$ are linearly independent, it follows that $X = \mathbf{R}^2$. This is indicative of the broader fact, which we leave to the reader to verify, that, aside from $\{0\}$ and $\mathbf{R}^2$, the only subalgebras of $\mathbf{R}^2$ are $\{0\} \times \mathbf{R}$, $\mathbf{R} \times \{0\}$ and the diagonal $D = \{(a, a) : a \in \mathbf{R}\}$.

**Theorem 16.5.7** STONE–WEIERSTRASS THEOREM Let $T$ be a completely regular Hausdorff space and let $X$ be a self-adjoint subalgebra of $C(T, \mathbf{F}, c)$. Then:

(a) For any $y \in C(T, \mathbf{F}, c)$, $y \in \operatorname{cl} X$ iff the following statements hold:

(1) For $s, t \in T$ such that $y(s) \neq y(t)$, there exists $x \in X$ such that $x(s) \neq x(t)$.

(2) For $t \in T$ such that $y(t) \neq 0$, there exists $x \in X$ such that $x(t) \neq 0$.

(b) $X$ is dense in $C(T, \mathbf{F}, c)$ iff $X$ is nonvanishing and separates points of $T$.

(c) Let $\mathbf{R}^n$ ($n \in \mathbf{N}$) carry the Euclidean topology. The $n$-variable polynomials in $C(\mathbf{R}^n, \mathbf{F}, c)$ are a dense subalgebra of $C(\mathbf{R}^n, \mathbf{F}, c)$.

**Proof.** We use the notation of the statement. (a) First, suppose that $\mathbf{F} = \mathbf{R}$. If $y \in \operatorname{cl} X$ and $y(s) \neq y(t)$, let $r > 0$ be such that $|y(s) - y(t)| > 3r$. Let $K = \{s, t\}$. As $y \in \operatorname{cl} X$, there exists $x \in X$ such that $p_K(x - y) < r$. Thus $|x(s) - x(t)| \geq |y(s) - y(t)| - |x(s) - y(s) - (x(t) - y(t))| > 3r - 2r = r > 0$ and we have proved (1). The proof of (2) is the same as (1) except that we take the compact set $K = \{t\}$.

Conversely, suppose that $y \in C(T, \mathbf{F}, c)$ satisfies (1) and (2). Let $s, t \in T$ and consider the multiplicative linear map $H : C(T, \mathbf{R}) \to \mathbf{R}^2$, $v \mapsto (v(s), v(t))$. We show next that $Hy \in H(X)$.

Since $H(X)$ is a subalgebra of $\mathbf{R}^2$, it is one of the subalgebras listed prior to the statement of the theorem. If $y(s) \neq 0$, it follows from (2) that there exists $x \in X$ such that $x(s) \neq 0$; therefore $H(X) \not\subset \{0\} \times \mathbf{R}$; in other words, $Hy \notin \{0\} \times \mathbf{R}$ implies that $H(X) \not\subset \{0\} \times \mathbf{R}$. Likewise, $Hy \notin \mathbf{R} \times \{0\}$ implies that $H(X) \not\subset \mathbf{R} \times \{0\}$. By (1), if $Hy = (y(s), y(t)) \notin D$, there is an element in $H(X)$ that does not belong to $D$, i.e., $Hy \notin D \Rightarrow H(X) \not\subset D$. Finally, if $Hy \neq (0,0)$ then using (2) again, $H(X) \not\subset \{(0,0)\}$. We conclude from this that $Hy \in H(X)$—in order that $Hy \notin H(X)$, $H(X)$ must be one of the proper subalgebras of $\mathbf{R}^2$; then either $Hy \notin \{0\} \times \mathbf{R}$, while $H(X) \subset \{0\} \times \mathbf{R}$, or $Hy \notin \mathbf{R} \times \{0\}$ while $H(X) \subset \mathbf{R} \times \{0\}$, or $Hy \notin D$ while $H(X) \subset D$, or $Hy \neq \{(0,0)\}$ while $H(X) \not\subset \{(0,0)\}$. As we have already observed, none of these situations can occur. Therefore there exists $x \in X$ such that $(x(s), x(t)) = (y(s), y(t))$. By Theorem 16.5.2 , the algebra $\operatorname{cl} X$ is a lattice; since $p_{\{s,t\}}(y - x) = 0$, it follows from the Kakutani-Stone theorem [16.5.5], that $y \in \operatorname{cl} X$.

Now let $\mathbf{F} = \mathbf{C}$. The argument that $y \in \operatorname{cl} X \Rightarrow$ (1) and (2) is the same as for $\mathbf{F} = \mathbf{R}$. Conversely, suppose that (1) and (2) hold for $y \in C(T, \mathbf{C})$ and that $s, t \in T$. We show that (1) and (2) are satisfied for $\operatorname{Re} y$ with respect to $\operatorname{Re} X$. If $\operatorname{Re} y(s) \neq \operatorname{Re} y(t)$ then by (1), there exists $x \in X$ such that $x(s) \neq x(t)$. By Theorem 16.5.4(a), there exist $u, v \in \operatorname{Re} X$ such that $x = u + iv$. Since $x(s) \neq x(t)$, it follows that $u(s) \neq u(t)$ or $v(s) \neq v(t)$. Hence $\operatorname{Re} X$ satisfies (1) with respect to $\operatorname{Re} y$. Similarly, it follows that $\operatorname{Re} X$ satisfies (2). By the argument for the real case and since $\operatorname{Re} X$ is a subalgebra of $C(T, \mathbf{R})$, it follows that $\operatorname{Re} y \in \operatorname{cl} \operatorname{Re} X$. The same argument proves that $\operatorname{Re} iy \in \operatorname{cl} \operatorname{Re} X$. Hence $y = \operatorname{Re} y + i \operatorname{Re}(-iy) \in \operatorname{cl} \operatorname{Re} X + i \operatorname{cl} \operatorname{Re} X = \operatorname{cl}(\operatorname{Re} X + i \operatorname{Re} X) = \operatorname{cl} X$ by Theorem 16.5.4(b).

(b) Suppose $\operatorname{cl} X = C(T, \mathbf{F}, c)$ and let $\beta T$ denote the Stone-Cech compactification of $T$. If $s, t \in T$, $s \neq t$, it follows from Urysohn's lemma that there exists $\beta y \in C(\beta T, \mathbf{F}, c)$ such that $\beta y(s) = 1$, $\beta y(t) = 0$ and $0 \leq \beta y(u) \leq 1$ for all $u \in \beta T$. Let $y = \beta y|_T$. Since $\operatorname{cl} X = C(T, \mathbf{F}, c)$, with $K = \{s, t\}$, there exists $x \in X$ such that $p_K(x - y) < 1/3$. Thus $|x(s) - y(s)| < 1/3$ and $|x(t) - y(t)| < 1/3$. These inequalities together with $y(s) = 1$, $y(t) = 0$ imply that $x(s) \neq x(t)$ and $x(s) \neq 0$. Thus $X$ separates points of $T$ and is nonvanishing on $T$.

Conversely, if the self-adjoint subalgebra $X$ separates points and is nonvanishing then, given distinct points $s, t \in T$, there exists $x, y \in X$ such that $x(s) \neq x(t)$ and $y(t) \neq 0$. By (a), it follows that any $z \in C(T, \mathbf{F})$ belongs to $\operatorname{cl} X$.

(c) We prove (c) by observing that polynomials in $C(\mathbf{R}^n, \mathbf{F})$ can be constructed that have different values at different points of $\mathbf{R}^n$ and are nonzero at a given point of $\mathbf{R}^n$. Since the polynomials are a self-adjoint subalgebra, they are dense in $C(\mathbf{R}^n, \mathbf{F}, c)$ by (b). ■

There are many other versions of the Stone–Weierstrass theorem. For functions taking values in an ultravalued field $(K, |\cdot|)$, we have:

**Theorem 16.5.8** NON-ARCHIMEDEAN STONE–WEIERSTRASS THEOREM For a complete ultravalued field $(K, |\cdot|)$ and Hausdorff space $T$, let $C(T, K)$ be the algebra of continuous functions mapping $T$ into $K$.

(a) Suppose $T$ is compact and consider $C(T, K, \|\cdot\|_\infty)$, $C(T, K)$ endowed with the sup norm $\|\cdot\|_\infty$. If a subalgebra $X$ of $C(T, K)$ contains constants and separates points *strongly*—$s \neq t \Rightarrow$ there exists $x \in C(T, K)$ such that $x(s) = 0$ and $x(t) = 1$—then $X$ is dense in $C(T, K, \|\cdot\|_\infty)$.

(b) It $T$ is locally compact and $C_\infty(T, K, \|\cdot\|_\infty)$ denotes $K$-valued functions that vanish at $\infty$ then a subalgebra $X$ that separates points strongly is dense in $C(T, K, \|\cdot\|_\infty)$.

Both results are due to Kaplansky [1950] with inspiration from Dieudonné [1944]; the proofs can also be found in Narici et al. [1971, p. 159f.] and Schikhof [1984, pp. 127 and 273]. For further material on the subject, see also Chernoff et al. [1968].

## 16.6 EXERCISES

16.101 A BANACH SPACE NOT STRICTLY CONVEX Let $T$ be a compact Hausdorff space with more than one point. Show that $C(T, \mathbf{F}, \|\cdot\|_\infty)$ is not strictly convex.

16.102 SUBSPACES NOT CHEBYSHEV Let $M$ be a subspace of a normed space $X$. Show that $M$ is not a Chebyshev space iff there exists $f \in M^\perp$ such that $\|f\| = 1$ and distinct $x, z \in X$ such that $f(y) = f(z) = \|y\| = \|z\|$ while $y - z \in M$.

16.103 BEST APPROXIMATIONS Let $K$ be a convex subset of a normed space $X$. Show that for any $x \in X$ the set $B_K(x) = \{y \in K : \|x - y\| = d(x, K)\}$ is convex and closed.

16.104 Show that a normed space $X$ has the unique extension property iff every maximal subspace $M$ of $X$ has the unique extension property.

16.105 CLARKSON'S RENORMING LEMMA Given a separable normed space $(X, \|\cdot\|)$, show that there exists an equivalent strictly convex norm.

16.201 $L_p[0,1]$ AND $\ell_p$ ARE UNIFORMLY CONVEX (a) Show that *Clarkson's inequalities* (1)–(3) [Clarkson 1936, p. 400] hold. Notice that (1) and (3) reduce to the parallelogram law for $p = 2$. For any $x, y \in L_p[0,1]$ or $\ell_p$ for $2 \le p < \infty$ and $\frac{1}{p} + \frac{1}{q} = 1$,

$$2\left(\|x\|_p^p + \|y\|_p^p\right) \le \|x+y\|_p^p + \|x-y\|_p^p \le 2^{p-1}\left(\|x\|_p^p + \|y\|_p^p\right) \tag{1}$$

$$2\left(\|x\|_p^p + \|y\|_p^p\right)^{q-1} \le \|x+y\|_p^q + \|x-y\|_p^q \tag{2}$$

$$\|x+y\|_p^p + \|x-y\|_p^p \le 2\left(\|x\|_p^q + \|y\|_p^q\right)^{p-1} \tag{3}$$

(b) Show that if $1 < p \le 2$, the reverse inequalities hold in (1)–(3).

(c) Show that $L_p[0,1]$ and $\ell_p$ are uniformly convex for $p > 1$.

(d) Show that $L_1[0,1]$ and $\ell_1$ are not strictly convex.

16.202 CLOSED SUBSPACES NEED NOT BE PROXIMINAL Let $(c_0, \|\cdot\|_\infty)$ be the Banach space of null sequences. Let $M = \{(a_n) \in c_0 : \sum_{n\in\mathbf{N}} 2^{-n} a_n = 0\}$. Show that $M$ is a closed infinite-dimensional subspace of $c_0$ and that if $x = (b_n) \notin M$ then there is no $m \in M$ such that $\|x - m\| = d(x, M)$.

16.203 APPROXIMATION IN HILBERT SPACE Let $M$ be a closed subspace of a Hilbert space $(X, \langle\cdot,\cdot\rangle)$.

(a) Prove the Beppo Levi inequality: If, $m, n \in M$ and $x \in X$ then

$$\|m - n\| \le \sqrt{\|x-m\|^2 - [d(x,M)]^2} + \sqrt{\|x-n\|^2 - [d(x,M)]^2}$$

(b) Show that $M$ is a Chebyshev subspace; thus, by Theorem 16.3.15, Hilbert spaces are strictly convex.

16.204 BEST APPROXIMATE SOLUTIONS The purpose of this exercise is to describe and characterize best approximate solutions for inconsistent linear equations. There are algorithms for finding such solutions (ascent and descent algorithms); the details can be found in Cheney [1966, Chap. 1].

Consider the linear equations

$$\sum_{j=1}^{n} a_{ij}x_j = b_i \qquad i = 1,\ldots,m \quad (m,n \in \mathbf{N})$$

where $a_{ij}$, $b_i$, and $x_j$ are real numbers (the $x_j$ are the solutions).

(a) Show that if $n > m$, there may or may not be solutions but they can never be unique.

(b) Show that if $n \leq m$, there may or may not be solutions and they may or may not be unique.

For the remainder of this exercise, we take $m > n$. If $r(i) = \sum_{j=1}^{n} a_{i,j}x_j - b_i$, let $x = (x_1,\ldots,x_n)$, $d(x) = \max_i r(i)$ and $D(x) = \max_i |r(i)|$. The *Chebyshev best-approximation solution* for an inconsistent set of equations is a solution which minimizes $D(x)$. It is also called a *minimax* solution.

(c) Why is there always a Chebyshev best-approximation solution?

(d) If a theory including algorithmic procedures is developed for finding a minimum of $d(x)$, how might this be used to find a minimum for $D(x)$?

16.205 Reflexive but Not Uniformly Convex We present M. M. Day's [Day 1941] class of examples of reflexive but not uniformly convex spaces. In fact, they cannot be equivalently renormed in a manner that would make them uniformly convex. Let $(X_n)$ be a sequence of Banach spaces. For $p > 1$, let $L_p((X_n)) = \{(x_n) : x_n \in X_n \text{ and } \sum_{n\in\mathbf{N}} \|x_n\|^p < \infty\}$. A norm on $L_p((X_n))$ is given by $\|(x_n)\| = \left(\sum_{n\in\mathbf{N}} \|x_n\|^p\right)^{1/p}$.

(a) Show that $L_p((X_n))$ is a Banach space.

(b) Show that $L_p((X_n))$ is strictly convex iff each $X_n$ is strictly convex.

(c) Show that $L_p((X_n))' = L_q((X_n'))$, where $1/p + 1/q = 1$.

(d) If each $X_n$ is reflexive, show that $L_p((X_n))$ is reflexive.

(e) Let $X_n = \{(b_{n1},\ldots,b_{nn}) : b_{nj} \in \mathbf{F}$, and $\|(b_{n1},\ldots,b_{nn})\| = \max_j |b_{nj}|$.

Show that there is no norm on $L_p((X_n))$ equivalent to the norm given above that will make $L_p((X_n))$ uniformly convex.

**Hints to Exercises**

16.101 Let $A$ and $B$ be closed, disjoint subsets of $T$. Let $x, y \in C(T, \mathbf{R})$ be such that $x(A) = \{0\}$, $x(B) = \{1\}$, $y(A) = \{1/2\}$, $y(B) = \{1\}$, $x(T) \subset [0,1]$ and $y(T) \subset [0,1]$. The midpoint of $[x, y]$ belongs to the surface of the closed unit ball of $C(T, \mathbf{F})$.

16.103 For $y, z \in B_K(x)$, note that $d(x, K) \leq \|x - [ay + (1-a)z]\|$ for any $a \in [0,1]$. Now add and subtract $ax$ in the last norm.

16.105 Let $\{x_n : n \in \mathbf{N}\}$ be a dense subset of $SU(X)$. For each $n \in \mathbf{N}$ choose $f_n \in SU(X')$ such that $f_n(x_n) = \|f_n\| = 1$ and, for $x \in X$, consider $\|x\|^* = \|x\| + \left(\sum_{n \in \mathbf{N}} \frac{|f_n(x)|^2}{2^n}\right)^{1/2}$.

16.201 Proofs of the inequalities (1) and (3) in (a) and (b) can be found in Clarkson [1936] and Köthe [1983, pp. 355f.].

(c) For $p \geq 2$, if $\|x\|_p = \|y\|_p = 1$,

$$\|x+y\|_p^p + \|x-y\|_p^p \leq 2^{p-1}\left(\|x\|_p^p + \|y\|_p^p\right)$$

Hence if $\|x - y\| \geq a$ then $\left\|\left(\frac{1}{2}\right)(x+y)\right\|_p \leq [1 - (a/2)^p]^{1/p}$. If $1 < p \leq 2$ then $\left\|\left(\frac{1}{2}\right)(x+y)\right\|_p^q + \|(x-y)/2\|_p^q \leq 2\left(\|x/2\|_p^p + \|y/2\|_p^p\right)^{q-1}$, so if $\|x-y\|_p \geq a$ then $\left\|\left(\frac{1}{2}\right)(x+y)\right\|_p \leq (1 - (a/2)^q)^{1/q}$.

16.202 Letting $a = \sum_{n \in \mathbf{N}} 2^{-n} b_n$, if $x_k = 2^k / (2^k - 1)((a_n) - (b_n))$ where $a_n = a$ for $n \leq k$, $a_n = 0$ for $n > k$, show that $x_k \in M$ and $\|x - x_k\| \to |a|$ from above. Then $d(x, M) \leq |a|$. If $m = (c_n) \in M$ is such that $\|x - m\| \leq |a|$, use the fact that sequences in $c_0$ converge to 0 to show that $\left|\sum_{n \in \mathbf{N}} 2^{-n} b_n\right| = \left|\sum_{n \in \mathbf{N}} (b_n - c_n)\right| \leq \sum_{n \in \mathbf{N}} 2^n |b_n - c_n| < |a|$ and arrive at a contradiction.

16.203 (a) Let $m, n \in M$ and $-1 \neq a \in \mathbf{R}$. Then $p = (m + an)/(1+a) \in M$. Since $\|x - p\|^2 \geq d(x, M)^2$ then $(1+a)^2 \|x - p\|^2 = \|(x - m) + a(x - n)\|^2 \geq (1+a)^2 d(x, M)^2$ and therefore $a[\|x - n\|^2 - d(x, M)^2 + 2a \operatorname{Re}\langle x - m, x - n\rangle - d(x, M)^2] + \|x - m\|^2 - d(x, M)^2 \geq 0$. Since $a$ is any real number not equal to $-1$, it follows that $(\operatorname{Re}\langle x - m, x - n\rangle - d(x, M)^2)^2 \leq (\|x - m\|^2 - d(x, M)^2)(\|x - n\|^2 - d(x, M)^2)$. Apply this to $\|m - n\|^2 = \|(x - n) - (x - m)\|^2 = \|x - m\|^2 + \|x - n\|^2 - 2\operatorname{Re}\langle x - m, x - n\rangle$.

(b) Suppose $m_n \in M$ are such that $d(x, M) = \lim_n \|x - m_n\| = c$. Apply the Beppo Levi inequality of (a) to show that $(m_n)$ is a Cauchy sequence.

16.204 (a) The system has a solution $(b_1, \ldots, b_m)$ iff $(b_1, \ldots, b_m)$ belongs to the linear span of $\{(a_{i,j}, \ldots, a_{m,j}) : j = 1, \ldots, n\}$. The solution is unique iff the set $\{(a_{i,j}, \ldots, a_n, j) : j = 1, \ldots, n\}$ is linearly independent.

(b) Same hint as (a).

(c) The situation is covered by Theorem 16.3.6(a). Identify the subspace $M$, the desirable norm, and then apply Theorem 16.3.6(a).

(d) Double the number of equations by multiplying the scalars $b_i$ and the coefficients $a_{ij}$ by $-1$.

16.205 (a) If $((x_i^n))_{i \in \mathbf{N}}$ is a Cauchy sequence in $L_p((X_n))$, show that $(x_i^n)$ is a Cauchy sequence for each $i$. Then, for each $i \in \mathbf{N}$, $\lim_n x_i^n = x_i \in X_i$ and $\lim_n (x_i^n) = (x_i) \in L_p((X_n))$.

(b) If $L_p((X_n))$ is strictly convex then each $X_i$ has a linearly isometric image in $L_p((X_n))$; consequently each $X_i$ is strictly convex. Conversely, suppose that

$$\left(\sum_{i \in \mathbf{N}} \|x_i + y_i\|^p\right)^{1/p} = \left(\sum_{i \in \mathbf{N}} \|x_i\|^p\right)^{1/p} + \left(\sum_{i \in \mathbf{N}} \|y_i\|^p\right)^{1/p}$$

Since $\|x_i + y_i\| \leq \|x_i\| + \|y_i\|$ for each $i \in \mathbf{N}$, it follows from

$$\begin{aligned} \left(\sum_{i \in \mathbf{N}} \|x_i + y_i\|^p\right)^{1/p} &\leq \left(\sum_{i \in \mathbf{N}} (\|x_i\| + \|y_i\|)^p\right)^{1/p} \\ &\leq \left(\sum_{i \in \mathbf{N}} \|x_i\|^p\right)^{1/p} + \left(\sum_{i \in \mathbf{N}} \|y_i\|^p\right)^{1/p} \\ &= \left(\sum_{i \in \mathbf{N}} \|x_i + y_i\|^p\right)^{1/p} \end{aligned}$$

that $\|x_i + y_i\| = \|x_i\| + \|y_i\|$ for all $i \in \mathbf{N}$. As $X_i$ is strictly convex for all $i \in \mathbf{N}$, there exist $t_i \geq 0$ such that $y_i = t_i x_i$. Since $\ell_p$ is strictly convex, there exists $t \geq 0$ such that $(\|y_i\|) = t(\|x_i\|)$. Thus $t = t_i$ for all $i \in \mathbf{N}$ and $(y_i) = t(x_i)$.

(c) Let $f \in L_p((X_n))'$ and $x_i \in X_i$ $(i \in \mathbf{N})$. Let $z_i = (0, \ldots, 0, x_i, 0, \ldots)$ and define $f_i \in X_i'$ by taking $f_i(x_i) = f(z_i)$. Clearly $f_i$ is locally bounded and $\|f_i\| \leq \|f\|$. Define $\hat{f} = (f_i)$ where $\hat{f}(x) = \sum_{i \in \mathbf{N}} f_i(x_i)$ with $x = (x_i) \in L_p((X_n))$. It is clear that $\hat{f}(x) = f(x)$ for all $x \in L_p((X_n))$. Using arguments similar to those of Example 15.7.2, one can show that $(f_i) \in L_q((X_i'))$ and, in fact, $\|f\| = \left\|\hat{f}\right\| = \left(\sum_{n \in \mathbf{N}} \|f_n\|^q\right)^{1/q}$.

Conversely, if $(f_i) \in L_q((X_n'))$ and $x = (x_i) \in L_p((X_n))$ then $|f(x)| \leq \sum_{i \in \mathbf{N}} |f_i(x_i)| \leq \sum_{i \in \mathbf{N}} \|f_i\| (\|x_i\|) \leq \left(\sum_{i \in \mathbf{N}} \|f_i\|^q\right)^{1/q} \left(\sum_{i \in \mathbf{N}} \|x_i\|^p\right)^{1/p}$ and $f \in L_p((X_i))'$ with $\|f\| \leq \left(\sum_{i \in \mathbf{N}} \|f_i\|^q\right)^{1/q}$.

(d) $L_q((X'_n))' = L_p((X''_n)) = L_p((X_n))$.

(e) Suppose that there is a norm $\|\cdot\|_1$ on $L_p((X_n))$ with respect to which $L_p((X_n))$ is uniformly convex. If $0 < a \leq 2$, define $b(a) = (1/2)\inf\{2 - \|x + x'\|_1\}$ where $x$,$x'$ belong to the unit ball of $(L_p((X_n)), \|\cdot\|_1)$ and satisfy $\|x - x'\|_1 \geq a$. Show that $b(a) > 0$ iff $(L_p((X_n)), \|\cdot\|_1)$ is uniformly convex. Assuming that $\|\cdot\|$ and $\|\cdot\|_1$ are equivalent, there exist $m, n > 0$ such that $m\|\cdot\| \leq \|\cdot\|_1 \leq n\|\cdot\|$. With no loss of generality, assume that $m = 1$ and $n > 1$. Restricting these norms to the images of $X_n$ in $L_p((X_n))$, it follows that

$$\begin{aligned}\{(b_{i1}, \ldots, b_{ii}) : \sup_{j\leq i} |b_{ij}| \leq 1/n\} &\subset \{(b_{i1} \ldots, b_{ii}) : \|(b_{i1}, \ldots, b_{ii})\|_1 \leq 1\} \\ &\subset K_i = \{(b_{i1}, \ldots, b_{ii} : \sup_{j\leq 1} |b_{ij}| \leq 1\}\end{aligned}$$

If $c = 1/n$, $(a_1c, \ldots, a_ic) \in K_i$, where $a_i = \pm 1$. If $x = (c, \ldots, c)$ and $x' = (c, \ldots, c, -c)$ then $x - x' = (0, \ldots, 0, 2c)$ while $(x + x')/2 = (c, \ldots, c, 0)$. Now $\|x - x'\| = 2c$, so $\|(x + x')/2\|_1 \leq 1 - b(2c)$. Thus

$$|| (c/(1 - b(2c)), \ldots, c/(1 - b(2c)), 0)_1|| \leq 1$$

If we repeat this process using $(c/(1 - b(2c), \ldots, c/(1 - b(2c), 0)$, we obtain

$$\left\|\left(\frac{c}{(1-b(2c)\left[1-b(\frac{2c}{1-b(2c)})\right]}, \ldots, \frac{c}{(1-b(2c)\left[1-b(\frac{2c}{1-b(2c)})\right]}, 0, 0\right)\right\|_1 \leq 1$$

But if $d \leq e$, clearly $b(d) \leq b(e)$ [from the definition of $b(\cdot)$]. Thus $b(2c) \leq b(2c/(1-b(2c)))$ or, alternatively, $1/(1-b(2c)) \leq 1/(1-b(2c/(1-b(2c)))$. Consequently, $\|(c/(1-b(2c))^2, \ldots, c/(1-b(2c))^2, 0, 0)\|_1 \leq 1$. If we continue this process down to the first entry, we obtain

$$\left\|\left(\frac{c}{(1 - b(2c))^{i-1}}, \frac{c}{(1 - b(2c))^{i-1}}, 0, \ldots, 0\right)\right\|_1 = \frac{c}{(1 - b(2c))^{i-1}} \leq 1$$

But $0 < 1 - b(2c) < 1$, so that, for sufficiently large $i$, we have arrived at a contradiction. In this argument the identification between $X_i$ and its natural isometric image in $L_p((X_n))$ has been used rather freely. The space $L_p((X_n))$ is not strictly convex because the $X_n$ are not strictly convex. Day [1941] showed that $L_p((X_n))$ could be constructed which is strictly convex, reflexive, and yet cannot be made uniformly convex with respect to any norm equivalent to the defined norm $\left(\sum_{n\in\mathbf{N}} \|x_n\|^p\right)^{1/p}$.

# Bibliography


[1] Abramovich, Y. A., and A. Kitover [2000] *Inverses of disjointness preserving operators,* Mem. Amer. Math. Soc. *143*, no. 679, American Mathematical Society, Providence, RI 02940-6248.

[2] Adasch, N. [1970] Tonnelierte Räume und zwei Sätze von Banach, Math. Ann. *186*, 209–214.

[3] Adasch, N., B. Ernst, and D. Keim [1978] *Topological vector spaces, the theory without convexity conditions*, Lecture Notes in Mathematics 639, Springer-Verlag.

[4] Agnew, R., and A. Morse [1938] Extensions of linear functionals with applications to limits, integrals, measures and densitites, Ann. Math. *39*, 20–30.

[5] Ajupov, S. A. [1974] The Krein–Milman theorem in locally $K$-convex spaces, Dokl. Akad. Nauk. Uzebekskoi SSR, Tashkent, no. 5, 3–5 (Russian).

[6] Akhieser, N. [1956] *Theory of approximation*, Ungar, New York.

[7] Akhieser, N., and M. Krein [1962] Some questions in the theory of moments, Amer. Math. Soc., Providence.

[8] Akilov, G. [1948] Necessary conditions for the extension of linear operations, C. R. (Doklady) Acad. Sci. URSS. N. S. *59*, 417–418 (Russian).

[9] Alaoglu, L. [1940] Weak topologies of normed linear spaces, Ann. of Math. *41*, 252–267 (Thesis from University of Chicago 1938).

[10] Albius, E., and M. Morillon [2001] Uniform smoothness entails Hahn–Banach, Quaest. Math. *24*, 425–439.

[11] Alfsen, E., and E. Effros [1972] Structure in real Banach spaces I–II, Ann. of Math. *96*, 98–128, 129–173.

[12] Alfsen, E. M. [1971] *Compact convex sets and boundary integrals*, Ergebnisse der Mathematik und ihrer Grenzgebiete, Vol. 57, Springer-Verlag, Berlin.

[13] Ameniya, I., and Y. Komura [1968] Über nicht-vollständige Montel-Räume, Math. Ann. *177*, 273–277.

[14] Amir, D. [1965] On isomorphisms of continuous function spaces, Israel J. Math. *3*, 205–210.

[15] Andenaes, P. R. [1970] Hahn–Banach extensions which are maximal on a given cone, Math. Ann. *188*, 90–96.

[16] Ando, T. [1973] Closed range theorems for convex sets and linear liftings, Pacific J. Math. *44*, 393–410.

[17] Araujo, J. [1998] Separating maps and linear isometries between some spaces of continuous functions, J. Math. Anal. Appl. *226,* 23–39

[18] Araujo, J. [1999a] **N**-compactness and automatic continuity in ultrametric spaces of bounded continuous functions, Proc. Amer. Math. Soc. *127,* 2489–2496

[19] Araujo, J. [1999b] Linear isometries between spaces of functions of bounded variation, Bull. Austral. Math. Soc. *59*, 335–341.

[20] Araujo, J. [2001] A new version of the non-Archimedean Banach–Stone theorem, *p-adic functional analysis* (Ioannina, 2000), 13–19, Lecture Notes in Pure and Appl. Math., 222, Dekker, New York.

[21] Araujo, J. [2002a] Realcompactness and spaces of vector-valued functions, Fund. Math. *172,* 27–40.

[22] Araujo, J. [2002b] **N**-compactness and weighted composition maps, Proc. Amer. Math. Soc. *130*, 1225–1234

[23] Araujo, J. [2004a] Realcompactness and Banach–Stone theorems, Bull. Belg. Math. Soc. Simon Stevin *11*, 247–258.

[24] Araujo, J. [2004b] Linear biseparating maps between spaces of vector-valued differentiable functions and automatic continuity, Adv. Math. *187*, 488–520.

[25] Araujo, J. [2006] The noncompact Banach–Stone theorem, J. Operator Theory *55*, 285–294.

[26] Araujo, J., E. Beckenstein and L. Narici [1995] Biseparating maps and homeomorphic realcompactifications, J. Math. Anal. Appl. *192*, 258–265.

[27] Araujo, J., E. Beckenstein and L. Narici [1996] When is a separating map biseparating? Arch. Math. *67,* 395—407.

[28] Araujo, J., and J. Font [1997] Linear isometries between subspaces of continuous functions, Trans. Amer. Math. Soc. *349*, 413–428.

[29] Araujo, J., and J. Font [2000] Linear isometries on subalgebras of uniformly continuous functions, Proc. Edinburgh Math. Soc. *43*, 139–147.

[30] Araujo, J., J. Font and S. Hernández [1996] A note on Holsztynski's theorem, Ann. New York Acad. Sci *788*, 9–12.

[31] Araujo, J., and K. Jarosz [2001] Isometries of spaces of unbounded continuous functions, Bull. Austral. Math. Soc. *63*, 475–484.

[32] Araujo, J., and K. Jarosz [2003a] Biseparating maps between operator algebras, J. Math. Anal. Appl. *282*, 48–55.

[33] Araujo, J., and K. Jarosz [2003b] Automatic continuity of biseparating maps, Studia Math. *155,* 231–239.

[34] Araujo, J., and J. Martínez-Maurica [1990] The non-Archimedean Banach–Stone theorem, *Proceedings of a conference on padic analysis* edited by F. Baldassari, S. Bosch and B. Dwork, Lecture Notes in Mathematics 1454, Springer, 64–79.

[35] Arendt, W. [1983] Spectral properties of Lamperti operators, Indiana Univ. Math. J. *32*, 199–215.

[36] Arens, R. [1947] Duality in linear spaces, Duke Math. J. *14*, 787–794.

[37] Arens, R., and J. Kelley [1947] Characterizations of the space of continuous functions over a compact Hausdorff space, Trans. Amer. Math. Soc. *62*, 499–508.

[38] Arias de Reyna, J. [1982] Normed barely Baire spaces, Israel J. Math. *42*, nos. 1–2, 33–36.

[39] Aronszajn, N., and P. Panitchpakdi [1956] Extensions of uniformly continuous transformations and hyperconvex metric spaces. Pac. J. Math. *6*, 405–439.

[40] Artin, E. [1967] *Algebraic numbers and functions*, Gordon and Breach, New York. (A reprint of 1951 Princeton lecture notes.)

[41] Ascoli, Guido [1932] Le curve limite di una varietà data di curve, Atti dell R. Acad. dei Lincee, Memorie Cl. Sci. Fis. Mat. Nat. *18*, 521–586

[42] Asimow, L., and A. Ellis [1980] *Convexity theory and its applications in functional analysis*, London Mathematical Society Monographs no. 16, Academic Press, New York.

[43] Asmar, N., E. Berkson and T. Gillespie [1991] Transference of strong type maximal inequalities by separation-preserving representations, Amer. J. Math. *113*, 47–74.

[44] Bachman, G. [1964] *Introduction to p-adic numbers and valuation theory*, Academic Press, New York.

[45] Bachman, G., and L. Narici [2000] *Functional analysis*, Dover Publications, 31 East 2nd St., Mineola, NY 11501, a reprint of the 1966 edition published by Academic Press, New York.

[46] Bachman, G., L. Narici and E. Beckenstein [2000] *Fourier and wavelet analysis,* Springer-Verlag, New York.

[47] Bachman, G., E. Beckenstein, L. Narici, and S. Warner [1975] Rings of continuous functions with values in a topological field, Trans. Amer. Math. Soc. *204*, 91–112.

[48] Balakrishnan, A. V. [1981] *Applied functional analysis*, 2nd ed. Springer-Verlag, Berlin.

[49] Banach, S. [1922] Sur les opérations dans les ensembles abstraits et leur application aux équations intégrales, Fund. Math. *3*, 133–181.

[50] Banach, S. [1929] Sur les fonctionelles linéaires I and II, Studia Math. *1*, 211–216, 223–239, reprinted in Banach 1979 below.

[51] Banach, S. [1932] *Théorie des opérations linéaires*, Monografje Matematyczne, Warszawa, reprinted by Chelsea, New York 1932 and in English translation as *Theory of linear operations* by North-Holland, Amsterdam-New York-Oxford-Tokyo, 1987, reprinted in 2009 by Dover Publications, 31 East 2nd St., Mineola, NY 11501; in this translation all Banach's footnotes and a majority of his references were removed so the history of the subject cannot be properly understood without looking at the Warsaw or New York version. For more on Banach, see http://banach.univ.gda.pl/e-index.html.

[52] Banach, S. [1979] *Oeuvres,* vol II, PWN-Éditions Scientifiques de Pologne, Warsaw.

[53] Banach, S., and S. Mazur [1933] Zur Theorie der linearen Dimension, Studia Math. *4*, 100–112, reprinted in Banach 1979 above.

[54] Banach, S., and H. Steinhaus [1927] Sur le principe de condensation de singularités, Fund. Math. *9*, 50–61.

[55] Bandyopadhyay, P., and A. Roy [2003] Nested sequences of balls, uniqueness of Hahn–Banach extensions and the Vlasov property, Rocky Mountain J. Math. *33*, 27–67.

[56] Bary, N. [1964] *A treatise on trigonometric series I*, Macmillan, New York.

[57] Batt, J., and E. Berg [1969] Linear bounded transformations on the space of continuous functions, J. Funct. Anal. *4*, 215–239.

[58] Beckenstein, E., and L. Narici [1987] A non–Archimedean Stone–Banach theorem, Proc. Amer. Math. Soc. *100*, 242–246.

[59] Beckenstein, E., L. Narici and A. Todd [1988] Automatic continuity of linear maps on spaces of continuous functions, Manuscripta Math. *62*, 257–275.

[60] Beckenstein, E., L. Narici, and C. Suffel [1977] *Topological Algebras*, North-Holland, Amsterdam.

[61] Behrends, E. [1978] On the Banach–Stone theorem, Math. Ann. *233*, 261–272.

[62] Behrends, E. [1979] *M-structure and the Banach–Stone theorem,* Lecture Notes in Mathematics *736*, Springer-Verlag, New York.

[63] Behrends, E. [1991] Points of symmetry of convex sets in the two-dimensional complex space—a counterexample to D. Yost's problem, Math. Ann. *290*, 463–471.

[64] Bell, J., and D. Fremlin [1972] A geometric form of the axiom of choice, Fund. Math. *77*, 167–170.

[65] Bennett, G., L. Dor, V. Goodman, W. Johnson and C. Newman [1977] On uncomplemented subspaces of $L_p$, $1 < p < 2$, Israel J. Math. *26*, 178–187.

[66] Berberian, S. [1974] *Lectures in functional analysis and operator theory,* Springer-Verlag, Berlin.

[67] Beurling, A., and A. Livingston [1962] A theorem on duality mappings in Banach spaces, Ark. Math. *4*, 405–411.

[68] Bierstedt, K., and J. Bonet [2003] Some aspects of the modern theory of Fréchet spaces, Rev. R. Acad. Cien. Serie A. Math. *97*, 159–188.

[69] Birkhoff, G. (junior) [1935] Orthogonality in linear metric spaces, Duke Math. J. *1*, 169–172.

[70] Birkhoff, G. (junior) [1936] A note on topological groups, Compositio Math. *3*, 427–430.

[71] Bishop, E., and R. Phelps [1961] A proof that every Banach space is subreflexive, Bull. Amer. Math. Soc. *67*, 97–98.

[72] Blair, Ch. E. [1977] The Baire category theorem implies the principle of dependent choices, Bull Acad. Polon. Sciences XXV, 933–934.

[73] Blatter, J., and G. Seever [1975] Interposition of semicontinuous functions by continuous functions, in *Analyse fonctionelle* (Proceedings of a conference in Rio de Janeiro, August 1972), L. Nachbin, ed., Hermann, Paris, pp. 27–51.

[74] Bohnenblust, H., and A. Sobczyk [1938] Extensions of functionals on complex linear spaces, Bull. Amer. Math. Soc. *44*, 91–93.

[75] Bohnenblust, H. [1942] A characterization of complex Hilbert spaces, Port. Math. *3*, 103–109.

[76] Bombal, F. [2003] Análisis funcional: una perspectiva histórica, Proceedings of the Seminar of Mathematical Analysis 2002–2003, Secretariado de Publicaciones, Unversidad de Sevilla, 81–117.

[77] Bonnesen, T., and W. Fenchel [1934] *Theorie der konvexen Körper*, Springer-Verlag, Berlin.

[78] Bonnice, W., and R. Silverman [1966] The Hahn–Banach theorem for finite-dimensional spaces, Trans. Amer. Math. Soc. *121*, 211–222.

[79] Bonnice, W., and R. Silverman [1967] The Hahn–Banach theorem and the least upper bound properties are equivalent, Proc. Amer. Math. Soc. *18*, 843–850.

[80] Bonsall, F., and I. Duncan [1971] Numerical ranges of operators on normed spaces and elements of normed algebras, London Math. Soc., Lec. notes 2, Cambridge Univ. Pr., New York.

[81] Bourbaki, N. [1938] Sur les espaces de Banach, C. R. Acad. Sci. Paris *206*, 1701–1704.

[82] Bourbaki, N. [1950] Sur certains espaces vectoriels topologiques, Ann. Inst. Fourier 2, 5–16.

[83] Bourbaki, N. [1953] *Espaces vectoriels topologiques*, Hermann, Paris, Chaps. I and II.

[84] Bourbaki, N. [1955] *Espaces vectoriels topologiques*, Hermann, Paris, Chaps. III and IV.

[85] Bourbaki, N. [1966a] *Elements of mathematics: General topology, Part 1*, Addison Wesley, Reading, Mass.

[86] Bourbaki, N. [1966b] *Elements of mathematics: General topology, Part 2*, Addison-Wesley, Reading, Mass.

[87] Brace, J. [1960] Almost uniform convergence versus pointwise convergence, Proc. Amer. Math. Soc. *6*, 986–987.

[88] Brace, J. [1962] Compactness in function spaces, Duke Math. J. *29*, 157–166.

[89] Brace, J. [1965] Convergence on filters and simple equicontinuity, III. J. Math. *9*, 286–296.

[90] Brace, J., G. Friend, and P. Richetta [1969] Locally convex topologies on function spaces, Duke Math. J., 709–714.

[91] Bridges, D. [1979] *Constructive functional analysis*, Pitman, London.

[92] Browder, F. [1965] Multivalued monotone nonlinear mappings and duality mappings in Banach spaces, Trans. Amer. Math. Soc. *118*, 338–351.

[93] Brown, L., and N-C. Wong [2004] Unbounded disjointness preserving linear functionals, Monats. fur Math. *141*, 21–32.

[94] Bushaw, D. [1963] *Elements of general topology*, Wiley, New York.

[95] Buskes, G. [1993] The Hahn–Banach theorem surveyed, Diss. Math. *327*, 1–49.

[96] Butzer, P., S. Gieseler, F. Kaufmann, R. Nessel, and E. Stark [1980] Eduard Helly [1884–1943), Eine nachträgliche Würdigung, Jahresber. Deutsch. Math.-Verein *82*, 128–151.

[97] Butzer, P., R. Nessel, and E. Stark [1984] Eduard Helly (1884–1943), in memoriam, Results in Mathematics *7*, 145–153.

[98] Cambern, M. [1966] A generalized Banach–Stone theorem, Proc. Amer. Math. Soc. *17*, 396–400.

[99] Cambern, M. [1967] On some isomorphisms with small bound, Proc. Amer. Math. Soc. *18*, 1062–1066.

[100] Cambern, M. [1970] Isomorphisms of $C_0(X)$ onto $C_0(Y)$, Pacific J. Math. *35*, 307–312.

[101] Cambern, M. [1975] On mappings of spaces of functions with values in a Banach space, Duke Math. J. *42*, 91–98.

[102] Cambern, M. [1977] The Banach–Stone property and the weak Banach–Stone property in three-dimensional spaces, Proc. Amer. Math. Soc. *67*, 55–61.

[103] Cao, J., I. Reilly and H. Xiong [2003] A latticle-valued Banach–Stone theorem, Acta Math. Hungar. *98 (1–2)*, 103–110.

[104] Carleson, L. [1966] On convergence and growth of partial sums of Fourier series, Acta Math. *116*, 135–157.

[105] Castillo, J., and M. González [1997] *Three-space problems in Banach space theory,* Lecture Notes in Math. *1363*, Springer, New York.

[106] Cech, E. [1937] On bicompact spaces, Ann. of Math. *38*, 823–844.

[107] Cederquist, J., and T. Coquand [2000] *Entailment relations and distributive lattices*, Logic Colloquium '98 (Prague), 127–139, Lect. Notes Log., 13, Assoc. Symbol. Logic, Urbana, IL.

[108] Cederquist, J., T. Coquand and S. Negri [1998] The Hahn–Banach theorem in type theory, in *Twenty-five years of constructive type theory* (Venice, 1995), Oxford Logic Guides, *36*, Oxford Univ. Press, New York, 57–72.

[109] Cengiz, B. [1973a] On extremely regular function spaces. Pacific J. Math. *49*, 335–338.

[110] Cengiz, B. [1973b] A generalization of the Banach–Stone Theorem, Proc. Amer. Math. Soc. *50*, 426–430.

[111] Cengiz, B. [1978] On topological isomorphisms of $C_0(X)$ and the cardinal number of $X$. Proc. Amer. Math. Soc. *72* 105–108.

[112] Cengiz, B. [1990] On isometries of completely regular spaces of continuous functions. Doğa Mat. *14*, 186–190.

[113] Chalmers, B., M. Ostrovskii, and B. Shekhtman [2003] Hahn–Banach operators: A review, J. Computation Anal. Appl. *3*, 11–24.

[114] Chen, Jin Xi, Zi Li Chen and Ngai-Ching Wong [2008] A Banach–Stone theorem for Riesz isomorphisms of Banach lattices, Proc. Amer. Math. Soc. *136*, 3869–3874.

[115] Cheney, E. [1966] *Introduction to approximation theory.* McGraw-Hill, New York.

[116] Chernoff, P., R. Rasala and W. Waterhouse [1968] The Stone–Weierstrass theorem for valuable fields, Pac. J. Math. *27*, 233–240.

[117] Choquet, G. [1958] Une classe régulière d'espaces de Baire, C. R. Acad. Sci. Paris *246*, 218–220.

[118] Choquet, G. [1969a] Outils topologiques et métriques de l'analyse mathématique, Cours rédigé par C. Mayer, C.D.U., Paris.

[119] Choquet, G. [1969b] *Lectures on analysis*, Vol. I: *Integration and topological vector spaces*, Benjamin/Cummings, Reading, Mass.

[120] Choquet, G. [1969c] *Lectures on analysis*, Vol. II: *Representation theory*, Benjamin/Cummings, Reading, Mass.

[121] Ciesielski, K. [2007] On Stefan Banach and some of his results, Banach J. Math. Anal *1*, 1–10.

[122] Civin P., and B. Yood [1957] Quasi-reflexive spaces, Proc. Amer. Math. Soc. *8*, 906–911.

[123] Clarkson, J. [1936] Uniformly convex spaces. Trans. Amer. Math. Soc. *40*, 396–314.

[124] Cohen, H. [1975] A bound two isomorphism between $C(X)$ Banach spaces, Trans. Amer. Math. Soc. *40*, 396–414.

[125] Coquand, T. [2006] Geometric Hahn–Banach theorem, Math. Proc. Cambridge Philos. Soc. *140*, no. 2, 313–315.

[126] Costara, C., and D. Popa [2001] Effective determination of all the Hahn–Banach extensions of some linear and continuous functionals, An. c Stiinc t. Univ. Ovidius Constanc ta Ser. Mat. *9*, 31–43.

[127] Cristescu, R. [1977] *Topological vector spaces.* Noordhoff, Leyden, The Netherlands.

[128] Curtain, R., and A. Pritchard [1977] *Functional analysis in modern applied mathematics.* Academic Press, New York.

[129] Danet, R-M., and N-C. Wong [2002a] Hahn–Banach–Kantorovich type theorems with the range space not necessarily (o)-complete, Taiwanese J. Math. *6*, 241–246.

[130] Danet, R-M., and N-C. Wong [2002b] Extension theorems without Dedekind completeness, Rendiconti del Circolo Mathematico di Palermo., Series II, Supp. 68, 381–387.

[131] Davie, A. M. [1973] The approximation problem for Banach spaces, Bull. Lond. Math. Soc. *5*, 261–266.

[132] Davie, A. M. [1975] The Banach approximation problem, J. Approx. Theory *13*, 392–394.

[133] Davis, W. [1977] A characterization of $P_1$-spaces, J. Approx Theory *21*, 315–318.

[134] Day, M. [1940] The spaces $L_p$ with $0 < p < 1$, Bull. Amer. Math. Soc. *46*, 816–823.

[135] Day, M. [1941] Reflexive Banach spaces not isomorphic to uniformly convex spaces, Bull. Amer. Math. Soc. *47*, 313–317.

[136] Day, M. [1955] Strict convexity and and smoothness of normed spaces, Proc . Amer. Math. Soc. *78*, 516–528..

[137] Day, M. [1973] *Normed linear spaces*, Third ed., Springer-Verlag, New York.

[138] Dedekind, R. [1932] *Gesammelte mathematische Werke*, 3 vols., Braunschweig (Vieweg), Vol. 2, p. 288.

[139] Delbaen, F., and W. Schachermayer [1994] A general version of the fundamental theorem of asset pricing, Math. Ann. *300*, 463–520.

[140] de Grande-de Kimpe, N. [1971] $c$-compactness in locally K-convex spaces, Indag. Math. *33*, 176–180.

[141] de Guzmán, M. [1966] Nota sobre la unicidad en el teorema de extensión de Hahn–Banach, Rev. Real Acad. Ciencias Exactas, Físicas y Nat. *60*, 577–584.

[142] de Wilde, M. [1967] Sur le théorème du graphe fermé, C. R. Acad. Sci. Paris, *265*, 376–379.

[143] de Wilde, M. [1969a] Réseaux dans les espaces linéaires à semi-normes, Mém. Soc. Roy. Sci. Liège *18*, Fasc. 2, 1–144.

[144] de Wilde, M. [1969b] Sur les sous-espaces de codimension finie d'un espace linéaire à semi-normes, Bull. Soc Roy. Sci, Liège *38*, 450–453.

[145] de Wilde, M. [1973] Various types of barreledness and increasing sequences of balanced and convex sets in locally convex spaces, in *Summer school on topological vector spaces*, L. Waelbroeck, ed., Lecture Notes in Mathematics, Vol. 331, Springer-Verlag, Berlin, pp. 211–217.

[146] de Wilde, M. [1978] *Closed graph theorems and webbed spaces*, Pitman, London.

[147] de Wilde, M., and C. Houet [1971] On increasing sequences of absolutely convex sets in locally convex spaces, Math. Ann. *192*, 257–261.

[148] Dierolf, P., S. Dierolf, and L. Drewnowski [1978] Remarks and examples concerning unordered Baire-like and ultrabarreled spaces, Coll. Math. *39*, 109–116.

[149] Diestel, J. [1975] *Geometry of Banach spaces*, Lecture Notes in Mathematics, Vol. 485, Springer-Verlag, New York.

[150] Diestel, J. [1984] *Sequences and series in Banach spaces*, Graduate Texts in Mathematics 92, Springer-Verlag, New York

[151] Dieudonné, J. [1942] La dualité dans les espaces espaces vectoriels topologiques, Ann. Soc. Ecole Norm. Sup. *59*, 107–139. Also Dieudonné 1981b, below, pp. 235–267.

[152] Dieudonné, J. [1944] Sur les fonctions continues $p$-adiques, Bull. Sci. Math. *68*, 79–95. Also Dieudonné 1981b, below, pp. 177–192.

[153] Dieudonné, J. [1952] Sur les propriétés de permanence de certains espaces vectoriels topologiques, Ann. Soc. Math. Polon. *25*, 50–55. Also Dieudonné 1981b, below, pp. 346–351.

[154] Dieudonné, J. [1953] Recent developments in the theory of locally convex spaces, Bull. Amer. Math. Soc. *59*, 495–512.

[155] Dieudonné, J. [1957] Denumerability conditions in locally convex vector spaces, Proc. Amer. Math. Soc. *8*, 367–372. Also Dieudonné 1981b, below, pp. 357–362.

[156] Dieudonné, J. [1969] *Foundations of modern analysis I*, Academic Press, New York.

[157] Dieudonné, J. [1981a] *History of functional analysis*, North-Holland, Amsterdam.

[158] Dieudonné, J. [1981b] *Choix d'oeuvres mathématiques I*, Hermann, Paris.

[159] Dieudonné, J., and L. Schwartz [1950] La dualité dans les espaces $(F)$ et $(LF)$, Ann. Inst. Fourier *1*, 61–101.

[160] Dirichlet, P. [1829] Sur la convergence des séries trigonométriques qui servent a représenter une function arbitraire entre des limites données, J. für d. reine u. angewandte Mathematik. *4*, 157–169.

[161] Diximier, J. [1948] Sur un théorème de Banach, Duke Math. J. *15*, 1057–1071.

[162] Diximier, J. [1964] *Les $C^*$-algèbres et leurs represéntations*, Gauthier-Villars, Paris.

[163] Dodu, J., and M. Morillon [1999] The Hahn–Banach property and the axiom of choice, MLQ Math. Log. Q. *45*, 299–314.

[164] Dugundji, J. [1966] *Topology*, Allyn and Bacon, Boston.

[165] Dunford, N., and J. Schwartz [1958] *Linear operators, part I: General theory*, Interscience, New York.

[166] Duren, P., Romberg, B. and A. Shields [1969] Linear functionals on $H^p$ spaces with $0 < p < 1$, J. Reine u. Angew. Math. *238*, 32-60.

[167] Eberlein, W. [1978] Weak compactness in Banach spaces, Proc. Nat. Acad. Sci. USA *33*, 51–53.

[168] Edwards, R. E. [1965] *Functional analysis*, Holt, Rinehart and Winston, New York.

[169] Eggleston, H. [1965] Sets of constant width in finite-dimensional Banach spaces, Israel J. Math. *3*, 163–172.

[170] Enflo, P. [1973] A counterexample to the approximation property in Banach spaces, Acta Math. *130*, 309–317.

[171] Engelking, R. [1977] *Outline of general topology*, Polska Akademia Nauk., Inst. Matematyczny, Warsaw.

[172] Farid, F., and K. Varadarajan [1994] Isometric shift operators on $C(X)$, Can. J. Math. *46*, 532–542.

[173] Feinberg, M., and R. Lavine [1983] Thermodynamics based on the Hahn–Banach theorem: The Clausius inequality, Arch. Rat. Mech and Anal. *82*, 203–293.

[174] Fichtenholz, G., and L. Kantorovitch [1934] Sur les opérations linéaires dans l'espace des fonctions bornées, Studia Math. *5*, 69–98.

[175] Fischer, E. [1907] Sur la convergence en moyenne, C. R. Acad. Sci. *144*, 1022–1024.

[176] Fréchet, M. [1904] Generalisation d' un théorème de Weierstrass, C. R. Acad. Sci. *139*, 848–850.

[177] Fréchet, M. [1905] Sur les opérations linéaires, Trans. Amer. Math. Soc. *6*, 134–140.

[178] Fréchet, M. [1906] Sur quelques points du calcul fonctionnel, Rend. Circ. Mat. Palermo *22*, 1–71. (Fréchet's thesis.)

[179] Fréchet, M. [1908] Essai de géométrie analytique à une infinité de coordonnées, Nouv. Ann. de Math. *8*, 97–116 and 289–317.

[180] Fréchet, M. [1925] Les espaces abstraits topologiquement affines, Acta Math. *47*, 25–52.

[181] Fuchsteiner, B. [1977a] Decomposition theorems, Manus. Math. *22*, 151–164.

[182] Fuchsteiner, B. [1977b] When does the Riesz representation theorem hold? Arch. Math. *28*, 173–181.

[183] Fuchsteiner, B. [1981] Generalized Hewitt–Nachbin spaces arising in statespace completions, in *Functional analysis, holomorphy and approximation theory,* S. Machado, ed., Lecture Notes in Mathematics, Vol. 843, Springer-Verlag, Berlin.

[184] Fuchsteiner, B., and König [1978] New versions of the Hahn–Banach theorem, *Proc. 2nd international conference on general Inequalities,* Math. Research Institute, Oberwolfach. Black Forest, July 30–August, 1978, E. Beckenbach, ed., International Series of Numerical Math., Vol. 47, Birkhäuser Verlag, Basel, pp. 255–266.

[185] Fuchssteiner, B., and W. Lusky [1981] Convex cones, *Proc. 2nd international conference on general inequalities,* Math. Research Institute, Oberwolfach Black Forest, July 30–August 5, 1978, E. Beckenbach, ed., International Series of Numerical Math. *47*, Birkhäuser Verlag, Basel, 255–266.

[186] Gajek, L., J. Jachymski, J., and D. Zagrodny [1995] Projections, extendability of operators and the Gateaux derivative of the norm, J. Appl. Anal. *1*, 29–38.

[187] Garabedian, P., and M. Schiffman [1954] On solution of partial differential equations by the Hahn–Banach theorem, Trans. Amer. Math. Soc. *76*, 288–299.

[188] Garnir, H., M. de Wilde, and J. Schmets [1968] *Analyse fonctionelle I,* Birkhäuser Verlag, Basel.

[189] Gau, H-L., J-S. Jeang and N-C. Wong [2002] A Banach–Stone theorem for separating linear bijections, Taiwanese J. Math. *6*, 399–403.

[190] Gau, H-L., J-S. Jeang and N-C. Wong [2003] Biseparating linear maps between continuous vector-valued function spaces, J. Australian Math. Soc., Series A, *74*, 101–111.

[191] Gelfand, I., and A. Kolmogorov. [1939] On rings of continuous functions on topological spaces, Dokl. Akad. Nauk SSSR *22*, 11–15.

[192] Giles, J. [1967] Classes of semi-inner product spaces, Trans. Amer. Math. Soc. *321*, 436–446.

[193] Giles, J. [1982] *Convex analysis with applications in differentiation of convex functions*, Pitman, London.

[194] Gillman, L., and M. Jerison [1960] *Rings of continuous functions*, van Nostrand, New York.

[195] Goldstine, H. [1938] Weakly complete Banach spaces, Duke Math. J. *4*, 125–131..

[196] Goodner, D. B. [1950] Projections in normed linear spaces, Trans. Amer. Math. Soc. *69*, 89–108.

[197] Goodner, D. B. [1960] Separable spaces with the extension property, J. Lond. Math. Soc. *35*, 239–240.

[198] Goodner, D. B. [1964] The closed convex hull of certain extreme points, Proc. Amer. Math. Soc. 15, 256–258.

[199] Goodner, D. B. [1965] On the continuous function space of a basically disconnected space, Proc. Amer. Math. Soc. *16*, 932–934.

[200] Goodner, D. B. [1970] Subspaces of $C(S)$ isometric to $m$, J. Lond. Math. Soc. (2), *3*, 488–492.

[201] Goodner, D. B. [1973] Spaces isomorphic to $P_1$ spaces, Indian J. Math. *15*, 73–76.

[202] Goodner, D. B. [1976] Banach spaces with the bound-one projection property, Indian J. Math. *18*, 113–115.

[203] Gowers, W. [1994] A solution to Banach's hyperplane problem, Bull. Lond. Math. Soc. *26*, 523–530.

[204] Gowers, W. [1996] A solution of the Schroeder–Bernstein problem for Banach spaces, Bull. Lond. Math. Soc. *28*, 297–304.

[205] Gowers, W., and B. Maurey [1993] The unconditional basic sequence problem, J. Amer. Math. Soc. *6*, 851–874.

[206] Granero, A., M. Jiménez-Sevilla and J. Moreno [2004] Intersections of closed balls and the geometry of Banach spaces, Extracta Math. *19*, 55–92.

[207] Gray, J., and M. Micallef [2008] About the cover: The work of Jesse Douglas on minimal surfaces, Bull. Amer. Math. Soc. *45,* 293–302.

[208] Gregory, D., and J. Shapiro [1970] Nonconvex linear topologies with the Hahn–Banach extension property, Proc. Amer. Math. Soc. *25*, 902–905.

[209] Grilliot, T. [1967] Extensions of algebra homomorphisms, Mich. Math. J. *14*, 107–116.

[210] Grothendieck, A. [1950a] Sur la complétion du dual d'un espace vectoriel localement convexe, C. R. Acad. Sci. Paris *230*, 605–606.

[211] Grothendieck, A. [1950b] Quelques résultats relatifs à la dualité dans les espaces $(F)$, C. R. Acad. Sci. Paris *230*, 1561–1563.

[212] Grothendieck, A. [1950c] Critères généraux de compacité dans les espaces $(LF)$, C. R. Acad. Sci. Paris *231*, 940–942.

[213] Grothendieck, A. [1951a] Quelques résultats sur les espaces vectoriels topologiques, C. R. Acad. Sci. Paris *233*, 839–841.

[214] Grothendieck, A. [1951b] Sur une notion de produit tensoriel topologique d'espaces vectoriels topologiques et une classe remarquable d'espaces vectoriels liées à cette notion, C. R. Acad. Sci. Paris *233*, 1556–1558.

[215] Grothendieck, A. [1952] Critères de compacité dans les espaces fonctionnels généraux, Amer. J. Math. *74*, 168–186.

[216] Grothendieck, A. [1953a] Sur les applications linéaires faiblement compactes d'espaces du type $C(K)$, Canad. J. Math. *5*, 129–173.

[217] Grothendieck, A. [1953b] Sur certains espaces fonctions holomorphes I, II, J. Reine Angew. Math. *192*, 35–64, 77–95.

[218] Grothendieck, A. [1953c] Sur les espaces de solutions d'une classe générale d'équations aux dérivées partielles, J. Anal. Math. *2*, 243–280.

[219] Grothendieck, A. [1954a] Sur les espaces $(F)$ et $(DF)$. Summa Brasil. Math. *3*, 57–123.

[220] Grothendieck, A. [1954b] *Espaces vectoriels topologiques*, Sociedade de Matemática de São Paulo, São Paulo.

[221] Grothendieck, A. [1954c] Resumé des résultats essentiels dans la théorie des produits tensoriels topologiques et des espaces nucléaires, Ann. Inst. Fourier *4*, 73–112.

[222] Grothendieck, A. [1955] Produits tensoriels topologiques et espaces nucléaires. Mem. Amer. Math. Soc. *16*, Providence.

[223] Grothendieck, A. [1973] *Topological vector spaces* [translation of Grothendieck 1954b], Gordon and Breach, New York.

[224] Grünbaum, B. [1960] Some applications of expanson constants, Pacific J. Math. *10*, 193–201.

[225] Grünbaum, B., and V. Klee [1967] Convexity and applications, Committee on the Undergraduate Program in Mathematics no. 16, Mathematical Association of America, Washington, D.C.

[226] Gutek, A., D. Hart, J. Jamison and M. Rajagopalan [1991], Shift operators on Banach spaces, J. Funct. Anal. *101*, 97–119.

[227] Hadamard, J. [1968] *Oeuvres*, 4 vols. CNRS, Paris.

[228] Hahn, H. [1922] Über Folgen linearer Operationen, Monatsh. Math. Phys. *32*, 1–88.

[229] Hahn, H. [1927] Über linearer Gleichungssysteme in linearer Räumen, J. Reine Angew. Math. *157*, 214–229.

[230] Halmos, P. [1974a] *Measure theory*, Graduate Texts in Mathematics 18, Springer-Verlag, New York, a reprint of the 1950 edition from van Nostrand, Princeton, N.J.

[231] Halmos, P. [1974b] *Finite-dimensional vector spaces*, 2nd ed., Undergraduate Texts in Mathematics, Springer-Verlag, New York. A reprint of the 1958 2nd edition published by D. Van Nostrand, Princeton, N.J.

[232] Halmos, P. [1974c] *Lectures on Boolean algebras*, Springer-Verlag, New York. A reprint of the 1963 edition from Van Nostrand, Princeton, N.J.

[233] Halpern, J. [1964] The independence of the axiom of choice from the Boolean prime ideal theorem, Fund. Math. *55*, 57–66.

[234] Hardy, G., and W. Rogosinski [1950] *Fourier series*, Cambridge Tracts 38, Cambridge.

[235] Harmand, P., D. Werner and W. Werner [1993] *M-ideals in Banach spaces and Banach algebras*, Lecture Notes in Math. 1547, Springer-Verlag, New York.

[236] Hasumi, M. [1958] The extension property of complex Banach spaces, Tohoku Math, J. (2), *10*, 135–142.

[237] Hausdorff, F. [1914, 1917] *Grundzüge der Mengenlehre*, Veit, Leipzig.

[238] Helly, E. [1912] Über linearer Funktionaloperationen, Österreich. Akad. Wiss. Math.-Natur. Kl. S.-B. IIa, *121*, 265–297.

[239] Helly, E. [1921] Über Systeme linearer Gleichungen mit unendlich vielen Unbekannten, Monatsh. für Math. Phys. *31*, 60–91.

[240] Helly, E. [1923] Über Mengen konvexer Körper mit gemeinschaftlichen Punkten, Jahresber. Deutsch. Math.-Verein *32*, 175–176.

[241] Helly, E. [1930] Über Systeme von abgeschlossenen Mengen mit gemeinschaftlichen Punkten, Monatsh. für Math. und Phys., *37*, 281–302.

[242] Helly, W. [1975] *Urban systems models*, Academic Press, New York.

[243] Hennefeld, J. [1979] $M$-ideals, HB-subspaces and compact operators, Indiana Univ. Math. J. *28*, 927–934.

[244] Henriksen, M. [1956] On the equivalence of the ring, lattice and semigroup of continuous functions, Studia Math. *28*, 289–294.

[245] Herman, R., and R. Whitley [1967] An example concerning reflexivity, Proc. Amer. Math. Soc. *7*, 959–960.

[246] Hernández, S., E. Beckenstein and L. Narici [1995] Banach–Stone theorems and separating maps, Manuscripta Math. *86*, 409–416.

[247] Herrero, P. [2003] El teorema de Hahn–Banach: historia, versiones analíticas, geometría y aplicaciones. La propiedad de extensión, Universidad de Murcia, Murcia, Spain.

[248] Hildebrandt, T. [1923] On uniform limitedness of sets of functional operations, Bull Amer. Math. Soc. *29*, 309–315.

[249] Hildebrandt, T. [1931] Linear functional transformation in general spaces, Bull Amer. Math. Soc. *37*, 185–212.

[250] Hille, E., and R. Phillips [1957] *Functional analysis and semigroups*, Amer. Math. Soc. Coll. Publ. *31*, New York.

[251] Hogbe-Nlend, H. [1971] *Théorie des bornologies et applications.* Lecture Notes in Mathematics, Vol. 213, Springer-Verlag, Berlin.

[252] Hogbe-Nlend, H. [1973] Techniques de bornologie en théorie des espaces vectoriels topologiques, in *Summer School on Topological Vector Spaces* (L. Waelbroeck, ed.), Lecture Notes in Mathematics, Vol. 331, Springer-Verlag, Berlin, pp. 84–162.

[253] Holbrook, J. [1975] Concerning the Hahn–Banach theorem, Proc. Amer. Math. Soc. *50*, 322–327.

[254] Holmes, R. [1972] *A course on optimization and best approximation*, Lecture Notes in Mathematics, Vol. 257, Springer-Verlag, Berlin.

[255] Holmes, R. [1975] *Geometric functional analysis and its applications*, Graduate Texts in Mathematics, Vol. 24, Springer-Verlag, Berlin.

[256] Holmes, R., B., Scranton and J. Ward [1975] Approximation from the space of compact operators and other $M$-ideals, Duke Math. J. *42*, 259–269.

[257] Holsztynski, W. [1966] Continuous mappings induced by isomegtries of spaces of continuous functions, Studia Math. *26*, 133–136.

[258] Horvath, J. [1966] *Topological vector spaces and distributions*, Vol. I, Addison-Wesley, Reading, Mass.

[259] Horvath, J. [1973] Locally convex spaces, in *Summer school on topological vector spaces*, L. Waelbroeck, ed., Lecture Notes in Mathematics, Vol. 331, Springer-Verlag, Berlin, pp. 41–83.

[260] Howe, R. [1983] Very basic Lie theory, Amer. Math. Monthly *90*, 589–623.

[261] Hu, S. [1949] Boundedness in a topological space, J. Math. Pures Appl. *28*, 287–320.

[262] Huijsmans, C. and B de Pagter [1994] Invertible disjointness preserving operators, Proc. Edinburgh Math. Soc.. *37*, 125–132.

[263] Hunt, R. [1968] *Orthogonal expansions and their continuous analogues*, Southern Illinois University Press, Carbondale, Illinois, 235–255.

[264] Husain, T. [1965] *The open mapping and closed graph theorems in topological vector spaces*, Oxford University Press, London.

[265] Husain, T. [1966] *Introduction to topological groups*, W. B. Saunders, Philadelphia.

[266] Husain, T., and S. Khaleelulla [1978] *Barreledness in topological and ordered vector spaces.* Lecture Notes in Mathematics, Vol. 692, Springer-Verlag, Berlin.

[267] Hustad, O. [1973] A note on complex $P_1$-spaces, Israel J. Math. *16*, 117–119.

[268] Hustad, O. [1974] Intersection properties of balls in complex Banach spaces whose duals are $L_1$ spaces, Acta Math. *132*, 283–313.

[269] Ingleton, A. [1952] The Hahn–Banach theorem for non-Archimedean fields, Proc. Camb. Philos. Soc. *48*, 41–45.

[270] Ioffe, A. [1981] A new proof of the equivalence of the Hahn–Banach extension and the least upper bounded properties. Proc. Amer. Math. Soc. *82*, 385–389.

[271] Ishihara, H. [1989] On the constructive Hahn–Banach theorem, Bull. London Math. Soc. *21*, 79–81.

[272] Iyahen, S. [1968] On certain classes of linear topological spaces, Proc. Lond. Math. Soc. (3) *18*, 285–307.

[273] Iyahen, S. [1971] On certain classes of linear topological spaces II, J. Lond. Math. Soc. (2), *3*, 609–617.

[274] Iyahen, S. [1972] Corrigendum. On certain classes of linear topological spaces II, J. Lond. Math. Soc. (2), *5*, 740.

[275] Jacobson, N. [1953] *Lectures in abstract algebras II: Linear algebra*, D. Van Nostrand, Princeton, N.J.; reprinted in 1975 as Graduate Texts in Mathematics, Vol. 31, by Springer-Verlag, Berlin.

[276] Jacobson, N. [1964] *Lectures in abstract algebra III: Theory of fields and Galois Theory*, D. Van Nostrand, Princeton, N.J.; reprinted in 1976 as Graduate Texts in Mathematics, Vol. 32, by Springer-Verlag, New York.

[277] James, R. C. [1947] Orthogonality and linear functionals in normed linear spaces, Trans. Amer. Math. Soc. *61*, 1947, 265–292.

[278] James, R. C. [1950] Bases and reflexivity of Banach spaces, Ann. of Math. *52*, 518–527.

[279] James, R. C. [1951] A non-reflexive Banach space isometric with its second conjugate space, Proc. Nat. Acad. Sci. USA *37*, 174–177.

[280] James, R. C. [1957] Reflexivity and the supremum of linear functionals, Ann. of Math. *66*, 159–169.

[281] James, R. C. [1964] Characterizations of reflexivity, Studia Math. *23*, 205–216.

[282] James, R. C. [1957] Reflexivity and the supremum of linear functionals, Ann. of Math. *66*, 159–169.

[283] James, R. C. [1972] Reflexivity and the sup of linear functionals, Israel J. Math. *13*, 289–300.

[284] James, R. C. [1974] A separable somewhat reflexive Banach space with nonseparable dual, Bull. Amer. Math. Soc. *80*, 738–743.

[285] Jameson, G. [1970] *Ordered linear spaces*, Lecture Notes in Mathematics, Vol. 141, Springer-Verlag, Berlin.

[286] Jamison, R. E., R. O'Brien, and P. Taylor [1976] On embedding a compact convex set into a locally convex topological vector space, Pac. J. Math. *64*, 193–205.

[287] Jarchow, H. [1981] *Locally convex spaces*, B. G. Teubner, Stutgart, West Germany.

[288] Jeang, J.-S., and Ngai-Ching Wong [1996] Weighted composition operators of $C_0(X)$'s, J. Math. Anal. Appl. *201*, 981–993.

[289] Jeang, J.-S., and N-C. Wong [1997] Into isometries of $C_0(X, E)$'s, J. Math. Anal. Appl. *207*, 286–290.

[290] Jeang, J.-S., and N-C. Wong [2003] On the Banach–Stone problem, Studia Math. *155*, no. 2, 95–105.

[291] Jerison, M. [1950] The space of bounded maps into a Banach space, Ann. of Math. *52*, 309–327.

[292] Jarosz, K. [1990] Automatic continuity and separating linear isomorphisms, Ann. of Math. *52*, 309–327.

[293] Johnson, W., H. Rosenthal and M. Zippin [1971] On bases, finite-dimensional decompositions and weaker structures in Banach spaces, Israel J. Math. *0*, 488–506.

[294] Jones, W. B. [1966] Duality and types of competeness in locally convex spaces, Can. Math. Bull. *33*, 1–6.

[295] Kadets, M. [1967] A proof of the topological equivalence of all separable infnite-dimensional Banach spaces, Funct. Anal. Appl. *1*, 53–62.

[296] Kakol, J. [1992] Simple construction of spaces without the Hahn–Banach extension property, Comment. Math. Univ. Carolin. *33*, 623–624.

[297] Kakutani, S. [1936] Über Metrisation der toplogiischen Gruppen, Proc. Imp. Acad. Math. Tokyo *12*, 82–84.

[298] Kakutani, S. [1939] Weak topology and regularity of Banach spaces, Proc. Imp. Acad. Math. Tokyo *15*, 169–173.

[299] Kakutani, S. [1941] Concrete representations of abstract $(M)$-spaces, Ann. of Math. *42*, 994–1024.

[300] Kalton, N. [1975] Basic sequences in $F$-spaces and their applications, Proc. Edinburgh Math. Soc. *19*, 151–167.

[301] Kalton, N. J. [1980] An $F$-space with trivial dual where the Krein–Milman theorem holds, Israel J. Math. *36*, 41–50.

[302] Kalton, N., and N. Peck [1980] A re-examination of the Roberts example of a compact convex set without extreme points, Math. Ann. *253*, 89–101.

[303] Kalton, N., N. Peck and J. Roberts [1984] *An F-space sampler*, Cambridge University Press, London Mathematical Society Lecture Notes 89, Cambridge.

[304] Kaplansky, I. [1947] Lattices of continuous functions, Bull. Amer. Math. Soc. *53*, 617–623.

[305] Kaplansky, I. [1950] The Weierstrass theorem in fields with valuations, Proc. Amer. Math. Soc. *1*, 356–357.

[306] Katznelson, Y. [1968] *An introduction to harmonic analysis*, Wiley, New York.

[307] Kaufman, R. [1966a] Interpolation of additive functionals, Studia Math. 27, 269–272.

[308] Kaufman, R. [1966b] A type of extension of Banach spaces, Acta Sci. Math. (Szeged) *27*, 163, 166.

[309] Kelley, J. [1950] The Tychonoff product theorem implies the axiom of choice, Fund. Math. Soc. *37*, 75–76.

[310] Kelley, J. [1951] Note on a theorem of Krein and Milman, J. Osaka Inst. Sci. Tech., Part I, *3*, 1–2.

[311] Kelley, J. [1952] Banach spaces with the extension property, Trans. Amer. Math. Soc. *72*, 323–326.

[312] Kelley, J. [1976] *General topology*, Graduate Texts in Mathematics, Vol. 27, by Springer-Verlag, New York., a reprint of the 1955 edition published by D. Van Nostrand, Princeton, N.J.

[313] Kelley, J., and I. Namioka [1976] *Linear topological spaces*, Graduate Texts in Mathematics 36, Springer-Verlag, New York, a reprint of the 1963 version published by D. Van Nostrand, Princeton, N. J.

[314] Klee, V. [1953] Convex bodies and periodic homeomorphisms in Hilbert space, Trans. Amer. Math. Soc. *74*, 10–43.

[315] Klee, V. [1963] Convexity, Proc. Symp. Pure Math. 7, V. Klee ed., American Mathematical Society, Providence, R.I.

[316] Koldunov, A. [1995] Hammerstein operators preserving disjointness, Proc. Amer. Math. Soc. *123*, 1083–1095.

[317] Kolmogorov, A. [1934] Zur Normierbarkeit eines allgemeinen topologischen Raumes, Studia Math. *5*, 29–33.

[318] Komura, Y. [1962] On linear topological spaces, Kuamoto J. Sci. A *5*, 148–157.

[319] Komura, Y. [1964] Some examples on linear topological spaces, Math. Ann. *153*, 150–162.

[320] König, D. [1922] Über konvexe Körper, Math. Z. *14*, 208–210.

[321] König, H. [1968] Über das von Neumannsche Minimax theorem, Arch. Math. *19*, 482–487.

[322] König, H. [1970] On certain applications of the Hahn–Banach and minimax theorems, Arch. Math *21*, 583–591.

[323] König, H. [1972a] Sublineare Funktionale, Arch. Math. *23*, 500–508.

[324] König, H. [1972b] Sublineare Funktionale, Matematisk Institut, Aarhus Universitet, Open House for Functional Analysts, March 20–28, 1972.

[325] König, H. [1978] Neue Methoden und Resultate aus Funktionalanalysis und konvexer Analysis, Oper. Res. Verf. *28*, 6–16.

[326] König, H. [1980] Der Hahn–Banach Satz von Rodé für unendlichstellige Operationen, Arch. Math. *35*, 292–304.

[327] König, H. [1982] On some basic theorems in convex analysis, in *Modern applied mathematics: Optimization and operations research*, B. Korte, ed., North-Holland, Amsterdam, pp. 106–144.

[328] König, H. [1987] On the abstract Hahn–Banach theorem due to Rodé, Aequ. Math *34*, 89–95.

[329] Köthe, G. [1950] Über zwei Sätze von Banach, Math. Z. *53*, 203–209.

[330] Köthe, G. [1965] General linear transformations of locally convex spaces, Math. Ann. *159*, 309–328.

[331] Köthe, G. [1968] Abbildungen von $(F)$-Räumen in $(LF)$-Räumen, Math. Ann. *178*, 1–3.

[332] Köthe, G. [1979] *Topological vector spaces II*, Grundlehren der Math. Wiss. 237, Springer-Verlag, Berlin-Heidelberg-New York.

[333] Köthe, G. [1983] *Topological vector spaces I*, Second printing, revised, Grundlehren der Math. Wiss. 159, Springer-Verlag, Berlin-Heidelberg-New York.

[334] Krabs, W. [1979] *Optimization and approximation*, Wiley, New York.

[335] Krein, M. [1938] The $L$-problem in an abstract linear normed space, in Akhieser and Krein 1962, pp. 175–204.

[336] Krein, M., and D. Milman [1940] On the extreme points of regularly convex sets, Studia Math. *9*, 133–138.

[337] Krein, M., and V. Smulian [1940] On regularly convex sets in the space conjugate to a Banach space, Annals of Math. *41*, 556–583.

[338] Krull, W. [1928] Galoissche Theorie unendlicher algebraischen Erweiterungen. Math. Ann. *100*, 678–698.

[339] Lacey, H., and J. Cohen [1969] On injection envelopes of Banach spaces, J. Funct. Anal. *4*, 11–30.

[340] Lamperti, J. [1958] On the isometries of certain function spaces, Pacific. J. Math. *8*, , 459–466.

[341] Landsberg, M. [1956] Lineare topologische Räume die nicht lokalconvex sind, Math. Z. *65*, 104–112.

[342] Larsen, R. [1973a] *Functional analysis*, Marcel Dekker, New York.

[343] Larsen, R. [1973b] *Banach algebras.* Marcel Dekker, New York.

[344] Lassonde, M. [1998] Uniqueness of Hahn–Banach extensions and liftings of linear dependences, Math. Scand. *53*, 97–113.

[345] Lau, K. [1956] A representation theorem for isometries of $C(X, E)$, Pacific J. Math. *60*, 229–232.

[346] Leigh, J. [1980] *Functional analysis and linear control theory*, Academic Press, New York.

[347] Levin, M., and S. Saxon [1971] A note on the inheritance of properties of locally convex spaces by subspaces of countable codimension, Proc. Amer. Math. Soc. *29*, 97–102.

[348] Lévy, P. [1922] *Leçons d'analyse fonctionnelle*, Gauthier-Villars, Paris.

[349] Lima, Å. [1983] Uniqueness of Hahn–Banach extensions and liftings of linear dependences, Math. Scand. *53*, 97–113.

[350] Lindenstrauss, J. [1962] On the extension property for compact operators, Bull. Amer. Math. Soc. *68*, 484–487.

[351] Lindenstrauss, J. [1963a] On a problem of Nachbin concerning extension of operators, Israel J. Math. *1*, 75–84.

[352] Lindenstrauss, J. [1963b] Some results on the extension of operators, Bull. Amer. Math. Soc. *69*, 582–586.

[353] Lindenstrauss, J. [1964a] On the extension of operators with a finite-dimensional range, Ill. J. Math. 8, 488–499.

[354] Lindenstrauss, J. [1964b] On the extension of operators with range in a $C(K)$ space, Proc. Amer. Math. Soc. *15*, 218–225.

[355] Lindenstrauss, J. [1964c] Extension of compact operators, Mem. Amer. Math. Soc. (Providence) *48*.

[356] Lindenstrauss, J. [1966] On extreme points in $\ell_1$, Israel J. Math. *5*, 153–156.

[357] Lindenstrauss, J. [1967] On complemented subspaces of $m$, Israel J. Math. *4*, 59–61.

[358] Lindenstrauss, J., and C. Stegall [1967] Examples of separable Banach spaces which do not contian $\ell_1$ and whose duals are non-separable, Studia Math. *54*, 81–105.

[359] Lindenstrauss, J., and L. Tzafriri [1971] On the complemented subspaces problem, Israel J. Math. *9*, 263–269.

[360] Lindenstrauss, J., and L. Tzafriri [1977] *Classical Banach spaces I*, Lectures Notes in Mathematics, Vol. 338/Ergebnisse der Mathematic und ihre Grenzgebiete, Vol. 92, Springer-Verlag, Berlin.

[361] Lorch, E. [1939] On a calculus of operators in reflexive vector spaces, Trans. Amer. Math. Soc. *45*, 217–234.

[362] Łoś, J. and C. Ryll-Nardzewski, [1951] *On the applications of Tychonoff's theorem in mathematical proofs*, Fund. Math. *38*, 233–237.

[363] Löwig, H. [1934] Komplexe euklidische Räume von beliebiger endlicher oder transfiniter Dimensionzahl, Acta Sci. Math (Szeged) *7*, 1–33.

[364] Lumer,G. [1961] Semi-inner product spaces, Trans. Amer. Math.Soc. *100*, 1961, 29–43.

[365] Luxemburg, W. [1962] Two applications of the method of construction by ultrapowers to analysis, Bull. Amer. Math. Soc. *68*, 416–419.

[366] Luxemburg, W. [1967a] *Beweis des Sätzes von Hahn–Banach*, Arch. Math (Basel) *18*, 271–272.

[367] Luxemburg, W. [1967b] Reduced powers of the real number system and equivalents of the Hahn–Banach theorem, Technical Report 2, Cal. Inst. Tech. It also appeared in *Applications of model theory to algebra, analysis and probability*, Holt, Rinehart and Winston, New York, 1969, 123–137.

[368] Luxemburg, W. and M. Väth [2001] The existence of non-trivial bounded functionals implies the Hahn–Banach theorem, Z. Anal. Anwendungen *20*, 267–279 (electronic publication).

[369] Luxemburg, W., and A. Zaanen [1971] *Riesz spaces*, Vol. I, North-Holland, Amsterdam.

[370] Mackey, G. [1945] On infinite-dimensional linear spaces, Trans. Amer. Math. Soc. *57*, 155–207.

[371] Mackey, G. [1946] On convex topological linear spaces, Trans. Amer. Math. Soc. *60*, 519–537.

[372] Maddox, I., and J. Roles [1969] Absolute convexity in certain topological linear spaces, Proc. Camb. Philos. Soc. *66*, 541–545.

[373] Mahler, K. [1973] *Introduction to p-adic numbers and their functions*, Cambridge, University Press, Cambridge.

[374] Mahowald, M. [1961] Barreled spaces and the closed graph theorem, J. Lond. Math. Soc. *36*, 108–110.

[375] Martineau, A. [1966a] Sur le théorème du graphe fermé, C. R. Acad. Sci. Paris *263*, 870–871.

[376] Martineau, A. [1966b] Sur le théorème du graphe fermé, Sém. Lelong. 7[e] année, no. 6, Faculté des Sciences de Paris.

[377] Martineau, A. [1968] Sur les théorèmes de S. Banach et L. Schwartz concernant le graphe fermé, Studia Math. *30*, 43–51.

[378] Mauldin, R. (ed.) [1981] *The Scottish book*, Birkhäuser, Boston, Mathematics from the Scottish Cafe, including selected papers presented at the Scottish Book Conference held at North Texas State University, Denton, Texas, May 1979. For a copy of the original with entries in Polish, Russian, English, French and German, see http://banach.univ.pl/e-scottish-book.html. For an English version, see http://www.math.uni.wroc.pl/wydzial/index/php.

[379] Mazur, S. [1930] Über die kleinste konvexe Menge, die eine gegebene kompakte Menge enthält, Studia Math. *2*, 7–9.

[380] Mazur, S. [1933] Über konvexe Mengen in linearen normierten Räumen, Studia Math. *4*, 70–84.

[381] Mazur, S., and W. Orlicz [1933] Über Folgen linearer Operatoren, Studia Math. *4*, 152–157.

[382] Mazur, S., and W. Orlicz [1953] Sur les espaces métriques linéaires II, Studia Math. *13*, 137–179.

[383] Mazur, S., and S. Ulam [1932] Sur les transformations isométriques d'espaces vectoriels normés, C. R. Acad. Sci. Paris *194*, 946–948.

[384] Megginson, R. [1998] *An introduction to Banach space theory*, Grad. Texts in Math. *183*, Springer, New York.

[385] Milman, D. [1938] On some criteria for the regularity of spaces of type (B), Dokl. Akad. Nauk SSSR N.S. *20*, 243–246 (Russian).

[386] Minkowski, H. [1896] *Geometrie der Zahlen*, Teubner, Leipzig.

[387] Minkowski, H. [1911] *Gesammelte Abhandlungen I–II*, Teubner, Leipzig-Berlin. For the definition of extreme point in $\mathbf{R}^3$, see *Theorie der konvexen Körper, ins besondere Begründung ihres Oberflächenbegriffs*, posthumously published in Vol. II, pp. 131–229.

[388] Mira, J. [1982] A unified approach to the extension problem for normed spaces, Boll. Un. Mat. Ital. *1*, 225–232.

[389] Moreno, J. [2007] Semicontinuous functions and convex sets in $C(K)$ spaces, J. Australian Math. Soc. *82*, 111–121.

[390] Moreno, J., P. Papini and R. Phelps [2005] Diametrically maximal and constant width sets in Banach spaces, Canadian J. Math., *58*, 820–842.

[391] Mukherjea, A., and K. Pothoven [1978] *Real and functional analysis*, Plenum, New York.

[392] Mulvey C., and J. Pelletier [1991] A globalization of the Hahn–Banach theorem, Adv. Math. *89*, 1–59.

[393] Murray, F. [1936] Linear transformations in $L_p$, $p > 1$, Trans. Amer. Math. Soc. *39*, 83–100.

[394] Murray, F. [1937] On complementary manifolds and projections in spaces $L_p$ and $\ell_p$, Trans. Amer. Math. Soc. *41*, 138–152.

[395] Myers, S. [1948] Banach spaces of continuous functions, Trans. Amer. Math. Soc. *41*, 138–152.

[396] Nachbin, L. [1949] On strictly minimal topological division rings, Ann. Math. Soc. *49*, 132–140.

[397] Nachbin, L. [1950] A theorem of the Hahn–Banach type for linear transformations, Trans. Amer. Math. Soc. 68, 28–46.

[398] Nachbin, L. [1954] Topological vector spaces of continuous functions, Proc. Nat. Acad. Sci. USA *40*, 471–474.

[399] Nachbin, L. [1960] Some problems in extending and lifting continuous linear transformations, Proc. International Symposium on Linear Spaces, Jerusalem, pp. 340–350.

[400] Nachbin, L. [1965] *Topology and order*, D. Van Nostrand, Princeton, N.J.: reprinted in 1976 by Robert Krieger, Melbourne, Fla.

[401] Nachbin, L. [1967] *Approximation theory*, D. Van Nostrand, Princeton, N.J.; reprinted in 1976 by Robert Krieger, Melbourne, FL.

[402] Nachbin, L. [1981] *Introduction to functional analysis: Banach spaces and differential calculus*, Marcel Dekker, New York.

[403] Naimark, M. [1960] *Normed rings*, Noordhoff, Groningen, The Netherlands.

[404] Nakano, H. [1971] On the Hahn–Banach theorem, Bull. Acad. Polon. Sci. *19*, 743–745.

[405] Narici, L. [2007] On the Hahn–Banach theorem, *Proceedings of the second international school of analysis in Andalucía 2004*, M. Velasco and A. Rodríguez eds., World Scientific Publishing, Singapore, 87–122. Also: Topology Atlas Preprint #554, http://at.yorku.ca/p/a/a/o58.htm.

[406] Narici, L. and E. Beckenstein [1981] Strange terrain—non-Archimedean spaces, Amer. Math. Monthly *88*, 667–676.

[407] Narici, L., E. Beckenstein and G. Bachman [1971] *Functional analysis and valuation theory*, Marcel Dekker, New York.

[408] Narici, L., E. Beckenstein and J. Arajuo [1994] Separating maps and rings of continuous functions, *p-adic functional analysis*, ed. by N. de Grande-de Kimpe, S. Navarro and W. Schikhof, Editorial de la Universidad de Santiago de Chile, Av. Lib. Bernardo O'Higgins 3363, Santiago, 69–82.

[409] Narici, L., and E. Beckenstein [1997] The Hahn–Banach theorem: The life and times, Topology Appl. *77* (2), 193–211.

[410] Narici, L., and E. Beckenstein [2004] Non-uniqueness of certain Hahn–Banach extensions, Vladikavkaz Mat. Zh. *6*, 26–28.

[411] Narici, L., and E. Beckenstein [2008] The Hahn–Banach theorem and the sad life of E. Helly, *Proceedings of the third international school of analysis in Andalucía 2007*, J. Delgado and T. Domínguez, eds., World Scientific Publishing, Singapore, 97–110.

[412] Neumann, M. [1994] Sandwich theorems for operators of convex type, J. Math. Anal. and Appl. *188*, 759–773.

[413] Neumann, M. and V. Velasco [1998] Network analysis, *Minimax theory and applications*, ed. by B. Ricceri and S. Simons, Kluwer, Dordrecht,167–189.

[414] O'Connor, J., and E. Robertson [2002] *Tibor Radó,* MacTutor History of Mathematics, http://www-history.mcs.st-andrews.ac.uk/Biographies/Rado.html.

[415] Oja, E. [1984] On the uniqueness of the norm preserving extension of a linear functional in the Hahn–Banach theorem, Iav. Akad. Nauk. Es. SSR *33*, 424–439 (Russian).

[416] Oja, E. [1997] HB-subspaces and Godun sets of subspaces in Banach spaces, Mathematika *44*, 120–132.

[417] Oja, E. and M. Põldvere [1999] Intersection of ball sequences and uniqueness of Hahn–Banach extensions, Proc. Royal Soc. Edinburgh *129A*, 1251–1262.

[418] Ono, T. [1953] A generalization of the Hahn–Banach theorem, Nagoya Math. J. *6*, 171–176.

[419] Ostrovskii, M. [2001] Hahn–Banach operators, Proc. Amer. Math. Soc., electronic publication.

[420] Oxtoby, J. [1961] Cartesian products of Baire spaces, Fund. Math. *49*, 157–266.

[421] Oxtoby, J. [1971] *Measure and category*, Graduate Texts in Mathematics, Vol. 2, Springer-Verlag, Berlin.

[422] Oxtoby, J., and S. Ulam [1938] On the equivalence of any set of first category to a set of measure 0, Fund. Math. *31*, 201–206.

[423] Páles, Z. [1992] Geometric versions of Rodé's theorem, Rad. Mat. *8*, 217–229.

[424] Park, S. [1993] A little generalization of the Hahn–Banach extension property, J. Korean Math. Soc. *30*, 139–150.

[425] Peano, G. [1888] *Calcolo geometrico secondo l'Ausdehnungslehre di H. Grassmann preceduto dalle operazioni della logica deduttiva*, Bocca, Turin; see also the reproduction in Formulario Matematico, vol. V, Edizioni Cremonese, Rome, 1960.

[426] Peressini, A. [1967] *Ordered topological vector spaces*, Harper & Row, New York.

[427] Pelczynski, A. [1958] On the isomorphism of the spaces $m$ and $M$, Bull Acad. Polon. Sci., Sér. Math. *6*, 695–696.

[428] Pelczynski, A. [1971] Any separable Banach space with the bounded approximation property is a complemented subspace of a Banach space with a basis, Studia Math. *40*, 239–243.

[429] Peller, V. [1976] An analog of J. von Neumann's inequality for the space $L^p$, Soviet Math. Dokl. *17*, 1976, 1594–1598, translated from Dokl. Akad. Nauk SSSR *231*, 1976.

[430] Pérez Carreras, P., and J. Bonet [1987] *Barrelled locally convex spaces*, North-Holland Math. Studies 131, North-Holland, New York.

[431] Pérez-García, C. [1992] The Hahn–Banach extension property in $p$-adic analysis, in *P-adic functional analysis*, Lecture Notes in Pure and Appl. Math. *137*, Marcel Dekker, New York, 127–140.

[432] Pérez-García, C., and W. Schikhof [2003] Finite-dimensional orthocomplemented subspaces in *P-adic normed spaces. Ultrametric functional analysis*, Contemp. Math. *319*, Amer. Math. Soc., Providence, RI, 281–298.

[433] Pettis, B. [1939] A proof that every uniformly convex space is reflexive. Ultrametric functional analysis, Duke Math. J. *5*, 249–253.

[434] Phelps, R. R. [1957] Subreflexive normed linear spaces, Arch. Math. (Basel) *8*, 444–450, with a correction in vol. 9, pp. 439–440.

[435] Phelps, R. R. [1960a] Uniqueness of Hahn–Banach extensions and unique best approximation, Trans. Amer. Math. Soc. *95*, 238–255.

[436] Phelps, R. [1960b] A representation for bounded convex sets, Proc. Amer. Math. Soc. *11*, 976–983.

[437] Phelps, R. R. [1966] *Lectures on Choquet's theorem*, D. Van Nostrand, Princeton, N.J.; now available through Springer-Verlag, Berlin.

[438] Pietsch, A. [1972] *Nuclear locally convex spaces*, Ergebnisse der Mathematif und ihre Grenzgebiete, Vol. 66, Springer-Verlag, New York.

[439] Pietsch, A. [2007] *History of Banach spaces and linear operators*, Birkhäuser Verlag, Boston-Basel-Berlin.

[440] Pincherle, S. [1954] *Opere scelte*, 2 vol., Cremonese, Roma.

[441] Pincus, D. [1972] Independence of the prime ideal theorem from the Hahn–Banach theorem, Bull. Amer. Math. Soc. *78*, 766–770.

[442] Pincus, D. [1974] The strength of the Hahn–Banach theorem, Proc. Victoria Symposium on Nonstandard Analysis, Lec. Notes Math. 369, Springer-Verlag, New York, 203–248.

[443] Przeworska-Rolewicz, D., and S. Rolewicz [2005] Historical remarks on bounded sets, European mathematicians in the last centuries, Wroclaw, 87–97.

[444] Prolla, J. [1977] *Approximation of vector-valued functions*, North-Holland, Amsterdam.

[445] Prolla, J. [1982] *Topics in functional analysis over valued division rings.* North-Holland Mathematics Studies, Vol. 77, North-Holland, Amsterdam.

[446] Pták, V. [1953] On complete topological linear spaces, Ceh. Mat. Zur. *3* (78), 301–364. (Russian with English summary.)

[447] Pták, V. [1956] On a theorem of Mazur and Orlicz, Studia Math. XV, 365–366.

[448] Pták, V. [1958] Completeness and the open mapping theorem, Bull. Soc. Math. Fr. *86*, 41–74.

[449] Putnam, D. [1968] An easier condition than total boundedness. Pi Mu Epsilon J. *4*, 370–372.

[450] Radó, T. [11930a] Some remarks on the problem of Plateau, Proc. Nat. Acad. Sci. USA *16,* 242–248.

[451] Radó, T. [11930b] On Plateau's Problem, Annals of Math. *31,* 457–469.

[452] Radó, T. [11930c] The problem of the least area and the problem of Plateau, Math. Z. *32*, 763–796.

[453] Radon, J. [1921] Mengen konvexer Körper, die einen gemeinsamen Punkt enthalten, Math. Ann. *83*, 113–115.

[454] Raikov, D. [1966] Double closed graph theorem for topological linear spaces, Siberian Math. J. *7*, 2, 287–300.

[455] Rickart, C. [1960] *General theory of Banach algebras*, D. Van Nostrand, Princeton, N. J.

[456] Riesz, F. [1907] Sur les systèmes orthogonaux de fonctions, C. R. Acad. Sci. *144*, 615–619.

[457] Riesz, F. [1910a] Sur certain systèmes d'équations fonctionelles et l'approximation des fonctions continues, Académie des Sciences, Paris, Comptes Rendus *150*, 674–677.

[458] Riesz, F. [1910b] *Untersuchungen über Systeme integrierbarer Funktionen,* Math. Ann. *69*, 449–497.

[459] Riesz, F. [1911] Sur certain systèmes singuliers d'équations intégrales, Ann. Sci. École Norm. Sup. *28*, 33–62.

[460] Riesz, F. [1913] Les systèmes d'équations linéaires à une infinité d'inconnues, Gauthier-Villars, Paris.

[461] Riesz, F. [1918] Über lineare Funktionalgleichungen, Acta Math. *41*, 71–98. Also in his complete works [1960 below] 1053–1080.

[462] Riesz, F. [1960] *Oeuvres complètes*, 2 vol., Akadémiai Kiadó, Budapest.

[463] Roberts, J. [1976] Pathological compact convex sets in the spaces $L_p$, $0 < p < 1$, The Altgeld Book 1975/76, University of Illinois.

[464] Roberts, J. [1977] A compact convex set with no extreme points, Studia Math. *60*, 255–266.

[465] Roberts, A., and D. Varberg [1973] *Convex functions*, Academic Press, New York.

[466] Robertson, W. [1958] Completions of topological vector spaces, Proc. Lond. Math. Soc. (3). *8*, 242–257.

[467] Roberson, W. [1972] On the closed graph theorem and spaces with webs, Proc. Lond. Math. Soc. *24*, 692–738.

[468] Robertson, A., and W. Robertson [1956] On the closed graph theorem, Proc. Glasgow Math. Assoc. *3*, 9–12.

[469] Robertson, A., and W. Robertson [1973] *Topological vector spaces*, 2nd ed., Cambridge Tracts no. 53, Cambridge University Press, Cambridge.

[470] Rodé, G. [1978] Eine abstrakte Version des Sätzes von Hahn–Banach, Arch. Math. *31*, 474–481.

[471] Rodríguez-Salinas Palero, B. [1971] Algunos problems y teoremas de extensión de aplicaciones lineales, Rev. Real Acad. Ci. Exact. Fis. Natur. Madrid *65*, 677–704.

[472] Rolewicz, S. [1978] On convex sets containing only points of support, Soc. Math. Polon. Comm. Math. Tomus Specialis Honorem, Panstwowe Wydawnictwo Naukowe, Warsaw, pp. 279–281.

[473] Rosell, J., [2007] El teorema de la acotación uniforme, Master's Thesis directed by B. Cascales, Universidad de Murcia, Murcia, Spain.

[474] Roy, N. [1987] Extreme points of convex sets in infinite dimensional spaces, Amer Math. Monthly *94*, 409–422.

[475] Royden, H. {1968] *Real analysis*, 2nd ed., Macmillan, New York.

[476] Saccoman, J. [2001] On the extension of linear operators, Int. J. Math. Math. Sci. *28*, 621–623.

[477] Saxon, S. [1972] Nuclear and product spaces, Baire-like spaces and the strongest locally convex topology, Math. Ann. *197*, 87–106.

[478] Saxon, S. [1974] Two characterizations of linear Baire spaces, Proc. Amer. Math. Soc. *45*, 204–208.

[479] Saxon, S., and M. Levin [1971] Every countable codimensional subspace of a barreled space is barreled, Proc. Amer. Math. Soc. *29*, 91–96.

[480] Schaefer, H. [1971] *Topological vector spaces*, Springer-Verlag, Graduate Texts in Math. 3, 3rd printing, corrected, Berlin-Heidelberg-New York.

[481] Schauder, J. [1927] Zur Theorie stetiger Abbildungen in Funktionalräumen, Math. Z. *26*, 47–65.

[482] Schauder, J. [1930a] Über linearer vollstetige Funktionaloperatoren, Studia Math. *2*, 183–196.

[483] Schauder, J. [1930c] Die Fixtpunktsatz in Funktionalräumen, Studia Math. *2*, 1–6.

[484] Schauder, J. [1930c] Über die Umkerhrung linearer, stetige Funktionaloperatoren, Studia Math. *2*, 171–180.

[485] Schikhof, W. [1984] *Ultrametric calculus*, Cambridge University Press, New York.

[486] Schikhof, W. [1986] Compact-like sets in non-Archimedean functional analysis, *Proceedings of the conference on p-adic analysis*, Hengelhoef, 137–147.

[487] Schikhof, W. [1999] *Banach spaces over non-Archimedean valued fields*, Report 9937, Math. Dep't., U. of Nijmegen.

[488] Schmets, J. [1976] *Espaces de fonctions continues*, Lecture Notes in Mathematics, Vol. 519, Springer-Verlag, Berlin.

[489] Schmets, J. [1983] *Spaces of vector-valued continuous functions*, Lecture Notes in Mathematics, Vol. 1003, Springer-Verlag, Berlin.

[490] Schmets, J., and M. de Wilde [1971] Caractérisation des espaces $C(S)$ ultrabornologiques, Bull. Soc. Roy. Sci Liège *40*, 119–120.

[491] Schmidt, E. [1908] *Über die Auflösung linearer Gleichungen mit unendlich vielen Unbekannten*, Rend. Palermo XXV, 53–77.

[492] Schreier, O. [1925] Abstrakte kontinuierliche Gruppen, Abh. Math. Sem. Univ. Hamb. *4*, 15–32

[493] Schwartz, L. [1966a] *Théorie des distributions*, rev. ed., Hermann, Paris.

[494] Schwartz, L. [1966b] Sur le théorème du graphe fermé, C. R. Acad. Sci Paris *263*, 602.

[495] Semadeni, Z. [1971] *Banach spaces of continuous functions*, Polish Scientific Publishers, Warsaw.

[496] Shapiro, J. [1970] Existence of linear functionals on $F$-spaces with bases, Duke Math. J. *37*, 639–645.

[497] Shirota, T. [1952] A generalization of a theorem of Kaplansky, Osaka Math. J. *4*, 121–132.

[498] Shirota, T. [1954] On locally convex vector spaces of continuous functions, Proc. Japan Acad. *30*, 294–298.

[499] Shuchat, A. [1972] Approximation of vector–valued continuous functions, Proc. Amer. Math. Soc. *31*, 97–103.

[500] Simons, S. [1965] The sequence spaces $\ell(p_v)$ and $m(p_v)$, Proc. Lond. Math. Soc. (3) *5*, 422–436.

[501] Simons, S. [1970a] Minimal sublinear functionals, Studia Math. *37*, 37–56.

[502] Simons, S. [1970b] Formes sousliéaires minimales, Sém. Choquet 1970/71, no. 23.

[503] Simons, S. [1975] Convergence theorems, Hahn–Banach and Choquet theorems, minimax theorem and James's theorem, in *Analyse fonctionelle* (Proceedings of a conference in Rio de Janeiro, August 1972), L. Nachbin, ed., Hermann, Paris, pp. 271–276.

[504] Simons, S. [2003] A new version of the Hahn–Banach theorem, Arch. Math. (Basel) *80*, 630–646.

[505] Simons, S. [2008] *From Hahn–Banach to monotonicity*, 2nd ed., Lecture Notes in Mathematics *1693*, Springer–Verlag, New York.

[506] Sims, B. [2001] A mathematical pilgrimage, Australian Math. Soc. Gazette *28*, 232–236.

[507] Singer, I. [1962] Choquet spaces and best approximation, Math. Ann. *148*, 330–340.

[508] Singer, I. [1970] *Best approximation in normed linear spaces by elements of linear subspaces*, Die Grundlehren der Mathematischen Wissenschaften, Vol. 171, Springer-Verlag, Berlin.

[509] Slowikowski, W. [1961a] On continuity of inverse operators, Bull. Amer. Math. Soc. *67*, 467–470.

[510] Slowikowski, W. [1961b] Quotient spaces and the open mapping theorem, Bull. Amer. Math. Soc. *67*, 498–500.

[511] Slowikowski, W. [1965] Extensions of sequentially continuous linear functionals in $LF$-spaces, Bull. Acad. Polon. Sci. Sér. Sci. Math. Astronom. Phys. *13*, 7, 455–458.

[512] Slowikowski, W. [1966] Extensions of sequentially continuous linear functionals in inductive sequences of $F$-spaces, Studia Math. *26*, 193–221.

[513] Smith, R., and J. Ward [1940] Über linear topologische Räume, Mat. Sbornik *7*, 425–444.

[514] Smulian, V. [1939] Mat. Sbornik *6*, 77–89, English summary 90–94.

[515] Smulian, V. [1978] $M$-ideal structure in Banach algebras, J. Func. Anal. *27*, 337–349.

[516] Sobczyk, A. [1941] Projection of the space $(m)$ on its subspace $(c_0)$, Bull. Amer. Math. Soc. *47*, 938–947.

[517] Sobczyk, A. [1944] On the extension of linear transformations, Trans. Amer. Math. Soc. *55*, 153–169.

[518] Sobolev, S. [1936] Méthode nouvelle à résoudre le problème de Cauchy pour les équations linéaires hyperboliques normales, Mat. Sb. (N.S.) *1*, 39–72.

[519] Soukhomlinov, G. A. [1938a] On the extension of linear functionals in complex and quaternion linear spaces, Matem. Sbornik *3*, 353–358 [Russian with German summary].

[520] Soukhomlinov, G. [1938b] Über Fortsetzung von linearen Funktionalen in linearen komplexen Räumen und linearen Quaternionräumen, Mat. Sb. (N.S.) *3* (45), 353–358. (In Russian; German summary in Zbl. Math. *19*, 169 (1938–1939).

[521] Springer, T. [1965] Une notion de compacité dans la théorie des espaces vectoriels topologiques, Indag. Math. *27*, 182–189.

[522] Steinitz, E. [1913/16] Bedingt konvergente Reihen und konvexe Systeme, J. Reine Angew. Math. *143*, 128–175, *144*, 1–40, *146*, 1–52.

[523] Stone, M. [1936] The theory of representations for Boolean algebras, Trans. Amer. Math. Soc. *40*, 37–111.

[524] Stone, M. [1937a] Algebraic characterization of special Boolean rings, Fund. Math. 29, 223–303.

[525] Stone, M. [1937b] Applications of the theory of Boolean rings to general topology, Trans. Amer. Math. Soc. *41*, 375–481.

[526] Stone, M. [1940] A general theory of spectra I, Proc. Nat. Acad. Sci. USA 26, 280–283.

[527] Stone, M. [1941] A general theory of spectra II, Proc. Nat. Acad. Sci. U.S.A. *27*, 83–87.

[528] Stone, M. [1949] Boundedness properties in function lattices, Can. J. Math. *1*, 176–186.

[529] Stout, E. [1971] *The theory of uniform algebras*, Bogden and Quigley, Tarrytown-on-Hudson, New York.

[530] Sullivan, F. [1977] Geometrical properties determined by the higher duals of a Banach space, Illinois J. Math. Math. *21*, 315–331.

[531] Sundaresan, K. [1973] Spaces of continuous functions into a Banach space, Studia Math. *48*, 15–22.

[532] Taylor, A. [1939] The extension of linear functionals, Duke Math. J. *5*, 538–547.

[533] Tihonov, A. [1930] Über die topologische Erweiterung von Räumen, Math. Ann. *111*, 767–776.

[534] Tihonov, A. [1935] Ein Fixtpunktsatz, Math. Ann. *102*, 544–561.

[535] To, T. O. [1971] The equivalence of the least upper bound property and the Hahn–Banach extension property in ordered vector spaces, Proc. Amer. Math. Soc. *30*, 287–296.

[536] Todd, A. [1980] Coverings of products of linear topological spaces, J. Austal Math. Soc. *29*, 281–290.

[537] Todd, A., and S. Saxon [1973] A property of locally convex Baire spaces, Math. Ann. *206*, 23–34.

[538] Treves, F. [1967] *Topological vector spaces, distributions and kernels*, Academic Press, New York.

[539] Valdivia, M. [1971a] A hereditary property in locally convex spaces, Ann. Inst. Fourier Grenoble, *21*, 1–2.

[540] Valdivia, M. [1971b] Absolutely convex sets in barreled spaces, Ann. Inst. Fourier Grenoble *21*, 3–13.

[541] Valdivia, M. [1971c] On *DF* spaces, Math. Ann. *191*, 38–43.

[542] Valdivia, M. [1971d] A class of bornological barreled spaces which are not ultrabornological, Math. Ann. *194*, 43–51.

[543] Valdivia, M. [1981] On suprabarreled spaces, In *Functional analysis, holomorphy and approximation theory* (Proceedings of a conference in Rio de Janeiro, August 1978), Lecture Notes in Mathematics, Vol. 843, Springer-Verlag, Berlin, pp. 572–580.

[544] Valentine, F. [1964] *Convex sets*, McGraw-Hill, New York.

[545] van Rooij, A. [1978] *Non-Archimedean functional analysis*, Marcel Dekker, New York.

[546] van Tiel, J. [1965] Espaces localement $K$-convexes I–III, Indag. Math. *27*, 249–289.

[547] van Tiel, J. [1984] *Convex analysis*, Wiley, New York.

[548] Volterra, V. [1887] Sopra le funzioni che dipendono da altre funzioni, Rend. Accad. Lincei *3*, 97–105; see also *Opere matematiche*, 5 vol., Acc. dei Lincei, 294–302, 1954–1962.

[549] von Neumann, J. [1935] On complete topological linear spaces, Trans. Amer. Math. Soc. *37*, 1–20.

[550] Waelbroeck, L. [1971] *Topological vector spaces and algebras*, Lecture Notes no. 230, Springer-Verlag, Berlin.

[551] Waksmundzka-Hajnos, M. [2006] Defending grandfather, Focus *11*, p. 16. An article by Banach's niece about his father. An expanded version is available at http://banach.univ.gda.pl/e-greczek.html

[552] Warner, S. [1968] Locally compact vector spaces and algebras over discrete fields, Trans. Amer. Math. Soc. 130, 463–493.

[553] Warner, S. [1969] Compact and finite-dimensional locally compact vector spaces, Ill. J. Math. *13*, 383–393.

[554] Webb, J. [1981] Subspaces of barreled spaces, Quaestiones Math. *4*, 323–324.

[555] Weber, H. *Lehrbuch der Algebra*, 2nd ed., 3 vol., Braunschweig, Vieweg, 1898–1908.

[556] Wehausen, J. [1938] Transformations in linear topological spaces, Duke Math. J. *4*, 157–169.

[557] Weir, M. [1975] *Hewitt–Nachbin spaces*, North-Holland Mathematics Studies, Vol. 17, North-Holland, Amsterdam.

[558] Weierstrass, K. [1885] Über die analytische Darstellbarkeit sognenannter willkürlicher Funktionen reller Argumente, Sitzungsber. Königl. Preuss. Akad. Wiss. vom. 9. Juli 1885, 633–639.

[559] Wheeler, R. [1972] The equicontinuous weak-$*$ topology and semireflexivity, Studia Math. *41*, 243–256.

[560] Whitley, R. [1966] Projecting $m$ onto $c_0$, Amer. Math. Monthly *73*, 285–286.

[561] Wiener, N. [1922] Limit in terms of continuous transformation, Bull. Soc. Math. Fr. *50*, 119–134.

[562] Wiener, N. [1923] Note on a paper of M. Banach, Fund. Math. *4*, 136–143.

[563] Wilansky, A. [1976] On a characterization of barreled spaces, Proc. Amer. Math. Soc. *57*, 375.

[564] Yaglom, I., and V. Boltyanskii [1961] *Convex figures*, translated by P. Kelly and L. Walton, Holt, Rinehart and Winston, New York.

[565] Zelazko, W. [1973] *Banach algebras*, Elsevier, New York/Polish Scientific Publishers, Warsaw.

[566] Zippin, M. [2003] Extensions of bounded linear operators, *Handbook of the geometry of Banach spaces*, vol. 2, 1703–1741.

# Index